AF517537

FLORA OF
SOUTH CENTRAL RAJASTHAN

FLORA OF SOUTH CENTRAL RAJASTHAN

B.L. Yadav
K.L. Meena

Lecturer in Botany
M.L.V. Government College
Bhilwara (India)

Published by:
Scientific Publishers
5A, New Pali Road, P.O. Box 91
Jodhpur 342 001, India

E-mail: info@scientificpub.com
Website: www.scientificpub.com

ISBN: 978-81-7233-683-7

Reprinted from Journal of Economic & Taxonomic Botany
Additional Series # 34 (2011)
ISSN: 0970-3306

Typeset: Rajesh Ojha
Printed in India

PREFACE

Rajasthan in the North Western part of India with an area of 342,274 sq. km. is the largest state and lies between 23°3' – 30°12' N latitude and 69°31' – 78°17' E longitude The desert belt encircling half of the world merges into India's tropical part through this state causing thereby enormous variation in soil, climate, vegetation and mineral aspects of this part of India The Aravalli ranges, one of the oldest mountain ranges of the world, divide the state into two distinct climatic zones. The area west to Aravallis is desert and semi desert characterized by the high temperature, high wind velocity, low rainfall and scrub jungle. While that of east and south east of Aravalli is a humid zone with mixed deciduous type of forests. The western part and the south eastern part of the state have extensively been worked out from the floristic point of view by several workers but the south central part of the state has largely been ignored. Therefore, the extensive floristic work on this part of the state for the last 10 years has culminated in the form of this flora.

The flora of south-central Rajasthan deals with the floristic composition of the area of Ajmer, Bhilwara, Chittorgarh and Rajsamand districts which is represented by the admixture of western dry and southern wet zone elements of Rajasthan. 686 species belonging to 416 genera and 117 families have been presented. Besides providing and up-to-date nomenclature and full synonymy of all the taxa, the elaborate description of species provided is based on personal observations of the authors. Special attention has been given to the information on nomenclature, identification and critical evaluation of species. Ecological notes along with phenology and ethnobotanical importance have been incorporated to almost all the taxa. The distribution of each species in the world as well as in India has also been provided. 187 good quality illustrations and more than 340 colored photographs have added to the value of the book. Moreover, 7 new taxa have been added to the flora of Rajasthan. It is hoped that this book will certainly serve as an excellent reference material and practical guide for teachers, researchers, students and environmentalist as well as for conservation of biodiversity and sustainable utilization of plant resources.

In addition, easy to use keys with an emphasis on field characteristics have been provided. Only those localities from where actual collections known the authors have been made, are mentioned. But of course this does not mean that there is no short coming in this work. Therefore, the valuable suggestions are welcomed.

We are beholden to Prof. T.N. Bhardwaja, ex Vice-Chancellor, Kota Open University, Kota, and Prof. C.B. Gena, ex Vice-Chancellor, Shri Ganga Singh University, Bikaner, for their valuable suggestions frank criticism, critical comments on the identity of specimens, encouragement and blessings throughout the course of this work.

We specially grateful to former Director, M. Sanjappa and Joint Director, Dr. M.S. Mondal of Botanical Survey of India, Kolkata for the confirmation of identity of various species. Help rendered by Joint Directors and Dr. S.L. Meena, Scientist 'D' of Arid Zone Regional Centre, Jodhpur (BSI) is gratefully acknowledged. Sincere thanks are due to all the member of Indian Association of Angiosperm Taxonomy (IAAT) and to Dr. M.L. Verma, former-lecturer in Botany, Government College, Ajmer for their help in identification of the plant specimens.

We are highly thankful to the forest officials of Ajmer, Bhilwara, Chittorgarh and Rajsamand districts for their help and co-operation during the plant collection of these areas.

We would be falling in our duty if we do not place on record thanks and appreciation to the tribals of the study area for providing the traditional knowledge about plants. Thanks are also due to Mewar Bhil Corps (MBC) Khairwara, for their cooperation during this work.

We record our deep sense of gratitude to Principal and Vice Principals of M.L.V. Government College, Bhilwara, for their blessings, encouragement and for providing necessary facilities for this work.

B.L. Yadav
K. L. Meena

CONTENTS

Plate 1-2

1. A view of *Malaji hills* near Jahazpur (Bhilwara)

2. Mixed deciduous forest near Mandal Garh (Bhilwara)

Plate 3-4

3. A rocky terrace near Badnor (Bhilwara) with clumps of dendroid *Euphorbia caducifolia* Haines mixed with *Prosopis juliflora* (Sw.) DC.

4. Stabilized sand dunes near Budha Puskar (Ajmer) showing mixed association of *Ephedra ciliata* Fisch. & May. ex C.A. Mey, *Acacia senegal* (L.) Willd., *Crotalaria burhia* Buch.-Ham. ex Benth., *Leptadenia pyrotechnica* (Forssk.) Decne, *Maytenus emarginatus* (Willd.) Ding Hou, *Prosopis cineraria* (L.) Druce.

Plate 5-6

5. A view of deciduous forest of *Acacia senegal* (L.) Willd., *Boswellia serrata* Roxb. ex Cocleb., *Lannea coromandelica* (Houtt.) Merrill and *Sterculia urens* Roxb. near village Gurala (Bhilwara).

6. *Phoenix sylvestris* (L.) Roxb. in pure stands near village Karoi (Bhilwara)

Plate 7-8

7. Grass land near Hamirgarh (Bhilwara) showing association of *Apluda mutica* L., *Aristida adscensionis* L. and trees of *Butea monosperma* (Lam.) Taub. in the background

8. Wetland near Mainal (Bhilwara) showing association of *Ipomoea aquatica* Forsk., *Nelumbo nucifera* Gartn.; and trees of *Anogeissus pendula* Edgew, *Mitragyna parviflora* (Roxb.) Korth. in the background.

Plate 9-16

9. *Annona squamosa* L.

10. *Cissampelos pareira* L.

11. *Cissampelos pareira* L.

12. *Tinospora cordifolia* (Willd.) Miers

13. *Nymphaea nauchali* Burm.

14. *Nymphaea pubescens* Willd.

14A. *Nymphaea pubescens* Willd.

15. *Nelumbo nucifera* Gaertn.

16. *Argemone mexicana* L.

Plate 17-32

17. *Argemone ochroleuca* Sweet.

18. *Lepidium sativum* L.

19. *Sisymbrium irio* L.

20. *Cleome burmannii* Wight & Arn.

21. *Cleome scaposa* D.C.

22. *Cleome simplicifolia* (Camb) Hook. & Thoms.

23. *Cleome viscosa* L.

24. *Capparis decidua* (Forsk.) Edgew.

25. *Capparis sepiaria* L. var. *sepiaria*

26. *Capparis zeylanica* L.

29. *Polygala umbonata* Craib

28. *Maerua oblongifolia* (Forsk.) Rich.

27. *Crateva adansonii* DC. subsp. *odora* (Buch.-Ham.) Jacobs

30. *Vaccaria pyramidata* Medic

31. *Portulaca oleracea* L.

32. *Tamarix ericoides* Rottl.

Plate 33-44

33. *Abutilon hirtum* (L.) Sweet

34. *Abutilon indicum* (L.) Sweet

35. *Abutilon ramosum* (Cav.) Guill. & Perr.

36. *Malvastrum coromandelianum* (L.) Garcke

37. *Sida acuta* Burm. f.

38. *Sida ovata* Forsk.

39. *Hibiscus caesius* Gracke

40. *Malva parviflora* L.

41. *Urena lobata* L.

42. *Bombax ceiba* L.

43. *Bombax insigne* Wallich

44. *Adansonia digitata* L.

Plate 45-56

45. *Helictres isora* L.

46. *Melhania futtyporensis* Munro ex Mast.

47. *Melochia corchorifolia* L.

48. *Pterospermum acerifolium* Willd.

49. *Sterculia urens* Roxb.

50. *Corchorus aestuans* L.

51. *Corchorus olitorius* L.

52. *Grewia tenax* (Forsk.) Fiori.

53. *Triumfetta rhomboidea* Jacq.

54. *Heptage benghalensis* (L.) Kurz

55. *Fagonia schweinfurthii* (Hadidi) Hadidi

56. *Tribulus rajasthanensis* Bhandari & Sharma

Plate 57-68

57. *Impatiens balsamina* L.

58. *Biophytum sensitivum* (L.) DC.

59. *Aegle marmelos* (L.) Corr.

60. *Murraya paniculata* (L.) Jack.

61. *Ailanthus excelsa* Roxb.

62. *Balanites aegyptiaca* Roxb. ex Cocls.

63. *Boswellia serrata* (L.) Del.

64. *Commiphora wightii* (Arn.) Bhandari

65. *C. wightii* (Arn.) Bhandari Fruiting branch

66. *Melia azedarach* L.

67. *Celastrus paniculata* Willd.

68. *Maytenus emarginatus* (Willd.) Ding Hou

Plate 69-80

69. *Ziziphus mauritiana* Lam.

70. *Ziziphus nummularia* (Burm.f.) Wight & Arn.

71. *Ziziphus xylopyrus* (Retz.) Willd.

72. *Ampelocissus latifolia* (Roxb.) Planch.

73. *Cayratia trifolia* (L.) Domin

74. *Cardiospermum halicacabum* L.

75. *Sapindus trifoliatus* L.

76. *Lannea coromandelica* (Houtt.) Merrill.

77. *Lannea coromandelica* (Houtt.) Merrill.

78. *Rhus sinuata* Thunb.

79. *Rhus sinuata* Thunb.

80. *Moringa oleifera* Lam.

Plate 81-92

81. *Abrus precatorius* L.

82. *Aeschynomene indica* L.

83. *Alhagi maurorum* Medic.

84. *Alysicarpus longifolius* (Rottl. ex Spreng.) Wight & Arn.

85. *Alysicarpus ovalifolius* (Schum.) Leonard

86. *Butea monosperma* (Lam.) Taub.

87. *Crotalaria burhia* Buch.-Ham. ex Benth.

88. *Crotalaria hebecarpa* (DC.) Rudd.

89. *Crotalaria medicaginea* Lam.

90. *Desmodium gangeticum* (L.) DC

91. *Indigofera caerulea* Roxb.

92. *Indigofera cordifolia* Heyne ex Roth.

Plate 93-104

93. *Indigofera hirsuta* L.

94. *Indigofera hochstetteri* Baker

95. *Indigofera linifolia* (L.f.) Retz. var. *linifolia*

96. *Indigofera linnaei* Ali

97. *Indigofera oblongifolia* Forsk.

98. *Indigofera trita* L.f. subsp. *trita*

99. *Lathyrus aphaca* L.

100. *Melilotus indica* (L.) All.

101. *Mucuna pruriens* (L.) DC.

102. *Rhynchosia capitata* (Heyne ex Roth) DC.

103. *Rhynchosia rothii* Roxb. ex Aitch.

104. *Sesbania bispinosa* (Jacq.) Wight

Plate 105-116

105. *Tephrosia strigosa* (Dalz.) Sant. & Mahesh.

106. *Tephrosia uniflora* Pers. ssp. *petrosa* (Blatt. & Hallb.) Gillett & Ali

107. *Vicia sativa* L.

108. *Chamaecrista absus* (L.) Irwin & Barneby

109. *Chamaecrista mimosoides* (L.) Greene

110. *Chamaecrista pumila* (Lam.) Singh

111. *Senna alexandrina* Gars. ex Miller

112. *Senna auriculata* (L.) Roxb.

113. *Senna italica* Mill. ssp. *italica*

114. *Senna tora* (L.) Roxb.

115. *Senna uniflora* (Mill.) Irwin & Barneby

116. *Tamarindus indica* L.

Plate 117-128

117. *Acacia auriculiformis* A. Cunn. ex Benth.

118. *Acacia catechu* (L. *f.*) Willd.

119. *Acacia farnesiana* (L.) Willd.

120. *Acacia leucophloea* (Roxb.) Willd.

121. *Acacia nilotica* (L.) Willd. ex Del. subsp. *astringens* (Schum. & Thonn.) Roxb.

122. *Acacia nilotica* (L.) Willd. ex Del. subsp. *cupressiformis* (J. L. Stewart) Ali & Faruqi

123. *Acacia nilotica* (L.) Willd. ex Del. subsp. *indica* (Benth.) Brenan

124. *Acacia senegal* (L.) Willd.

125. *Acacia tortilis* (Forsk.) Hayne

126. *Dichrostachys cinerea* (L.) Wight & Arn.

127. *Mimosa hamata* Willd.

128. *Mimosa himalayana* Gamble

Plate 129-140

129. *Mimosa himalayana* Gamble

130. *Pithocellobium dulce* (Roxb.) Benth.

131. *Prosopis cineraria* (L.) Druce

132. *Terminalia arjuna* (Roxb. ex DC.) Wight & Arn.

133. *Terminalia bellirica* (Gaertn.) Roxb.

134. *Terminalia elliptica* Willd

135. *Syzygium heyneanum* (Duthie) Wall. ex Gamble

136. *Ammania auriculata* Willd.

137. *Ammania auriculata* Willd.

138. *Ludwigia adscendens* (L.) Hara

139. *Ludwigia hyssopifolia* (G. Don) Excell

140. *Citrullus colocynthis* (L.) Schrad.

Plate 141-152

141. *Cucumis prophetarum* L.

142. *Dactyliandra welwitschii* Hook.

143. *Melothria maderas-patana* (L.) Cogn.

144. *Momordica balsamina* L.

145. *Momordica dioica* Roxb. ex Willd.

146. *Opuntia elatior* Mill.

147. *Corbiconia decumbens* (Forsk.) Exell.

148. *Glinus lotoides* L.

149. *Glinus oppositifolius* (L.) DC.

150. *Alangium salvifolium* (L.f.) Wangerin

151. *Borreria articularis* (L.f.) F.N. Will.

152. *Mitragyna parvifolia* (Roxb.) Korth.

Plate 153-164

153. *Acanthospermum hispidum* DC.

154. *Amberboa ramosa* (Roxb.) Jafri

155. *Blainvillea acmella* (L.) Philipson

156. *Blumea mollis* (D. Don) Merr.

157. *Caesulia axillaris* Roxb.

158. *Dicoma tomentosa* Cass.

159. *Echinops echinatus* Roxb.

160. *Glossocardia bosvallea* (L. f.) DC.

161. *Gnaphalium pulvinatum* Delile

162. *Grangea maders-patana* (L.) Poir.

163. *Launaea procumbens* (Roxb.)

164. *Pulicaria angustifolia* DC.

Plate 165-176

165. *Pulicaria crispa* (Forssk.) Benth. & Hook.

166. *Sonchus asper* (L.) Hill.

167. *Sphaeranthus indicus* L.

168. *Synedrella nodiflora* (L.) Gaertn.

169. *Tridax procumbens* L.

170. *Xanthium indicum* Koenig

171. *Dyerophytum indicum* (Gibs. ex Wt.) O. Ktze

172. *Plumbago zeylanica* L.

173. *Anagallis arvensis* L.

174. *Madhuca indica* J.F. Gmelin

175. *Madhuca indica* J. F. Gmelin

176. *Manilkara hexandra* (Roxb.) Dub.

Plate 177-188

177. *Mimusops elengi* L.

178. *Diospyros melanoxylon* Roxb.

179. *Diospyros montana* Roxb.

180. *Jasminum multiflorum* (Burm. f.) Andrews

181. *Nyctanhes arbor-tristis* L.

182. *Salvadora oleoides* Decne.

183. *Salvadora persica* L.

184. *Carisa congesta* Wight

185. *Catharanthus pusillus* (Murr.) G. Don

186. *Thevetia peruviana* (Pers.) Merr.

187. *Wrightia arborea* (Dennst.) Mabb.

188. *Wrightia tinctoria* (Roxb.) R. Br.

Plate 189-202

189. *Calotropis gigantea* (L.) R. Br.

190. *Calotropis procera* (Ait.) Ait. f.ssp. *hamiltonii* (Wight) Ali

191. *Ceropegia bulbosa* Roxb. var. *bulbosa*

193. *Leptadenia reticulata* (Retz.) Wight & Arn.

194. *Pergularia daemia* (Forsk.) Chiov.

196. *Tylophora hirsuta* (Wall.) Wight & Arn.

192. *Leptadenia pyrotechnica* (Forsk.) Decne.

195. *Sarcostemma acidum* (Roxb.) Voigt, in asso. with *E. caducifolia*

198. *Cryptolepis buchananii* Roem. & Schult.

199. *Cryptostegia grandiflora* R. Br.

197. *Wattakaka volubilis* (L. f.) Stapf

200. *Enicostema axillare* (Lam.) Raynal

201. *Nymphoides hydrophylla* (Lour.) O. Kuntze

202. *Nymphoides indica* (L.) Kuntze

Plate 203-214

203. *Arnebia hispidissima* (Lehm.) DC.

204. *Heliotropium curassavicum* L.

205. *Heliotropium indicum* L.

206. *Heliotropium supinum* L.

207. *Cordia crenata* Delile

208. *Cordia gharaf* (Forsk.) Ehrenb.ex Asch.

209. *Ehretia laevis* Roxb.

210. *Convolvulus prostratus* Forsk.

211. *Convolvulus prostratus* Forsk. var. *prostatus*

212. *Cressa cretica* L.

213. *Evolvulus alsinoides* (L.) L.

214. *Ipomoea aquatica* Forsk.

Plate 215-226

215. *Ipomoea carnea* Jacq. subsp. *fistulosa* (Mart. ex Choisy) Austin

216. *Ipomoea mauritiana* Jacq.

217. *Ipomoea nil* (L.) Roth. var. *nil*

218. *Ipomoea pes-tigridis* L.

219. *Ipomea sindica* Stapf.

220. *Merremia aegyptia* (L.) Urban

221. *Merremia emarginata* (Brum.f) Hall. f.

222. *Rivea hypocrateriformis* (Desr.) Choisy

223. *Cuscuta chinensis* Lam.

224. *Cuscuta reflexa* Roxb.

225. *Datura fastuosa* L.

226. *Datura ferox* L.

Plate 227-238

227. *Nicotiana plumbaginifolia* Viv.

228. *Physalis minima* L.

229. *Solanum incanum* L.

230. *Solanum sisymbriifolium* Lam.

231. *Solanum virginianum* L.

232. *Withania coagulans* (Stocks) Dunal

233. *Withania somnifera* (L.) Dunal

234. *Antirrhinum orontium* L.

235. *Bacopa monnieri* (L.) Wettst.

236. *Craterostigma plantagineum* Hochst

237. *Kickxia ramosissima* (Wall.) Janchen

238. *Limnophila indica* (L.) Druce

Plate 239-250

239. *Lindernia ciliata* (Colsm.) Penn.

240. *Scoparia dulcis* L.

241. *Striga angustifolia* (D. Don.) Sald.

242. *Striga gesnerioides* (Willd.) Vatke

243. *Verbascum chinense* (L.) Sant.

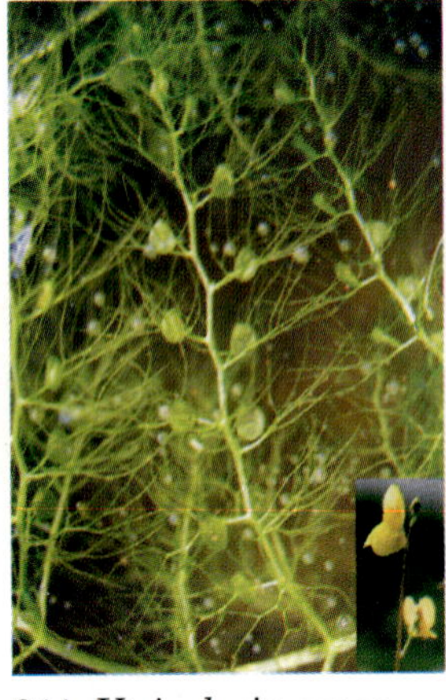

244. *Utricularia aurea* Lour.

245. *Utricularia stellaris* L. f.

246. *Jacaranda mimosifolia* D. Don

247. *Millingtonia hortensis* L.f.

248. *Tecomella undulata* (Sm.) Seem.

249. *Pedalium murex* L.

250. *Sesamum indicum* L.

Plate 251-262

251. *Martynia annua* L.

252. *Barleria acanthoides* Vahl

253. *Barleria cristata* L. var. *cristata*

254. *Barleria cristata* L. var. *dichotoma* (Roxb.) Prain

255. *Eranthemum purpurascens* Wight ex Nees

256. *Gantelbua urens* (Heyne ex Roth) Bremek.

257. *Haplanthodes neilgherrensis* (Wight) Majumdar

258. *Hemigraphis hirta* (Vahl) T. Anders

259. *Justicia trinervia* Vahl.

260. *Peristrophe paniculata* (Forsk.) Brum.

261. *Clerodendrum multiflorum* (Brum. f.) Kuntze.

262. *Gmelina arborea* Roxb.

Plate 263-274

263. *Lantana wightiana* Wall. ex Gamble

264. *Phyla nodiflora* (L.) Greene

265. *Tectona grandis* L.f.

266. *Vitex negundo* L.

267. *Anisomeles indica* (L.) O. Kuntze

268. *Lavandula bipinnata* (Roth) O. Kuntze

269. *Leonotis nepetifolia* (L.) R. Br.

270. *Leucas aspera* (Willd.) Link

271. *Salvia aegyptiaca* L.

272. *Plantago exigua* Juss. ex Murr.

273. *Achyranthes aspera* L. var. *argentea* (Thw.) Hook. *f.*

274. *Aerva javanica* (Burm. f.) Juss. ex Schult. var. *javanica*

Plate 275-286

275. *Alternanthera pungens* Kunth

276. *Amaranthus lividus* L.

277. *Celosia argentea* L.

278. *Digera muricata* (L.) Mart.

279. *Chenopodium album* L.

280. *Chenopodium botrys* L.

281. *Chenopodium murale* L.

282. *Basella rubra* L.

283. *Persicaria barbata* (L.) Hara

284. *Polygonum plebeium* R. Br. var. *plebeium*

285. *Aristolochia bracteolata* Lam.

286. *Dendrophthoe falcata* (L.f.) Etting

Plate 287-298

287. *Santalum album* L.

288. *Acalypha ciliata* Forsk.

289. *Baliospermum montanum* (Willd.) Muell.- Arg.

290. *Bridelia retusa* (L.) Spreng.

291. *Chrozophora rottleri* (Geis.) A. Juss. ex Spreng.

292. *Croton bonplandianum* Baill.

293. *Euphorbia dracunculoides* Lam.

294. *Euphobia elegans* Spreng.

295. *Euphorbia heterophylla* L.

296. *Euphorbia neriifolia* L.

297. *Ficus palmata* Forsk. sub. sp. *virgata* (Roxb.) Browitez

298. *Ceratophyllum demersum* L.

Plate 299-310

299. *Ottelia alismoides* (L.) Pers.

300. *Crinum defixum* Ker-Gawl

301. *Agave americana* L.

302. *Curculigo orchioides* Gaertn.

303. *Dioscorea bulbifera* L. var. *bulbifera*

304. *Asparagus racemosus* Willd.

305. *Gloriosa superba* L.

306. *Monochoria vaginalis* (Burm. f.) Presl.

307. *Commelina forskalaei* Vahl

308. *Cyanotis cristata* (L.) D. Don

309. *Murdania nudiflora* (L.) Brenan

310. *Typha angustifolia* L.

Plate 311-322

311. *Lemna perpusilla* Torrey

312. *Limnophyton obtusifolium* (L.) Miq.

313. *Sagittaria guayanensis* Kunth subsp. *lappula* (D. Don) Bogin

314. *Eriocaulon cinereum* R. Br.

315. *Cyperus alulatus* Kern

316. *Cyperus compressus* L.

317. *Cyperus difformis* L.

318. *Cyperus iria* L.

319. *Cyperus rotundus* L.

320. *Fimbristylis bisumbellata* (Forsk.) Bubani

321. *Kyllinga bulbosa* Beauv.

322. *Schoenoplectus articulatus* (L.) Palla

Plate 323-334

323. *Schoenoplectus supinus* (L.) Palla

324. *Apluda mutica* L.

325. *Aristida adscensionis* L. var. *adscensionis*

326. *Arundo donax* L.

327. *Brachiaria reptans* (L.) Gard. & Hubb.

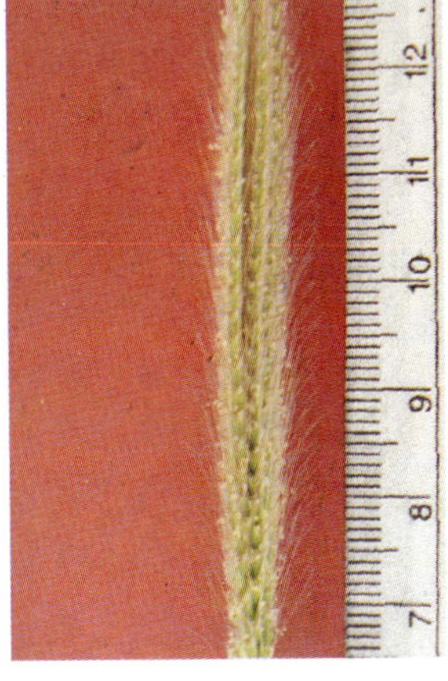

328. *Chloris virgata* Sw.

329. *Coix lacryma-jobi* L.

330. *Dactyloctenium scindicum* Boiss.

331. *Echinochloa colona* (L.) Link

332. *Eragrostis cilianensis* (All.) Vignolo-Lut. ex. F.T. Hubbard

333. *Oryza rufipogon* Griff.

334. *Phalaris minor* Retz.

Plate 335-341

335. *Rottboellia exaltata* L.

336. *Saccharum spontaneum* L.

337. *Setaria glauca* (L.) P. Beauv.

337. *Setaria glauca* (L.) P. Beauv.

338. *Setaria italica* (L.) P. Beauv.

339. *Sorghum halepense* (L.) Pers.

340. *Tetrapogon tenellus* (Koen. ex Roxb.) Chiov.

341. *Tragus Roxburghii* Panigrahi

341A. *Tragus Roxburghii* Panigrahi

INTRODUCTION

Geographically, Rajasthan is situated in the north-western part of India between 23°3' and 30°12' N. latitude and 69°30' and 78°17' E. longitude. It occupies an area of about 3, 42, 274 sq. km i.e. nearly 11 percent of the total area of India. The elevation of land surface varies from 214 to 1375 m. In shape, it is an irregular rhomb with north-south and east-west diagonals, the former about 784 km. and the later 850 km long. The western and north-western boundaries are marked by Pakistan. In the north and north-east, it is bounded by Punjab, Haryana and Uttar Pradesh, in the east and South-East by Madhya Pradesh and in the South-West by Gujarat States of India. Phyto-geographically, the state of Rajasthan forms the eastern extremity of the great arid and semi-arid belt of the world, the great Sahara desert belt passes through the western part of Rajasthan State. The major part of eastern and South-Eastern regions forms the western part of the Ganegetic plains, the southern region forms a part of *Malva* plateau and south-eastern region a part of Deccan plateau.

The region covered in the present flora lies in the southern central part of Rajasthan state between 24°15' - 26°58' N latitude and 73°28' - 75°49' E longitude and covers an area of about 33652 sq. km (Fig. 1). It includes the area of districts of Ajmer, Rajsamand, Bhilwara and Chittorgarh. The area which is roughly triangular in shape is approximately 100 - 532 meter above the mean sea level. Its north-western boundary is made by Nagaur and Pali districts. On southward side it is bounded by Udaipur district while southern eastern boundary is made by Kota as well as Bundi districts of Rajasthan and Neemuch district of Madhya Pradesh state. Its north eastern part is shared by Tonk and Jaipur districts.

Aravalli ranges, one of the oldest mountain ranges of the world are main physiographic factor of the area. Entire area is represented by hilly tracts, plains, sand dunes and plateau region. Most of the area is the part of *Banas basin* while on the south-eastern side it is the part of *upper mal* region. Ajmer district and part of Rajsamand from the northern part while Bhilwara, Chittorgarh and part of Rajsamand form the central and southern part of the area under present investigations.

PHYSIOGRAPHY

The area is almost triangular in shape. It is generally a level plain interspersed with low hills which run in the north westerly directions throughout the area under investigation. Puskar, Kishangarh and Mangliyawas area of Ajmer district is sandy.

The Aravalli ranges run across the entire area and come into prominence near Kumbhalgarh (Rajsamand), Chittorgarh, Mandalgarh (Bhilwara) and Ajmer city. The second

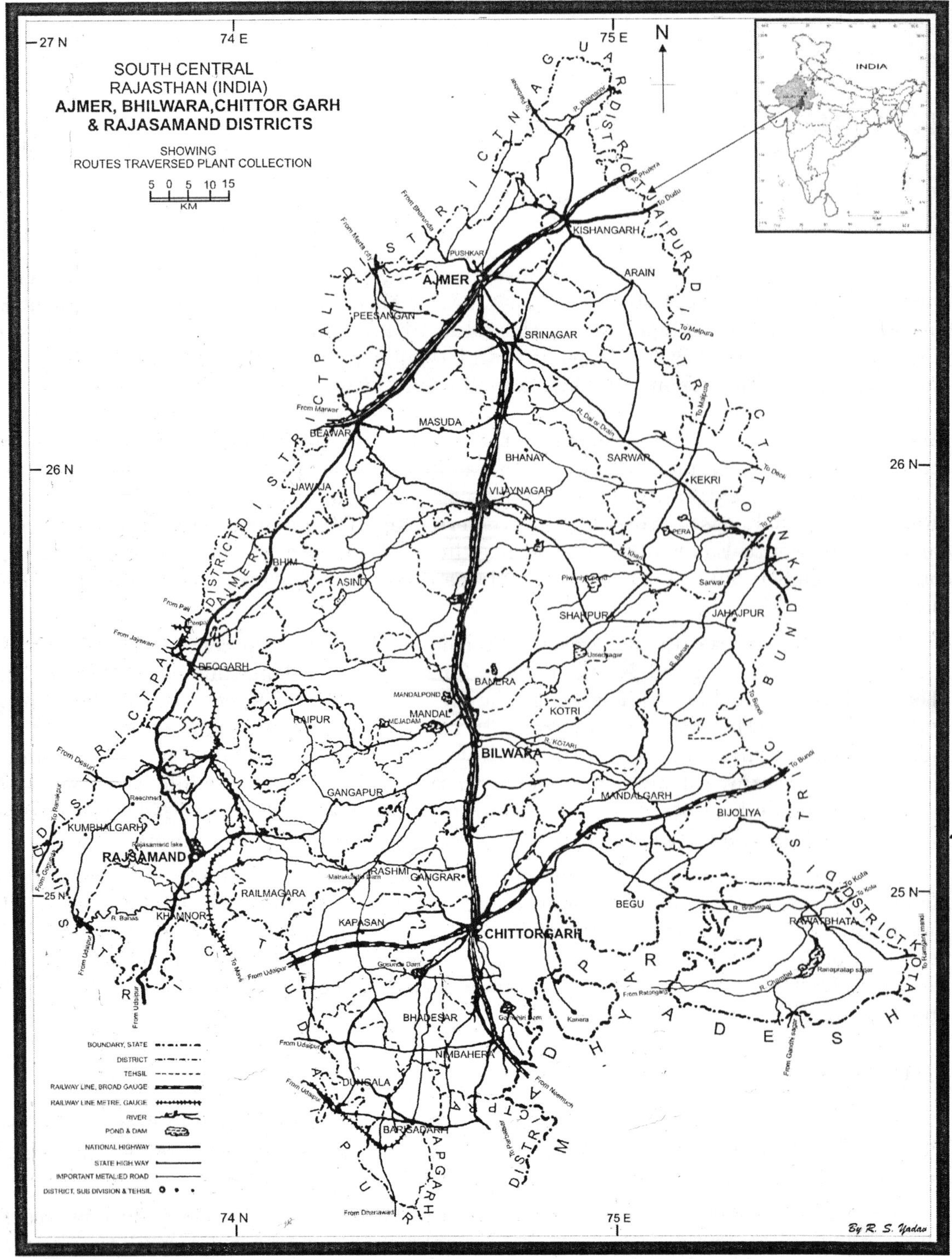
SOUTH CENTRAL
RAJASTHAN (INDIA)
AJMER, BHILWARA, CHITTOR GARH
& RAJASAMAND DISTRICTS
SHOWING
ROUTES TRAVERSED PLANT COLLECTION
5 0 5 10 15
KM
INDIA
N
27 N
26 N
25 N
74 E
75 E
AJMER
KISHANGARH
PUSHKAR
ARAIN
PEESANGAN
SRINAGAR
MASUDA
BEAWAR
BHANAY
SARWAR
KEKRI
VIJAYNAGAR
JAWAJA
BHIM
ASIND
SHAHPURA
JAHAJPUR
DEOGARH
BANERA
MANDAL
KOTRI
RAIPUR
BILWARA
GANGAPUR
MANDALGARH
BIJOLIYA
KUMBHALGARH
RAJSAMAND
RAILMAGARA
RASHMI
GANGRAR
BEGU
KHAMNOR
KAPASAN
CHITTORGARH
RAWATBHATA
BHADESAR
NIMBAHERA
DUNGALA
BARISADARI
BOUNDARY, STATE
DISTRICT
TEHSIL
RAILWAY LINE, BROAD GAUGE
RAILWAY LINE METRE GAUGE
RIVER
POND & DAM
NATIONAL HIGHWAY
STATE HIGH WAY
IMPORTANT METALIED ROAD
DISTRICT, SUB DIVISION & TEHSIL
By R. S. Yadav

Fig. 1.

highest point of Aravalli is Jarga hills (914 m) near Kumbhalgarh in Rajsamand district. The Nag pahar at a distance of about 5 km west of Ajmer city, Taragarh (Ajmer), Chittor fort, 5 km east of Chittor city, Mandal garh fort west of Mandal garh city and Kumbhal garh fort (Rajsamand) attain a scarcely inferior elevation. Range of hills between Ajmer and Nasirabad marks the dividing watershed of the Indian subcontinent. The rains which falls on the southern side of Nasirabad finds its way through the tributaries of Banas which join Chambal and ultimately falls into the bay of Bengal, which falls on other side is discharged by the Luni into the Gulf of Kutch.

The area of Rajsamand district consists of an elevated plateau on the north and fertile plains on the east. The southern and western portions are mostly covered with rocks, hills and dense forests. The Aravalli ranges enter this district in Bhim tehsil from Ajmer and attain a height of 914 m at Jarga near Sayra.

The area of Bhilwara and Chittorgarh districts is intersected by Aravalli ranges at several places which are predominated in the Chittorgarh, Mandal garh (Bhilwara) and in Jahazpur tehsil (Bhilwara). Bijoliya tehsil area of Bhilwara (South east of Bhilwara) is a plateau region.

There are several rivers which flow through the study area. All rivers are mere revolute in hot weather but become torrent in the rains. The Banas, one of the main river of study area originates in Khamnor hills of Aravalli about 5 km from the fort of Kumbhalgarh (Rajsamand). After flowing through the area of Rajsamand and Railmagara tehsils of Rajsamnad district, the river passes to Rashmi tehsil of Chittorgarh district. It enters Bhilwara near village Doodiya. This river approaches the hills of Mandalgarh and is joined by the river Berach on the right bank and Kothari on the left bank. It then flows through Jahazpur tehsil of Bhilwara and enters Ajmer district near village Jeetpura in Kekri tehsil. Thus, the entire area taken for present investigation comes under Basin region. The famous Shrinathji temple at Nathdwara is located on the right bank of Banas in Rajsamand district. The Khari river rises in the hills near village Birjal (Ajmer) and after forming the boundary between Bhilwara and Ajmer districts falls into Banas. The river Sagarmati rises near the Bisala tank in Ajmer and meets river Saraswati at the extreme north of village Govindgarh (Ajmer tehsil). River Rupnagar flows in the major part of Kishangarh tehsil of Ajmer.

Berach, Gambhiri, Chambal, Gunjali, Bamani, Kaldi and Wagon are the rivers which flow through south and the central part of the study area. The river Berach rises from the hills near Udaipur and fills the Udaisagar lake and then in the form of overflow passes through the portions of Kapasan, Bhadesar, Gangrar and Chittorgarh tehsils. It then joins rives Gambhiri near Chittorgarh town and falls into the river Banas near village Bigod in Mandalgarh tehsil of Bhilwara district. The gambhiri which originates in Madhya Pradesh flows through Nimbahera and Chittorgarh tehsils and joins the river Berach at village Bherda (Chittorgarh tehsil). The Chambal (the only perennial river) and Gunjali rivers have their sources in the plains of M.P. state and flow through the portions of tehsil Bhaisrorgarh. Bamani river is the tributary of Chambal. It rises in the hills near Begun and joins the Chambal near Bhaisrorgarh.

Other important rivers which flow in the central part of the study area are Mansi, Kothori, Nagadi and Chandrabhaga. Mansi has its origin in Mandal tehsil and passing through the portions of Asind and Hurda tehsils, it joins the Khari river in Shahpura tehsil. Khari river originates near Deogarh in Rajsamnd district and is joined by Mansi river in

Shahpura tehsil. Rivers which flow in the south western part of the study area (particularly in Rajsamand district) are Chandrabhaga, Gomati, Jamri etc.

There are natural lakes namely Pushkar and Budhapuskar near Ajmer city and Sargaon and Karantia near Beawar town (Ajmer district). Another large lake called Rajsamand is situated at Rajnagar (Rajsamand district). Among important tanks whose water is utilized for irrigation as well as for domestic use are - Bisala, Dilwar, Foysagar, Jawaja, Kalinjar, Makreda, Phoolsagar, Ramsar (Ajmer district); Bari, Bassi, Kapasan, Magrol, Nagawali, Singoli, (Chittorgarh district); Karera, Mandal, Meza, Gurla, Gowta, Guwardi, Naharsagar, Piwnia, Ummedsagar (Bhilwara).

CLIMATE

There are great variations in the climatic conditions as we move from northwest sides to southwards side of the area under investigations. The area has a hot dry summer and cold bracing winter. Winter extends from December to February while the summer season extends from March to the end of June. The rainy season is comparatively short in this region and lasts only till mid September, about 90% of the annual rainfall is received during the period from June to September. July and August being the rainiest months.

The maximum temperature was 44°C and the minimum 1.0°C at Ajmer centre during 1980. During this period the mean temperature remained 22.5°C. Climate of northwestern and north eastern part of study area is more or less dry while that of central south eastern and south western part which include Bhilwara, Rajasamand and Chittorgarh districts is moderate and generally healthy.

This part lies in the belt having an average minimum temperature ranging from 15°C to 20°C and an average maximum temperature 25°C to 30°C. However, during recent year temperature of the area has gone beyond 45°C during summers and dips up to 5 - 7°C during winters. Generally, Chittorgarh remains the coolest during winters in area of present investigations.

The rainfall generally increases from northwest to the southeast directions. The southern part which includes Chittorgarh gets maximum rains while northern part (Ajmer) gets minimum rains. The normal annual rainfall in the northern part was recorded 41.53 cm in 1980. On an average there are 26 rainy days in a year. Normal annual rainfall in north western part is 49.50 cm and normal rainy days in the area are 28 a year. The normal annual rainfall of central and southern part is 74.54 cm and 75.28 cm respectively. The normal rainy days in these areas are 32 and 34 a year respectively.

GEOLOGY AND MINERALS

The geology of most part of the area is that of Aravalli system. However, Vindhyan system, banded gneissic complex and Bundelkhand gneiss, Delhi system and Raialo series are also found.

The geology of northern part which includes Ajmer district is that of Aravalli ranges which extends throughout the district except for the northwestern part which is flat, plain and mostly covered by sand dunes. The oldest rocks exposed in the district are pre-Aravalli Banded Gneissic complex over 2,500 million years old and comprise schist, genisses and amphibelite. These are the rocks of Delhi super group comprising of quartzite, conglomerate, amphibolite and mica schist (Alwar group) and Calc-schists, delomitic marble, cal-gneisses, amphibolic and phyllite (Ajabgarh group). The post Delhi intrusivcs, granite and alkaline rocks i.e. nepheline, syenites and its associated rocks are seen near Kishangarh.

The rocks of north western part (Rajsamand district) consists for the most part of schist belonging to the Aravalli system. Other rocks are composed of Banded gnessic complex and Bundelkhand gneiss, Delhi system, Raialo series and ultra-basic intrusive of pre-cambrian age.

The geological formation of central part of study area consists of Banded gnessic-complex and Bundelkhand gneiss, Vindhyan, Aravalli and Delhi system while the geology of southern part is that of Deccan traps and Aravalli system in most part. Besides, some part of the area is covered by Vindhyan system, Banded gneissic complex and Bundelkhand gneiss and the rest by Raialo series.

The area is endowed with a large number of minerals which add several lakhs of rupees to the state's treasury. The main minerals found in the area are:

Muscovite mica associated in the pegmatite veins is mined near Dadiya (Kishangarh tehsil, Ajmer district), Basundri (Kekri tehsil, Ajmer) mines near sarwar and arain. There are several mica mines in Raipur and Sahara tehsils of Bhilwara district. Mica is also available at village Bhagwasa in Jahazpur tehsil (Bhilwara). Emerald of the gem variety is quarried from Rajgarh and Bubani-mahami mines (Ajmer), Deogarh and Amet tehsils (Rajsamand).

There are soap stone quarries at Jahazpur, Mandalgarh and Kotri tehsils of Bhilwara district, various places of Rajsamand and Ajmer distrcits.

There are important quarries of marble in Rajsamand and Ajmer districts. Kishangarh (Ajmer) and Rajsamand are the important centers of white marble. Rajsamand is very well known for its marble production, as the largest producing as well as the largest single unit in the whole country. A mine for pink marble is at Devimata near Jawar mines (Rajsamand). Sand stone is found in large quantity at various places in Mandalgarh and Bijoliya tehsils of Bhilwara. Iron ore is available near villages Gangrar (Chittorgarh), Asind and Banera (Bhilwara).

Other minerals found in the area are Asbestos, Felspar, Garnet, Lime stone, Vermiculite, Clay, Kyanite, Calcite, Lead-zink ore, Manganese ore etc.

SOIL

The soil in the upland contains large amount of well – sorted finer silt, sand and coarse cobbles, while the lowlands / plains have soil with higher percentage of clay, silt and sand. On the basis of their occurrence, the soils of the area may be divided into following groups.

1. *Red and Yellow soils :* These soils occur in the western part of the area. They are poorer in carbonate and humus with the pH ranging from 5.5 to 8.5. These soils have good water holding capacity due to considerable percentage of silt and clay proportion.
2. *Ferruginous red soil :* At some open land these soils are observed in the western and southern part of the area. This soil has low content of lime, potash, iron oxide, phosphorous, nitrogen and humus.

3. *Mixed red and black soil* : The eastern part of Bhilwara and Chittorgarh districts are covered with this soil. pH of this soil ranges from neutral to alkaline.
4. *Medium Black soil:* South eastern part of Bhilwara and Chittorgarh districts is covered mainly with black soil but at some patches grey brown to dark brown coloured, medium heavy soil and yellow and light red soils have also been found.
5. *Sandy soil:* This type of soil is found in the area of Kishangarh, Pushkar and Mangliyawas of Ajmer district. This soil is characterized by low organic content and poor water holding capacity.

Flora and Fauna

The northern part of the area which includes Ajmer representing central Rajasthan comprises of plant species common to the arid tracts of India. The central part of the study area is represented by the elements of western dry zone and southern wet zone of Rajasthan while species found in southern part (Chittorgarh and south Rajsamnad) are common to humid zone.

Due to geographical position and limited rainfall the flora of north western and north eastern part of study area is not rich. The forest crop is mainly of xerophytic species common to the arid tracts of India. This area is completely outside the timberline i.e. teek, sal and sisso zone of the tropical and subtropical India. The main species found in the area are Dhokra (*Anogeissus pendula*), Kumta (*Acacia senegal*), Salar (*Boswellia serrata*), Khejri (*Prosopis cineraria*), Khair (*Acaca catechu*), Ber (*Ziziphus nummularia*), Koulassi (*Dichrostachys cinerea*), Khejra (*Acacia lecucophloea*), Tambolia (*Ehretia laevis*), Thor (*Euphorbia caducifolia*), Danssara (*Rhus sinuata*), Khirni (*Wrightia arborea*), Kankran (*Flacourtia indica*), Ker (*Capparis decidua*), Aak (*Calotropis procera*) etc.

Northern dry mixed deciduous type of forests are more common in south western part of the study area. Other types of forests found in this area are subsidiary edaphic type of dry tropical forests. Among the species found are Aam (*Mangifera indica*), Babul (*A. nilotica*), Dhak (*Butea monosperma*), Khair (*A. catechu*), Gular (*F. racemosa*), Salar (*Boswellia serrata*), Hingota (*Balanites aegyptiaca*), Semal (*Bombax ceiba*), Timaroo (*Diospyros melanoxylon*), Bans (*Bambusa arundinacea*). Among the small shrubs available are *Calotropis procera*, *Cassia auriculata*, *Carissa congesta* etc.

The central part of the study area is represented by mixed deciduous forests. The main species found are *Acacia catechu*, *Acacia senegal*, *Anogeissus pendula*, *Boswellia serrata*, *Calotropis procera*, *Capparis decidua*, *Prosopis cineraria*, *Dichrostachys cinerea*, *Ehretia laevis*, *Euphorbia caducifolia*, *Flacourtia indica*, *A. lecucophloea*, *Rhus sinuata*, *Wrightia arborea*, *Ziziphus nummularia*.

South and south eastern part is a humid zone and comprises of teak forests in the hills of Barisadari while subsidiary edaphic type of dry tropical forest are common in Bhaisrorgarh, Gangrar, Chittorgarh, Nimbahera and Kapasan. Some of the main species of the area are *Sterculia urens*, *Madhuca indica*, *Tectona grandis*, *Lannea coromandelica*, *Butea monosperma*, *Acacia luecophloa*, *Erythrina suberosa*, *Anogeissus latifolia*, *A. pendula*, *Bridelia retusa* etc.

Deer, Neel gai, Wild pigs, chital, Fox, Panther, Boar, Hares, Grey partridges, Sand grouse, Black buck, Snakes constitute the fauna of the area.

PREVIOUS WORK AND PRESENT STUDY

The work on the flora and present day Botany has been pioneered by Linnaeus with his classic publication Genera Plantarum (1737). In India, the work on floral exploration had been initiated by Roxburgh (1820-1824) who published "Flora Indica". Later on Hooker and Thomson (1855) published an Introductory essay in the flora Indica in which they concluded that the vegetation of the plains of Marwar, which are continuous with the great sandy desert tract of Africa is not known in detail

The earliest recorded information on botanical explorations in Rajasthan is about Jacquemont's journey in 1832 from Delhi to Bombay via Ajmer and Neemuch. Systematic work on phytodiversity of Indian desert was taken up in 1868 by Sir George King, he made collections in Rajasthan which resulted in his papers "Famine Food of Marwar (1869)", "Notes on vegetable products used as food during famine in Rajputana (1870)" and "Sketch of the Flora of Rajputana (1878)". Brandis (1874) published "The forest flora of the North-West and Central India". It was a result of the plant explorations made by him during 1869 and 1870 in Rajputana forests. Duthie (1886) also made an exploratory tour of Rajputana forests. In 1890, Macadam published a list of plants from Mt. Abu. Adams (1899) listed about 50 species of plants in his book "The Western Rajputana States". Blatter and Hallberg (1918-1921) presented "The Flora of the Indian Desert" (Jodhpur and Jaisalmer), which for about 30 years remained the only authentic systematic account of the plants of Rajasthan. Parker (1918) recorded a number of plants from Jaipur. Duthie (1903-1929) in his "Flora of Upper Gangetic Plains", included the major part of eastern and south-eastern Rajasthan.

After the publication of Duthie's "Flora of the Upper Gangetic Plains and of the adjacent Siwalik and Sub-Himalayan tract 1903-29", the floristic contributions from eastern Rajasthan were negligible till 1950. Since then, however, a large number of papers have been published on the vegetation of the Aravallis and the parts in the east of it (Raizada & Sharma, 1962, Vyas & Ramdeo, 1965, Vyas, 1967, Bhandari, 1965, Gupta, 1965a, Maheshwari & Singh, 1974, Majumdar, 1969, 76, 77, Jain, 1970, Sharma, 1978, Sharma *et al.*, 1980, Singh, 1977, 1980, Parmar & Singh, 1982, Yadav & Meena, 2008, 2009, Meena & Yadav, 2006, 2009, Meena, 2010).

The publication of the "Flora of N.E. Rajasthan" (Sharma & Tiagi, 1979), "Flora of Banswara" (Singh, 1983) and "Flora of Tonk District" (Shetty & Pandey, 1983) have further added to our knowledge of the Flora and Floral composition of eastern Rajasthan.

A large number of publications dealing with the floral composition of Rajasthan have been reviewed by Jain (1970), Majumdar (1977), Bhandari (1978), Shetty & Pandey (1979) and Sharma (1980).

The aridity and other factors influencing the vegetation in the west of Aravallis have attracted the attention of the whole nation and also the UNESCO. As a result, a number of lists of plants and accounts on the vegetation of Rajasthan desert have been published. The notable contributions are : Ramchandra Rao (1941), Griffith (1946), Sankhala (1951), Sarup (1951, 1952a, 1954, 1957, 1958a, 1958b), Krishnaswamy & Gupta (1952), Biswas & Rolla (1953), Bhandari (1954, 1961a, 1961b, 1963, 1964a, 1964b, 1965, 1967), Bakshi (1954), Bhandari & Sarup (1957), Nair & Joshi (1957) Joshi (1958, 1961), Sarup & Puri (1960a, 1960b) Puri & Jain (1961, 1962), Puri *et al.* (1964), Bhandari & Singh (1964), Gupta & Bhandari (1965), Kanodia & Gupta (1969), Guhabakshi (1969), Bhandari & Verdcourt (1970), Bhandari & Sharma (1977b) Dillon *et al.* (1974), etc. More recently, Puri *et al.* (1964),

Bhandari (1978), Dawre *et al.* (1979), Parmar & Singh (1982), Pandey *et al.* (1983), Pandey (1984), Pandey & Shetty (1985), Pandey *et al.* (1985), Sharma *et al.* (1990).

The most remarkable work on the phytodiversity of Indian desert was done by Bhandari (1978) who published a comprehensive "Flora of Indian desert" containing 594 species with keys and detail description of species. He revised the flora in 1990 with the addition of 90 more plant species from this region of Rajasthan. Subsequently, Shetty & Pandey (1978, 1979, 1983), Bhandari & Kumar (1995) further added to the flora of Indian desert. Phyto-diversity of Thar-desert has been reported by Gena (2006).

The floristic composition of various protected areas of Rajasthan has been investigated by several workers. Prasad *et al.* (1996) have presented "Illustrated flora of Keoladeo National Park, Bhartpur, Rajasthan. The biodiversity of desert National Park Rajasthan has been worked out by Singh & Singh (2006) which includes almost unexplored plants and animals, due to its inaccessible approach and being on the western border line. Biodiversity of Sariska Tiger Project in the North-East Rajasthan has compiled by Yadav (2006) while that of Ranthambhore National Park has been investigated by Singh & Srivastava (2007).

Sharma & Tiagi (1979) have recorded 627 species grouped into 343 genera belonging to 95 families of angiosperms in their Flora of North Eastern Rajasthan. While, Shetty & Singh (1987-93), in the Flora of Rajasthan included 1911 species, 780 genera and 154 families in which 1 species was of Gymnosperm. Recently, Sharma (2002) in his Flora of Rajasthan (Hadoti Plateau) included 1187 species, 628 genera & 139 families. More recently, Tiagi & Aery (2007) recorded 1378 plant species (including cultivated as well as ornamental) belonging to 126 families in their Flora of Rajasthan (South and South East Rajasthan).

However, the floristic composition of south central part of the state which forms the geographical heartland of Rajasthan has sadly been neglected except some brief reports made by Sharma (1958) and Sharma *et al.* (1990), therefore, to fill up this gap the present work has been taken up.

BIOLOGICAL SPECTRUM

A community consists of many species with different kinds of population fluctuations and interaction with one another. The physiognomic forms and structural studies of land communities have been discussed in more rewarding manner by Whittaker (1975). For the classification of the vegetation of different regions of the world, different systems have been proposed by several ecologists (Humboldt, 1805; Drude, 1897, 1913; Raunkiaer, 1911, 1934; Braum - Blanquet, 1932; Cain, 1950; Cabrera, 1952). But Raunkiaer's life form system (1934) has a universal applicability. The plant climate of a region, according to Raunkiaer (1934), is characterised by the life forms which is the biological spectrum of that region may or may not exceed the percentage of the same life form in the normal spectrum (Spectrum of the whole flora of the earth). Thus, the biological spectrum of different regions may be used to compare the widely separated plant communities in terms of their climatic adaptability.

Several workers have studied the biological spectrum of the Indian desert (Das & Sarup, 1951; Mertia & Bhandari, 1978 and Charan *et al.*, 1978). Pandey *et al.* (1985) have reviewed biological spectrum of Rajasthan desert and pointed out that the phytoclimate of the desert is of "thero-cryptophytic". Pandey & Parmer (1993) have analyzed the biological spectrum of Rajasthan as a whole and have again advocated "thero-cryptophytic" phytoclimate of the state. According to Mehar-Homji (1962b), the biological spectrum reveals "therophytic phytoclimate" for arid zones and for intensively cultivated areas. More recently, Singh & Singh (2006) have analysed the biological spectrum of the desert national park and advocated " Thero-cryptophytic" phytoclimate in desert national park, Rajasthan.

Sen (1973) has recognised leaflessness as a characteristic physiognomic feature of the Indian arid zone vegetation. Leaflessness in the arid zone perennials is either permanent, seasonal or rhythmic. In permanent leaflessness, green leaves are borne only during seedling stage, as in *Leptadenia pyrotechnica*, *Calligonum polygonoides*, *Sarcostemma acidum* etc. Seasonal leaflessness is the development of green foliage during rainy season after which majority of arid zone shrubs become partially or wholly leafless under moisture stress, as in *Crotalaria burhia*, *Convolvulus microphyllus*, *Citrullus colocynthis* etc. The rhythmic types are those where leaflessness is not related to any season or time of year. In this situation they have to face acute moisture scarcity as shown by *Capparis decidua* and *Euphorbia caducifolia*.

A thorough survey of literature indicates that no comprehensive studies have been made on the floristics of south central Rajasthan, hence, to present the flora of this region would be worthwhile as the area represents the connecting link between western dry and eastern humid zone of Rajasthan.

MATERIAL AND METHODS

To achieve the objectives of the present work, intensive and extensive botanical exploration tours were conducted in the area under investigation for 10 years in different seasons so as to collect more and more plant species in flowering and fruiting state for better understanding of flora and floral composition of this area. Excursions were undertaken two to four times a week, one of which was usually of longer duration. The special attention was paid on the collection of ephemerals, which complete their life – cycle within a few days and disappear with the onset of drier conditions. Field trips were arranged in such a way so as to cover all the localities at more or less regular intervals. All the specimens collected were serially numbered. The field notes including habit, habitat, colour of flower, associations and other pertinent features which generally cannot be studied from the pressed specimens were recorded. Local names and uses of almost all the species were also recorded. Efforts were made to identify the plants from the fresh material, those which could not be satisfactorily identified in the field or in the laboratory at Bhilwara, were preserved and later checked and authenticated at Herbaria of Central National Herbarium, Howrah and Botanical Survey of India, Jodhpur.

Floristic survey

The rainy season was the best time and attempts were made all these years to visit as many places as possible during this favourable season. The other time suitable for collection was near about October-December when plants of angiosperm were in their full bloom. Besides these, the aquatic plants could be studied as by this time the turbulent water of the rainy season settled and allow the aquatic vegetation to come up. Flowering stages of some trees also could be collected in this season. Many trees, which do not bear flowers and foliage together, blossom in spring were also recorded.

Long before the spring season, the collected water in the reservoirs is consumed for irrigational purposes or otherwise evaporated, the land exposed exhibits some vegetation if not already utilized for the winter crop, which too is full of weeds.

Ethnobotanical survey

To achieve the goal, intensive and extensive ethnobotanical survey and interviews were conducted with the tribal people and local *Vaids* who are in touch with forest environments. Generally tribals, who know about the medicinal plants, do not want to give all the informations because they believe that if medicinal properties of a plant are disclosed, then that plant will lost medicinal value for ever. They also believe that this knowledge is exclusively for their community and not for outsiders. For these reasons the informations collected from the tribals is an important aspect of ethnobotanical study. The peoples who can provide informations about medicinal plants include experience medicine men and women, elders, birth attendants, *bhopa*, woodcutters, shepherds and headmen of the community. These have been consulted and interviewed for authenticity, collected informations were cross examined at different places through different informants and in different seasons.

Each of the plant species recorded has been collected with the help of the informants and photographed. Species were identified with the help of reputed flora (Sharma & Tiagi, 1979, Shetty & Singh, 1987-1993, Bhandari, 1995). Voucher specimens were deposited in the Herbarium, Department of Botany, MLV Government College, Bhilwara.

VEGETATION

A general survey of the vegetation was conducted for the following different areas –

(a) Urban areas

(b) Hilly areas

(c) Agriculture fields

(d) Plain areas.

Urban areas

In urban areas, trees are mainly recorded at road sides, in gardens, open grounds and near houses. Different cities of the area were surveyed at different times of the year. The tree flora of urban area is represented by *Acacia leucophloea, Azadirachta indica, Bombax ceiba, Cassia roxburgii, C. siamea, Dalbergia sissoo, Ficus benghalensis, F. religiosa, Leucaena latisiliqua, Polyalthia longifolia, Pongamia pinnata, Prosopis juliflora* and *Tamarinds indica.*

Hilly areas

A survey of various hills of Ajmer, Rajsamand, Bhilwara and Chittorgarh distrcts was conducted for tree flora of hilly areas. The majority of hills are almost Barran, however, at places vegetation comprises of trees like, *Anogeissus pendula, Boswellia serrata, Maytenus emarginata, Sterculia urens, Wrightii tinctoria* etc. Shrub species like *Dichrostachys cinerea, Grewia tenax* have also been observed.

Agricultural fields

The important trees of agriculture fields are *Acacia leucophloea, Acacia nilotica, Azadirachta indica, Butea monosperma, Mangifera indica, Moringa oleifera, Phoenix sylvestris, Prosopis cineraria, Psidium guajava* L., *Ziziphus mauritiana*. Shrubs like *Capparis sepiaria, Clerodendrum phlomidis, Euphorbia caducifolia, Kigelia africana, Lantana camara* and *Pedilanthus tithymaloides* are commonly growing as hedge shrubs along the boundries of agricultural fields.

Plain areas

The localities of plain areas are represented by trees like, *Acacia leucophloea, Prosopis juliflora, Holoptelea integrifolia, Butea monosperma, Acacia senegal, Wrightii tinctoria*, and *Phoenix sylvestris*. Among the shrubs and bushes, *Capparis decidua, Lantana camara, Dichrostachys cinerea, Capparis sepiaria, Grewia flavescens* and *Ziziphus nummularia* are the common taxa in these areas.

VEGETATION TYPES

Seven main series of natural vegetation have been categorized in the area under investigation:

1. ***Anogeissus pendula* Series:** This is the most wide ranging series found almost all over Aravalli ranges of this area. Other taxa of this series are *Acacia leucophloea, A. nilotica, Balanites aegyptiaca, Butea monosperma, Dichrostachys cinerea* etc.
2. ***Butea monosperma- Acacia leucophloea* series:** This series occurs in the south western part, northern part of Bhilwara and Chittorgarh districts. Common taxa of this series are: *Azadirachta indica, Balanites aegyptiaca, Butea monosperma. Calotropis gigantea, Capparis decidua, Prosopis cineraria, Lantana camera* etc.
3. ***Maytenus emarginata - Acacia senegal* Series:** This is the most characteristic of hill fringes and foothills occurring in western part of Bhilwara and Ajmer districts. This series contains major taxa viz. *Anogeissus pendula, Cordia dichotoma, Dichrostachys cinerea, Diospyros montana, Grewia flavescens, Indigofera oblongifolia, Rhus sinuata, Wrightia tinctoria,* etc.
4. ***Prosopis cineraria – Crotalaria burhia* Series:** This is characteristic of the vast area. The vegetation is extremely sparse with scattered trees and bushes. This commonly occurs in all areas of Ajmer, Bhilwara and Rajsamand districts. Other plant taxa in this series are *Acacia tortilis* ssp. *raddiana, Aerva javanica, Capparis decidua, Calotropis procera, Ziziphus nummularia* etc.
5. ***Acacia leucophloea – Capparis decidua* Series:** This series occurs in Ajmer, Bhilwara and Rajsamand districts and chiefly contains the taxa like - *Acacia nilotica, Aegle marmelos, Anogeissus pendula, Balanites aegyptiaca, Butea monosperma, Heliotropium indicum, H. subulatum, Prosopis juliflora* etc.
6. ***Boswellia serrata - Lannea coromandelica* series :** This is the most prevalent type of natural vegetation in the southern part of the area and contains large number of species in growing stock, the common ones of which are *Boswellia serrata, Lannea coromandelica, Wrightia tinctoria* and *Butea monosperma* etc.
7. ***Tectona grandis Series***: This type of series restricted particularly to the southern part (Bari Sadari) of the study area. The common associates of *Tectona* are *Boswellia serrata, Anogiessus latifolia, Lannea coromandelica, Acacia catechue, Soymida febrifuga.*

8. ***Grasslands*** - These occurs almost throughout area. Different types of grasses are found depending upon the soil types, along with isolated trees. Some common grasses are as *Apluda mutica, Cenchrus ciliaris, Cymbopogon martinii, Dichanthium annulatum, Heteropogon contortus, Sehima nervosum, Sorghum halepense* etc.

In the present work *Calotropis acia, Cleome burmannii, Polygala umbonata, Senna uniflora* and *Solanum siysmbriifolium* have been recorded as new species from Rajasthan, as these have not been described by any of the earlier workers.

The vegetation of the area has been classified as "mixed deciduous type. This area lies in the south central part of aravalli ranges which is assumed to be of ecological significance as it shares the flora of western arid and south-east humid zone of Rajasthan. Hill tops are practically barren while on the slopes, where some sand and moisture accumulate, thick growth of plants such as *Acacia leucophloea, A. senegal, Balanites aegyptiaca, Maytenus emarginata, Prosopis cineraria, Anogeissus pendula, Wrightia tinctoria* etc. has been recorded. Among the perennials and shrubs, *Capparis decidua, Barleria prionitis, Leptadenia pyrotechnica, Crotalaria burhia, Ziziphus nummularia, Dichrostachys cinerea, Rhus sinuata* are common. A thick carpet of annuals was recorded during the favourable rainy season. These are ephemerals which complete their life cycle before the onset of drier conditions. Some perennials also sprout during this favourable season.

The most disastrous role in the area of present investigation is played by the erratic rainfall. The average annual rainfall can be compared with some parts of the gangetic plain but it is rendered ineffective due to the long intermittent dry spells of drought often as long as two to three weeks. Amongst the annuals, only those species occur which are able to complete their life cycle within a very short moist period. Amongst the perennial species only those which are suitably adapted to these conditions, occur.

The extreme variations in the diurnal as well as seasonal temperature makes it very difficult for the vegetation to successfully adapt itself to above conditions. Therefore, the plants capable of either adapting themselves to high or low temperatures and discouraging conditions of rainfall, are found. This perhaps is the reason for a very low percentage of tree species.

Whatever vegetation grows in spite of such difficult and discouraging conditions of climate, gets a detrimental treatment at the hands of biotic factors. Tree species are excessively lopped for fodder, fuel and other purposes by men. The domestic animals like, camel, goat and sheep are the worst enemies of the vegetation.

Ecological Classification of Vegetation

For the sake of convenience of study, the vegetation of the area has been classified on the basis of habitats as vegetation of (i) waste lands (ii) Hilly tracts (iii) Grasslands & cultivated areas (iv) Aquatic habitats.

(I) Wasteland vegetation

The vegetation of the wasteland areas is semi-xerophytic with sparse tree layer. The common trees of this habitat are: *Acacia leucophloea, A. nilotica, Balanites aegyptiaca, Butea monosperma, Mangifera indica, Maytenus emarginata, Anogeissus pendula, Holoptelea integrifolia, Prosopis cineraria, Phoenix sylvestris, Salvadora persica, Ziziphus mauritiana* etc. The progressive regeneration of most of the trees, both by seeds and coppice, is negligible due to great biotic pressure on them. *Prosopis juliflora* is the only species which shows highest survival percentage under adverse climatic and biotic conditions.

Besides these, a large number of annual and perennial herbs are encountered in the waste lands. With the advent of the winter season a considerable number of compositaceous meadow herbs, in association with *Acalypha ciliata, Achyranthes aspera, Aregmone mexicana, Leucas cephalotes, Solanum nigrum* etc. appear and bloom.

(II) Vegetation of Hilly Tracts

The hilly regions of the area are comprising of subtropical deciduous type of vegetation with species like *Anogeissus pendula, Acacia leucophloea, Boswellia serrata, Balanites aegyptiaca, Diospyros montana, Wrightia tinctoria* which support some seasonal (annual) vegetation during rains. Bushes like *Adhatoda zeylanica, Plumbago zeylanica* and shrubs like *Barleria prionitis* and *Grewia flavescens* are commonly observed at higher elevations. The middle zone is dominated by *Anogeissus pendula, Maytenus emarginata*, associated with *Lannea coromandelica, Wrightia tinctoria* etc. At the basal zone *Acacia nilotica, Dichrostachys cinerea, Butea monosperma, Prosopis cineraria, Phoenix sylvestris* etc. grow almost in equal proportions.

(III) Grasslands and Cultivated areas

Grasslands

The grasslands, managed for grass production and subsequent grazing as well as to check the exposure of soil to the action of rains and high velocity winds, are few. They are usually managed on the outskirts of the forests, on naked hillocks, in the degraded forests, protected forests, waste lands etc. The grasslands of such habitats are dominated by tall grasses like *Apluda mutica, Aristida adscensionis, Cenchrus ciliaris, Chloris virgata, Cymbopogon martinii, Dichanthium annulatum, Heteropogon contortus, Pennisetum hohenackeri, Sehima nervosum* etc.

Vegetation of areas under cultivation (Weeds)

A large number of weeds grow with the crop plants. Most of them are well equipped for dissemination by wind, water, main and animals. The highest weed density may be observed during the month of January and February of winter season. The typical weeds of the winter crops are; *Ageratum conyzoides, Anagallis arvensis, Aregmone mexicana, Asphodelus tenuifolius, Chenopodium album, Cynodon dactylon, Euphorbia dracunculoides, Fumaria indica, Lathyrus aphaca, Lepidium sativum, Oxalis corniculata* etc. *Asphodelus tenuifolius* is a common geophytic herb of these areas.

(iv) Vegetation of aquatic habitats

The area under investigation is devoid of any permanent streams and rivers except chambal which flows through small part of Chittorgarh. There are seasonal rivers whose water may stay beyond the rainy season in temporary tanks, ponds, puddles, anicuts etc. The water is gradually utilized for irrigating the fields and the bottom area thus exposed of these reservoirs is utilized for agriculture. Therefore, a permanent aquatic vegetation is absent except some deep ditches inside the area of the reservoirs. The aquatic vegetation may be free floating, submerged or anchored on marginal belts.

Free floating and submerged vegetation consists of members of the family Hydrocharitaceae such as *Hydrilla verticillata*, *Vallisneria spiralis* and members of family Potamogetonaceae - *Potamogeton perfoliatus*. Species of *Lemna* and *Wolffia* form scum on water and in places devoid of nitrogenous matter. *Sagittaria guayanensis*, *Vallisneria spiralis*, are the common rooted water plants. The common plants of marshy habitats are: *Aeschynomene indica, Caesulia axillaris, Cyperus rodundus, Ipomoea aquatica, Scirpus roylei, Typha angustata, Echinochloa colonum* and *Phoenix sylvestris* (an arborescent species).

The details of family-wise break up, number of genera and species under each family have been given in table 1 which indicate that 685 species under 415 genera belonging to 116 families occur in the area under investigation. Of these, 541 species under 325 genera and 91 families are of dicots and 144 species under 90 genera and 25 families belong to monocots.

Table 1. Proportional relationship of dicotyledonous and monocotyledonous taxa in south central Rajasthan.

Group	Families		Genera		Species	
	No.	%	No.	%	No.	%
Dicots	91	78.45	325	78.31	541	78.98
Monocots	25	21.55	90	21.69	144	21.02
Total	116	100%	415	100%	685	100%

The ratio of the total number of genera to species is 1:1.65 which is rather low in comparison to a corresponding ratio for whole of India (1:7), Gangetic plain region (1:2.2) and whole of Rajasthan 1 : 2.4 (Shetty and Singh, 1987-93) and higher than the Rajasthan desert (Bhandari 1:1.19), but is more or less in conformity with the ratio for Delhi state 1:1.63 (Maheshwari, 1963).

The only Gymnosperm collected from the area was *Ephdra ciliata* Fisch. Mayer ex. C. A. Mayar. of family Gnetaceae.

It is also interesting to note from Table 1 that the percentage occurrence of the indigenous genera (78.31%) and species (78.98%) of dicotyledons are more or less identical. Similarly, this is also evident in the monocotyledonous genera and species.

The present study indicates that the dicots have been represented by higher number of families, genera and species in comparison to the monocotyledons. Poaceae has been recorded as the largest family of monocotyledons in this area, which is represented by 67 species, on the other hand Papilionaceae has been recorded to be the largest family of dicots with 53 plant species. Families of monocotyledons like Agavaceae, Alismataceae, Arecaceae, Musaceae and Pandanaceae are poorly represented with 1 genus and 1 species each. Among-the dicotyledons, Aristolochiaceae, Balanitaceae, Balsaminaceae, Basellaceae, Casuarinaceae, Ceratophyllaceae, Dioscoreaceae, Elatinaceae, EriocaAulaceae, Fumariaceae, Lentibulariaceae, Loranthaceae, Malpighiaceae, Moringaceae, Nelumbonaceae, Pandanaceae, Potamogetonaceae, Santalaceae, Simaroubaceae, Trapaceae, Typhaceae, Urticaceae and Violaceae are the families represented in this area with 1 genus and 1 species.

Table 2 shows a comparison of ten dominant families occuring in south central Rajasthan with that of W. Rajasthan, Delhi state, Gangetic plain region, Rajasthan, Rajasthan desert, Saurashtra and Rajasthan (South & South-east) given in order of the frequency of species. There is a general resemblance of the flora of these areas, as far as the ten dominant families are concerned, although the frequency of species amongst these ten families varies in different areas. However, the grasses and Legumes occupy the formost place in all these regions.

Families Asteraceae and Cyperaceae take up the next positions but the former has more species in all parts of Rajasthan and Delhi state, whereas Cyperaceae is more dominant in Gangetic plains. So far as family Acanthaceae is concerned, it occupies seventh position in the area under investigation, and Gangetic plains. But in W. Rajasthan it stands at 8th and it occupies the 5th position in Delhi, Saurashtra and South & South east Rajasthan.

Vegetation of area under investigation is of mixed deciduous type dominated by *Anogeissus pendula* Edgew., associated with *Acacia leucophloea* (Roxb.) Willd., *A. nilotica* (L.) Wild. ex Del., *Acacia senegal* (L.) Willd., *Butea monosperma* (Lam.) Taub., *Dichrostachys cinerea* (L.) Wight & Arn., *Wrightia tinctorea* (Roxb.) R. Br. etc. with some stunted, thorny or prickly shrubs and perennial herbs capable of drought resistance. The ephemerals come up during rainy season, complete their life cycle before the advent of winters and subsequently major part of the area is transformed into desolate and barren. Ephemerals, which constitute the bulk of vegetation of the area, appear suddenly above the ground just after showers and complete their life cycle in an incredibly short time. They dry up as soon as the soil gets dry or perennate by underground stems.

BIOTIC FACTORS

A large number of plants and animals have little future, unless immediate steps are not taken to arrest the causes leading to biological impoverishment. It is estimated by the IUCN that about 20,000-25,000 of the world's vascular plants are dangerously rare or under threat. Further a list of threatened and endemic plants of the countries has been compiled by the IUCN (1976).

Biotic factors play an important role in the denudation of forest and scantier vegetation of this region. It is believed that the geological events brought about warmer and drier climates in the adjoining western districts of Rajasthan and consequently the people took to nomadic life.

Vast areas of the forest were used for fire wood, much beyond their natural reparative power. The herds of goat and sheep of the nomadic tribes used up whatever natural vegetation existed. It has now been recognized that though in the desert, aridity originated by geological events but the man and his hungry herds of grazing animals have also contributed a great deal in accentuating these conditions down the ages. In the area under investigation, a large number of herds of goat, sheep and camel enter during winter and summer seasons from *Marwar* region of the state.

Table 2. Synopsis of the indigenous flora-ten dominant families of south central Rajasthan and adjoining regions (Family Leguminosae here includes the taxa of all three families Fabaceae, Caesalpiniaceae & Mimosaceae taken together)

S. N.	South central Rajasthan (Present work)	W. Rajasthan (M.M. Bhandari, 1978)	Delhi (Maheshwari, 1963)	Gangetic plains (Hooker, 1907)	Rajasthan (Jain, 1970)	Rajasthan desert (Blatt. E Hallb. 1919)	Saurashtra (Santapau, 1953)	Rajasthan (South & South-east) (Tiagi & Aery, 2007)
1.	Leguminosae	Poaceae (Gramineae)	Poaceae (Gramineae)	Poaceae (Gramineae)	Poaceae (Gramineae)	Poaceae (Gramineae)	Leguminosae	Leguminosae
2.	Poaceae (Gramineae)	Leguminosae	Leguminosae	Leguminosae	Leguminosae	Leguminosae	Poaceae (Gramineae)	Poaceae (Gramineae)
3.	Asteraceae	Asteraceae	Asteraceae	Cyperaceae	Asteraceae (Compositae)	Asteraceae (Compositae)	Asteraceae (Compositae)	Asteraceae (Compositae)
4.	Cyperaceae	Cyperaceae	Cyperaceae	Asteraceae (Compositae)	Cyperaceae	Cyperaceae	Malvaceae	Cyperaceae
5.	Malvaceae	Convolvu-laceae	Acanthaceae	Scrophu-laraceae	Acanthaceae	Convolvu-laceae	Acanthaceae & Convolvulaceae	Acanthaceae
6.	Euphor-biaceae	Malvaceae	Euphorb-iaceae	Malvaceae	Euphorb-iaceae	Malvaceae	Cyperaceae	Euphorbi-aceae
7.	Acanthaceae	Euphorbi-aceae	Convolvu-laceae	Acanthaceae	Malvaceae	Euphorb-iaceae	Rubiaceae & Solanaceae	Lamiaceae (Labiatae)
8.	Convolvu-laceae	Acanthaceae	Malvaceae	Euphorbi-aceae	Lamiaceae (Labiatae)	Acanthaceae	Cucurbitaceae & Lamiaceae	Malvaceae
9.	Amaran-thaceae	Cucurbit-aceae	Amarant-haceae	Convolv-ulaceae	Convolv-ulaceae	Cucurbit-aceae	Scrophular-aceae	Rubiaceae
10.	Lamiaceae	Amaranth-aceae	Scrophu-laraceae	Lamiaceae (Labiatae)	Scrophula-raceae	Amarant-haceae	Amaranthaceae, Tiliaceae & Asclepiadaceae	Convolv-ulaceae

These create the major loss to the natural biodiversity of the area of present study. The devastating propensities of goat and camel have been beautifully summarized in a Marwari proverb:

"*Oont Chhode Akaro, Bakri Chhode Kankro*".

Meaning thereby that the camel will eat everything except *Calotropis* spp. but the goat will devour even that, leaving only the pebbles. Therefore, continuous grazing has almost completely prevented natural regeneration.

Besides these, some exotic plants such as *Lantana camara*, *Parthenium hysterophorus* and *Prosopis juliflora* also affect the local biodiversity Some of these have allelopathic effects while others are fast growing and compete with the native angiospermic flora. These three plant species have changed the scenario of vegetation of area under investigation.

As a consequence of environmental degradation, certain plants such as *Commiphora wightii* which were once abundant in the south central area are now under threat of extinction.

BIOLOGICAL SPECTRUM

A community consists of many species with different kinds of population fluctuations and interaction with one-another. The physiognomic forms and structural studies of land communities have been discussed in more rewarding manner by Whittaker (1975). Based on the Raunkiaer (1934) and modified by Clapham (1935) and Cain (1950), the percentage distribution of species among the life forms of the flora of south central region (i.e. biological spectrum) is grouped into the following categories.

Table 3. Biological spectrum of Flora of South central Rajasthan

S.No.	Life forms	Category	Percentage
1.	Phanerophytes	(Ph.)	16.00
2.	Nanophanerophytes	(N.)	20.32
3.	Chamaephytes	(Ch.)	04.32
4.	Hemicryptophytes	(H.)	02.52
5.	Geophytes	(Ge.)	05.76
6.	Hydrophytes	(HH.)	07.02
7.	Therophytes	(Th.)	42.99
8.	Parasites	(P.)	01.07

ECONOMIC AND MEDICINAL VALUE OF PLANTS

FOOD

Southern part of the area under investgation is tribal dominated. *Bheel* and *Meena* are the main tribes of this area. The tribal agriculture in Rajasthan is not a traditional one, having originated in the later half of the last century. Before this, tribals like *Bheel* were satisfying their hunger from the flash of the game animals and wild plant food, these people passing through the stage of slash and burn cultivation have eventually come to depend on settled and skilled agriculture. The antiquity of the cultivated crops in this region, such as Maize, Jowar is associated with Banas culture of "Ahar" (Udaipur district) estimated 2100 : 1300 BC, today a famous theme for maize is:

"*Makki ko roto khani, Aur mewar chhorke ni jani.*"

The maize continues to be the staple food of all the tribes in this region. Difficulties in agriculture are many. The undulatory and rocky topography, inadequate irrigation, fragmented and small land holdings are a few of them.

Besides cash crops, an inclination towards raising fruit crops (viz., *Carica papaya, Citrus medica, Mangifera indica, Musa paradisiaca, Psidium guajava* L., *Syzygium cumini* &

Ziziphus mauritiana in orchards for subsidiary income has been noticed. As regards animal food and food habit of tribals need a special mention. They are extremely fond of the flash of game animals, birds and fishes that are sought by the all tribes.

In general the tribals are non-vegetarian but religious sects that have come up, like the "*Bhagat*" Bheels have given up the use of liquor and animal flesh.

WILD FAMINE FOOD PLANTS

Different plants and plant parts used during famine are given below:

1. *Acacia leucophloea,* family - Mimosaceae - stem bark and crown of stem, are powdered for making bread.
2. *Aegle marmelos,* family - Rutaceae - fruits supplement diet during famine.
3. *Bauhinia racemosa,* family - Caesalpiniaceae - the powdered stem bark is used for augmenting flour during famine.
4. *Ficus benghalensis,* family - Moraceae - receptacles mixed with maize or wheat flour are rolled into bread in scarcity.
5. *Ficus racemosa,* family - Moraceae - during scarcity, flour is supplemented with the powdered receptacles or its stem bark.
6. *Ficus religiosa,* family - Moraceae - receptacles used as in *Ficus benghalensis.*
7. *Madhuca indica,* family - Sapotaceae - flowers and fruits collected and stored in storage bins. In scarcity they are taken out, baked and eaten. During famine the stem bark is boiled to make a sort of broth called *rab* and consumed.
8. *Mucuna pruriens,* family - Fabaceae - Fruits consumed in scarcity. *Terminalia bellirica,* family - Combretaceae - seeds used during acute scarcity condition. Over eating of course, causes intoxication.
9. *Ziziphus nummularia,* family - Rhamnaceae - Fruits are dried and stored in sufficient quantity, removing the seeds they are also powdered and pressed into laddoos, used in scarcity.
10. *Cyperus rotundus,* family - Cyperaceae - bulbous tubers are edible.
11. *Phoenix sylvestris,* family - Arecaceae - the upper meristematic part of stem is eaten raw. The ripe fruits fresh as well as dried are consumed.
12. *Urgenia indica,* family - Liliaeae - the bulb is consumed during scarcity.

TIMBER YIELDING PLANTS

(a) Hard wood timbers: *Acacia nilotica* ssp. *indica* is the prized timber of the study area. The other hard-wood species are *Acacia leucophloea, A. catechu, Albizia lebbek, Anogeissus latifolia, A. pendula, Dalbergia latifolia, D. sissoo, Diospyros melanoxylon, Madhuca indica, Prosopis cineraria, Salvadora oleoides, S. persica* and *Tamarindus indica.*

(b) Soft-wood timbers: The major softwood timber species witnessed in study area are *Ailanthus excelsa, Bombax ceiba, Boswellia serrata, Butea monosperma, Holoptelea integrifolia, Mangifera indica, Mitragyna parvifolia, Phoenix sylvestris, Syzygium cumini, Ziziphus mauritiana* etc.

(c) Bamboo (*Bambusa arundinacea*) is in great demand for various uses in rural as well as urban areas for making houses, huts, agricultural practices. Importance wise it is next to *Acacia* wood in the area.

AGRICULTURAL IMPLEMENTS

Ploughing and sowing

Acacia leucophloea, Acacia nilotica, Bambusa sp., *Butea monosperma, Diospyros melanoxylon, Holoptelea integrifolia, Madhuca indica, Phoenix sylvestris, Tectona grandis, Wrightia tinctoria* and *Ziziphus mauritiana* are common taxa used for agricultural equipment.

FIRE WOOD AND CHARCOAL

Wood of *Acacia nilotica* ssp. *indica, Anogeissus latifolia, A. pendula, Butea monosperma, Diospyros melanoxylon, Phoenix sylvestris, Prosopis cineraria, Salvadora oleoides, S. persica, Sterculia urens, Syzygium cumini, Tamarindus indica* etc. are used as fire wood. Recently introduced species *Prosopis juliflora* is widely used in fire wood as well as for producing charcoal.

AROMATIC PLANTS

Rhizomes of *Curcuma aromatica*, heartwood and underground portions of *Santalum album*, leaves of *Mentha arvensis* L., *Murraya koenigii*, Flowers of *Jasminum multiflorum, J. sambac, Rosa gallica* L., *Tagetes erecta* L., *T. patula* L. etc. yield aromatic compounds.

DYE YIELDING PLANTS

Acacia catechu, Acacia leucophloea, Acacia nilotica ssp. indica, Arnebia hispidissima, Butea monosperma, Cocculus pendulus, Erythrina suberosa, Indigofera tinctoria, Lawsonia inermis, Nyctanthes arbor-tristis, Striga gesnerioides, Wrightia tinctoria etc.

TANNIN YIELDING PLANTS

Acacia catechu, Acacia nilotica ssp. *indica, Albizia lebbek, Anogeissus pendula, Bauhinia racemosa, Cassia fistula, Lannea coromandelica, Phyllanthus emblica, Tamarindus indica, Ziziphus nummularia* and *Z. xylopyrus*.

GUM-RESIN YIELDING PLANTS

Acacia catechu, Acacia nilotica ssp. indica, Acacia senegal, Anogeissus latifolia, Azadirachta indica, Boswellia serrata, Butea monosperma, Lannea coromandelica, Prosopis cineraria, Sorghum bicolor (L.) Moench and *Sterculia urens*.

FIBER YIELDING PLANTS

Abelmoschus moschatus, Abutilon indicum, Agave americana, Bombax ceiba, Butea monosperma, Crotalaria juncea, Hibiscus cannabinus, Helicteres isora, Holoptelea integrifolia, Leptadenia pyrotechnica, Phoenix sylvestris and *Sacchrum bengalense*.

VICES IN CULTURE

In the life of a member of the tribal world, intoxication is not only essential but of paramount importance, the only exceptions being the adherents of the "*Bhagat movement*" who have rejected this alongwith animal flesh.

Liquor

Liquor prepared from *Mahuwa* is the most prevalent intoxicant. The flowers of *Madhuca indica* - Fresh or dried, comprise the raw material of the local liquor called *Mahuri* or *Mahudi* or *Daru*.

Technique

In a covered earthen pot, the flowers and fruits of *Madhuca indica* are allowed to ferment in water for about a week till the characteristic odour is detected. The pot may at times be burried in the ground. When ready, the mixture is distilled.

In its simple form, the apparatus comprises of two pitchers each with a clay pot having a hole, placed upside down on its mouth. The pitcher holding the fermented flowers is placed over stones and allow to boil, the other is immersed in water. Using a hollow bamboo stem, connection is established through the holes of the inverted pots (Instead of the bamboo pipe, flower stalks or leaf petioles of *Nymphaea* or a hollowed and dried fruit of *Lagenaria siceraria* or PVC tube is being used). The apparatus is made air tight by plastering with clay.

Two and a half to three Kg. flowers are said to yield three to four bottles of *Mahuri*. For better quality production, sometimes double distillation is done. Plant ingredients as bark of *Acacia leucophloea, A. nilotica and Ziziphus nummularia* are added to flowers of *Madhuca indica*. This improve flavour, promote early fermentation, enhance the degree of intoxicity and prevent spoilage.

Jaggery Liquor

The liquor is also prepared by the fermentation and distillation of *Jaggery* ("*Gur*"). This liquor is coloured and considered inferior to *mahuri* which is of a dirty colour, and is sweet in taste.

Ganja and Bhang

Bhang (dried leaves, flowering shoots of male and female plants of *Cannabis sativa* L.) and *Ganja* (unfertilized inflorescences of female plants *C. sativa*) are more commonly used. The former is taken with sweets or in drink, the latter is smoked with tobacco.

The plants occur naturally as well as in cultivation.

Tobacco

Throughout the tribal and rural population the tobacco (*Nicotiana tabacum* L.) smoking habit is rampant. Men, women and even the children alike are fond of tobacco. Tobacco is smoked in the following ways:

1. Bidi - Tobacco is rolled up in the leaves of *Diospyros melanoxylon.*
2. Chilam - It is a smoking pipe made of clay or wood. When chilam is not available fresh leaves of *Calotropis procera* are fashioned into disposable *Chilams.* After filling them with tobacco they are smoked.
3. The tobacco is mixed with lime and chewed.

The customary offering of tobacco is a sign of hospitality or brotherhood.

Masticatories

1. *Tamarindus indica* - The seeds of this tree are kept in mouth and chewed for hours especially by children and womenfolk.
2. Paan - Chewing of paan or the leaves of *Piper betle* smeared with lime and catechu (product of *Acacia catechu* wood) with nuts of *Areca catechu* is also popular among the tribals. The two types of paan are chewed, (i) Mitha paan (ii) Tobacco paan. The nuts of *Areca catechu* may be chewed alone. Perhaps the use of paan in ceremonies is a result of Hindu contact and influence.

HERBS AND HEALING

Rajasthan which can even be called as a rich storehouse of medicinal plants as it covers varied geographical areas, phytogeographical regions like mixed deciduous forests, hills of Aravalli and dry Desert areas. The history of medicine and surgery dates back perhaps to the origin of the human race itself. The references of curative properties of some herbs in 'Rigveda' (3500-1800 BC) though in brief, seem to be the earliest record of use of plants in medicine. But the tribals treat various ailments through regular practices using plants from their surrounding environment. The area of present investigation is inhabitat by different tribal communities. These communities and villagers have strong belief in their traditional healing practices. A large number of plants or plant parts are being used in the treatment of various diseases by them.

This traditional knowledge is being transferred from generation to generation only through oral folklore. Therefore, an attempt has been made to document various plants, plant parts used as medicine. The observations recorded durring the present study are presented below:-

(a) **Abortifacients:** Roots of *Achyranthes aspera, Boerhavia diffusa, Gloriosa superba,* bark of *Bombax ceiba*, fruits of *Carica papaya*, seeds of *Trigonella foenum-graecum.*

(b) **Antiasthmatics:** Whole plant of *Triumfetta rhomboidea*, Roots of *Boerhavia diffusa, Curculigo orchioides, Vitex negundo,* stem, leaf and seed of *Baliospermum montanum*, bark and leaf of *Acacia catechu*, leaf of *Adhatoda zeylanica*, *Tylophora indica*, flowers of *Madhuca indica,* fruits of *Solanum virginianum* and seeds of *Achyranthes aspera.*

(c) **Antidiabetic:** Whole plants of *Catharanthus roseus*, roots of *Helicteres isora*, flowers of *Pongamia pinnata*, seeds of *Syzygium cumini*.

(d) **Antidysenteric:** Roots of *Calotropis procera, Cissampelos pareira* var. *hirsuta*, bark of *Murraya koenigii*, rhizomes of *Curculigo orchioides*, fruits of *Helicteres isora*, leaf of *Alangium salvifolium*, seeds of *Terminalia bellirica* and gum of *Acacia nilotica, Butea monosperma*.

(e) **Antirheumatic:** Roots of *Achyranthes aspera, Gloriosa superba, Urena lobata*, root and leaf of *Boerhavia diffusa*, leaf of *Madhuca indica, Ricinus communis*.

(f) **Antisterility:** Roots of *Chlorophytum tuberosum*.

(g) **Boils:** Roots of *Abutilon indicum, Asparagus racemossus*, fruits of *Ziziphus nummularia*.

(h) **Cold:** Roots of *Abrus precatorius*, leaves of *Annona squamosa, Eucalyptus camaldulensis, Ziziphus nummularia*, flowers of *Butea monosperma, Madhuca indica*.

(i) **Cough:** Whole plant of *Euphorbia hirta*, roots of *Echinops echinatus, Tylophora hirsuta*, leaves of *Adhatoda zeylanica, Barleria prionitis, Hygrophila auriculata*. flowers of *Madhuca indica*, fruits of *Solanum virginianum, Tribulus terrestris* and seeds of *Achyranthes aspera*.

(j) **Fever:** Whole plants of *Enicostema hyssopifolia axillare, Evolvulus alsinoides, Leucas aspera*, leaves of *Peristrophe paniculata*, stem of *Tinospora cordifolia*, flowers of *Vitex negundo* and fruits of *Tamarindus indica*.

(k) **Jaundice:** Whole plants of *Trianthema portulacastrum*, rhizome of *Curculigo orchioides*, stem of *Cuscuta reflexa, Tinospora cordifolia*.

(l) **Ophthalmic Disease:** Leaves of *Nicotiana tabacum*.

(m) **Piles:** Whole plant of *Leucas aspera*, leaves of *Chenopodium murale, Launaea procumbens*, fruits of *L. aspera, Plantago ovata*.

(n) **Skin Disease:** Whole plants of *Euphorbia hirta*, stem of *Tinospora cordifolia*, leaves of *Plumbago zeylanica*, seed oil of *Madhuca indica*.

(o) **Snake and scorpion bites:** Root of *Annona squamosa*, fruit pulp of *Aegle marmelos, Cassia fistula*, pedicels and thalamus of fruits of *Cucurbita moschata*, whole plant of *Polycarpaea corymbosa* etc.

(p) **Stomachache:** Gum of *Acacia catechu*, leaves of *Pergularia daemia*, flowers of *Capparis decidua*, fruits of *Helicteres isora*.

(q) **Toothache:** Roots of *Barleria prionitis*,

(r) **Wounds:** Whole plant of *Striga gesnerioides, Tridex procumbens*, bark of *Azadirachta indica*, leaves of *Acacia nilotica, Azadirachta indica, Lepidagathis trinervis, Madhuca indica, Maytenus emarginata* and latex of *Croton bonplandianum*.

KEY TO THE FAMILIES

1a. Flowers arranged in cones. Ovules naked, born on the surface of megasporophylls. Stigma absent. .... .Gnetaceae

1b. Flowers not arranged in cones. Ovules enclosed in megasporophylls. Stigma present.................... 2

2a. Leaves reticulate veined. Flowers 4 - 5-merous. Cotyledons 2. 3

3a. Perianth 2 to several seriate, differentiated into calyx and corolla. 4

4a. Petals all free. 5

5a. Ovary superior. 6

6a. Stamens numerous. 7

7a. Stamens united by filaments in a tube, cup or sometimes several groups. 8

8a. Anthers 1-celled. Stamens polyadelphous.................... Bombacaceae

8b. Anthers 2-celled. Stamens not as above. 9

9a. Terrestrial shrubs. Fruits winged, samaroid. ....Malpighiaceae

9b. Aquatic herbs. Fruits with 2 spines.................... .Trapaceae

7b. Stamens free. 10

10a. Aquatic plants. Leaves rotundate-peltate. Carpels embedded in the fleshy receptacle....11

11a. Leaf base cordate or sagittate. Filaments digitate at base. Carpels connate in many cell ball resembling a syncarpous ovary. .... Nymphaeaceae

11b. Leaf base retuse. Filaments linear. Carpels free embedded in the cavities of spongy receptacle.................... ...Nelumbonaceae

10b. Terrestrial plants. Leaves and carpels not as above.................... 12

12a. Flowers 3-merous.................... .Annonaceae

12b. Flowers 4-5-merous.................... 13

13a. Carpels many, free.................... .Rosaceae

13b. Carpel 1, if more, then united. 14

14a. Flowers unisexual.................... .Euphorbiaceae

14b. Flowers bisexual. 15

15a. Ovary 1-celled. 16

16a. Ovary usually borne on a gynophore. Fruit an indehiscent berry..Capparidaceae

16b. Ovary sessile, Fruit a dehiscent capsule or legume. 17

17a. Fruit a legume.................... *Mimosaceae*

17b. Fruit a capsule. 18

18a. Capsule upto 1 cm long, open at the top. .... .Resedaceae

18b. Capsule more than 1 cm long, dehiscent along valves. 19

19a. Capsule 2-valved. .... .Cleomaceae

19b. Capsule 3-5-valved. 20

20a. Leaves entire or dentate. Capsules 2-3 cm longy. Plants with milky or yellow sap. .... Papaveraceae

20b. Leaves entire. Capsules upto 0.5 cm long. Plants with watery sap. .... .Portulacaceae

15b. Ovary 2 or more celled. 21

21a. Leaves gland dotted. .... .Rutaceae

21b. Leaves not gland dotted. 21

22a. Sepals connate. Petals wrinkled. Stamens arising from the calyx-tube. .... .Lythraceae

22b Sepals free. Petals not wrinkled. Stamens arising from the receptacle. .... Tiliaceae

6b. Stamens as many a s or twice the number of petals or sometimes. 23

23a. Fruit a legume, sometimes lomentum or indehiscent. 24

24a. Corolla actinimorphic. Petals valvate. .... .Mimosaceae

24b. Corolla zygomorphic. Petals imbricate. 25

25a. Corolla papilionaceous, odd petal outermost. Stamens 1 or 2-adelphous. .... Papilionaceae

25b. Corolla not papilionaceous, odd petal innermost. Stamens free .... .Caesalpiniaceae

23b. Fruits various, but not a legume or lomentum. 26

26a. Leaves scale-like, appressed and reduced. .... Tamaricaceae

26b. Leaves well developed. 27

27a. Flowers irregular. 28

28a. Placentation parietal. 29

29a. Leaves 2-3 pinnatisect. Sepals 2. Stamens 6, into 2 bundles. .... .Fumariaceae

29b. Leaves shallowly serrate. Sepals 5. Stamens 5, free .... Violaceae

28b. Placentation axile. 30

30a. Sepals 3, posterior one spurred. Stamens 5. Capsule 5-celled, 5-valved, many-seeded. Seeds estrophiolate .... Balsaminaceae

30b. Sepals 5, not spurred. Stamens 8. Capsule 2-celled, 2-valved, 2-seeded. Seeds strophiolate. .... .Polygalaceae

27b. Flowers regular. 31

31a. Leaves simple 32

32a. Stamens usually tetradynamous. Corolla cruciform. Placentation parietal. .... .Brassicaceae

32b. Stamens, corolla and placentation not as above. 33

33a. Ovary usually 1-celled, atleast when ripe 34

34a. Trees, usually with milky or resin passages in stem. .... .Anacardiaceae

34b. Scandent or twining shrubs or erect to prostrate herbs, without milky resinous sap. 35

35a. Scandent or twining shrubs. Flowers unisexual. Fruit drupaceous, 1-seeded. .... Menispermaceae

35b. Erect or prostrate herbs. Flowers bisexual. Fruit capsule, few to many seeded. 36

36a. Succulent herbs. Flowers surrounded by a whorl of leaves. Sepals or calyx lobes-2. Capsules circumsessile. .... .Portulacaceae

36b. Plants not succulent. Flowers not surrounded by whorl of leaves. Sepals 5. Capsule valvular. .... .Caryophyllaceae

33b. Ovary 2 or more celled. 37

37a. Flowers 1-sexual. .... .Euphorbiaceae

37b. Flowers 2-sexual or intermixed with 1-sexual flowers 38

38a. Plants climbing with tendrils. .... .Vitaceae

38b. Plants erect or prostrate, without tendrils. 39

39a. Leaves opposite. 40

40a. Stipules present. Sepals imbricate. Stamens inserted below the ovary. Style 2-5. .... .Elatinaceae

40b. Stipules absent. Sepals valvate. Stamens inserted on the calyx tube. Style 1. .... .Lythraceae

39b. Leaves alternate. 41

41a. Torus small, not expanded. ....Tiliaceae

41b. Torus thickened or expanded into a fleshy disc 42

42a. Stamens opposite to the petals. Calyx lobes valvate. .... .Rhamnaceae

42b. Stamens alternate with petals. Calyx lobes imbricate .... .Celastraceae

31b. Leaves compound. 43

43a. Ovary 1-celled. .... Moringaceae

43b. Ovary 2 or more celled. 44

44a. Plants climbing with tendrils. (except *Sapindus*) 45

45a. Leaves biternately compound. Flowers irregular. Fruit an inflated capsule. ....Sapindaceae

45b. Leaves ternately or pinnately compound. Flowers regular. Fruit a berry. .... .Vitaceae

44b. Plants erect or prostrate, without tendrils. 46

46a. Herbs or shrublets. 47

47a. Style 1. Fruit schizocarpic with spiny cocci. .... .Zygophyllaceae

47b. Style 5. Fruit capsular without spines. .... .Oxalidaceae

46b. Trees or shrubs 48

48a. Leaves covered with aromatic glandular dots. .... .Rutaceae

48b. Leaves without aromatic glandular dots. 49

49a. Staminal filaments united into a tube or stamens 1-adelphous. 50

50a. Anther 1-celled. .... .Malvaceae

50b. Anther 2-celled. 51

51a. Leaves simple. Hypogynous disc absent. .... .Sterculiaceae

51b. Leaves pinnately compound. Hypogynous disc present. .... .Meliaceae

49b. Staminal filament free or distinct. 52

52a. Flowers polygamous. Fruit samaroid, 1-seeded.. ...... .Simaroubaceae
52b. Flowers bisexual. Fruit a drupe, containing 3, 1-seeded pyrenes...53
53a. Stem with flaky bark. Branches without spines. Fruits fleshy. .... .Burseraceae
53b. Stem with yellowish white bark. Branches with axillary spines. Fruits stony. .... .Balanitaceae
5b. Ovary inferior. 54
54a. Stamens as many as the petals or less. 55
55a. Plants climbers with tendrils. Flowers unisexual.......... .Cucurbitaceae
55b. Plants without tendrils. Flowers bisexual. 56
56a. Plants herbaceous. 57
57a. Petioles of leaves inflated below the apex. Fruits drupaceous, 2-spined, 1-seeded. . .Trapaceae
57b. Petioles not as above. Fruit a capsule, spineless, many seeded. .... .Onagraceae
56b. Plants woody. 58
58a. Flowers white. Fruits wingless. .... .Alangiaceae
58b. Flowers pinkish. Fruits winged. .... .Combretaceae
54b. Stamens more than or twice the number of petals. 59
59a. Succulent plants with jointed phylloclades. Leaves reduced to spines or bristles. .... .Cactaceae
59b. Woody plants with normal stem. Leaves well developed.......... .Myrtaceae
4b. Petals all united, at least the base..........60
60a. Ovary inferior or halfinferior. 61
61a. Leaves exstipulate. Flowers in involucral heads. Anthers syngenesious. Ovary 1-celled. .... .Asteraceae
61b. Leaves stipulate, stipules usually inter or intrapetiolar. Flowers not in involucral heads. Anthers free. Ovary 2 or more celled.......... .Rubiaceae
60b. Ovary superior. 62
62a. Plants either insectivorous with bladder like insect-traps or parasite. 63
63a. Aquatic, leafy, insectivorous herbs with minute bladder-like insect-traps. Flowers irregular. .... .Lentibulariaceae
63b. Terrestrial. Stem parasite, with yellow, leafless much branched climbing stem. Flowers regular.......... .Cuscutaceae
62b. Plants not as above. 64
64a. Placentation basal..........65
65a. Trees. Fruit an indehiscent berry or drupe. .... Salvadoraceae
65b. Herbs. Fruit a circumsessile capsule. .... .Plantaginaceae
64b. Placentation various not basal. 66
66a. Stamens more than the corolla lobes. 67
67a. Flowers bisexual. Corolla campanulate, usually 8 - lobed. Fruit 1 - 2-seeded. .... .Sapotaceae
67b. Flowers unisexual. Corolla unceolate, usually 4-lobed. Fruit 4-8 seeded. .... .Ebenaceae

66b. Stamens as many as the corolla-lobes or fewer. 68

68a. Stamens opposite to the corolla lobes. 69

69a. Calyx with stalked glands. Corolla long tubular. Style 5. .... Plumbaginaceae

69b. Calyx without stalked glands. Corolla rotate. Style 1. .... Primulaceae

68b. Stamens alternate with the corolla lobe. 70

70a. Flowers regular. 71

71a. Leaves opposite. 72

72a. Plants with milky sap. Fruit follicular (except *Carissa* & *Thevetia*). 73

73a. Pollens distict not in pollinia. Styles 2, united from the base forming a common style with a stigma. Gynostegium absent. .... Apocynaceae

73b. Pollens in pollinia. Styles 2, distinct in lower region, united in stagmatic region and bearing a single stigma. Anthers united with the stigma to form a gynostegium 74

74a. Pollen masses granular. Pollen carrier spathulate. .... Periplocaceae

74b. Pollen masses waxy. Pollen carrier slender .... Asclepiadaceae

72b. Plants without milky sap. Fruit not follicular. 75

75a. Stamens 2. .... Oleaceae

75b. Stamens 4-5 76

76a. Anthers connate, forming a cone. Fruit of 4-nutlets. .... Boraginaceae

76b. Anthers free. Fruit capsular. 77

77a. Shrubs. Corolla-lobes imbricate. Placentation axile. .... Buddlejaceae

77b. Herbs. Corolla-lobes contorted. Placentation usually parietal .... Gentianaceae

71b. Leaves alternate. 78

78a. Plants aquatic. Ovary 1-celled. .... Menyanthaceae

78b. Plants terrestrial (except *Ipomoea aquatica*). Ovary usually 2-4-celled. 79

78a. Ovules many in each cell of ovary. 80

80a. Anthers sagittate. Styles 1. Carpels not placed obliquely. Placenta not swollen. .... Hydrophyllaceae

80b. Anthers not sagittate. Styles 2. Carpels placed obliquely. Placenta swollen. .... Solanaceae

78b. Ovules 1-2 in each cell of ovary. 81

81a. Trees. Flowers mostly in panicled cymes. Fruit a drupe. .... Ehretiaceae

81b. Erect, prostrate or twining herbs or shrubs. Fruit a nut of 2-4 nutles or dry hard berry. .... Convolvulaceae

70b. Flowers irregular. 82

82a. Leaves compound. Fruit elongated. Seeds prominently winged (except *Jacaranda*). .... Bignoniaceae

82b. Leaves simple (except *Vitex*) or pinnately lobed. Fruit not elongated. Seeds not winged 83

83a. Stem nodes swollen. Bracts conspicuous. Fruits usually elastically dehiscing. Seeds attached on retinaculae. .... .Acanthaceae

83b. Stem nodes not conspicuously swollen. Bracts minute or absent. Fruits not elastically dehiscing. Seeds without retinaculae. 84

84a. Inflorescence verticillaster. Ovary 4-lobed, separating into 4, 1-seeded nutlets. Style gynobasic. .... Lamiaceae

84b. Inflorescence otherwise. Ovary without lobed, not separating into nutlets, dehiscing variously or indehiscent. Style terminal. 85

85a. Plants viscid hairy. Flowers with glands at the base of padicels. 86

86a. Flowers in terminal racemes. Fruits beetle shaped, 2-horned. .... .Martyniaceae

86b. Flowers solitary axillary. Fruits not as above. .... .Pedaliaceae

85b. Plants not viscid hairy. Flowers without glands at the base of padicels. 87

87a. Ovary cells 1 or 2-ovuled. Fruits usually drupaceous. .... .Verbenaceae

87b. Ovary cells more than 2-ovuled. Fruits usually capsular. 88

88a. Ovary 1-celled. Placentation parietal. .... .Gentianaceae

88b. Ovary 2-celled. Placentation axile. .... .Scrophulariaceae

3b. Perienth 1-seriate, not differentiated into calyx and corolla or absent. 89

89a. Aquatic plants. .... .Ceratophyllaceae

89b. Terrestrial plants. 90

90a. Branches jointed. Leaves reduced to many-toothed sheaths surrounding the nodes. Flowers arranged in cone-like spikes. Fruits crowded into a cone. .... .Casuarinaceae

90b. Branches not jointed. Leaves normal. Flowers not in cone-like spikes. Fruits not as above. 91

91a. Ovary inferior. 92

92a. Seeds without distinct seed-coat, attached to the pericarp of the fruit. 93

93a. Partial stem parasites. Ovary 1-ovuled. .... .Loranthaceae

93b. Partial root-parasites. Ovary 2 to 3-ovuled. .... .Santalaceae

92b. Seeds with distinct seed-coat, not attached to the pericarp of the fruit. 94

94a. Trees. Stamen 10. .... .Combretaceae

94b. Herbs. Stamens 5 or less. .... .Chenopodiaceae

91. Ovary superior. 95

95a. Ovary 2-6 celled. 96

96a. Ovules many in each cell of ovary. 97

97. Style 1. Sepals 4. Stamens 4. .... .Lythraceae

97b. Styles 3-5. Sepals 5. Stamens 5. 98

98a. Leaves unequal in each pair at the nodes, base dilated into a pauch. Tepals gamophyllous.. .... .Aizoaceae

98b. Leaves equal in size, opposite or whorled. Tepals polyphyllous.. ....Molluginaceae

96b. Ovules usually 1-2 in each cell of ovary. 99

99a. Perianth zygomorphic. Ovary inferior, 6-celled........................ .Aristolochiaceae

99b. Perianth actinomorphic. Ovary superior, 3-celled...100

100a. Leaves pinnately compound. Flowers 2-sexual. ....Sapindaceae

100b. Leaves usually simple or lobed. Flowers 1-sexual...................Euphorbiaceae

95b. Ovary 1-celled. 101

101a. Stipules usually ocreate, persistent or rarely deciduous and leaving a circular scar.. .Polygonaceae

101b. Stipules absent if present not ocreate..102

102a. Perianth-segment united into a long tube, petaloid.............................. .Nyctaginaceae

102b. Perianth-segment free or united at the base, sepaloid. 103

103a. Herbs or herbaceous twiners or undershrubs. 104

104a. Twining fleshy herbs. Utricles enclosed in fleshy perianth.................... .Basellaceae

104b. Erect or prostrate herbs or undershrubs. Fruits not enclosed in the perianth.....105

105a. Bracts and tepals scarious. Stamens usually connate below........ . Amaranthaceae

105b. Bracts (if present) and tepals not scarious. Stamens free..................................106

106a. Flowers bisexual. 107

107a. Plants erect herbs. Perianth sepaloid. Anthers erect. Fruits not enclosed in the perianth...Chenopodiaceae

107b. Plants twining herbs. Perianth petaloid. Anthers versatile. Fruits enclosed in fleshy perianth... .Basellaceae

106b. Flowers unisexual..108

108a. Leaves simple, stipulate. Perianth segment 4. Stamens 4. ....Urticaceae

108b. Leaves pinnately compound. Perianth segment 5. Stamens 5...Cannabinaceae

103b. Trees. 109

109a. Fruits samaroid/winged.. .Ulmaceae

109b. Fruits not winged...110

110a. Inflorescence a catkin or hypanthodium. Flowers unisexual, spicate or crowded on or within a fleshy receptacle. Stipules present. ....Moraceae

110b. Inflorescence otherwise. Flowers bisexual, in axillary clusters or in short branched racemes. Stipules absent.. .Flacourtiaceae

2b. Leaves usually parallel-viened. Flowers 3-merous. Catyledon 1. 111

111a. Free floating. Plant body thalloid..Lemnaceae

111b. Terrestrial or aquatic. Plant body differentiated into root, stem and leaves. 112

112a. Monoecious tree-like erect plants with stem formed by imbricate base of petioles. .... .Musaceae

112b. Plants not as above. 113

113a. Inflorescence of heads or of spikelets composed of flowers (florets) in axils of glumes or between 2 glumes. (lemma and palea)............114

114a. Flowers in involucral heads. Perianth present, usually 2-seriate. Ovary 3-celled.. Eriocaulaceae

114b. Flowers in spikelets. Perianth absent or represented by bristles, scales or lodicules. Ovary 1-celled...115

115a. Stem terete, noded, usually internodes hollow. Leaf-sheath splitting (open). Flowers enclosed between 2-glumes (lemma & palea). Fruit caryopsis. ....Poaceae

115b. Stem angled, usually nodless. Leaf-sheath not splitting (closed). Flowers subtended by a single glume. Fruit an achene.. .Cyperaceae

113b. Inflorescence various, but not of heads or spikelets............................116

116a. Climbing herbs or shrubs...117

117a. Leaf tip circinate or leaves reduced to spinescent scales, with needle like cladodes in their axils or leaves like grasses. Flowers 2-sexual...118

117b. Leaves cauline, cordate but not as above. Flowers 1-sexual.................................Dioscoreaceae

118a. Herbs. Roots fibrous. Leaves not spinescent, apex with long tendril. Flowers large showy. Perianth lobes with crisply waved margin. Fruit a capsule. ....Colchicaceae

118b. Shrubs. Roots tuberous. Leaves spinescent, apex without tendril. Flowers small, not showy. Perianth lobes not as above. Fruits globose..Liliaceae

116b. Plants of various habits but not as above...119

117a. Flowers 3-merous..120

120a. Stem bulbous, geophytes. Leaves ensiform. Capsule oblong ellipsoid. ....Hyacinathaceae

120b. Stem globose, aerial. Leaves linear-lanceolate. Capsule obovoid-oblong........Asphodelaceae

119b. Flowers not as above..121

121a. Perianth absent or represented by hairs or bristles. 122

122a. Dioecious woody plants, usually supported by stilt roots. Leaves with spinous margins..Pandanaceae

122b. Monoceious herbaceous plants, without stilt roots. Leaves not spinaceous. .... Typhaceae

121b. Perianth present. 1 or 2-seriate, sometimes differentiated into calyx and corolla..........123

123a. Annuals. Leaves denticulate, auriculate. .... .Najadaceae

123b. Perennials. Leaves entire, not auriculate...124

124a. Ovary superior. 125

125a. Trees or shrubs. Stem with leaf-scars. Leaves with pinnately divided. Inflorescence spadix.. ....Arecaceae

125b. Herbs or shrubs. Stem without leaf-scars. Leaves simple. Inflorescence otherwise: 126

126a. Carpels 3 - many (except Alismataceae), free..127

127a. Perianth segment 6, arranged in two whorl of 3 each, differentiated into calyx and corolla. 128

128a. Inflorescence umbels. Friut dehiscent, of 6-7 follicles. Ovules and seeds many in each carpel. .... Limnocharitaceae

128b. Inflorescence racemes or whorled of racemes. Fruit indehiscent, of 3 or more achenes. Ovules and seeds solitary in each carpel.................. .Alismataceae

127b. Perianth segment 1-4, uniseriate, sepaloid or petaloid not differentiated into calyx and corolla. 129

129a. Tepals 1-3, white. Stamens 6 or more. .... .Aponogetonaceae

129b. Tepals 4, green. Stamens 4. .... Potamogetonaceae

126b. Carpels solitary or if more, then united. 130

130a. Perianth not distinguished into calyx and corolla. .... Agavaceae

130b. Perianth distinguished into calyx and corolla. 131

131a. Terrestrial. Flowers in cymes or panicles. .... .Commelinaceae

131b. Aquatic. Flowers in racemes or spikes. 132

121a. Tepals scale like, not concuously petaloid. .... Juncaceae

132b. Tepals not scale like, concuously petaloid .... .Pontederiaceae

124b. Ovary inferior. 133

133a. Submerged aquatic herbs without tubers or bulbs. Flowers mostly unisexual. Ovary 1-celled. .... Hydrocharitaceae

133b. Terrestrial or semiaquatic herbs with tuber or bulbs. Flowers bisexual. Ovary 2 or more celled. 134

134a. Flowers yellow. Scape very short. Leaves rigid, plicate. ....Hypoxidaceae

134b. Flowers white. Scape stout. Leaves fleshy not plicate. ....Amaryllidaceae

SYSTEMATIC ENUMERATION OF PLANTS

ANNONACEAE

KEY TO THE GENERA

1a. Ripe carpels united, forming many-loculed syncarpus fruits .... *Annona*

1b. Ripe carpels free, forming apocarpus fruits. 2

2a. Mostly climbing shrubs. Peduncle flattened and indurated, hooked .... *Artabotrys*

2b. Trees, never climbing. Peduncle neither flattened and indurated nor hooked.. *Polyalthia*

ANNONA L.

Annona squamosa L., Sp. Pl. 537. 1753; DC., Prodr. 1: 85. 1824; Hook. *f.* & Thoms., in Hook. *f.*, Fl. Brit. Ind. 1: 78. 1872; Cooke, Fl. Pres. Bomb. 1: 15. 1901; Duthie, Fl. Gangetic Plain 1: 23. 1903; Parmar, in Shetty & Singh Fl. Rajasthan 1: 58. 1987; Raju, in sharma *et al.* Fl. India. 1: 207 - 208. 1993. (Fig. 1 & Plate 9)

Vern.: *Sitaphal.*

Evergreen shrub or small tree, 3 - 6 m high. Branches sparsely hairy when young, glabrous with age. Leaves elliptic to oblong-obovate, 8 - 14 x 2.5 - 5 cm, acute to obtuse with cuneate at base, glabrous on both sides, young leaves sparsely hairy. Petiole 0.6 - 1.5 cm long, glabrous. Flowers axillary, solitary or a few together, leaf-opposed; pedicels 1-3 cm long. Bract and bracteole minute, pilose. Pedicel 0.5 - 0.7 cm long, glabrous. Sepals broadly ovate, 1.5 - 3 x 3 - 4 mm, basally connate, pubescent outside. Outer petals narrowly oblong, 1.5 - 3 x 0.3 - 0.5 cm, pale yellow with deep purple spot inside at base, obtuse, triquetrous, basally concave within, sparsely pilose to glabrous outside, puberulous inside; inner petals frequently absent. Stamens numerous ca 1 mm long, narrow cuneate, filament short, locules equal, connective-tip subtruncate. Carpels many basally connate, ovary dorsally pilose, 1-ovuled, style short, stigma narrow conical. Fruit 5 - 10 x 5 - 7.5 cm, yellowish green, glaucous, tuberculate, monocarps loose and easily separable in ripe fruit, pulp soft, smooth, pure white or yellow tinged. Seeds many, shiny, arillate, dark brown to black.

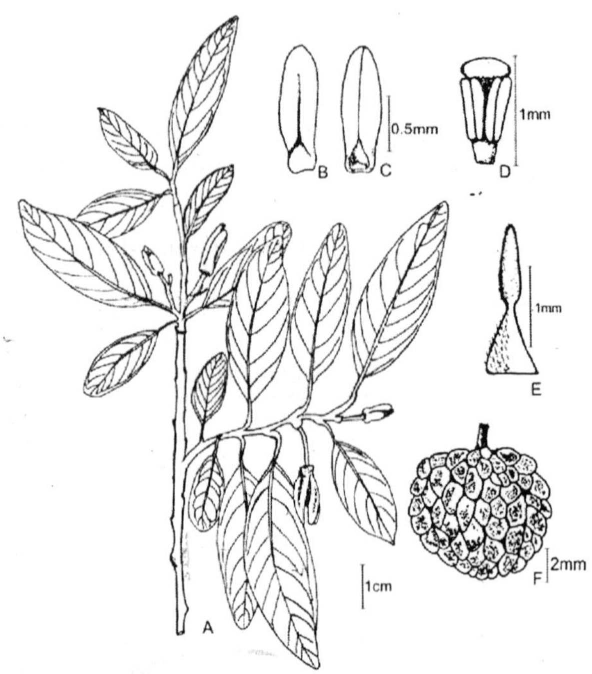

Fig. 1. *Annona squamosa* L. : A. Flowering twig, B. & C. Outer petal (B. Outer face; C. Inner face); D. Stamen, E. Part of carpel, F. Fruit.

Field note: Common in scrub forest as well as widely cultivated.

Fl. & Fr.: May - January.

Specimen examined: Gurla 85, Chittorgarh Fort 3527 (MLVGCB Herb).

Distribution: Widely cultivated in Old and New World tropics.

Uses: The Custard apple is indigenous to tropical America. Fruit are edible. The pulp is said to be rich in Vitamin C. Seeds are strong irritant to eyes and may cause blindness. Leaves and unripe fruits are insecticidal. Root is a strong purgative.

ARTABOTRYS R. Br.

Artabotrys hexapetalus (L.f.) Bhandari in Baileya 12(4): 147. 1964; Parmar, in Shetty & Singh Fl. Rajasthan 1: 59. 1987; Raju, in Sharma *et al*. Fl. India. 1: 251-252. 1993. *Annona hexapetala* L.f., Suppl. 270. 1781. *Anona uncinata* Lam., Encycl. Meth. 2: 127. 1786. *Artabotrys odoratissimus* R. Br. in Edwards Bot. reg. 5: t. 423. 1819 (non Blume 1825); Hook. *f.* & Thoms. in Hook. *f.*, Fl. Brit. Ind. 1: 54. 1872; Cooke, Fl. Pres. Bomb. 1: 10. 1901. *Uvaria odoratissima* Roxb., Fl. Ind. 2: 666. 1824. *Artabotrys uncinatus* (Lam.) Merr. in Philipp. J. Sci. (Bot.). 7: 234. 1912. (Fig. 2)

Vern.: *Nag-champa.*

Large evergreen shrub, climbing by the hooked peduncles or semi-scandent. Branches glabrous, appressed hairs on young branches. Leaves oblong-elliptic to oblanceolate, 10 - 22 x 2.8 - 6 cm, acuminate, subcoriaceous, glossy above, glabrous on both sides lateral nerves 6 - 18 pairs, juvenile leaves covered with appressed golden-brown hairs. Petiole ca 5 - 7 mm long, glabrous. Flowers fragrant, ca 2 cm across, on hooked peduncles, solitary or in pairs. Pedicel glabrous. Sepals 3, basally connate, ovate, 7-10 x 4 - 7 mm, acute, glabrate, apex recurved. Petals 6 (3+3), ovate-lanceolate, 2-2.5 x 0.5-1.3 cm, yellowish green to bright yellow, pubescent on both sides, narrow and concave at base, expanded above the concave part, subacute, flattened. Outer petals with a minute ligule like outgrowth just above the concave part, inner petals with a knee like projection above the concave part, concealing the stamens and carpels. Receptacle flat. Stamens many spirally arranged, 1.5 mm long, cuneate, filament shorter than anthers, locules unequal, connective-tip expanded triangular, apiculate. Ovary 5 mm long, glabrous, style curved short, stigma blunt, papillate. Fruit of 1 - 10 baccate monocarps, 2.5 - 3.5 cm across, sessile, obovoid, apiculate, yellow and fragrant when ripe, glabrous. Seeds 1 - 2, deeply grooved one side, brown.

Field note: Occasionally cultivated in gardens for its fragrant flowers. Usually its growth is kept in check by regular pruning to make it a dense shrub, otherwise it becomes a hook-climber.

Fl. & Fr.: Almost throughout the year (except winter

Specimen examined: Rawatbhata 4096 (MLVGCB Herb).

Distribution: Widely cultivated in Old World tropics and subtropics.

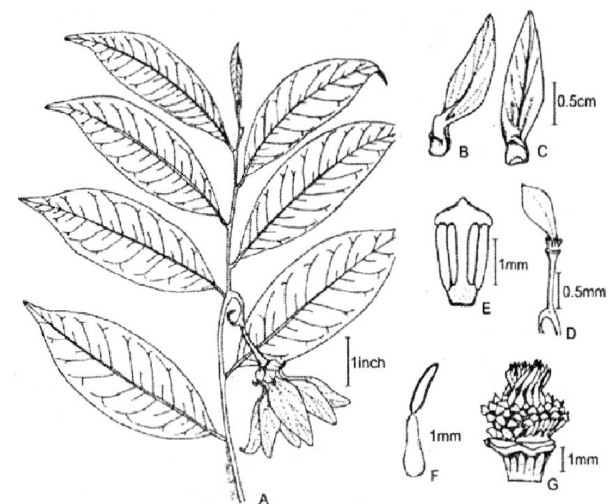

Fig. 2. *Artabotrys hexapetalus* (L.f) Bhandari: A. Flowering twig, B. Inner petal, C. Outer petal, D. Fruit, E. Stamen, F. Carpel, G. Receptacle with stamen & Carpel.

POLYALTHIA Blume

Polyalthia longifolia (Sonner.) Thwait., Enum. Pl. Zeyl. 398. 1864; Hook. *f.* & Thoms. in Hook. *f.*, Fl. Brit. Ind. 1: 62. 1872; Cooke, Fl. Pres. Bomb. 1: 13. 1901; Parmar, in Shetty & Singh Fl. Rajasthan 1: 64. 1987; Raju, in

Sharma *et al.* Fl. India. 1: 274 - 275. 1993. *Guatteria longifolia* Wall., Cat. 6442. 1832; Wight & Arn., Prodr. 10. 1834. *Uvaria longifolia* Sonner., Voy. aux Indes. 2: 233. 1782. (Fig. 3)

Vern.: *Ashok, Ashapal.*

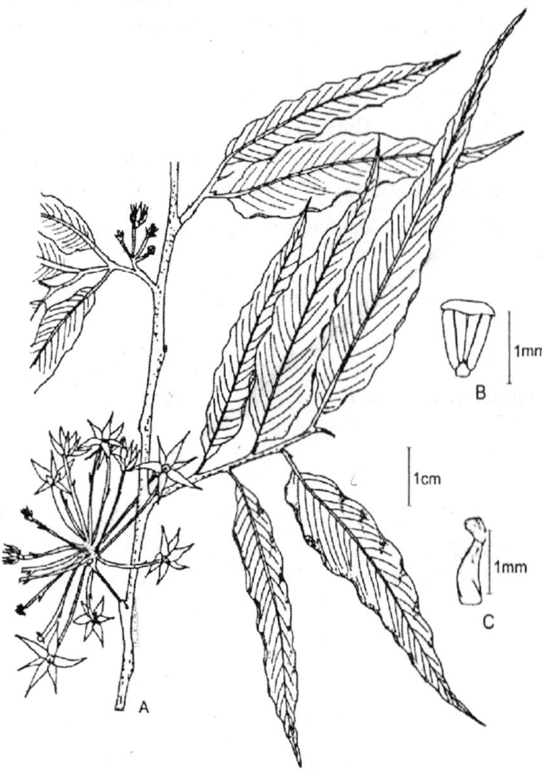

Fig. 3. *Polyalthia longifolia* (Sonnerat) Thw. : A. Flowering twig, B. Stamen, C. Carpel.

Evergreen tree with straight trunk, upto 20 m tall. Young branches spreading, pendulous, pilose, becoming glabrous. Leaves oblong-lanceolate to narrow-lanceolate, 21.5 - 30.0 x 3.7 - 5.6 cm, tapering to a fine point, margins undulate, glossy above, glabrous on both sides (juvenile leaves tomentose). Petiole 4-8 mm long, glabrous. Flowers greenish yellow, 2.5 - 3 mm long, numerous in subulate clusters on 3 - 10 mm long peduncle. Pedicels slender, 2 - 3 cm long, pubescent. Bract minute, caducous, bracteole semi-amplexicaule, densely hairy. Sepals 3, broadly deltiod, 2 x 2 - 3 mm, obtuse to subacute, free, pubescent on both sides. Petals narrow lanceolate or linear, outer 6 - 7 x 2 mm, slightly smaller than the inner, tapering into a fine point, slightly undulate, pubescent on both sides, pale green to yellowish green, inner 15 x ca 2 - 5 mm. Receptacle convex. Stamens numerous, 1 mm long, anther subsessile, locules unequal, connective-tip subtruncate. Carpels free, ca 1.5 mm long, ovary glabrous except few hairs in the upper part, stigma subsessile, papillate. Fruit of 10-15 baccate monocarps on upto 37 mm long woody stalk. Mature monocarp ca 20 mm long, ovoid, rounded at both ends, purplish black, 1-seeded, stipe ca 12 mm long, stout, glabrous. Seeds, ovoid, smooth or slightly grooved, pale brown.

Field note : It is commonly cultivated as an avenue tree along roadsides as well as in gardens.

Fl. & Fr.: March - May.

Specimen examined: Bhilwara 607, Mandalgarh 918 (MLVGCB Herb).

Distribution: Indigenous to Sri Lanka, cultivated in India, Malaya, Pakistan and Tropical East Africa

Notes: The Ashapal/Ashok tree is indigenous to Sri Lanka, elsewhere it is cultivated. The cultivar *pendula* (weeping variety) is columnar with short drooping branches, sometimes the lower branches touching the ground. The tree is supposed to have several medicinal properties. The bark is febrifuge and the inner bark yields a useful fibre. Berries are relished by bats, reportedly eaten by human being during scarcity. The tall, straight trunks were used as mast in the olden days of sailing ships.

MENISPERMACEAE

KEY TO THE GENERA

1a. Leaves peltate. Bracts enlarged in fruit. Stamens united in a column or disc. Carpel 1.. *Cissampelos*

1b. Leaves not peltate. Bracts not enlarged in fruit. Stamens free. Carpels 3 or more............2

2a. Leaves 3-5-nerved, not deeply cordate at the base. Carpels 3 to many. .... *Cocculus*

2b. Leaves 7-nerved, deeply cordate at the base. Carpels 3. .... *Tinospora*

CISSAMPELOS L.

Cissampelos pareira L. var. ***hirsuta*** (Buch. - Ham. ex DC.) Forman, in Kew Bull. 22: 356. 1968; Duthie, Fl. Gangetic Plain 1: 33. 1903; Parmar, in Shetty & Singh Fl. Rajasthan 1: 60. 1987; Gangopadhyay, in sharma *et al.* Fl. India. 1: 317. 1993. *C. hirsuta* Buch. - Ham. ex DC., Syst. Nat. 1: 535. 1817; *C. pareira* L., Sp. Pl. 1031. 1753, *p.p.,* quoad B; Hook. *f.,* Fl. Brit. India 1: 103. 1872. (Plate 10 & 11)

Slender tomentose climber, compact, thicket forming. Leaves peltate, 3 - 12 x 3.5 - 11 cm, triangularly broad-ovate, or orbicular, obtuse, mucronate, base cordate or truncate, ± tomentose on both sides, ultimately becoming glabrous above and glaucous below, palmetly 5 - 7 - nerved; petiole 2 - 10 cm long, pubescent. Inflorescence pendulose, subcorymbose cymes upto 6 cm long. Flowers minute, pedicels filiform. Male flowers in pedunculate branched cymes, clustered in the axil of a small leaf; sepals 4, obovate-oblong, hairy outside; petals 4, united to form a 4-toothed cup, hairy outside; stamens 4, filaments united, column short, anthers connate, encircling the top of the column. Female flowers clustered in the axils of orbicular, 1 - 1.5 mm, hoary imbricate bracts, on 5-10 cm long racemes; sepal 1, ovate-oblong, pubescent outside; petal 1, obtriangular subreniform; carpel 1, densely hairy; style shortly 3-fid. Drupe 5 - 7 x 2.5 - 4 mm, subglobose, compressed, hairy-pubescent, red when fresh, black when dry, endocarp transversely ribbed, tuberculate. Seeds horseshoe-shaped.

Field note:Common along hedges and at the foothills.

Fl. & Fr.: July - October

Specimen examined: Keli 3504, Mainal 3841 (MLVGCB Herb).

Distribution: India and Pakistan.

Notes: The type variety *pareira*, appears to be confined to America and has more cordate and less hairy leaves with somewhat shorter petioles than variety *hirsuta*.

Uses: It is very common in hedges at the foothills and up to c. 2300 m. It is a source of alkaloids like Seeprine, Bebeerine and Cissampeline. The leaves and roots are used as a cure for dyspepsia, diarrhoea, dropsy and in snake-bite. Stem yields a strong fibre.

COCCULUS DC.

KEY TO THE SPECIES

1a. Leaves densely greyish-tomentose or puberulous, generally ovate-oblong, Female inflorescence 0.5- 2.5 cm long. .... *C. hirsutus*

1b. Leaves glabrous or slightly puberulous, linear- oblong or trapezoid. Female inflorescence 0.7-1.3 cm long. ....*C. pendulus*

Cocculus hirsutus (L.) Diels in Engler, Pflanzenr. IV. 94: 236. 1910; Parmar, in Shetty & Singh Fl. Rajasthan 1 : 61. 1987; Gangopadhy-aym in Sharma *et al.* Fl. India. 1: 318 - 319. 1993. *Menispermum hirsutum* L., Sp. Pl. 341. 1753. *Cocculus villosus* DC., Syst. 1:525. 1817. Prodr. 1: 98. 1824; Hook. *f.* & Thoms. in Hook. *f., l.c.* 1: 101. 1817; Duthie, Fl. Gangetic Plain 1: 28. 1903. *Cebatha hirsuta* (L.) O. Ktze., Rev. Gen. 1: 9. 1891. *Menispermum villosum* Lam., Encycl. 4: 97. 1797. (Fig. 4)

Vern.: *Jal-jamni.*

Climbing undershrub, often densely pubescent-tomentose. Branches densely tomentose. Leaves 3 - 8 x 2.0 - 3.5 cm (basal larger), ovate or ovate- cordate, sometimes 3-5-lobed, base subcordate, cuneate or truncate, apex obtuse or mucronate, densely tomentose when young, later subglabrous; basal nerves 3 - 5, lower ones main in branches, 3 -5 lobed; upper ones usually entire; petiole 0.5-2.5 cm long. Male flowers: sepals hairy, the outer 3, oblong-lanceolate, 1.5 - 2.5 x 0.5-0.8 mm, inner 3 broadly ovate, 1.5 - 2.5 x 1.7 - 2 mm, greenish. Petals ovate-oblong, 0.5 - 1.5 x 0.3-6 mm, sparsely pubescent to glabrous; stamens 6, free, 0.7 - 1 mm long, enclosed by petals. Female flowers: 1 - 3 in axillary fascicles or rarely racemes; bracts minute, linear; carpels 0.7 - 1 mm long, styles terete. Drupe dark purple, 4 - 8 x 3 - 5 mm, endocarp annular or ribbed with a prominent dorsal crest, perforated.

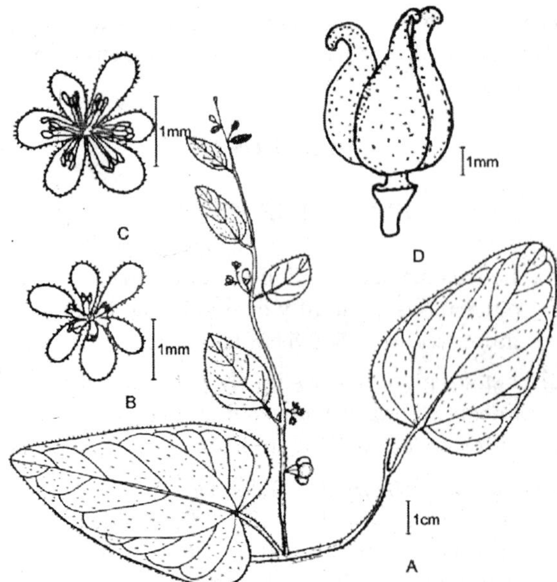

Fig. 4. *Cocculus hirstus* (L.) Diels: A. Male flowering twig, B. Male flower, C. Female flower, D. Carpel.

Field note: Common climber in hedges and also under growths in forests.

Fl. & Fr.: February - April.

Specimen examined: Karoi 816, Mandalgarh 1022 (MLVGCB Herb).

Distribution: Tropical Africa, India and Pakistan.

Uses: The juice of the ripe fruit yields a permanent bluish-purple ink and the roots as well as the leaves are used in native medicines and as a tonic.

Cocculus pendulus (J. R. & G. Forst.) Diels, in Engler, Pflanzenr. 46: 237. 1910; Bhandari, Fl. Indian Desert 28 - 29. 1978; Parmar, in Shetty & Singh Fl. Rajasthan 1: 61. 1987; Gangopadhyay, in Sharma *et al.* Fl. India. 1: 320. 1993. *Epibatrium pendulum* J.R. & G. Forst., Char. Gen. 108. t. 54. 1776. *Menispermum leaeba* DC. Prodr. Syst. Nat. 1: 98. 1817; Hook .f. & Thoms. in Hook. *f.*, Fl. Brit. India 1: 102. 1872; Duthie, Fl. Gangetic Plain 1: 29. 1903. *Cocculus leaeba* DC., Syst. Nat. 1: 529. 1817; Hook .f. & Thoms. in Hook. *f.*, *l.c.* 1: 102. 1872. *C. cebatha* DC., Syst. 1: 526. 1818; BH., 26 (1) 219. 1918.

Vern.: *Pilwan*.

Scandent shrub with long slender, puberulous branchlets. Stem up to 15 cm in diam., branches slender, puberulous, ca 5 - 6 m long. Leaves 1.6 - 5 x 0.5-2 cm, oblong-lanceolate, ovate or trapezoid, base truncate, cuneate, rounded or trilobed-hastate, apex obtuse, mucronate or emarginate, generally glabrous or slightly puberulous on both sides, basal nerves 3-5. Flowers minute. Male flowers: in axillary panicles, 0.5 - 20 mm long, with peduncles up to 1.5 cm long, pedicels 0.5 mm or absent; sepals ovate-elliptic, fleshy or membranous, thickened at the base, the outer 3, smaller, 1 - 1.5 x 0.4 - 0.7 mm, sparsely puberulous, the inner 3 larger, puberulous to glabrous; petals ovate-obovate, 0.8 - 2 x 0.5 - 1 mm; stamens 0.8 - 1.5 mm long. Female flowers: 1-2 on short axillary peduncles, 0.7 - 1.3 cm long; sepals and petals as in males; carpels c. 1 mm long. Drupe reddish when fresh but turning black when dried, reniform, compressed, 4 - 7 x 4 - 5 mm; endocarp ribbed on the lateral faces and without a prominent crest, not perforated in the centre.

Field note: Common among hedges and at the foothills.

Fl. & Fr.: Almost throughout the year.

Specimen examined: Pansal 902 (MLVGCB Herb).

Distribution: Tropical and subtropical India, Pakistan and Africa

Uses: Juice of the leaves has the property of coagulating water into a green jelly like substance which is taken with sugar as a tonic. Roots and the leaves are used in rheumatic pains.

TINOSPORA Miers.

Tinospora cordifolia (Willd.) Miers in An. & Mag. Hat. Hist. Ser. 2.7: 38.1851; Hook. *f.*, Fl. Brit. India 184. 1855 & in Fl. Brit. India 1: 97. 1872; Duthie, Fl. Gangetic Plain 1: 26. 1903; Bhandari, Fl. Indian Desert 29. 1978; Parmar, in Shetty & Singh Fl. Rajasthan 1: 62. 1987; Pramanik & Gangopadhyay, in Sharma *et al.* Fl. India. 1: 347 - 348. 1993. *Menispermum cordifolium* Willd., Sp. Pl. 4:

826. 1806. *Cocculus cordifolius* (Willd.) DC., Syst. 1: 517. 1818. (Plate 12)

Vern.: *Nim-giloe.*

Tall deciduous climber upto 12 m high, entirely glabrous. Stem ca 6.5 cm in diam., smooth, shining, lenticellate, bark light grey, papery, branches pubescent. Leaves 4 - 15 x 4 - 13 cm, broadly ovate or orbicular, deeply cordate at the base, 5 - 7-nerved, subacuminate, young pubescent above, tomentose beneath; petiole 2 - 12.3 cm long, thickened and twisted at the base. Flowers greenish yellow in racemes 7 - 14 cm long, pedicels slender, usually solitary in the female, clustered in the male. *Male flowers*: sepals 6 in 2 series, the outer ones small, ovate-oblong, obtuse, concave, the inner large suborbicular, membranous, flat, 2.5 - 4 x 2 - 3 mm; petals subequal, 2 - 3 x 1.2 - 1.8 mm, clawed, membranous, smaller than the sepals; stamens 6, filaments free, 2.5-3 mm long, thickened at the apex. Female flowers: sepals and petals as in the male flowers; staminodes 6, ca 1.5 mm long; carpels 3, ellipsoid, stigma capitate. Drupes 1 - 3, ovoid, radiating from a ca 2 mm long carpophore, 6 - 9 x 4 - 5 mm, smooth, crimson, endocarp tubercled.

Field note: Common along hedges, other trees and at the foothills.

Fl. & Fr.: March - June.

Specimen examined: Bhilwara 417, Hamirgarh 913 (MLVGCB Herb).

Distribution: Naturalized and cultivated throughout tropical and subtropical regions of Pakistan, India, Burma and Sri Lanka.

Uses: It contains a bitter substance, Berberine, also yields Giloin and Giloinin. The watery extract *"Sat Giloe"* is sold in the market and is often used as a febrifuge under the name of *"Indian quinine"*. An infusion is prepared from the stem and the root which is considered to be a valuable tonic and a cure for intermittent fever and dyspepsia. The stem juice is kept overnight and then used in fever, jaundice, diabetes & general disability.Juice is applied to cure skin disease. 1/4th cup of stem juice mixed with 1 glass of water and is taken for 7 days to cure leucorrhoea.

NYMPHAEACEAE

NYMPHAEA L.

KEY TO THE SPECIES

1a. Leaves entire or distanly toothed, lower surface glabrous. .... *N.nauchali*

1b. Leaves irregularly and sharply serrate, lower surface pubescent. .... *N. pubescens*

Nymphaea nauchali Burm. f., Fl. Indica 120. 1768; Bhandari, Fl. Indian Desert 29. 1978; Parmar, in Shetty & Singh Fl. Rajasthan 1: 64. 1987; Mitra, in Sharma *et al*. Fl. India. 3: 430 - 431. 1993. *N. stellata* Willd. Sp. PL. 2: 1153. 1799; Hook. *f.* & in Hook. *f.,* Fl. Brit. India 1: 114. 1872; Duthie, Fl. Gangetic Plain 1: 34. 1903; Subran., Aquatic Angiosp. 6. f. 1. 1962. (Fig. 5 & Plate 13)

Vern.: *Kumudani.*

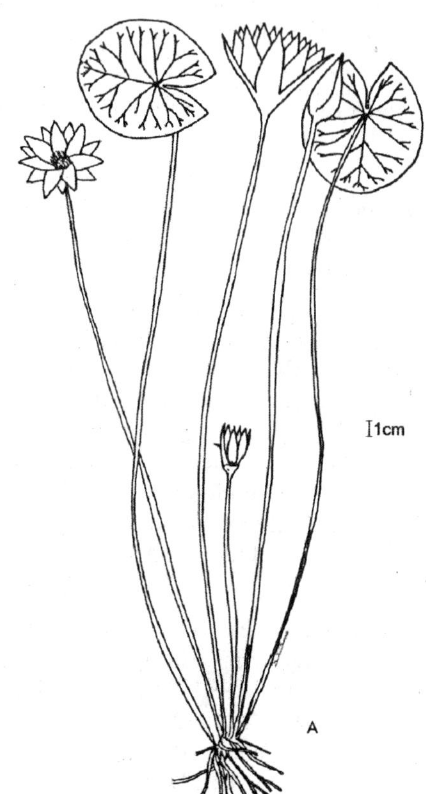

Fig. 5. *Nymphaea nauchali* Burm. f. A. Habit.

Perennial aquatic herb. Rhizome erect, unbranched. Leaf blade elliptic-orbicular to orbicular, 7 - 15 cm in diam., papery, abaxially glabrous, peltate a few mm from base of sinus, base cordate, basal lobes parallel to spreading, margin subentire to deeply crenate. Flower slightly emergent, 3 - 15 cm in diam. Calyx insertion on receptacle circular; sepals lanceolate to oblong-lanceolate, 2.5 - 8 cm, slightly veined, persistent. Petals 10 - 30, white tinged with purple, blue, or purple-red, linear-oblong to lanceolate, 4.5 - 5 cm, transition to stamens gradual. Filament of inner stamens somewhat as wide as anther; connective apically appendaged. Carpels only partially united, walls between locules of ovary double. Stigma rays 10 - 30; carpellary appendages triangular-tapered. Fruit globose, 1.5 - 4.5 cm in diam. Seeds ellipsoid-globose, 0.5 - 1.3 mm, with longitudinal rows of hairs.

Field note: Common during rainy season in ponds, ditches and tanks in association with *Nymphoides, Najas* etc.

Fl. & Fr.: July - December

Specimen examined: Mandal 614, Mainal 940 (MLVGCB Herb).

Distribution: Afghanistan, Bangladesh, India, Indonesia, Myanmar, Nepal, New Guinea, Pakistan, Philippines, Sri Lanka, Thailand, Vietnam and Australia.

Notes: In FRPS and a number of other works this species is named as *Nymphaea stellata* . An examination of the type of *N. nauchali* by Verdcourt (Kew Bull. 44: 179. 1989) indicated that the name should be applied to this species, not to *N. pubescens*. Much work remains to be done to improve our understanding of this wide-ranging and highly variable taxon and its relationship to related taxa in Africa

Nymphaea pubescens Willd., Sp. Pl. 2: 1154. 1799; Bhandari, Fl. Indian Desert 28 - 30. 1978; Parmar, in Shetty & Singh Fl. Rajasthan 1: 64. 1987; Mitra, in Sharma *et al*. Fl. India. 1: 251 - 252. 1993. *N. lotus auct. non* L. (1753); Hook. *f.* & Thoms., in Fl. Brit. India. 1: 114. 1872; Duthie, Fl. Gangetic Plain 1: 33. 1903. *N. lotus* L. var. *pubescens* (Willd.) Hook. & Thoms., in Fl. Brit. India. 1: 114. 1872. (Fig. 6 & Plate 14)

Vern.: *Kamal*.

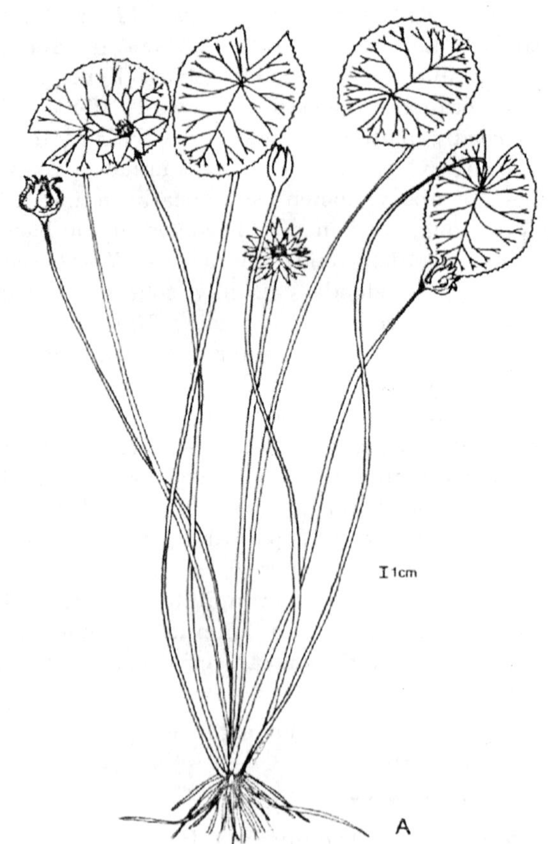

Fig. 6. *Nymphaea pubescens* Willd. A. Habit

Perennial, aquatic herb. Rhizome erect, producing slender stolons. Leaf blade ovate-elliptic to suborbicular, 15 - 26 cm, papery, abaxially densely pubescent, peltate more than 5 mm from base of sinus, base deeply cordate and basal lobes subparallel, margin dentate and teeth acute to subspinose. Flower emergent, 5 - 8 cm in diam. Calyx insertion on receptacle circular; sepals oblong, 5 - 8 cm, conspicuously veined, caducous or decaying after anthesis. Petals 12 - 14 white, red, or pink, oblong, 5 - 9 cm, transition to stamens abrupt. Filament of inner stamens only slightly wider than anther; connective apically unappendaged. Carpels completely united, walls between locules of ovary single. Stigma rays 12 - 20, carpellary appendages linear. Fruit ovoid to subglobose, 3.5 - 5 cm. Seeds ellipsoid to globose, 1-2 mm, with longitudinal ridges.

Field note: Common during rainy season. in association with *Nymphoides, Najas* etc. in ponds.

Fl. & Fr.: August - October.

Specimen examined: Mandal 604, Mainal 1012 (MLVGCB Herb).

Distribution: India, Bangladesh, Indonesia, Myanmar, New Guinea, Pakistan, Philippines, Sikkim, Sri Lanka, Thailand and Vietnam.

Notes: The relationship between the Asiatic *Nymphaea lotus* var. *pubescens,* often treated as *N. pubescens,* and the var. *lotus* of Africa is in need of further study. Cultivated forms of var. *pubescens* with red flowers, originating from India, are known from Guangdong Province and perhaps elsewhere in S. China. Although they are reportedly sterile, they have sometimes been segregated as *N. rubra* Roxb.ex Andrews.

Uses: Stems are eaten as vegitable. Fruits are also edible.

NELUMBONACEAE

NELUMBO Adans.

Nelumbo nucifera Gaertn., Fruct. Sem. Pl. 1: 73. t. 19., f. 2. 1788; Parmar, in Shetty & Singh Fl. Rajasthan 1: 64. 1987; Mitra, in Sharma *et al.* Fl. India. 1: 441 - 442. 1993. *Nelumbium speciosum* Willd., Sp. Pl. 2: 1258. 1799; Hook. *f.* & Thoms., in Fl. Brit. India 1: 116. 1872; Duthie, Fl. Gangetic Plain 1: 35. 1903. (Plate 15)

Varn.: *Kamal.*

Perennial aquatic herb with milky juice. Petiole 1 - 2 m, terete, fistulous, glabrous or papillae hard and scattered; leaf blade abaxially blue-green, orbicular, 25 - 90 cm in diam., papery, glabrous, glaucous, water-repellent, margin entire. Flowers 10 - 23 cm in diam.; peduncles longer than petioles, glabrous or sparsely spinulate. Tepals caducous, pink or white, oblong-elliptic to obovate, 5 - 10 × 3 - 5 cm. Stamens slightly longer than receptacle; filament slender; anther linear, 1 - 2 mm; connective appendage clavate, to 7 mm, incurved. Receptacle accrescent, turbinate, 5 - 10 cm in diam. Fruit oblong to ovoid, 1.0 - 2.0 × 7 - 15 cm, glabrous; pericarp thick, hardened.

Field note: Common in pond; also cultivated in garden ponds for ornamental purposes.

Fl. & Fr.: June - August

Specimen examined: Mainal 84, Mandal 1575 (MLVGCB Herb).

Distribution: India, Indonesia (Java), Bhutan, Japan, Korea, Malaysia, Myanmar, Nepal, New Guinea, Pakistan, Philippines, Russia (Far East), Sri Lanka, Thailand, Vietnam; SW Asia, Australia.

This species is cultivated for its edible rhizomes and seeds.

PAPAVERACEAE

ARGEMONE L.

KEY TO THE SPECIES

1a. Flowers bright yellow. Stigma shallowly dissected, style absent in fruit...... *A. mexicana*

1b. Flowers white turning pale-yellowish with age. Stigma deeply dissected, style 1 - 2 mm long in fruit................................ *A. ochroleuca*

Argemone mexicana L., Sp. P1. 508. 1753; Hook. *f.* & Thoms in Hook. *f.,* Fl. Brit. India 1 : 117. 1872; Duthie, Fl. Gangetic Plain 1: 36. 1903; BH., 26 (1): 219. 1918; Bhandari, Fl. Indian Desert 30. 1978; Parmar, in Shetty & Singh Fl. Rajasthan 1: 64. 1987; Debnath & Nayar, in Sharma & Balakrishnan Fl. India. 2: 2 - 3. 1993. (Fig. 7 & Plate 16)

forma : ***mexicana***

Vern.: *Pili Kantali, Satyanashi.*

Prickly glabrous annual herb, 30 - 125 cm tall, branched. Leaves alternate, elliptic-oblong, pinnatifid, semiamplexicaul, sinuate-lobulated; variegated green and white, 5 - 20 cm long, 2 - 8 cm broad, ultimate segments, dentate, prickly on the margin, midrib and the veins beneath; Flowers sessile, 3 - 8 cm in diam., subtended by 2

- 3 foliaceous bracts. Sepals 8 - 12 mm long, 5 - 7 mm broad, with an acute, terete horn below the apex, very sparsely prickly outside, concave, imbricate, caducous. Petals 4 - 6, obovate, 2.5 - 3.5 x 1.5 - 2.5 cm, narrowed below, bright yellow, imbricate, more or less crumpled in bud. Stamens indefinite, 8 - 12 mm long; anther ca 2 mm long, curved after flowering. Ovary ovate, 8 - 10 x 3 - 5 mm broad covered with long soft spines; stigma red, 3 - 6 lobed; lobes usually broad. Capsule, oblong or elliptic-oblong, 2.5 - 4 x 1.2 - 2 cm, with rounded ribs, covered with sharp erect prickles; valves 3 - 6. Seeds many, blackish brown to brown, rounded, 1.5 - 2 mm in diam, with fine, conspicuous tuberculae.

Fig. 7. *Argemone mexicana* L. A. Habit

Field note: Common in fallow-fields, wasteplaces, weed of cultivatd fields.

Fl. & Fr.: February - May.

Specimen examined: Bhilwara 602, Shahpura 1187 (MLVGCB Herb).

Distribution: Native of West Indies and Mexico, but naturalized in most of the warm countries of the world as a weed.

Uses: Yellow juice of the plant is medicinally used in cutaneous infection. Seeds are said to be poisonous, have narcotic properties and yields a oil which has been in use amongst West Indian practitioners as an aperient. It exercises a soothing influence when applied externally in case of headache and also in herpetic and other forms of skin disease. A pale yellow clear limpid oil, obtained from the seeds, is used in lamps and medicinally in ulcers and eruptions

Argemone ochroleuca Sweet, Brit. Fl. Gard. 3: t. 242. 1828; Venkatesh. in Curr. Sci. 31: 250. 1962; Duthie, Fl. Gangetic Plain 1: 36. 1903; BH. 26 (1) 219. 1918; Maheshwari, Fl. Delhi 55. 1963; Parmar, in Shetty & Singh Fl. Rajasthan 1: 67. 1987; Debnath & Nayar, in Sharma & Balakrishnan Fl. India. 2: 5. 1993. *Argemone mexicana* Sweet var. *ochroleuca* (Sweet) Lindl. Bot. Reg. 1343. 1830. (Fig. 8 & Plate 17)

Vern.: *Dholi satyanashi*.

A prickly glaucous herb, usually 35 - 40 cm tall; young stem whitish purple or violaceous, fading with age; prickles 6 - 9 mm long, with broader base. Leaves sessile, amplexicaul, basal leaves forming a rosette, upper alternate, variegated with white, 10 - 30 x 4-10 cm, sinuate-pinnatifid; segments incised dentate, with sharp yellowish spines on the margin, midrib and the veins beneath; midrib and the veins on the lower side purplish blue; Flowers solitary axillary, white or creamish white, becoming yellowish on fading, 2.5 - 3 cm across, sessile or on a short pedicel, subtended by 2-3 foliaceous bracts; bracts 2 - 4 x ca 5 mm. Sepals 3, obovate oblong, with horn as long as the body, 8 - 10 x 5 - 7 mm. Petals 6, obovate, 2.8 - 3 x ca 1.8 cm. Stamens 7 - 10 mm long; anthers oblong, recurved. Ovary ovate, 8 - 10 x 5 mm, covered with long pointed spines. Stigma deeply dissected, 5 lobed, reddish brown with an obscure style. Capsule oblong, 3 - 4 x 1.5 - 2 cm, (with a prominent, 1-2 mm long style) covered with sharp, erect prickles. Seeds black, rounded, 1.5-2 mm in diam., with fine reticulations.

Field note: A common weed of cultivated fields and waste lands.

Fl. & Fr.: February - June.

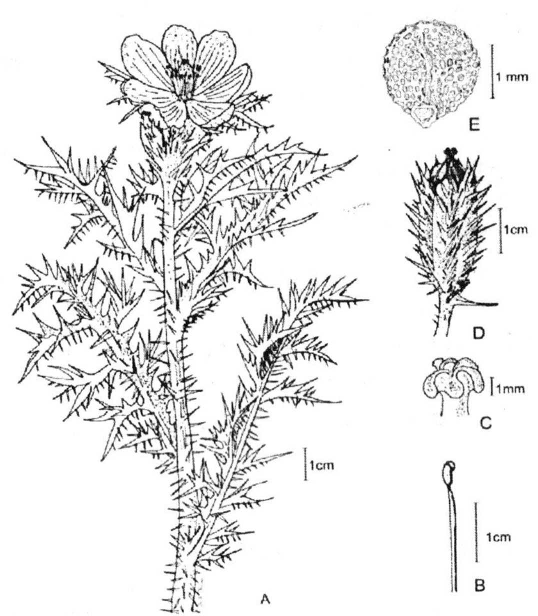

Fig. 8. *Argemone ochroleuca* Sweet: A Habit, B. Stamen, C. Stigma, D. Fruit, E. Seed

Specimen examined: Bijoliya 84, Nandrai 1029 (MLVGCB Herb).

Distribution: Mexico; naturalized in many warm countries.

FUMARIACEAE

FUMARIA L.

Fumaria indica (Haussk.) Pugsley in Journ. Linn. Soc. Bot. 44: 313. 1919; Blatter *et al.* in J. Bomb. Nat. Hist. Soc. 34 (3): 6676. 1933; Santapau, Fl. Saur. 1: 9. 1962; BH. 26 (1): 219. 1918; Bhandari, Fl. Indian Desert 31. 1978; Parmar, in Shetty & Singh Fl. Rajasthan 1: 64. 1987; Ellis & Balakrishnan, in Sharma & Balakrishnan Fl. India. 2: 84. 1993. *F. vaillantii* Loisel. var. *indica* Haussk. in Flora 56: 443. 1873. *F. parviflora* Lam. subsp. *vaillantii auct. plur.* non *F. vaillantii* Loisel. 1809; Hook. *f.* & Thoms. in Fl.Ind. 1: 258. 1855; Hook. *f.*, Fl. Brit. India 1: 128. 1872; Duthie, Fl. Gangetic Plain 1: 37. 1903. (Fig. 9)

Vern.: *Pitpapra.*

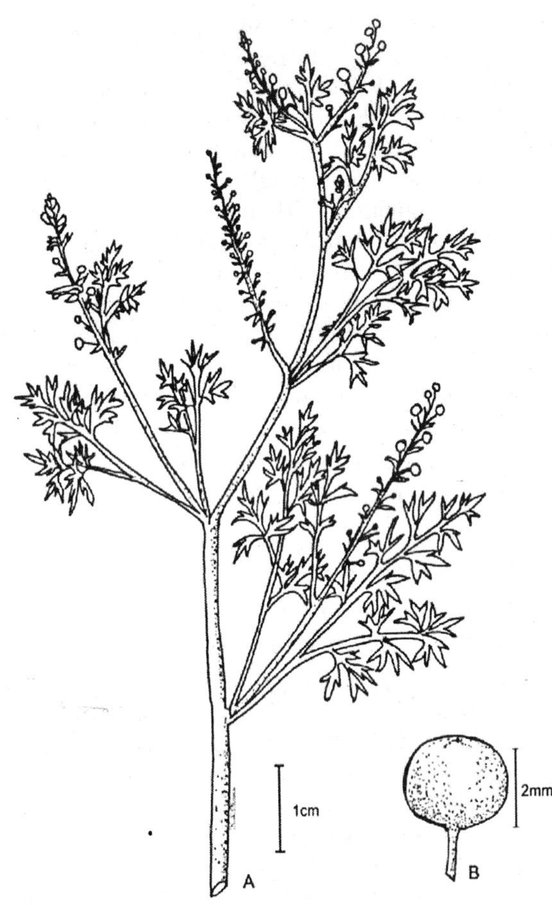

Fig. 9. *Fumaria indica* (Haussk.) Pygsley: A. Habit, B. Fruit.

Small, delicate, much branched, leafy, diffuse herb, 5 - 30 cm long, glaucous, glabrous. Leaves 3 - 10 x 1.5 - 5 cm, long-stalked to subsessile, much dissected, 2 - 3 -pinnatisect or decompound, with 3 - 5 pairs of lateral pinnae and a terminal one; pinnae long-petioluled to subsessile; 1 - 3 x 0.5 - 2.5 cm, often ternately lobed with each lobe deeply, finely lobuled into linear ultimate segments (lobules) up to 5 x 1 mm, flat to slightly channelled. Racemes shortly peduncled to subsessile, leaf-opposed, 6-12-flowered, 10 - 20 mm long; bracts as long as or slightly longer than pedicels; linear, submembranous, whitish. Pedicel erect, 1.5 - 2 mm long, with midian ridge and somewhat expanded margins, slightly broader below the fruit or at its apex. Flowers 5 - 6 mm long, usually white or pale pinkish. Sepals minute, less than 1 x 0.5 mm, sometimes obsolete or inconspicuous, laciniate-dentate, whitish, membranous. Upper petal with very short suborbicular, slightly oblong and obscurely downcurved spur, ca 1.5 mm long (or about ½ of the lamina of upper petal). Fruit ca 2 mm in diam., suborbicular, with usually rounded apex when mature (slightly or minutely apiculate when young), slightly or obscurely keeled (on the margins corresponding to pedicel margins), keel more distinct in slightly young fruits, somewhat rugose when dried and with 2 obscure apical pits, 1-seeded; seed 1 - 1.5 mm in diam., brownish.

Field note: Common weed during winter season in cultivatd fields and wasteplaces.

Fl. & Fr.: March - June.

Specimen examined: Gurla 1519, Karoi 54 (MLVGCB Herb).

Distribution: India, Pakistan, Afghanistan & C. Asia; introduced elsewhere.

Notes: Very similar to *Fumaria parviflora* Lam. but leaf segments flat or almost flat, very much like those in *Fumaria vaillantii* Lois., from which it differs by its bracts which are as long as or slightly longer than pedicels. It is said to be a hybrid between the 2 species. *Fumaria parviflora* Lam. and *Fumaria asepala* Boiss., both recorded by Blatter & Fernandez (l.c. 675-676) from Waziristan are probably mis-identification of this species. These species do not occur in the present study area, but closely resemble *Fumaria indica* in having bracts as long as the pedicels.

Uses: Used as fodder and in local medicines as blood purifier.

BRASSICACEAE

KEY TO THE GENERA

1a. Fruits beaked, beak one or few seeded, seeds in 2 row in each loculus. .... *Eruca*

1b. Fruits not beaked or rarely with seedless stylar beak.. 2

2a. Fruits terete, angular, inflated or compressed parallel to septum, replum equaling the width of fruit... 3

3a. Cotyledons (Radicle) mostly accumbent.................................... ....*Rorippa*

3b. Cotyledons (Radicle) mostly incumbent............................ ...... *Sisymbrium*

2b. Fruits compressed at right angles to the septum, replum much narrower than the width of fruit...4

4a. Fruits indehiscent or breaking into 1- seeded halves.. *Coronopus*

3b. Fruits dehiscent, liberating the seeds free......5

5a. Seeds 1 in each locule........................*Lepidium*

5b. Seeds 2-many in each locule. ....*Cochlearia*

COCHLEARIA L.

Cochlearia cochlearioides (Roth) Santapau & Maheshw. in journ. Bombay nat. Hist. Soc. 54: 804. 1957; Parmar, in Shetty & Singh Fl. Rajasthan 1: 72. 1987; Bhaumik in Sharma & Balakrishnan Fl. India. 2: 191. 1993. *Alyssum cochearioides* Roth, Nov. Pl. Sp. 322. 1721. *Cochlearia flava* Buch.-Ham. ex Roxb., Fl. Ind. 1: 45. 1820; Hook. *f.* & Anders. in Hook. *f.*, Fl. Brit. India 1: 145. 1872; Duthie, Fl. Gangetic Plain 1: 41. 1903.

Annual, erect herbs diffusely branched, glabrous. Roots long, fusiform, sparsely branched. Leaves 2 - 7.0 x 0.5 - 2.5 cm, lyrately pinnatifid; cauline ones 0.8 - 2.0 x 0.4 - 0.6 cm, elliptic-oblong or obovate-oblong, entire or pinnatified. Racemes numerous, 3 - 15 cm long, Flowers small, yellow, slyle short, thick. Sepals oblong, ca 1.5 x 0.75 mm. Fruit ellipsoid, 3 - 5 seeded 5 - 6 x 2 - 3 mm, smooth, valves slightly swollen, membranous, without nerves; style ca 1 mm long. Seeds many, rugose, with filiform funicles, dark brown to blakish.

Field note: Occasional in river beds and other damp places.

Fl.: December - February; *Fr.:* February - April.

Specimen examined: Karoi 422, Nandrai 1223 (MLVGCB Herb).

Distribution: India: Upper and lower Gangetic plains and Gujarat.

CORONOPUS Zinn nom. cons.

Coronopus didymus (L.) Smith, Fl. Brit. 2: 691. 1800; Mahesh., Illus. Fl. Delhi f. 5. 1966; Bhandari, Fl. Indian Desert 31 - 32. 1978; Parmar, in Shetty & Singh Fl. Rajasthan 1: 72. 1987; Bhaumik, in Sharma & Balakrishnan Fl. India. 2: 192. 1993. *Lepidium didymum* L. Syst. 2: 433. 1754. *Senebiera pinnatifida* DC. in Mem. Soc. Hist. Nat. Paris 144, t. 9. 1799. *Senebiera didyma* (L.) Pers. Syn. Pl. 2: 185. 1807. (Fig. 10)

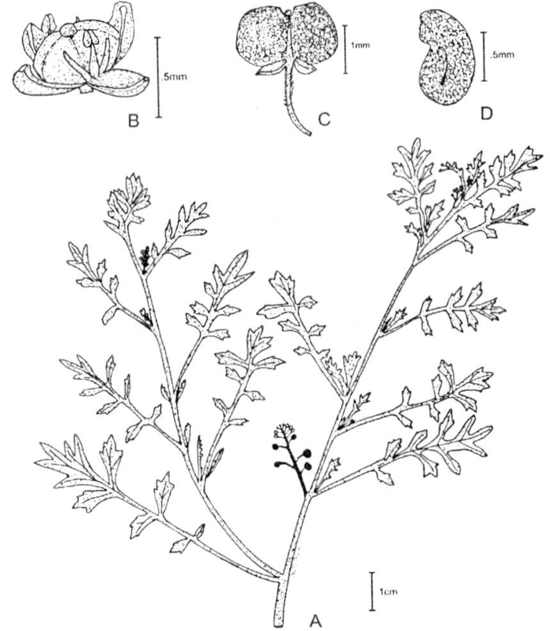

Fig. 10. *Coronopus didymus* (L.) Smith: A. Habit, B. Flower, C. Fruit, D. Seed

Annual or biennial, 10 - 40 cm long, diffuse or procumbent, often somewhat foetid herb, branched mostly from below, glabrous or hairy. Basal and lower leaves rosulate, pinnatisect, stalked, deeply pinnatipartite, 5 - 10 x 1.5 - 2 cm, often glabrous, segments usually pinnatifid; upper leaves similar or pinnatifid and only 3 - 5-partite, 1.5 - 3 cm long, sessile or subsessile; lobes sinuate toothed, often only on one side. Racemes 30 - 60-flowered, dense, up to 8 cm long (usually much shorter) in fruit. Flowers minute, ca l mm across; pedicel short, 3 - 4 mm long in fruit, filiform. Petals present or O. Stamens only 2 or 4, but only 2 fertile. Siliculae shorter than the pedicel, ca 1.5 x 2.5 mm, broader than long, bilobed; valves globose, reticulately rugose, glabrous; septum very narrow, inconspicuous, ca 0.3 mm broad; seed ca 1 - 1.5 mm long, reniform, flattened, reddish-brown.

Field note: Common weed during winter season in cultivatd fields and wasteplaces.

Fl. & Fr.: January- April.

Specimen examined: Bijoliya 152, Kotri 1205 (MLVGCB Herb).

Distribution: Native of S. America, but widely introduced almost throughout the world.

Uses: A good fodder.

ERUCA Mill.

Eruca sativa Mill., Gard. Dict., ed. 8, no.1. 1768; Hook. *f.* & T. Anders. in Hook. *f.*, Fl. Brit. India 1: 158. 1872; Duthie, Fl. Gangetic Plain 1: 46. 1903; Parmar, in Shetty & Singh Fl. Rajasthan 1: 73. 1987; Hajra, in Chowdhery & Bhaumik in Sharma & Balakrishnan Fl. India. 2: 143. 1993. *Eruca vesicaria* (L.) Cav. var. *sativa* (Mill.) Thell. in Hegi, III. Fl. Mittleur. 4 (1): 201. 1918; Bhandari, Fl. Indian Desert 32. 1978. *Brassica eruca* L. Sp. Pl. 667. 1753. *Brassica erucoids* Roxb., Fl. Ind. 3: 117. 1832. (Fig. 11).

Vern.: *Tara Mira*.

Annual, somewhat foetid herb, 25 - 50 cm. tall, erect, simple sometimes branched, hispid rarely subglabrous. Leaves lyrate-pinnatifid, fleshy; lower subrosulate, shortly stalked, 4 - 5-partite, usually 7 - 15 x 3 - 5 cm; terminal lobe somewhat larger, usually ovate, obtuse, irregularly toothed; lateral lobes ovate-oblong; upper leaves 1 - 3-partite, smaller, subsessile or sessile. Racemes 15 - 50-flowered, lax, up to 30 cm long in fruit. Flowers large, 15 - 20 mm across, whitish to dull yellow or dirty white; pedicel 3 - 5 mm long in fruit, becoming somewhat thick and appressed or subappressed to the axis. Sepals 8 - 10 x 1.8 - 2 mm, oblong, hairy, often pale-violet. Petals 15 - 22 x 5 - 6.5 mm, obovate, long clawed, brown or violet veined, apex truncate. Stamens 9 - 13, 11 - 15 mm long; anthers 2.5 - 3. 5mm long. Fruit 15 -

30 x 3.5 - 4.5 mm, ovate-oblong to oblong-ellipsoid (including 6-8 mm long, sabre-shaped, compressed, seedless beak, with 5 parallel veins); valves with a prominent mid-vein only. Seeds 6 - 12 in each locule, ovoid or subglobose, 1.5 -2 mm in diam., pale brown.

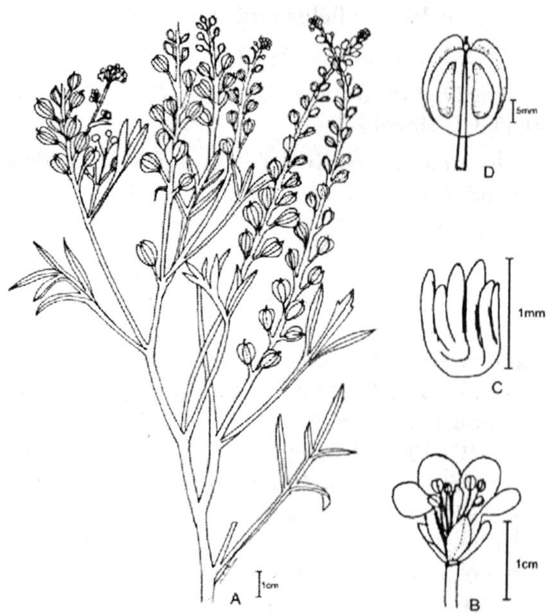

Fig. 11. *Eruca sativa* Mill.:
A. Habit, B. Flower, C. Embryo, D. Fruit

Field note: Widely cultivated for oil and fodder.

Fl. & Fr.: January- April.

Specimen examined: Jahajpur 1137, Sahada 611 (MLVGCB Herb).

Distribution: Europe, N. Africa, S.W. Asia; widely introduced elsewhere.

Notes: *'Tara Mira'* is widely cultivated within this area for fodder and its seed oil, valued medicinally and other purposes. It is a very variable species and *E. cappadocica* Reut., originally distinguished by its long beak and size of seed, is only a form of it.

LEPIDIUM L.

Lepidium sativum L. Sp. P1. 644. 1753; Hook. *f.* & T. Anderson, in Fl. Brit. India 1: 159. 1872; Parmar, in Shetty & Singh Fl. Rajasthan 1: 75. 1987; Bhaumik in Sharma & Balakrishnan Fl. India. 2: 206 - 207. 1993; Duthie, Fl. Gangetic Plain 1: 48. 1903. (Plate 18)

Vern.: *Ahlya*.

Annual, herb, upto 50 cm tall, erect, branched, glabrous, rarely pilose. Lower leaves long-petioled, irregularly pinnatisect or lyrate-pinnate, 4-10 x 2.5 - 3.5 cm, stalked to subsessile; uppermost leaves linear, sessile. Racemes much branched, each 20-30-flowered, ebracteate. Flowers small, ca 3 mm across, white or pinkish; pedicel 2 - 4 mm long in fruit, not thickened, ± ascending or subappressed. Sepals ca 1.5 x 1 mm, oblong, obtuse, sparsely pubescent. Petals spathulate, ca 2.5 x 2.75 mm, narrowed below, apex rounded. Stamens 6, ca 1.8 - 2.2 mm long; anthers minute. Fruits 4.5 - 6 x 3.5 - 5 mm, broadly elliptic to suborbicular, glabrous, narrowly winged and distinctly notched at the apex with short style included with-in; septum ca 1 mm broad. Seed ca 3 x 1 mm, brown.

Field note: Common weed of cultivated areas.

Fl. & Fr.: January- April.

Specimen examined: Karoi 55, Sahara 921 (MLVGCB Herb).

Distribution: Native of Egypt & W. Asia; introduced and naturalized elsewhere or cultivated throughout the world.

Uses: Fresh leaves are used as salad.

RORIPPA Scop.

Rorippa indica (L.) Hiern, Cat. Afr. Pl. Welw. 1: 26 add. et corr. 1896 et 2: 481, errata 1899; Parmar, in Shetty & Singh Fl. Rajasthan 1: 77. 1987; Bhaumik in Sharma & Balakrishnan Fl. India. 2: 129. f. 22. 1993. *Sisymbrium indicum* L., Mant. Pl. 1: 93. 1767. *Nasturtium indicum* (L.). DC., Syst. Nat. 2: 199. 1821; Hook .f. & T. Anderson in Hook. *f.*, Fl. Brit. India 1: 134. 1872; Duthie, Fl. Gangetic Plain 1: 39. 1903.

Annual herb, 50 - 60 cm tall, sparsely hairy to glabrous, often much branched above. Lower leaves deeply lyrate-pinnatifid to pinnatipartite with 1 - 4 segments on either side, 6 - 10 x 2.5 cm;

upper ones sessile, ovate-lanceolate to lanceolate-oblong, entire, lyrate-pinnatifid or dentate above base. Racemes elongated, 25 - 35-flowered, up to 30 cm long in fruit. Flowers bright yellow, ca 2.5 mm across; pedicels up to 5 mm long in fruit, ascending or subspreading. Sepals 2 - 2.5 mm long. Petals usually about as long as the sepals, sometimes reduced or suppressed. Stamens ca 1.5 - 2 mm long. Stigma subsessile. Fruits usually 10 - 15 x ca 1.5 mm, slender but often slightly up-curved, blunt apexed with ca 1 mm long. Seeds sub-biseriate or biseriate, minute, ca 0.5 mm, rugose, spherical, usually about 100 or more in each fruit.

Field note: Rarè in moist, open areas.

Fl. & Fr.: December - March.

Specimen examined: Patola Mahadev 1306, Shahpura 1352 (MLVGCB Herb).

Distribution: Himalaya, India, Pakistan, Indonesia and Africa

Notes: Sometimes nodal thread-like roots are present below in addition to the main root.

SISYMBRIUM L.

Sisymbrium irio L. Sp. Pl. 659. 1753; Hook. *f.* & T. Anderson in, Fl. Brit. India 1: 150. 1872; Duthie, Fl. Gangetic Plain 1: 42. 1903; Bhandari, Fl. Indian Desert 35. 1978; Parmar, in Shetty & Singh Fl. Rajasthan 1: 77-78. 1987; Hajra, Chowdhery & Das, in Sharma & Balakrishnan Fl. India. 2: 243 - 244. 1993. *S. irioides* Boiss. in Ann. Sci. Nat. ser. 2, 17: 76. 1842. (Plate 19)

Vern.: *Asalio.*

Annual, 15 - 70 cm tall, erect, branched, glabrous or sparsely hairy with simple hairs. Basal and lower leaves petioled, numerous in rosette, pinnately lobed with a hastate terminal lobe larger than the laterals, 2 - 6-partite, very variable in size; cauline leaves similar but only 1 - 3-partite. Racemes 40 - 80-flowered, up to 30 cm long in fruit with siliquae overtopping young flowers and buds. Flowers ca 3 mm across, yellow; pedicels 10 - 15 mm long in fruit, filiform, ascending or sub-spreading, rarely spreading. Sepals 2 - 2.5 mm long. Petals 3 - 4 x ca 1 mm, usually slightly longer than the sepals. Stamens ca 2 - 3 mm long; anthers ca 0.5 mm long. Fruit 30 - 45 x 1 mm, linear, often slightly upcurved; valved submembranous, convex, obscurely to distinctly torulose, faintly 3-veined but the mid-vein distinct, glabrous; style inconspicuous, thickened, with a depressed, sub-bilobed stigma; septum hyaline. Seeds 20 - 40 in each locule, ca 1 mm long, oblong-ellipsoid, yellowish-brown.

Field note: A common species found from hills to plains in this area, very variable in leaf, flower and pedicel size but rocket like leaves, smaller flowers and fruits on thin ascending pedicels distinguishes it.

Fl. & Fr.: January- March.

Specimen examined: Bijoliya 832, Karoi 1693 (MLVGCB Herb).

Distribution: Europe, most of Asia and N. Africa

Uses: The whole green plant is used as fodder to increase milk production in cattles. Seeds are used medicinally.

CLEOMACEAE

CLEOME L.

KEY TO THE SPECIES

1a. Leaves simple. 2

2a. Leaves broadly ovate, elliptic or suborbiculr. broad. ....*C. scaposa*

2b. Leaves obovate, lanceolate, linear or oblong-elliptic. Stamens 8 - 12. ....*C. simplicifola*

1b. Leaves compound, 3 - 9 foliolate.................3

3a. Androgynophore mostly absent, if present under 3 mm long. ....*C. burmannii*

3b. Androgynophore conspicuous, 5 mm or more. 4

4a. Flowers white or dirty white to pinkish. stamens 6. .... *C. gynandra*

4b.Flowers yellow, stamens more than 10 ....*C. viscosa*

Cleome burmannii Wight & Arn., Prodr. 1: 22. 1834; Hook. *f.* & Thomson in Hook. *f.*, Fl.

Brit. India 1: 170. 1872; Sundara Raghwan in Sharma & Balakrishnan, in Fl. India 2: 304 - 305. 1993. Yadav & Meena, Rheedea 19 (1 & 2): 65 - 66. 2009. (Fig. 12 & Plate 20)

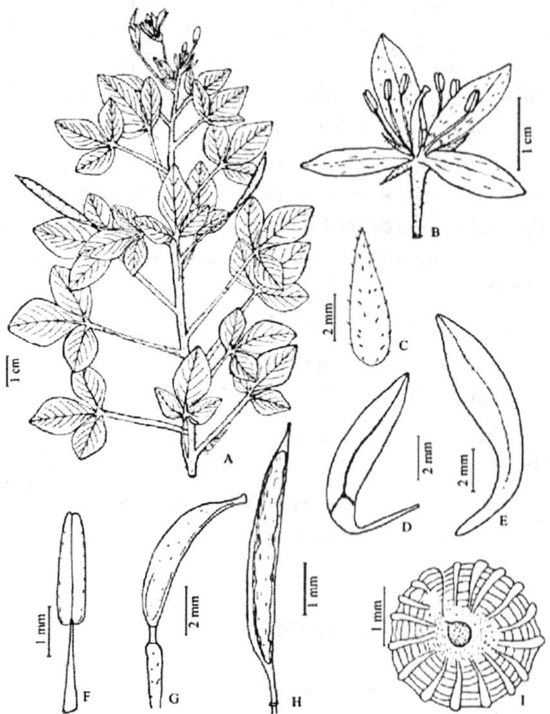

Fig. 12. *Cleome burmannii* Wights & Arn.: A. Flowering and fruiting branch, B. Flower, C. Calyx outer surface, D. & E. Petals; F. Stamen,G. Ovary, H. Fruit, I. Seed

Herbaceous, annual, erect and scattered soft prickles with fibrous root, branched, 15 - 50 cm tall; stem erect, much branched, somewhat flattened, ribbed, clothed with scattered soft 0.5 - 2.5 mm long prickles. Leaves alternate 1 - 7 cm long, palmately compound with 3 leaflets, petiolate, petiole 2.5 - 4.0 cm long, upper smaller, leaf 1 - 7 cm long, leaflets with serrulate margins, ovate to elliptic- lanceolate, cuneate and webbed at base, acute, subacute or subrounded at apex, faintly crenate- serrulate along margins, 2.0 - 6.0 x 0.4 - 1.8 cm, nerve pair on middle leaflet 7 and on lateral leaf lets 5, leaflets glabrous, petiolules up to 1 mm long. Flower purple, axillary, solitary, zygomorphic by displacements of petals to one side, tetramerous, pedicilate, pedicels 5 - 18 mm glandular hairy. Calyx 4, green, hairs present on outer surface, 0.3 to 0.4 cm long not much longer than the tube, polysepalous, lanceolate, acuminate. Petals 4, purple, clawed, elliptic-oblong, 8 - 11 mm long, polypetalous. Stamen filament pink, 8-11 mm long, dorsifixed, dithecous, free, dehiscing by longitudinal slits. Gynophore short, 0.5 - 1.5 mm, elongating up to 4 - 7 mm after fertilisation. Ovary bicarpellery unilocular, syncarpous, placentation parietal; stigma sessile, capitate. Capsules linear, terete, torulose, glabrous, 3.5 - 4.5 cm long, 1 - 1.5 mm thick, narrowed at both ends, beaked. Seeds 10 - 15, 1.2 - 1.4 mm across, with concentric ribs and closed cleft, with elaiosome.

Field note: Extremely rare (endangered) in waste lands, sites of old cultivation, road sides, gardens etc. on moist sandy soil & wet places.

Fl. & Fr. : May - November

Distribution: India (Rajasthan, Tamil Nadu and Kerala).

Specimen examined: Pansal 51, Triveni 1576 (MLVGCBHerb).

Notes: Extremely rare. Besides type, the only other historic collection from Peninsular India is Koenig's specimen from Ramanathapuram in Tamil Nadu. Venkobarao had collected it from Shenkotta along Kerala- Tamil Nadu border in 1914. Ellis in 1962 had collected it from Tanjavur district of Tamil Nadu (Sharma and Balakrishnan 1993). Reports of its collection from Gujarat (Shah, 1978), Kerala (Manilal and Sivarajan, 1983) and reports from other parts of Tamil Nadu are misidentification of *C. rutidosperma* DC. (Sundara Raghvan, 1984).

Cleome gynandra L. Sp. Pl. 2: 671. 1753; Bhandari, Fl. Indian Desert 38-39. 1978; Parmar, in Shetty & Singh Fl. Rajasthan 1: 80 - 81. 1987; Sundara Raghwan in Sharma & Balakrishnan, Fl. India 2: 309-310. 1993. *Gynandropsis pentaphylla* (L.) DC. Prodr. 1: 238. 1824; Hook. *f.* & Thomson in Fl. Brit. India 1: 171. 1872; Duthie, Fl. Gangetic Plain 1: 51. 1903. *G. gynandra* (L.) Briq. in Ann. Cons. Jard. Bot. Geneve 17: 382. 1914. (Fig. 13)

var. ***gynandra***

Vern.: *Dhoro bagro, Hulhul.*

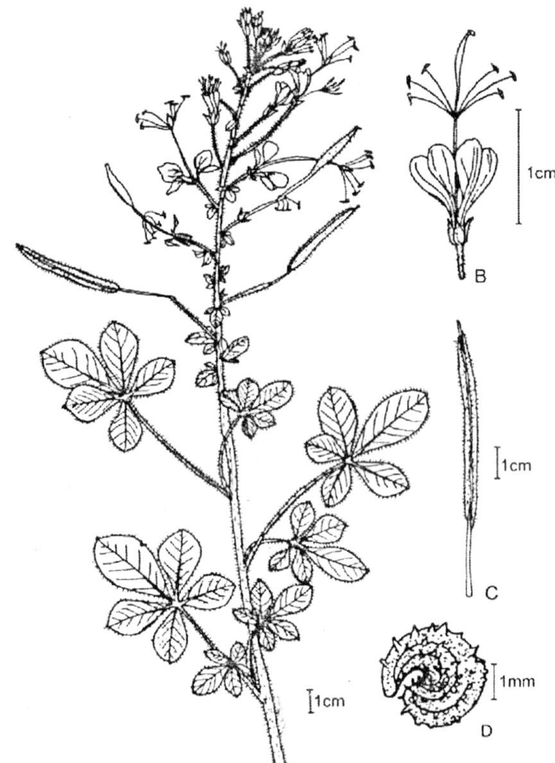

Fig. 13. *Cleome gynandra* L. A. Flowering twig, B. Flower, C. Fruit, D. Seed

Annual herb, erect or suberect, 10 - 110 cm tall, strongly foetid. Stem glandular, pubescent. Leaves petiolate, 3 - 5-foliolate, very variable in size; leaflets 7 x 6.5 cm, oblong-elliptic, oblanceolate or rhombate, cuneate at base, usually acute or slightly acuminate, finely crenate to subentire; petiole 3 - 10 cm long. Inflorescence lax, many flowered corymbose raceme, showy, up to 30 cm long, bracteate, bracts much smaller than the leaves. Flowers 1 - 2 cm across, mostly white or pinkish; pedicels generally 1 - 2 cm long. Sepals 4, ovate to obovate, elliptic or oblanceolate, 1.5 - 5 x 0.5 - 1.5 mm, glandular. Petals 4, obovate-oblanceolate, long clawed, 3 - 20 x 1 -10 mm. Androphore 10 - 25 mm long. Stamens 6, filaments subequal, 1.4 - 2 cm long, stiff and somewhate spreading conspicuously from a long androphore, filaments often pinkish or lilac, anther linear, ca 2mm long. Gynophore upto 2cm long, usually shorter in flowers but increasing in fruit. Ovary green, style 1 - 2 mm long, stigma discoi, depressed at apex, purple. Fruit capsular, 3-12 cm long, 0.2 - 0.8 cm broad, linear, narrowed towards both the ends, glandular-pubescent, finely and obliquely striated, many seeded. Seeds 1 - 1.8 mm in diam., finely longitudinally striated, with slightly cristate transverse ridge, dark-brown or black.

Field note: Throughout the area at road sides and waste lands. It is sometimes attackd by *Cuscuta* spp.

Fl. & Fr.: September to December.

Specimen examined: Pansal 81, Bhilwara 1627 (MLVGCB Herb).

Distribution: Pantropical

Uses: Leaves used in treating rheumatism. Leaf juice is used to treat headache by applying to forehead and pouring into nostrils. Seed paste is used in killing lice. Boiled seeds are used in curing cough.

Cleome scaposa DC., Prodr. 1: 239. 1824; Parmar, in Shetty & Singh Fl. Rajasthan 1: 80 - 82. 1987; Sundara Raghwan in Sharma & Balakrishnan, Fl. India 2: 314. 1993. *C. gracilis* Edgew. in J. Asiat. Soc. Beng. 16: 111. 1847; Bhandari, Fl. Indian Desert 39. 1978. *C. papillosa* Steud. Nom. ed. 2 (1): 382. 1840 (nom. nud.); Hook. *f.* & Thomson in Fl. Brit. India 1: 168. 1872; BH. 26 (1): 221. 1918. (Fig. 14 & Plate 21)

Annual, erect, upto 60 cm tall, slender, sparsely branched from below, hairy with glandular or eglandular hairs. Leaves simple, broadly ovate, elliptic or suborbicular, 1.3 - 3 x 0.5 - 2.4 cm, scarbid, lateral nerves 2 - 4 pairs; petiole 5 - 30 mm long, ciliate hairy; upper leaves almost subsessile. Inflorescence rather lax, racemose, elongated, increasing up to 20 cm in fruit, often bracteate below. Flowers 3 - 5 mm across, actinomorphic, white turning yellowish rarely pinkish; pedicel 2.5 - 4.5 mm long increasing up to 10 mm in fruit, filiform, ultimately spreading, slender. Bracts foliaceous at lower flowers, absent at upper flowers. Sepals about 2 mm long, lanceolate or elliptic, 1.5-2 mm, glandular pubescent. Petals 2.5 - 3.5 x 3 - 4 mm, ovate or obovate, glabrous, yellowish. Stamens 6, about as long as the petals, not exserted. ovary sessile, linear-oblong. Capsule linear, 20 - 30 x 0.8 - 1.2 mm, slender, sessile, often spreading and arcuately curved, glabrous or slightly glanular, many seeded. Seeds about 0.7 mm in diam., glabrous, minutely granulate, brown-black.

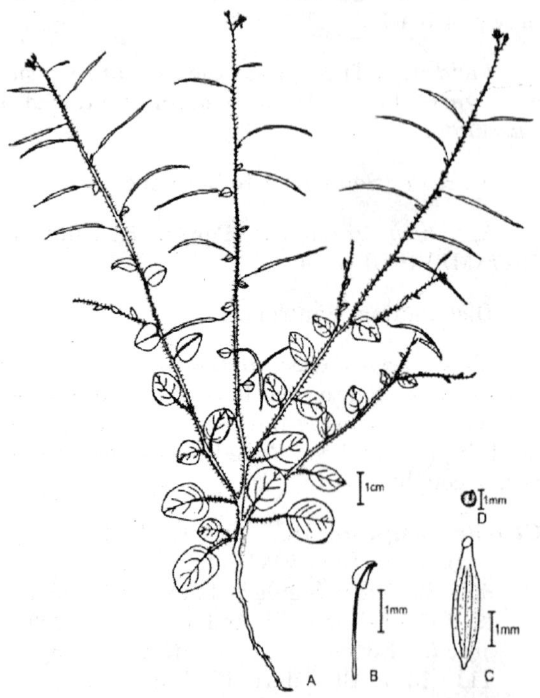

Fig. 14. *Cleome scaposa* DC. :
A. Habit, B. Stamen, C. Fruit, D. Seed

Field note: Rare, found in gravelly soils.

Fl. & Fr.: August- December

Specimen examined: Mezadam 1558, Bhilwara 1748 (MLVGCB Herb).

Distribution: North and Tropical Africa, Egypt, Arabia, West Pakistan and India.

Cleome simplicifolia (Camb.) Hook. *f.* & Thomson in Fl. Brit. India 1: 169. 1872; Duthie, Fl. Gangetic Plain 1: 49. 1903; Parmar, in Shetty & Singh Fl. Rajasthan 1: 82-83. 1987; Sundara Raghwan in Sharma & Balakrishnan, Fl. India 2: 314 - 315. 1993. *Polanisia simplicifolia* Camb. in Jacq. Voy. Bot. 20, t. 20, 1844. *P. burtporensis* Munro in Wight, *l.c.* 3 (4): 5. t. 1072. 1846. *Cleome asperrima* Blatt. in J. Asiat. Soc. Bengal n.s. 26: 340. 1937, *syn. nov.* (Fig. 15 & Plate 22)

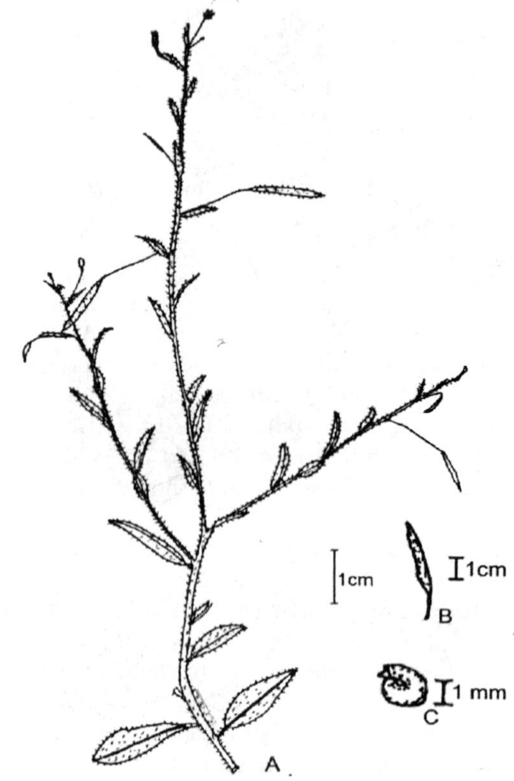

Fig. 15. *Cleome simplicifolia* (Camb.) Hook.f. & Thoms: A. Habit, B. Fruit, C. Seed

Erect herb, upto 60 cm, branched. Stem scabrid trigose. Leaves obovate, lanceolate, linear-oblong or oblong-elliptic, tapering at base, obtuse or subacute at apex, densely strigose with scaly hairs; petioles scabrid at base, upto 2 cm long but almost absent upwards. Flowers 4 - 5 mm across, pink or purple, solitary in axils of leafy bracts forming lax few-flowered racemes. Pedicels upto 1.8 cm long, filiform. Sepals triangular or ovate, acute at apex, 1.8 - 3.5 x 0.8 - 1.5 mm, scabrid. Petals oblong, obovate, oblanceolate or spathulate, indistinctly clawed at base, rounded with a mucro at apex. Stamens 8 - 16; filament glabrous, 3 - 5 mm long, slender throughout or slightly thickened at apex. Ovary cylindric. Capsule striate, subtorulose. Seeds 4 - 15, reniform, 1.5 - 1.8 mm, yellowish brown, turning black after drying, elaiosome white.

Field note: Rare, in waste lands.

Fl. & Fr.: Rainy season.

Specimen examined: Bari Sadari 3511 (MLVGCB Herb).

Disribution : India, Endemic.

Cleome viscosa L. Sp. pl. 1:672. 1753; Hook. *f.* & Thomson in Fl. Brit. India 1: 170. 1872; BH. 26 (1) : 221. 1918; Duthie, Fl. Gangetic Plain 1: 50. 1903; Bhandari, Fl. Indian Desert 40 - 41. 1978; Parmar, in Shetty & Singh Fl. Rajasthan 1: 82-83. 1987; Sundara Raghwan, in Sharma & Balakrishnan, Fl. India 2: 309 - 317. 1993. *Cleome icosandra* L. Sp. Pl. 1 : 672. 1753. *Polanisia viscosa* (L.) DC. Prodr. 1: 242. 1824. *P. icosandra* Wt. & Arn. Prodr. Fl. Ind. 22. 1834; Merrill., Enum. Phil. Fl. Pl. 21: 209. 1923. (Fig. 16 & Plate 23)

var. ***viscosa.***

Vern.: *Bagro.*

Strict or considerably branched annual 2-16 dm tall, densely glandular-pubescent throughout, unarmed; petioles 1-6 cm long, exstipulate; leaflets 3-5, oblanceolate-elliptic, acute to obtuse at apex, cuneate to attenuate at base, the central leaflets 1-6 x 0.5 - 3 cm, entire, densely glandular-ciliate, petiolules 1-3 mm long. Flowers relatively few, solitary in axils of upper leaves; pedicels 5-23 mm long; sepals free, narrowly elliptic to oblanceolate-elliptic, acuminate, 5-10 x 1-4 mm, deciduous; petals pale to deep yellow or orange-yellow, obovate to spatulate, rounded at apex, gradually attenuate to base, 6 - 14 x 3-5 mm; disk obsolete. Stamens 12 - 18, included, the filaments 5 - 8 mm long, 4 - 10 of the adaxial shorter than others and slightly clavate; anthers 2 - 3 mm long. Ovary cylindric, densely glandular-pubescent, sessile; style 1- 3.5 mm long at anthesis, to 4 - 5 mm long in fruit; stigmas capitate. Fruits linear-cylindric, attenuate toward apex, sessile, 3 - 10 x 2 - 5 cm, erect, glandular-pubescent, teret, linear-oblong, tapering at both ends, beaked. Seeds obovoid-reniform, 1.0 - 1.8 mm across, dark reddish brown, strikingly rugulose transversely with about 30 rugae, these again minutely transversely rugulose; closed at cleft with very narrow opening, glabrous.

Field note: Often locally abundant as a naturalized weed in cultivated fields and along roadsides.

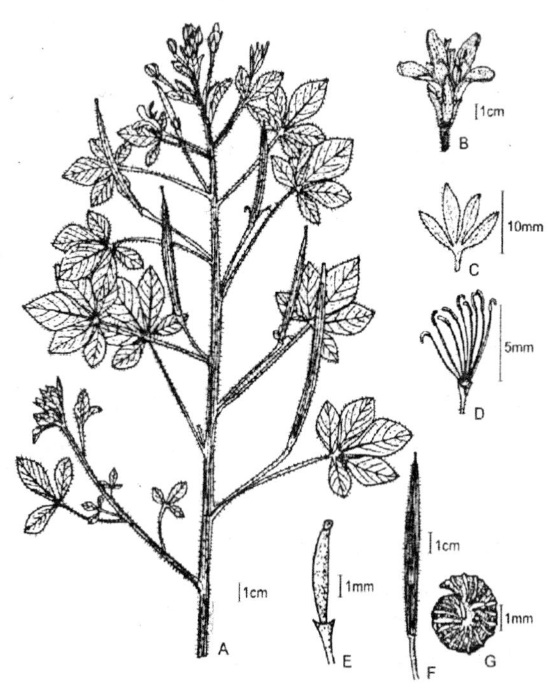

Fig. 16. *Cleome viscosa* L. : A. Flowering and fruiting branch, B. Flower, C. Calyx, D. Stamens, E. Ovary, F. Fruit, G. Seed

Fl. & Fr.: Rainy season.

Specimen examined: Mandal 80 (MLVGCB Herb).

Distribution: Old World, probably indigenous of Asia, now pantropical.

Uses: The paste of whole plant is applied externally for healing ulcer & wounds. One fourth tsp leaf juice is used in fever and as ear drops to reduce earache.Leaf extract has strong antifungal properties.

The seeds are carminative, anthelmintic, effective in diarrhoea, kidney pain and liver disorders. The poultice of seeds relieve chronic joint pains.

CAPPARIDACEAE

KEY TO THE GENERA

1a. Climber or straggling shrubs. Sepals fused below forming a distinct tube ....*Maerua*

1b. Shrubs or trees. Sepals distant, inserted on the edges of a distinct disc. 2

2a. Unarmed tree. Leaves trifoliolate. ....*Crateva*

2b. Armed shrubs. Leaves simple. ....*Capparis*

CAPPARIS L.

KEY TO THE SPECIES

1a. Plant leafless, if present, minute (only on young twigs), early caducous. Flowers red. ....*C. decidua*

1b. Plants prominently leafy. Flowers white or purplish. 2

2a. Leaves cuneate, rounded; flowers in umbels, ripe berries dark purple to black. ...*C. sepiaria*

2b. Leaves ovate to lanceolate, flowers supra-axillary, ripe berries reddish-orange to purple ....*C. zeylanica*

Capparis decidua (Forssk.) Edgew. in J. Linn. Soc. 6: 184. 1862; Mahesh, Illu. Fl. Delhi f. 12. 1966; Bhandari, Fl. Indian Desert 37. 1978; Parmar, in Shetty & Singh Fl. Rajasthan 1: 85. 1987; Sundara Raghwan in Sharma & Balakrishnan, Fl. India 2: 265. 1993. *Sodada decidua* Forssk. Fl. Aegypt. - Arab. 81. 1775. *Capparis aphylla* Roth, Nov. Sp. Pl. Ind. Or. 238. 1821; Duthie, Fl. Gangetic Plain 1: 53. 1903. (Fig. 17 & Plate 24)

Vern.: *Kair, Ker.*

Low shrub to small tree, leafless green, up to 5 m high; branches crooked, spiny. Leaves present on young twigs, caducous, linear, 4 - 20 x 1 - 3 mm, often spine-tipped, subsessile; stipular spines 1 - 6 mm long, straight or slightly curved, yellow or brown. Inflorescence few to many flowered, ebracteate racemes. Flowers showy, coral-pink or brick-red, orange-red or yellowish, 1 - 2.5 cm across; pedicel 1 - 1.5 cm, slender. Sepals petaloid, unequal, 5 - 10 x 3 - 5 mm, ovate-oblong, upper one distinctly saccate, floccose-ciliate margins. Petals about as long as the sepals, puberulous, upper pair slightly larger and hidden in the saccate sepal. Stamens generally 10 - 16, filaments 10 - 35 mm long, reddish. Gynophore 10 - 15 mm long, slender, glabrous; ovary glabrous, globose, beaked. Fruit globose, 10 - 20 mm across, slightly beaked, glabrous smooth, deep red when ripe, pericarp thin. Seeds reniform, 2 - 5 x 2 - 4 mm.

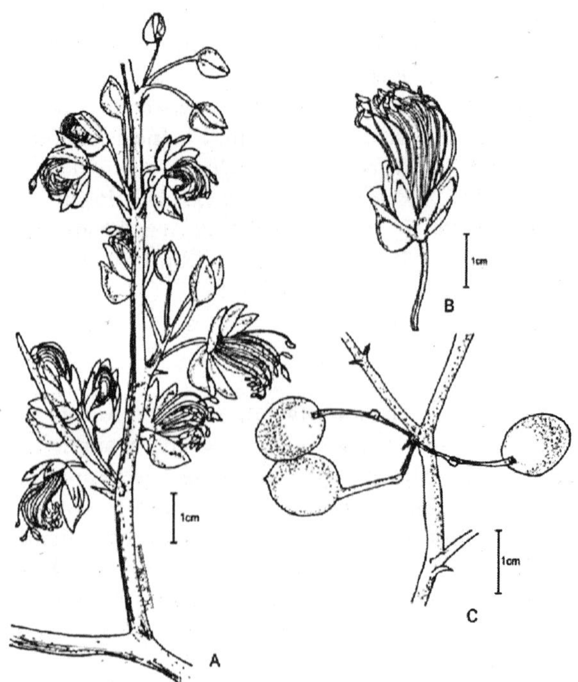

Fig. 17. *Capparis decidua* (Forsk.) Edgew: A. Flowering twig, B. Flower, C. Fruiting twig

Field note: One of the common shrubs of arid plains, flowering abundantly during March to April.

Fl.: February - May; *Fr.*: June - November.

Specimen examined: Pissangan 4102, Shahpura 1129 (MLVGCB Herb).

Distribution: N. and Tropical Africa,Arabia, eastward to India.

Uses: The wood is hard, bitter and resistant to attacks of white ants; it is used for making knees of boats. The young fruits and flower buds are pickled. Flower buds are eaten to relieve stomachache. Root paste is applied on the scorpion bite. 1-2 tsp powdered coal of stem with 1 cup of water is taken for the treatment of bone fracture.

Capparis sepiaria L., Syst. Nat. ed. 2. 1071. 1759; Hook. *f.* & Thoms. in Fl. Brit. Ind. 1: 177. 1872; Duthie, Fl. Gangetic Plain 1: 53. 1903; Cooke, Fl. Pres. Bomb. 1: 48. 1903; Mahesh, Illu. Fl. Delhi f. 13. 1966; Parmar, in Shetty & Singh Fl. Rajasthan 1: 85 - 86. 1987; Sundara Raghwan in Sharma & Balakrishnan, Fl. India 2: 289 - 290. 1993. *Capparis incanescens* DC., Prodr. 1: 247. 1824. *C. sepiaria* L. var. *incanescens* (DC) Hook. *f.* & Thoms, *l.c.* 177. *C. glauca* Wall. ex Hook. *f.* & Thomson, *1. c.* 180. (Fig. 18 & Plate 25)

var. ***sepiaria***

Vern.: *Heens, Khantar.*

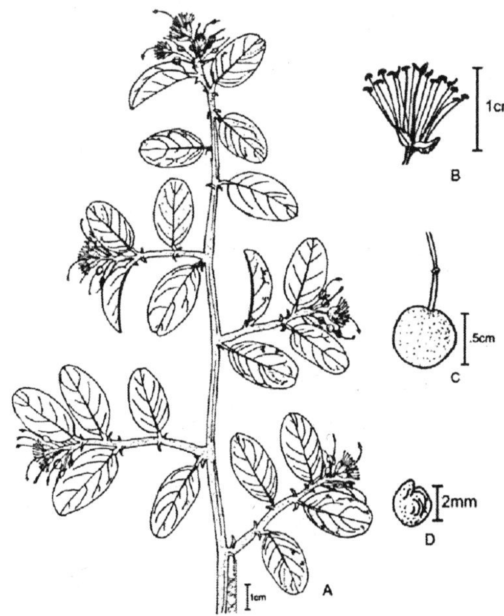

Fig. 18. *Capparis sepiaria* L. var. *sepiaria*
A. Flowering branch, B. Flower, C. Fruit, D. Seed

Much branched shrub, 2 - 5 m tall, more or less pubescent with ashy grey simple hairs. Twigs zig-zag. Stipular spines 2 - 5 mm long, hooked, often brown-black. Leaves elliptic-oblong to ovate or suborbicular, 1.5 - 5 x 1 - 2 cm, acute, softly pubescent to glabrous, lateral nerves 4 - 6 pairs; petiole 2 - 5 mm long, pubescent. Flowers usually in corymbose terminal clusters of 10 - 30, small, 5 - 10 mm across, white; pedicels 1 - 2 cm long, not thickened in fruit, pubescent. Sepals 3 - 6 x 2.5-3.5 mm, subequal. Petals oblong-spathulate, 3.5 - 8 x 2-3 mm, more or less hairy at the base. Stamens many, 5 - 10 mm long. Gynophore 5 - 10 mm long, slightly thickened in fruit; ovary ovoid, glabrous, ovules few. Fruit globose, 8 - 12 mm in diam., crustaceous, dark-brown, smooth, 1-2 seeded. Seeds ca 6 mm, brown.

Field note: Common in the scrub forest and hedges of crop fields.

Fl.: February -April; *Fr.*: April - June.

Specimen examined: Pansal 1233, Hamirgarh 1756 (MLVGCB Herb).

Distribution: India, Sri Lanka, Pakistan, Malesia, Indochina, China, Australia and tropical Africa

Uses: It is used as a hedge plant.

Capparis zeylanica L., Sp. Pl. ed. 2. 720. 1762; Dunn in Kew Bull. 62. 1916; Hook. *f.* & Thoms., in Hook. *f.* Fl. Brit. Ind. 1: 174. 1872; Parmar, in Shetty & Singh, Fl. Rajasthan 1: 86. 1987; Sundara Raghwan in Sharma & Balakrishnan, Fl. India 2: 298 - 299. 1993. *C. horrida* L., *f.* Suppl. 264. 1781; Hook. *f.* Thoms. *l.c.* 174. 1872; Duthie, Fl. Gangetic Plain 1: 53. 1903. (Plate 26)

Vern.: *Heens, Kanthar.*

Climbing or scrambling shrub, 2 - 5 m long, densely armed with 4 - 6 mm long recurved stipules, young parts clothed with scarlet-brown rufous tomentum; hairs irregularly branched and soon falling off. Leaves simple, elliptic-oblong, rarely ovate or obovate, subcordate or cuneate at base, acute, 2.5 - 12 x 2-7 cm, lateral nerves 4 - 8 pairs; petiole up to 1.5 cm long; stipule recurved, ca 4 mm long, yellowish-brown. Flowers showy, pink or purple, supra axillary rows of 2 or sometimes 1 or 3 flowers, 1.5 - 2.5 cm across, the upper most largest and the oldest; pedicels 1.5 - 3.5 cm long, tomentose. Sepals 8 - 12 x 5 - 8 mm, subequal, tomentose. Petals 12 - 18 x 5 - 12 mm, subentire, hairy on the outer surface. Stamens indefinite (30 - 50), filaments creamy white turning to pink, about twice as long as the petals. Gynophore glabrous, 3 - 4.5 cm long, thickened in fruit. Fruit globular or ellipsoid, pustulate, pericarp thick and strong, smooth, reddish orange to purple. Seeds many, embedded in scarlet pulp. 5-7 x 4 -6 mm.

Field note: Along hedges, roadsides, scrub forests.

Fl. & Fr.: March-June.

Specimen examined: Keli 3510, Nimbahera 4176 (MLVGCB Herb).

Distribution: India, Sri Lanka, East Pakistan, eastward to Malesia and Indochina.

Notes: The leaves are highly variable, sometimes narrowly lanceolate with a pair of lateral basal lobes, hence appearing hastate.

Uses: It is used as a hedge plant. The fresh leaves are boiled in water and bath is taken to cure swelling in body.

CRATEVA L.
(Crataeva)

Crateva adansonii DC. subsp. ***odora*** (Buch. - Ham) Jacobs in Blumea 12: 198. 1964; Mahesh, Illu. Fl. Delhi f. 14. 1966; Parmar, in Shetty & Singh Fl. Rajasthan 1: 87 - 88. 1987; Sundara Raghwan in Sharma & Balakrishnan Fl. India 2: 322 - 324. 1993. *C. odora* Buch. - Ham. in Trans. Linn. Soc. 15: 118. 1827. *C. roxburghii* R. Br. in Denh. & Clapp. Narr. Trav. Disc. Afr. App. 224. 1826. *C. religiosa* Forst. f. var. *roxburghii* (R. Br.) Hook. *f.* & Thoms. in Hook. *f.* Fl. Brit. Ind. 1: 172. 1872; Duthie, Fl. Gangetic Plain 1: 52. 1903. *C. nurvala* sensu Blatter in J. Bombay Nat. Hist. Soc. 35: 293. 1931, non Buch.-Ham. (Fig. 19 & Plate 27)

Vern.: *Barna*.

Unarmed, deciduous tree, 6 - 15 m tall, glabrous. Leaves 3-foliolate, petiolate; leaflet shortly petiolulate, elliptic, elliptic-lanceolate, ovate-lanceolate, more or less oblique, acuminate, narrowed towards the base, 5 - 12 x 2.5 - 6 cm, herbaceous to subcoriaceous, often turning brownish when dry; petiole 4 - 8 cm long. Inflorescence corymbs 12 - 30-flowered, rachis 2 - 3 cm long, often growing through with a few axillary flowers. Flowers 3- 6 cm across, subirregular, greenish-white turning yellowish after opening, appearing before or just after the flush of new leaves, pedicellate, ebracteate; pedicels 2 - 4 cm long. Sepals 4 - 5 mm long, ovate-deltoid, greenish-white, distant. Petals ovate, or suborbicular, reticulately veined, eventually 20 - 30 x 10 -17 mm (incl. 4-8 mm long claw); 2 adaxial ones generally somewhat larger. Stamens usually 20 - 25, exceeding the petals (3-5 cm long), inserted at base of the gynophore. Gynophore generally 3 - 5 cm long, slender, thickened in fruit, often deciduous inhibiting the development of fruit. Fruit globose, 3 - 5 cm in diam., yellowish. Seeds broadly ellipsoid, 5 - 6 x 2 mm, smooth, brown.

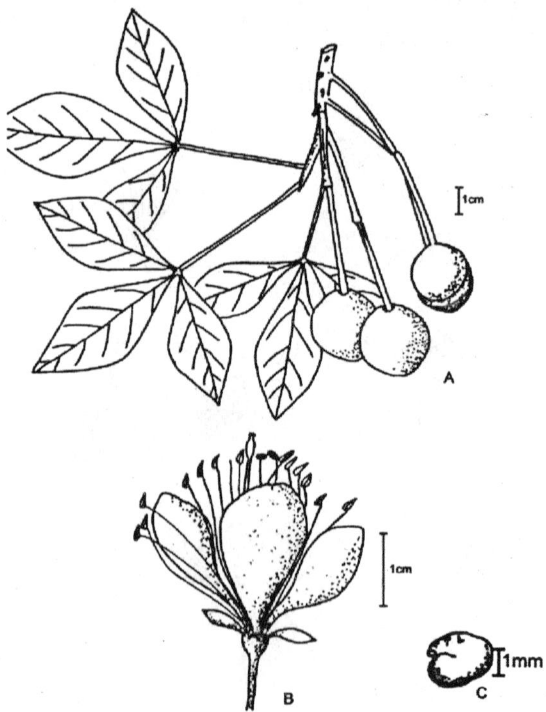

Fig. 19. *Crateva adansonii* DC. subsp. *odora* (Buch.-Ham.) Jocobs.: A. Fruiting branch, B. Flower, C. Seed.

Field note: Common in open forests or scrub forests.

Fl. : April - May; *Fr.* : May - August.

Specimen examined: Luhari kala 1757, Rawaliya 3702 (MLVGCB Herb).

Distribution: India, Pakistan, Myanmar and Sri Lanka.

Notes: It is distinguished from *Crateva religiosa* Forst. f. by its herbaceous, more or less coriaceous leaflets turning brownish when dry (not thin papery and greenish) and smaller fruits. The two have fewer (less than 10) pairs of nerves on their leaflets, which distinguishes them from *Crateva nurvala* Ham. (which usually has more than 10 pairs of nerves). Also, the fruit is ovoid in *C. nurvala*.

Uses: Bark crushed and the filterate is used to increase lactation by tribal ladies. Bark extrct with pepper used to cure chest pain and menstrual disorders.

MAERUA Forssk.

Maerua oblongifolia (Forsskal) A. Rich. in Guill. & Perr. Fl. Seneg. Tent. 1: 32. t. 5. 1831; Hook. *f.* & Thoms., in Hook. *f.*, Fl. Brit. Ind. 1: 171. 1872; Blatter, in Journ. Bomb. Nat. Hist. Soc. 31: 900.1927; Bhandari, Fl. Indian Desert 42. 1978; Sundara Raghwan, in Sharma & Balakrishnan Fl. India 2: 331. 1993. *Capparis oblongifolia* Forssk. Fl. Aegypt. Arab. 99. 1775. *Niebuhria arenaria* DC. Prodr. 1: 244. 1824. *Maerua arenaria* (DC.) Hook. *f.* & Thoms. in Hook. *f.*, Fl. Brit. India 1: 171. 1872; Parmar, in Shetty & Singh Fl. Rajasthan 1: 87 - 89. 1987. (Fig. 20 & Plate 28)

var. ***scabra*** Hook. *f.* & Thoms *l.c.* 1: 171. 1872.

Vern.: *Orapa*.

Scandent, climbing, up to 4 m long; branches glabrous to scabrous, divaricate; hairs short, simple. Leaves elliptic-oblong or lanceolate, 2 - 5 x 0.7 - 2.5 cm, entire, obtuse, often slightly mucronate, lateral veins 5 - 6 pairs; petiole 2 - 6 mm long. Flowers usually in dense, corymbose racemes, greenish-white, pedicellate. Sepals 4, ovate-elliptic, acute or slightly acuminate, margin white-pubescent; calyx-tube 3 - 8 mm long; calyx lobes elliptic-oblong, 6 - 14 x 3 - 7 mm. Petals 4, on cup-shaped disc, ovate-lanceolate to obovate, acuminate, 3 - 8 x 2-5 mm, margin a somewhat undulate. Stamens 20 - 26, filaments radiating, up to 2 cm long, inserted on ca 5 mm long torus. Gynophore 1.5 - 2.5 cm long. Ovary cylindrical, glabrous, 2 - 4 mm long. Fruit cylindrical, 3 - 8 x 1 - 1.5 cm, torulose or irregularly many knotted, pale brown, often somewhat twisted. Seeds 5 - 8 mm in diam., globose, tuberculate.

Fig. 20. *Maerua oblongifolia* (Forssk.) A. Rich. var. *scabra* Hook. *f.* & Thoms: A. Flowering and fruiting twig, B. Flower.

Field note: Common in semiarid sandy tracts and scrub forests often associated with *Capparis decidua, Euphorbia caducifolia*, etc.

Fl. & Fr.: January- May.

Specimen examined: On way to Bijoliya 1565, Gurla 1758 (MLVGCB Herb).

Distribution: India, Thialand, Sri Lanka, Pakistan, Arabia, Africa and Sri Lanka.

Notes: A very variable species with regards to hairiness and size of leaves. This species has thinner leaves, more or less ovate-elliptic with a mid rib and conspicuous secondary veins. The resemblance in the two species is so close that Pax & Hoffm. (*l.c.*) gave the range of Distribution of *M. arenaria* as India and Arabia, while Elffers *et al.* (*l.c.*) considered the Arabian plants to belong to *M. oblongifolia*.

Uses: The leaves serve as fodder for goat and camels.

RESEDACEAE

OLIGOMERIS Cambess., *nom. cons.*

Oligomeris linifolia (Vahl) Macbr., in Contr. Gray Herb. n.s. 53: 13. 1918; Raizada, in Ind. For. 92: 318. 1966; Mahesh Illus. Fl. Delhi f. 15. 1966; Parmar, in Shetty & Singh Fl. Rajasthan 1: 87-89. 1987; Balakrishnan & Premanath, in Sharma & Balakrishnan Fl. India 2: 337-338. 1993. *Reseda subulata* Del., Fl. Egypt 15. 1813, *nom. nud. R. linifolia* Vahl., in Hornem., Hort. Hafin. 2: 501. 1815. *Oligomeris glaucescens* Camb., in Jacq., Voy. Bot. 4: 24, t. 25. 1844; Hook. *f.* & Thoms. Fl. Brit. Ind. 1: 181. 1872; Duthie, Fl. Gangetic Plain 1: 54. 1903. (Fig. 21)

Vern.: *Bui.*

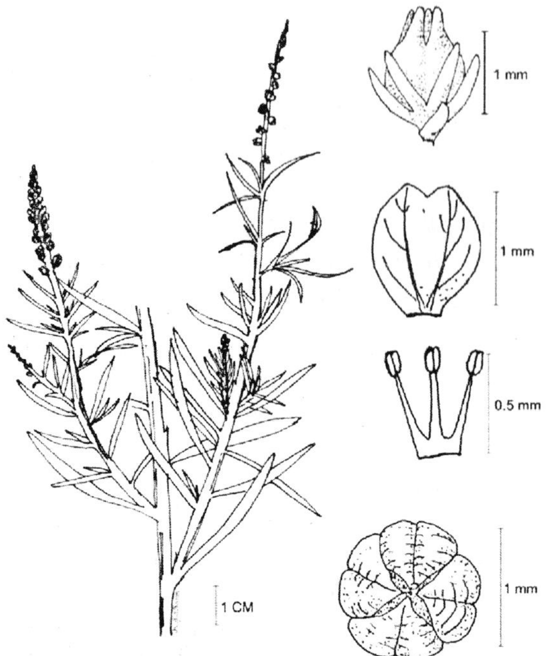

Fig. 21. *Oligomeris linifolia* (Vahl) Macbr.: A. Flowering and fruiting branch, B. Flower, C. Petal, D. Stamen, E. Seed.

Annual herbs glabrous, glaucous, 7 - 35 cm tall. Stem erect to ascending, ribbed. Leaves 2.5 - 7.5 x 0.2 - 0.3 cm, glabrous to scabrid, in clusters, narrowly linear, with 1 - 2 minute basal lobes, ca 1 mm long. Flowers bisexual, in dense, 2 - 25 cm long spikes, greenish; bracts 1 - 1.5 mm long, linear to lanceolate, persistent. Sepals 4, subequal, 1 -1.5 mm long, connate at the base. Petals 2, united more than half way up or free. Stamens 3 - 4, sometimes abortive; filaments united at the base, ca 1.5 mm long. Capsule erect, subglobose, ca 3 mm broad; teeth acute, ca 1 mm long, membranous. Seeds less than 1 mm long, greenish-black, glaucous.

Field note: Rarely observed in the area. A common desert spring annual of the plains and the foothills. Often found in waste and saline places. Those growing near moist areas tend to be less branched and with numerous slender leaves and shorter spikes. The lobes at the bases of leaves may be stipular in nature.

Fl. & Fr.: January-April.

Specimen examined: Bhilwara 86 (MLVGCB Herb).

Distribution: South - west U.S.A., N. Africa, Syria, Iran, India, Pakistan and West-wards to Spain and the Canary Islands.

VIOLACEAE

HYBANTHUS Jacq.

Hybanthus enneaspermus (L.) F. Mueller, Fragm. 10: 81. 1876; Parmar, in Shetty & Singh Fl. Rajasthan 1:91. 1987; Banerjee & Pramanik in Sharma & Balakrishnan Fl. India 2: 343 - 345. 1993. *Viola enneasperma* L., Sp. Pl. 2: 937. 1753. *V. suffruticosa* L., Sp. Pl. 2: 937. 1753. *Ionidium suffrutico-sum* (L.) Roemer & Schultes, Syst. Veg. 5: 394. 1819; Hook. *f.* & Thoms. in Hook. *f.*, Fl. Brit. Ind. 1: 185. 1872. (Fig. 22)

Vern.: *Ratanpurus.*

Annual or perennial herb, Tap root long, slender. Stem, procumbent, 10 - 20 cm tall, rarely to 40 cm. Young branches canaliculate, puberulous. Leaves alternate, sessile or subsessile; stipules greenish, triangular or subulate, very small, ciliate; leaf blade linear-lanceolate, linear-oblan-ceolate, or narrowly spatulate, unequal in size, 0.5 - 3.5 × 0.2-0.5 cm, midvein slightly raised abaxially, lateral veins inconspicuous, base cuneate, decurrent, margin entire, sparsely 1-3-serrulate, apex acute, rarely acuminate, mucronulate. Flowers axillary, solitary in leaf axils, blue-purple; pedicels not exceeding leaves, 2-bracteolate near flower; bracteoles opposite, lanceolate or subtriangular, membranous, ciliate. Sepals subequal, lanceolate, 2-3 mm, margin membranous, white, ciliate. Petals irregular, upper 2 smallest, oblong, apex acute, lateral 2 ovate-orbicular, apex slightly recurved, lower 1 largest, 6-7 mm, ca 2 × as large as others, anterior half broadest, 2 lateral sides incurved, more deeply colored, shortly spurred. Stamens free, filaments short, 2 with curved calcarlike (spurlike) appendages, anthers longitudinally dehiscent, appendages at top of connectives yellow-brown, membranous. Ovary subglobose; styles clavate, base dilated, slightly curved in upper narrow part; stigma slightly incurved. Capsule pendulous, globose. Seeds creamy-yellow, ovoid-ellipsoidal, conspicuously longitudinally striate.

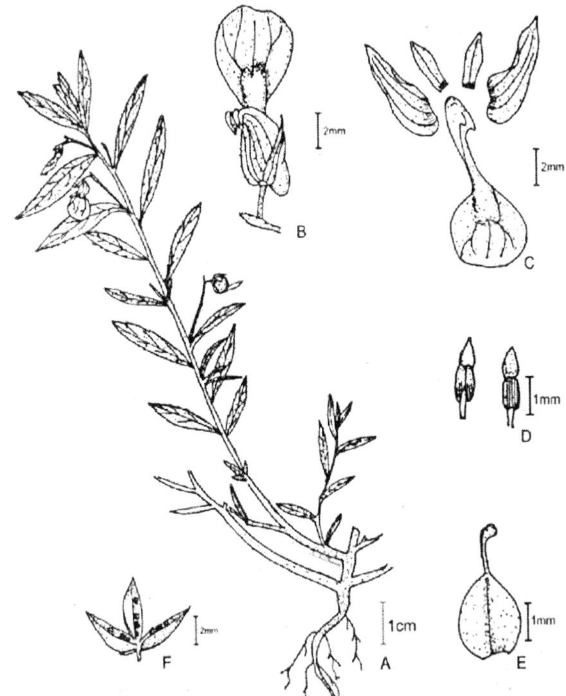

Fig. 22. *Hybanthus enneaspermus* (L.) Muell. : A. Habit, B. Flower, C. Petals, D.Stamens, E. Pistil, F. Dehisced fruit.

Field note: Common in grass lands.

Fl.: June - October; *Fr.*: Aug - December.

Specimen examined: Gurla 86, Ladpura 1331 (MLVGCB Herb).

Distribution: Tropical Asia, Africa, Australia and America

Uses: Decoction or powder of the whole plant is taken to improve memory and vitality. It is also used as a remedy in congestion, asthma, fever and leprosy. Shampoo made from this plant removes dandruff. (Sharma & Balakrishnan, 1993).

FLACOURTIACEAE

FLACOURTIA L. Herit.

Flacourtia indica (Burm. f.) Merr., Interpr. Rumph. Herb. Amb. 377. 1917; Sleumer in Fl. Males. I, 5: 76. 1954; Parmar, in Shetty & Singh Fl. Rajasthan 1:94. 1987; Mitra in Sharma & Balakrishnan Fl. India 2: 401 - 402. 1993. *Gmelina indica* Burm. f. Fl. Ind. 132, t. 39, f. 5. 1768. *Flacourtia sepiaria* Roxb., Pl. Corom. 1: 48, t. 68. 1796; Hook. *f.* & Thoms. in Hook.f., Fl. Brit. Ind. 1: 194. 1872. *F. ramontchi* L' Herit., Strip. Nov. 3: 59, t. 30 & 30 B. 1786; Hook. *f.* & Thoms. Fl. Brit. Ind. 1: 193. 1872, pro maj. parte excl.

var. ***latifolia.***

Vern.: *Kanker, Kanju.*

A shrub or small tree, up to 6 m tall, dioecious, polymorphous, deciduous, upto 10 m tall. Spines axillary. Younger branches pubescent. Leaves usually clustered towards apices on older branches, usually cuneate and 3-5-veined at base, acute, obtuse or sometimes emarginate at apex, 1.5 - 5 x 1 - 3 cm. Flowers ca 4 mm across, yellowish green, solitary, or few in axillary, bracteate, glabrous, racemose clusters or tomentose panicles up to 3 cm long. Sepals 4 - 5, slightly connate at base, ovate, obtuse, subglabrous outside, hairy inside. *Male flowers*: Stamens numerous; filaments long; anthers short, oblong, opening by slits. *Female flowers*: Ovary on a disc, glabrous; stigmas 5 - 10. Fruit ellipsoid to subglobose, with 6 - 7 pyrenes, 5 - 10 mm in diam. Seeds trigonous, 2 - 2.5 mm long.

Field note: Occurs throughout the area.

Fl. & Fr.: January- May.

Specimen examined: Nayasamaliya 407, Ladpura 1427 (MLVGCB Herb).

Distribution: Widespread in S.E. Asia and Tropical Africa

Notes: Following Sleumer (*l.c.*) *F. indica* is accepted here in a wider sense. *F. ramontchii* is usually a small tree with larger leaves, distinct paniculate inflorescence and found in dry deciduous and mixed forests. *F. indica* (*F. spiaria*) on the other hand is usually a twiggy shrub with smaller leaves, few flowers in axillary clusters and found in dry thickets and scrub forests (Sharma & Balakrishnan, 1993).

Uses: Fruits are edible and eaten raw, these are also used to cure jaundice and enlarged spleen.

POLYGALACEAE

POLYGALA L.

KEY TO THE SPECIES

1a. Leaves usually petiolate; petiole 5 - 15 mm long ....*P. umbonata*

1b. Leaves usually sessile or subsessile; petiole up to 2 mm long. 2

2a. Leaves mucronate. Wing sepals broader above middle, obtuse at apex. Capsule glabrous with ciliate margins.........*P. arvensis*

2b. Leaves obtuse or acute. Wing sepals broader below middle, acute-acuminat at apex. Capsule pubescent with long apical and marginal hairs...............................*P. erioptera*

Polygala arvensis Willd. Sp. P1. 3. 2: 876. 1802; Roxb., Fl. Ind. 3: 218. 1832; Burtt, in Notes Roy. Bot. Gard. Edinb. 32: 404. 1973; Parmar, in Shetty & Singh Fl. Rajasthan 1: 95. 1987; Banerjee, in Sharma & Balakrishnan Fl. India 2: 460. f. 87. 1993. *P. chinensis auct. plur.*, non L. 1753; Bennett, in Hook. *f.* & Thoms., in Hook. *f.*, Fl. Brit. Ind. 1: 204. 1872; Duthie, Fl. Gangetic Plain 1: 62. 1903; Mukherjee, in Bull. Bot. Soc. Bengal 22: 38. 1960; Mahesh Illus. Fl. Delhi f. 16. 1966. *P. chinensis* L. var. *linarifolia* sensu Mukerjee, in Bull. Bot. Soc. Bengal 12: 40. 1958.

Vern.: *Miradu, Gabhainsini.*

Annual, pubescent, procumbent or erect, 5 - 30 cm tall. Leaves obovate, linear-lanceolate, elliptic or linear, mucronate, 4 - 55 x 5 - 20 mm, margin recurved. Racemes lateral, shorter than the leaves, dense, truncate, 10 - 15 mm long. Bracts persistent, shorter than the pedicel. Sepals persistent; outer sepals equal, acute-acuminate, glabrous, broadly hyaline margined; inner sepals ovate-elliptic, acute-acuminate, oblique, falcate 3 - 5 x 1.5 - 2.5 mm, slightly enlarged in fruit, covering ca 1/2 of the fruit, ciliate except at apex, 3 - 4 nerved, veinlets anastomosing. Lateral petals longer than keel; keel 4 mm long, crested. Capsule oblong-orbicular, emarginate, 3.5 - 5 x 3 - 4 mm, glabrous, wing ca 0 - 0.25 mm, ciliate. Seeds oblong, pilose; strophiole 3-appendiculate, appendages minute.

Field note: Found throughout the area.

Fl. & Fr.: August - October

Specimen examined: Naya Samaliya 407, Ladpura 1427 (MLVGCB Herb).

Distribution: Tropical Asia and Australia.

Polygala erioptera DC. Prod. 1:326. 1824; Boiss., Fl. Or. 1: 469.1867; Bennett, in Hook. *f.* & Thoms. in Hook. *f.* Fl. Brit. Ind. 1: 203. 1872; Duthie, Fl. Gangetic Plain 1: 62. 1903; Mukherjee, in Bull. Bot. Soc. Bengal 12: 47. 1960; Bhandari, Fl. Indian Desert 44 - 45. 1978; Parmar, in Shetty & Singh Fl. Rajasthan 1: 96. 1987; Banerjee in Sharma *et al.* Fl. India 2: 467 - 469. 1993. *P. vahliana* DC. Prodr. 1: 326. 1824. *P. erioptera* DC. var. *vahliana* (DC.) Chodat. Mem Soc. Phys. Hist. Nat. Genev. 31: 2. 345. 1893.

Annual or perennial herb, erect, pubescent, 5 - 32 cm tall. Leaves subsessile, oblong to linear, obtuse or acute, 2 - 45 x 0.75 - 8 mm, recurved. Racemes lateral, lax, shorter than leaves, 3 - 27 mm. Flowers purple, solitary or in up to 3.5 cm long condensed leaf-opposed or extra-axillary racemes. Bracts persistent, shorter than pedicel. Sepals persistent; outer sepals ovate-lanceolate to ovate-elliptic, acute-mucronate at apex, concave,

1.5 - 2.5 x 0.7 - 1.2 mm, unequal; upper larger than the lower ones, broadly hyaline margined; inner sepals ovate, obtuse, 3 - 5 x 1.5-3 mm, enlarged in fruit, oblique, pubescent, 3-nerved, veinlets anastomosing. Lateral petals shorter or equalling the keel; keel 2.5 - 4.5 mm, crested. Staminal sheath free or slightly adnate to middle lobe of petal, ca 1.5 mm long; free portion of filaments ca 1.5 mm long. Ovary ovoid, ca 1 mm long; style curved, broadened at middle, up to 3 mm long; stigma capitate. Capsule elliptic-oblong, obliquely notched, pubescent, 3.5 - 5 x 2 - 2.5 mm. Seed ca 3 x 1.5 mm, pilose, strophiole 3-appendiculate, appendages minute.

Field note: Found throughout the area. Sepals open twice, once at the time of pollination and again at the time of dehiscence of capsule.

Fl. & Fr.: throughout the year.

Specimen examined: Thago ka khera 87, Jahajpur 1363 (MLVGCB Herb).

Distribution: Asia, Tropical Africa

Polygala umbonata Craib in Bull. Misc. Inf. Kew 1916: 260. 1916; Mukherjee in Bull. Bot. Soc. Bengal 12: 35. 1958. *P. triphylla* sensu Haines in Bot. Bihar Orissa 42. 1925, non Burm. f. 1798, nec D. Don 1825. (Plate 29)

Annual herb, erect, 15 - 20 cm tall. Stem grooved, glabrous. Leaves lanceolate, ovate - lanceolate,tapering at base, mucronate at apex, 4.5 - 6 x 2 - 3 cm, setulose along margins, membranous, glabrous above, sparsely pubescent and pale beneath; petioles 8 - 10 mm long. Flowers 4 - 5 mm long, in axillary or terminal glabrous racemes; pedicels 1 - 1.2 mm long; bracts persistent, longer than pedicels, narrow, deciduous. Sepals 5, outer sepals 1 - 1.5 mm long, solitary, ovate, ca 2 mm long; wing sepals obovate, cuneate at base, rounded at apex, 3.7 - 4 x 3 - 3.5 mm. Petals 3 - lobed, ca 4 mm long; lateral petals acute; keel petals with 2 - lobed rounded crests. Stamens 2.5 - 3 mm long; filaments coherent below, glabrous; free portion 1 - 2 mm long, hyaline. Ovary ovoid; style flate, bent at middle, 2.5 3 mm long; stigma minute. Capsule obovate or orbicular, 3 - 3.5 mm across, narrowly winged, emarginated at apex. Seeds oblong, 1-2 mm long, black, pubescent, minutely and copiously tubercled; caruncle 3-lobed, gelatinous; strophiole black, shiny, hollow, about ane-third as long as seed, narrower than seed, curved slightly at apex, ending in an orbicular sunken membranous lamella.

Field note: Rare in the area.

Fl. & Fr.: July - October

Specimen examined: Bhilwara 405 (MLVGCB Herb).

Distribution: Myanmar, Vietnam, Laos and Combodia.

CARYOPHYLLACEAE

KEY TO THE GENERA

1a. Leaves stipulate. 2

2a. Leaves apparently whorled, Styles 5, free from base, capsules 5-valved. ....*Spergula*

2b. Leaves pseudo-whorled. Style 3, fused, capsules 3-valved.......................*Polycarpaea*

1b. Leaves exstipulate..3

3a. Sepals free. Flowers white, styles usually 3..*Stellaria*

3b. Sepals united. Flowers pinksh to purplish, styles usually 3.................................*Vaccaria*

POLYCARPAEA Lam.

Polycarpaea corymbosa (L.) Lam, Tabl. Encycl. 2: 129. 1797; Wight, *l.c.* 2 (4): 6. t. 712. 1843; Edgew. & Hook. *f.,* in Hook.f., Fl. Brit. Ind. 1: 245. 1872; Duthie, Fl. Gangetic Plain 1: 68. 1903; Bhandari, Fl. Indian Desert 46. 1978; Parmar, in Shetty & Singh Fl. Rajasthan 1: 98-99. 1987; Majumdar, in Sharma *et. al.* Fl. India 2: 549 - 551. 1993. *Achyranthes corymbosa* L. Sp. Pl. 205. 1753.

var.: ***corymbosa***

Vern.: *Rati khard.*

Annual or short-lived perennial exect herb. 15-35 cm tall, mostly branching above middle, occasionally branched at base, white pilose. Stipules ovate-lanceolate, 2-4 mm, scarious, margin hyaline, apex acute. Leaves apparently verticillate; leaf blade linear or acicular, 1.5 - 2 cm × ca 1 mm, erect or uniformly recurved from erect base, apex acute to needlelike. Bracts lanceolate, longer than pedicel, hyaline, membranous. Pedicel slender, white pilose. Sepals lanceolate, (2-) 2.7 - 3.5 × 0.5 - 1 mm, membranous, hyaline, apex acuminate. Petals milky white or reddish, broadly ovate, ca 1/2 as long as sepals, apex obtuse. Stamens usually 5, shorter than petals. Style short, 1/3 - 1/2 as long as ovary. Capsule brown, ovoid, ca 1/2 as long as calyx, 5 - 13-seeded. Seeds brown, flat, ca 0.5 × 0.3 mm, obscurely reticulate.

Field note: Mostly a weed of open, cultivated or open sandy as well as moist sandy soils, less often in grassy area.

Fl. & Fr.: Aug - October

Specimen examined: Pansal 312, Vindhiya Bhata 1337 (MLVGCB Herb).

Notes: An elegant silvery tinged herb, often turning reddish. This species is used medicinally.

SPERGULA L.

Spergula fallax (Lowe) E. H. Krause in Sturm, Deutschl. Fl. ed. 2. 5: 21. 1901; Lowe in Hooker's J. Bot. Kew Gard. Misc. 8: 289. 1856; Bhandari, Fl. Indian Desert 47. 1978; Parmar, in Shetty & Singh Fl. Rajasthan 1: 100. 1987; Majumdar, in Sharma *et al*. Fl. India 2: 578 - 580. 1993. *Spergularia fallax* Lowe in Hook. J. Bot. Kew Journ. Bot. 8: 289. 1856. *Spergula vernalis* auct. non Willd. 1809; Bhatt *et al*.. in Bull. Bot. Surv. India 12: 264. 1970; Blatt in J. Bombay Nat. Hist. Soc. 68: 492. 1971. *S. pentandra sensu* Edgew. & Hook. *f*. in Hook.f., Fl. Brit. Ind. 1: 243. 1872, pro parte, non L. 1753; Duthie, Fl. Gangetic Plain 1: 67. 1903.

Vern.: *Hiyaroo bagdo*

Plants erect ascending annual herb. Stems branched from base, branches cylindrical, glabrous. Leaves spuriously whorled, 1.5 - 4.0 x 0.04 - 0.1 cm, linear, flate, obtuse. Stipules ovate, acute. Inflorescence few to many flowered; bracts ovate, 1mm long, acute, scarious; Pedicels upto 1.75 cm long, deflexed in fruits. Sepals white margined, lanceolate, acute or acuminate, usually glandular pubescent outside. Petals white. Ovary ellipsoid; style 3. Capsule ovoid- subglobose, 3-valved. Seed granulate, polished, black with hyaline wings.

Field note: The plant is a winter crop weed of cultivated fields, and is used as fodder.

Fl. & Fr.: Decembr- March.

Specimen examined: Karoi 57, Raipur 1557 (MLVGCB Herb).

Distribution: Cosmopolitian weed.

STELLARIA L.

Stellaria media (L.) Villars, Hist. Pl. Dauphiné. 3: 615. 1789, p.p.; Edgew. & Hook. *f.*, in Hook. *f.*, Fl. Brit. Ind. 1: 230. 1872; Duthie, Fl. Gangetic Plain 1: 66. 1903; Bhandari, Fl. Indian Desert 47. 1978; Parmar, in Shetty & Singh Fl. Rajasthan 1: 101. 1987; Majumdar, in Sharma et al. Fl. India 2: 585. 1993; Mahesh., Illus. Fl. Delhi f. 17. 1966. *Alsine media* L. Sp. Pl. 272. 1753.

Annual herb. Stem decumbent or ascending, pale purplish, 10 - 30 cm tall, sparsely branched at base, with 1(or 2) lines of hairs. Basal leaves long petiolate, distal leaves sessile or shortly petiolate; leaf blade broadly ovate to ovate-orbicular, 0.8 - 2.5 × 0.5 - 1.5 cm, base narrowed or cordate, apex acuminate or acute. Flowers in sparse terminal or axillary cymes. Pedicel 0.7 - 1.4 cm, elongate and nodding after anthesis, slender, with 1 line of hairs. Sepals 5, ovate-lanceolate or ovate-oblong, ca 2 - 2.5 or 4 mm, outside glandular pubescent, margin broadly membranous, apex slightly obtuse or nearly rounded. Petals oblong, shorter than or subequaling sepals, 2-cleft nearly to base; lobes nearly linear. Stamens 3-5, shorter than petals. Styles 3, linear. Capsule ovoid, slightly longer than persistent sepals, 6-valved. Seeds numerous, red-brown, ovoid to compressed globose, 1-1.2 mm in diam., semiglobose-tuberculate or curved reticulate.

Field note: A common weed of gardens and cultivated fields.

Fl. Fr.: January- April.

Specimen examined: Karoi 1139, Sahara 1169 (MLVGCB Herb).

Distribution: Afghanistan, Bhutan, India, Japan, Korea, Pakistan, Russia, Sikkim; Europe.

VACCARIA Wolf

Vaccaria pyramidata Medic. Philos. Bot. 1: 96. 1789; Parmar, in Shetty & Singh Fl. Rajasthan 1: 101. 1987; Majumdar, in Sharma *et al.* Fl. India 2: 593. 1993; *Saponaria vaccaria* L. Sp. Pl. 409. 1753; Edgew. & Hook. *f.*, in Hook. *f.*, Fl. Brit. Ind. 1: 217. 1874; Duthie, Fl. Gangetic Plain 1: 64. 1903. (Plate 30)

Vern.: *Sabuni*.

Erect, annual herb, roboust, 15 - 60 cm high, branched, glabrous; taproot slender. Leaves cordate-lanceolate, acute at apex, 2.5 - 7.7 x 0.8 - 1.8 cm, glabrous, entire; lower leaves somewhat petiolate, the rest sessile. Flowers erect, in lux corymbose dichasial cymes, pink; pedicels slender. Calyx tube inflated, *ca* 1.2 cm across, with 5 sharp angles or wings; teeth 5, triangular. Petals rose-coloured; limb cuneate or rounded or somewhate emarginated; coronal scales absent. Capsule ovoid-globose, 4-valved, 4-locular below, included within the calyx. Seeds globose, *ca* 2mm in diam., brown-black, granulate.

Field note:Common weed of cultivated fields.

Fl. & Fr.: December - March.

Specimen examined: Karoi 154, Ladpura 1157 (MLVGCB Herb).

Distribution: Temperate Asia, Europe, N. America and Australia.

Uses: Used as a fodder.

PORTULACACEAE

PORTULACA L.

KEY TO THE SPECIES

1a. Nodal appendages present. ....*P. quadrifida*

1b. Nodal appendages absent.............................2

2a. Sepals distinctly carinate; axillary hairs not conspicuous, upto 1 mm long, deciduous... *P. oleracea*

2b.Sepals not carinate, at the most only apically keeled; axillary hairs prominent, more than 2 mm long, mostly persistant. ....*P. pilosa*

Portulaca oleracea L. Sp. Pl. 445. 1753; Dyer in Hook. Fl. Brit. India, l; 246. 1874; Duthie, Fl. Gangetic Plain 1: 69. 1903; BH 26 1: 224. 1918; Bhandari, Fl. Indian Desert 49. 1978; Parmar, in Shetty & Singh Fl. Rajasthan 1: 102. 1987; Rao in Sharma *et al.*Fl. India 3: 4 - 5. 1993. (Plate 31)

Vern.: *Lunaki*.

Annual or perennial, prostrate or erect, ca 25 cm tall, succulent, glabrous, green or purplish green herb. Branches and stem 3 - 20 mm rarely up to 50 mm long internodes. Leaves alternate or subopposite, closely crowded below the flowers, spathulate or obovate-oblong to linear-oblong, attenuate at the base, sub-sessile, obtuse or truncate, 3 - 25 x 1.5-8 mm, thick, fleshy, glabrous, glistening white below, green or purplish-green above; stipular appendages usually absent. Inflorescence usually in the forks of branches, cymose, with clusters of 3 - 6 flowers subtended by 4-leaved involucre, rarely flowers solitary and terminal. Flowers sessile, yellow, 5 - 8 mm across, bracteate; bracts membranous, ovate, ca 3 mm long, acuminate, white or somewhat purplish. Sepals subequal, basally united into a short, 2 - 3 mm long tube, keeled; lobes 2 - 3 mm long, slightly hooded, margin broad membranous, acute, deciduous. Petals 5, deliquescent, slightly united at the base, obovate, 5 - 6 x 2.5 - 3 mm, yellow, emarginate with mucronulate notch. Stamens 7 - 12, basally somewhat united and adnate to petals, filaments ca 2.5 mm long, sensitive to touch, anthers ovoid. Ovary ca 2 mm long, half embedded in calyx tube, ovoid; style 1.5 - 2 mm long, stigmas 4 - 5, sticky, ca 1 mm long. Capsule many-seeded, 6-8 x 3-4 mm. Seeds shining black, ca 0.5 - 0.8 mm, reniform, testa tuberculate.

Field note: A weed of wastelands throughout. Flowers open in morning. This is a much variable species, 2-3 forms of this species can be seen easily.

Fl. & Fr.: July - February

Specimen examined: Bhilwara 88, Mandal 1420 (MLVGCB Herb).

Distribution: Cosmopolitan weed. Probably native of South-West parts of United States and now widely distributed in warm temperate, tropical and subtropical regions throughout the world.

Uses: The extract of stem is applied on skin against burning sensation and prickly heat. The purslane constitutes a useful article of diet in scurvy and diseases of lungs, liver and kidney. Leaves are slightly acidic and used as refrigerant, anti-scorbutic, astringent in dysuria, irritation of bladder, haematuria, haemoptysis and gonorrhoea. Seeds are used in preparation of Unani and Ayurvedic medicines, as a demalcent, astringent, diuretic and vermifuge.

Portulaca pilosa L., Sp. Pl. 445. 1753; Grees., in Blumea 17: 295; 1969 & in Steenis, Fl. Males. 1. 7: 131. 1971 (quoad subsp. *pilosa* 'rase' *pilosa*); Sivarajan, in J. Bombay Nat. Hist. Soc. 78: 1981 (quoad subsp. *pilosa*); Raizada, Suppl. Duthie, Fl. Gangetic Plain 22. 1976; Parmar, in Shetty & Singh Fl. Rajasthan 1: 103. 1987; Rao, in Sharma *et al*. Fl. India 3: 6. 1993. *P. tuberose* Roxb. Fl. India. 2: 464. 1832; Dyer, in Hook. *f.* Fl. Brit. India, l: 247. 1874; *P. parvula* auct. non A. Gray; Murthy & Singh, in Proc. Nat. Inst. Sci. India 27: 14. 1961; Sharma in Bull. Bot. Surv. India 15: 136. 1976. *P. suffruticosa* Wall. ex Wt. & Arn. Prodr. 356. 1834; Thiselton Dyer, in Hook. *f. l. c.* l: 249. 1874.

Vern.: *Lunki.*

Annual, prostrate or ascending, somewhat fleshy herb. Stem and branches slender, terete, with 3-6 mm long internodes, glabrous. Leaves alternate, sessile, somewhat fleshy, linear to linear-lanceolate, 5 - 8 x 1.5 - 3 mm, acute, entire; stipular hairs copious, 3 - 6 mm long, twisted. Inflorescence a terminal, sessile cluster of 2 - 6 flowers surrounded by long pale hairs and 8 leaved- involucre. Flowers pink, *ca* 10 mm across; bracts minute, elliptic-ovate, acute. Sepals free, *ca* 3.5 mm long, triangular-ovate, acute, glabrous. Petals 5, slightly united at the base, obovate, 5 - 6 x 4 mm, obtuse, pink. Stamens 20, filaments filiform, 3 mm long, basally united and adnate to petals; anthers minute. Carpels 4, syncarpous; ovary ovoid; style deeply 4-fid; stigmas 4, elongated, 1.5 mm long. Capsule ovoid, 5 - 6 x 4 mm, circumscissile in the middle, enclosed by withered corolla and hairs, crustaceous, wall parallel-veined. Seeds numerous, reniform, dark brown to black, papillate.

Field note: Native of tropical and subtropical America It is rare and probably a casual introduction in the area.

Fl. & Fr.: March-October.

Specimen examined: Bhilwara 409, Mandal 1759 (MLVGCB-Herb).

Distribution: Cuba, West Indies, Puerto Rico, Flordia, Mexico, Guatemala, Venezuela, Brazil, Uruguaya, Rio de Janiero, West Pakistan.

Uses: Leaves are intensely bitter in taste and have been used as a diuretic, stomachic and emmenagogue.

Portulaca quadrifida L. Mant. 1: 73. 1767; DC. Prodr. 3: 353. 1828; Boiss. Fl. Or. 1: 757. 1867; Dyer in Hook. *f.,* Fl. Brit. India 1: 247. 1874; Cooke, Fl. Bomb. Pres. 1: 72. 1908; Gamble, Fl. Madras 1: 47. 1935; Maheshwari, Fl. Delhi, 70. 1963; Bhandari, Fl. Indian Desert 49 - 50. 1978; Parmar, in Shetty & Singh Fl. Rajasthan 1: 103 - 104. 1987; Rao in Sharma *et al.* Fl. India 3: 6. 1993. *P. meridiana* L. f. Suppl. Sp. Pl. 248. 1781. *P. quadrifida* L. var. *meridiana* (L.) DC. Prodr. 3: 354. 1828.

Vern.: *Nunya ka sag, Lunki.*

Diffuse annual, prostrate, glabrous, somewhat fleshy, rooting at nodes. Tap root somewhat woody, nodal adventitious roots fibrous, up to 3 cm long. Stem and branches filiform, internodes 2-25 mm long. Leaves opposite, sessile to subsessile, ovate-oblong to elliptic-oblong or ovate to ovate-lanceolate, 3 - 10 x 2 - 3 mm, acute or somewhat obtuse, fleshy; stipular hairs copious, silvery white, 2 - 10 mm long. Flowers solitary terminal, yellow, 6-8 mm across, surrounded by a 4-leaved involucre and silvery hairs. Sepals equal, membranous, basally united into a minute tube half embedded in the floral axis, ovate-oblong, 3 - 4.5 x 2 - 2.5 mm, obtuse. Petals 4, united at the base, pandurate, 3 - 4 x 1.5 - 2 mm, obtuse, yellow. Stamens 8, adnate to petals; filaments filiform, 1.5 - 2.5 mm long. Carpels 4, syncarpous; ovary ca 2 mm, half sunk in floral axis; style c. 2 mm long, quadrifid, with 4 linear, reclinate stigmas. Capsule 3.5 - 4.5 x 2.5-3 mm, oblong-conical. Seeds black, ca 1 mm, reniform, somewhat papillose.

Field note: Common weed of moist as well as rocky and gravelly habitats.

Fl. & Fr.: August-March.

Specimen examined: Karoi 1184 (MLVGCB Herb).

Distribution: India, Pakistan, Tropical Africa and Asia.

Uses: Fresh sap of whole plant is used in skin diseases.

TAMARICACEAE

TAMARIX L.

Tamarix ericoides Rottl. & Willd. in Gesel. Naturf. Fr. Berl. Neue Schr. 4: 214. t. 4. 1803; Dyer, in Hook. *f.* Fl. Brit. India 1: 249. 1874; Duthie, Fl. Gangetic Plain 1: 72. 1903; Bhandari, Fl. Indian Desert 53. 1978; Parmar, in Shetty & Singh Fl. Rajasthan 1: 106. 1987; Shetty & Pandey in Sharma *et al.* Fl. India 3: 25. 1993. *Myricaria vaginata* Desv. in Ann. Sci. Nat. ser. 1, 4: 350. 1825. *Trichaurus ericoides* (Rottl. & Willd.) Arn. ex Wight & Arn., Prodr. 40. 1834, *p.p.* (Plate 32)

Tall shrub tree with blackish bark, 2 - 4 m high, glabrous to subglabrous, somewhat papillose. Leaves vaginate in the lower part and pseudo-vaginate in upper part, subamplexicaul in the upper part, ovate, lanceolate, accumintate or acute at apex, 1 - 7 mm long with punctate salt glands. Racemes mostly vernal and aestival both, simple 4 - 19 (-20) cm x 8 - 10 mm; rachis papilose; bracts 2.25 - 5 mm long, ca 2 mm broad, semiamplexicaul or amplexicaul, acute to acuminate, triangular, somewhat denticulate at the margins. Pedicel 2 mm long, flowers 6 mm across, pink or pale pink. Calyx 5-lobed, sepals 2 - 3.5 x 1.2 - 2.2 mm, ovate, minutely and irregularly denticulate, outer 2 more acute than the inner three. Petals 5, free, elliptic to obovate 4 - 6.5 (-6) x 4 - 4.5 mm, very finely and irregularly denticulate. Stamens 10, alternating with sepals and petals, antipetalous stamens somewhat shorter than antisepalous stamens, 5-longer stamens antisepalous, inserted hypodiscally with filaments 3.5 - 5.5 mm long, alternating with 5 shorter antipetalous ones inserted peridiscally with filaments 2.5 - 4.1 mm long. Disc cupshaped, fleshy, forming somewhat membranous torus. Ovary ovate, ca 1 mm long, capsule 1.25 cm x 4 mm broad. Seeds 1 - 1.5 mm long; coma 5 - 8.5 mm long.

Field note: Common speciceis in the river bed at Jahajpur.

Fl. & Fr.: Almost throughout the year.

Specimen examined: Jahajpur 412, 1760 (MLVGCB Herb).

Distribution: Sri Lanka, Bengladesh, Pakistan, India (Bengal, W. Peninsula, Bihar, Orissa, Karnataka, Saurashtra, Rajasthan).

Uses: Leaves are used in the treatment of cough and cold in children. The trunk as well as twigs are used as firewood.

ELATINACEAE

BERGIA L.

Bergia ammannioides Roxb. [Hort. Beng. 34. 1814, nom nud.] ex Roth, November P1. Sp. 219. 1821; Roxb., Fl. Ind. 2: 457. 1832, 'ammanoides'; Dyer in Hook. *f.*, Fl. Brit. Ind. 1: 251. 1874; Duthie, Fl. Gangetic Plain 1: 72. 1903; Cooke, Fl. Bomb. Pres., 1: 77.1908; Maheshwari, Fl. Delhi 72, 1963; FPB. 1: 77; BH. 26 1: 225. 1918; Bhandari, Fl. Indian Desert 54. 1978; Parmar, in Shetty & Singh Fl. Rajasthan 1: 107-108. 1987; Bhattacharya in Sharma *et al.* Fl. India 3: 33 - 34. 1993. *Elatine ammannioides* (Roxb. ex Roth) Wight & Arn. Prodr. 41. 1834. *Bergia ammannioides* Roxb. ex Roth var. *pentandra* Wight, III. Ind. Bot. 54, t. 25a. 1840.

Vern.: *Jal bhangra.*

Annual, 8 - 35 cm tall, erect or somewhat decumbent, branched herb of moist places. Stem and branches mostly pinkish, glandular pubescent, rarely sub-glabrous. Leaves opposite, subsessile or shortly petiolate, oblong-obovate, elliptic oblong, lanceolate or oblanceolate, 4 - 25 x 2 - 15 mm, apex acute, margin minutely and distinctly serrate, glandular pubescent or almost glabrous; stipules triangular lanceolate, hairy with serrate margins, acute, 2 - 4 mm long. Inflorescence dense or loose axillary fasciculate cyme of 6 - 12 or more flowers. Flowers reddish, usually 5-merous, 0.5 - 2 mm long, ca 2 mm across; pedicels slender, 1 - 3 mm long, glandular pubescent. Sepals 5, free, 1.5 - 3.2 mm long, linear-lanceolate to ovate-oblong, margins ciliolate, glandular pubescent. Petals 5, lanceolate or oblanceolate, 1 - 2.5 mm long, obtuse or sub-acute, reddish pink. Stamens usually 5, rarely more, subfiliform, ca 1 mm long. Ovary subglobose, sulcate, glabrous. Carpels 5, united; styles 5, free, erect or shortly recurved, 0.5 - 1 mm long; stigmas thick, capitate. Capsule 0.2 - 4 mm long, reddish, ovoid or subglobose, margins of the valves incurved. Seeds minute, with rounded ends, dark brown, shining, somewhat reticulate.

Field note: Commonly occurs in sandy marshy places.

Fl. & Fr.: November-June.

Specimen examined: Bhilwara 1197 (MLVGCB Herb).

Distribution: Tropical Asia from Iran to S. China, Philippines, Australia and Tropical Africa

MALVACEAE

KEY TO THE TRIBES

1a. Styles twice as many athe number ofs carpels... ....Ureneae

1a. Style as many as carpels or styles undivided and stigma entire 2

2a. Fruit a capsule..............................Hibisceae

2b. Fruit a schizocarp................................ 3

3a. Stigmas decurrent on the adaxial side of style ....Malveae

3b. Stigmas apical or nearly so, capitate, discoid or obliquely truncate ....Abutileae

KEY TO THE GENERA IN TRIBES

Tribe 1. ***ABUTILEAE*** Endl.

KEY TO THE GENERA

1a. Seeds 2 or more in each mericarp......*Abutilon*

1b. Seeds one in each mericarp.............................2

2a. Epicalyx present............................*Malvastrum*

2b. Epicalyx absent..*Sida*

ABUTILON Mill.

KEY TO THE SPECIES

1a. Carpels 8-12......................... ....*A. ramosum*

1b. Carples more than 15............................ 2

2a. Branches, petiole and pedicels densely covered with patent, long, simple hairs, minute stellate hairs and viscid by short glandular hairs, Stipules lancolate. Corolla orange with purple centre.....................................*A. hirtum*

2b. Branches, petiole and pedicels densely covered with stellate hairs, and rarely with some slender, simple hairs. Glandular hairs absent. Stipules linear. Corolla yellow or pale yellow, without purple centre ....*A. indicum*

Abutilon hirtum (Lam.) Sweet, Hort. Brit. ed. 1. 53. 1826; Wight & Arn., Prodr. 56. 1834. *Sida hirta* Lam., Encycl. 1: 7. 1983; Parmar, in Shetty & Singh Fl. Rajasthan 1: 116. 1987; Paul, in Sharma *et al.* Fl. India 3: 264 - 265. 1993. *S. graveolens* Roxb. [Hort. Beng. 50. 1814, nom. nud.] ex Hornem., Hort. Hafn. Suppl. 77. 1819; Roxb., Fl. Ind. 3: 179. 1932. *Abutilon graveolens* [Roxb. ex Hornem.] Wight & Arn. ex Cat. 13. 1833 & Wight & Arn., Prodr. 56. 1834; Masters, in Hook. Fl. Brit. India 1: 327. 1874, incl. var. *hirtum*; Duthie, Fl. Gangetic Plain 1: 83. 1903. (Fig. 23 & Plate 33)

var.: ***hirtum***

Erect, annual herb or undershrub, 0.5 - 2 m tall. Stem, petiole and pedicel clammy pubescent and usually long, dense simple, spreading hairy. Leaves with 3 - 15 cm long petiole; stipules 5-10 mm long, 1.5 - 2 mm broad, falcate, lanceolate, eventually reflexed; blade 2.5 - 12 x 3.5 - 13 cm, densely stellate, glutinous hairy on both sides mixed with simple hairs, more so on the nerves beneath, usually broadly ovate or ovate, cordate at base, acute to acuminate at apex, crenulate-dentate or serrate, yellowish, yellowish green to green. Flowers axillary, solitary; pedicel 1.5 - 3 cm and in fruit up to 4 cm long, articulate above the middle. Calyx 6 - 9 mm long, in fruit up to 12 mm, usually fused to the middle; lobes ovate, acuminate-cuspidate, glutinous, simple and stellate pubescent without, sericeous within. Corolla 3 - 3.5 cm across, orange-yellow, yellow; petals 1.5 x 1 cm, densely simple hairy, occasionally stellate on the covering margins, more towards apex, broadly obovate, claw ciliate. Staminal column 5 - 7 mm, yellow or dark purple, stellate pubescent at the base, upper tubular part glabrous. Schizocarps 1 - 2 cm across, globular, indented at apex; mericarps 20 - 25 (-30), separating early, 10 - 15 mm high, radially 7 - 10 mm, broadly ovate-reniform, shortly acuminate, sometimes rounded, dorsally stellate-tomentose, 3-seeded. Seeds reniform, ca 2 - 2.5 mm across, with minute stellate hairs, sometimes more so at the hilum, brownish black.

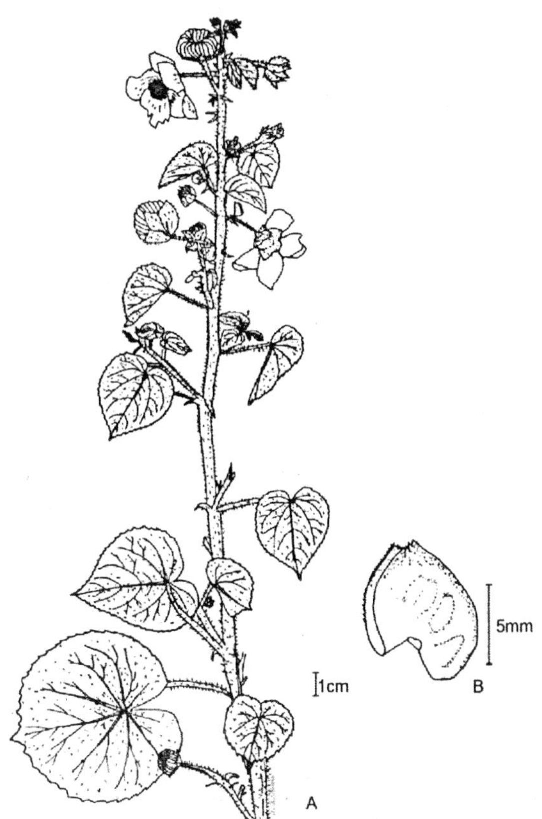

Fig. 23. *Abutilon hirtum* (Lam.) Sweet.: A. Flowering and fruiting branch, B. Mericarp

Field note: Rare in waste places and along roadsides

Fl. & Fr.: November - December.

Specimen examined: Bangreda 3741 (MLVGCB Herb).

Distribution: Semiarid tropics of the old world, introduced in tropical America, India (Throughout in scrub forests up to 600 m)

Abutilon indicum (L.) Sweet, Hort. Brit. ed. 1. 54. 1826; Masters, in Hook. *f.* Fl. Brit. India 1: 326. 1874; Duthie, Fl. Gangetic Plain 1: 83. 1903; BH. 26 1: 226. 1918; Santapau, Fl. Saur. 1: 40. 1962; Bhandari, Fl. Indian Desert 57. 1978; Parmar, in Shetty & Singh Fl. Rajasthan 1: 116 - 117. 1987; Paul, in Sharma *et al.* Fl. India 3: 266 - 267. 1993. *Sida indica* L. in Torner, Cent. Pl. 2: 26. 1756. (Plate 34)

Subsp. ***indicum.***

Vern.: *Dabi dablya, Tara kunchi.*

Undershrub to shrub, 50 cm to 3 m tall, stellate pubescent, intermingled with small, simple, spreading hairs atleast on the young parts. Branches usually green sometimes purplish. Leaves 2 - 18.5 x 1.5 - 16 cm broad, cordate at base, acute to long acuminate at apex, serrate to crenate, 7 - 9-nerved, sometimes 3-angular, stellate pubescent, velutinous and cinereous on both sides, sometimes green above, petiole 2 - 18 cm long, stellate pubescent mixed with sparse, weak, simple, spreading hairs; stipules linear, 2 - 5 mm long. Flowers axillary, solitary; pedicel 1.5 - 8 cm long, jointed near the apex, accrescent and geniculate. Calyx 5 - 8 mm across, not or slightly accresent in fruit, ultimately reflexed, fused at the base or nearly to the middle, pubescent on both sides; lobes lanceolate to ovate, acute to acuminate. Corolla orange-yellow or yellow, 1.5 - 2.5 cm across; petals obovate, 1 - 1.5 cm long and broad, truncate, rounded or emarginate at apex, claw hairy on the margin. Staminal column 5 - 7 mm long, basal part conical, stellate pubescent, upper portion tubular, glabrous. Schizocarps, cylindric truncate, globular, flat or slightly indented at apex, 1.5 - 2 cm across, stellate pubescent; mericarps 15 - 27, 10 - 18 x 7 - 9 mm, acute or somewhat acuminate at the back, erect at maturity and not spreading stellately, dehisce after separating from the central axis. Seeds 2 -3 mm in diam., reniform, minutely stellate-hairy or glabrescent, brownish black.

Field note: Common in wasteplaces, fields and gardens.

Fl. & Fr.: September-April.

Specimen examined: Mandal 92, Bhilwara 1761 (MLVGCB Herb).

Distribution: Tropics and subtropics of New and Old Worlds.

Uses: The root, leaves and bark are mucilaginous and are used to soothen and to protect the mucous membranes of the respiratory & urinary systems. A decoction of the root is given for chest problems such as bronchitis. It is a tonic for the skin due to mucilaginous nature. Root powder is applied to wounds and is also used for treatment of boils and ulcers. Seeds are laxative and are useful in killing thread worms. Affected rectum may be treated by inhailing the smoke of the powdered seeds. Plant is also used in the urinary tract infection.

Abutilon ramosum (Cay.) Guill. & Perr. in Guill. *et al.* Fl. Senegamb. Tent. 1: 68. 1831; Masters, in Hook. Fl. Brit. India 1: 328. 1874; Duthie, Fl. Gangetic Plain 1: 83. 1903; Bhandari, Fl. Indian Desert 58. 1978; Parmar, in Shetty & Singh Fl. Rajasthan 1: 119. 1987; Paul, in Sharma *et al.* Fl. India 3: 266 - 271. 1993. *Sida ramosa* Cav. Diss. 1: 28. t. 6. f. 1. 1785. (Plate 35)

Erect, much branched, perennial, 1 - 3 m tall herb or undershrub. Stem ash-coloured. Branches stellulate tomentose mixed with simple, spreading weak hairs. Leaves with 3 - 10 cm long pubescent petiole; stipules 6 - 10 mm long, filiform-linear, hairy; blade 2.5 - 13 x 2 - 11 cm, chartaceous, broadly ovate, acuminate, cordate at base, coarsely crenate or serrate up to the base, stellulate pubescent to glabrescent on both sides, sometimes 3-angular. Flowers axillary or terminal, solitary or paired or divided above dichotomously (once or twice, each into two) as in cyme; peduncle 1.5 - 4.5 cm long, tomentose and mixed with a few simple, weak, spreading hairs; pedicel in solitary 3 - 4 cm long, articulate above the middle, in cyme 3-6 mm long, articulate in the middle or above or without articulation. Calyx 5 - 7 mm long, fused up to the middle; lobes deltoid-ovate, acuminate-cuspidate, 3 - 5. mm broad, pubescent, viscid. Corolla yellow, 1.5 cm across; petals 5 - 12 mm long, glabrous. Staminal column very short (ca 2.5 mm across), antheriferous towards tip, stellate-hairy. Schizocarps cylindric, 1 x 1 cm, stellate pubescent, slightly viscid; mericarps ca 8 - 10, each with ca 2 mm long, somewhat spreading awns, 3 - 4 mm broad. Seeds 2 to 4 per mericarp, reniform, minutely stellate-puberulous, glabrescent, brown-black, 1.5 - 2 mm across.

Field note: Common species during rainy season. This species is different from other *Abutilon* sp. due to fewer number of carpels, which are awned.

Fl. & Fr.: September - December.

Specimen examined: Mandal 92, Asind 1769 (MLVGCB Herb).

Distribution: Tropical Africa, India and Pakistan.

MALVASTRUM A. Gray, nom. cons.

Malvastrum coromandelianum (L.) Garcke in Bonplandia. 5: 297. 1857; Parmar, in Shetty & Singh Fl. Rajasthan 1: 126 - 127. 1987; Paul, in Sharma *et al.* Fl. India 3: 266 - 277 - 279. 1993. *Malva coromandeliana* L. Sp. Pl. 687. 1753. *M. tricuspidata* R. Br. in Aiton f. Hort. Kew. ed. 2. 4: 210. 1912. *Malvastrum tricuspidatum* (R.Br.) A.Gray, Pl. Wright 1: 16. 1852; Masters, in Hook. Fl. Brit. India 1: 321. 1874; Duthie, Fl. Gangetic Plain 1: 79. 1903; Mahesh. Illus. Fl. Delhi f.26. 1966. (Fig. 24 & Plate 36)

Annual herb or undershrub, up to 1 m high; branches with appressed, strigose, stellate, 4-rayed or simple hairs. Leaves 1 - 7 x 0.5 - 4 cm, upper surface with simple appressed hairs, lower with simple and 4 rayed stellate hairs particularly on the nerves, acute or obtuse, coarsely serrate, 3 - 5-nerved; petiole 0.5 - 5 cm long, stellate hairy; stipules linear-lanceolate, acuminate, 2 - 10 mm long, entire or denticulate, hairy on the margins and outer surface. Flowers axillary, solitary or fascicles, upto 6 cm long spikes. Bracts 4 - 10 x 1.5 - 2.5 mm, ovate, bifid or biparted with acute or acuminate segments, densely covered with simple and minute stellate hairs outside, glabrous inside, caducous. Epicalyx segments 3 - 6 x 1 - 1.5 mm, linear to lanceolate, strigose hairy. Calyx usually fused below the middle, 5 - 8 mm long, strigose hairy outside; lobes deltoid or triangular, acuminate, 3 - 4 x 2 - 5 mm. Petals 7 - 10 x, 4 - 6 mm, obliquely obovate, emarginate at apex, hairy at base. Staminal tube 2 - 4 mm long, conical at base, glabrous. Schizocarps globular; mericarps 8 - 15, ca 2 mm across, dorsally 1 mm broad, deeply curved, with a short, 1 - 1.5 mm long apical awn and 2, 0.5 mm long dorsal awns near the middle hispid above. Seed 1 in each mericarp, ca 1 mm across, reniform, brown-black, glabrous.

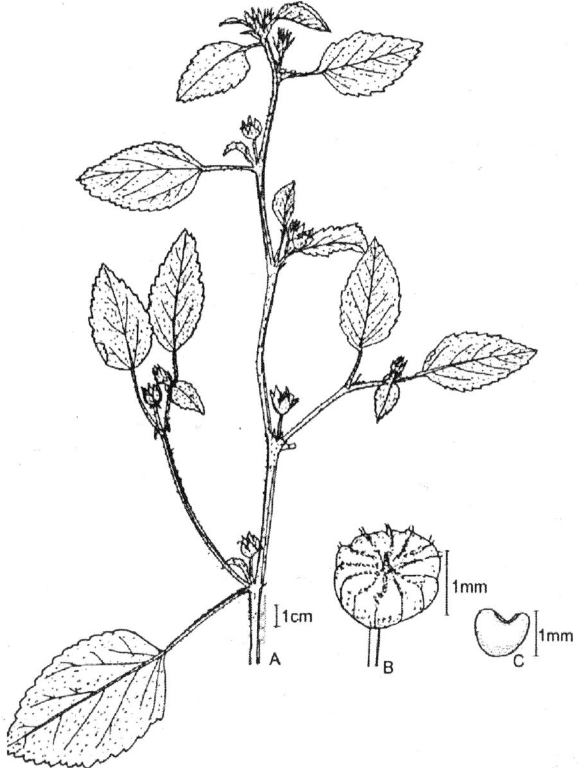

Fig. 24. *Malvastrum coromandelianum* (L.) Garcke: A. Flowering and fruiting branch, B. Fruit, C. Seed

Field note: Widely distributed throughout the area.

Fl. & Fr.: Almost throughout the year.

Specimen examined: Bhilwara 1142, Mandalgarh 1628 (MLVGCB Herb).

Distribution: Tropical regions of both New and Old Worlds.

Notes: It shows a wide range of variations in the shape and size of leaves often in the same plant. This species is often confused with the *Sida acuta*, but can be easily distinguished from the latter by its epicalyx and strigose indumentums on stems, petioles and pedicels.

SIDA L.

KEY TO THE SPECIES

1a. Mericarps 5...................................*S. cordata*

1b. Mericarps 6 - 12...2

2a. Stipuls of each pair dissimilar, one linear to lanceolate and other linear to filiform; leaves sparsely hairy, soon glabrescent.........*S. acuta*

2b. Stipuls of each pair not dissimilar, densely pubescent...3

3a. Mericarps indehiscent...........................*S. ovata*

3b. Mericarps dehiscent...4

4a. Mericarps with 2 awns, awns retrorsely hairy..*S. cordifolia*

4b. Mericarps with or without awns, awns never retrorsely hairy......................... *S. rhombifolia*

Sida acuta Burm., *f.* Fl. Indica 147. 1768; Roxb., Fl. Ind. 3: 171. 1832; *emend.* K. Schum. Fl. Bros. 12. 326. 1891; Wight. *l.c.* 1: t. 95. 1838; Duthie, Fl. Gangetic Plain 1: 81. 1903; Bhandari, Fl. Indian Desert 66. 1978; Parmar, in Shetty & Singh Fl. Rajasthan 1: 130 - 131. 1987; Paul, in Sharma *et al.* Fl. India 3: 281. 1993. *S. lanceolata* Retz., Obs. Bot. 4: 119. 1786; Roxb., Hort. Beng. 50. 1814 & Fl. Ind. 3: 175. 1832. *S. carpinifolia* auct. non. L. f. 1785; Masters, in Hook. Fl. Brit. India 1: 323. 1874. (Plate 37)

Sub-shrub or herb, erect or ascending, 0.5 - 2 m tall. Stem pilose or subglabrous, with minute stellate hairs mixed with some simple hairs. Leaves distichous; stipules filiform, 6 - 12 x 1 - 1.5 mm, often longer than petiole, usually persistent; petiole 2 - 6 mm, sparsely pilose; leaf blade ovate, oblong, lanceolate, or linear-lanceolate, 1 - 9 × 0.4 - 2.5 cm, both surfaces glabrous or sparsely stellate pilose, rarely with simple hairs adaxially, base obtuse, margin dentate, sometimes partly entire toward its base, apex acute or acuminate. Flowers solitary or paired, axillary, sometimes congested at stem apex. Pedicel 4-12 mm, pilose, articulate at middle. Calyx shallowly cup-shaped, connate in basal 1/2, ca 6 mm, mostly glabrous, margins often ciliate, lobes 5, caudate. Corolla yellow, less often white or yellow-orange, 8 - 10 mm in diam.; petals obovate, 6 - 7 mm, ciliate, base attenuate, apex rounded, sparsely glandular hairy outside. Staminal column ca 4 mm long with simple and glandular hairs, antheriferous towards apex. Ovary 1.5 mm across, ovoid or globular. Schizocarp nearly globose; mericarps 6 - 10, segmentiform-tetrahedral, ca 3.5 - 4 mm, basally transversely ridged, side walls reticulate-veined, glabrous, apex beaked, extending into 2 awns, apically dehiscent. Seeds trigonous, ca 2 mm, glabrous except around hilum, dark brown.

Field note: Common in scrub forests, along roadsides, waste lands, etc.

Fl. & Fr.: September - November.

Specimen examined: Mainal 89, Bhilwara 1762 (MLVGCB Herb).

Distribution: India, Bhutan, Cambodia, Laos, Nepal, Thailand, Vietnam.

Notes: The glabrous calyx is usually diagnostic for *Sida acuta*. *Sida bodinieri* Gand. was described as glabrous and if correctly placed in *Sida* would most likely belong under *S. acuta*. *Sida chanetii* Gand. was related by the original author to *S. acuta* but it was based on a collection from near Beijing, well north of any other records for *Sida*, and so its identity needs confirmation.

Uses: The bark is a good fiber source, and the roots are used medicinally.

Sida cordata (Burm.f.) Borss. in Blumea. 14 1: 182. 1966; Bhandari, Fl. Indian Desert 67. 1978; Parmar, in Shetty & Singh Fl. Rajasthan 1: 131. 1987; Paul, in Sharma *et al.* Fl. India 3: 283. 1993. *Melochia cordata* Burm.*f.* Fl. Ind. 143. 1768. *Sida veronicifolia* Lamk. Encycl. 1: 5. 1783. *S. humilis* Cav., Diss. 5, t. 134, f. 2. 1788; Mast., in Hook. Fl. Brit. India 1: 322. 1874, incl. var. *veronicifolia*. *S. beddomei* Jacob in J. Bombay Nat. Hist. Soc. 47: 50, 1950.

Prostrate or semiprostrate herb, up to 80 cm tall. Profusely branched at the base, all parts stellate pubescent, mixed with simple spreading hairs on branches, petioles, pedicels and calyx. Leaves 0.5 - 8 x 0.3 - 5.5 cm, narrowly to broadly ovate, lanceolate in terminal branches, crenate to serrate, cordate or rounded at base, acuminate at apex, lower surface stellate pubescent, upper one with stellate and simple, strigose hairs; stipules filiform, 1 - 4 mm long, linear, filiform, hairy; petiole 1.5 - 30 mm long. Flowers axillary, solitary or pseudoraceme by reduction of leaves in the terminal parts; pedicel 1 - 2.5 cm long, in fruit up to 3.5 cm, jointed slightly above the middle. Calyx fused to the middle, campanulate, 5-fid; lobes, 4 -6 x 0.2 mm, deltoid to triangular, acuminate. Corolla pale yellow, 7 - 8 mm across; petals ca 5 x 4 mm, obovate, ciliate at base. Staminal column 2 - 3 mm long, basal portion conical, tubular part short, glabrous or with some short simple hairy. Schizocarps ca 4 x 3 mm, depressed globose, enclosed in persistent calyx, brownish black, 3 - 4 mm across, pubescent at the top, beaked; mericarps 5, ca 4 x2 mm, slightly longitudinally keeled on the back, tetrahedral with rounded angles, awnless, dehiscent, membranous, smooth, disintegrated at the radial surfaces, 2.5 mm long. Seeds brown, ca 2 x 1 mm, glabrous, dispersed by withering of wall, brownish black.

Field note: The plants are erect but at fruiting stage they become trailing or procumbent.

Fl. & Fr.: September - December

Specimen examined: Bhilwara 1782 (MLVGCB Herb).

Notes: This is a highly variable species in indumentum and inflorescence. Variation in the inflorescence sometimes gives the illusion of the presence of at least 2 taxa. It has been observed in the herbarium specimens that there is continuous variation in the inflorescence hence the separation of taxa on the basis of this character is not advisable. However, discontinuous variation has been observed in the colour of branches and pedicels, and in the stiffness of pedicels on the basis of which 2 taxa may be easily delimited.

Uses: Stem yields good fibre.

Sida cordifolia L., Sp. Pl. 684. 1753; Mast., in Hook. *f.,* l: 324. 1874; Duthie, Fl. Gangetic Plain 1: 82. 1903; Bhandari, Fl. Indian Desert 67. 1978; Parmar, in Shetty & Singh Fl. Rajasthan 1: 131. 1987; Paul, in Sharma *et al*. Fl. India 3: 285. 1993.

Vern.: *Bal.*

Erect, velvety undershrub, up to 1 m tall. Branches, petiole and pedicel stellately pubescent mixed with simple, spreading hairs. Leaves stipulate, stipule 2 - 10 mm long, filiform, stellate hairy; petiole 4 - 5 mm long, stellate tomentose; blade 0.5 - 6 x 0.4 - 5 cm, ovate to oblong or orbicular or lanceolate, rounded or slightly cordate at base, 7 -nerved at base, acute at apex, crenate to serrate, on both surface stellate pubescent, also mixed with simple hairs on the nerves beneath, velvety. Flowers axillary, solitary or paired or more in terminal branches; pedicel 2 - 10 mm, in fruit up to 2 cm long, jointed near the top. Calyx 5 - 10 x 5 - 8 mm, densely stellate tomentose outside and mixed with simple, spreading hairs, fused to the middle; lobes triangular or deltoid, acute to acuminate. Corolla yellow, ca 15 mm across; petals obliquely obovate, truncate at apex; ciliate at base. Staminal column ca 2.5 - 3 mm long, hairy or glabrous. Ovary conical, stellate-hairy. Schizocarps discoid, 5 - 8 mm across, stellate pubescent in the upper half, dehiscent; mericarps 8 - 10, reticulate, radially 3-3.5 mm long, ca 2.5 mm broad, dorsally 2 - 2.6 mm broad, slightly grooved; awns 2, divergent, 3 - 5 mm long, retrorsely hairy. Seeds flattened reniform, ca 2 mm long and broad, glabrous except hilum, brown to dark brown.

Field note: Common in waste lands, particularly in xeric habitats, species having an unpleasant smell.

Fl. & Fr.: September - December.

Specimen examined: Bhilwara 155, 1763 (MLVGCB Herb).

Distribution: Pantropical.

Sida ovata Forsk., Fl. Aegypt. -Arab. 124. 1775; Andrews, Fl. Pl. Anglo-Egypt. Sud. 2: 41. 1952; Santapau. Fl. Saur. 1: 36. 1962; Bhandari, Fl. Indian Desert 73. 1978; Parmar, in Shetty & Singh Fl. Rajasthan 1: 131 - 132. 1987; Paul, in Sharma *et al.* Fl. India 3: 288 - 189. 1993. *S. grewioides* Guill. & Perr. in Guill. *et al.* Fl. Seneg. Tent. 1: 71. 1830; Mast., in Hook. *f.,* l: 323. 1874; Duthie, Fl. Gangetic Plain 1: 81. 1903 (Plate 38).

Vern.: *Dabi.*

Erect, perennial herb, up to 30 cm tall, all parts stellate pubescent. Leaves with up to 1 cm long, filiform or linear stipule; petiole 0.5 - 1.5 cm long; blade 1. 5 - 5 x 1 - 4 cm, ovate-oblong, coarsely serrate-crenate, entire towards the base, obtuse or somewhat truncate, stellate pubescent on both sides, darker above, cinereous below, penninerved, at base mostly 5 nerved. Flowers white, axillary, solitary or paired; pedicel 5 - 8 mm, in fruit up to 10 mm, rarely up to 15 mm long, with or without articulation at the top. Calyx 5 lobed, ca 4 mm across, 4.6 mm long, slightly angular; lobes triangular-deltoid, acuminate, 2.3 x 3-4 mm. Corolla 1 - 1.5 cm across, yellow or yellowish brown; petals 7 - 24 mm long, 4 - 12 mm broad, oblique, unguiculate, sparsely covered with bulbous based hairs, claw hairy, glabrous outside. Staminal tube up to 37 mm long, simple and stellate-hairy. Carpels 7 - 8; locules, style branches and stigmas as many as carpels. Schizocarps depressed globose, 3 - 5 mm in diam., incompletely enclosed by calyx; mericarps 2 - 3 mm long, radially 2 - 2.5 mm broad, dorsally 1 - 1.5 mm broad, reticulate towards the margin, not so towards the centre, glabrous except awns, indehiscent; awns 2, 1 mm long, connivent, inflexed. Seeds 1.5 - 2 mm long, rounded reniform, slightly longitudinally grooved on the dorsal surface, hilum with few short stellate hairs, brown to dark brown.

Field note: A common species in fields and gardens. The calyx remain attached to the mature fruits and slightly opened at the apex.

Fl. & Fr.: September - December.

Specimen examined: Bhilwara 90, (MLVGCB Herb).

Distribution: In drier parts of Africa, Arabia, India, Pakistan and Persia.

Sida rhombifolia L., Sp. Pl. 2: 684. 1753; emend. Mast., in Hook. *f.*, Fl. Brit. India l: 323. 1874; Duthie, Fl. Gangetic Plain 1: 181. 1903; Parmar, in Shetty & Singh Fl. Rajasthan 1: 132. 1987; Paul, in Sharma *et al.* Fl. India 3: 289. 1993. *S. orientalis* Cav. Diss. 1: 21. t. 12. f. 1. 1785.

subsp.: ***rhombifolia***

Annual or perennial herb or undershrub, erect or prostrate, many branched, to ca 1.5 m tall. Branchlets stellate. Stipules spinelike, 3 - 5 mm; petiole 2 - 7 mm, stellate puberulent; leaf blade rhombic to oblong-lanceolate or obovate, rarely linear-lanceolate, 0.5 - 8 × 0.3 - 5 cm, abaxially gray-white stellate pilose, adaxially sparsely stellate pilose to subglabrous, base broadly cuneate, margin dentate, apex obtuse to acute. Flowers solitary, axillary. Pedicel 1 - 2.5 cm, densely stellate tomentose, articulate above middle. Calyx cup-shaped, 4 - 6 x 3 - 5 mm, abaxially stellate pubescent, lobes triangular, apices acute. Corolla ca 1 cm in diam.; petals yellow, obovate, ca 15 mm, base attenuate, apex rounded. Filament tube 4 - 5 mm, glabrous. Style branches 8-10. Schizocarps semiglobose to broadly turbinate, 6-7 mm in diam.; mericarps 7-10, 2.5 - 3 mm excluding awn, shallowly grooved to near base, eventually dehiscent, side walls usually thin, not veined, stellate puberulent, apex usually 2-awned, awns to 1.5 mm. Seeds reniform, ca 2 mm, glabrous, blackish.

Field note: Common in waste lands, particularly in xeric habitats.

Fl. & Fr.: September - December.

Specimen examined: Bhilwara 95, Mainal 3668 (MLVGCB Herb).

Distribution: Pantropical.

Notes: *Sida rhombifolia* is the type species of the genus *Sida*. The taxonomy of this species/ species complex is controversial and is discussed by Verdcourt (Kew Bull. 59: 233-239. 2005). The lectotype of *S. rhombifolia* almost certainly came from Jamaica, not India as suggested by some authors, and is a match with neither African nor Asian material. Verdcourt recognized six varieties for East African material, differing most obviously in mericarp morphology: most notably the degree of dehiscence, which varies from completely indehiscent to dehiscing by an apical slit to breaking into two valves, the degree of sculpturing, and the presence or absent of awns. Chinese material appears to have a comparable range of variation, though many collections lack fully mature mericarps, and more detailed studies could lead to the recognition of more, comparable taxa.

Hu (Fl. China, Malvaceae [Fam. 153], 20-21. 1955) recognized two varieties, var. *rhombifolia* and var. *corynocarpa* (Wallich ex Masters) S. Y. Hu (Fl. China, Malvaceae [Fam. 153], 20. 1955), differing only in minor quantitative characters. The validity of var. *corynocarpa* is open to question, as Masters (Fl. Brit. India 1: 324. 1874) only mentioned the supposed basionym "*Sida corynocarpa*" in a note under *S. rhombifolia* var. *retusa* and therefore cannot be said to have accepted it as a species. Plants with prostrate stems from Taiwan, Japan (Ryukyu Islands), and the Philippines have been placed in *S. rhombifolia* subsp. *insularis*. Their status needs more detailed investigation. Borssum Waalkes included *S. alnifolia* L. within *S. rhombifolia* as var. *retusa*. Hu 12784, from Hong Kong, has 10 indehiscent mericarps per flower, each with a single awn, suggesting a relationship to *Sida rhombifolia* var. *maderensis* (Lowe) Lowe (*S. maderensis* Lowe; *S. unicornis* Marais). That taxon has rather more strongly sculptured mericarps than Hu 12784, which might represent a distinct taxon.

Uses: The entire plant is used medicinally.

Tribe 2. ***HIBISCEAE*** Endl.

KEY TO THE GENERA

1a. Trees ....*Thespesia*

1b. Herbs or shrubs 2

2a. Calyx irregularly 2 - 3-lobed, spathaceous, deciduous........................... ....*Abelmoschus*

2b. Calyx regularly 5-lobed, not spathaceous, persistent.. .3

3a. Capsules winged..................................*Fioria*

3b. Capsules not winged...................... ..... *Hibiscus*

ABELMOSCHUS Medicus

KEY TO THE SPECIES

1a. Epicalyx segments 4 - 6, linear to lanceolate and caducous before the expansion of the corolla ....*A. ficulneus*

1b. Epicalyx segments 6 or more, linear to narrowly lanceolate....................................... .2

2a. Capsules shorter or equaling the epicalyx....................................... ......*A. crinitus*

2b. Capsules much longer than the epicalyx.......3

3a. Epicalyx as long as calyx. Mature capsules more than 6 cm long, truncate at the base; peduncles very short as compared to the length of capsules...........................*A. esculentus*

3b. Epicalyx shorter than calyx. Mature capsules 3-6 cm long, narrow at the base; peduncles nearly as long as capsules ....*A. moschatus*

Abelmoschus crinitus Wall., Pl. As. Rar. 1: 39. t. 44. 1830; Borssum in Blumea 14: 103. 1966; Parmar, in Shetty & Singh Fl. Rajasthan 1: 111. 1987; Paul, in Sharma *et al.* Fl. India 3: 302. 1993. *Bamia crinita* Wall. [Cat. No. 1920. 1829, nom. nud.]. *Hibiscus crinitus* (Wall.) G. Don, Gen. Hist. 1: 380. 1831. *H. cancellatus* Roxb., [Hort. Beng. 51. 1814, nom. nud.] Fl. India 3: 201. 1832; Mast., in Hook. *f.* Fl. Brit. India l: 342. 1874; Duthie, Fl. Gangetic Plain 1: 92. 1903. *Abelmoschus cancellatus* (Roxb.) Voigt, Hort. Sub. Calc. 119. 1845.

Vern.: *Marbhindi.*

Erect herb, 0.5 - 1.5 m tall. Roots usually tuberous. Branches, petioles and pedicels simple and stellate hairy, usually hispid or hirsute. Leaves 5 - 8 cm across, orbicular or broadly ovate, deeply cordate at base, 5 -7 nervrd, simple or stellate usually hirsute, angular or 5 - 7 lobed, coarsely serrate to dentate; stipules 1 - 3 cm long, linear to filiform; petiole 0.5 - 24 cm long. Flowers axillary, usually solitary; pedicel c. 2 cm long, in fruit up to 3 cm. Epicalyx segments 10 - 16, 2 - 5 cm long, 0.5 -1 mm broad, linear to filiform, hairy. Calyx 2 - 2.5 cm long, densely puberulous to tomentose. Corolla 6 - 7 cm across, yellow with a purple spot in the centre; petals 4 - 9 cm long, 2 - 4 cm broad, obovoid, glabrous. Staminal column ca 2 cm long. Capsule 2 - 5 cm long, 2 - 3.5 cm across, ovoid-globular, simple and stellate pubescent, shortly acuminate or rounded, hirsute. Seeds 3-5 mm long, globular-reniform, black, rusty tomentose or rarely glabrous.

Field note: This species shows great variations in the degree of incision of the leaves and the density of the indumentums. The capsules are more or less enclosed by the epicalyx segments.

Fl. & Fr.: July- December

Specimen examined: Mezadam 94 (MLVGCB Herb).

Distribution: South Asia, China, Malesia (Java) and Philippines.

Notes: The tuber like swollen tap root enables the species to withstand periodic burnung of the vegetation (Borssum Waalkes in Blumea 14: 103. 1966)

Abelmoschus esculentus (L.) Moench., Meth. Pl. 617. 1794. *Hibiscus esculentus* L., Sp. Pl. 2: 696. 1753; Mast., in Hook. *f.*, Fl. Brit. India l: 343. 1874; Duthie, Fl. Gangetic Plain 1: 93. 1903; Parmar, in Shetty & Singh Fl. Rajasthan 1: 111-112. 1987.

Annual herb, 1-2 m tall, most parts with very sparse prickly hairs. Stem often hollow. Stipule filiform, 7-10 mm, sparsely hirsute; petiole 7-30 cm, long hirsute, often with pubescent adaxial groove; leaf blade palmately 3 - 7-lobed, 8 - 30 cm in diam., lobes broad to narrow, sparsely hirsute on both surfaces, margin sparsely dentate and emarginate. Flowers solitary, axillary. Pedicel 0.5 - 3cm, sparsely strigose. Epicalyx lobes 7-10 (-12), filiform, 5 - 18 × 1-2.5 mm, sparsely hirsute. Calyx campanulate, 2 - 3 cm, densely stellate puberulent. Corolla yellow or white with dark purple center, 5 - 7 cm in diam.; petals obovate,

3.5-5 × 3-4 cm. Filament tube 2 - 2.5 cm. Capsule cylindric, 10-25 × 1.5-2 cm, long beaked, sparsely strigose. Seeds dark brown or gray, globose to reniform, 5 - 15 per locule, 4 - 5 mm, striate, minutely warty.

Field note: Commonly cultivated for its fruits as vegetables.

Fl. & Fr.: July-December

Specimen examined: Kotri 1377, Rajgarh 1765 (MLVGCB Herb).

Distribution: Throughout the tropical and warm temperate regions around the world.

Uses: The young fruits are used as vegetable in S. Asia

Abelmoschus ficulneus (L.) Wight & Arnott ex Wight, Cat. No. 14. 1833. Wight & Arn., Prodr. 53. 1834; Parmar, in Shetty & Singh Fl. Rajasthan 1: 112. 1987; Paul, in Sharma *et al.* Fl. India 3: 304. 1993. *Hibiscus ficulneus* Linn. Mast., in Hook. Fl. Brit. India l: 340. 1874; Duthie, Fl. Gangetic Plain 1: 91. 1903.

Vern.: *Malbhindi.*

Herb or undershrub, 0.5 - 1.5 m tall. Stem usually simple, and short hairy, sometimes prickly, usually thick and fistular. Leaves 2 - 12 cm long and broad, orbicular, cordate at base, serrate, scabrous on both sides, 3 - 5-parted, lobes 2 - 8 x 1 - 4 cm, obovate to spathulate, rounded or obtuse at apex; stipules 5 - 12 mm long, linear or filiform, hirsute; petiole 1.5 - 20 cm long, simple hairy. Flowers axillary, solitary or in terminal racemes by the reduction of leaves; pedicel 1 - 1.5 cm long, in fruit up to 3.5 cm, densely simple hairy, glabrescent. Epicalyx segments 5 - 6, 5 - 10 x 1 - 1.5 cm, linear to lanceolate, acute, hirsute, caducous before expansion of corolla. Calyx in bud lageniform with 3 mm long linear lobes. Corolla ca 2 cm across, white, becoming pink, with a dark purple spot in the centre; petals obovate, 2 - 3 x 1.5 - 2.5 cm, obovate, rounded at apex. Staminal tube ca 1.5 cm long, glabrous. Capsule 2 - 4.5 cm long, 1.5 - 2 cm across, 5 angular, obtuse, rarely short acuminate, ovoid, simple, hairy, glabrescent, valves membranous, hirsute inside with short simple hairs. Seeds ca 3 mm across, globular, stellate-tomentose in concentric rings, black, glabrous to stellate pilose.

Field note: Common in waste lands. This species can be easily recognized by its flask-shaped calyx in buds and blunt, hairy capsules. The leaves resemble those of *Ficus carica* L., hence the specific epithet.

Fl. & Fr.: August - December

Specimen examined: Shambhupura 1351, Mandalgarh 1766 (MLVGCB Herb).

Distribution: N. Australia, S. Asia, Malesia, E. Africa and Madagascar

Uses: Seeds yield an oil and are also used for flavouring coffee. Stem yields an excellent fibre (Sharma *et al.*, 1993).

Abelmoschus moschatus Medic., Malv. 46. 1787; Bhandari, Fl. Indian Desert 55. 1978; Parmar, in Shetty & Singh Fl. Rajasthan 1: 113. 1987; Paul, in Sharma *et al.* Fl. India 3: 308. 1993. *Hibiscus abelmoschus* L., Sp. Pl. 696. 1753; Mast., in Hook. *f.*, 1: 342. 1874; Duthie, Fl. Gangetic Plain 1: 92. 1903.

Herb or undershrub, erect 0.5 - 3 m tall. Stem simple and retrorse hairy, hispid to prickly. Leaves extremely variable in shape and size, 4 - 18 x 3 - 20 cm, orbicular or broadly ovate, cordate at base, not lobed or 3 - 7 lobed or parted, hispid on both sides; lobes deltoid to oblong-lanceolate, serrate to dentate or crenate or rarely entire; stipules linear or filiform, 5 - 12 mm long, simple hairy; petiole 2 - 20 cm long, hispid. Flowers axillary, solitary, pedicel 2 - 8 cm long, in fruit up to 19 cm. Epicalyx segments 6 - 10, free, 8 - 20 x 1 - 2.5 mm, appressed, linear to lanceolate. Calyx 1.5 - 3 cm long, stellate-tomentose outside, sericeous inside. Corolla 10 cm across, yellow with a deep purple spot at the base; petals 5 - 10 x 1.5 - 5.5 cm, fleshy, ciliate at base. Staminal tube 1.5 - 2 cm long. Capsule 4 - 8 cm long, 2 - 3.5 cm across, ovoid to globose, acuminate with a rostrum of ca 5 mm, densely simple hairy, usually hispid. Seeds 3 - 4 mm across, ovoid-reniform, usually glabrous or rarely stellate pubescent, black, often musk scented.

Field note: Occurs throughout the area in cultivated as well as waste lands.

Fl. & Fr.: August - December.

Specimen examined: Asind 1778 (MLVGCB Herb).

Distribution: Native of the Old World tropics.

FIORIA Mattei

Fioria vitifolia (L.) Mattei in Bot. Ort. Bot. Palermo n.s. 2: 71. 1916; Paul, in Sharma *et al.* Fl. India 3: 311. 1993. *Hibiscus vitifolius* L., Sp. Pl. 696. 1753; Mast in Hook. *f.* 1: 338. 1874; Parmar, in Shetty & Singh Fl. Rajasthan 1: 124. 1987; *H. obtusifolius* Willd., Sp. Pl. 3: 829. 1801. *H. vitifolius* L. var. *genuine* f. *indica* Hochr. in Ann. Cons. Jard. Bot. Geneve 4: 169. 1900.

Vern.: *Markappa.*

Herb or undershrub, up to 2 m tall, usually densely stellate tomentose all over. Leaves 2.5 - 15 x 2 - 12 cm, subcordate-rounded at base, acute at apex, serrate to crenate, not lobed or shallowly 3 - 7-lobed, broadly ovate to orbicular, densely pubescent on both surfaces; petiole 2 - 13 cm long; stipules filiform, 2 - 5 mm long. Flowers axillary, solitary; pedicel 1.5 - 6 cm long, accerescent, articulate; epicalyx segments 7 - 12, linear, 6 - 12 x 0.5 - 1 mm, simple hairy on the margin, erect, in fruit spreading to generally reflexed. Calyx fused below the middle, 5-lobed, lobes 5 - 15 x 6 - 10 mm, 5 nerved, reticulately veined, simple and 2 rayed hairy within; lobes ovate to deltoid, acute. Corolla 4 - 6 cm across, pale yellow to yellow with a purple centre; petals 2.5 - 5 x 1 - 3 cm, outside with simple and 2 rayed hairs towards the top, glabrescent, obovate. Staminal column 1 - 2 cm long, regularly antheriferous all over, glabrous. Ovary ovoid, obtuse, 5-angular, pubescent 3 - 4 mm long; style arms upto 4 mm long, glandular hairy; stigmas clavate, hairy. Capsule globular, 1.5 - 2 cm across, shorter than calyx, shortly beaked, beaks ca 3 mm long, 5-winged, wings bristly, splitting laterally on the keels during dehiscence. Seeds 2 - 4 in each cell, reniform 2 - 3 mm across, regularly verruculose, glabrous, brownish-black.

Field note: Common in dry deciduous forest.

Fl. & Fr.: August - November

Specimen examined: Karoi 1783, Bijoliya 1764 (MLVGCB Herb).

Distribution: Tropical and subtropical regions of the Old World.

Notes: The species was treated previously by Masters (1874) and Borssum Waalkes (1966) under *Hibiscus*. Mattei (l.c.) transferred it to *Fioria* based mainly on fruit with conspicuous, scarious, strongly veined wings - a unique character absent in the genus *Hibiscus*.

HIBISCUS L. nom. cons.

KEY TO THE SECTIONS

1a. Epicalyx segment forked at apex, linear or oblanceolate; calyx lobes prominently 3-veined; stem with prickles or bristles or both ...... *Fucaria*

1b. Epicalyx segment not forked at apex; calyx lobes without prominently 3- veined; stems without prickles or bristles or both 2

2a. Seeds hairy ...... *Hibiscus*

2b. Seeds glabrous or tuberculate 3

3a. Calyx more or less inflated particularly in fruit, lobes many-veined; mostly shrubs, sometimes herbs ...... *Trionum*

3b. Calyx not inflated, lobes 3-veined; herbs or undershrubs...... *Ketmia*

Section 1. ***FUCARIA*** DC.

Hibiscus cannabinus L., Syst. Nat. ed. 10. 2: 1149. 1759; Mast in Hook. *f.,* 1: 339. 1874; Duthie, Fl. Gangetic Plain 1: 90. 1903; FPN, 1: 116; BH, 26. 1: 228. 1918; Bhandari, Fl. Indian Desert 60. 1978; Parmar, in Shetty & Singh Fl. Rajasthan 1: 136. 1987; Paul, in Sharma *et al.* Fl. India 3: 324. 1993.

Vern.: *Ambari.*

Annual or perennial, erect, robust herb to 2.5 - 4 m tall, glabrous, sparsely spiny. Stipules silk like, 3 - 6 mm; petiole 3 - 15 cm, sparsely spiny; leaf blade dimorphic; blades on proximal part of stem cordate, those on distal part of stem palmately 3-7-lobed, lobes lanceolate, 3-12 × 0.4 - 2 cm, base cordate or nearly rounded, margin serrate, glabrous on both surfaces, apex acuminate; basal veins 5 - 7, glandular near base of midrib. Flowers solitary, axillary or in racemes. Epicalyx lobes 7 - 8, filiform, free, spreading or reflexed, upto 1.5 cm long, sparsely spiny. Calyx nearly campanulate, connate for ca 1/2 length, ca 3 cm, spiny and white tomentose; lobes 5, free almost to the base, lanceolate, 1 - 2 cm long, caudate, distinctly 3-nerved with a large nectary on midvein outside. Corolla yellow with crimson center; petals oblong-obovate, 4 - 6 x 1 - 3 cm, spreading, glabrous. Staminal column 1 - 2.5 cm, glabrous. Styles 5, glabrous. Capsule globose, ca 2 x 1.5 cm, dehiscing into 5-valves, densely spiny out side, glossy inside, apex shortly beaked. Seeds 7 x 4 mm, reniform, dotted with minute brownish scale-like tufted hairs, brownish.

Field note: Cultivated.

Fl. & Fr.: September- October

Specimen examined: Mezadam 1348 (MLVGCB Herb).

Distribution: Tropical and subtropical Africa. Cultivated in most tropical countries.

Notes: *Hibiscus cannabinus* is cultivated as a fiber crop in association with *Zeya mays* L. as well as *Arachis hypogaea* L. etc.. Rough fibres obtained from the stem are used to prepare ropes as well as strings. The stem is used as a fuel and children also smoke as biri or cigarettes.

Section 2. ***HIBISCUS*** L. nom. cons.

Hibiscus micranthus L. f., Suppl. 308. 1781; Mast., in Hook.f., Fl. Brit. Ind. 1: 335. 1874; Cooke, Fl. Bomb. Pres. 1: 113. 1958 (reprint. ed); Duthie, Fl. Gangetic Plain 1: 89. 1903; Parmar, in Shetty & Singh Fl. Rajasthan 1: 122. 1987; Paul, in Sharma *et al.* Fl. India 3: 330. 1993. *Hibiscus ovalifolius* sensu Boisser, Fl. Or. 1: 839. 1867, auct. plur. non (Forssk.) Vahl, 1790.

Vern.: *Okda.*

Undershrub, sparsely to profusely branched, upto 2.6 m high. Branches and leaves scabrous to bristly hairy. Leaves subsessile to long petiolate; blade 1.5 - 4.5 x 0.5 - 2 cm, elliptic, ovate or ovate-rhomboid, scabrous to bristly or sparsely hairy on both sides; stipules 0.3 - 1.3 cm long, filiform. Flowers axillary, solitary; pedicel 0.3 - 2 cm long, with or without joint. Epicalyx segments, 6 - 8, 2 - 4 mm long, filiform, stiff. Calyx 5 mm long, 5 lobed; lobes lanceolate or triangular. Corolla 1 - 1.5 cm across, white or pink; petals 5, reflexed, stellate hairy outside. Staminal tube very short, up to 5 mm long. Capsule globose, 7 - 9 mm across, dehiscing into 5-valves, smooth out side, glossy inside. Seeds reniform, cottony, hairs upto 8 mm long, black.

Field note: Common on hills, particularly in association of *Euphorbia caducifolia* Haines. The flowers change colour from pink to white during the day. Dehisced capsules are persistent

Fl. & Fr.: August - November

Specimen examined: Pansal 1307, Nayasamaliya 1402 (MLVGCB Herb).

Distribution: India, Pakistan, Arabia and Tropical Africa

Section 3. ***KETMIA*** DC.

Hibiscus caesius Garcke, in oestr. Bot. Zeit. 850. 1849; Rakshit & Kundu in Bull. Bot. Surv. India 12: 173. 1972; Bhandari, Fl. Indian Desert 60. 1978; Pandey *et al.*, in Bull Bot. Surv. India 24: 133. 1982; Parmar, in Shetty & Singh Fl. Rajasthan 1: 121. 1987; Paul, in Sharma *et al.* Fl. India 3: 334. 1993. *H. gibsonii* Stocks ex Harv. in Harv. & Sond. Fl. Capensis 2: 587. 1962; Mast., in Hook. *f.*, Fl. Brit. Ind. 1: 339. 1874; Duthie, Fl. Gangetic Plain 1: 91. 1903. (Plate 39)

Erect, 1 - 1.5 m tall herb or undershrub. Branches with scattered bristles. Leaves palmately 3 - 5-partite; leaflets 1.5 - 7.5 x 0.5 - 3 cm, commonly elliptic, serrate, upper surface almost glabrous or simple or stellate hairy on the veins, lower surface with scattered stellate hairs; stipules, 3 - 10 mm long, filiform with scattered, simple, slightly stiff hairs; petiole 1 - 8 cm long, with a few scattered stiff hairs. Flowers axillary, solitary, large; pedicel up to 12 cm long with scattered, tubercled simple, stiff hairs, more so near the top, articulate, slightly geniculate. Epicalyx segments commonly 10, 1.5 - 3.5 cm x 1 mm, needle like subspiny, terete, hairy, straight or slightly curved, radiate. Calyx fused at the base, shorter than or equal to epicalyx segments; lobes lanceolate, long acuminate, 3-nerved, sparsely simple hairy. Corolla large, 3 - 5.5 x 5 - 8 cm across, yellow or rarely white with purple centre; petals obovate, sparsely stellate hairy outside, glabrescent, claw ciliate margin. Staminal column 1.5 - 2 cm long, purple, antheriferous below the middle. Capsule ovoid, 7-13 x 6 - 9 mm, enclosed within the calyx, hairy outside, glabrous inside, dehiscing into 5 valves, beaked. Seeds 3 - 4 per cell, dark brown, 3 mm across, reniform, minutely pilose.

Field note: Common in moist, shady places and in wetlands.

Fl. & Fr.: August - October

Specimen examined: Hamirgarh 91 (MLVGCB Herb).

Distribution: Tropical and subtropical regions of the Old hemisphere.

Uses: The seed oil is considered nonedible but is used industrially and in oil-lamps.

Section 4. ***TRIONUM*** DC.

Hibiscus trionum L., Sp. Pl. 697. 1753; Mast., in Hook. *f.,* Fl. Brit. Ind. 1: 334. 1874; Duthie, Fl. Gangetic Plain 1: 88. 1903; Rakshit & Kundu, in Bull. Bot. Surv. India 12: 168. 1972; Parmar, in Shetty & Singh Fl. Rajasthan 1: 123. 1987; Paul, in Sharma *et al.* Fl. India 3: 341. 1993.

Vern.: *Mar-ambari.*

Annual, usually erect or somewhat spreading or straggling, hispid, 25 - 60 cm tall herb. Leaves 2 - 7.5 cm long, 3 - 5-partite, central lobe longest, each lobe toothed to pinnately partite, lower leaves sometimes not lobed, nearly glabrous or sparsely stiff hairy, more so beneath; stipules 2 - 7 x 1 mm, linear, stiff hairy, more so beneath; petiole ca 1.5 - 4 cm long, minutely stellate pubescent and sparsely mixed with spreading stellate hairs. Flowers axillary, solitary; pedicel 2 - 4.5 cm long, stiff hairy, more so above, articulate near the top. Epicalyx segments 7 - 12, 5 - 15 cm long, linear, margin with long simple, spreading, very stiff hairs, tubercled at base. Calyx twice the length of epicalyx, fused above the middle, with green-purplish raised veins, stellate pubescent, spreading and stiff hairy on nerves, very much inflated and enclosing the fruit; lobes deltoid, acuminate, ca 1 cm broad, minutely stellate and long, simple appressed hairy within towards the margin. Corolla white-pale yellow with a purple mouth, 1.5-3 cm across; petals 1 - 2 x 0.5 cm, glabrous. Staminal column 3-8 mm long, purplish; filaments ca 2 mm long, purplish; anthers yellow. Capsule 0.5-2 x 0.6-1.5 cm, oblong, obtuse, black, very hispid. Seeds many, ca 2 mm long and broad, black tuberculate.

Field note: Rarely found near ponds in sandy soils.

Fl. & Fr.: September - November

Specimen examined: Karoi 311 (MLVGCB Herb).

Distribution: Common in hotter parts of the Old World from Southern Europe to S. Africa, Madagascar, Asia and Australia, naturalized in America

Uses: An infusion of flowers is used in itching.

THESPESIA Sol. ex Correa, nom. cons.

Thespesia populnea (L.) Sol. & Corr. Ann. Mus. Hist. Nat. Paris. 9: 290, t. 8. f. 1: 1807; Mast., in Hook. *f.,* Fl. Brit. India 1: 345. 1874; Parmar, in Shetty & Singh Fl. Rajasthan 1: 134. 1987; Paul, in Sharma *et al.* Fl. India 3: 352. 1993. *Hibiscus populneus* L. Sp. Pl. 694. 1753. *Thespesia macrophyla* Blume, Bijdr. 2: 73. 1825.

Vern.: *Rampipal, Paraspipal.*

A medium sized tree, 4 - 8 m tall. Branches green, peltate hairy, glabrescent. Leaves ovate, short acuminate, entire, 5 - 15 x 3 - 10 cm, peltate hairy to almost completely glabrous, young parts green; stipules linear-lanceolate, somewhat ensiform, 5 -8 x ca 2 mm broad, peltate hairy, caducous; petiole 5 - 10 cm long, sparsely peltate hairy, glabrescent. Flowers axillary, solitary; pedicel jointed at base, 3 mm across, 1 - 5 cm long, accrescent in fruit, peltate hairy, erect; bracteoles 2, bracteate joints present, subulate to linear-lanceolate, opposite, 2 mm broad. Epicalyx segments 3, spirally arranged, up to 0.5 - 2 cm x 2 - 4 mm, lanceolate, peltate hairy, caducous. Calyx cupular, 1 - 1.5 x 1.5 - 2 cm across, entire or minutely 5 toothed; peltate hairy without, dense, simple, appressed hairy within, coriaceous, persistent. Corolla pale yellow with a crimson centre, turning reddish in the evening, 3 - 6 cm across; petals 6 - 7 cm x ca 5 cm, peltate hairy outside on the covering margin, claw fleshy, ciliate. Staminal column 1 cm long, glabrous. Young fruits and buds when cut exude yellow gum. Fruit globose, 2 - 3 cm across, often with a mucro above, indehiscent, easily pressed by finger, smooth or somewhat 5-angular, peltate hairy, more so above, glabrescent, 5-celled, each cell 2 - 4 seeded. Seeds obovoid, angular, ca 1 cm x 6 mm, rounded above, acute below, covered by dense soft hairs, long and somewhat loose hairy at base.

Field note: Commonly planted near temple, along road sides.

Fl. : October-December; *Fr.*: December - April

Specimen examined: Bhilwara 1158, Pur 1783 (MLVGCB Herb).

Distribution: Common in tropical countries. India, Myanmar, Indochina, Malesia, Australia and Africa

Tribe 3. ***MALVEAE*** A Gray

MALVA L.

KEY TO THE SPECIES

1a. Epicalyx segments ovate or ovate-oblong .... *M. sylvestris*

1b. Epicalyx segments linear to lanceolate 2

2a. Dorsal surface of mericarps smooth or finely ridged, with angles not keeled .... *M. verticillata*

2b. Dorsal surface of mericarps distinctly reticulated with keeled angles. .. *M. parviflora*

Malva parviflora L., in Hoejer, Demonstr. Pl. Hort. Ups. 18. 1753 & Amoen. Acad. 3: 416,1756 & Sp. Pl. ed. 2: 969. 1763; Mast., in Hook. *f.*, Fl. Brit. India l: 321. 1874; Duthie, Fl. Gangetic Plain 1: 79. 1903; Mahesh., Illus. Fl. Delhi f. 25. 1966; Bhandari, Fl. Indian Desert 62. 1978; Parmar, in Shetty & Singh Fl. Rajasthan 1: 125. 1987; Paul, in Sharma *et al.* Fl. India 3: 361. 1993.(Fig. 25 & Plate 40)

Vern.: *Golio.*

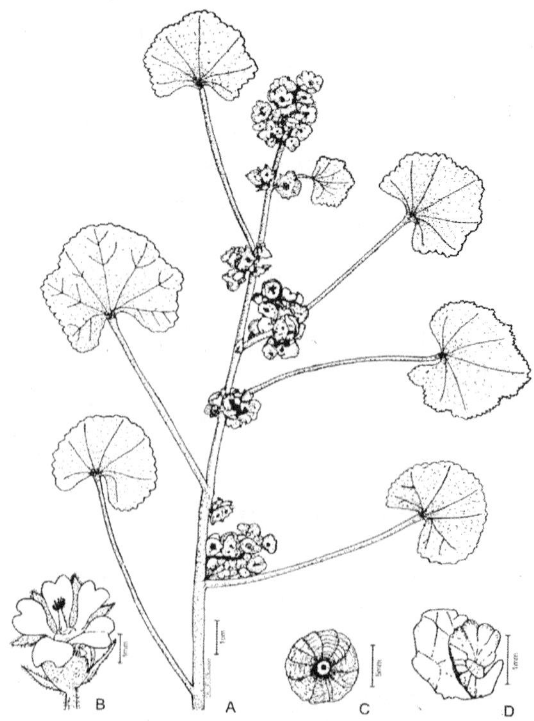

Fig. 25. *Malva parviflora* L. A. Fruiting branch, B. Flower, C. Fruit, D. Mericrip.

Annual, prostrate-procumbent herb. Stem up to 50 cm long, sparsely stellate pubescent to glabrescent. Leaves orbicular, cordate at base, 2 - 4.5 x 2 - 7 cm, upper surface with generally sparsely simple or 2-rayed hairs or glabrous, lower surface more hairy, simple, 2 rayed and stellate, often 3 - 7-lobed, crenate-serrate; stipules lanceolate to ovate, 1 - 3 x 1 - 2 mm, ciliate margined; petiole 1 - 25 cm long, with a line of simple and stellate hairs on upper side. Flowers 2 - 6 in axillary fascicled, compact, rarely solitary; pedicel 3 - 5 mm long, usually not visible, stellate pubescent, in fruit a few up to 10 mm long, glabrescent. Epicalyx segments linear, ca 6 mm long, 1 mm broad, stellate pubescent. Calyx cupular, ca 5 x 3 mm, divided to the middle, ovate, acute, densely stellate pubescent, free nearly to the middle, accrescent in fruit, c. 1 cm long. Petals bluish-white, pinkish at the tips, 3 - 7 x 2 - 3 mm, usually shorter than or equal to sepals, occasionally longer, slightly notched above, glabrous, oblong, slightly narrow at base. Staminal column glabrous, ca 3 mm long, antheriferous towards apex. Fruit discoid, 5 - 8 mm across; mericarps 10, dorsally with raised reticulate venation, radially transversely ribbed, 1.5 - 2 mm across in all direction margin winged. Seed 1.8 - 2 mm in diam., brown, black, glabrous.

Field note: A common weed of cultivated fields, gardens as well as waste places.

Fl. & Fr.: December - Feberuary

Specimen examined: Karoi 58 (MLVGCB Herb).

Distribution: Europe, N. Africa and S. & W-Asia,.

Notes: In the Linnaean Herbarium (L.) there are 2 sheets of this species viz. 870.16 and 810.17 bearing specific epithet *parviflora* in the handwriting of Linnaeus. The specimen on the former sheet does not match with this description. While the specimen on the latter sheet bears fasciculate, subsessile flowers and fruits and is in agreement with his description. This is considered as holotype.

Malva sylvestris L., Sp. Pl. 689. 1753; Mast., in Hook. *f.*, Fl. Brit. India 1: 321. 1874; Duthie, Fl. Gangetic Plain 1: 79. 1903; Parmar, in Shetty & Singh Fl. Rajasthan 1: 126. 1987; Paul, in Sharma *et al*. Fl. India 3: 363. 1993.

Usually biennial, up to 1.25 m tall, pubescent to glabrescent herb. Leaves 1.5 - 8 x 1 - 11 cm, 3-7-lobed, 3 - 7- nerved at base, reniform to suborbicular, cordate, lobes semicircular to oblong, obtuse at apex, crenate, glabrous or sparsely simple hairs beneath; stipules lanceolate, scarious, ca 5 x 1.5 mm; petiole 1.5 - 7 cm long, pilose. Flowers axillary 1 - 4 in fascicles; pedicel 0.5 - 2.5 cm long, unequal in a fascicle, pubescent with simple and stellate fascicles; pedicels 0.5 - 2.5 cm, unequal in a fascicle, pubescent with simple and stellate hairs, ultimately glabrous. Epicalyx segments 2 - 5 x 1.5 - 2 mm, ovate, ciliate by simple hairs, glabrous. Calyx free to the middle, 3-7 x 2-4 mm, glabrescent; lobes broadly triangular, acute, densely hirsute outside with stellate and a few simple hairs, persistent. Corolla dark purple or violet; petals 1.5-2.5 x 0.5-1 cm, obovate, emarginate, glabrous, claw ciliate. Staminal column ca 4 mm long, stellate pubescent; filaments ca 2 mm long. Fruit glabrous, 4 - 10 mm across, discoid, glabrous; mericarps 10 - 12, reticulate. Seeds 1.5 - 2 mm long, reniform, glabrous, brownish black.

Field note: Common weed of cultivated fields, gardens as well as waste places.

Fl. & Fr.: February - April

Specimen examined: Karoi 59 (MLVGCB Herb).

Distribution: W. Europe, N. Africa and Asia.

Malva verticillata L., Sp. Pl. 689. 1753; Mast., in Hook. *f.*, Fl. Brit. India 1: 320. 1874; Duthie, Fl. Gangetic Plain 1: 78. 1903; Parmar, in Shetty & Singh Fl. Rajasthan 1: 126. 1987; Paul, in Sharma *et al*. Fl. India 3: 363 - 364. 1993. *M. neigherrensis* Wight, Icon. Pl. Ind. Orient. t. 950. 1845.

Annual or perennial herb, 0.5 - 2.5 m tall. Stem green to purplish stellate hairy, glabrescent. Leaves usually orbicular, 3 - 12 x 2 - 10 cm, suborbicular, cordate or subtruncate at base, upper surface with sparse simple or 2-rayed hairs, glabrescent, lower with stellate simple or 2-rayed hairs, more hairy than above coarsely, crenate-dentate, usually 5-6-lobed; stipule lanceolate to deltoid, 3 - 5 x 2 - 4 mm, acuminate, glabrous except long ciliate margin; petiole 2 - 15 cm long, densely tomentose above, elsewhere glabrescent except the groove. Flowers axillary, 2-many, loose or compact; pedicel visible or not, sparsely stellate hairy, 2 - 15 mm long. Epicalyx segments linear or linear-lanceolate 3 - 4 x 1 - 1.5 mm, glabrescent, margin ciliate. Calyx free to halfway down; sepals 5 - 6 x 3 - 4 mm, very sparsely stellate hairy without finely reticulately veined accrescent in fruit, up to 12 x 8 mm; lobes deltoid with long ciliate margins, acute at apex. Corolla purplish; petals 7 - 8 mm long, purplish glabrous clawed, slightly notched. Staminal column 3 - 4.5 mm long glabrous or simple hairy above. Ovary discoid, carpels 10 - 12; styles free; stigmas decumbent on adaxial side of styles. Fruit enclosed in accrescent calyx, 5 - 7 mm across; mericarps 10 - 12, glabrous, rugose on the margins, centrally grooved or smooth on the dorsal side, radially rugose, ca 2 mm across in all directions. Seeds dark brown-black, ca 1 - 1.5 mm across, glabrous.

Field note: Rare, found as a spring season weed.

Fl. & Fr.: December - April.

Specimen examined: Karoi 1784 (MLVGCB Herb).

Distribution: Asia, Europe, Egypt and S. Africa

Tribe 4. ***URENEAE*** Benth. & Hook.f.

KEY TO THE GENERA

1a. Leaves with nectaries on midrib beneath...*Urena*

1b. Leaves without nectaries........... ....*Pavonia*

PAVONIA Cav. nom. cons.

KEY TO THE SPECIES

1a. Mericarps smooth ....*P. odorata*

1b. Mericarps winged ....*P. zeylanica*

Pavonia odorata Willd., Sp. Pl. 3: 837. 1800. Mast., in Hook. *f.,* Fl. Brit. India l: 331. 1874; Duthie, Fl. Gangetic Plain 1: 78. 1903; Bhandari, Fl. Indian Desert 62. 1978; Parmar, in Shetty & Singh Fl. Rajasthan 1: 128 - 129. 1987; Paul, in Sharma *et al.* Fl. India 3: 373 - 374. 1993. *Hibiscus odoratus* Roxb. [Hort. Ben. 50. 1814. nom. nud.] ex. Wight & Arn., Prodr. 47. 1934.

Erect odorous herb, upto 40 cm high, Stem, petioles and pedicels covered with simple glandular hairs. Leaves 2 - 6 x 1.5 - 7 cm, cordate-ovate, 3 - 5 angled, 4 - 6 cm long, 5 - 7 cm broad. Flowers axillary, solitary or fascicled in the terminal branches, pedicels 1.5 - 4 cm, accrescent up to 6 cm long, jointed above the middle, yellow with purple branches. Epicalyx segments 10 - 12, free, 0.5 - 1.5 cm long, linear, ciliate, persistent. Calyx 5-lobed, ca 3 mm across, lobes connate at base, ca 4 x 1.5 mm, ovate-lanceolate, scattered simple hairy on both surfaces. Corolla pink, 1 - 2 times longer than calyx, glabrous. Staminal column shorter than petals, glabrous. Ovary globose; stigma capitate. Fruit globular; mericarps 5, ca 4 x 2 mm, more or less reniform, unarmed, smooth, wingless. Seeds 2 mm long, reniform, minutely papillose, brownish-black.

Field note: Occasionally occurring in gardens and waste lands.

Fl. & Fr.: October - November

Specimen examined: Gurla 60 (MLVGCB Herb).

Distribution: India, Pakistan, Burma, Sri Lanka and East Tropical Africa

Uses: Leaves are edible. Roots are reported to be used for making a perfume called '*Hina*'.(Paul 1993).

Pavonia zeylanica (L.) Cav, Diss. 3: 34. 48. f. 2: 1787; Mast. Hook. *f.,* Fl. Brit. India 1: 331. 1874; Mahesh. Illus. Fl. Delhi f.28. 1966; Raizada, Suppl. Duthie, Fl. Gangetic Plain 32. 1976; Bhandari, Fl. Indian Desert 65. 1978; Parmar, in Shetty & Singh Fl. Rajasthan 1: 129. 1987; Paul, in Sharma *et al.* Fl. India 3: 377- 378. 1993. *Hibiscus zeylanicus* L. Sp. Pl. 697. 1753.

Perennial profusely branching towards tip, hispid, undershrub, upto 1m high. Stem, petiole and pedicel with short, simple and stellate hairs mixed with long, spreading, simple, sometimes tubercled hairs. Leaves 1 - 4 x 0.8 - 3.5 cm, upper leaves lanceolate to ovate, entire, lower ones usually 3-fid to partite, shallowly cordate or truncate at base, upper surface usually simple hairy and lower stellate; lobes oblong or obovate, entire or irregularly toothed, middle one longest; stipules 2 mm long, linear, hairy, deciduous; petiole 0.5 - 8 cm long. Flowers axillary, solitary; pedicel 1 - 4 cm long, articulate near the apex; epicalyx segments 8 - 10, 6 - 10 mm long, free, filiform, ciliate. Calyx 2 - 4 x 1.5 mm, fused at the base, stellate pubescent outside, glabrous inside; lobes lanceolate, acute. Corolla ca 15 mm long, pale-pink; petals 10 x 5 mm glabrous. Staminal column 8 mm long, antheriferous throughout. Ovary 1.5 mm across, more or less ovoid, glabrous; style-branches 10; stigmas capitate. Fruit 5 mm across, pubescent, globular; mericarps 5, winged, 4 x 3 mm, radial glabrous, smooth or somewhat wrinkled. Seeds 1.5 x 1 mm, reniform, dark brown, minutely pubescent.

Field note: Common in open fields and among hedges of gardens and fields.

Fl. : September - November; *Fr.*: October - December

Specimen examined: Atun 304, Ajmer 1127 (MLVGCB Herb).

Distribution: Sri Lanka, India, Pakistan, S.E. Arabia, Tropical Africa and Mauritius.

URENA L.

Urena lobata L. Sp. Pl. 692. 1753; Mast. Hook. *f.*, Fl. Brit. India 1: 329. 1874; Duthie, Fl. Gangetic Plain 1: 85. 1903; Mahesh., Illus. Fl. Delhi f. 27. 1966; Parmar, in Shetty & Singh Fl. Rajasthan 1: 134 - 135. 1987; Paul, in Sharma *et al.* Fl. India 3: 380- 381. 1993. (Fig. 26 & Plate 41)

subsp. ***lobata*** var. ***lobata***

Erect, annual to perennial herb or under-shrub or shrub, 0.5 to 2 m tall, stellate pubescent. Leaves extremely variable in size and shape, 1 - 12 x 0.5 - 12.5 cm, ovate to orbicular, truncate or shallowly cordate at base, acute at apex, serrate to crenate, rarely subentire, near the base with a nectary on the midrib beneath, sometimes also present on 2 adjacent nerves, stellate pubescent on both sides more, densely beneath, scabrulous above, velutinous below, unlobed, angular or shallowly lobed; lobes, rounded or acute; stipules 2 - 4 mm long, linear to lanceolate, acute; petiole 0.5 - 12 cm long. Flowers axillary, usually solitary or in fascicles of 2 - 3; pedicel 2 - 3 mm long, in fruit up to 5 mm, stellate tomentose; epicalyx segments closely enveloping calyx and shortly adnate to it, 3 - 10 x 1 - 3 mm, linear-lanceolate, minute stellate-hairy outside, often fimbriate sericeous by appressed, simple and stellate hairs towards apex inside. Calyx equalling the epicalyx, 5-parted, tubular to campanulate; lobes 4 - 6 x 1.5 - 2 mm, ovate to deltoid, acute to acuminate, nerves somewhate prominent with nectaries at base, hairs similar to epicalyx. Corolla 2 - 3 cm across, pink with purple centre; petals 1 - 1.5 x 0.5 cm, obovate, rounded at apex. Fruit 5 - 8 mm across, stellate pubescent, globular, glochidate spiny; mericarps 4 - 5 mm long, dorsally 3 mm broad, radially 4 mm broad and reticulate, shortly awned. Seeds 2 - 3 mm across, reniform or so, brown, pubescent, glabrescent.

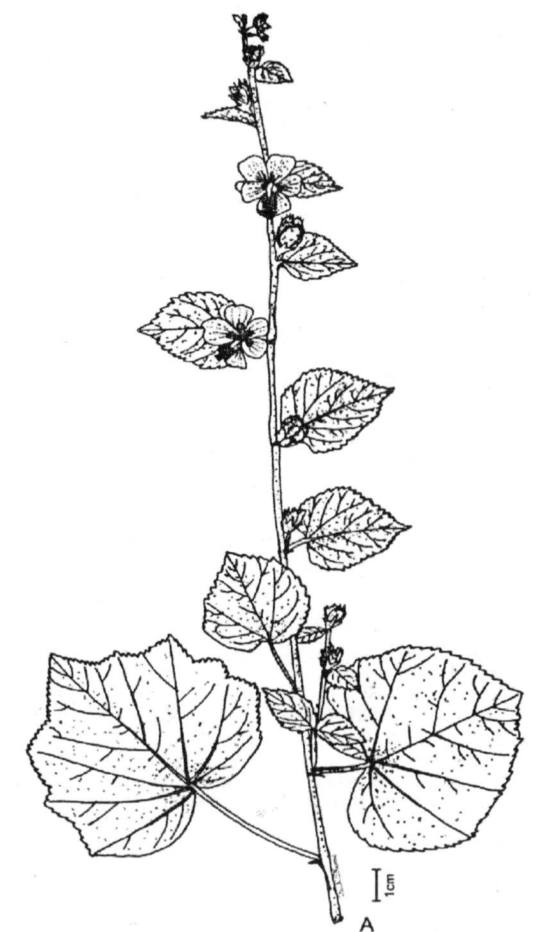

Fig. 26. *Urena lobata* L. ssp. *lobata*:
A. Flowering and fruiting twig

Field note:Common in waste places and forest edges.

Fl.: September - November, *Fr.* : November - January

Specimen examined: Triveni 96 (MLVGCB Herb).

Distribution: Tropical regions of both the hemispheres.

Uses: It is an important fibre plant. Fibres are said to be long lasting than jute. In some countries like Cuba, Madagascar, Nigeria and Brazil etc. it is cultivated for making coffee sacks.

BOMBACACEAE

BOMBAX L. nom. cons.

KEY TO THE SPECIES

1a. Trees covered with stout hard conical prickles. Flowers 10 - 12 cm long. Capsule 10 - 13 cm long. Seeds dark brown................... ......*B. ceiba*

1b. Trees without prickles. Flowers 12 - 17 cm long. Capsule 18 - 22 cm long. Seeds black....................................... ....*B. insigne*

Bombax ceiba L., Sp. Pl. 511. 1753 pro parte.; Robyns, in Taxon 10: 160. 1961 & in Bull. Jard. Bot. Brux. 83: 88. t. 3. 1963; Parmar, in Shetty & Singh Fl. Rajasthan 1: 137. 1987; Nayar & Biswas, in Sharma *et al.* Fl. India 3: 398. 1993. *B. malabaricum* DC., Prodr. 1: 479. 1824; Mast., in Hook. Fl. Brit. India 1: 349. 1874; Duthie, Fl. Gangetic Plain 1: 98. 1903; *Salmalia malabarica* (DC.) Schott in Schott & Endl., Melet. Bot. 35. 1832. *Gossampinus malabarica* (DC.) Merr. in Lingn. sci. J. 5: 126. 1927. (Plate 42)

Vern.: *Simbal, Simal, Hemro.*

Tall tree, trunk usually unbranched up to considerable height. Bark grey, covered with hard small conical prickles, usually disappearing with age. Petiole 12 - 30 cm long, pulvinate at the base; stipules triangular, 5 - 10 x 4 mm, margin hairy, caducous. Leaves digitate, 5 - 7-foliolate; leaflets glabrous, entire, elliptic-lanceolate, acuminate, attenuate at base, more or less leathery, unequal, 5 - 24 x 7 - 10 cm; petiolule 1 - 3 cm long. Inflorescence many fascicles of 1 - 4, flowers borne at or near the end of branches. Flowers large, showy, red; pedicel thick, glabrous or tufted puberulous, 1 - 2.5 cm long. Calyx 3-lobed, cup-shaped, 3.5 - 4 cm long, smooth outside, densely silky within, falling with corolla and stamens. Petals 5, bright red or white, twisted in bud, stellate tomentose outside, sparsely pubescent inside, elliptic-oblong, fleshy, usually recurved, 8 - 18 x 3.5 - 5 cm. Stamens 65 - 80, 3 - 7.5 cm long in 6 bundles in 2 series, the central bundle with 15 stamens, of which 5 longer (4 - 5 cm) and 10 shorter (3 - 3.5 cm) and 5 bundle in the outer series, each with 10 stamens (4.5 - 5 cm long); staminal tube short; filaments 3.5 - 6 cm long, pink, somewhat tomentose, flattened at base; anthers long, afterward twisted, violet. Ovary conical, green, covered with silky hairs, 0.5-1.2 cm long; style simple, 5.9 - 6.5 cm long; stigmas 5, filiform, 5 - 6 mm long. Capsule 10 - 18 cm long; oblong, woody, 5 valved, valves sikly inside. Seeds brown, smooth, obovid, 6 mm long, embeded in silky white wool.

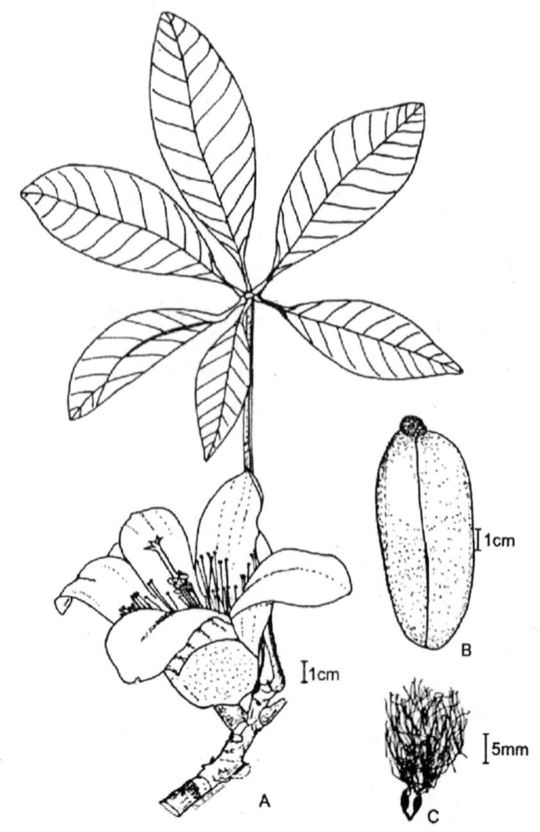

Fig. 27. *Bombax ceiba* L.: A. Flowering and fruiting twig, B. Fruit, C. Seed

Field note: Commonly found wild as well as cultivated along roadsides and gardens.

Fl. & Fr.: December-March.

Specimen examined: Bhilwara 1195, Ajmer 1578 (MLVGCB Herb).

Distribution: India, Sri Lanka, S,E.Asia, China, Australia and China

Uses: The young roots are dried in shade and cooked as vegetable. The vegetable is used increase the number of sperms in semen. The silk cotton produced from the fruits, is used for stuffing beds, pillows and quilts.

Bombax insigne Wallich, Pl. Asiat. Rar. 1: 71, t. 79, 80. 1830; Mast., in Hook. Fl. Brit. India 1: 349. 1874; Nayar & Biswas, in Sharma *et al.* Fl. India 3: 398. 1993. *Salmalia insignis* (Wall.) Schott & Engl. Melet. Bot. 35. 1832. *B. insigne* Wall. vars. *andamanica, polystemon & wightii* Prain in J. Asiat. Soc. Beng. 65: 1893. (Plate 43)

Vern.: *Simbal, Semal, Hemro.*

Tree upto 20 m high; trunk marked with large leafscars, not spiny; branchlets spiny or not. Petiole longer than leaves, pubescent. Leaves digitate, glabrous, deciduous; leaflets 7-9, petiolules 1.2 - 1.6 cm; blades obovate, cuspidate-acuminate at apex, 10-12 × 4-5 cm, nearly leathery, abaxially villous only on midrib and lateral veins, base tapering, apex shortly acuminate. Flowers nearly terminal, 1.2 - 1.7 cm long, upto 22 cm in diam.; pedicel rod-shaped, ca 1.9 cm, robust. Calyx urceolate or globose, 3 - 5 cm, abaxially nearly glabrous, adaxially very densely woolly, apex obscurely lobed. Petals orange or yellow, oblong or filiform, boat-shaped, incurved, 1.0 - 1.5 × 0.3 cm, fleshy, abaxially sericeous, adaxially glabrous, obtuse. Stamens numerous in 5 bundles of 50 or more each opposite petals; filaments 0.8 - 1.0 cm long, unequal. Ovary ovoid, tomentose; stigmas 5-fid. Capsule cylindric, 5-valved, 18 - 30 × 3.5 - 6 cm, curved at apex, glabrous. Seeds obovate, smooth, black, embedded in dense creamy silk fibers.

Field note: Commonly occurring wild as well as cultivated along roadsides and in gardens.

Fl.: January- March; *Fr.* April - May

Specimen examined: Bhiwara 1207, Bigod 1224 (MLVGCB Herb).

Distribution: India (Maharashtra, Rajasthan, Karnataka, Tamilnadu and Andaman & Nicobar Islands), Maynmar.

CULTIVATED SPECIES

ADANSONIA L.

Adansonia digitata L., Sp. Pl. 1190. 1753; Mast. Hook. *f.,* Fl. Brit. India 1: 348. 1874; Cook, Fl. Bombay 1: 119. 1901; Shah, Fl. Gujarat 1: 118. 1978; Parmar, in Shetty & Singh Fl. Rajasthan 1: 137. 1987; Nayar & Biswas in Sharma *et al.*Fl. India 3: 404. 1993. (Plate 44)

Vern.: *Kalp virksha.*

Tall tree, upto 20 - 25 m high, deciduous; trunk short, thick up to 10 m in diam., abruptly tapering into thick branches, branches spreading, pubescent when young; bark usually smooth, grey. Leaves usually gathered near tips of branchlets, digitate, 3-9-foliolate; petiole 10-30 cm long, pulvinate at the base, pubescent when young, some-times glabrous; leaflets 3 -9, 2.5 - 10 x 5.5 cm, tufted pubescent when young, entire, oblong to obovate, cuneate and decurrent at base, gradually caudate-acuminate at apex. Flowers axillary, pendulous, showy, large, white; peduncles ca 90 cm long, tomentose; bracteoles 2; flower buds globose to ovoid-elliptical, acute. Calyx triangular, acute at apex, sericeous inside, tomentose outside. Petals 5, 7 - 9 x 6 - 8 cm, obovate to flabelliform, rounded at apex, unguiculate, white becoming brown when dry, glabrous or sparsely hairy outside, adnate to the base of staminal tube. Staminal tube 5 - 7 cm long, cylindrical to conical, glabrous, divided above into numerous, slender filament as long as tube; anthers 2 mm long, reniform. Ovary 5 - 10-loculed, hirsute, ovules many in each locule; styles 1 - 1.5 cm long exserted, hirsute at base; stigmas 5 - 10- lobed, lobes oblong, radiating. Capsule 20 - 40 cm long; oblong to oblong-ovoid, woody, longitudinally sulcate, acute at base, rounded at apex, indehiscent, 6 - 12 cm in diam., pale brown. Seeds many, dark brown, embedded in farinaceous pulp.

Field note: Rare, often planted near temples due to its religious importance.

Fl.: April - May; *Fr.*: June - December.

Specimen examined: Mangliyawas 3201 (MLVGCB Herb).

Distribution: Native of tropical Africa, Cultivated in various parts of inida.

Uses: The fruit pulp is used as a remedy in dysentery.

STERCULIACEAE

KEY TO THE GENERA

1a. Trees or shrubs 2

1b. Herbs or undershurbs.................................. 5

2a. Flowers unisexual or uni and bisexual on the same individual (plants polygamous). Petals absent. Carpels free or becoming free in fruit.................................. ......*Sterculia*

2b. Flowers bisexual. Petals present. Carpels mostly united 3

3a. Staminal column free from carpels, anthers alternating with staminodes along the margin of the copular column at its apex. ....*Guazuma*

3b. Staminal column adnate to gynophore forming stalk, dialated above into a cup; fertile anthers alternating with staminodes towards margin. 4

4a. Anther locules parallel with apiculate connective. Style simple, clavate; stigma 5-grooved. Seeds winged in the upper part ....*Pterospermum*

4b. Anther locules divaricate or confluent into one. Styles 5, subulate, free or more or less united, thickened and stigmatic at apex. Seeds without wings.........................*Helicteres*

5a. Petals clawed, style solitaty. Capsules 2-valved ....*Waltheria*

5b. Petals not clawed, style 5 or 5-fid. Capsules 5-valved 6

6a. Tomentose undershurbs, flowers orange-yellow, staminodes present...............*Melhania*

6b. Glabrous herbs, flowers pink or white, staminodes absent...................... ......*Melochia*

GUAZUMA Mill.

Guazuma ulmifolia Lam., Encycl. 3: 52. 1789. Bailey, Stand Cycl. Hort. 1: 1415. 1947; Robyns, in Ann. Miss. Bot. Gard. 51: 103.1964; Pandey, in Shetty & Singh, Fl. Rajasthan 1: 139 - 140. 1987; Malick in Sharma *et al.*Fl. India 3: 424. 1993. *Theobroma guazuma* L. Sp. Pl. 782. 1753. *Guazuma tomentosa* Kunth, in H.B. & K., November Gen. Sp. 5: 320. 1823; Mast. Hook. *f.*, Fl. Brit. India 1: 375. 1874. (Fig. 28)

Vern.: *Rudraksh*.

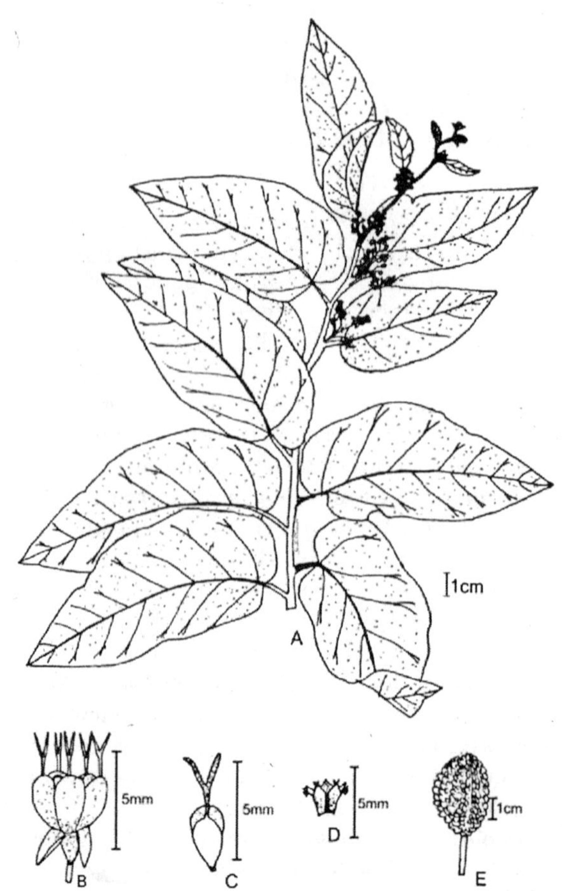

Fig. 28. *Guazuma ulmifolia* Lam. A. Flowering twig, B. Flower, C. Patal, D. Stamen, E Fruit

A medium sized tree, up to 8 - 15 m high. Leaves ovate, or ovate-oblong or lanceolate, 5 - 17.5 x 2 - 9 cm, oblique, acuminate, somewhat scabrous above and velvety below; petiole 0.5 - 2 cm long. Inflorescence axillary or terminal, many-flowered, thyrsiform cyme. Flowers small, ca 8 mm across, yellow; pedicel 3 - 6 mm long. Sepals 3 - 5, elliptic-lanceolate or ovate, basally connate, 2 - 3 x. 1.5 mm, tomentose outside, nearly glabrous within, reflexed. Petals obovate, 3 x 2 mm, appendage 3 - 4 mm long, puberulous outside. Staminal cup ca 3 x 2 mm, bearing the stamens and staminodes, fimbriate, acute. Ovary, sessile, ca 1 mm long, globose with fine tubercules; style 5, ca 2 mm long, connate at base. Capsule globose or oblong-ellipsoid, up to 3 cm in diam., black, tuberculate. Seeds many 1 mm in diam., globose.

Field note: Commonly planted as Rudrakhsa tree in gardens, near temples etc. Occasionally seen in the forests, particularly in scrub vegetation.

Fl.: March-May; *Fr.*: September-December.

Specimen examined: Harni Mahadev 1583, Bhilwara 1795 (MLVGCB Herb).

Distribution: A native of tropical America and Java, cultivated in India, Burma and Pakistan.

Uses: The fruit resembling black mulberry contains a small quantity of sweet edible pulp and is said to be valuable for fattening cattles. Mucilaginous sap from the inner bark of the tree is used for clearing sugar in W. Indies. Bark is reported to be used in treatment of elephantiasis, skin and chest diseases in Mauritius and W. Indies.

HELICTERES L.

Helicteres isora L., Sp. Pl. 963. 1753; Dalz. & Gibs.,Bomb. Fl. 22.1861; Roxb., Fl. Ind.ed. Carey 3: 143. 1832; Mast., in Hook. *f.*, Fl. Brit. India 1: 365. 1874; Duthie, Fl. Gangetic Plain 1: 102. 1903; Cooke,Fl. Bomb. Pres. 1: 136. 1958; Bhandari, Fl. Indian Desert. 70. 1978; Pandey, in Shetty & Singh, Fl. Rajasthan 1: 140. 1987; Malick, in Sharma *et al.* Fl. India 3: 426. 1993. (Fig. 29 & Plate 45)

Vern.: *Maror phali.*

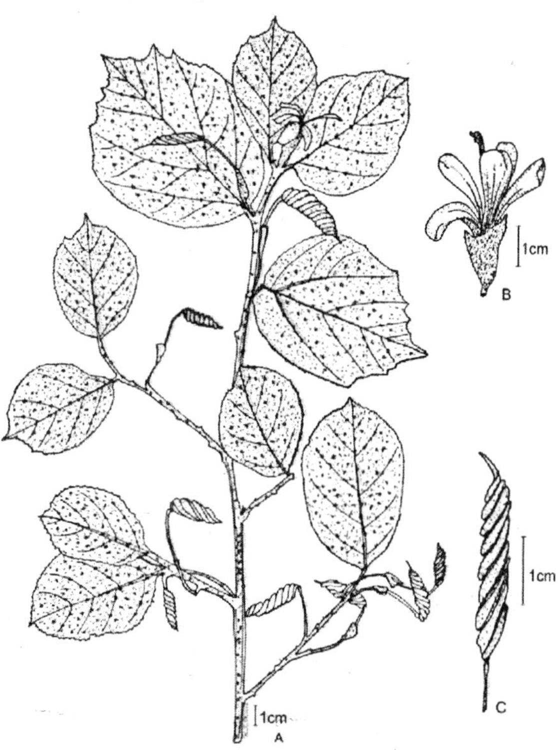

Fig. 29. *Helicters isora* L. :
A. Fruiting branch, B. Flower, C. Fruit

A large shrub. Branches rough with scattered stellate hairs. Leaves oblong, obovate or orbicular, 5 - 23 x 4 - 17 cm, irregularly toothed, often triangular towards apex, oblique and cordate or rounded at the base, shortly acuminate, scabrous above; petiole 1 - 8 mm long, tomentose; stipules linear, 3 - 24 mm long, tomentose. Inflorescence 2 - 6-flowered axillary cyme. Flowers 3 - 3.5 cm long, red, fading to lead-grey; pedicel 3 - 6 mm long, tomentose; bracts linear, 3-5 mm long, tomentose. Calyx bilabiate, 1 - 2 cm long, toothed, teeth triangular, gibbous, laterally pubescent. Petals unequal, 2 - 5 cm long, upper 3 longer and narrower than the lower 2 reflexed, claw auriculate. Stamens 15, 10 fertile in 5 pairs, each pair alternating with scaly bifid 5 staminodes, exserted, 3 - 4 cm long, abruptly reflexed; anthers ovate, arranged in a ring around the ovary. Ovary ovoid or conical, 5-grooved, tomentose, raised on a 3 - 4 cm long, curved, tomentose gynophore; style curved, 5-grooved at the base. Fruit cylindric, 4 - 8 x 0.5 - 1 cm in diam., long beaked, splitting into 5 spirally twisted densely tomentose follicles. Seeds many, angular, wrinkled tomentose.

Field note: Common in dry deciduous forests.

Fl. & Fr.: July-December.

Specimen examined: Pansal 1511 (MLVGCB Herb).

Distribution: India, Sri Lanka, Pakistan, Malayan Peninsula and Australia.

Uses: It is a good source of strong fibres obtained from the bark. The fruit is said to be medicinally important in the treatment of gastric and intestinal disorders, 2 tsp powder of fruit is taken with water twice a day for 3 days.

MELHANIA Forssk.

Melhania futteyporensis Munro ex Mast., in Hook. *f.,* Fl. Brit. Ind. 1: 373. 1874; Duthie, Fl. Gangetic Plain 1: 106. 1903; Malick in Sharma *et al.* Fl. India 3: 438. 1993. *M. tomentosa* Stocks ex Masters var. *major* Blatt. & Hallb. in J. Bombay Nat. Hist. Soc. 26: 228. 1918. *M. futteyporensis* Munro ex Mast. var. *major* (Blatt. & Hallb.) Santapau in J. Bombay Nat. Hist. Soc. 56: 278. 1959; Bhandari, Fl. Indian Desert. 71. 1978; Pandey, in Shetty & Singh, Fl. Rajasthan 1: 141. 1987. (Plate 46)

A small undershrub, 40-70 cm tall, coarsely pubescent. Leaves 5 - 7-costate, lanceolate, elliptic-lanceolate, ovate or ovate-lanceolate, unlobed or occasionally obscurely 3-lobed, 2 - 15 x 1 - 8 cm, velvety on both sides, serrate-crenate; petiole 1 - 3 cm long, hairy; stipules filiform, 4-8 mm long, brown. Inflorescence in 2 - 4 axillary and terminal, peduncled cyme; peduncle 1 - 5 cm long, pubescent. Flowers yellow, 1.5 - 2 cm across; pedicel 5 - 10 mm long, pubescent, elongated in fruit; involucral bracts ovate, 0.8 - 2 x 0.5 - 1 cm, cuspidate. Sepals 5, ovate-oblong, 0.8 - 14 x 3 - 4 mm, acuminate or cuspidate, pubescent outside, persistent. Petals oblique, obovate, ca 1.5 x 1 - 1.8 cm, obovate, membranous, veined. Stamens 8 - 24 mm long, staminodes ca 1 - 1.5 cm long. Ovary subglobose, 3 mm in diam., villous. Capsule oblong-subglobose, 0.6 - 1.5 x 0.5 cm in diam., obtuse or rarely minutely beaked, villous. Seeds 4 in each locule, angular, muricate, black.

Field note: Common in gravel, sandy clay or sandy soils in plains and low hills.

Fl. & Fr.: August-December.

Specimen examined: Pansal 1623 (MLVGCB Herb).

Distribution: India, Pakistan Arabia and Tropical Africa

MELOCHIA L.

Melochia corchorifolia L., Sp. Pl. 2: 675. 1753; Mast., in Hook. *f.,* Fl. Brit. Ind. 1: 374. 1874; Duthie, Fl. Gangetic Plain 1: 107. 1903; Robyns, in Ann. Miss. Bot. Gard. 51: 83. 1964; Mahesh., Illus. Fl. Delhi f. 30. 1966; Pandey, in Shetty & Singh, Fl. Rajasthan 1: 142. 1987; Malick, in Sharma *et al.* Fl. India 3: 441. 1993. (Plate 47)

Herb or undershrub, less than 1 m tall, erect or decumbent. Branches yellow-brown, sparsely stellate puberulent. Stipules linear, 2 - 4 mm; petiole upto 3.5 cm long; leaf blade 2.5 - 10 × 1 - 7 cm, variable in shape, ovate, oblong-ovate or lanceolate, rarely obscurely 3-lobed, thinly papery, abaxially sparsely stellate puberulent, adaxially nearly glabrous, basal veins 5, base rounded or cordate, margin dentate, apex acute or obtuse. Inflorescence a dense cyme or glomerule, terminal or axillary. Flowers surrounded by 4 - 5 bracteoles. Epicalyx lobes 4, linear, hairy. Calyx campanulate, 5-lobed, 1 - 2 mm long, abaxially villous and hispid, adaxially glabrous, lobes triangular. Petals 5, white or pink, drying reddish, oblong, 3 - 4 mm long, narrowed at base. Stamens 5, connate at base, opposite petals; anther cells divergent. Ovary sessile, ca 1 mm long, 5 - loculed hairy; styles 5, filiform. Capsule depressed subglobose, 5-angular, 3 - 6 mm in diam., hispid. Seeds 1 or 2 per cell, brown-black, ovoid, slightly triangular, 2 - 3 mm long.

Field note: Common in open wet habitats and as an undergrowth in the forests.

Fl. & Fr.: August- December

Specimen examined: Mainal 1426, Mandalgarh 1085 (MLVGCB Herb).

Distribution: Pantotropical.

PTEROSPERMUM Schreb. nom. cons.

Pterospermum acerifolium (L.) Willd., Sp. Pl. 3: 729. 1801; Mast., in Hook. *f.,* Fl. Brit. Ind. 1: 368. 1874; Duthie, Fl. Gangetic Plain 1: 103. 1903; Pandey, in Shetty & Singh, Fl. Rajasthan 1: 143. 1987; Malick, in Sharma *et al.* Fl. India 3: 449. 1993. *Pentapetes acerifolia* L. Sp. Pl. 698. 1753. (Plate 48)

Vern.: *Kanak-champa.*

Large tree, up to 30 m tall, with grey bark and rusty pubescent young parts. Leaves oblong, broadly obovate to ovate, orbicular or rectangular, 20 - 40 x 8 - 35 cm, cordate, often peltate, margin wavy to distantly coarse toothed or irregularly lobed; silvery to rusty pubescent beneath, glabrescent and dark green above; petiole 5 - 15 cm long, densely tomentose at the base; stipules pinnatifid, caducous. Flowers mostly solitary, 10 - 15 cm long and across, white, fragrant; pedicel ca 2 cm long; bracts laciniate. Sepals 5, linear-lanceolate, united at the base into short tube, 8 - 12 x 0.6 - 0.9 cm, obtuse, rusty pubescent outside, thick, reflexed, deciduous. Petals 5, white, linear-oblong or obovate, 6 - 9.5 x 1 cm, reflexed, obtuse. Fertile stamens 5-9 cm long, staminodes equalling the petals; anthers 1-1.5 cm long. Ovary 1 - 3 cm long on 1-1.3 cm long gynophore; stigmas clubshaped. Carpels 5; ovary oblong, pentagonal, 5-loculed, densely rusty tomentose. Capsule 10 - 20 x 3-6 cm, woody, 5 -valved, rusty brown-glabrescent. Seeds 1-2 x 1-1.5 cm, compressed, obliquely oval, wings brown, membranous.

Field note: Planted as a roadside tree or garden ornamental.

Fl. & Fr.: March- August.

Specimen examined: Bhilwara 1555 (MLVGCB Herb).

Distribution: Often cultivated, throughout warmer regions. Probably a native of North East India; Bangla Desh (Chittagong), Burma, Malayasia, Pakistan and N. America

Uses: The indumentum from the lower side of leaf is said to be used to prevent bleeding from wounds and as tinder. A good tonic is prepared from the flowers which is also a cure for inflammation, ulcers, tumours, blood troubles and leprosy. Bark and leaves are used in small pox. Flowers kept in clothes impart a pleasant perfume and keep away insects.

STERCULIA L.

Sterculia urens Roxb., Pl. Corom. 1: 25, t. 24. 1795 & Fl. Ind. 3: 145. 1832; Mast., in Hook. *f.,* Fl. Brit. Ind. 1: 355. 1874; Duthie, Fl. Gangetic Plain 1: 100. 1903; Bhandari, Fl. Indian Desert. 72. 1978; Pandey, in Shetty & Singh, Fl. Rajasthan 1: 144. 1987; Malick in Sharma *et al.* Fl. India 3: 470 - 471. 1993. (Plate 49)

Vern.: *Kadda, Kadaya.*

Deciduous tree more or less pubescent; trunk strait; bark white, smooth, outer papery peeling off, inner fibrous. Leaves simple, alternate, entire or palmately lobed or digitately compound 3 - 5 lobed; cordate at base, lobes 5 or obscurely 2 or more, caudate, acuminate at apex, entire, 11 - 30 cm in diam., crowded at the ends of the branchlets, stipules caducous. Inflorescence an axillary or terminal panicle or raceme. Flowers, small, yellow to olive yellow, 5 - 9 mm across, numerous, unisexual and bisexual on the same plant (plants polygamous) or only unisexual (monoecious), terminal mostly pistillate and precocious, usually actinomorphic, glandular pubescent, terminal panicles, appearing before the leaves at the ends of branchelets; bracts lanceolate, deciduous. Calyx ca 5 mm long campanulate, hoary, tube about as long as the lobes, acute, glandular hairy inside at base. Male flowers: stamens 10-15, monadelphous into a short or long column. Female flowers: ovaries 2 mm in diam. on ca 3 mm long gynandrophore with sterile anthers at base; styles as long as ovary; stigmas radiating. Petals absent. Carpels (3 - 6), syncarpous, becoming free in fruit; ovary usually sessile, locules as many as carpels, 2-many ovuled; style with 4 - 5-lobed or peltate stigma. Fruit an etaerio of stellately spreading, ventrally dehiscent, woody follicles, follicles 5, usually oblong or ovoid-oblong, densely pubescent, often mixed with stinging hairs. Seeds 3 - 6 in each follicle, ca 7 x 5 mm, oblong, dark brown or black, glossy.

Field note: Common at the top of hills in mixed deciduous forests.

Fl.: December - February; *Fr* .:March - May.

Specimen examined: Gurala 1374, Bijoliya 1567 (MLVGCB Herb).

Distribution: A genus of ca 100 species, distributed in tropics and warm temperate regions.

Uses: Paste of leaves is applied on cuts & gums. Decoction prepared from fresh powder of dried tuber with jaggery is taken twice a day for 5-7 days in asthma and bronchitis. "*Ladoos*" prepared from dried tuber powder are given to women after delivery as tonic.

WALTHERIA L.

Waltheria indica L., Sp. Pl. 673. 1753. excl. syn. Hort. Cliff., pro parte, quoad descrip et spec.; Mast., in Hook. *f.*, Fl. Brit. Ind. 1: 374. 1874; Duthie, Fl. Gangetic Plain 1: 107. 1903; Robyns, in Ann. Miss. Bot. Gard. 51: 74. 1964; Bhandari, Fl. Indian Desert. 72. 1978; Pandey, in Shetty & Singh, Fl. Rajasthan 1: 145. 1987; Malick, in Sharma *et al.* Fl. India 3: 473. 1993. *W. americana* L., Sp. Pl. 673. 1753. *W. angustifolia* L. Syst. Nat. ed. 10. 1140. 1759.

Erect or ascending undershrub, up to 1 m tall, multi-branched. Branchlets densely puberulent. Petiole 0.5 - 2.5 cm, densely pubescent; leaf blade ovate or long elliptic-ovate, 2.5-6.5 × 1.5 - 4.5 cm, both surfaces densely puberulent, base rounded or shallowly cordate, margin serrulate, apex obtuse. Inflorescence cymose, capitate, axillary; peduncle nearly absent to 1.5 cm. Epicalyx lobes narrowly lanceolate, ca 4 mm. Calyx tube-shaped, 5-lobed, 3 - 4 mm, lobes triangular, much longer than tube. Petals ca 4 mm, yellowish, spatulate, apex truncate, slightly longer than sepals. Filament tube enclosing ovary. Ovary unilocular, sessile, puberulent; style inserted obliquely; stigma penicillate. Capsule obovoid, ca 3 mm, hairy, 1-seeded, enclosed by persistent calyx. Seeds 2mm long, obovate, very small, smooth.

Field note: Common in waste lands and dry thickets

Fl. & Fr.: August - November

Specimen examined: Hamirgarh 1421, Mezadam 1749 (MLVGCB Herb).

Distribution: Pantropical.

TILIACEAE

KEY TO THE GENERA

1a. Basal serrations of lamina mostly glandular. Ovary with uncinate hairs. Capsule echinate with hooked prickles .... *Triumfetta*

1b. Basal serrations of lamina eglandular. Ovary glabrous or with simple or stellate hairs. Capsule not as above 2

2a. Shrubs. Petals clawed, thick, pitted or glandular at the base. Fruit entire or lobed, 1 - 4-pyrened drupe .... *Grewia*

2b. Herbs, rarely undershrubs. Petals not clawed, neither pitted nor glandular at the base. Fruit an elongated or subglobose-globose capsule .... *Corchorus*

CORCHORUS L.
(*Antichorus* L.)

KEY TO THE SPECIES

1a. Leaf blade without setose or filiform appendages at the base 2

1b. Leaf blade with setose or filiform appendages at the base 5

2a. Perennial, prostrate. Carpels 4. Capsule 4-loculed .... *C. depressus*

2b. Annual, erect or suberect. Carpels 3. Capsule 3-loculed 3

3a. Sepals neither hooded nor awned at the apex. Capsule winged................ *C. fascicularis*

3b. Sepals hooded, awned at the apex. Capsule 3-winged 4

4a. Stamens 12 - 30, capsules 6- angled prominently winged. ....*C. aestuans.*

4b. Stamens 12 - 30, capsules 6- angled not prominently winged. ....*C. tridens*

5a. Capsule subglobose-globose, beakless and depressed at the apex. ....*C. capsularis*

5b. Capsule elongated, beaked at the apex...*C. olitorius*

Corchorus aestuans L., Syst. Nat. ed. 10. 1079. 1759; Mahesh., Illus. Fl. Delhi f. 35. 1966; Bhandari, Fl. Indian Desert. 73. 1978; Singh, in Shetty & Singh Fl. Rajasthan 1: 146. 1987; Daniel & Chandrabose, in Sharma *et al.* Fl. India 3: 485. 1993. *C. acutangulus* auct. non Forssk. 1755; Lam. Encycl. 2: 104. 1786; Wight, Ic. 3 (1) : 1. t. 739. 1844; Mast., in Hook. *f.*, Fl. Brit. Ind. 1: 398. 1874; Duthie, Fl. Gangetic Plain 1: 121. 1903. *C. fuscus* Roxb., Fl. Ind. 2: 582. 1832. (Plate 50)

Vern.: *Chonch.*

Erect or spreading, 10 - 60 cm high, much branched, annual herb. Stem and branches purple, pilose. Leaves 3 - 5-costate, costae hairy to almost glabrous, lanceolate to ovate, 2 - 10 x, 1 - 5 cm, oblique or obtuse at the base, serrate, basal serratures prolonged into setaceous appendages or not, glabrous except the hairy to glabrescent costae, acute; petiole 0.5 - 4 cm long, purple, pilose; stipules setaceous, 5 - 10 mm long, pilose, purplish green. Cyme antiphyllous, very shortly pedunculate, mostly 2, rarely 3-flowered. Flowers golden yellow, ca 1 cm across, pedicel ca 2 mm long, jointed; bracts akin to stipules, 4 - 6 mm long, purple. Sepals linear-oblong, 3 - 4 mm long, hooded and awned at the apex, purple dotted outside, green within, glabrous. Petals obovate, 3 - 5 x 3 mm, obtuse, hairy at the base. Stamens 12 - 30, filaments ca 3 mm long. Carpels 3; ovary oblong-cylindric, ca 1.5 - 2 mm long, minutely hairy, 3-loculed, truncate above; style 3-fid, ca 1.5 mm long, stigmas bilobed. Capsules solitary or paired, with 3 fairly stout wings, 10 - 30 x 4 - 6 mm, 6-angled, truncate and terminating in 3, bifid, radiating, 3 - 7 mm long beaks at apex, 3-loculed, locules transversely septate or aseptate. Seeds numerous, blackish-brown, angular, rough, obliquely truncate at both ends.

Field note: Common annual in waste lands, plains and hills.

Fl. & Fr.: August-October.

Specimen examined: Bhilwara 1309, Jahajpur 1746 (MLVGCB Herb).

Distribution: India, Sri Lanka, Pakistan, Bangla Desh, Burma, Malaya, Indo-China, Australia, tropical Africa, West Indies and Central America

Uses: A coarse fibre is occasionally extracted from this plant in parts of India which is of no commercial value. The roots and leaves are said to cure gonorrhoea and used in making an injection for urethral discharge. The seeds are stomachic and used in pneumonia.

Corchorus capsularis L. Sp. Pl. 529. 1753; Mast., in Hook. *f.*, Fl. Brit. Ind. 1: 397. 1874; Duthie, Fl. Gangetic Plain 1: 119. 1903. Maheshwari, Fl. Delhi 89. 1963; Singh, in Shetty & Singh Fl. Rajasthan 1: 146 - 147. 1987; Daniel & Chandrabose, in Sharma *et al.* Fl. India 3: 485- 486. 1993.

Vern.: *Ambari, Jute.*

Large glabrous, annual, herb up to 2.5 m tall. Leaves 3-5-costate, ovate-lanceolate or linear-lanceolate, 5 - 15 x 1.5 - 8 cm, acute or acuminate, coarsely serrate, basal serratures backwardly prolonged into setaceous appendages; petiole up to 5 cm long; stipules linear, 6 - 10 mm long. Cyme 1 or 2-flowered, axillary or antiphyllous. Flowers yellow, 8 - 10 mm across, pedicellate; bracts linear-ovate, 2 - 3 x 1.5 - 2 mm. Sepals linear oblong, 4-5 mm long. Petals obovate, 3 - 5 x 2.5 - 3 mm, notched at the apex. Stamens 20 - 30, filaments ca 3 mm long. Carpels 5; ovary subglobose, 5-loculed, glabrous, truncate; style minute. Capsule globose, 10-12 mm in diam., beakless and depressed at apex, scabrous, ridged, tuberculate or muricate, 5-loculed, locules aseptate. Seeds cuneiform, ca 2 mm long, brown, glabrous.

Field note: Cultivated as a fibre crop.

Fl. & Fr.: September-October.

Specimen examined: Mainal 1359 (MLVGCB Herb).

Distribution: India, Srilanka, Pakistan, Bangla Desh, Burma and Malayan Peninsula.

Uses: The plant is extensively cultivated in India and Bangla Desh as well as in other tropical countries including Pakistan for the most valuable fibres of remarkable strength, extracted from the bast by retting and called jute or golden fibre. The fibre is used for making gunny bags, ropes, carpets, rugs, rough clothes and many other similar articles of daily use. The part, left after the fibre has been extracted, is used in the paper industry and in preparation of alcohol. An infusion of leaves is a demulcent, stomachic, carminative, laxative, stimulant and used to increase appetite. It is also given in dysentry, fever, dyspepsia and disorders of the liver. Decoction of roots and unripe fruits is used in diarrhoea. Leaves contain glucoside capsularin which is related to corchorin and corchoritin, extracted from seeds having action similar to digitalis group of genins and used in cardiac diseases.

Corchorus depressus (L.) Vicary, in J. Asiat. Soc. Bengal 16: 1160. 1847; Stocks, in Proc. Linn. Soc. 1: 367. 1848; Maheshwari, Fl. Delhi 88. 1963; Mahesh., Illus. Fl. Delhi f. 32. 1966; Bhandari, Fl. Indian Desert. 73. 1978; Singh, in Shetty & Singh Fl. Rajasthan 1: 147. 1987; Daniel & Chandrabose, in Sharma *et al.* Fl. India 3: 486. 1993. *Antichorus depressus* L. Mant. Pl. 1 : 64. 1767. *Corchorus antichorus* Raeusch. Nom. Bot. ed. 3. 158. 1797; Mast., in Hook. *f.*, Fl. Brit. Ind. 1: 398. 1874; Duthie, Fl. Gangetic Plain 1: 121. 1903.

Vern.: *Munderi, Boh phali, Bahu phali, Cham ghas.*

A mat-forming, prostrate, much branched woody perennial. Stem diffusely branched from woody rootstock, young branches sparsely hairy, older ones glabrous. Leaves 3-costate, elliptic to broadly elliptic, 1 - 4 x 0.7 - 1.5 cm, glabrous except the scattered hairy costae, plicate in bud, crenate-serrate, without basal setose appendages, obtuse; petiole 1.5 - 2.5 cm long, filiform, minutely hairy; stipules linear-falcate, ca 3 mm long. Cyme mostly 2 - 4-flowered, antiphyllous, peduncle minute. Flowers yellow, 6 - 8 mm across, pedicel ca 1 mm long; bracts linear-lanceolate, ca 1.5 mm long. Sepals yellowish-green, linear-oblong, 3 - 4 mm long, spreading, acute. Petals obovate-spathulate, as long as sepals, obtuse. Stamens 8 - 10, filaments c. 4 mm long, erect. Carpels 4; ovary ca 1 mm long, oblong-cylindric, 4-loculed; style equalling the ovary, stigma 4-lobed. Capsule oblong-cylindric, straight or curved, glabrescent, 7 - 22 mm long, with ca 1.5 - 2 mm long beak, 4-loculed, locules transversely septate. Seeds blackish-grey, ca 1.5 mm long, roughly triangular, obliquely truncate, black.

Field note: Common in sandy clay and saline or gravelly areas in arid and semi-arid regions throughout the area. Size of leaves and fruits is very much reduced in saline and rocky soils.

Fl. & Fr.: February-November.

Specimen examined: Mezadam 97 (MLVGCB Herb).

Distribution: Central and North West India and Pakistan to north and tropical Africa and Cape Verde Islands.

Uses: Plant sold by herbalists and Ayurvedic doctors under the name *Munderi.* Leaves are used as an emollient and cooling agent. Mucilage is used for the treatment of gonorrhoea and applied as a poultice for healing wounds. Decoction of seeds and leaves with milk and sugar is a good tonic.

Corchorus fascicularis Lam., Encyc. 2: 104. 1786; Mast., in Hook.f., Fl. Brit. Ind. 1: 398. 1874; Duthie, Fl. Gangetic Plain 1: 121. 1903; Bhandari, Fl. Indian Desert. 74. 1978; Singh, in Shetty & Singh, Fl. Rajasthan 1: 147. 1987; Daniel & Chandrabose, in Sharma *et al.* Fl. India 3: 486 - 487. 1993. *C. brachycarpus* Guill. *et al.*, Fl. Seneg. Tent. 89. 1831.

Vern.: *Chonchi.*

Annual, suberect, 50 - 60 cm tall herb. Stem woody, with scaly bark, glabrous. Leaves 3-costate, elliptic-oblong or lanceolate to narrowly ovate, 1.5 - 8 x 0.5 - 2 cm, glabrous, serrate, basal serratures not prolonged into filiform appendages, obtuse; petiole 3 - 12 mm long, hirsute; stipules subulate-filiform, ca 5 mm long. Cyme fascicle of 2 - 5 (-8) flowers, antiphyllous, very shortly pedunculate. Flowers yellow, 5 - 6 mm across; bracts minute. Sepals linear-oblong, ca 1.5 - 2.5 mm long, apiculate. Petals oblong-obovate, as long as sepals. Stamens as many as or twice the number of petals, filaments ca 1.5 - 2 mm long. Carpels 3; ovary oblong-ovate to linear, 3-loculed, hairy; style short, stigma capitate. Capsules 2 - 5(-8) in each fascicle, sessile, 1 - 1.5 cm long, with 1 - 2 mm long entire beak, somewhat triangular, pubescent, 3-loculed, locules without transverse septa. Seeds wedge shaped, ca 1.5 mm long, angular, black, obliquely truncate at both ends.

Field note: Widespread in hotter parts of India but not common in South central Rajasthan.

Fl. & Fr.: August-September.

Specimen examined : Bhilwara 305, Paroli 1799 (MLVGCB Herb).

Distribution: India (Rajasthan, Utter Pradesh, Madhya Pradesh, Bihar, Peninsula), Sri Lanka, Pakistan, Burma, North Australia, Arabia and tropical Africa

Uses: Fibre extracted from this plant is used for making ropes. Plant on grinding forms a mucilaginous jelly like substance which is used in seminal and general weakness as a tonic and is valued as an astringent and restorative.

Corchorus olitorius L., Sp. Pl. 529. 1753; Mast., in Hook. *f.*, Fl. Brit. Ind. 1: 397. 1874; Duthie, Fl. Gangetic Plain 1: 120. 1903; Mahesh., Illus. Fl. Delhi f. 33. 1966; Bhandari, Fl. Indian Desert. 74. 1978; Singh, in Shetty & Singh Fl. Rajasthan 1: 148. 1987; Daniel & Chandrabose, in Sharma *et al.* Fl. India 3: 487. 1993. *Corchorus decemangularis* Roxb., Fl. Ind. 2: 582. 1832. (Plate 51)

Vern.: *Chamghas.*

An erect, subglabrous, annual or biennial, up to 3 m tall (in cultivation). Stem basally woody, branched. Leaves 3 - 5-costate, lanceolate to ovate-lanceolate, 2.2 - 9 x 1.8 - 4.2 cm, glabrous except the scattered hairy veins, serrate, basal serratures prolonged into filiform deflexed appendages, acute or acuminate; petiole 1 - 3.2 cm long, hairy; stipules subulate, 8 - 12 mm long. Cyme 1 or 2-flowered, antiphyllous, shortly pedunculate. Flowers yellow, 12 - 15 mm across; subsessile; bracts subulate, 4-5 mm long. Sepals linear-oblong, 5 - 7 x 1.5 - 3 mm, keeled, caudate, somewhat bullate outside. Petals oblanceolate, 7 - 8 x 2.5 - 4 mm, hairy at the base, obtuse. Stamens numerous, somewhat united at the base, filaments 6 - 7 mm long; anthers subglobose. Carpels 5; ovary cylindric, subsulcate, 5-loculed, thinly hairy; style short, stigma 5-lobed, minutely papillate. Capsules 1 or 2 together, 2 - 7.2 cm long, with 4 - 5 mm long entire beak, ca 4 - 5 mm across, 10-angled, glabrous, 5-loculed, locules transversely septate. Seeds greenish-black, triangular, ovate, ca 2 x 1.5 mm long.

Field note: Common in plains and hills, as a weed of cultivated field and fallow lands in moist shady places.

Fl. & Fr.: August - October

Specimen examined: Kodukota 1143, Hamirgarh 1227 (MLVGCB Herb).

Distribution: A native of the Indo-Pakistan subcontinent, now widespread throughout the world in tropical countries and north Australia, either by cultivation or as escape.

Uses: This is another species of *Corchorus* which is extensively cultivated in most tropical countries especially in India and Bangla Desh for the commercial jute fibres obtained from the sclerenchymatous lignified secondary phloem elements. The fibre and other parts are used for the same purpose as *Corchorus capsularis* L. Plant is used in West Africa and Sudan as a pot herb. Leaves are useful in chronic cystitis, gonorrhoea and dysuria and their infusion is given as tonic and febrifuge. Seeds have purgative properties and the fruit contains ascorbic acid (Vit. C.).

Corchorus tridens L., Mant. Alt. 566. 1767; Mast., in Hook.f., Fl. Brit. Ind. 1: 398. 1874; Duthie, Fl. Gangetic Plain 1: 121. 1903; Maheshwari, Fl. Delhi 88. 1963; Mahesh., Illus. Fl. Delhi f. 36. 1966; Bhandari, Fl. Indian Desert. 74. 1978; Singh, in Shetty & Singh Fl. Rajasthan 1: 148. 1987; Daniel & Chandrabose, in Sharma *et al.* Fl. India 3: 488. 1993. *C. trilocularis* auct. non L. Mant. Pl. 77. 1767; Burm. f., Fl. Ind. 125. t. 37, f. 2. 1768. *C. burmanii* DC., Prodr. 1: 505. 1824.

Vern.: *Kawava torai, Kag nasha*

Erect or suberect, 30 - 60 cm tall, annual herb. Stem and branches glabrous. Leaves 3 - 4-costate, oblong-lanceolate or linear-lanceolate or somewhat elliptic-obovate, 1.5 - 12 x 0.6 -2.2 cm, glabrous except the sparsely hairy veins, crenate-serrate, basal serratures prolonged into filiform spreading appendages or not, acute; petiole 5 - 18 mm long, hairy; stipules setaceous, 2 - 4 mm long. Cyme 1 - 4-flowered, antiphyllous, short peduncled. Flowers yellow, ca 1 cm across, subsessile; bracts subulate, 3 - 4 mm long. Sepals linear-oblong, 4 - 5 mm long, ca 2 mm broad, apiculate. Petals oblong, 3 - 4 x 2 mm, obtuse. Stamens 10 - 15; filaments 4 - 5 mm long. Carpels 3; ovary cylindric, sulcate, 3-loculed, hairy; style small, stigma sparsely papillate. Capsules 1 - 3 together, erect, cylindrical, 1.5 - 4 cm long, ca 2 mm in diam., not ribbed, glabrous, terminated by 3 spreading bifid tips, 3-loculed, locules aseptate. Seeds angular, obliquely truncate at both ends black.

Field note: Common throughout the area in cultivated fields and waste lands, on sandy alluvial moist soils, particularly after rains.

Fl. & Fr.: July-November.

Specimen examined: Kodukota 1128, Kakroliyaghati 1520 (MLVGCB Herb).

Distribution: Widespread in tropical and subtropical countries of Africa, Asia and in North Australia.

Uses: Young leaves and shoots are used as pot herb in India, Pakistan, tropical Africa and are sometimes dried for the purpose to be used in soup. It is commonly eaten by goats and camels. Stem yields a good fibre which is used in north Nigeria and elsewhere for fishing lines.

GREWIA L.

(*Microcos* L.)

KEY TO THE SPECIES

1a. Leaves glabrous or nearly so, peduncles capillary, 1 - 2 flowered........................*G. tenax*

1b. Leaves stellately hairy or tomentose at least on lower surface, peduncles solitary, 2 - 3 flowered..... 2

2a. Leaves oblong, ovate-oblong or ovate-elliptic; peduncles much shorter than or as long as petioles; petals yellow ....*G. flavescens*

2b. Leaves narrowly lanceolate to ovate, obtuse; peduncles much longer than petioles; petals white...... .... *G. damine*

Grewia damine Gaertn., Fruct. Sem. Pl. 2: 113. t. 106. f. 1790; Dunn, in Gamble, Fl. Pres. Madras 118. 1915; Singh, in Shetty & Singh Fl. Rajasthan 1: 151. 1987; Daniel & Chandrabose, in Sharma *et al* .Fl. India 3: 496. 1993. *G. salvifolia* Heyne ex Roth. November Pl. Sp. 239. 1821; Mast., in Hook. *f.*, Fl. Brit. Ind. 1: 386. 1874, *p. p.*; Duthie, Fl. Gangetic Plain 1: 116. 1903.

Vern.: *Gangechi*

A shrub or small sized tree, 2 - 4 m tall. Stem and branches rugose, bark ash-grey, young shoots pale hairy. Leaves with 2 - 6 mm long, upwardly dilated, densely hairy petiole; lamina with scattered stellate hairs above, hoary tomentose and whitish-grey beneath, narrowly lanceolate to ovate, 1.2 - 9 x 0.8 - 3.5 cm, 3-costate, unequal sided, obtuse-oblique at the base, minutely serrate, obtuse; stipules falcate, 8 - 10 (-15) mm long, villous. Cyme 2 - 3-flowered, axillary, peduncles solitary, rarely 2 or 3, longer than petiole. Flowers yellow, ca 1.5 cm across; pedicel 8 - 10 mm long, pale hairy; bracts linear-lanceolate, ca 5 mm long. Sepals linear-oblong, 12 - 13 x ca 3 mm, stellate hairy outside. Petals oblong-obovate, claw ca 1.5 mm long, with white gland, hairy, limb 3.5 - 6 x ca 2 mm broad, notched at the apex. Stamens ca 40 in 2 whorls, filaments 3 - 4.5 mm long. Ovary 1 - 2-lobed, densely hairy; style clavate, 6 - 7 mm long, stigma 4-lobed. Drupe slightly succulent, globose, rarely 2-lobed, 6 - 10 mm across, almost glabrous.

Field note: Common in the dry arid plains and low hills.

Fl. & Fr.: July - September.

Specimen examined: Pansal 98, Bijoliya 1629 (MLVGCB Herb).

Distribution: India (Punjab, Madhya Pradesh, Peninsula), Pakistan, Nepal and tropical Africa.

Uses: Root is dried to make powder. In case of bone fracture, after 3 days of fracture 1/2 cup of decoction is given once a day for 3 days to animals. In human beings 1-2 tsp decoction is taken once a day for 3-7 days. Subacidic, slightly succulent fruits are edible. Branches afford fine walking sticks.

Grewia flavescens A. L. Juss. in Ann. Mus. Natl. Hist. Nat. 4 : 91. 1804; Singh, in Shetty & Singh Fl. Rajasthan 1: 152. 1987; Daniel & Chandrabose, in Sharma *et al.*Fl. India 3: 498. 1993. *Grewia pilosa auct. plur.* non Lam. 1789; Mast., in Hook. *f.*, Fl. Brit. Ind. 1: 388. 1874, *pro parte;* Duthie, Fl. Gangetic Plain 1: 111. 1903.

Vern.: *Pirigangechi*

An erect shrub or small tree, up to 6 m tall. Stem with grey bark. Leaves 1.5 - 13 x 1 - 7 cm, ovate-elliptic, oblong or ovate-oblong, subcordate or rounded at base, usually acute, rarely obtuse or obovate at the apex, serrate, scabrous above, tomentose beneath, 3-nerved; petioles upto 7 mm long. Flowers in short axillary or extra-axillary, pedunculate, umbillate cymes, peduncles 1 - 3 together, yellow. Sepals linear-lanceolate, acute, tomentose outside, glabrous inside, 1.2 - 1.7 cm long. Petals yellow, spathulate or linear oblong, usually 2-fid; glands ca 3 mm long, oblong, 5 - 10 mm long, Stamens 1 cm long. Ovary ca 2 x 0.5 mm, subglobose or ovoid, pilose; stigma 2- fid. Drupes usually 2-4-lobed, 0.7 - 1 x 0.7 - 1.5 cm, globose when not lobed, yellowish brown, stellate-tomentose.

Field note: Common on the hills throughout the area.

Fl. & Fr.: August - November.

Specimen examined: Pur 1088, Mandalgarh 1804 (MLVGCB Herb).

Distribution: Tropical Africa and India.

Grewia tenax (Forsk.) Fiori in Bos. Piant. Legn. Eritr. 246. 1909 & in Agric. Colon 5. Suppl. 23. 1912; Naray & Rao in J. Indian Bot. Soc. 29: 179. 1950; Maheshwari, Fl. Delhi 87. 1963; Mahesh., Illus. Fl. Delhi f. 31. 1966; Bhandari, Fl. Indian Desert. 82. 1978; Singh, in Shetty & Singh Fl. Rajasthan 1: 155. 1987; Daniel & Chandrabose, in Sharma *et al.* Fl. India 3: 511. 1993. *Chadara tenax* Forsk. Fl, Aegypt. - Arab. 105. 1775. *Grewia populifolia* Vahl, Symb. Bot. 1 : 33. 1790; Mast., in Hook. *f.*, Fl. Brit. Ind. 1: 385. 1874; Duthie, Fl. Gangetic Plain 1: 111. 1903. *Chadara betulaefolia* Juss. in Ann. Mus. Natl. Hist. Nat. 4 : 92. t. 2. 1804. (Plate 52)

Vern.: *Gangechi, Kango*

Suberect to erect shrub, up to 3 m tall. Stem with ash-grey bark, young twigs stellate hairy. Leaves 3 - 5-costate, almost glabrous to sparsely or densely stellate hairy on both sides, ovate-elliptic or obovate to almost orbicular, 0.5 - 4.5 x 0.4 - 2.5 cm, sharply serrate, cuneate at the base, acute to obtuse, rarely emarginate at the apex; petiole 2 - 24 mm long, hairy; stipules linear-lanceolate, caducous. Flowers solitary or rarely paired, on solitary, antiphyllous, 1.5 cm long, hairy peduncle, white, rarely yellowish-white, 2 - 2.5 cm across; pedicel ca half as long as peduncle, stellate tomentose. Sepals linear-oblong, 1.2 - 1.8 x 2 - 4 mm, stellate hairy outside. Petals linear, 1.1 - 1.4 cm x ca 2 mm, claw much smaller than limb, with somewhat reniform densely ciliate gland, bilobed at the apex. Androgynophore ca 0.5 mm long. Stamens numerous; filaments slightly shorter than style or subequal. Torus angular. Ovary 4-lobed, glabrous, rarely stellate hairy; style ca 8 - 10 mm long, clavate, glabrous, rarely sparsely stellate hairy; stigma somewhat cupular. Drupe usually 2 - 4-lobed, lobes ca 5 - 10 mm in diam., glabrous, rarely with sprinkled stellate hairs, orange yellow with reddish tinge.

Field note: Common in semi-arid plains and hills throughout the area.

Fl. & Fr.: August - October.

Specimen examined: Gurla 1206, Beawar 4161 (MLVGCB Herb).

Distribution: Tropical and North Africa to Iran, Pakistan, India and Sri Lanka.

Notes: A highly variable species, especially in size, shape and hairiness of leaves and petiole in addition to size of floral parts. It includes 2 taxa which can be separated from each other by sharp discontinuity of petiole length. However, We have observed discontinuity of the petiole size even in one and the same plant hence, it is not possible to recognize the two taxa even on the basis of this character.

Uses: The plant is commonly used as fodder.

TRIUMFETTA L.

(Bartramia L.)

KEY TO THE SPECIES

1a. Spines glabrous, stamens mostly 15, rarely fewer but not less than 10. Spines on the capsule glabrous.................... ... *T. rhomboidea*

1b. Spines pubescent or tomentose, stamens mostly 5, rarely 10. Spines on the capsule with a line of spreading hairs on the dorsal side.. ...... *T. pentandra*

Triumfetta pentandra A. Rich. in Guill. *et al.*, Fl. Seneg. Tent. 1: 93. t. 19. 1831; Bhandari, Fl. Indian Desert. 78. 1978; Singh, in Shetty & Singh Fl. Rajasthan 1: 157. 1987; Daniel & Chandrabose, in Sharma *et al.* Fl. India 3: 519. 1993. *T. neglecta* Wight & Arn., Prodr. 75. 1834; Mast., in Hook. *f.,* Fl. Brit. Ind. 1: 396. 1874; Duthie, Fl. Gangetic Plain 1: 119. 1903. *T. rhomboidea* Jacq. var. *pentandra* (A. Rich.) J. L. Ellis in Bull. Bot. Surv. India 24: 209. 1983.

An erect, branched, 20 - 60 cm tall, annual herb. Stem subscabrous, stellate hairy. Leaves 5-costate, basal rhomboid-ovate, palmately 3-lobed, terminal ones ovate-lanceolate, unlobed, 3 - 10 x 3 - 8 cm, sparingly simple hairy above, stellate pubescent below, acuminate-acute, cuneate at the base, unequally coarsely serrate; petiole 1 - 5.5 cm long, scabrous, densely hairy; stipules 5 - 6 mm long, subulate, ciliate and hispid-glandular along margins. Cymes antiph-yllous, short peduncled, arranged racemosely due to abortion of upper leaves. Flowers yellow, ca 4 - 5 mm across, almost sessile; bracts linear, ca 2 - 3 mm long, hairy. Sepals lorate, ca 2.5 mm long, cucullate and awned at the apex, sparsely stellate hairy outside. Petals spathulate, as long as sepals, glabrous at the base, obtuse. Stamens mostly 5, rarely up to 10; filaments hairy outside at the base. Carpels 2; ovary ovoid, 2-loculed, hairy. Capsule ovoid, densely tomentose, 4.5 - 6 x 3 - 4 mm in diam. including ca 1.5-2 mm long, ascending, uncinate spines with a line of spreading hairs on the upper side. Seeds 4, somewhat trigonous, smooth, brown.

Field note: Common in waste lands, open forests and along the roads. Flowers open during mid-day.

Fl. & Fr.: July - October.

Specimen examined: Gangapur 1192, Kot 1384 (MLVGCB Herb).

Distribution: Pakistan, India, Sri Lanka, Cape Verde Islands and Tropical Africa

Triumfetta rhomboidea Jacq., Enum. Syst. Pl. 22. 1760; Mast., in Hook.f., Fl. Brit. Ind. 1: 395. 1874; Duthie, Fl. Gangetic Plain 1: 118. 1903; Cooke. Fl. Bomb. Pres. (reprint. ed.) 1: 156. 1958; Santapau in Bull. Bot. Surv. Ind. 3: 21. 1961; Maheshwari, Fl. Delhi 89. 1963; Singh, in Shetty & Singh Fl. Rajasthan 1: 157. 1987; Daniel & Chandrabose, in Sharma *et al.* Fl. India 3: 519. 1993; *T. bartramia* L., Syst. Nat. ed. 10, 2: 1044. 1759, nom. illegit. *Bartramia indica* L. Sp. Pl. 378. 1753, non *Triumfetta indica* Lam. 1792. *T. angulata* Lam., Encycl. 3: 421. 1792; Wight & Arn., Prodr. 74. 1834. Nat. ed. 10, 2: 1044. 1759. *T. trilocularis* Roxb., Fl. Ind. 2: 462. 1832. *T. tungarensis* Billore, in J. Econ. Tax. Bot. 3: 621. 1982. (Plate 53)

Vern.: *Lapta.*

Much branched, 0.5 - 2 m tall herb or undershrub. Leaves 3 - 7-costate, variable in shape, generally rhomboid-ovate, 3 - 9.5 x 2.5 - 7.5 cm, stellate or simple hairy to glabrescent, palmately 3-lobed or not, acute at the apex, cuneate-cordate at base, irregularly serrate; petiole up to 5 cm long, pubescent, subulate, short in terminal leaves; stipules subulate, ca 3 - 4 mm long, subulate, pubescent. Cymes terminal or antiphyllous. Flowers golden yellow, ca 5 - 6 mm across, subsessile. Buds oblong, constricted in the middle. Sepals ca 5 mm, lorate, strongly cucullate and awned at the apex, hairy outside. Petals oblong-obovate, equal to or slightly shorter than the sepals, hairy at the base. Stamens 8 - 15, rarely fewer, filaments as long as sepals, papillose. Carpels 2 - 3, syncarpous; ovary subglobose, 2 -• 3-loculed, with uncinate hairs; style subulate, stigma 2 - 3-lobed. Capsule globose or subglobose, indehiscent, tomentose, ca 5 - 6 mm in diam. including 1.5 - 2 mm long, glabrous, uncinate spines.

Field note: Common in waste lands, open forests and along the roads. Flowers open during mid-day.

Fl. & Fr.: August-November.

Specimen examined: Jalindri 302, Gangapur 1802 (MLVGCB Herb).

Distribution: Throughout the tropics.

Notes: This species is common in this area. It is confused with *Triumfetta pentandra* A. Rich to which it resembles in leaf shape. The two can be easily distinguished with the help of key characters.

Uses: Leaves, flowers and fruits are demulcent, astringent and given in gonorrhoea. Bark and fresh leaves are used in diarrhoea and dysentry. Seeds are said to possess anthelmintic and aperient properties.

MALPIGHIACEAE

HIPTAGE Gaertn.

Hiptage benghalensis (L.) Kurz in J. Asiat. Soc. Bengal. 43. 2: 136. 1874; Pandey, in Shetty & Singh, Fl. Rajasthan 1: 160. 1987; Srivastava, in Hajra *et al.* Fl. India 4: 14. 1997. *Banisteria benghalensis* L., Sp. Pl. 427. 1753. *Hiptage madablota* Gaertn. Fruct. 2: 169. t. 116. 1791; Hook. *f.*, Fl. Brit. Ind. 1: 418. 1874; Duthie, Fl. Gangetic Plain 1: 124. 1903. (Plate 54)

Vern.: *Madhavilata.*

Large woody climbing shrub; young shoots sparsely pubescent and often with minute medifixed hairs. Leaves elliptic, 9 - 19 x 4 - 9 cm, acute or acuminate, pale and sparsely pubescent beneath, shiny above; petiole up to 1 cm long. Axillary and terminal racemes up to 20 cm long, erect, softly tomentose. Pedicels 1.5 - 2 cm across. Flowers 1.5 - 2 cm across. Bract 3 - 4.5 mm long, ovate-lanceolate, acuminate, pubescent, deciduous; bracteoles 2, ca 3 mm long, linear lanceolate, present halfway up the pedicel; pedicel 1.5-2.5 cm long. Calyx lobe ovate, elliptic or oblong, 8 - 10 x 2 - 4 mm, fulvous-serceous outside; gland oblong, elliptic or lanceolate, 4 - 5 mm long, half decurrent on pedicels. Petals alternating with the sepals, suborbicular, ca 1 cm long, clawed, creamish white, the posterior one the largest and with a yellow centre, appressed pubescent hairs to the outside, glabrous to sparsely pubescent within. Stamens unequal; filament of longest stamen l - 1.2 cm long, twice as long or longer than the shorter ones, slightly dilated and connate at the base; anthers ovate, ca 2 mm long. Ovary pubescent; style 1.3 - 1.5 cm long, slightly curved, lower part with appressed hairs. Samaras variable; middle wing oblanceolate, oblong, 2 - 3 x 0.8 - 1.5 cm; dorsal crista ca 11 mm long. Seeds sub-globose.

Field note: Rare, in the area. Cultivated in gardens as ornamental. A strong woody climber with showy fragrant flowers.

Fl. & Fr.: March - April.

Specimen examined: Bhilwara 99 (MLVGCB Herb).

Distribution: Native of India and Malaya; W. China, Java, Burma, Philippines, Sri Lanka and the sub-Himalayan tract from the Indus eastward.

Uses: Leaves are medicinal, used in rheumatism and skin diseases. It also have insecticidal properties and used in the treatment of scabies.

ZYGOPHYLLACEAE

KEY TO THE GENERA

1a. Leaves 1 to 3-foliate, Fruit capsule, seeds albuminous ....*Fagonia*

1b. Leaves simple, Fruit a schizocarp, seeds exalbuminous......................................*Tribulus*

FAGONIA Tourn. ex L.

Fagonia schweinfurthii (Hadidi) Hadidi ex Ghafoor, in Jafri & El Gadi, Fl. Libya 38 : 31. 1977; Bhandari, Fl. Indian Desert. 80. 1978; Pandey, in Shetty & Singh, Fl. Rajasthan 1:

163. 1987; Singh & Singh, in Hajra *et al.* Fl. India 4: 44. 1997. *F. indica* Burm. f. var. *schweinfurthii* Hadidi, in Rech. f. Fl. Iran 98 : 6. t. 6. 1972. (Plate 55)

Vern.: *Dhamso.*

Annual to biennial, up to 50 cm tall, glandular to glabrous shrublet. Stem basally woody, branches prostrate to erect, terete, striate, with up to 1 - 3 cm long internodes. Lower leaves trifoliolate, petiole, up to 10 - 12 mm long, terminal leaves unifoliolate, with up to 5 mm long petiole; leaflets linear-oblong, up to 3.5 cm long, acute, central the largest; stipular spines aculeate, shorter than leaves and internodes, some times much reduced, patent, not reflexed. Flowers pinkish-purple, ca 8 - 10 mm across, pedicel up to 8 mm long, glandular to glabrous. Sepals ovate, 3 - 4 mm long, sparsely glandular hairy outside, deciduous or semi-persistent. Petals obovate, 4 - 6 x 2 - 3 mm, obtuse. Stamens with 3 - 5 mm long filaments. Capsule ca 4 x 4 mm, pubescent, pedicel ca twice as long as fruit.

Field note: Common in gravelly and sandy habitats.

Fl. & Fr.: Almost throughout the year.

Specimen examined: Atun 100, Asind 1670 (MLVGCB Herb).

Distribution: India, Afghanistan, Iraq, Pakistan, Iran, Aden, Eritrea, Aethiopia, Sudan, Somalia and Kenya.

Notes: In Indian Floras *F. schweinfurthii* Hadidi and *F. indica* Burm. f. are erroneously identified either as *F. cretica* L. or *F. arabica* L. Examination of the type of *F. cretica* L. has revealed that it does not occur in Indian subcontinent and is probably restricted to mediterranian region (Hadidi, 1972).

TRIBULUS Tourn. ex L.

KEY TO THE SPECIES

1a. Mericarps with 2 major divergent spines and 20 - 30 minor spines, basal pairs always absent. ....*T. rajasthanensis*

1b. Mericarps dorsally tuberculate, spines usually 4. ....*T. terrestris*

Tribulus rajasthanensis Bhandari & Sharma in Bot. Notis. 129: 367. f. 1. 1977; Bhandari, Fl. Indian Desert. 83. 1978; Pandey, in Shetty & Singh, Fl. Rajasthan 1: 166. 1987; Singh & Singh, in Hajra *et al.* Fl. India 4: 53. f. 17. 1997. (Plate 56)

Vern.: *Gokharu.*

Annual or biennial, prostrate, densely appressed whitish silky pubescent herb. Stem hirsute to sericeous, branches many from woody root stock, spreading up to 50 cm long, hirsute. Leaves paripinnate, 2 - 5 cm long; stipules lanceolate to falcate, 3 - 7 x 1 - 2 mm; leaflets 4 - 7 pairs, elliptic-oblong to ovate- oblong, pubescent above, villous beneath, 4 - 12 x 3 - 6. Stipules 5 - 7 x 1 - 2 mm, subulate, falcate, persistent. Flowers yellow, 10 - 15 mm across; pedicel up to 20 mm long. Sepals 5 - 7 x 1 - 2 mm, linear-lanceolate, acute, pubescent. Petals obovate, 5 - 8 x 2 - 5 mm, cuneate at base. Stamens 10, filaments ca 3.5 mm long, intrastaminal glands broadly triangular, free; extrastaminal glands round, truncate or emarginate; anthers oblong, versatile. Ovary ca 2 mm across, ovoid, hirsute; style ca 1 - 1.5 mm long; stigmas 5 - rayed, decurrent. Fruit up to ca 1 cm in diam., 4 - 8 mm long, mericarps 6 - 7 x 4 - 5 mm, densely hispid; major 2 long patent spines, ca 5 mm long; minor spines 20 - 30, unequal, 1 - 3 mm long.

Field note: Occasionaly, found in gravelly plains.

Fl. & Fr.: Almost throughout the year.

Specimen examined: Atun 156, Bhilwara 1785 (MLVGCB Herb).

Distribution: India (Rajasthan and Gujrat) and Pakistan.

Uses: Fruits are taken as a remedy for impotency.

Tribulus terrestris L., Sp. Pl. 387. 1753; Edgew. & Hook. *f.* in Hook. *f.,* Fl. Brit. Ind. 1: 423. 1874, p.p.; Duthie, Fl. Gangetic Plain 1: 127. 1903; Cooke, Fl. Bomb. Pres. (reprint. ed.) 1: 170. 1958; Bhandari, Fl. Indian Desert. 84. 1978; Pandey, in Shetty & Singh, Fl. Rajasthan 1: 166. 1987; Singh & Singh, in Hajra *et al.* Fl. India 4: 55. 1997.

Vern.: *Gokharu, Gokharu ki kanti.*

Annual or biennial, prostrate, densely appressed whitish silky pubescent herb. Stem hirsute to sericeous; branches spreading. Leaves paripinnate, 2.5 - 7 cm long; stipules subulate to falcate, 2 - 5 mm long; leaflets 4 - 7(-8) pairs ovate to elliptic-oblong, oblique, acute or obtuse, 4 - 11 x 2 - 8 mm. Flowers yellow, 1 - 1.5 cm across; pedicel up to 1.5 cm long. Sepals ovate-lanceolate, 2 - 6 x 1.5 - 2 mm, acute. Petals obovate, 2.5 - 8 x 1 - 4 mm, obtuse. Stamens 10; filaments ca 3 - 5 mm long; anthers, oblong, versatile. Ovary ca 1.5 mm across, ovoid, hirsute; style ca 1 - 1.5 mm long; stigmas decurrent. Fruit up to ca 10 - 15 mm in diam., 4 - 8 mm long; mericarps 4 - 8 x 5 - 7 mm, densely crested and tuberculate on dorsal side, densely hairy to glabrescent, with 2 long patent and 2 short downwardly directed spines (rarely all or lower 2 reduced to tubercles).

Field note: Commonly, found in gravelly plains, as well as in sandy soils of barren lands and cultivated fields as a weed.

Fl. & Fr.: Almost throughout the year.

Specimen examined: Atun 1420, Masuda 1573 (MLVGCB Herb).

Distribution: Tropical and subtropical countries in Asia, Africa, S. Europe, North Australia and introduced in new world tropics.

Notes: The fruit spines creat a problem to cattles feet so the name Gokharu ki kanti is popular among local peoples. A highly variable species in leaf, flower size and fruit characters. Many authors recognize subspecies and varieties on the basis of presence or absence of spines, tubercles and indumentum on the dorsal side of mericarps but these characters do not seem constant and therefore we prefer to recognize one variable species without any formal infra-specific taxa.

Uses: Fruit powder is used 3 times daily for cough, asthma & backache. Fruit powder & root extract is taken orally in stone problems. Plant extract is applied on wounds. It is used as a diuretic, demulcent, tonic, aphrodisiac and aperient. Fruit is used in painful micturition, urinary diseases, impotency, cough and heart diseases.

BALSAMINACEAE

IMPATIENS L.

Impatiens balsamina L., Sp. Pl. 938. 1753; Hook. *f.*, Fl. Brit. Ind. 1 : 453. 1874 & in Rec. Bot. Surv. Ind. 4: 7. 1904; Cooke, Fl. Pres. Bomb. 1: 184. 1901; Babu, Herb. Fl. Dehradun 105. 1977; Pandey, in Shetty & Singh, Fl. Rajasthan 1: 172. 1987; Vivekananthan *et al.*, in Hajra *et al.* Fl. India 4: 123 - 124. 1997. (Plate 57)

Vern.: *Gulmendi.*

Erect, annual, 45 - 80 cm tall, flaccid, pubescent. Stem simple or branched, glabrous or pubescent. Leaves alternate, distinct, lanceolate-elliptic, 3 - 15 x 1 - 3 cm, serrate along margins. Flowers white, orange, pink-red or purple, axillary, solitary or in fascicles of 2 - 3, 1 - 1.5 cm long; pedicel up to 1 - 2 cm long, glabrous or pubescent. Lateral sepals ca 2 - 3 mm long, ovate-lanceolate, sparsely ciliate; Lip cymbiform, 1.2 - 1.8 cm long, petaloid with yellowish blotch; spur filiform, incurved, 1.5 - 3.5 cm long. Standard 1.2 cm long, with a crest terminating into a mucro. Wings obovate, 2 - 2.5 cm long, deeply notched; auricles ovate-rounded. Capsule broadly ovoid to fusiform, 1.2 - 2 cm long, densely tomentose, pendulous. Seeds globose, reticulate.

Field note: Common in moist, shady localities, also cultivated as an ornamental.

Fl. & Fr.: Late August - September.

Specimen examined: Bijoliya 1521, Bhilwara 1810 (MLVGCB Herb).

Distribution: India, Pakistan, Sri Lanka, Malaya and also in China; introduced in Turkey and S. Europe.

Uses: The *'garden balsam'* or *'touch-me-not'* is variable in the size of the plant, pubescence and colour of the flower. Several varieties of this species are known.

Notes: It is an extremely variable species particularly with regard to size of leaves and flowers.

OXALIDACEAE

KEY TO THE GENERA

1a. Leaves fascicled at top of stem, even-pinnate ....*Biophytum*

1b. Leaves not fascicled at top of stem, 3-foliolate ....*Oxalis*

BIOPHYTUM DC.

Biophytum sensitivum (L.) DC. Prodr. 1: 690. 1824; Edgew.& Hook. *f.* in Hook. *f.*, Fl. Brit. Ind. 1: 436. 1874; Duthie, Fl. Gangetic Plain 1: 131. 1903; Cooke, Fl. Bomb. Pres. (reprint. ed.) 1 : 170. 1958; Pandey, in Shetty & Singh, Fl. Rajasthan 1: 166. 1987; Manna, in Hajra *et al.* Fl. India 4: 238. 1997. *Oxalis sensitiva* L., Sp. Pl. 1: 434. 1753. (Fig. 30 & Plate 58)

var.: ***sensitivum***

Vern.: *Chhui-mui.*

Annual 5 - 35 cm tall, base woody. Stem simple, slender to robust, hispid especially apically. Leaves 3 - 13 cm; rachis slender, moderately hispid; leaflets 6 - 15 pairs; leaflet opposite; blades oblong to obovate-oblong, 3 - 15 × 2 - 7 mm, usually glabrous, occasionally sparsely covered with trichomes, base almost symmetric. Umbels several flowered; peduncle 2 - 7 cm, various lengths, up to 14 cm long, appressed-strigose and patently glandular-hairy; bracts several, lanceolate, ca 2 mm, densely crowded at apex of peduncle, cluster ca 3 mm. Pedicel ca 1 mm at anthesis but to 3 mm in fruit. Sepals ovate-lanceolate, acute, 4 - 6 x 0.5 - 1.5 mm, with glandular septate trichomes. Petals yellow, 5 - 7 x 1 - 2 mm, lanceolate. Anthers 1.5 - 2.5 mm long. Ovary 0.5 - 0.75 mm, apical ciliate; ovules 2 - 5 per locule; style 0.5 - 1 mm; stigma flattened. Capsule ellipsoid-obovoid, 4-5 × 3-4 mm, pubescent. Seeds prominently ridged.

Field note: Rare, in the forest and near shallow riverbeds.

Fl. & Fr.: September - December.

Specimen examined: Nimbahera 1625 (MLVGCB Herb).

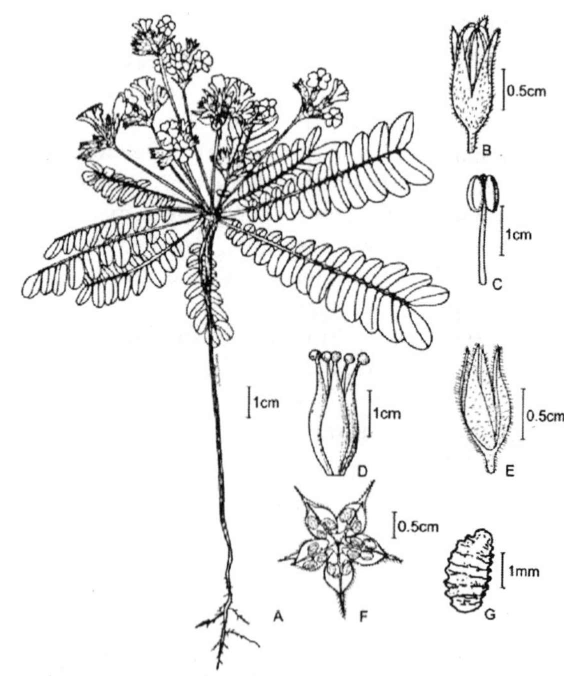

Fig. 30. *Biophytum sensitivum* (L.) DC. A. Habit, B. Flower bud, C. Stamen, D. Ovary, E. Fruit, F. Dehiscing fruit, G. Seed

Distribution: India, Indonesia, Malaysia, Nepal, Philippines, Sri Lanka, Thailand, Vietnam and tropical Africa

Uses: This species is used medicinally.

OXALIS L.

KEY TO THE SPECIES

1a. Ascending or creeping herb. Flowers yellow ....*O. corniculata*

1b. Acaulescent herb with underground bulbous root-stocks. Flowers purplish..................................*O. dehradunensis*

Oxalis corniculata L., Sp. Pl. 435. 1753; Wight, Ic. 1: t. 18. 1838; Edgew. & Hook. *f.* in Hook. *f.*, Fl. Brit. Ind. 1: 436. 1874; Duthie, Fl. Gangetic Plain 1: 130. 1903; Bhandari, Fl. Indian Desert. 87. 1978; Pandey, in Shetty & Singh, Fl. Rajasthan 1: 170. 1987; Manna, in Hajra *et al.* Fl. India 4: 242. f. 60. 1997. (Fig. 31).

Vern.: *Khatti booti.*

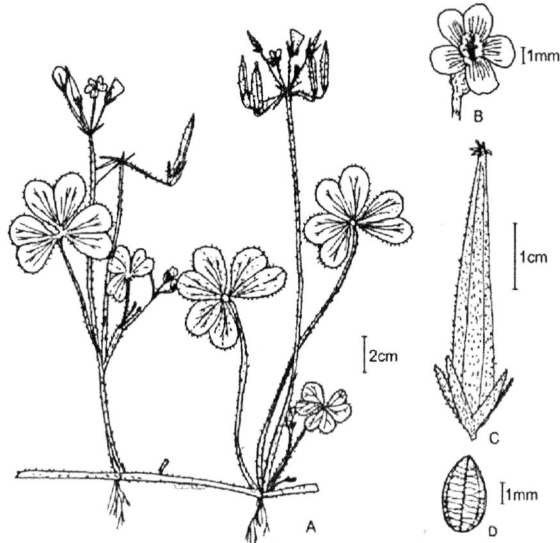

Fig. 31. *Oxalis corniculata* L.
A. Habit, B. Flower, C. Fruit, D. Seed

Plants creeping, pubescent, rooting at the nodes. Leaflets 4 - 15 x 8 - 32 mm, obcordate, pilose-tomentose. Flowers solitary or in 2 - 5 flowered axillary umbels. Pedicel 5 - 15 mm long, deflexed in fruit. Bracts 3 - 5, linear. Sepals 5, linear lanceolate, pilose. Petals 5, yellow. Stamens 10, united; Long ones ca 5 mm long; short ones ca 4 mm long, filaments glabrous; anthers dehiscing longitudinally. Ovary ellipsoid to cylindric, ca 2 x 1 mm, puberulous; stigma small, sometimes flattened and minutely bifid; ovules 5 - 11 per locule in a row. Capsule 9 - 20 x 2 - 4 mm, usually linear-cylindric, pubescent, shortly beaked, 5-angled. Seeds 5 - 11 per locule, flattened-ovoid, 1.5 mm long, brown, transversely ribbed, acute.

Field note: A common weed of gardens, lawns, banks of water streams, fields and other moist and shady places.

Fl. & Fr.: Almost throughout the year.

Specimen examined: Jahajpur 603, Potla 1624 (MLVGCB Herb).

Distribution: Cosmopolitan weed.

Notes: A very variable species in form and size. Common during the spring season.

Uses: Leaves occasionally used as vegetable. Juice of the plant mixed with onion is used to remove warts. Fresh leaves are chewed for bitter taste.

Oxalis dehradunensis Raizada in Suppl. Duthie, Fl. Gangetic Plain 37. 1976; Pandey, in Shetty & Singh, Fl. Rajasthan 1: 171. 1987; Manna, in Hajra *et al*. Fl. India 4: 246 - 249. f. 62. 1997. *O. richardiana* Babu, Herb. Fl. Dehradun 104. 1977. *O. latifolia* auct. plur. non H.B.K. 1821; Calder in Rec. Bot. Surv. India 6 (8): 335. 1919; Fyson, Fl. S. Indian Hill Stat. 2: t. 51. 1932.

Glabrous or sparsely pubescent, acaulescent herb, with bulbous root stock, perennial. Bulbils clustered or solitary and arising from the stolons. Leaves 2 - 6, 3-foliolate; petioles 10 - 20 cm long; leaflets 15 - 40 x 30 - 70 mm, triangular in outline, cuneate at base, shallowly notched at apex, with oblique lobes, epunctate except two glands below the apex, glabrous. Scapes 1 - 4, up to 30 cm long, glabrous to sparsely pubescent, slender. Flowers many, upto 9 mm long, umbellate or subumbellate, purple with a green-viend base. Pedicel 1 - 1.8 cm long, glabrous. Sepals 5, elliptic oblong, equal, ca 4 x 1.5 mm, glabrous, tip with a reddish brown callus. Corolla infundibuliform, 1 - 1.5 cm long; petals 5, pink. Stamens 10, alternately long and short; longer filament inapendiculate, ca 4 mm long; short ones ca 2 mm long, both filaments strigose. Ovary elongate, 5-lobed, ca 2.5 mm long, glabrous; styles 5, distinct, glabrous; stigmas 5, discoid with wavy margins; ovules 4 in each locule.

Field note: Commonly found as a weed of cultivated fields and irrigated ditches.

Fl. & Fr.: July - November.

Specimen examined: Mezadam 1554 (MLVGCB Herb).

Distribution: A native of South and Central America

RUTACEAE

KEY TO THE GENERA

1a. Plants unarmed .... *Murraya*

1b. Plants armed with spines 2

2a. Leaves simple ....*Citrus*

2b. Leaves compound.................................... 3

3a. Leaves alternate, leaflets 3. Style short, stigma capitate or obtuse........................*Aegle*

3b. Leaves spirally arranged, leaflets 3 - 9. Style absent, stigma fusiform ....*Limonia*

AEGLE Corr. nom. cons.

Aegle marmelos (L.) Corr., in Trans. Linn. Soc. 5: 223. 1800; Cooke, Fl. Bomb. Pres. 1: 203. 1901; Hook. *f.,* Fl. Brit. Ind. 1: 516. 1875; Duthie, Fl. Gangetic Plain 1: 143. 1903; Parmar, in Shetty & Singh Fl. Rajasthan 1: 172. 1987; Nair & Nair, in Hajra *et al.* Fl. India 4: 264. 1997. *Crataeva marmelos* L. Sp. Pl. 444. 1753. *Aegle marmelos* (L.) Correa var. *mahurensis* Zate, in Indian J. For. 5: 35. 1982. (Plate 59)

Vern: *Billa, Bel.*

Tree, upto 12 m tall; branches with spines ca 4 cm long, single or paired. Leaves petiolate, alternate, pinnately trifoliate, occassionally 5-foliate, sparsely white pubescent; leaflets 3 - 5, the lower 2 subsessibie, ovate-lanceolate, subcrenulate, obtuse. Flowers greenish-white, fragrant. Pedicels pubescent. Calyx cupular with 5 small deltate or sub-orbicular teeth, pubescent, deciduous. Petals 4 or 5, 13 mm, oavte-oblong, fleshy, spreading. Stamens ca 50, free or in fascicles, filaments subulate, ca 7 mm long, glandular, glabrous; anthers linear-oblong, apiculate, ca 8 mm long. Disc cylindric, ca 1 mm high, greenish, glabrous. Ovary ovoid-oblong, 3.5 - 5 mm long, faintly ridged, glabrous, greenish, 8 - 22-locular, locules with many biseriate ovules; style very short, glabrous; stigma cylindric or bluntly conical, longitudinally grooved, light greenish, often silky. Fruit 5 - 10 cm (or more) in diam., 8 - 20 -locular; greenish-yellow or greenish, globose to pyriform; rind 3 mm thick; pericarp hard, woody, grey or yellowish, depressed above, many seeded, aromatic. Seeds flat, oblong, densely woolly, embedded in transparent sticky gum; testa white, woolly-pubescent.

Field note: Common in deciduous and scrub forests; also planted in gardens.

Fl.: March-April. *Fr.* : September - December.

Specimen examined: Ladpura 1734, Pansal 1811 (MLVGCB Herb).

Distribution: India, Pakistan, Burma and China.

Uses: The *'bengal quince'* is frequently cultivated in the plains for its fruits. The pulp of fresh fruits can be used for making jams or drinks (sherbet). Dried pulp is used medicinally for the treatment of diarrhoea and dysentery. Also used in paints. Leaves are offered to *"God Shiva"*

CITRUS L.

KEY TO THE SPECIES

1a. Petiole narrowly to broadly winged...................................... *C. aurantifolia*

1b. Petiole narrowly marginate............... *C. limon*

Citrus aurantifolia (Christm.) Swingle in J. Wash. Acad. Sci. 3: 465. 1913 & in Citrus Indust. 1: 374. 1967; Parmar, in Shetty & Singh Fl. Rajasthan 1: 175. 1987; Nair & Nair, in Hajra *et al.* Fl. India 4: 278 - 279. 1997. *Limmonia aurantifolia* Chrim. & Panz., Pflanzenr. Syst. 1: 618. 1777. *Citrus acida* Roxb., Fl. Ind. 3: 390. 1832. *C. medica* L. var. *acida* (Roxb.) Hook. *f.,* Fl. India 1: 515. 1875.

Vern.: *Kaghzi nimboo.*

Shrub or small spiny bush or tree, 2-3 m tall, spines short and sharp. Leaflets small; petioles narrowly winged; wings spathulate to obcordate, upto ca 15 x 13 mm, ariculate above with the base of blade; leaflets blade ovate elliptic to oblong, rounded at base, obtusely acute at apex, crenate along margins, 5 - 7.5 x 3.5 - 5 cm. Inflorescence lax, axillary racemes, 2 - 7-flowered. Flowers 3 - 10, 20 - 25 mm across, perfect and staminate, white. Calyx cupular, 4- or 5-lobed, greenish-white; lobes deltoid, acute, glandular, fleshy. Petals 4 or 5, oblong, acute, glandular, white. Stamens 20 - 28; filaments polyadelphous, white; anthers oblong, yellowish. Ovary depressed -

globose, greenish; style cylindric, white; stigma capitate. Fruit ovoid or rounded, 25 - 50 mm in diam., smooth, pitted, glandular; endocarp segments 9 - 12, strongly adherent, centre solid, pulp-vesicles slender, greenish-yellow when ripe, pulp greenish; juice abundent, very acidic. Seeds ovoid.

Field note: Cultivated for fruits throughout the area.

Fl.: Winter months. *Fr.*: April-September.

Specimen examined: Pansal 1751 (MLVGCB Herb).

Distribution: Probably native of Malesia. Cultivated in many parts of sub-tropical countries.

Uses: Fruits are used for refreshing drinks, pickles, etc.

Citrus limon (L.) Burm. f., Fl. Ind. 173. 1768; Maheshwari, Fl. Delhi, 95.1963; Hook. *f.*, Fl. Brit. Ind. 1: 516. 1875; Parmar, in Shetty & Singh Fl. Rajasthan 1: 175. 1987; Nair & Nair, in Hajra *et al.* Fl. India 4 : 282. 1997. *C. limonum* Risso, Ann. Mus. Natl. Hist. Nat. 20: 201. 1813. *C. aurantium* L. var. *limonum* Wight & Arn., Prodr. 98. 1834. *C. medica* L. var. *limonum* (Wight & Arn.) Hook. *f.*, Fl. Brit. India 1: 515. 1875.

Vern.: *Jatti khatti, Moto limbu.*

Spinous shrub or small tree, 3-6 m tall. Leaves unifoliate, petiole marginate or narrowly winged, often articulate at apex with base of blade; leaflet blades ovate or elliptic-oblong, obtuse to rounded at base, subacute or obtuse at apex, glandular serrulate along margins, 5.5 - 11.5 x 2-5 cm. Flowers shortly pedicellate, 5-merous, bisexual or male. Calyx urceolate; sepals suborbicular, minute, glabrous. Petals ovate, oblong, 15 - 20 x 3 - 5 mm, white, tinged purple. Stamens 20 - 40; filaments monoadelphous or irregularly polyadelphous at base, free above; anthers oblong, apiculate. Ovary subcylindric, 5 - 7 x 3 -4 mm; style thick, caducous; stigma globose. Fruit ovoid-oblong, upto 6 -8 cm across, yellow when ripe; pulp abundant and strongly acidic. Seeds ovoid, 5 - 10 x 5 - 6 mm.

Field note: Cultivated for fruits.

Fl. & Fr. : August - March

Specimen examined: Mandalgarh 1580, Bhilwara 1830 (MLVGCB Herb).

Distribution: Native of S. E. Asia: cultivated throughout tropical and subtropical countries.

Uses: Used as a rootstock for oranges and grapefruits. Also commonly used for pickles.

LIMONIA L.

Limonia acidissima L., Sp. Pl. ed. 2: 554. 1762; Airy Shaw in Misc. Inform. 1939: 293. 1939; Stone & Nicolson in Taxon 27: 551. 1978; Nair & Nair, in Hajra *et al.* Fl. India 4: 294. 1997. *Schinus limonia* L., Sp. Pl. 389. 1753; *Feronia elephantum* Correa in Trans. Linn. Soc. 5: 225. 1800; Wight, Ic. 1: t. 15. 1838; Hook. *f.*, Fl. Brit. Ind. 1: 516. 1875; Duthie, Fl. Gangetic Plain 1: 143. 1903. *Hesperuthusa acidissima* (L.) M. Roemer, Syn. Mon. Hesper. 1: 38. 1846. *Feronia limonia* (L.) Swingle in J. Wash. Acad. Sci. 4: 328. 1914 & Citrus Ind. 1: 416. 1967; Bhandari, Fl. Ind. Des. 92. 1978; Parmar, in Shetty & Singh Fl. Rajasthan 1: 175. 1987. *Limonia elephantum* (Correa) Panigrahi in Taxon 26 : 567. 1977.

Vern.: *Kaith, Kaitha.*

Tree, upto 9 m tall, semideciduous. Branches sligltly zig-zag when young, arms with axillary sharp spines, ca 3.5 cm. Leaves alternate, imperi-pinnate, upto 11 cm long; leaflets 5-7, opposite, ca 3 x 2 cm, obovate, subsessile, entire to obscurely crenulate, retuse, glandular-punctate. Petiole and rachis narrowly winged. Inflorescence axillary and terminal, many flowered panicles, up to 5 cm long. Flowers dull red, small, reddish in bud, turning to pale yellowish, bisexual, fragrant; pedicels 5 mm long, glabrescent. Calyx 5-toothed; teeth lobes small, deltoid, ca 0.5 mm long, puberulent, ciliate near tip. Petals 5, ovate-oblong, acute. Stamens 10 or 12; filaments subulate, hairy at the base, glabrous above; anthers linear-oblong, acute, ca 4 mm long. Disc minute. Ovary globose, incompletely 4 - 6-locular with many ovules in several series; style very short, thick, fleshy; stigma oblong-fusiform. Berry globose, 5 - 9 cm in diam., wooly, 1- locular, with sweet-sour mucilaginous pulp; Seeds numerous, oblong, brown, woolly.

Field note: Common in waste lands, also planted in gardens.

Fl. & Fr.: March - December.

Specimen examined: Mainal 1226 (MLVGCB Herb).

Distribution: India, Pakistan, Java and Sri Lanka.

Uses: The pulp of the fruit is used in the treatment of diarrhoea and in preparation of chutney.

MURRAYA J. G. Koenig ex L., nom. cons.

KEY TO THE SPECIES

1a. Leaves with 15 - 27 leaflets, aromatic; inflorescence with many flowers; berries subglobose purplish to black when ripe.. ......*M. koenigii*

1b. Leaves with 7 leaflets, non-aromatic; inflorescence with 3 - 6-flowers; berries ovoid-ellipsoid, reddish when ripe. ......*M. paniculata*

Murraya koenigii (L.) Spreng., Syst. Veg. 2: 315; Hook. *f.,* Fl. Brit. Ind. 1: 503. 1875; Duthie, Fl. Gangetic Plain 1: 137. 1903; Parmar, in Shetty & Singh Fl. Rajasthan 1: 175. 1987; Nair & Nair, in Hajra *et al.* Fl. India 4: 351. 1997. *Bergera koenigii* L. Mant. Pl. Alt. 563, 1771. *Chalcas koenigii* (L.) Kurz, in J. Asiat. Soc. Bengal, Pt. 2, Nat. Hist. 44: 132. 1875.

Vern.: *Meetha neem, Curry patta.*

Small tree or large shrub, up to 6 m tall, branches pubescent. Bark green to greysh. Leaves imparipinnate, up to 50 cm long; petiole and rachis densely white-puberulent; leaflets alternate, 9 - 27, 2 - 8 x 1 - 3 cm, ovate to ovate-lanceolate or orbicular, oblique, acute to obtuse, apex retuse, margin irregularly crenulate, dark green above, paler beneath, foetid-scented. Flowers white, in terminal, branched cymes, many, scented; pedicels slender, short (ca 4mm). Calyx saucer shaped; sepals 5, fused at base, deltate, puberulent abaxially, glandular. Petals 5, 6 - 8 x 1 - 1.5 mm, linear, obtuse, glandular. Stamens 10, filaments subulate, 5 - 7 mm long, glabrous; anthers dorsifixed, ellipsoid, short, pale greenish. Disc annular, slightly conical. Ovary oblong-ovoid, 2 locular, ovules 1 (-2) in each locule; style short; stigma captitate, glandular. Berry ovoid, 9 x 10 mm, purplish to black when ripe. Seeds 1 or 2, ovoid-oblong.

Field note: Planted near houses or in gardens.

Fl. : April-June. *Fr.*: June-August

Specimen examined: Mandalgarh 1815, Bhilwara 1253 (MLVGCB Herb).

Distribution: India, Pakistan, China and Sri Lanka.

Uses: Leaves are used as a condiment to flavour curries.

Murraya paniculata (L.) Jack., in Malay Misc. 1: 31. 1820; Bailey, Man. Cult. Pl. 609, 1949; Parmar, in Shetty & Singh Fl. Rajasthan 1: 176. 1987; Nair & Nair, in Hajra *et al.* Fl. India 4: 352. f. 92. 1997. *Chalcas paniculata* L. Mant. 1: 68, 1767. *Murraya exotica* L. Mant. Pl. 2: 543, 1771; Hook. *f.,* Fl. Brit. Ind. 1: 503. 1875; Cooke, Fl. Bomb. Pres. 1: 193. 1958; Duthie, Fl. Gangetic Plain 1: 137. 1903. (Plate 60).

Vern.: *Kamni.*

Large shrub or small tree up to 10 m high. Branches sparsely pubescent, with an ash coloured bark. Leaves imparipinnate, upto 20 cm long; leaflets glabrous, 3 - 5, sometimes 7, 3.5 - 8 x 2 - 3.5 mm, elliptic, oblique; apex drawn out, obtuse, sometimes notched, undulate. Flowers fragrant, white, in few-flowered panicles. Pedicels slender, upto 10 mm long, glandular, glabrous. Calyx lobe 5, small, 0.5 - 1 mm long, deltate, acute, glandular, glabrous. Petals 5, 12 - 15 x 3 - 4 mm, oblong-lanceolate, sparsely pubescent, glandular. Stamens 10; filaments subulate above, dialated below, 5 - 10 mm, glabrous; anthers ellipsoid-oblong, ca 1 mm long,yellowish. Disk annular, lobulate, ca 1 mm high, ca 1.5 mm broad, glabrous. Ovary 2 - 3-locular, each locule with one ovule, elongate; style cylindric, 4 - 8 mm long, glabrous; stigma capitate, 2 - 3- lobed. Berry ovoid-ellipsoid, 7 - 13 x 6 - 8 mm, beaked at the top, deep orange to red when ripe. Seeds 1 or 2, ellipsoid.

Field note: Cultivated for its fragrant flowers and red berries.

Fl. & Fr.: March-September

Specimen examined: Bhilwara 1552 (MLVGCB Herb).

Distribution: Sub-himalayan tracts of India-Pakistan, China and the Pacific Islands.

SIMAROUBACEAE

AILANTHUS Desf., nom. cons.

Ailanthus excelsa Roxb., Pl. Cor. 1: 24, t. 23. 1795; Bennett in Hook. *f.,* Fl. Brit. Ind. 1: 518. 1875; Duthie, Fl. Gangetic Plain 1: 144. 1903; Bhandari, Fl. Indian Desert. 88. 1978; Parmar, in Shetty & Singh Fl. Rajasthan 1: 176. 1987; Basak in Hajra *et al*. Fl. India 4: 410-411. 1997. *Pongelion wightii* Tieghem in Ann. Sci. Nat. Bot. 1: 277. 1906. (Plate 61)

Vern.: *Ardu*.

A large polygamous deciduous lofty tree, 10 - 20 m high . Bark smooth, ash coloured or greenish white. Leaves 20 - 40 cm long, abruptly pinnate; leaflets 4 - 16 x 3 - 8 cm, tomentose when young, irregularly glandular-toothed, 6 - 14 pairs, alternate or subopposite, variable in shape, unequal at base, glabrous; petiolules 2 - 3 cm long. Inflorescence a large lax panicle; bract minute, covered with soft hairs. Flowers small, many, slightly yellow tinged. Bisexual flower: Calyx 5 lobed, ovate, triangular, ca 2 mm long. Petals 5 glabrous, ovate-lanceolate, much larger than the calyx lobes. Stamens 10, inserted below the disc; filaments glabrous, shorter than the petals; anthers ca 1 mm long; stigma short. Male flower similar to bisexual flower but without rudimentary ovary. Ovary sparsely hairy; styles free or connate; stigma curling. Samara 3 - 7.5 x 1 - 1.5 cm, linear, twisted at the top, red coloured, acute. Seed one in the centre of samara.

Field note: Common in waste lands, also planted in gardens as wll as road sides.

Fl. & Fr.: February-March.

Specimen examined: Bhilwara 102, Gurla 1767 (MLVGCB Herb).

Distribution: Australia, China, India, Pakistan.

Uses: Bark contains quassionoids (Such as ailanthone & quassin) alkaloids, flavons & tannins. The anticancer properties of quassionoids widely extensively known. 1 tsp bark powder is taken daily in abdominal pain. Wood is used as firewood and in making toys. Green leaves are lopped for fodder.

BALANITACEAE

BALANITES Delile nom. cons.

Balanites aegyptiaca (L.) Del. Fl. Aegypt. 77, 221. t. 28. f. 1: 1813; Wight, Ic. 1: t. 274. 1840; Bhandari, Fl. Indian Desert. 88. 1978; Parmar, in Shetty & Singh Fl. Rajasthan 1: 176 - 177. 1987. *Ximenia aegyptica* L. Sp. Pl. 1194. 1753. *Balanites roxburghii* Planch., in Ann. Sc. Nat. ser. 4. 2: 258. 1854; Bennett, in Hook. *f.,* Fl. Brit. Ind. 1: 522. 1875; Singh & Singh, in Hajra *et al*. Fl. India 4: 40. 1997. *B. aegyptica* Delile var. *roxburghii* (Planch.) Duthie, Fl. Gangetic Plain 1: 145. 1903. (Plate 62)

Vern.: *Hingora, Hingoriya, Hingota.*

A moderate sized spiny tree, deciduous, 3 - 5 m high; bark pale brown or whitish. Stem pubescent when young, branches ending in strong sharp thorns. Spines 2.5 - 6 cm long, axillary, straight, very strong, often bearing leaves and flower. Leaves 2-foliolate; leaflets 1 - 6 x 1 - 3 cm, elliptic-oblong, obovate-oblong or oblanceolate, rounded or cuneate at base, obtuse or acute or rarely slightly mucronate; petioles 3 - 5 mm long. Flowers axillary, fascicled cymes, 4 - 12 mm in diam., pale or greenish yellow, fragrant; pedicels 5 - 15 mm long, pubescent. Sepals 5, 3 mm long, pubescent out side, silky within. Petals 5, elliptic-oblong, glabrous outside, silky-villous inside. Stamens 10, 2 - 3 mm long; filaments 1.5 - 2 mm long, glabrous. Disc cupular, 10-lobed, glandular margins. Ovary globose, ca 2 mm across, silky-villous; style short, conical. Fruits ovoid or subspherical, 5-grooved, 3 - 6 x 2 - 4 cm, yellowish-green when ripe, stony. Seeds solitary, stony with fibrous testa.

Field note: Common in scrub forests, waste lands, along roadsides. It propogates by root and is found in gregarious.

Fl.: December - March; *Fr.*: March - July.

Specimen examined: Atun 938, Nimbahera 101 (MLVGCB Herb).

Distribution: India and Tropical Africa.

Uses: About 150 gm of dried fruits are grinded with water to make a paste which is applied locally twice a day for 4 - 5 days to cure skin eruptions. Fruits are used in washing clothes. Wood is used in making household, agricultural equipments as well as firewood. Leaves are used as fodder.

BURSERACEAE

KEY TO THE GENERA

1a. Medium- sized tree. Spines absent. Leaves imparipinnate. Fruits capsular dehiscent .... *Boswellia*

1b. Shrub. Spines present. Leaves simple. Fruits drupaceous indehiscent .. *Commiphora*

BOSWELLIA Roxb. ex Colebr.

Boswellia serrata Roxb. ex Colebr. in Asiat. Res. 9: 379. t. 5. 1807; Bennet, in Hook. *f.*, Fl. Brit. India 1: 528. 1875; Duthie, Fl. Gangetic Plain 1: 147. 1903; Bhandari, Fl. Indian Desert. 89. 1978; Parmar, in Shetty & Singh Fl. Rajasthan 1: 177. 1987; Chithra & Henry, in Hajra *et al.* Fl. India 4: 432 - 433. 1997. *B. serrata* Roxb. ex Colebr. var *glabra* (Roxb.) A. W. Bennet in Hook. *f.,* Fl. Brit. India 1: 528. 1875. *Chloroxylon dupada* Ham. Journ. Mysore, 1 : 184. 1807 nom. subnud. *Boswellia glabra* Roxb. Pl. Cor. t. 207. 1811. (Plate 63)

Vern.: *Salar.*

Deciduous tree, with grayish-white bark peeling off in thin flakes, 10-15 m high. Branches with annular swelling at intervals, pubescent. Leaves crowded at the end of branches, imparipinnate, leaflets 8-25, variable in size, shape and degree of pubescent, ovate-lanceolate, oblong-ovate or elliptic, crenate-serrate, oblique, cuneate, rounded at base, acute, obtuse or retuse, mucronate at apex, dark green, pubescent. Flowers in 8-20 cm long, pink or white with pinkish tinge, 5- merous, in axillary, pubescent racemes born at the end of branches, fragrant. Calyx tube broadly campanulate, 1.5 - 3 x 2.5 - 3 mm; calyx-lobes triangular-ovate, pubescent, persistent. Petals ovate-oblong, 5 - 7 x 2.5 - 4 mm, shortly clawed, obtuse, inflexed at apex, pubescent outsid except margins. Stamens 2.5 - 5 mm long, glabrous, pubescent; anther lobes oblong. Pistil ca 4 mm long; stigma capitate. Fruits 3 - gonous, 1.2 - 2.4 x 1 - 1.4 cm, reticulately veined, glabrous, retuse at one end, dehiscent, 3 -valved. Seed 3, compressed, margins winged.

Field note: Common in forests, particularly on hills.

Fl.: January - May; September - December. *Fr.*: April - August; November - March.

Specimen examined: Gurla 601, Jogniya Mata 2206 (MLVGCB Herb).

Distribution: India (Endemic)

Uses: Bark is sweet, cooling and tonic. It is good in vitiated conditions of Pitta, cough and asthama. It is useful in fever, urethrorrhea, diaphoresis, convulsions, chronic laryngitis, jaundice and arthritis. Gum is sold in market under the name *"Lobhan"*. which is put on fire for fragrance during acts of devotion. The fragrant gum-resin is used for lighting by tribals. Leaves are used as fodder and wood is used in paper manufecturing.

COMMIPHORA Jacq.

Commiphora wightii (Arn.) Bhandari, in Bull. Bot. Surv. Ind. 6: 327. 1964 & Bhandari, Fl. Indian Desert. 95. f. 34. 1978; Parmar, in Shetty & Singh Fl. Rajasthan 1: 178. 1987; Chitra & Henry, in Hajra *et al.* Fl. India 4: 445. 1997. *Balsamodendron wightii* Arn. in Ann. Nat. Hist. 3: 86. 1839. *Commiphora mukul* (Hook. ex Stocks) Engl. in DC. Monogr. Phan. 4: 12. 1883. *Balsamodendron mukul* Hook. ex Stocks in Kew J. Bot. 1: 259. t. 8. 1849; Bennet in Hook. *f.,* Fl. Brit. India 1 : 529. 1875. (Plate 64 & 65)

Vern.: *Guggul, Gugal.*

Shrub up to 4 m tall; branches aromatic, thorny and knotty with papery bark. Leaves sessile, alternate or fascicled, 1-3-foliolate; leaflets glabrous, the terminal sessile or subsessile, obovate, serrate (sometimes serrate only towards the apex), 1-5 x 0.5-2.5 cm; lateral when present sessile, serrate or entire, less than half the size of the terminal leaflet. Plants dimorphic, one having bisexual and male flowers and the other having female flowers with staminodes. Bisexual and male flowers solitary or in 2 - 3 flowered fascicles, red or pinkish white. Bracts 2, opposite, glandular hairy. Calyx fused basally with the disc tubular or urceolate, 1 - 2 mm long; lobes usually triangular, valvate, glandular hairy outside. Petals 4-5, broadly linear, reflexed, acute, 3 - 5 x ca 1 mm, reflexed at apex. Stamens 8, very rarely 10, 3 - 5 mm long, free, alternately short and long, included, sometimes equalling the petals; shorter stamens inserted alternately in deeper. Disc 8 - 10 lobed, conspicuous. Ovary 2-loculed, oblong, ovoid, attenuate into style; stigma obscurely 2-lobed. Fruit up to 1 cm long, red when ripe, marked with 2 white longitudinal lines (or grooves), mucronate; mesocarp yellow, rarely orange, 4 lined and fused at the base; epicarp dehiscing from the base upwards on maturation. In female flowers, sepals 2 mm long, petals 3 - 4 x 1 mm; staminodes 8, alternately short and long, 1 - 1.5 mm long; ovary and fruit the same as in bisexual flowers.

Field note: Very rare and threatened, found in gravelly and rocky habitats growing in association with *Euphorbia caducifolia.*

Fl. & Fr.: December - July.

Specimen examined: Mangaliyawas 362, Ajmer 1606 (MLVGCB Herb).

Distribution: India, Arabia, Baluchistan and Pakistan.

Uses: Decoction of plant or gum resin in warm water is used for gargling against pyorrhoea and spongy gums. Gugal is also used in *Dhoop* and to cure bronchitis, nasal catarrh and laryngitis.

Notes: As stated earlier 3 types of flowers are met within this taxon. Of these, the functionally female flowers are borne on a separate plant, and have staminodes that apparently look like well developed stamens but they contain only sterile pollen grains. This was determined by acetocarmine and cotton/blue-lactophenol tests. The bagged flowers did not produce any fruits. This excludes the possibility of apomixis. As normal fruits are met with in these plants, it is inferred that pollen grains from bisexual or male flowers, present on separate plants are needed for the production of normal fruits.

It is the source of Guggul or Indian Bdellium, a gum-resin that exudes from the branches. It is largely used as an incense, in medicine, perfumery, and as a substitute for African Bdellium. It is also used to adulterate myrrh.

MELIACEAE

KEY TO THE GENERA

1a. Leaves entire. Fruit a capsule. Seeds winged. Sarcotesta corky or woody................. *Soymida*

1b. Leaves serrate or crenate. Fruit a drupe. Seeds not winged. Sarcotesta fleshy...............2

2a. Leaves unipinnate. Stigma trifid. Drupe oblong, 1-seeded, endocarp thin... *Azadirachta*

2b. Leaves imparipinnate, Stigma capitate. Drupe globose, 3 - 6-seeded, endocarp stony.. *Melia*

AZADIRACHTA A. Juss.

Azadirachta indica A. Juss., in Mem. Mus. Hist. Nat. Par. 19: 221. t. 13. f. 5. 1830; Bhandari, Fl. Indian Desert. 90. 1978; Parmar, in Shetty & Singh Fl. Rajasthan 1: 179. 1987; Jain & Bennet, in Hajra *et al.* Fl. India 4: 478. 1997. *Melia azadirachta* L. Sp. Pl. 385. 1753; Hiern, in Hook. *f.,* Fl. Brit. India 1: 544. 1875; Duthie, Fl. Gangetic Plain 1: 150. 1903. *M. indica* Brandis, Forest Fl. N. W. India 67. 1874.

Vern.: *Neem, Limbro.*

Tall, deciduous tree, up to 15 m tall; Bark greyish-black, slightly longitudinally furrowed. Branches glabrous. Leaves imparipinnate, 10 - 45 cm long, pulvinus at the base; leaflets alternate-opposite, 2.5-7 x 1-2 cm, ovate, subsessile, serrate, glabrous, acuminate. Flowers white, fragrant, in 10 - 22 cm long, axillary panicles, sweet-scented. Bracts caducous. Sepals obovate, 1.5 - 4 mm long; lobes ovate-suborbicular, puberulous outside, glabrous inside, imbricate. Petals 5 - 6 mm long, obvoate to oblong, puberulous outside, white, margin ciliate. Staminal tube ca 4 - 5 mm long, dilated at apex, puberulous, 10-striate, 10-toothed; teeth 2-lobed; anthers oblong, basifixed. Ovary globose, 3-celled, glabrous; style linear, ca 2.5 mm long; stigma trifid, with a light coloured rim at base. Drupe oblong, smooth, 1.3 - 2 cm long, greenish-yellow, 1-seeded. Seeds ellipsoid, glabrous, whitish-brown, hard.

Field note: Common in waste lands, usually planted along roads and near habitations.

Fl. & Fr.: Feberuary - May.

Specimen examined: Bhilwara 289 (MLVGCB Herb).

Distribution: A native of India and China, cultivated and naturalized throughout India, Malaysia and Pakistan.

Uses: It yields good timber. All parts of the plant are of medicinal importance. Gum is extracted from the bark. Leaves are used as a poultice for boils; dried leaves serve in place of mothballs. Fruit are edible and oil extracted from the seeds is used as a purgative and anthelmintic.

MELIA L.

Melia azedarach L., Sp. Pl. 1: 384. 1753; Wight, Icon. Pl. Ind. Orient. t. 160. 1839; Hiern in Hook. *f.*, Fl. Brit. India 1: 544. 1875; Duthie, Fl. Gangetic Plain 1: 151. 1903; Parmar, in Shetty & Singh Fl. Rajasthan 1: 179 - 180. 1987; Jain & Bennet, in Hajra *et al.* Fl. India 4: 494. 1997 (Plate 66).

Vern.: *Bakain.*

Tree, up to 12 m tall; young shoots tomentose. Leaves 2-pinnate, up to 90 cm long; leaflets opposite, elliptic, 2.5 - 5 cm x 5 - 19 mm, serrate to sub-serrate, acuminate, often oblique, subsessile. Flowers lilac, sweet-scented, in axillary panicles; pedicel 2-7 mm long, puberu-lous. Calyx 5 lobed; lobes ca 2 mm long, acute, pubescent out side. Petals 5, 7 - 9 mm long, spathulate to lanceolate, ciliate, imbricate in bud. Staminal tube 6-24 mm long, cylindrical, expanded at the base and apex, 10-striate, with 20 teeth at the apex; anthers sessile, 1 between each pair of teeth. Disc glabrous, fused with the ovary base. Ovary usually 5-locular, glabrous; style 4 - 5 mm long, clavate at apex; stigma 10-lobed, capitate. Drupe 1.5 - 2 cm long, ellipsoid or globose, 3 - 6 seeded, fleshy and yellow when ripe.

Field note: It is a fast growing tree of the plains and foot-hills, cultivated along road-sides and in villages.

Fl. & Fr.: March - June.

Specimen examined: Pur 103 (MLVGCB Herb).

Distribution: Wild in W. Himalaya, up to 1700 m. Cultivated and naturalized in parts of Iran, China, Burma, Turkey, India & W. Pakistan.

Uses: Fruits are eaten by goats and sheep and the stony endocarps are used as beads. Decoction of 1/2 kg fresh leaves is given to buffalo once a day for dewormification.

SOYMIDA A Juss.

Soymida febrifuga (Roxb.) A. Juss. Mem. Mus. Hist. Nat. Paris 19: 251. t. 22. f. 26. 1830; Beddome, Fl. Sylv. S. India t. 8. 1869; Hiern in Hook. *f.*, Fl. Brit. India 1: 567. 1875; Duthie, Fl. Gangetic Plain 1: 153. 1903; Parmar, in Shetty & Singh Fl. Rajasthan 1: 180. 1987; Jain & Bennet in Hajra *et al.* Fl. India 4: 501. 1997. *Swietenia febrifuga* Roxb. Pl. Cor. t. 17. 1795.

Vern.: *Rohan.*

Tree up to 22 m tall; bark thick bluish grey or brown. Branches with leaf scars. Leaves paripinnate, 25 - 50 cm long; rachis 6 - 12 cm long; leaflets 3 - 6 pairs, 3 - 20 x 1 - 12 cm, subopposite or alternate, ovate-oblong, rounded, inequilateral at base, entire or coarsely crenate-serrate along margins, obtuse at apex, coriaceous; petiolules up to 1 cm long. Flowers in long terminal and axillary panicles, greenish white, ca 1 cm across.

Calyx 5-lobed; lobes ovate, pubescent, imbricate. Petals 5, obovate, notched at apex, ca 5 mm long, pubescent outside. Staminal tube cup shaped, urceolate, less than half as long as petals; anthers 10. Ovary glabrous.. Capsule 5-locular, smooth, ovoid, reddish purple, woody. Seeds many, ca 5 cm long, winged at one end.

Field note: Occasionally found in mixed forests on the hill slopes of the area.

Fl. : Feberuary - April; *Fr.*: March - July.

Specimen examined: Gurla 104 (MLVGCB Herb).

Distribution: N. & S. Tropical America and W. Indies.

CELASTRACEAE

KEY TO THE GENERA

1a. Unarmed climbing shrub. Stamens arising form the edge of the disc. Ovary seated on the disc. Seeds completely enveloped by arils .... *Celastrus*

1b. Armed shrub or a small tree. Stamens arising below the disc. Ovary confluent with the disc. Seeds partly enveloped by arils........ *Maytenus*

CELASTRUS L.

Celastrus paniculata Willd., Sp. Pl. 1: 1125. 1797; Roxb., Fl. Ind. ed. Wall. 1: 621. 1824; Lawson, in Hook. *f.,* Fl. Brit. India 1: 617. 1875; Duthie, Fl. Gangetic Plain 1: 158. 1903; Parmar, in Shetty & Singh Fl. Rajasthan 1: 182. 1987; Ramamurthy, in Singh *et al.* Fl. India 5: 87. f. 18. 2000. (Fig. 32 & Plate 67)

Vern.: *Mal kangni.*

A large woody climber, bark corky; young shoots marked with lenticels. Leaves very variable, elliptic, broadly ovate, suborbicular, obovate or oblong-ovate, glabrous, 5 - 15 x 2 - 10 cm, base cuneate, obtuse or rounded; apex acute, acuminate or obtuse, rarely emarginate. Panicles terminal, large, 4-8 cm long, drooping, pubescent, branching into compound cymes. Male flowers: minute, pale-green; calyx lobes suborbicular, toothed, 1 x 2 mm; petals oblong or obovate-oblong, obtuse, entire, 2.5 - 3 x 1 - 1.5 mm; disc cupular, sometimes flat, lobes obscure or slightly triangular; stamens ca 3 mm long, filaments subulate; anthers ovoid, obtuse. Female flowers: sepals, petals and the disc similar to those of the male flowers, stamens sterile; carpel 2 - 2.5 mm long; ovary globose; style columnar; stigma 3-lobed. Capsules up to 10 - 8 mm, subglobose, trivalved, 3 - 6-seeded. Seeds ellipsoid, 3 - 5 x 2 - 4 mm, smooth, yellowish-brown, each enclosed in a red fleshy aril.

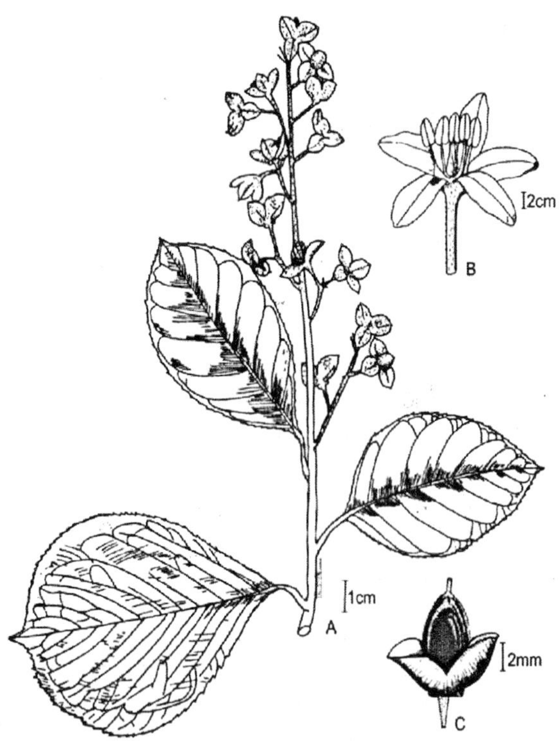

Fig. 32. *Celastrus paniculata* Willd. A. Fruiting branch (dehisced fruit), B. Flower C. Fruit

Field note: Occasionally found in mixed forests of the area.

Fl. & Fr.: April-June.

Specimen examined: Bari Sadri 365, Kanera 907 (MLVGCB Herb).

Distribution: India, S. China, S.E. Asia to Austraila.

Uses: Leaves and the seeds are used in medicine. Seed oil is used as a hair tonic.

MAYTENUS Molina

Maytenus emarginata (Willd.) Ding Hou in Van Steenis, Fl. Males. I. 6 (2): 241. 1962; Bhandari, Fl. Indian Desert. 91. 1978; Parmar, in Shetty & Singh Fl. Rajasthan 1: 182 - 183. 1987; Ramamurthy in Singh *et al.* Fl. India 5: 120. 2000. *Celastrus emarginatus* Willd. Sp. Pl. 1. 2: 1128. 1798. *C. montana* Roth, in Roem. & Schult. Syst. Veg. 5: 427. 1819; Wight, Ic. 2 (1): 7. t. 382. 1840. *Gymnosporia montana* (Roth) Benth. Pl. Austr. 1: 400. 1863, pro comb.; Lawson in Hook. *f.,* Fl. Brit. India 1: 621. 1875. *G. emarginata* (Willd.) Thwaites, Enum. Pl. Zeyl. 409. 1864; M. Lawson, in Fl. Brit. India 1: 621. 1875. *Celastrus senegalensis* Duthie, Fl. Gangetic Plain 1: 159. 1903. *Maytenus senegalensis auct.* non Exell 1952; Mahesh., in Bull. Bot. Surv. Ind. 3(1): 93. 1961. (Fig. 33 & Plate 68)

Vern.: *Kankedo*.

A tall shrub or a small compact tree, 3-5 m high; young branches often spiny, purple, bearing leaves and flowers on the spines. Leaves coriaceous, glabrous, very variable, obovate, spathulate, obtuse, entire or crenulate, decurrent at the base, 3.5 - 14.5 x 2 - 9 cm; petioles ca 2 - 12 mm long, nerves 5 - 6 pairs. Flowers small in axillary, dichotomous or fasciculate cymes on short branchlets, often forming terminal, elongate panicles. Pedicels filiform; bracts small, deltoid, fimbricate. Calyx lobes broadly elliptic-oblong, ciliate. Petals oblong or ovate-oblong, 2 - 3 x 1 mm. Stamens 2 - 3 mm long, inserted beneath the margin of the disc. Ovary glabrous, partially immersed, 3-loculed, orbicular; style deeply 2 - 3-cleft. Disc fleshy, many lobed. Capsules globose, 10 - 12 x 8 - 9 mm, bivalved, reddish-purple. Seed ellipsoid, 3 x 2 mm, chestnut brown, glabrous, rugose, arillode orange.

Field note: Common in open fields. Leaves are very variable in size.

Fl. & Fr.: October - March.

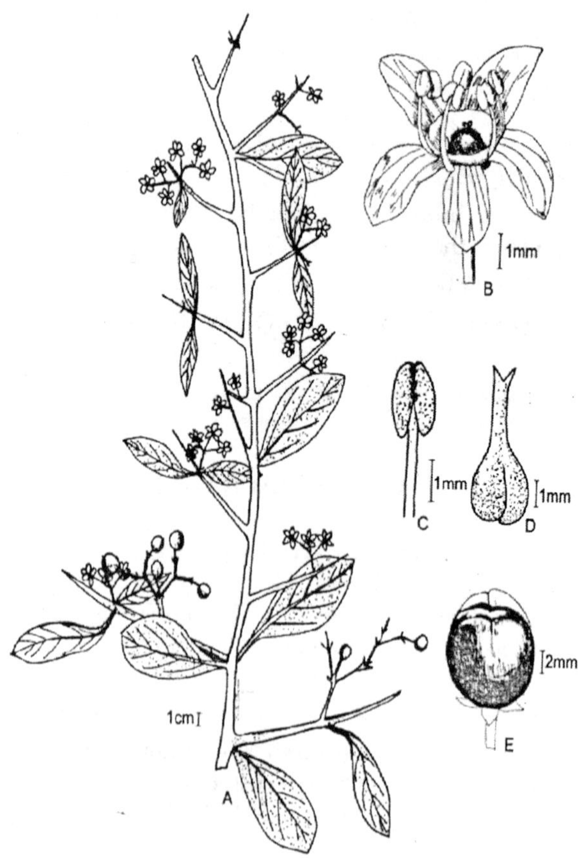

Fig. 33. *Maytenus emarginata* (Willd.) Ding Hou. A. Flowering and fruiting twig, B. Flower, C. Stamen, D. Ovary, E. Fruit

Specimen examined: Hamirgarh 941, Adharshila 1692 (MLVGCB Herb).

Distribution: Tropical and subtropical Africa and Asia.

Uses: The root bark is used locally to treat dysentery. An oil of some medicinal value is extracted from the seeds and known as Oleum Nigrum or Black oil. An ointment is prepared by mixing of ash of burnt leaves and butter for healing of wounds. Fruits are sweet, digestible and used to treat ulcers, piles & corneal opacities.

RHAMNACEAE

ZIZIPHUS Mill.

KEY TO THE SPECIES

1a. Trees. Flowers in pedunculate cymes. Drupes more than 1.5 cm in diam............................. 2

1b. Diffuse, much branched shrubs. Flowers in sessile cymes. Drupes less than 1.2 cm in diam.. *Z. nummularia*

2a. Flowers in sessile cymes or cymose fascicles. Styles 2, connate or free above the middle. Fruits with fleshy edible pulp... *Z. mauritiana*

2b. Flowers in pedunculate cymes. Styles 3, distint. Fruits woody, hard, not edible, ... *.Z. xylopyrus*

Ziziphus mauritiana Lam., Encycl. 3: 319. 1789; Mahesh. Illu. Fl. Delhi f. 44. 1966; Bhandari, Fl. Indian Desert. 99. 1978; Parmar, in Shetty & Singh Fl. Rajasthan 1: 182-185. 1987; Bhandari & Bhansali, in Singh *et al*. Fl. India 5: 233. 2000. *Z. jujuba* (L.) Gaertner, Fruct. 1: 203. 1788 (non Mill. 1768); Lawson, in Hook. *f.*, Fl. Brit. India 1: 632. 1875; Duthie, Fl. Gangetic Plain 1: 163. 1903. *Rhamnus jujuba* L. Sp. Pl. 194. 1753. (Plate 69)

Vern.: *Ber, Bor*.

A medium size evergreen tree, upto 4 m high, bark dark grey or nearly black, with deep vertical cracks, reddish and fibrous inside; younger parts rusty tomentose; stipular prickles in pairs, 1.5 - 2 cm long, straight or one of them recurved; nodes slightly enlarged around the leaf-scars. Leaves 2 - 9.8 x 1 - 4.3 cm, alternate, elliptic oblong or sub-orbicular, rounded at both ends, mucronate at apex, serrulate or entire, pubescent above, tomentose to densely brown or silky-white beneath, 3-nerved at base; petiole 1 - 1.5 cm long; stipules spinescent. Flowers in axillary clusters or in short peduncled cymes; pedicels 2 - 5 mm in flowers, greenish yellow. Calyx-lobes glabrous within, tomentose outside, tube campanulate. Petals 5, 1 - 1.5 mm long, spathulate, deflexed within calyx-lobes. Stamens enclosed in the petals. Disc distinctly 10-grooved. Ovary half sunken in disc; Styles 2, connate for half of the length. Fruit ovoid, 1 - seeded, 1.5 - 3.5 cm long, often pointed at top, generally yellowish green, reddish in colour when ripe, fleshy and juicy. Seeds ca 1 cm across, compressed, light brown; stony 1 to 2 celled.

Field note: Very common in outskrits of agricultural fields, scrub forests, also planted for edible fruits.

Fl. & Fr.: September - March.

Specimen examined: Swarupganj 1422, Karoi 1836 (MLVGCB Herb).

Distribution: Tropical and subtropical Africa and Asia.

Uses: Fruits are edible. Wood is used for agricultural purposes. Branches are also used for fencing the fields.

Ziziphus nummularia (Burm. f.) Wight & Arn., Prodr. Fl. Penins. Ind. Or. 162. 1834; Lawson, in Hook. *f.*, Fl. Brit. Ind. 1: 633. 1875; Mahesh., Fl. Delhi 102. 1963; Mahesh., Illu. Fl. Delhi f. 45. 1966; Bhandari, Fl. Indian Desert. 99. 1978; Pandey, in Shetty & Singh, Fl. Rajasthan 1: 185. 1987; Bhandari & Bhansali, in Singh *et al*. Fl. India 5: 233. 2000. *Rhamnus nummularia* Burm. f., Fl. India 61. 1768. *Ziziphus rotundifolia* Lam. Encycl. 3: 319. 1789; Duthie, Fl. Gangetic Plain 1: 164. 1903; Duthie, Fl. Gangetic Plain 1: 164. 1903. (Fig. 34; Plate 70)

var.: ***nummularia***

Vern.: *Jari, Jhar-beri.*

A bushy much branched shrub, branches divaricating, flexuous, tomentose, young branches puberulous, grey, spines in unequal pairs, bigger straight ca 1 cm long, smaller recurved, whitish tomentose when young. Leaves 1 - 2 x 0.5 - 2 cm, orbicular or ovate-orbicular to elliptic, dark green and densely pubescent above, tomentose on both surface, entire or serrate, apex obtuse apiculate or mucronate, base round to sub cordate, lateral nerves prominent, pedicel 2 - 3 mm long. Flowers 3 - 4 mm in diam.; pedicels 2 - 5 mm long, tomentose. Calyx pubescent, lobes ca 1.5 - 2.2 mm long, ovate lanceolate, acute, keeled nearly to the base, campanulate, thickened at the apex. Petals ca 1-1.5 mm long, obovate-spathulate, margin convolute, rounded or truncate at the apex. Stamen included ca 0.8-1.2 mm long. Disc slightly 10-lobed, pitted. Ovary bicarpellary, bilocular; styles 2-cleft, united to above the middle. Drupe globose, 0.8 cm in diam., reddish brown-black when ripe, 2 celled, 2 seeded. Seeds compressed.

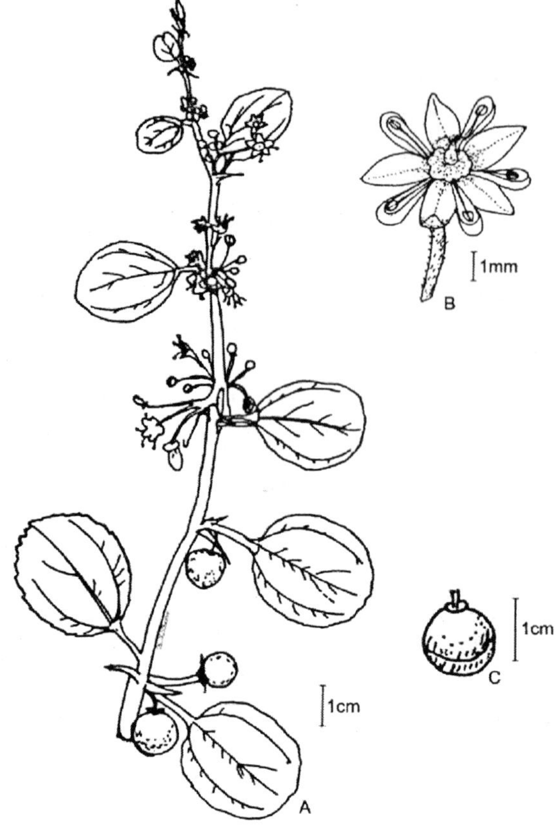

Fig. 34. *Ziziphus nummularia* (Burm.f.) Wight & Arn. A. Flowering and fruiting branch, B. Flower, C. Fruit.

Field note: Very common species in sandy to gravelly soils, in waste lands and outskrits of forests. Often forms pure formations.

Fl. & Fr.: March-June.

Specimen examined: Gopalpura 1829, Aroli 1803 (MLVGCB Herb).

Distribution: Palestine, Iraq, Iran, Afghanistan, Pakistan and India.

Uses: The branches are often used for fencing the fields, leaves are threshed out and used as fodder. Fruits edible. Leaves also applied in scabies and boils. Fruits boiled in water and water is used to take bath for curing sun strock.

Paste of leaves is used for healing cuts, boils and cutaneous diseases. In case of diarrhoea in goats, stem bark of *Bueta monosperma*, stem bark of *Z. nummularia*, and root bark of *Dichrostachys cinerea* are mixed and grind to make juice. 1/4th cup of this juice + 3/4th cup of water is given once a day to cure the diarrhoea.

Ziziphus xylopyrus (Retz.) Willd., Sp. Pl. 1: 1104. 1797, *'xylopyra'*; Lawson, in Hook. *f.*, Fl. Brit. Ind. 1: 634. 1875; Duthie, Fl. Gangetic Plain 1: 165. 1903; Pandey, in Shetty & Singh, Fl. Rajasthan 1: 187. 1987; Bhandari & Bhansali, in Singh *et al.* Fl. India 5: 243. 2000. *Rhamnus xylopyrus* Retz. Obs. Bot. 2: 11. 1781. (Plate 71)

Vern.: *Gut-bor.*

Large straggling shrub or small tree, 7 - 12 m tall; young shoots rusty tomentose; spines in pairs on younger branches, one straight, the other curved; nodes swollen at the leaf scars. Leaves broadly elliptic or orbicular, rarely ovate, 2-8 x 1-7 cm, obliquely subcordate, serrate, mucro at apex; petioles 2-7 mm long, fulvous tomentose. Inflorescence dense, axillary, dichotomous cymes; peduncles 2.5-4 mm long. Flowers 4-6 mm across, yellowish green; pedicels 3-4 mm long, tomentose. Calyx lobes 2-2.5 mm long, keeled up to the middle, thickend at apex, glabrous within. Petals 1.5-2 mm long. Stamens equal to petals. Ovary tricarpellary, 3-celled, rarely 2 or 4-celled, globose, nearly hidden by the disc; style cleft up to half to two third of the length. Drups globose. 1.8-2.5 cm in diam., woody, slightly rugose; stone thin and hard. Seeds 1 2 x 1-1.5 mm, compressed.

Field note: Common in deciduous forests.

Fl. & Fr.: April - December.

Specimen examined: Bari sadari 927; Manjawat 1928 (MLVGCB Herb).

Distribution: India, Nepal and Sri Lanka.

Uses: Plant yields hard and durable wood.

VITACEAE

KEY TO THE GENERA

1a. Leaves simple. Inflorescence tendril bearing. Flowers 5-merous .... *Ampelocissus*

1b. Leaves pinnately trifoliate. Inflorescence not tendril bearing. Flowers 4-merous ... *Cayratia*

AMPELOCISSUS Planch. nom. cons.

Ampelocissus latifolia (Roxb.) Planch. in Journ. Vigne Amer. 8: 374. 1884 & in DC., Monogr. Phan. 5: 370. 1887; Parmar, in Shetty & Singh Fl. Rajasthan 1: 188. 1987; Shetty & Singh, in Singh *et al*. Fl. India 5: 256. f. 58. 2000 . *Vitis latifolia* Roxb., Hort. Beng. 18. 1814 & Fl. India 2: 474. 1824; M. Lawson, in Hook. *f*., Fl. Brit. Ind. 1: 652. 1875; Duthie, Fl. Gangetic Plain 1: 171. 1903. (Plate 72)

Vern.: *Samp ki mosi*.

Weak herbaceous climbing shrub, with a tuberous root stock. Stem and branches hollow more or less glabrous, young parts puberulous. Leaves simple orbicular or broadly cordate 7 - 25 x 8 - 24 cm, 3 - 7 lobed, lobes acute, serrate-dentate, glabrous on both sides, prominently nerved beneath; petiole 5 - 15 cm long, stipules minute, deciduous. Inflorescence a compact thyrsoid cyme; peduncle 5 - 10 cm long, glabrous, ending in a long bifurcate tendril. Flowers numerous, deep reddish, pedicels ca 1 mm long. Clayx truncate or obscurely 5 toothed, entire, glabrous. Petals 5, oblong. Stamens ca 1.2 mm long; anthers globose. Ovary 10-lobed at apex, sunken in the disc; style absent; stigma subsessile, cup shaped. Berry globose, black, 6 - 9 x 4 - 5 mm, 2 seeded, rarely 3 seeded. Seeds elliptical, margin rugose; adaxial surface with a broad longitudinal ridge and a groove on either side; abaxial surface with an oblong-elliptic chalazal knot from which prominent fissures radiate towards margin.

Field note: Common among hedges and forest under growth. The local people call this species as *samp ki mosi*, and afraid of because they believe that there is possibility of snake in near by area of this species.

Fl. & Fr.: July - December

Specimen examined: Bari sadari 2150, Shakkargarh 1837 (MLVGCB Herb).

Distribution: India, Pakistan, Bangladesh & Nepal.

Uses: Roots are astringent.

CARATIA Juss.

Cayratia trifolia (L.) Domin, in Biblioth. Bot. 89: 370. 1927; Parmar, in Shetty & Singh Fl. Rajasthan 1: 188. 1987; Shetty & Singh, in Singh *et al*. Fl. India 5: 275. 2000. *Vitis trifolia* L. Sp. Pl. 203. 1953; Duthie, Fl. Gangetic Plain 1: 174. 1903. *Cissus carnosa* Lam., Encycl. 1: 31. 1789. *C. cinerea* DC., Prodr. In 631. 1820. *Vitis carnosa* (Lam.) Wall. ex M. Lawson, in Fl. Brit. India 1: 654. 1875. *Cissus trifolia* (L.) K. Schum. in Schum. & Hollr., Fl. Kaiser Wilhelmsl. 71. 1889. *Cayratia carnosa* (Lam.) Gagnep. in Not. syst. (Paris) 1: 348. 1911. *C. trifolia* (L.) Domin var. *cinerea* (Lam.) Babu, Herb. Fl. Dehra Dun 108. 1977. (Plate 73)

Vern.: *Amabel*.

A weak herbaceous climber, woody at base. Branches slender, terete, sub-fleshy, pubescent to glabrous; tendril leaf opposed much branched, wiry, not spirally coiled, ending in adhesive disc. Leaves pinnately trifoliate; leaflets ovate or obovate acute or obtuse, crenate-serrate, terminal leaflet cuneate at base, lateral leaflets slightly oblique or obtuse, pubescent on both side, somewhat fleshy and thick, 3-8 x 2-7 cm, lateral ones smaller, petiolulate, petiolule 2-8 mm long, stipule oblong ca 3 mm long. Inflorescence axillary or pseudoter-minal, leaf opposed dichotomous cymes. Peduncle 2.5 - 8 cm long, pubescent. Flower green ca 2.5 x 2 mm, pedicels ca 1.5-3 mm. Calyx obscure, sub-truncate, pubescent. Petals 4, greenish white, distinct, spreading, hooded at apex ca 2 x 1.2 mm, oblong, slightly pubescent at side. Stamens ca 1 mm long, anthers oblong. Disc 4 lobed, adnate to the base of ovary, style stout, subulate, ca 1 mm long; stigma capitate. Berry black, obovoid, globose, 3-4 seeded ca 1 cm across. Seeds triangular, obovoid, ca 6 x 4.5 mm, rounded at the back with linear tubercle, wedged shaped at the face.

Field note: Common among hedges and outskrits of forest.

Fl. & Fr.: July-September.

Specimen examined: Atun 1551, Pansal 1812 (MLVGCB Herb).

Distribution: Throughout tropical parts of India, Pakistan, Burma, Sri Lanka and Malacca.

Uses: Roots, leaves and seeds of the plant are used in native medicines due to its very acrid properties. The fruits are edible, pleasently acidic in taste.

SAPINDACEAE

KEY TO THE GENERA

1a. Climbing. Tendrillar herbs. Leaves biternate. Disc unilateral. Flowers in cymes or fascicles.................................. *Cardiospermum*

1b. Fruit indehiscent, saponaceous. Seeds globose, exarillate, Disc annular. Flowers in panicles or racemes. .... *Sapindus*

CARDIOSPERMUM L.

Cardiospermum halicacabum L. Sp. Pl. 366. 1753; Wight, *l.c.* 2(2): 8. t. 508. 1841; Hiern, in Hook. *f.,* Fl. Brit. Ind. 1: 670. 1875; Duthie, Fl. Gangetic Plain 1: 178. 1903. Mahesh. Illu. Fl. Delhi f. 47. 1966; Bhandari, Fl. Indian Desert. 93-99. 1978; Parmar, in Shetty & Singh Fl. Rajasthan 1: 191. 1987; Pant, in Singh *et al.* Fl. India 5: 356 - 357. 2000. *C. microcarpum* Kunth, Nov. Gen. *et* Sp. Pl. 5: 104. 1821. (Plate 74)

Vern.: *Lataphatkari.*

Annual, sometimes perennial herb, up to 3.5 m tall. Stem and branches slender, grooved, glabrous to sparsely pubescent. Leaves 51 - 7 x 0.4-1.5 cm; petiole 1.5-3 cm long; leaflets opposite, 1-2 cm long, pinnatisect, terminal leaflet 4 - 6 cm long, acute to acuminate, narrowed at base. Flowers white in 3-7 flowered, extra-axillary cymes; peduncle 5-15 cm long, sparsely pubescent; bracts 2, subulate, 1 mm long; pedicel 0.5-1.5 cm long, lowest pair modified into circinately coiled tendrils. Sepals free, concave, persistent; 2 outer ones orbicular, ca 2 mm long, ciliate; 2 inner ones oblong-ovate, 3 - 4 mm long, glabrous. Petals obovate, ca 3 mm long, 2 lower ones with a crested woolly scale, 2 upper ones with a large leafy scale. Disc reduced to 2 glands inside the lower petals. Stamens free; filaments compressed, 2 mm long, hairy; anthers 0.5 mm long; pistillode present in male flowers. Ovary obovoid, 2 - 3 mm long, hirsute; style short, pubescent; stigma 3-fid; staminodes 8 in the female flower. Capsule sub-globose or broadly pyriform 3 - 4 cm wide pubescent. Seeds 3, 1 in each locule, globose, ca 6 mm in diam., black; aril white, 5 mm broad, smooth.

Field note: Common among hedges and outskrits of forests.

Fl. & Fr.: August- January

Specimen examined: Hamirgarh 1579, Manpura 1796 (MLVGCB Herb).

Distribution: Africa, America, Malacca, Sri Lanka, India and Pakistan.

Uses: Leaves and seeds are reported to be used in indigenous system of medicines.

SAPINDUS L.

Sapindus trifoliatus L., Sp. Pl. 367. 1753; Hiern in Fl. Brit. India 1: 682. 1875; Pant, in Singh *et al.* Fl. India 5: 382. 2000. *S. laurifolius* Vahl, Symb. Bot. 3:54.1754; Duthie, Fl. Gangetic Plain 1: 180.1903, pro parte (excl. syn. *S. emarginatus* Vahl; Gamble, Fl. Pres. Madras 1: 178. 1956. (Repr. ed.); Parmar, in Shetty & Singh Fl. Rajasthan 1: 192. 1987. (Plate 75)

Vern.: *Aritha.*

Medium sized tree up to 5 - 10 m tall. Bark longitudinally fissurred. Leaves shortly petiolulate, alternate, pinnate; leaflets 2 pairs, rarely 3 pairs, opposite, 7 - 17.5 x 2.5 - 10 cm, lanceolate or elliptic-lanceolate, entire, shining above, sparsely hairy beneath. Flowers white, in 9 - 12 cm long, terminal, rusty pubescent panicles. Calyx pubescent outwards, glabrous within. Petals clawed, narrowly lanceolate, with 2 scales at base, softly hairy within. Disc concave, margins fleshy, hirsute. Stamens 8; anthers oblong. Fruits 2-3 lobed, coriaceous, fulvous hairy when young, wrinkled when ripe. Seeds black.

Field note: Common in deciduous forests also planted along roads as well as boundries of agriculture fields.

Fl. & Fr.: September - March.

Specimen examined: Bojunda 510, Chittor garh 1650 (MLVGCB Herb).

Distribution: India, Pakistan and Sri Lanka.

Uses: Fruits are saponaceous are used as a substitute of soap.

ANACARDIACEAE

KEY TO THE GENERA

1a. Leaves simple. Carpels solitary ..... .*Mangifera*

1b. Leaves compound, imparipinnate. Carpels united 2

2a. Flowers 4-merous. Drupe 12 mm or more in length, red ....*Lannea*

2b. Flowers 5-merous. Drupe less than 7 mm long, yellowish-brown ....*Rhus*

LANNEA A. Rich. nom. cons.

Lannea coromandelica (Houtt.) Merrill, J. Arn. 19: 353. 1938; Parmar, in Shetty & Singh Fl. Rajasthan 1: 194. 1987; Chandra & Mukherjee, in Singh *et al*. Fl. India 5: 463. 2000. *Dialium coromandelicum* Houtt., Nat. Hist. 2: 39. t. 5. f. 2. 1774. *Odina wodier* Roxb., Fl. Ind. 2: 293. 1832; Hook. *f.*, Fl. Brit. Ind. 2: 29. 1876. Duthie, Fl. Gangetic Plain 1: 188. 1903. (Fig. 35 & Plate 76 & 77)

Vern.: *Godal*.

Small to medium sized tree, 10 m or more tall. Bark dark brownish grey with longitudinal furrows. Leaves 20 - 40 cm, imparipinnate; leaflets 3 - 9 in number, subsessile, 9 - 15 x 5 - 7 cm, ovate to ovate-oblong, acuminate, base obliquely rounded, entire, glabrous above, rusty tomentose on the under surface. Bracts ca 1.2 mm long, broad ovate. Flowers unisexual, greenish, the male in compound and female in simple racemes. Calyx lobes 4, ca 1.2 mm long, broad ovate obtuse, ciliate. Petals 4, 2 mm long, oblong, green yellow. Male flowers: Stamens 8, as long as the petals and-inserted below the annular disk; anthers ca 1.2 mm long, basifixed; filaments glabrous. Female flowers: Ovary oblong; styles 3 - 4, stout. Drupe 10-12 mm long, oblong, compressed, red when ripe, with persistent sepals and styles.

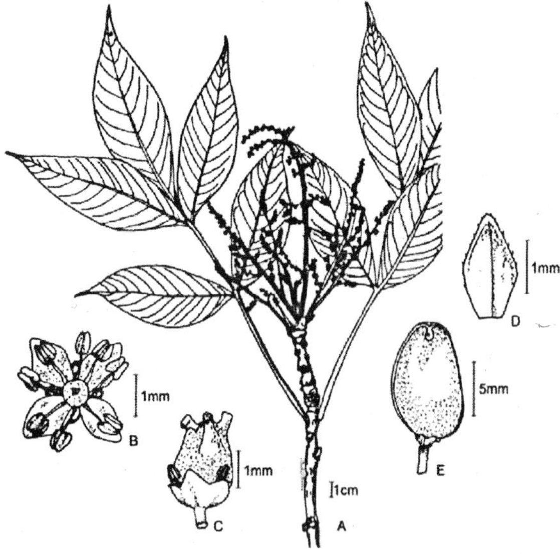

Fig. 35. *Lannea coromandelica* (Houtt.) Merill. A. Male flowering branch, B. Male flower, C. Female flower, D. Petal, E. Fruit

Field note: Common in deciduous forests. The species found leafless when in flower.

Fl.: January - June; *Fr.*: February - July.

Specimen examined: Gurla 1745, Singoli 1856 (MLVGCB Herb).

Distribution: The sub-himalayan tract to the Indus, Burma, Assam, Sri Lanka and the Andaman Islands.

Uses: The gum exuded from bark is used in medicines. Wood is used as timber. Leaves as fodder.

MANGIFERA L.

Mangifera indica L., Sp. Pl. 200. 1753; Hook. *f.*, Fl. Brit. Ind. 2: 13. 1876; Duthie, Fl. Gangetic Plain 1: 189. 1903; Parmar, in Shetty & Singh Fl. Rajasthan 1: 194-195. 1987; Chandra & Mukherjee, in Singh *et al*. Fl. India 5: 466 - 468. f. 100. 2000.

Vern.: *Aam*.

Tree up to 25 m tall. Leaves 11 - 30 x 3.5 - 8 cm, oblong, lanceolate, acuminate, coriaceous, shiny and dark green on upper surface; petiole 1.5 - 6 cm. Flowering panicles erect, conspicuous, longer than the leaves, pubescent. Flowers male and bisexual, 5 mm across; pedicel 1.5 - 3 mm long. Calyx 5-lobed, lobes 2 x 1 mm, ovate, pubescent on the outside. Petals 5, imbricate, elliptic, 3 x 1.2 mm, inner surface prominently 3-nerved. Stamens 5, 1 fertile, 2.5 mm long, rest sterile, small, teeth like; Disc 1.5 mm in diam., distinctly 5-lobed. Ovary 1 mm in diam., obliquely ovoid, 1-loculed, 1-ovuled; style 2 mm long, lateral. Drupe ovoid in outline, compressed, 3.5 - 20 cm long. Mesocarp fleshy. Endocarp (stone) hard and fibrous.

Field note: Planted in orchards, along the roads etc. for edible fruits.

Fl.: January - March; *Fr.*: March - July.

Specimen examined: Singoli 1419, Shahpura 1839 (MLVGCB Herb).

Distribution: Native of Burma. Widely cultivated in the tropics.

Uses: The tree is well known for its delicious fruits. Unripe fruits are used in making chutneys, jellies, pickles etc. Wood is used as timber.

RHUS L.

Rhus sinuata Thunb., in Prodr. Pl. Cap. 52. 1794. *R. mysurensis* G. Don, gen. syst. 2: 74. 1832; Heyne ex Wight & Arn., Prodr. 172. 1834; Hook. *f.,* Fl. Brit. Ind. 2: 9. 1876; Duthie, Fl. Gangetic Plain 1: 186. 1903; Mahesh. Illu. Fl. Delhi f. 48. 1966; Bhandari, Fl. Indian Desert 94. 1978; Parmar, in Shetty & Singh Fl. Rajasthan 1: 195. 1987; Chandra & Mukherjee, in Singh *et al.* Fl. India 5: 494. 2000. (Fig. 36 & Plate 78 & 79)

Vern.: *Dansaro*.

A dioecious shrub up to 2 m tall. Branchlets spinescent; young branches pubescent. Leaves trifoliate; petiole 8 - 18 mm long, pubescent. Leaflets 1 - 3 x 1 - 1.5 cm, the terminal the largest, obovate-cuneate, pubescent, the upper half irregularly lobed, lower entire; upper surface of a darker colour. Panicle slender, 3 - 10 cm long, terminal and axillary. Flowers unisexual, yellowish. Bracts ca 1 mm long, lanceolate, pubescent. Pedicel 2 - 4 mm long, pubescent. Calyx lobe ovate. Petals 5, ca 1 mm long, oblong-obtuse, glabrous, sometimes apex acutish. Disc faintly 5-lobed. Stamens 1 mm long, included, abortive in female flowers. Filaments slender; anthers suborbicular. Ovary ovoid, less than 1 mm long, glabrous. Styles 3, spreading. Drupe subglobose, compressed, 4 - 5 mm broad, yellowish-brown, epicarp crustaceous, shiny.

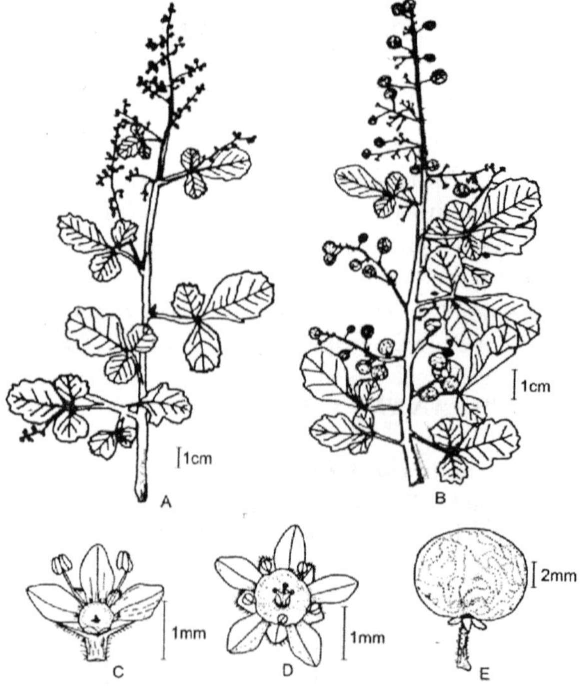

Fig. 36. *Rhus sinuata* Thunb. A. Male flowering twig, B. Female flowering twig, C. Male flower, D. Female flower, E. Fruit

Field note: A rare plant occasionally found in dry rocky areas.

Fl.: June - August; *Fr.*: July - October.

Specimen examined: Gurla 1582 (MLVGCB Herb).

Distribution: Pakistan and Rajasthan (India) and Sind.

Uses: Its bark is used for tanning. Fruits are eaten to increase lactation as well as to improve the digestion. Fruits are alos sold in market.

MORINGACEAE

MORINGA Adans.

KEY TO THE SPECIES

1a. Leaves mostly 2 - pinnate; leaflets 2 - 5 x 0.8 - 3 cm .... *M. cancanensis*

1b. Leaves mostly 2 - 3 (-4) pinnate; leaflets 0.5 - 2.5 x 0.3 - 1.5 cm.............................. *M. oleifera*

Moringa concanensis Nimmo ex Dalz. & Gibs., Bomb. Fl. 311. 1861; Stewart & Brandis, For. Fl. 130. 1874; Hook. f., Fl. Brit. Indaia 2: 45. 1876; Duthie, Fl. Gangetic Plain 1: 93. 1903; Bhandari, Fl. Indian Desert 102. 1978; Parmar, in Shetty & Singh Fl. Rajasthan 1: 196. 1987; Uniyal, in Singh *et al*. Fl. India 5: 515. 2000.

Vern. : *Sainjna*

A small tree with thick bark, glabrous, except younger parts and inflorescence. Leaves bipinnate (very rarely tripinnate); rachis articulated; leaflets 4 - 6 pairs with an odd one, broadly elliptic to orbicular, entire, obtuse at both ends, often emarginate at the apex. Inflorescence lax, panicles up to 45 cm long, tomentose. Flowers small; bracts minute, caducous. Calyx 5-lobed, minutely tomentose. Petals yellow, with red or pink veins, oblong or oblong-spathulate, the lower ca 9 mm long. Perfect stamens 5, alternating with 4-5 much smaller staminodes. Fruits linear, 20 - 30 cm long, sharply 3-angled. Seeds triangular, with a very thin, hyaline, wing at each angle, 1.5 - 2 cm long, white or pale yellow.

Field note: Not common and perhaps confined to certain localities in study area.

Fl. & Fr.: November- April.

Specimen examined: Gurla 277, Ajoliya ka khera 5531 (MLVGCB Herb).

Distribution: India (Rajasthan & M.P.), and Pakistan.

Moringa oleifera Lam., Encyl. Meth. 1: 398. 1785; Bhandari, Fl. Indian Desert 95. 1978; Parmar, in Shetty & Singh Fl. Rajasthan 1: 196. 1987; Uniyal, in Singh *et al*. Fl. India 5: 516. 2000. *M. pterygosperma* Gaertn. Fruct. 2: 314. 1791; Hook. *f.*, Fl. Brit. Ind. 2: 45. 1876; Duthie, Fl. Gangetic Plain 1: 192. 1903. (Fig. 37 & Plate 80)

Vern.: *Sahanjna, Sainjna.*

Large tree, up to 7 m high, younger parts tomentose. Leaves up to 50 cm long, alternate; rachis thickened and articulated at the base; leaflets; lateral leaflets elliptic, terminal obovate, slightly larger than lateral leaflets. Flowers white or yellowish, in large puberulouse, effuse panicles. Calyx lobes slightly unequal, petaloid, imbricate, linear - lanceolate 1.3 - 1.5 x 0.4 cm, reflexed, with prominent yellow streaks in the centre, entire, obtuse. Petals spathulate, unequal. Stamens 5, fertile alternating with 5 - 7 sterile stamens. Fruit 9-ribbed pendulous pod, 30 - 45 cm long. Seeds embedded in the pits of the valves, 3 angled, 3 - winged, blackish, rounded.

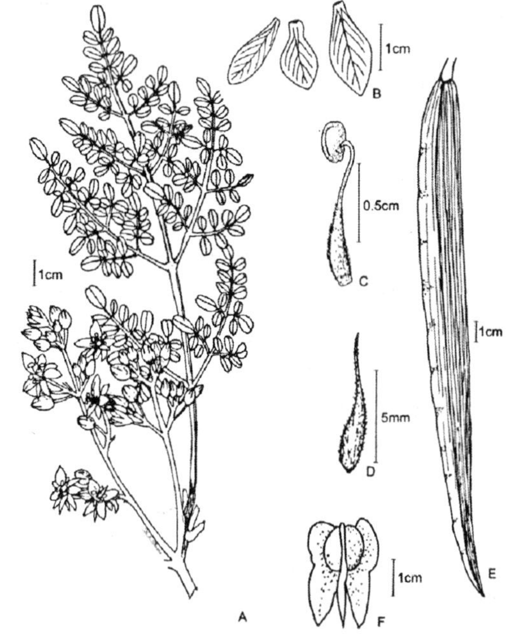

Fig. 37. *Moringa oleifera* Lam. A. Flowering twig, B. Petals, C. Stamen, D. Ovary, E. Fruit, F. Seed

Field note: Mostly cultivated in gardens, along roadsides and occasionally found in forests.

Fl. & Fr.: January- April.

Specimen examined: Bhilwara 1794 (MLVGCB Herb).

Distribution: India and many other tropical countries.

Uses: Flowers and immature fruits are used in curries and are said to be a good rubefacient. Branches are often lopped for fodder.

FABACEAE

KEY TO THE GENERA

1a. Trees 2

2a. Fruit indehiscent, sometimes inflated and bladder - like. 3

3a. Leaflets (mostly) alternate, rarely only 1 leaflet........ *Dalbergia*

3b. Leaflets opposite or a few at the same time alternate, or leaflets 3........ *Pongamia*

2b. Fruit dehiscent by a valve or valves (sometimes only at the apex). 4

4a. Branches armed with prickles. Flowers red. Fruit not silky .... *Erythrina*

4b. Branches unarmed. Flowers red tinged with orange. Fruit silky. .... *Butea*

1b. Shrubs or herbs (erect, prostrate or climbing) 5

5a. Inflorescence axis ending in spine, hairs biramous, Vexillary filament free, leaflets not toothed, anthers apiculate........ *Indigofera*

5b. Inflorescence, hairs and anthers not as above 6

6a. Plants spiny or prickly. 7

7a. Fruits breaking up longitudinally into 2- equal halves........ *Sesbania*

7b. Fruits usually breaking up into 1-seeded segments........ 8

8a. Keel petals obtuse or beaked, incurved. Filaments sometimes connate into 2 lateral bundles .... *Aeschynomene*

8b. Keel-petals often obliquely truncate at the apex. Filaments connate into a single sheath split adaxially........ *Alhagi*

6b. Plants not spiny or prickly 9

9a. Plants climbing or prostrate........ 10

10a. Fruits clothed with stinging hairs. ... *Mucuna*

10b. Fruits not clothed with stinging hairs. 11

11a. Flowers showy, dark purple........ *Clitoria*

11b. Flowers not as above 12

12a. Leaflets gland-dotted below .... *Rhynchosia*

12b. Leaflets not gland-dotted below .... *Abrus*

9b. Plants erect........ 13

13a. Leaf paripinnately compound or palmately foliolate 14

14a. Stamens diadelphous, 5 + 5, fruit jointed 15

15a. Leaf palmately 3-foliolate. Fruit 1 -6 jointed, echinate........ *Zornia*

15b. Leaf 2 - 8-foliate, Fruit 3 - 6 jointed, papilose .... *Smithia*

14b. Stamens diadelphous (9)+1, fruit not jointed 16

16a. Style tapering apically, glabrous or bearded at apex, all round or on lower surface only; staminal sheath oblique at the apex .... *Vicia*

16b. Style flattened or dorsiventrally compressed towards apex, glabrous or bearded along the upper surface; staminal sheath truncate at the apex........ *Lathyrus*

13b. Leaf imparipinnately or palmately compound or unifoliolate or simple. 17

17a. Fruit jointed 18

18a. Joints of pods turgid........ *Alysicarpus*

18b. Joints of pods flattened........ *Desmodium*

17b. Fruit not jointed 19

19a. Anthers dimorphic 20

20a. Stamens diadelphous (9)+1........ *Phaseolus*

20b. Stamens monadelphous. .... *Crotalaria*

19b. Anthers monomorphic........ 21

21a. Leaflets entire .... *Tephrosia*

21b. Leaflets toohed 22

22a. Leaf trifoliolate........ 23

23a. Style with tenuous lower part obsolete or long, upper portion barbate on the inner side .... *Vigna*

23b. Style not as above........ 24

24a. Fruit spirally coiled........ *Medicago*

24b. Fruit not spirally coiled 25

22b. Leaves pinnate; leaflets more than 3. 25

25a. Fruit more than 3 times as long as the calyx, if less, then the inflorescence rachis not elongated......................... *Trigonella*

25b. Fruit less than 3 times as long as the calyx, straight, inflorescence axis elongated .... *Melilotus*

ABRUS Adans

Abrus precatorius L., Syst. Nat. ed. 12: 472. 1767; Baker, in Hook. *f.,* Fl. Brit. Ind. 2: 175. 1876 ; Duthie, Fl. Gangetic Plain 1: 262. 1903; Mahesh., Illus. Fl. Delhi f. 71. 1966; Kew Bull. 24: 240. 1970; Bhandari, Fl. Indian Desert 97. 1978; Singh, in Shetty & Singh Fl. Rajasthan 1: 201. 1987; Sanjappa, Legumes India 74. 1992. *Glycine abrus* L. Sp. Pl. 753. 1753. (Plate 81)

Vern.: *Chirmi, Chirmu, Ratti.*

Perennial climber, much branched; branches slender. Stipules 3.5 - 6 mm long, linear, deciduous. Leaves paripnnate, 5 - 10 cm long. Petiole 0.6 - 1.2 cm long. Leaflets 9 - 20 pairs, opposite, petiolule less than 1 mm, 0.8 - 2.2 x 3.5 - 6 cm, oblong, tip rounded, apiculate, glabrous above, sparsely hairy below. Racemes axillary, pedunculate, shorter or equalling the leaves. Pedicels short. Calyx-lobes 2.5 mm long, glabrous or sparsely silky, teeth very short. Corolla pink or white with a pink tinge, vexillum 9-10 mm long. Stamens 9. Pod 2 - 4.5 x 1 - 1.3 cm, oblong, turgid, appressed with sharp deflexed beak, 3 - 5 seeded. Seeds ovoid, 7-8 mm long, scarlet with a black spot at the hilum, sometimes white with a black spot or uniformly black or white.

Field note: Common in clumps of shrubs in forests, hedges, also found climbing on *Euphorbia caducifolia.*

Fl. & Fr.: July-December.

Specimen examined: Mandalgarh 1403, Pansal 1853 (MLVGCB Herb).

Distribution: Pakistan, throughout India, Sri Lanka and Tropical Africa; introduced widely in the new and the old world.

Notes: According to the sub specific classification proposed by Verdcourt (Kew Bull. 24: 240. 1970), our specimens belong to *Abrus precatorius* ssp. *precatorius*.

Uses: Seeds are commonly used by jewellers as weight. They are poisonous and are put to many medicinal uses. Fresh leaves are chewed in case of mouth ulcer. These are also chewed by the professional singer to keep their throat and voice fit, because its sweet test releave hoarseness of voice, strengthen gums. Seeds are poisonous. 1-2 seeds are taken for 2 days for abortion.

AESCHYNOMENE L.

Aeschynomene indica L., Sp. Pl. 713. 1753; Wight, Ic. 2 (1): 9. t. 405. 1840; Baker, in Hook. *f.,* Fl. Brit. Ind. 2: 151. 1876; Duthie, Fl. Gangetic Plain 1: 270. 1903; Bhandari, Fl. Indian Desert 97. 1978; Singh, in Shetty & Singh Fl. Rajasthan 1: 202. 1987; Sanjappa, Legumes India 75. 1992. *Hedysarum neli-tali* Roxb., Fl. Ind. 3: 365. 1832. *Smithia aspera* Roxb., Fl. Ind. 3: 343. 1832. *Aeschynomene kashmiriana* Camb., in Jacq. Voy. Bot. 40. t. 48. 1844. (Fig. 38 & Plate 82)

Vern.: *Chhui-mui.*

Annual undershrub, 30 - 90 cm tall, stem glabrous, branches terete with prickle-like glandular excretions. Leaf 3.5 - 9.0 cm long, rachis furnished with glandular excretions looking like prickles. Leaflets 21 - 71, alternate, subsessile, overlapping, 2.5 - 6.5 x 1.2-2.0 mm, linear, obtuse, mucronate, glabrous. Inflorescence a 1 - 4-flowered axillary raceme, peduncle ca 2.5 cm long, peduncle and pedicel glandular, viscid, pedicel 5 - 9 mm long. Bracts ca 2 mm long. Calyx 6 - 7 mm long, bilabiate, glabrous, upper lip bifid, lower minutely 3-toothed. Corolla 9-10 mm long, yellow, vexillum veined with purple. Fruit 2.5 - 4.5 cm x 3-5 mm, slightly curved, 6 - 10 jointed, glabrous or slightly hairy, readily separating into 1-seeded parts.

Field note: Common in marshy areas, along the edges of streams and ponds etc.

Fl. & Fr.: August- November.

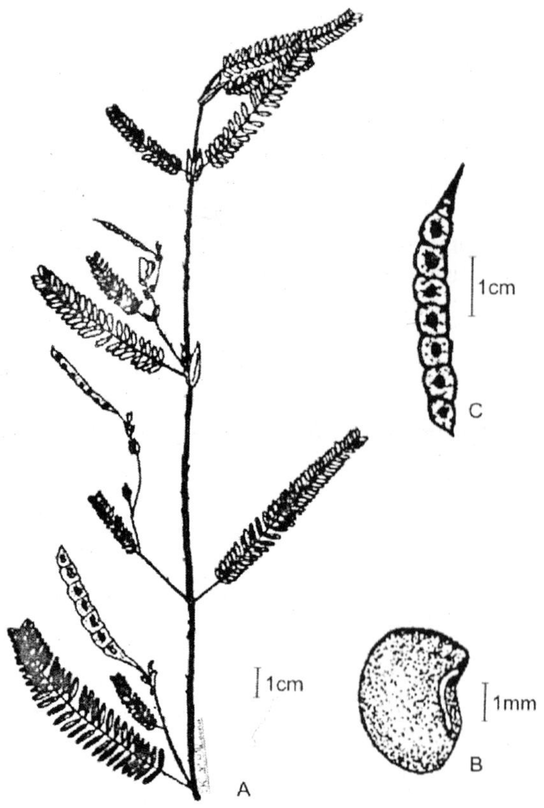

Fig. 38. *Aeschynomene indica* L. A. Flowering & fruiting branch, B. Seed, C. Fruit.

Specimen examined: Vindhyabhata 137, Mainal 1793 (MLVGCB Herb).

Distribution: India, Pakistan, Sri Lanka, Burma, Siam, Japan, Tropical Africa, Australia and N. America

ALHAGI Gagnebin

Alhagi maurorum Medic. in Vorles, Churpf. Phys. Oek. Ges. 2: 397. 1787; Baker, in Hook. *f.*, Fl. Brit. Ind. 2: 145. 1876; Bhandari, Fl. Indian Desert 97 - 98. 1978; Singh, in Shetty & Singh Fl. Rajasthan 1: 202. 1987; Sanjappa, Legumes India 76. 1992. *A. camelorum* Fisch. ex DC. Prodr. 2: 352. 1825; Duthie, Fl. Gangetic Plain 1: 279. 1903. *Hedysarum alhagi* L. Sp. Pl. ed. 1. 2: 745. 1753 p.p. quoad typum in Herb. L. ex Tataria. *H. pseudalhagi* M. Bieb. Fl. Taur.-Cauc. 2 : 174. 1808. *Alhagi pseudalhagi* (M. Bieb.) Fisch. Cat. Jard. Gorenk. ed. 2: 72. 1812; Bhandari, Fl. Indian Desert 105. 1978. (Fig. 39 & Plate 83)

Vern.: *Janasa*.

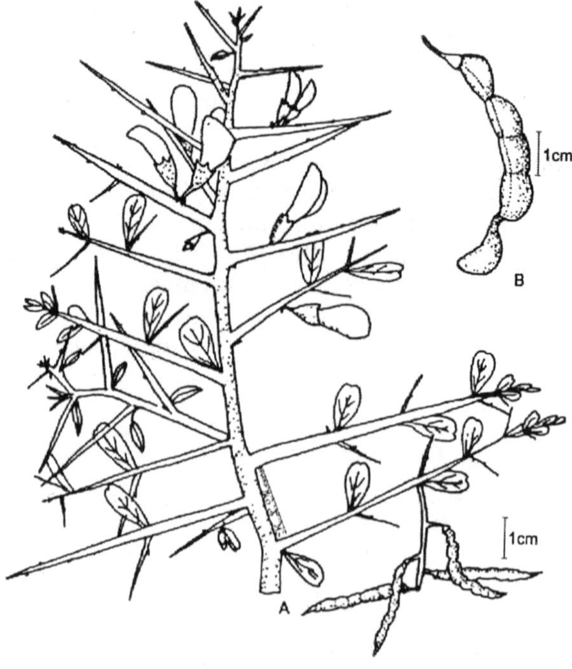

Fig. 39. *Alhagi maurorum* Medic.
A. Flowering & fruiting branch, B. Fruit

Undershrub, 60-100 cm tall, glabrous or pubescent. Leaf simple, 10-25 x 3 - 8 mm, obovate or elliptic-oblong, glabrous or pubescent, entire, apiculate; petiole ca 2 mm; stipules minute. Inflorescence lateral axillary racemes 1 - 5 cm long, ending in spine. Flowers red. Pedicel 1 - 3 mm, with 1 - 2 minute bracteoles. Calyx 2-2.5 mm, glabrous, teeth almost obsolete to triangular. Corolla 6 - 9 mm long, pink or reddish - violet. Fruit 19 - 34 x 2 - 3 mm, glabrous, more or less constricted between the seeds, 1 - seeded.

Field note: Common on alkaline soils.

Fl. & Fr.: March - July.

Specimen examined: Karoi 1524 (MLVGCB Herb).

Distribution: India, Eygpt, Baluchistan, Iran, Afghanistan, Pakistan, Russia, Turkey, Iraq, Syria, Palestine, Cyprus and N. Africa

Notes: Here Meikle (Fl. Cyprus, fined.) has been followed, who has interpreted *Alhagi maurorum* Medic to be conspecific with *Alhagi pseudalhagi* (M. Bieb.) Desv. contrary to other workers who regarded it as nomen confusum.

Uses: It is used as fodder. Sometimes it is also used for making tatties in summers. Medicinally the leaves are used in the traetment of asthama.

ALYSICARPUS Desv. nom. cons.

KEY TO THE SPECIES

1a. Pod not moniliform .... *A. ovalifolius*

1b. Pod moniliform.. 2

2a. Veins obscure on mature pod (one seeded part rounded turgid); inflorescence with 10 or less flowers .... *A. monilifer*

2b. Veins not obscure on mature pod (1 seeded part not rounded); inflorescence with more than 10 flowers .... *A. longifolius*

Alysicarpus longifolius (Rottl. ex Spreng.) Wight & Arn. Prodr. 233. 1834; Wight, Ic. 1: t. 251. 1840; Baker, in Hook. *f.,* Fl. Brit. Ind. 2: 159.1876; Duthie, Fl. Gangetic Plain 1: 278. 1903; Bhandari, Fl. Indian Desert 99. 1978; Singh, in Shetty & Singh Fl. Rajasthan 1: 205. 1987; Sanjappa, Legumes India 79. 1992. *Hedysarum longifolium* Rottl. ex Spreng. Syst. Veg. 3: 319. 1826. (Plate 84)

Vern.: *Ganoulo.*

Stem 1.2 - 1.5 m tall, slightly striate, glabrous. Stipules 9 - 20 mm long, scarious, very acute. Leaves unifoliolate; petioles 3 - 10 mm long; leaflets 5 - 15.0 x 0.9 - 1.9 cm, oblong or lanceolate, obtuse or subacute, base subcordate, glabrous above, appressly hairy on the nerves beneath; petiolules very short, stipels linear, acute. Inflorescence dense spicate raceme, 15 - 30 cm long. Bracts often longer than 1.3 cm, ovate, acuminate, silky outside, caducous. Pedicel ca 2.5 mm long. Calyx 6 - 16 mm long, striate, pubescent, teeth longer than the tube. Pods 0.9 - 1.3 cm x 2 - 2.5 mm, shortly stalked, slightly moniliform, terete-compressed, 4-6 jointed, minutely pubescent, reticulately veined, veins not obscure.

Field note: Common in cultivated fields.

Fl. & Fr.: September - November.

Specimen examined: Bhilwara 1634, Gangapur 1855 (MLVGCB Herb).

Distribution: Pakistan, India (Saurashtra, Madhya Pradesh, Bombay, Madras).

Uses: Mature plants are given to goats to increase milk production, whole plants are sold in market as fodder.

Alysicarpus monilifer (L.) DC., Prodr. 2: 353. 1825; Baker, in Hook. *f.,* Fl. Brit. Ind. 2: 157. 1876; Duthie, Fl. Gangetic Plain 1: 276. 1903; Bhandari, Fl. Indian Desert 107. 1978; Singh, in Shetty & Singh Fl. Rajasthan 1: 206. 1987; Sanjappa, Legumes India 79. 1992. *Hedysarum moniliferum* L. Mant. : 102. 1767.

Prostrate, much branched, branches pilose, hairs long and spreading. Stipules 3 - 8 mm long, scarious. Leaf unifoliolate, petiole ca 4 - 8 mm long, petiolules very short, stipels minute filiform. Leaflets 4 - 25 mm long, 4 - 15 mm broad, oblong-elliptic, rounded and mucronate at the apex, glabrous above, slightly hairy on the nerves below. Inflorescence 4 - 10-flowered short axillary raceme. Pedicel ca 1 mm long. Calyx 2.5 - 3.5 mm long, pilose, teeth narrow, acute. Pod 1.2 - 2.5 cm long, moniliform, joints 4 - 8, turgid, subglobose, finely downy with minute hooked hairs. Seeds subglobose or oblong, brownish.

Field note: Commonly found in waste lands.

Fl. & Fr.: August- November.

Specimen examined: Bhilwara 1857 (MLVGCB Herb).

Distribution: Pakistan, India (Punjab, Uttar Pradesh, Madhya Pradesh, Bengal, Madras & Bombay) and Ethiopia.

Alysicarpus ovalifolius (Schum.) J. Leon., in Bull. Jard. Bot. State Brux. 24: 88. 1954; Sanj. & Blatt., in J. Bombay Nat. Hist. Soc. 75. 250. 1978; Sanjappa, Legumes India 79. 1992. *Hedyasrum ovalifolium* Schum., Beskr. Guin. Pl. 359. 1827. *Desmodium ovalifolium* (Schum.) Walp. Rep. 1: 737. 1842. (Plate 85)

Annual herb, erect-spreading, 20 - 60 cm tall. Stem pubescent. Leaf unifoliolate, leaflets 1 - 10 x 0.6-3 cm, elliptic or oblong to narrowly lanceolate, acute to emarginate, mucronate, puberulous, petiole 2 - 8 mm long, petiolule 1-1.5 mm long, stipules 0.5 - 2.0 cm long. Inflorescence terminal or leaf-opposed or paniculate, peduncle 3-4 cm long, rachis 3 - 11 cm long, pedicel 1-2 mm long. Bracts 4 - 5 mm long. Calyx 4.5-6 mm long, pubescent, teeth 3 - 4 mm long. Corolla orange-buff to pink or reddish violet rarely whitish, Vexillum 4-6 mm long. Fruit 1.8-2.5 cm long, 2- 8-seeded, not constricted, each 1-seeded part 2.5- 4 x 2.0-2.5 mm, puberulous. Seeds brown, ellipsoid, slightly compressed with reddish-black patches.

Field note: Common in waste lands.

Fl. & Fr.: September - November.

Specimen examined: Nimbahera 2152 (MLVGCB Herb).

Distribution: India (Punjab, Uttar Pradesh, Madhya Pradesh, Bombay, Madras, Saurashtra), Pakistan, Bengal, Tropical Africa, Madagascar, Afghanistan, China, Indonesia (Java).

BUTEA Roxb. ex Willd. nom. cons.

Butea monosperma (Lam.) Taub., in Engler & Prantl, Pfianzenfam. 3 (3): 366. 1894; Mahesh., Illus. Fl. Delhi f. 74. 1966; Bhandari, Fl. Indian Desert 101. 1978; Singh, in Shetty & Singh Fl. Rajasthan 1: 211. 1987; Sanjappa, Legumes India 99. 1992. *Erythrina monosperma* Lam., Encycl. 2: 391. 1785. *Butea frondosa* Roxb. ex Willd. Sp. Pl. 3: 917. 1802; Baker, in Hook. *f.,* Fl. Brit. Ind. 2 : 194. 1876; Duthie, Fl. Gangetic Plain 1: 240. 1903. (Plate 86)

Vern.: *Dhak, Plash, Keshu, Khankhara, Sura.*

Deciduous tree, young branches tomentose. Bark ash coloured. Leaf pinnately trifoliolate; petiole 10 - 22.5 cm long, terminal leaflet 12.5 - 20 x 11 - 17.5 cm, broadly ovate, obtuse or retuse, glabrescent; lateral leaflet obliquely ovate, obtuse. slightly smaller than the terminal, glabrescent; petiolules 5 - 7.5 mm long, stipels small; stipules small, deciduous. Inflorescence an axillary or terminal raceme. 10 - 17.5 cm long, flowers fascicled at the nodes. Bracts and bracteoles small, velvety. Pedicel 1.2 -2.5 cm long, velvety. Calyx ca 1.2 cm long, velvety within and without, teeth deltoid, the upper 2 teeth united. Corolla bright red, tinged with orange, silky pubescent outside. Vexillum 3.7 - 5.0 cm long. Fruit 9 - 20 x 1.5-5.0 cm, silky, 1 seed at the apex, oblong, stalk 1.2-1.8 cm long. Seeds solitary.

Field note: 'Flame of the Forest' (*B. monosperma*) is commonly occurring in the area.

Fl. & Fr.: March-April.

Specimen examined: Atun 1180, Gangapur 1332 (MLVGCB Herb).

Distribution: Widely distributed in India, Pakistan, Sri Lanka.

Uses: The gum which is known as *kamarkas,* it is used in the form of *'laddu'* and is an essential medicine after delivery, it keeps waist in shape. The flower juice is given to children in fever and cold. Seeds are crushed on red stones and given to newly born child in case of diarrhoea. It is one of the most important hosts for the lac insect. Leaves are used for mking Pattal and Donas. The twigs are used in making roof of houses. Wood is used in making toys as well as firewood.

Notes: The taxon is usually described under the combination of *Butea frondosa* Koening ex Roxb. (1792), but its specific epithet is to be accepted from Lamarck's combination (*Erythrina monosperma*). Some of the modern authors attribute this combination to O. Kuntze (1891 Ic.) who is reported not to have made it (Sanjappa Fl. Saur. 1962 in Bull. Bot. Surv. India 3: 12. 1961). The present combination thus should be attributed to Taubert (1894 I.c.) who made it and published for the first time.

CLITORIA L.

Clitoria ternatea L. Sp. Pl. 753. 1753; Baker, in Hook. *f.,* Fl. Brit. Ind. 2 : 208. 1876; Duthie, Fl. Gangetic Plain 1: 230. 1903; Bhandari, Fl. Indian Desert 102. 1978; Singh, in Shetty & Singh Fl. Rajasthan 1: 213. 1987 Sanjappa, Legumes India 114. 1992.

var.: ***ternatea***

Vern.: *Gokran, Supli.*

A woody climber, perennial. Leaves 5 - 9.5 cm long, unipinnate, leaflets usually 5-7, 1.3-6.5 x 1-4 cm, elliptic-oblong, slightly emarginate or obtuse, short stalked, base acute or obtuse, petiolules up to 2 mm long, stipules linear, 4-10 mm long, apex acute. Inflorescence solitary axillary. Flowers bright blue. Bracts 4 - 5 mm long, bracteoles ca 7 mm long, orbicular. Pedicels 6 - 9 mm long. Calyx 12 - 16 mm long, 2 upper teeth joined near the base. Corolla rich blue in colour with greenish veins on the vexillum, vexillum 2.5 - 5.0 x 3.5 cm, Stamens 10, diadelphous, 9+1, dimorphic the vexillary stamen free, Ovary hairy, style simple, hairy along the inner side. Fruit up to 5 - 12 cm long, beaked 8 - 10-seeded, glabrous or sparsely pubescent. Seeds 6 - 10, yellowish-brown.

Field note: Common among hedges, particularly near habitations, also cultivated as an ornamental.

Fl. & Fr.: August - October.

Specimen examined: Meza 598, Bhilwara 910 (MLVGCB Herb).

Distribution: Cosmopolitan in tropics, widely cultivated as an ornamental plant; the true distribution now difficult to trace in view of cultivation.

Uses: Root is reported to have medicinal properties.

CROTALARIA L.

KEY TO THE SPECIES

1a. Leaves compound, 3-foliate ..... *C. medicaginea*

1b. Leaves simple 2

2a. Racemes all lateral, leaf opposed, few, often only 1-flowered ..... *C. hebecarpa*

2b. Racemes terminal, many -flowered, rarely opposite the leaves 3

3a. Pods glabrous ..... *C. spectabilis*

3b. Pods hairy 4

4a. Bush, low undershrubs, pods 3-4 seeded ..... *C. burhia*

4b. Herbs, pods 10 or more seeded ..... *C. juncea*

Crotalaria burhia Buch.- Ham. ex Benth. in Hook Lond. J. Bot. 2: 474. 1843; Baker, in Hook. *f.*, Fl. Brit. Ind. 2: 66. 1876; Duthie, Fl. Gangetic Plain 1: 202. 1903; Bhandari, Fl. Indian Desert 111. 1978; Singh, in Shetty & Singh Fl. Rajasthan 1: 216. 1987; Sanjappa, Legumes India 117. 1992. (Plate 87)

Vern.: *Shaneo*.

Undershrub, ca 30-60 cm tall; branches numerous, hoary with dense appressed pubescence. Leaves few, deciduous, simple, ca 0.6-2.5 cm x 3 - 10 mm, oblong-lanceolate, obtuse, pubescent on both sides. Inflorescence a 6-12-flowered, elongated raceme. Flowers yellow; pedicels very short; bracteoles 2. Calyx ca 7-9 mm long, pubescent, teeth lanceolate, acute. Corolla yellow, as long as calyx or slightly exserted. Stamens 10, monadelphous. Style slightly bearded at the top; ovary glabrous except hairy margins. Fruit ca 8 - 9 x 4 mm or less wide, hairy, 3 - 4-seeded. Seeds dark brown, notched at hilum, black.

Field note: Common in dry sandy habitats, in association with *Leptadenia pyrotechnica* (Forssk.) Decne., etc.

Fl. & Fr.: August - March.

Specimen examined: Hamirgarh 605, Mezadam 605s (MLVGCB Herb).

Distribution: India, Pakistan and Afghanistan.

Crotalaria hebecarpa (DC.) Rudd. in Phytologia 54 : 28. 1983; Sanjappa, Legumes India 121. 1992. *Hallia hirta* Willd. Sp. Pl. 3: 1169. 1802, non *Crotalaria hirta* Willd. 1803, nec Leg. 1816, nec Roth 1821. *Goniogyna hebecarpa* DC. Ann. Sci. Nat. Paris 4: 92. 1825. *Goniogyna leiocarpa* DC. Ann. Sci. Nat. Paris 4: 92. 1825, non *Crotalaria leiocarpa* Vog. 1843. *Heylandia hebecarpa* DC. Ann. Sci. Nat. Paris 4: 92. 1825, based on same collection as *Goniogyna hebecarpa* DC. *Heylandia hebecarpa* DC. Prod. 2: 213. Nov. 1825, non *Crotalaria leiocarpa* Vog. 1843. *H. latebrosa* DC. Mem. Legum. 6 : 201. Nov. 1826; Baker, in Hook. *f.*, Fl. Brit. Ind. 2 : 65. 1876. *Crotalaria uniflora* Koen ex Roxb., Fl. Ind. 3 : 271. 1832, non Baker 1871. *Goniogyna hirta* (Willd.) Ali in Taxon. 16 : 463. 1967; Bhandari, Fl. Indian Desert 114. 1978; Singh, in Shetty & Singh Fl. Rajasthan 1: 234. 1987. (Fig. 40 & Plate 88)

Vern.: *Gorakhi*.

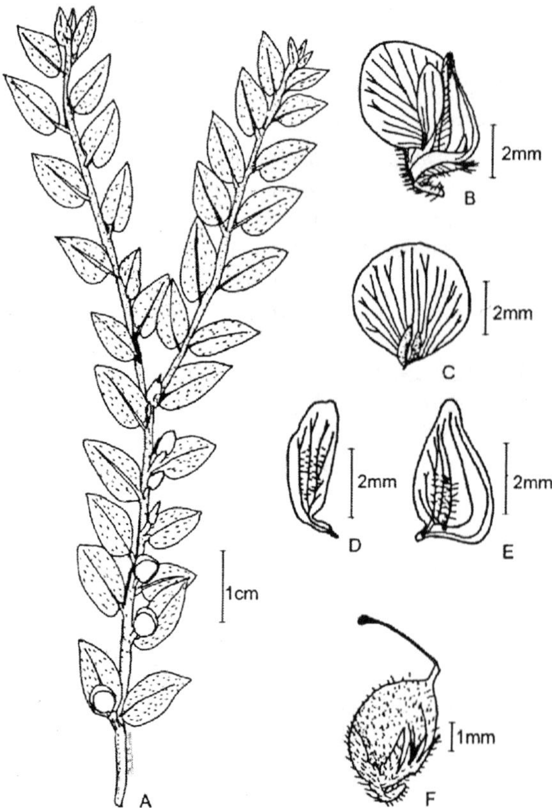

Fig. 40. *Crotalaria hebecarpa* (DC.) Rudd. A. Fruiting branch, B. Flower, C. Vexilum, D. Wing, E. Keel, F. Fruit

Prostrate annual herb, branches pilose, hairs long. Leaf simple, 0.6-1.7 x 0.4-0.8 cm, subsessile, ovate, apex subacute, obliquely cordate at the base, pilose on both sides. Flower solitary axillary, subsessile. Calyx pubescent, tube short, 2 upper teeth partly fused, lower 3 free. Corolla yellow. Vexillum ca 3-6.5 mm long. Anthers dimorphic. Fruit ca 3-4.5 mm long, subglabrous to pubescent, flattened, 2-seeded. Seeds reniform, brown.

Field note: Common in dry habitats as well as in grasslands.

Fl. & Fr.: February - June.

Specimen examined: Bhilwara 1147, Bansi 2153 (MLVGCB Herb).

Distribution: India, Pakistan and Sri Lanka.

Crotalaria juncea L., Sp. Pl. 714. 1753. Baker, in Hook. *f.,* Fl. Brit. Ind. 1 : 79. 1876; Duthie, Fl. Gangetic Plain 1: 206. 1903; Cooke, Fl. Bomb. Pres. (reprint ed.) 1: 320. 1958; Singh, in Shetty & Singh Fl. Rajasthan 1: 276. 1987; Sanjappa, Legumes India 122. 1992. *C. benghlensis* Lam., Encycl. 2: 196. 1786. *C. tenuifolia* Roxb. Fl. India 2: 263. 1832.

Vern.: *Sann, Hann.*

Annual, ca 60 - 250 cm tall; many ascending branches, pubescent. Leaf simple, ca 2.5 - 10.5 cm x. 6 - 20 mm, linear or oblong, obtuse or subacute, apiculate, pubescent on both sides, hairs appressed, silky; petiole ca 1.2 - 2.5 mm long; stipules almost absent. Inflorescence an erect terminal and lateral raceme, up to 30 cm long, 12 - 20-flowered. Pedicel ca 3 - 7 mm long. Bract minute; bracteoles 2, below the calyx. Calyx ca 1.8 - 2.0 cm long, pubescent, teeth linear-lanceolate. Corolla bright yellow. Vexillum ovate-oblong, slightly exserted. Fruit ca 2.5 - 3.2 cm long, sessile, pubescent, 10-15-seeded. Seeds subreniform, yellowish brown.

Field note: Commonly cultivated for fibres, which provide the Sunn Hemp of commerce. Young plants serve as green manuar and are destroyed in fields to increase soil fertility.

Fl. & Fr.: August - October.

Specimen examined: Kotri 2160 (MLVGCB Herb).

Distribution: Doubtfully wild; cultivated extensively for fibre in India, Pakistan, Burma, Malay Isles, Australia, Russia, introduced in Tropical Africa

Crotalaria medicaginea Lam. Encycl. Meth. Bot. 2: 201. 1786; Baker, in Hook. *f.,* Fl. Brit. Ind. 2: 81. 1876; Duthie, Fl. Gangetic Plain 1: 207. 1903; Bhandari, Fl. Indian Desert 111. 1978; Singh, in Shetty & Singh Fl. Rajasthan 1: 218. 1987; Sanjappa, Legumes India 124. 1992. *C. neglecta* Wight & Arn. Prodr. 192. 1834. *C. luxurians* Benth. in Hook. Lond. J. Bot. 2 : 578. 1843. *C. medicaginea* Lam. var. *neglecta* (Wight & Arn.) Baker & var. *luxurians* (Benth.) Baker, in Hook. *f., l.c.* 2 : 81. 1876. (Fig. 41 & Plate 89)

Vern.: *Gugrio.*

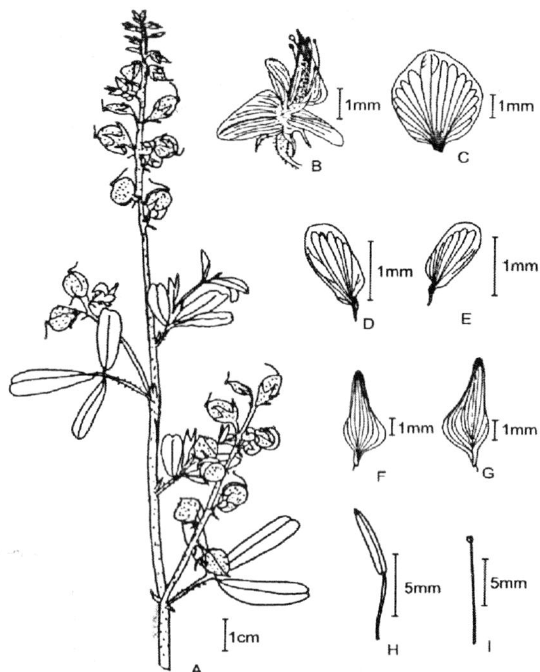

Fig. 41. *Crotalaria medicaginea* Lamk.
A. Flowering & fruiting branch, B. Flower, C. Vexillum, D. & E. Wings, F. & G. Keel, H. Fertile staman, I. Staminode.

Annual herb, ca 70 cm tall. Stem branched, sparsely pubescent, hairs appressed. Leaf trifoliolate, petiole ca 2.5 - 5.5 mm long; leaflets ca 6 - 10 x 2.5 - 4.5 mm, oblanceolate, obtuse to subtruncate, apiculate, glabrous above, pubescent below; petiolules short; stipules minute, filiform. Inflorescence terminal and leaf-opposed, 2-15 flowered raceme, flowers yllow. Bract minute, linear, subulate. Calyx ca 2.5 mm long, pubescent, teeth acute, longer than the tube. Corolla yellow. Vexillum ca 5-6 mm long, pubescent externally. Fruit subglobose, ca 4.0 mm or less long, pubescent, beaked, 2-seeded. Seeds 2, 1.5 - 1.8 mm long, suborbicular, compressed, glabrous. brown.

Field note: Common in waste lands, corners of fields etc.

Fl. & Fr.: August - September.

Specimen examined: Bhinay 1849 (MLVGCB Herb).

Distribution: India, Malay Isles, Afghanistan, Pakistan, Punjab, China and Australia.

Uses: The fresh plants aused as good fodder.

Crotalaria spectabilis Roth, Nov. Pl. Sp. 341. 1821; Ali, in Biologia 12 (2): 26. 1966; Singh in Shetty & Singh Fl. Rajasthan 1: 221. 1987; Sanjappa, Legumes India 130. 1992. *C. sericea* Retz. Obs. Bot. 5: 26. 1789 non L. 1768; Baker, in Hook. *f.,* Fl. Brit. Ind. 2: 75. 1876; Duthie, Fl. Gangetic Plain 1: 205. 1903; Cooke, Fl. Pres. Bombay 1: 319. 1958 (Repr.). *C. leschenenaulti* DC. Prodr. 2: 125. 1825; Baker, in Hook. *f.,* op. cit. 76.

Vern: *Marhan.*

An undershrub, 1 - 2 m tall, branches subglabrous. Leaf 4 - 15 x 1 - 7 cm, variable in breadth, oblong-lanceolate, acute or subacute, mucronate, glabrous above, finely pubescent below; petiole ca 1.5 - 3.5 mm long; stipules large, leafy. Inflorescence a 20 - 50-flowered, terminal raceme, ca 25 - 45 cm long. Bract 12 - 20 mm long, ovate, foliaceous; bracteoles 2, 1 - 2 mm long, situated below the middle of the pedicel. Calyx ca 11 - 14 mm long, subglabrous, teeth ca 8 - 9 mm long, upper teeth large, triangular. Corolla yellow with a purplish tinge. Fruit ca 2.5 - 5.5 cm long, linear-oblong, glabrous, 20 - 30-seeded; stipitate.

Field note: Rare in waste lands.

Fl. & Fr.: September - December.

Specimen examined: Bijoliya 1191, Mandalgarh 1938 (MLVGCB Herb).

Distribution: Pakistan, India, Sikkim, Nepal and Burma.

DALBERGIA L. f. nom. cons.

KEY TO THE SPECIES

1a. Leaflets 3 or 4 pairs, apex rounded or emarginate, leaf rachis straight .... .*D. latifolia*

1b. Leaflets 2 pairs, apex abruptly acuminate, leaf rachis zig zag ....*D. sissoo*

Dalbergia latifolia Roxb., Pl. Cor. 2: 7. t. 113. 1798; Baker, in Hook. *f.,* Fl. Brit. Ind. 2: 231.

1876; Duthie, Fl. Gangetic Plain 1: 263. 1903; Singh, in Shetty & Singh Fl. Rajasthan 1: 223. 1987; Sanjappa, Legumes India 137. 1992.

Vern.: *Shisham.*

Large tree, 10 -15 m high; branches glabrous. Leaf 10 - 15 cm long; leaflets 2.5 - 8.0 x 2.0 - 7.2 cm, elliptic-oblong, orbicular or elliptic-obovate, obtuse, rarely emarginate, glabrous; petiolule long. Inflorescence an axillary panicle. Flowers creamy white in axillary, lax panicles; pedicel as long as the calyx. Calyx teeth obtuse, shorter than the cup. Corolla whitish, almost twice as long as the cup. Stamens 9. Fruit oblong-lanceolate, reticulated veined, 1 - 4-seeded.

Field note: Occasional in mixed deciduous forests.

Fl. & Fr.: July - February.

Specimen examined: Bijoliya 609, Gurla 1185 (MLVGCB Herb).

Distribution: Native of India, Nepal, Sikkim; introduced in some parts of the Punjab.

Uses: The black heart wood with purple streaks is much valued for furniture. Foliage is used as fodder.

Dalbergia sissoo Roxb., Fl. Ind. 3: 223. 1832; Baker, in Hook. *f.,* Fl.Brit.Ind. 2: 231. 1876; Parker, For. Fl. Punj. ed. 3: 166. 1956; Baker, in Hook. *f., l. c.* 231. 1876; Duthie, Fl. Gangetic Plain 1: 264. 1903; Bhandari, Fl. Indian Desert 104. 1978; Singh in Shetty & Singh, Fl. Rajasthan 1: 224. 1987; Sanjappa, Legumes India 140. 1992.

Vern.: *Shisham.*

Tree upto 10 m high with rough bark and mainly longitudinal furrows, young branch pubescent. Leaf imparipinnate, rachis ca 3.7 - 7.5 cm long; leaflets 3 - 5, ca 3.5 -6.5 cm long, broadly ovate or suborbicular, acuminate, glabrescent, petiolule ca 5 - 8 mm long; stipules ca 5 mm long. Inflorescence an axillary panicle, composed of several short spikes with sessile to subsessile flowers. Bract small, pubescent, caducous. Calyx ca 5 mm long, teeth ciliate, unequal, shorter than the tube. Corolla yellowish white. Stamens 9, monadelphous, tube slit on the upper side only, anthers uniform. Ovary pubescent, 2 - 4-ovulate, style glabrous, stigma capitate. Fruit ca 3.7 - 10 cm x 7.0 -13 mm, strap-shaped, glabrous, 1 - 4-seeded. Seed flattened.

Field note: Very widely planted in the plains along the roadsides, fields and in the forest.

Fl. & Fr.: March - May.

Specimen examined: Kishangarh 80 (MLVGCB Herb).

Distribution: India, Pakistan, Afghanistan, Persia and Iraq.

Uses: The wood which is hard, heavy and durable, is very important commonly used for furniture, carts, boats, wheels etc.

DESMODIUM Desv. nom. cons.

Desmodium gangeticum (L.) DC. Prodr. 2 : 327. 1825; Baker, in Hook. *f.,* Fl. Brit. Ind. 2: 168. 1876; Duthie, Fl. Gangetic Plain 1: 286. 1903; Cooke, Fl. Bomb. Pres. (reprint ed.) 1: 379. 1958; Singh, in Shetty & Singh Fl. Rajasthan 1: 226. 1987; Sanjappa, Legumes India 153. 1992. *Hedysarum gangeticum* L. Sp. Pl. 746. 1753. *H. maculatum* L. *l.c.* 746. 1753. *Desmodium gangeticum* (L.) DC. var. *maculatum* (L.) Baker in Hook. *f., l.c.* 2 : 168. 1876; Duthie, *l.c.* 1: 287. 1903. (Plate 90)

Vern.: *Salvan, Salparni.*

Undershrub, erect, 60 - 120 cm tall, branches pubescent. Leaves unifoliolate, petiole 1 - 2.5 cm long, leaflets, 4 - 12.5 x 2.5 - 7.5 cm, ovate-oblong, rarely elliptic-oblong or lanceolate, acute or sub-acuminate, rounded at base, glabrous above, pubescent below; stipules 6 - 8.5 mm long, scarious; petiolule ca 1.5 mm long, stipules ca 3 mm long, lanceolate, scarious. Inflorescence terminal or axillary raceme, 15 - 30 cm long. Flowers clustered in slender. Bracts 1.5 - 3.5 mm long, triangular subulate. Pedicel 4 - 6.5 mm long, pubescent. Calyx ca 2.5 mm long, pubescent, teeth longer than the tube. Corolla 3.8 - 4.5 mm long, violet or white. Fruit 1.2 - 4.5 cm x 2.5 mm, subfalcate, upper suture slightly indented, lower deeply indented between the seeds, urticles 4 - 8, longer than broad, indehiscent, pubescent, hairs hooked, minute. Seeds suborbicular, smooth, brown.

Field note: Common in open forests, Waste lands etc.

Fl. & Fr.: August - November.

Specimen examined: Mainal 103 (MLVGCB Herb).

Distribution: India (Punjab, Kumaon, Bengal, Madras, Rajasthan, Madhya Pradesh, Bombay), Pakistan, Sri Lanka, Burma, Tropical Africa, Malaysia, Philippines and China.

ERYTHRINA L.

Erythrina suberosa Roxb., Fl. Ind. 3: 253. 1832; Baker, in Hook. *f.,* Fl. Brit. Ind. 2: 189. 1876; Duthie, Fl. Gangetic Plain 1: 238. 1903; Cooke, Fl. Bomb. Pres. (reprint ed.) 1: 392. 1958; Roxb. Ic. no. 108. 1971; Singh, in Shetty & Singh Fl. Rajasthan 1: 231. 1987; Sanjappa, Legumes India 173. 1992.

Vern.: *Dhoro khankharo.*

Erect tree, 10 - 17 m tall, branches armed with prickles. 3 - 6 mm long, which fall off after the third year. Leaf sometimes armed, trifoliolate, petiole 10 - 20 cm long, leaflets 7.5 - 15 x 12. 0 cm, the terminal rhomboid ovate, the lateral obliquely deltoid, glabrous above, densely pubescent below; stipels glanduliform; stipules linear-lanceolate, caducous. Racemes 14, near the tip of branches, 5 - 10 cm long, peduncle short. Flowers in scarlet fascicles of 1 - 3. Pedicel 12 - 13 mm long; bracts caducous, bracteoles minute. Calyx 6-10 mm long, pubescent, becoming bilabiate. Corolla scarlet. Vexillum 3.7 - 5.0 cm x 12-16 mm. Fruit stipitate, 12.5 - 15.0 cm long, torulose, 2 - 5-seeded. Seeds oblong.

Field note: Occasional in forests, along cultivated fields.

Fl. & Fr.: March - June.

Specimen examined: Rawaliya 2111 (MLVGCB Herb).

Distribution: Pakistan: India, commonly cultivated.

Uses: Wood is soft and light but tough, fairly durable. Flowers are given as an antidot of poisons.

INDIGOFERA L.

KEY TO THE SPECIES

1a. Leaves simple 2

2a. Leaves ovate-cordate.................... *I. cordifolia*

2b. Leaves not cordate 3

3a. Fruits 1-seeded............................... *I. linifolia*

3b. Fruits 6 -8 seeded...................... *I. oblongifolia*

1b. Leaves compound 4

4a. Leaflets 3 5

5a. Fruits quit flat and papery .... *I. hochstetteri*

5b. Fruits turgid...*I. trita*

4b. Leaflets more than 3 5

5a. Fruit tetragonal with stiff spreading hairs ....*I. hirsuta*

5b. Fruit not as above 6

6a. Leaflets alternate. Flowers in axillary, sessile or shortly peduncled heads. ... *I. linnaei*

6b. Leaflets opposite. Flowers in axillary racemes 7

7a. Fruits torulose. Fruit 10 - 20 seeded..............8

7b. Fruits not torulose. Fruit 1-6 seeded .... *I. argentea*

8a. Racemes 12 - 20 flowered. Fruits slightly curved, 8-10 seeded .... *I. tinctoria*

8b. Racemes 6 - 12 flowered. Fruits curved, 1-4 seeded. .... *I. caerulea*

Indigofera argentea Burm. f. Fl. Ind. 171. 1768; Cooke, Fl. Bomb. Pres. 1: 317.1902; Bhandari, Fl. Indian Desert 106. 1978; Singh, in Shetty & Singh Fl. Rajasthan 1: 237. 1987; Sanjappa, Legumes India 184. 1992 & in Hajra *et al.* Fl. India Fasc. 21: 12. 1995. *I. semitrijuga* Forssk. var. *tetrasperma* DC. Prodr. 2: 230. 1825. *I. semitrijuga auct.* non Forsk. 1775; Baker, in Hook. *f.,* Fl. Brit. Ind. 2: 998. 1876.

Undershrub, perennial, up to 60 cm tall, branches numerous, pubescent, hairs appressed as well as glandular. Leaf imparipinnately compound, 0.9 - 2.6 cm long, petiole ca 4.5 - 10 mm long. Leaflets 7-11, lateral leaflets subsessile, ca 3.5 - 4.5 mm long, broadly obovate, obtuse, densely appressed pubescent on both sides. Inflorescence a pedunculate raceme. Pedicel very short. Calyx 2 - 2.5 mm long, pubescent, teeth ca 1 - 1.2 mm long. Vexillum ca 5 - 6 mm long. Fruit ca 9 - 13 mm long, pubescent, hairs appressed as well as glandular, slightly torulose, generally 4 - 6 rarely 8 - seeded.

Field note: Rigid, woody herb, common in waste lands.

Fl. & Fr.: August - January

Specimen examined: Asind 158 (MLVGCB Herb).

Distribution: India (Rajasthan), Persia, Arabia, Pakistan, Egypt, Sudan, Somalia, Ethiopia, Libya and Sudan.

Indigofera caerulea Roxb., Fl. Ind. 3: 377. 1832; Wight, Ic. 2 (2): 5. t. 366. 1940; Bhandari, Fl. Indian Desert 107. 1978; Singh, in Shetty & Singh Fl. Rajasthan 1: 238. 1987; Sanjappa, Legumes India 186. 1992. *I. argentea* L. Mant. 2: 273. 1771; Baker, in Hook. *f.*, Fl. Brit. Ind. 2: 98. 1876. *I. argentea* L. var. *caerulea* (Roxb.) Baker, in Hook. *f.*, Fl. Brit. Ind. 2 : 99. 1876. (Plate 91)

Herb, up to 1 m tall, branches whitish, hoary. Leaf imparipinnately compound, petiole up to 2 cm long, rachis up to 8 cm long; leaflets 5-9 rarely 11, up to 32 mm long and 23 mm broad, oblong-obovate, densely pubescent below, glabrous above; stipels absent, petiolule ca 2 mm long; stipules ca 1-3 mm long. Inflorescence a raceme, sessile, up to 4 cm long. Bract ca 1 mm long; pedicel ca 1 mm long. Calyx ca 1.5-2 mm long, densely pubescent, teeth equal to the cup. Corolla golden brown. Vexillum ca 4.5 mm long. Fruit curved, torulose, indehiscent, with 3-4 segments, each ca 3-4 mm long, ca 2-3 mm wide, upper suture ca 1.5 mm wide, pubescent.

Field note: Common in waste land, hillocks and dry habitats.

Fl. & Fr.: August-November.

Specimen examined: Kodukota 1347, Tikar 1423 (MLVGCB Herb).

Distribution: India (Gujarat, Rajasthan), Pakistan, Arabia, Somali Republic and Sudan.

Indigofera cordifolia Heyne ex Roth, Nov. Pl. Sp. 357. 1821; Baker, in Hook. *f.*, Fl. Brit. Ind. 2: 93. 1876; Duthie, Fl. Gangetic Plain 1: 250. 1903; Cooke, Fl. Bomb. Pres. 1: 311. 1902; Bhandari, Fl. Indian Desert 117. 1978; Singh, in Shetty & Singh Fl. Rajasthan 1: 239. 1987; Sanjappa, Legumes India 187. 1992 & in Hajra *et al.* Fl. India Fasc. 21: 45. 1995. (Fig. 42 & Plate 92)

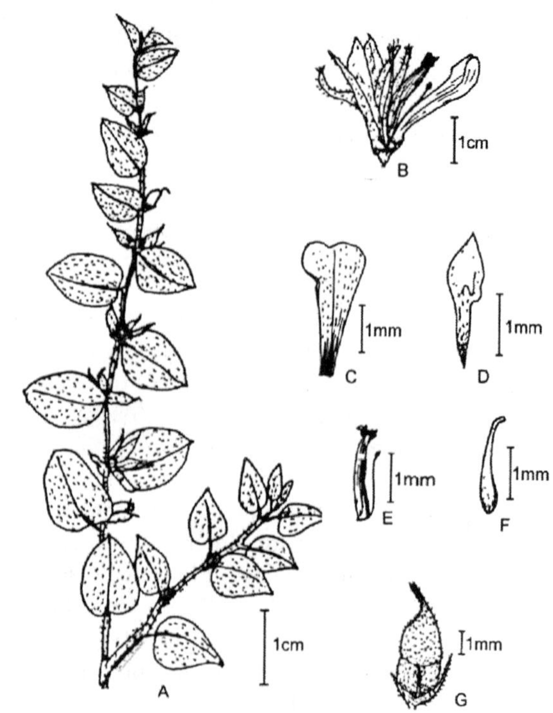

Fig. 42. *Indigofera cordifolia* Heyne ex Roth: A. Habit, B. Flower, C. Vexillum, D. Keel, E. Stamens, F. Gynoecium, G. Fruit.

Prostrate or diffuse annual herb, branched, stem pubescent. Leaf simple, ca 6 - 16 mm long, ca 4 - 13 mm broad, subsessile, broadly ovate, cordate, obtuse, mucronate, pubescent on both sides; stipules minute, setaceous. Inflorescence a subsessile, 4-8-flowered head. Calyx ca 3 - 3.5 mm long, pubescent externally, tube very short, teeth linear, densely pubescent. Corolla bright red, shorter than calyx. Fruit ca 4 - 5 mm long, oblong, densely pubescent, 2-seeded, rarely 1-seeded. Seeds pitted, inner side of the pericarp dotted.

Field note: Common in rocky or gravelly habitats.

Fl. & Fr.: August- December.

Specimen examined: Bhim 1415 (MLVGCB Herb).

Distribution: India, Pakistan, Afghanistan, Indonesia, Australia, Ethiopia, Sudan and Africa.

Indigofera hirsuta L., Sp. Pl. 751. 1753; Baker, in Hook. *f.,* Fl. Brit. Ind. 2: 98. 1876; Duthie, Fl. Gangetic Plain 1: 255. 1903; Mahesh. Illu. Fl. Delhi f. 59. 1966; Singh, in Shetty & Singh Fl. Rajasthan 1: 239. 1987; Sanjappa, Legumes India 191. 1992 & in Hajra *et al.* Fl. India Fasc. 21: 81. f. 31. 1995. *I. astragalina* DC., Prodr. 2: 228. 1825. (Fig. 43 & Plate 93)

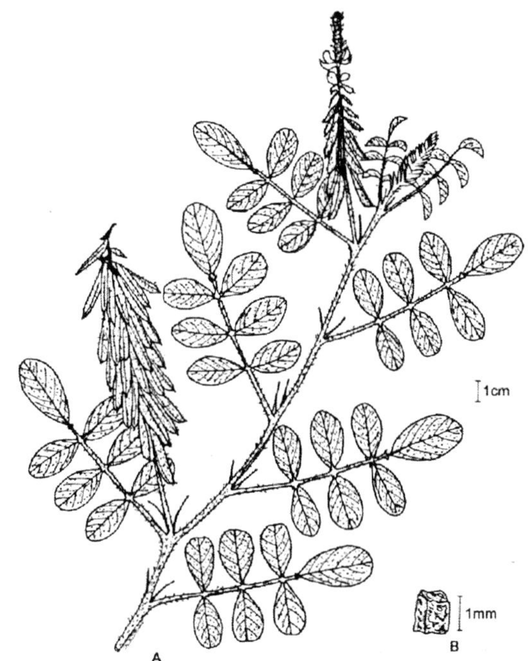

Fig. 43. *Indigofera hirsuta* L.
A. Flowering and fruiting twig, B. Seed

Erect or spreading annual, up to 1.5 m tall. Stem pilose, hairs stiff, mostly brown. Leaf imparipinnately compound, petiole up to 2 cm long, pilose; rachis up to 7 cm long, pilose; leaflets 5 - 11, up to 40 x 25 mm, pilose on both sides; petiolules ca 2 mm long; stipules up to 1 cm long. Inflorescence a raceme, peduncle 25 mm long, hirsute. Bracts upto 5 mm long, caducous. Pedicel ca 1 mm long. Calyx ca 4 mm long, brown hirsute, teeth linear setaceous, cup very small. Vexillum ca 4 - 4.5 mm long. Fruit tetragonal, ca 10 - 20 mm long, ca 3 mm wide, hirsute, up to 6-seeded.

Field note: Common in waste lands and open forests. It is highly polymorphic species due to variability in length of peduncle, size of pod, number of seeds, hair colour, leaflets number etc.

Fl. & Fr.: August - November.

Specimen examined: Bhilwara 105, Bhilwara 1423 (MLVGCB Herb).

Distribution: India, Pakistan, Sri Lanka, Burma, Indochina, China, Australia, New Guinea, Tropical Africa, Madagascar, S. Africa and introduced in West Indies.

Indigofera hochstetteri Baker in Oliver. Fl. Trop. Afr. 2: 101. 1871; Bhandari, Fl. Indian Desert 107. 1978; Singh in Shetty & Singh Fl. Rajasthan 1: 241. 1987; Sanjappa, Legumes India 191. 1992 & in Hajra *et al.* Fl. India Fasc. 21: 83. 1995. *I. anabaptista* Steud. ex Baker, in Hook. *f.,* Fl. Brit. India 2: 102. 1876; Duthie, Fl. Gangetic Plain 1: 251. 1903. (Fig. 44 & Plate 94)

Vern.: *Bagdo.*

Annual herb or undershrub, prostrate with numerous nearly glabrous branches. Leaf imparipinnate, ca 2.5 - 3.8 cm long, petiole ca 6-13 mm. Leaflets 3-5, opposite, ca. 1 - 2.5 cm x 0.4 - 0.8 mm, obovate-oblong or oblanceolate, apiculate, pubescent on both sides; stipules ca. 4.5 mm long. Inflorescence a short axillary raceme, up to 9 cm long. Flowers red. Calyx 2 - 3 mm long, pubescent, teeth longer than the tube. Corolla slightly shorter, equal or slightly longer than the calyx; standerd obovate-mucronate, keel densely hairy outside. Anthers ca 0.5 mm long. Ovary linear, sparsely hairy on dorsal side; style glabrous. Fruit 0.8 - 2.0 cm x 3.5 mm, flat, appressly pubescent, with a raised longitudinal ridge along the centre of each flattened face and a transverse ridge between the seeds, 5-9-seeded. Seeds discoid, brownish.

Field note: Commonly found in area during rainy season.

Fl. & Fr.: August - November.

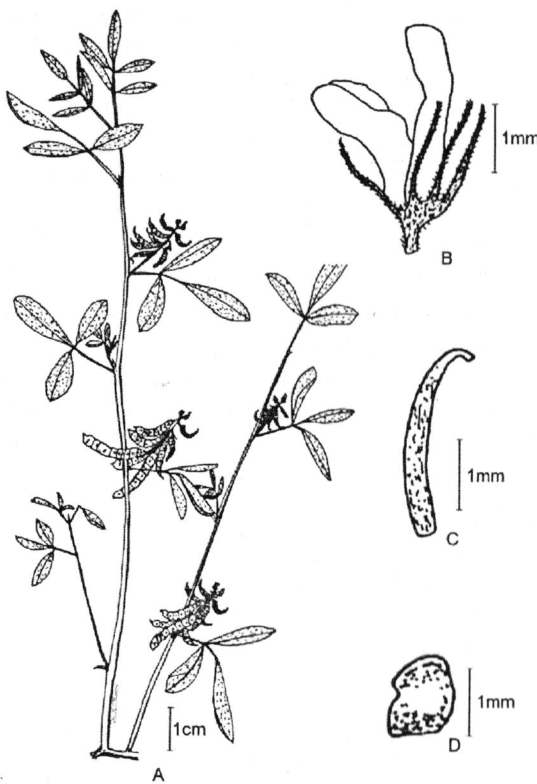

Fig. 44. *Indigofera hochstetteri* Baker.
A. Flowering & fruiting branch, B. Flower, C. Ovary D. Seed

Specimen examined: Atun 1338, Asind 1423 (MLVGCB Herb).

Distribution: India, Pakistan, Afghanistan, Arabia, Somali Republic, Ethiopia, Eritrea, Sudan, Nigeria, Kenya, Uganda, Tanganyika, Congo.

Uses : Tribals of Asind tehsil use seeds to mix with flour and consume as a food.

Indigofera linifolia (L.) Retz. Obs. 4: 29. 1786; Baker, in Hook. *f.,* Fl. Brit. Ind. 2: 92. 1876, incl. var. *campbelli* Baker; Duthie, Fl. Gangetic Plain 1: 249. 1903; Cooke, Fl. Bomb. Pres. 1: 310. 1902; Bhandari, Fl. Indian Desert 108. 1978; Singh, in Shetty & Singh , Fl. Rajasthan 1: 242. 1987; Sanjappa, Legumes India 192. 1992 & in Hajra *et al.* Fl. India Fasc. 21: 93. f. 36. 1995. *Hedysarum linifolium* L.f. Suppl. 331. 1781. *Sphaeridiophorum linifolium* (L.) Desv. in J. Bot. 3: 125. t. 6. f. 35. 1813. *Indigofera roxburghii* Tausch in Flora 14: 171. 1831. *Sphaeridiophorum abyssinicum* Jaub. & Spach, III. Pl. Orient. t. 494. 1857. (Fig. 45 & Plate 95)

Var: ***linifolia.***

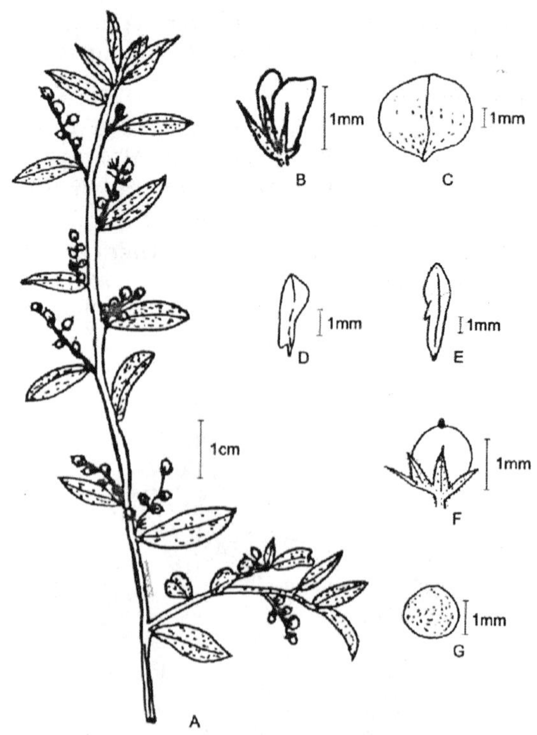

Fig. 45. *Indigofera linifolia* (L.) Retz. var. *linifolia* A. Flowering & fruiting twig, B. Flower, C. Vexillum, D. Wing, E. Keel, F. Fruit, G. Seed

Annual 20 - 50 cm long, much branched, prostrate; branches pubescent, hairs appressed. Leaf simple, ca 1.2 - 2.6 cm x 1.5 - 2.6 mm; leaflets 5 - 9, 4 - 8 x 2 - 5 mm, ovate to obovate, rounded at base, rounded to emarginate at apex, appressly pubescent on both sides; sessile to subsessile; stipules minute, setaceous. Inflorescence a sessile or shortly peduncled axillary raceme. Flowers pinkish-red; bracts, triangular, scarious. Calyx ca 2.5 - 2.6 mm, pubescent, teeth linear, longer than the tube. Corolla bright red. Vexillum ca 5-8 mm long, elliptic-ovate, hairy on the back; wing and keel glabrous. Anthers ca 0.3 mm log. Ovary globose, rounded to emarginate at apex. Fruit ca 1.5 - 2.0 mm long, globose, apiculate, appressly pubescent, 1-seeded. Seeds 1 mm, globose, dark brown.

Field note: Common in open plains and near cultivated fields.

Fl. & Fr.: Almost throughout the year.

Specimen examined: Bhilwara 106 (MLVGCB Herb).

Distribution: India, Pakistan, Bangla Desh, Afghanistan, Sri Lanka, Burma, Indochina, Siam, Indonesia, New Guinea, Australia, China, Eritrea, Ethiopia and Sudan.

Indigofera linnaei Ali in Bot. Notiser. 111: 549. 1958 & in Nasir & Ali, Fl. West Pakistan 100: 1977; Bhandari, Fl. Indian Desert 108. 1978; Singh, in Shetty & Singh Fl. Rajasthan 1: 242. 1987; Sanjappa, Legumes India 192. 1992 & in Hajra *et al*. Fl. India Fasc. 21: 96. 1995. *Hedysarum prostratum* L. Mant. Pl. 1: 102. 1967; Brum. f., Fl. India 168. t. 1. 1768. *Indigofera enneaphylla* Mant. Pl. 2: 272. 1771, nom. illegit; Baker, in Hook. *f.*, Fl. Brit. Ind. 2: 94. 1876; Duthie, Fl. Gangetic Plain 1: 250. 1903. *I. prostrata* (L.) Domin, Bibl. Bot. 89: 187. 1926, non Willd. 1802. (Plate 96)

Stem trailing prostrate, much branched, pubescent, hairs appressed, white. Leaf imparipinnately compound, leaflets alternate, 5 - 9 per leaf, 4 - 8 x 2 - 5 mm, ovate-obovate, rounded at base, sessile, obtuse to emarginate, pubescent on both sides; stipules ca 3 mm long, lateral, free, pubescent. Flowers in dense axillary spicate racemes, pinkish-red; peduncle 0 - 2 cm long. Bracts ca 2 mm long. Calyx ca 3 - 4 mm long, pubescent outer side, teeth longer than the cup. Corolla bright red; vexillum ca 4 mm long, ovate-spathulate, rounded or mucronate at apex, strigose on back; wings minutely ciliate along margins. Ovary 2 - 3 mm long; style glabrous. Fruit 4 - 6 x 1.5 - 2 mm, turgid, torulose, pubescent, dehiscent, apiculate, 1 - 3-seeded. Seeds truncate, greenish brown.

Field note: Common in open plains, near ponds, cultivated fields etc.

Fl. & Fr.: Almost throughout the year.

Specimen examined: Potla 1581, Bhilwara 1523 (MLVGCB Herb).

Distribution: Pakistan. India, Sri Lanka, Burma, Indochina, Indonesia and Australia.

Uses: Root is used to make a paste, 1 tsp paste + 1 glass of water is given to children 1-2 times a day to cure diarrhoea.

Indigofera oblongifolia Forsk., Fl. Aegypt. Arab. 137. 1775; Duthie, Fl. Gangetic Plain 1: 253. 1903; Bhandari, Fl. Indian Desert 109. 1978; Singh, in Shetty & Singh, Fl. Rajasthan 1: 243. 1987; Sanjappa, Legumes India 193. 1992 & in Hajra *et al*. Fl. India Fasc. 21: 108. 1995. *I. paucifolia* Delile, Fl. Egypt. 107. t. 37. f. 2. 1813; Baker, in Hook. *f.*, Fl. Brit. Ind. 2: 97. 1876. *I. argentea* auct. non Burm. f., 1768, nec L. 1771; Buch.-Ham. ex Roxb., Fl. Ind. 3: 374. 1832. *Bremontiera ammoxylon* DC. var. *burmanni* DC. Prodr. 2: 353. 1825. (Fig. 46 & Plate 97)

Vern.: *Jojru*.

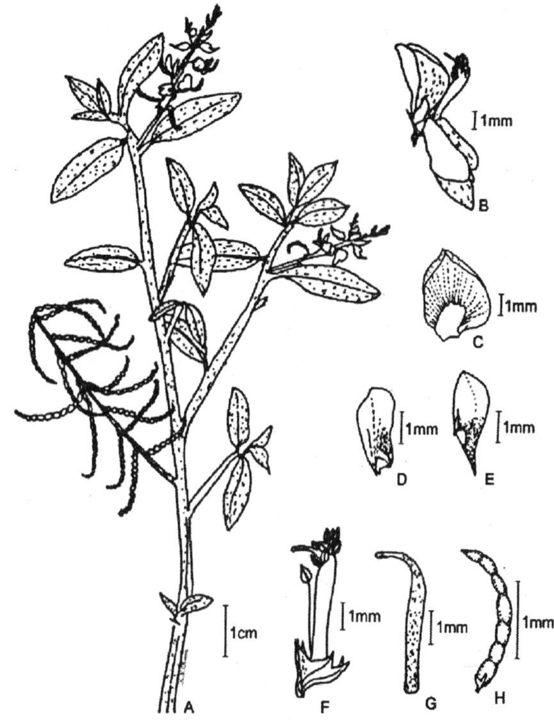

Fig. 46. *Indigofera oblongifolia* Forsk.
A. Flowering and fruiting twig, B. Flower, C. Vexillum, D. Wing, E. Keel, F. Sepals & Stamens, G. Pistil, H. Fruit.

Shrub with many stout branches, pubescent. Leaf imparipinnately compound, rarely unifoliolate, generally 3 - 5-foliolate, leaflets alternate, ca 1.5-2.6 cm x 6 - 13 mm, oblanceolate or elliptic-oblong, pubescent on both sides; petiole ca 6 - 13 mm long; stipules ca 3 - 3.5 mm long. Inflorescence a raceme up to 10 cm long. Flowers red; pedicel short. Bracts hairy. Calyx ca 2 - 2.5 mm long, pubescent outside, teeth ca 1.25 mm long. Corolla red; vexillum ca 5-10 mm long,. pubescent externally; wings fleshy, glabrous; keel ciliate along the margins, scabrous towards the tip. Fruit 1.2 - 2.0 cm long, slightly curved, torulose, hoary when young, hairs appressed, 6 - 8-seeded. Seeds 4 - gonous, oblong, smooth, reddish brown.

Field note: Rare in grasslands.

Fl. & Fr.: September - March.

Specimen examined: Bhilwara 107 (MLVGCB Herb).

Distribution: India, Pakistan, Sri Lanka, Java, Jordan, Yemen, Bahrein, Egypt, Sudan, Senegal, Angola and Nigeria.

Uses: Plants are used as fodder.

Indigofera tinctoria L. Sp. Pl. 751. 1753; Baker, in Hook. *f.,* Fl. Brit. Ind. 2: 99. 1876; Duthie, Fl. Gangetic Plain 1: 254. 1903; Cooke, Fl. Bomb. Pres. (reprint ed.) 1: 339. 1958; Bhandari, Fl. Indian Desert 109. 1978; Singh, in Shetty & Singh Fl. Rajasthan 1: 244. 1987; Sanjappa, Legumes India 195. 1992 & in Hajra *et al*. Fl. India Fasc. 21: 133. f. 51. 1995.

Vern.: *Leel*, *Neel*.

Shrub, 60 - 120 cm tall, branches pubescent, hairs appressed. Leaves 1.2 - 12.5 cm long; leaflets 1 - 2 x 0.6 - 1 cm, obovate or oblanceolate, glabrous to subglabrous above, appressly pubescent below; petiolule ca 1 mm long; stipules small. Inflorescence a subsessile raceme, ca 5 - 12 cm long. Calyx ca 1 - 1.5 mm long, pubescent, outside, teeth as long as the tube. Corolla pink; vexillum ca 4 - 4.5 mm long, ovate, adpressed hairy on the back. Fruit 1.8 - 3.8 cm x 2.5 mm, straight or slightly curved, glabrous, 10 - 12-seeded. Seeds truncate at both ends, brownish-black.

Field note: Rare, found in sandy gravelly habitats.

Fl. & Fr.: September - November.

Specimen examined: Tikar 1584, Jahajpur 1842 (MLVGCB Herb).

Distribution: Pakistan, India, Sri Lanka, Burma, Indonesia, Philippines, Siam, Indo China, Malaya, T. Africa and introduced in Tropical America, Formerly cultivated throughout the tropics for Indigo.

Indigofera trita L. *f.* Suppl. 335. 1781; Wight, Ic. 1: t. 315. 1840; Baker, in Hook. *f.,* Fl. Brit. Ind. 2: 96. 1876; Duthie, Fl. Gangetic Plain 1: 252. 1903; Singh, in Shetty & Singh Fl. Rajasthan 1: 245. 1987; Sanjappa, Legumes India 196. 1992 & Hajra *et al*. Fl. India Fasc. 21: 143. f. 54. 1995. *I. hedysaroides* Lam. Encycl. 3: 250. 1789. (Plate 98)

subsp. ***trita***

Erect woody undershrub, clothed with fine appressed hairs, 60 - 100 cm tall. Branches white canescent. Leaf trifoliolate, upto 5 cm long; leaflets 2 - 3.5 x 1 - 4 cm long, obovate or oblong, pubescent on both sides; stipels linear. Infloresc-ence axis 1 - 2 cm long, 6 - 12-flowered. Calyx white canescent; teeth setaceous. Standard obovate, obtuse-mucronate; keel densely hairy outside. Anthers 0.5 mm long. Ovary linear, glabrous. Fruit 2 - 3 x 0.2 - 0.3 cm. spreading or deflexed, 6 - 10-seeded. Seeds oblong, smooth.

Field note: Common in open forests and waste lands.

Fl. & Fr.: September - February.

Specimen examined: Bari Sadari 4218 (MLVGCB Herb).

Distribution: Pakistan, India, Burma, Sri Lanka, Australia and Indonesia.

LATHYRUS L.

KEY TO THE SPECIES

1a. Leaves without leaflets, stipules foliaceous, stem not winged, flowers yellow ....*L. aphaca*

1b. Leaveswith a pair of leaflets, stipules not foliaceous, stem winged, flowers purple ....*L. sativus*

Lathyrus aphaca L. Sp. Pl. 729. 1753; Baker, in Hook. *f.* Fl. Brit. Ind. 2: 179. 1876; Duthie, Fl. Gangetic Plain 1: 260. 1903; Bhandari, Fl. Indian Desert 110. 1978; Singh, in Shetty & Singh Fl. Rajasthan 1: 246. 1987; Sanjappa, Legumes India 200. 1992. (Fig. 47 & Plate 99)

Vern.: *Chapti-matri, Arak matri.*

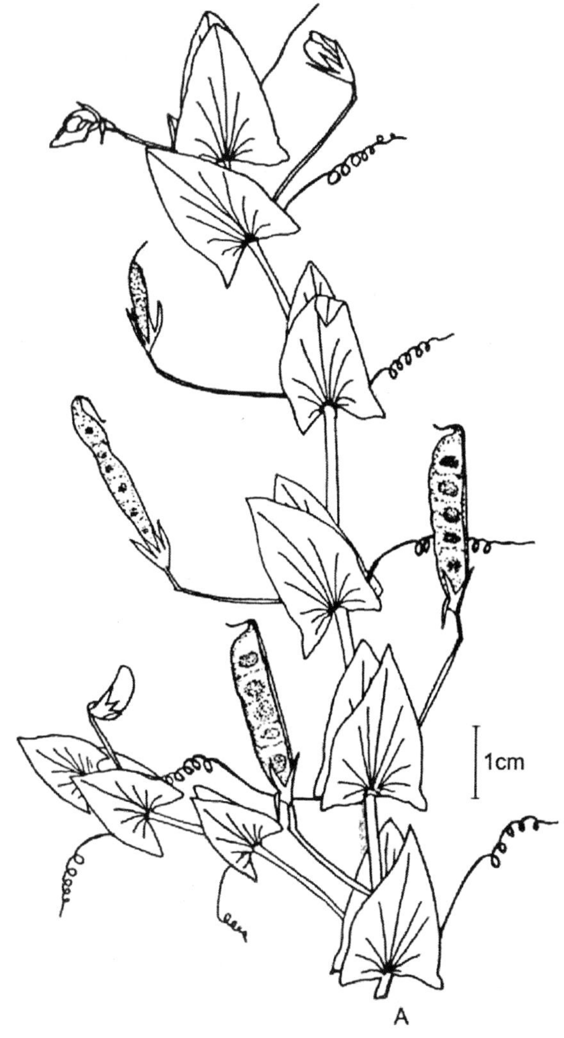

Fig. 47. *Lathyrus aphaca* L.
A. Flowering & fruiting twig.

Trailing or scrambling annual herb. Stem not winged, glabrous. Stipules foliaceous, 5-30 mm long, broadly ovate, hastate, rest of the leaf reduced to a tendril. Inflorescence a 1-2-flowered, axillary raceme. Flowers yellow; peduncle 1 - 4 times as long as the stipules. Calyx 3 - 9 mm long, teeth equal to 3 times as long as the tube. Corolla bright to pale yellow; vexillum 7 - 13 mm long. Fruit 18 - 35 x 4-6 mm, linear-lanceolate, glabrous, 4 - 6-seeded.

Field note: Common weed of cultivated fields.

Fl. & Fr.: January-March.

Specimen examined: Kachhola 1312, Kanera 1682 (MLVGCB Herb).

Distribution: Pakistan, India, Europe, N.Africa, S.W. & C.Asia, often cultivated.

Uses: The fresh plants are given to cattle as "*Leelo*" during winter for increasing milk production.

Lathyrus sativus L., Sp. Pl. 730. 1753; Baker, in Hook. *f.,* Fl. Brit. Ind. 2: 179. 1876; Duthie, Fl. Gangetic Plain 1: 260. 1903; Bhandari, Fl. Indian Desert 110. 1978; Singh, in Shetty & Singh Fl. Rajasthan 1: 246. 1987; Sanjappa, Legumes India 202. 1992.

Vern.: *Patri vatri.*

Glabrous to subglabrous, annual herb; stem winged. Leaf paripinnately compound, leaflets 2, 5 - 100 x 1.5 - 11 mm, narrowly linear- lanceolate, stipules lanceolate, semisagittate; median and upper leaves with mostly 3-sect tendrils. Flowers violet or bluish-purple; peduncle 1-flowered, 3.0 - 6.0 cm. Calyx 7 - 10 mm long, teeth subequal, 1 ½-3 times as long as the tube. Corolla red, blue or white, 12 - 24 mm long. Fruit 2.5 - 3.3 cm x 9-12 mm, upper suture broadly winged; wings 1 - 2.5 mm wide, 3 - 5-seeded.

Field note: Common weeds cultivated fields.

Fl. & Fr.: January-March.

Specimen examined: Bhilwara 1636 (MLVGCB Herb).

Distribution; Pakistan; Kashmir; India; Europe; N.Africa; S.W. Asia; Russia.

Uses : Cultivated as a fodder crop; human consumption of seeds is considered injurious and may even result in paralysis.

MEDICAGO L.

KEY TO THE SPECIES

1a. Corolla yellow. Fruit echinulate, spines on the margins alone .... *M. polymorpha*

1b. Corolla blue or purple. Fruit not echinulate, glabrous to pubescent .... *M. sativa*

Medicago polymorpha L. Sp. Pl. 779. 1753; Mahesh. Illus. Fl. Deihi f. 51. 1966; Singh in Shetty & Singh Fl. Rajasthan 1: 2468. 1987; Sanjappa, Legumes India 210. 1992. *M. denticulate* Willd. Sp. Pl. 3: 1414. 1802; Baker, in Hook. *f.,* Fl. Brit. Ind. 2: 90. 1876; Duthie, Fl. Gangetic Plain 1: 211. 1903. (Fig. 48)

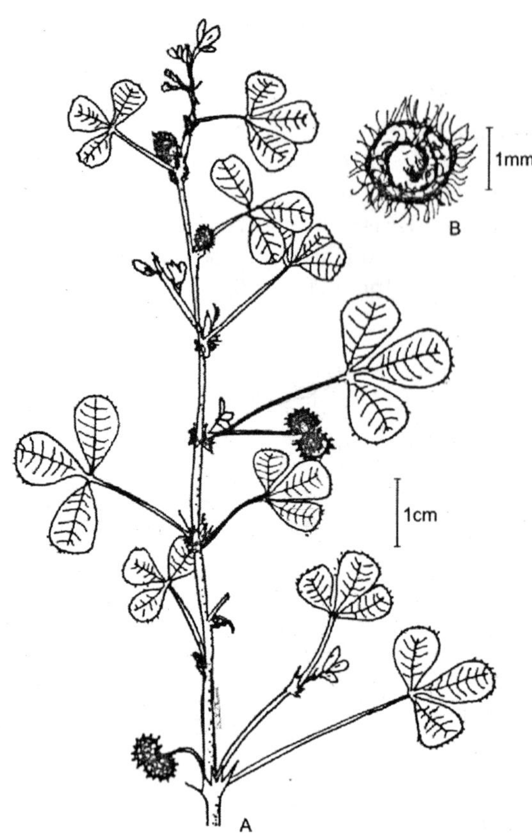

Fig. 48. *Medicago polymorpha* L.
A. Flowering & fruiting twig, B. Fruit

Annual herb, somewhat spreading. Leaves 3-foliate; petiole 1.2 - 4 cm long; leaflets 0.5 - 2.0 x 0.5 - 1.8 cm, obovate, obtuse, truncate to retuse, slightly toothed; stipules laciniate. Inflorescence a 2-8-flowered peduncled raceme, peduncle 5 - 15 mm long. Pedicel ca 5 mm long. Calyx 2.5 mm long, teeth subequal to the tube. Corolla ca 4 mm long, yellow. Fruit 2 - 12 mm high, with 2 - 4 (-6) spirals, glabrous, spines hooked in 2 rows, oriented almost parallel to the surface of the disc. Seeds dark brown.

Field note: Common weed of cultivated fields.

Fl. & Fr.: January-March.

Specimen examined: Gadoli 1035, Mandal 1833 (MLVGCB Herb).

Distribution: India and Pakistan; widely distributed throughout the world, except for tropical regions and desert.

Uses: Often used as fodder.

Medicago sativa L. Sp. Pl. 778. 1753; Baker, in Hook. *f.* Fl. Brit. Ind. 2: 90. 1876; Bhandari, Fl. Indian Desert 111. 1978; Singh, in Shetty & Singh Fl. Rajasthan 1: 277. 1987; Sanjappa, Legumes India 210. 1992.

Vern.: *Rizka.*

Erect perennial herb, 30 - 60 cm, pubescent to subglabrous. Leaflets 3, 1.2 - 2.0 x 0.3 - 0.5 cm, obovate-oblanceolate, dentate at apex, appressed pubescent; entire or dentate at base. Inflorescence a peduncled raceme, peduncle much longer than petiole. Flowers pink or purple. Calyx teeth as long as the tube. Corolla 6 - 12 mm long, violet to pale lavender. Fruit falcate or in a loose spiral of 11 - 4 turns, glabrous to appressed pilose, 10 - 20-seeded. Seeds yellow.

Field note: Common weed of cultivated fields.

Fl. & Fr.: March - June.

Specimen examined: Mandal 1526, Sareri 1904 (MLVGCB Herb).

Distribution: Pakistan, India, Central Asia to Europe, Orient and N. Africa; widely cultivated.

Uses: *Rizka* is widely cultivated as fodder. 1/2 cup of extract is taken at night for 3 days to cure night blindness.

MELILOTUS Mill.

Melilotus indica (L.) All. Fl. Pedem. 1: 308. 1785; Duthie, Fl. Gangetic Plain 1: 208. 1903; Cooke, Fl. Bomb. Pres. (reprint ed.) 1: 325. 1958; Mahesh. Illus. Fl. Deihi f. 53. 1966; Singh, in Shetty & Singh Fl. Rajasthan 1: 249. 1987; Sanjappa, Legumes India 211. 1992. *Trifolium indica* L. Sp. Pl. 765. 1753. *Melilotus parviflorus* Desf. Fl. Atlant. 2: 192. 1799; Baker, in Hook. *f.,* Fl. Brit. Ind. 2: 89. 1876. (Plate 100)

Vern.: *Ran-methi.*

Erect annual herb, 15 - 60 cm tall. Stem pubescent. Petiole 1.5 - 3.8 cm long; leaflets 0.4 - 2.0 x 0.2 - 0.7 cm, oblong or lanceolate, dentate, obtuse, retuse or truncate, glabrous to subglabrous. Inflorescence a 10 - 16-flowered raceme; peduncle up to 3 cm long. Bracts ca 0.5 mm long; pedicel ca 1 mm long, reflexed after flowering. Calyx ca 1.5 mm long. Corolla 2 - 3 mm long, yellow. Fruit 1.5 - 4 mm long, ellipsoid-oblong, tapering at both ends, with prominent veins, 1-seeded. Seeds brown, compressed.

Field note: Common weed of cultivated fields.

Fl. & Fr.: January-March.

Specimen examined: Ganeshpura 1163, Lakhola 1863 (MLVGCB Herb).

Distribution: India, Pakistan, Orient, Europe also introduced in warm temperate regions.

MUCUNA Adans. nom. cons.

Mucuna pruriens (L.) DC., Prodr. 2: 405. 1825; Baker, in Hook. *f.,* Fl. Brit. Ind. 2: 187. 1876; Duthie, Fl. Gangetic Plain 1: 208. 1903; Cooke, Fl. Bomb. Pres. (reprint ed.) 389. 1958; Mahesh. Illus. Fl. Deihi f. 73. 1966; Singh, in Shetty & Singh Fl. Rajasthan 1: 251. 1987; Sanjappa, Legumes India 217. 1992. *Dolichos pruriens* L. in Stickman, Diss. Herb. Amb. 23. 1754 & Syst. Nat. ed. 10. 1162. 1759. *Carpopogon pruriens* (L.) Roxb., Hort Beng. 34. 1814 nom. nud. & Fl. Ind. 3: 283. 1832. *Mucuna prurita* Hook., Bot. Misc. 2: 384. 1831. (Plate 101)

Vern.: *Kemach ki phali.*

Climbing herb; young branches densely pubescent, becoming glabrous. Leaf trifoliolate; petiole 2 - 40 cm long, leaflet 4.8 - 19 x 3.5 - 16.5 cm, obovate, elliptic, rhomboid or ovate, lateral leaflets very oblique, acute or acuminate, pubescent on both sides; petiolule 2-3 mm long; stipules ca 5 mm long. Inflorescence an axillary raceme, 15.0-.30 cm long, flowers single or 2-3 together. Bracts 12.5 mm long; pedicel 2.5 - 5.0 mm long. Calyx 7.5 - 9 mm long, silky, teeth equal or longer than the tube. Corolla purple; vexillum 1.5 cm long; keel 2.5 - 3.8 cm long. Fruit 5 - 6.3 cm long, not winged or plaited, with a longitudinal rib running the length of each valve, pubescent, hairs brown and irritating, 5-6-seeded. Seeds smooth, dark brown.

Field note: Common in scrub forests and boundries of cultivated fields.

Fl. & Fr.: January - March.

Specimen examined: Putholi 335, Bijolioya 1877 (MLVGCB Herb).

Distribution: India, Pakistan, Sri Lanka, Burma, Indochina, Indonesia and Philippines.

Uses: The bristles are used as a vermifuge, generally taken in syrup. Fruit hairs causes irritation on skin, which is get rid off with cattle dung.

PHASEOLUS L.

Phaseolus vulgaris L., Sp.P1. 723. 1753; Baker, in Hook. *f.,* Fl. Brit. Ind. 2: 200. 1876; Bailey, Man. Calt. Pl. 574. 1949; Cooke, Fl. Bomb. Pres. (reprint ed.) 1: 403. 1958; Singh, in Shetty & Singh Fl. Rajasthan 1: 277. 1987; Sanjappa, Legumes India 228. 1992.

Vern.: *Arak-mung.*

Annual, climber or suberect. Stem pubescent to glabrescent. Leaf trifoliolate; petiole 4 - 9 cm long; leaflets 4.5 - 15 x 2.5 - 6.5 cm, ovate to ovate-rhombic, acuminate, lateral leaflets oblique; petiolule 1.5 - 2.5 mm long; stipules 4 mm long. Inflorescence 1 - 3-flowered, peduncle upto 5 cm long. Bracts 3 mm long; bracteoles 5 - 6 mm long. Pedicel 2 - 10 mm long. Calyx pubescent, tube 1 - 3 mm long, teeth 1 mm long, upper 2 joined to form an emarginate lip. Corolla white, yellowish purple or pale-pink; vexillum 1 - 1.9 cm long, glabrous; keel 2.2 cm long, spirally incurved. Fruit 10 - 15 x 0.5 - 1.3 cm, linear-lanceolate, beaked, pubescent or glabrous, 5 - 12-seeded.

Field note: Common in scrub forests and boundries of cultivated fields.

Fl. & Fr.: August-Oct.

Specimen examined: Atun 2161 (MLVGCB Herb).

Distribution: It is of American origin, now widely cultivated in tropical and temperate regions.

PONGAMIA Vent. nom. cons.

Pongamia pinnata (L.) Pierre, Fl. For. Cochinch. 5: t. 385. 1899; Singh, in Shetty & Singh Fl. Rajasthan 1: 277. 1987; Sanjappa, Legumes India 230. 1992. *Cytisus pinnatus* L. Sp. Pl. 741. 1753. *Pongamia glabra* Vent. Jard. Malm. 1: 28. t. 1803; Wight. Ic. I: t. 59. 1838; Baker, in Hook. *f.*, Fl. Brit. Ind. 2: 240. 1876; Duthie, Fl. Gangetic Plain 1: 267. 1903. (Fig. 49)

Vern.: *Karnj, Kanjio.*

Tree with spreading branches, twigs glabrous. Leaf imparipinnate, rachis 5 - 15 cm long; leaflets 5 - 9, ca 5 - 10 cm long, elliptic or ovate-oblong, rounded at base, acute-acuminate, glabrous; rachis 5 - 15 cm long; petiolule ca 5 - 8.0 mm long; stipels absent; stipules small, caducous. Raceme shorter than the leaf. Flowers pinkish white; pedicel ca 7 - 10 mm long; bracts ca 2.5 mm long. Calyx ca 2.5 - 4.0 mm long, pubescent, obscurely toothed. Corolla white tinged with violet or pinkish; vexillum ca 1 - 1.3 cm long. Ovary subsessile; style incurved, glabrous. Fruit 2.5 - 6.0 x 1 - 3.5 cm, obliquely oblong, woody, indehiscent, usually 1-seeded. Seeds 1 - 2, reniform.

Field note: Common in waste lands, also planted along road sides, leaves are infected by insects causing galls. It is a beautiful roadside tree.

Fl. & Fr.: April-May.

Specimen examined: Bhilwara 574, Bijoliya 1562 (MLVGCB Herb).

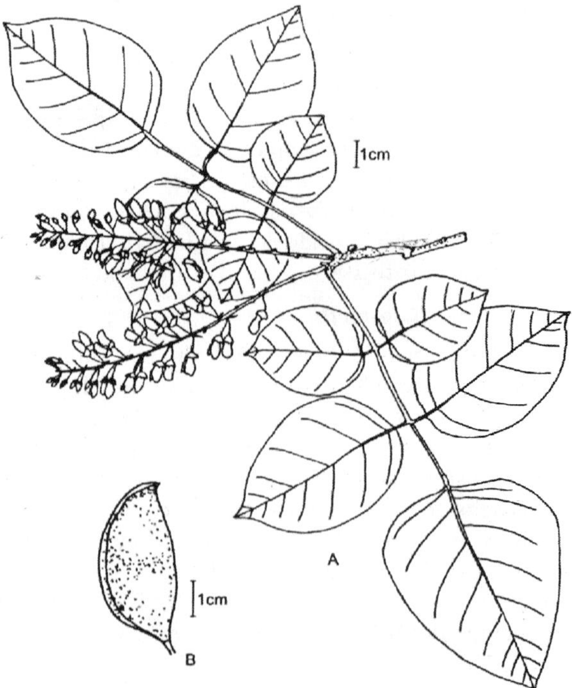

Fig. 49. *A. Pongamia pinnata* (L.) Pierre, A. Flowering twig, B. Fruit

Distribution: India, Pakistan, Sri Lanka, Burma, Malaya, N. Australia, Polynesia and Mascarenes.

Uses: Seeds are crushed and paste is applied in hairs to kill lices.

RHYNCHOSIA Lour. nom. cons.

KEY TO THE SPECIES

1a. Plants trailing, flowers in capitate racemes, pods sub orbicular, seeds strophiolate .... *R. capitata*

1b. Plants twining, flowers in lax racemes, pods elliptic - lanceolate, seeds estrophiolate......... 2

2a. Leaflets 6 - 10.5 cm long. Vexillum 1 cm or more long. .... *R. rothii*

2b. Leaflets 0.4 - 3 cm long. Vexillum 4 - 6.5 mm long.. 3

3a. Leaflets pubescent above, punctate below (not silky), glands orange to black in colour, fruits glabrescent .... *R. minima*

3b. Leaflets silky on both sides, glands yellow, sometimes almost contiguous, fruits thinly pubescent................................ *R. pulverulenta*

Rhynchosia capitata (Heyne ex Roth) DC. Prodr. 2: 386. 1825; Duthie, Fl. Gangetic Plain : 222. 1903; Singh, in Shetty & Singh Fl. Rajasthan 1: 254. 1987; Sanjappa, Legumes India 237. 1992. *Glycine capitata* Heyne ex Roth, Nov. Pl. Sp. 346. 1821 non *Nomismia capitata* (Heyne ex Roth) Wight & Arn. Prodr. 237. 1834. *Rhynchosia aurea* auct. non (Willd.) DC. 1828; Baker, in Hook. *f.*, Fl. Brit. Ind. 2 221. 1876; Bhandari, Fl. Indian Desert 124. 1978. (Plate 102)

Prostrate or climbing; stem pilose, branched. stipules 1.5 - 2.0 mm long, cordate, pilose. Leaf alternate, trifoliolate compound, petiole 2.5 - 12.5 cm long, pilose; petiolule up to 1.5 cm long, pilose; lamina 1.5-4.3 x 1.8-4.6 cm, rhomboid, sometimes oblique, entire, obtuse, scantily pubescent. Inflorescence 4-6-flowered, axillary umbellate. Flowers yellow; peduncle 0.5-3.5 cm, pilose. Bracts ca 2.5 mm long, cordate, pilose; pedicel ca 3-4 mm long, pilose. Calyx 10-13 mm long, pilose, lower tooth longest 7 - 9 mm long; vexillum 10 - 12 mm long, glabrous externally. Fruit 10.13 x 9-12 mm broad, mucronate transversely striated, pilose. Seeds 2, dark brown, reniform.

Field note: Common among agricultural fields, the flowers dried and fruits, remain buried on the soil.

Fl. & Fr.: August - October.

Specimen examined: Mandal 107, Atun 1873 (MLVGCB Herb).

Distribution: Pakistan, India and Sri Lanka.

Rhynchosia minima (L.) DC. Prodr. 2: 385. 1825; Baker, in Hook. *f.* F1. Brit. Ind. 2: 223. 1876; Duthie, Fl. Gangetic Plain 1: 222. 1903; Bhandari, Fl. Indian Desert 124. 1978; Singh, in Shetty & Singh Fl. Rajasthan 1: 255. 1987; Sanjappa, Legumes India 239. 1992. *Dolichos minimus* L., Sp. Pl. 726. 1753. *Rhynchosia medicaginea* DC. Prodr. 2: 386. 1825. *Rhynchosia laxiflora* Camb. in Jacq. Voy. Bot. 44. t. 54. 1844. *R. minima* (L.) DC. var. *laxiflora* (Camb.) Baker, in Hook. *f. l.c.* 2: 223. 1876.

Vern.: *Chiri-motio.*

Perennial, climbing or prostrate; stem glabrous to pubescent. Leaf trifoliolate, petiole 6 - 20 mm long; leaflets 1.0 - 2.5 cm long and broad, rhomboid-ovate or rhomboid-obovate, mostly acute apiculate, pubescent above, pubescent on the veins and glandular punctate below, glands orange to black; stipules up to 3.5 mm long. Inflorescence a lax or 6 - 12 in axillary raceme. Flowers reddish-yellow; pedicel very short. Calyx 3.5 - 4.5 mm long, pubescent, teeth longer than the tube. Corolla yellow; vexillum 5 - 6 mm long, pubescent externally. Fruit 1.2 - 2 cm x 6 -6.5 mm, glabrous, 2-seeded. Seeds ovate, pale green or mottled, estrophiolate.

Field note: Common along hedges, bushes and shrubs in waste lands, open forests and cultivated fields.

Fl. & Fr.: January-March.

Specimen examined: Mezadam 336, Karoi 1635 (MLVGCB Herb).

Distribution: India, Sri Lanka, Pakistan, Lower Burma, Tropical Africa, West Indies, U.S.A, Australia.

Rhynchosia pulverulenta Stocks in Hook. J. Bot. Kew Misc. 4: 147. 1852; Verde. in Kew Bull. 25: 105. 1971; Bhandari, Fl. Indian Desert 114. 1978; Singh, in Shetty & Singh Fl. Rajasthan 1: 255. 1987; Sanjappa, Legumes India 240. 1992. *R. memnonia* auct. non (Delile) DC. Baker in Hook. *f.*, F1. Brit. Ind. 2: 223. 1876. *R. rhombifolia* Blatter & Hallb. in Journ. Bombay Nat. Hist. Soc. 26: 242. 1918, non (Willd.) DC. 1825.

Woody herb, erect or scrambling, branches velvety tomentose. Leaf trifoliolate; petiole 0.5 - 3.0 cm long, terminal leaflet 0.4 - 3.0 x 0.6 - 4.0 cm, transversely elliptic to rhomboid, acute to obtuse; lateral leaflets smaller, obliquely ovate, velvety pubescent and gland dotted on both surfaces, specially on the lower side, glands yellow; stipules 2.5 - 4.0 mm long, pubescent. Inflorescence 1 - 10-flowered lax raceme. Flowers yellow. Bracts 1.5 - 2.0 mm long. Pedicel ca 0.5-3.0 mm long. Calyx pubescent and gland dotted, tube 1 - 2 mm long, teeth subequal, 1.5 - 4.5 mm long. Corolla pale yellow; vexillum 4 - 6.5 mm long, densely pubescent externally. Fruit 1.5 - 2.0 cm x 5.5-6.0 mm, densely pubescent and glandular punctate, 2-seeded.

Field note: Rare, among hedges, bushes and shrubs in waste lands, open forests and cultivated fields.

Specimen examined: Bijoliya 1178 (MLVGCB Herb).

Distribution: India, Pakistan, Arabia, S. W. Africa, Somali Republic, Ethiopia, Sudan, Tanganyika and Kenya.

Notes: Recent studies conducted in Karachi (Qureshi & Faruqi, Pak. J. Bot.2: 65. 1970) support this arrangement. More than 50 attempts to cross *Rhynchosia minima* (L.) DC. and *Rhynchosia pulverulenta* Stocks were not successful because of the non-germination of pollen on the stigma of the other species.

Rhynchosia rothii Benth. ex Aitch. Cat. Pl. Punjan & Sind. 50. 1869; Verdc., in Kew Bull. 25: 91. 1971; Singh, in Shetty & Singh Fl. Rajasthan 1: 255. 1987; Sanjappa, Legumes India 240. 1992. *R. sericea* Span., in Linnaea 15: 195. 1841, non Gill. ex Hook. & Arn. 1833; Baker, in Hook. *f.,* F1. Brit. Ind. 2: 225. 1876; Duthie, Fl. Gangetic Plain 1: 222. 1903. (Plate 103)

Stem woody, twining, pubescent, hairs spreading, glandular. Leaf pinnately trifoliolate, petiole 5-10 cm long, leaflet 6.0 - 10.5 x 3 - 8 cm, terminal ones rhomboid or suborbicular; the lateral leaflets very oblique, ovate or ovate-triangular, margin sinuate, acute-acuminate, pubescent on both sides, gland dotted on the lower side; petiolule ca 2.5 mm long; stipules 10 - 12 mm long. Inflorescence an axillary, pedunculate raceme, 20 - 30 cm long; Flowers yellow with purple dots; bracts much longer than pedicel. Calyx ca 7.5 mm long, pubescent and gland dotted; upper teeth partly connate, the lower longer than the cup; vexillum pubescent and glandular at the back. Fruit 2.5 - 3.0 cm x 7.5 mm, pubescent, 2 -seeded. Seeds ellipsoid-oblong, mottled

Field note: Rare, among hedges, bushes and shrubs in waste lands, open forests and cultivated fields.

Fl. & Fr.: July-September.

Specimen examined: Thago ka khera 108, Bijoliya 1879 (MLVGCB Herb).

Distribution: India, Pakistan, Upper Burma, Java and Thailand.

SESBANIA Adans. emend. Scop. nom. cons.

Sesbania bispinosa (Jacq.) Wight in U.S. Dept. Agr. Bur. Pl. Ind. Bull. 137: 15. 1909; Gillett, in Kew Bull. 17: 129. 1963; Bhandari, Fl. Indian Desert 116. 1978; Singh, in Shetty & Singh Fl. Rajasthan 1: 257. 1987; Sanjappa, Legumes India 242. 1992. *Aeschynomene bispinosa* Jacq. *l.c.* Pl. Rar. 3: 13. t. 564. 1792. *Coronilla aculeata* Willd. Sp. Pl. 3: 1147. 1802. ('*Sesban aculeatus*') nom. illegit; Baker, in Hook. *f.* F1. Brit. Ind. 2: 114. 1876; Duthie, Fl. Gangetic Plain 1: 243. 1903. *Sesbania aculeata* (Willd.) Poir. in Lam. Encycl. 7: 128. 1806. (Plate 104)

Vern: *Dadhan*.

Annual, erect, 1 - 3 m tall. Stem glabrous or sparsely pubescent when young, sparsely aculeate. Leaf paripinnate; rachis prickly 20 - 35 cm long, aculeate; leaflets 15 - 110, 0.4 - 1.2 x 0.2 - 0.4 cm, linear-oblong, obtuse and mucronate, sparsely pilose on margins and midrib below; stipules 6-10 mm long, membranous. Inflorescence a raceme, 3 - 12-flowered, up to 13 cm long. Flowers yellow; pedicel 6 - 11 mm long, filiform. Calyx ca 4 mm long, membranous teeth ca 1 mm long, margin pubescent. Corolla pale yellow, violet flecked; vexillum ca 12 - 13 mm long (including ca 3 mm long claw). Fruit ca 15 - 25 cm x ca 2 - 3 mm, linear, slightly falcate, torulose, beaked, glabrous, 35 - 40-seeded. Seeds 3 - 3.5 mm long, oblong, brown.

Field note: Common in cultivated fields, also at swampy places and thrives well in saline and alkaline soils.

Fl. & Fr.: August - October.

Specimen examined: Kishnawaton ki Kheri 1418, Kachhola 1884 (MLVGCB Herb).

Distribution: Pakistan, India, China, S.E. Asia, Mozambique, Rhodesia, Botswana; South West Africa, Lesotho, S. Africa, Madagascar, Mauritius, Kenya, Tanganyika and Jamaica

Uses : It is also grown as a green manure crop.

SMITHIA Ait. nom. cons.

Smithia conferta J. F Smith in Rees, Cyclop. 33. no. 2. 1819; Singh, in Shetty & Singh Fl. Rajasthan 1: 260. 1987; Sanjappa, Legumes India 246. 1992. *S. geminiflora* Roth, Nov. Pl. Sp. 352. 1821; Baker, in Hook. *f.* F1. Brit. Ind. 2: 149. 1876; Duthie, Fl. Gangetic Plain 1: 269. 1903. *S. geminiflora* Roth. var. *conferta* (Smith) Baker, in Hook. *f., l. c.* 2: 149. 1876; Duthie, *l. c.* 1: 270. 1903.

Annual prostrate or decumbent herb, 30 - 60 cm high, with straggling smooth branches. Leaf pinnately compound; leaflets 2 - 8 pairs, opposite, subsessile, 1 - 3 cm long, linear-oblong, bristly on the margins and midrib beneath. Flowers solitary or 2 - 4, in the upper axils of leaves at the end of branches so as to form a congested dense head, yellow. Bracts and calyx bristly on the back and apex. Calyx - lips equal, 5 - 6.5 mm long, lower lip long hairy; upper lip glabrous or hairy at apex. Corolla 8 - 10 mm long, yellow. Fruits 3 - 6 jointed; joints turgid, pipillose. Seeds 1 - 1.5 mm long, subglobose, glabrous, brown.

Field note: Common, in low-lying areas with grasses and also in ditches.

Fl. & Fr.: September - January

Specimen examined: Munjawa 1995, Hamirgarh 1905 (MLVGCB Herb).

Distribution: India, Nepal, China, Sri Lanka, Cambodia, Laos, Thailand, Vietnam, Indonesia, Philippines, Australia.

TEPHROSIA Pers. nom. cons.

KEY TO THE SPECIES

1a. Leaf simple....................................... *T. strigosa*

1b. Leaf compound 2

2a. Inflorescence not an elongated raceme.. *T. uniflora*

2b. Inflorescence an elongated raceme................ 3

3a. Fruit velvety tomentose .... *T. villosa*

3b. Fruit not velvety tomentose.......................4

4a. Fruit falcate or circinately coiled .... *T. falciformis*

4b. Fruit not as above, Seeds 5-9 per fruit, loosely packed, stem erect or ascending......... 5

5a. Leaflets 9-21, Flowers purple or bright pink.................................. .*T. purpurea*

5b. Leaflets 19-25, Flowers white......... *T. candida*

Tephrosia candida DC. Prodr. 2: 1825; Baker, in Hook. *f.* F1. Brit. Ind. 2: 111. 1876; Duthie, Fl. Gangetic Plain 1: 244. 1903; Singh in Shetty & Singh, Fl. Rajasthan 1: 263. 1987; Sanjappa, Legumes India 256. 1992. *Robinia candida* Roxb., Hort. Beng. 56. 1814.

Vern.: *Dhori-biyani.*

Small shrub, with grooved branches clothed with grey velvety pubescence, Leaflets 19 - 25, 5 - 8 cm long, oblong elliptic, cuspidate, rusted above and pale silky beneath. Stipules setaceous. Flowers in 15 - 25 cm long, terminal and lateral racemes, 1.5 cm across. Calyx 4 - 5 mm long silky, Corolla 1 - 1.5 cm white exserted, standard silky outside. Style flattened. Pods 7 - 10 cm long, slightly recurved, clothed with brown silky hairs. Seeds ellipsoid, compressed, strophiolate.

Field note: Rare, in waste lands.

Fl. & Fr.: August - October.

Specimen examined: Thago ka Khera 109, Atun 1862 (MLVGCB Herb).

Distribution: India, SE Asia and Malaysia.

Tephrosia falciformis Romaswami, in J. Asiat. Soc. Bengal. (n.s.). 12: 125. 1916; Raizada, Suppl. Duthie, Fl. Gangetic Plain 58. 1976; Bhandari, Fl. Indian Desert 129. 1978; Singh, in Shetty & Singh Fl. Rajasthan 1: 263. 1987; Sanjappa, Legumes India 256. 1992. (Fig. 50)

Vern.: *Rati-biyani*

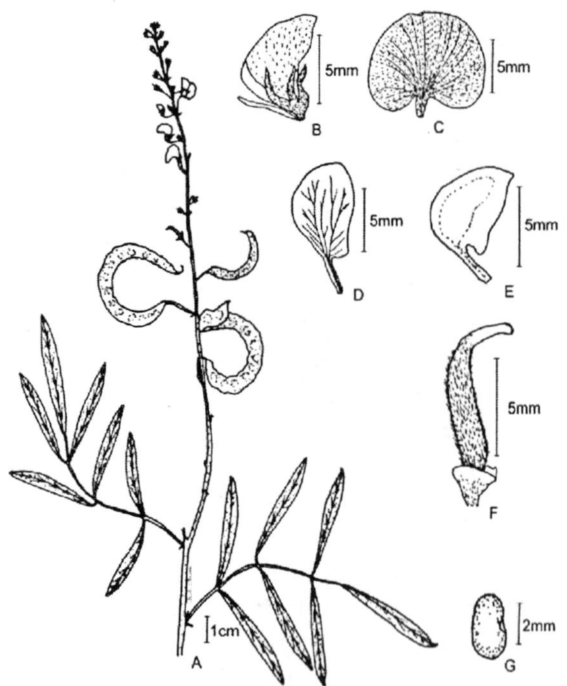

Fig. 50. *Tephrosia falciformis* Ramaswamy: A. Habit, B. Flower, C. Vexillum, D. Wing, E. Keel, F. Gynecelium, G. Seed.

Perennial shrub, ca 50 - 60 cm tall; young branches white-silky. Stipules linear, ca 3 mm long. Leaves imparipinnate, 5.0 - 10.5 cm long. Leaflets 5 - 11, 2.5 - 4.5 cm x 4.2 - 9.5 mm, narrowly oblanceolate, apex acute and mucronate, argenteocanescent on both surfaces. Inflorescence a terminal raceme, ca 9 - 30 cm long. Bracts ca 2.0 mm long. Pedicels ca 2.5 - 10.5 mm long. Calyx ca 4.0 mm long, densely pilose, teeth ca 2.0 mm long. Corolla purple; vexillum ca 7.5 - 9 mm long, orbicular-cordate, apex emarginate, pilose externally; wing ca 6.3 - 9 mm long, glabrous; keel ca 5 - 8 mm long. Ovary sericeous; stigma penicillate. Fruit falcate to circinate, flat, ca 3.2 - 9.0 cm x 6.5 mm, sparsely hairy, 3 - 5-seeded. Seeds obscurely reniform.

Field note: Rare, found in sandy soils.

Fl. & Fr.: August - November.

Specimen examined: Jhag (Kishangarh) 3756, Makrera 82 (Ajmer) (MLVGCB Herb).

Distribution: India, Pakistan.

Tephrosia purpurea (L.) Pers. Syn. Pl. 2: 329. 1807; Baker, in Hook. *f.,* Fl. Brit .Ind. 2: 112. 1876; Duthie, Fl. Gangetic Plain 1: 245. 1903; Cooke, Fl. Bomb. Pres. (reprint ed.) 1: 346. 1958; Bhandari, Fl. Indian Desert 131. 1978; Singh, in Shetty & Singh Fl. Rajasthan 1: 264. 1987; Sanjappa, Legumes India 258. 1992. *Cracca purpurea* L. Sp. Pl. 2: 752. 1753. *Galega purpurea* (L.) L. Sp. Pl. ed. 2. 2: 1063. 1763. *Tephrosia leptostachya* DC. Prodr. 2: 251. t. 2. 1825. *Galega lanceaefolia* Roxb., Fl. Ind. 3: 387. 1832. *G. diffusa* Roxb. *l.c.* 3 : 387. 1832. *Tephrosia diffusa* (Roxb.) Wight & Arn. Prodr. 213. 1834. *T. purpurea* (L.) Pers. var. *pubescens* Baker in Oliver, Fl. Trop. Africa 2: 125. 1871 & in Hook. *f., l.c.* 2: 113. 1876. *T. wallichii* Grah. ex Fawcett & Rendle, J. Bot. 55: 35. 1917. *T. hamiltonii* Drumm. ex Gamble, Fl. Madras 320. 1918.

Annual or perennial, erect up to 1 m tall, branches glabrous or sparsely pubescent. Leaf imparipinnate, 5 - 10 cm long, petiole ca 6 -10 cm long; leaflets 9 - 21, 1.5 - 2.4 x 0.5 - 1.8 cm, elliptic-oblong, obtuse or truncate, mucronate, glabrous above, appressed silky below; stipules narrowly triangular. Inflorescence a leaf-opposed raceme, some flowers in the upper leaf-axils also. Flowers purplish pink; bracts ca 4 mm long; pedicel ca 4 mm long. Calyx pubescent, cup ca 1.5 mm long, teeth ca 2 - 3 mm long. Corolla reddish purple or bright pink; vexillum ca 6 - 9 mm long, pubescent externally. Ovary pubescent, style flattened, glabrescent, stigma penicillate to almost glabrous. Fruit ca 2-6 x 0.4-0.6 cm, linear, subturgid, slightly curved, pubescent, 5-9-seeded. Seeds cylindrical, smooth, dark brown.

Field note: Common in waste lands, along roadsides, in open lands, fields etc.. Appreciable variation is met within the number of hairs on the stigma.

Fl. & Fr.: August - November.

Specimen examined: Karoi 237, Mangrop 1872 (MLVGCB Herb).

Distribution: Southern Asia, Australia, Tropical Africa, southwards to Napal; introduced in Tropical America

Tephrosia strigosa (Dalz.) Sant. & Maheshw., in J. Bomb. Nat. Hist. Soc. 54 (3): 805. 1957; Bhandari, Fl. Indian Desert 131. 1978; Singh, in Shetty & Singh Fl. Rajasthan 1: 265. 1987; Sanjappa, Legumes India 259. 1992. *Macronyx strigosus* Dalz. in Hook. Kew J. Bot. 2: 35. 1850. *Tephrosia tenuis* Wall. ex Dalz. & Gibs. Bombay Fl. 61. 1861; Baker, in Hook. *f.*, Fl. Brit. Ind. 2 : 111. 1876; Duthie, Fl. Gangetic Plain 1: 244. 1903. (Plate 105)

Vern.: *Gino- biyano.*

Annual, erect or decumbent, pubescent, hairs appressed. Leaf simple, alternate, subsessile, 4-8 x 0.3-0.6 cm, linear, acute, apiculate, sparsely pubescent above, silky below; petiole ca 2-2.5 mm long; stipules ca 2.5 - 3.5 mm long. Flowers 1-2 in the axil of the leaf, violet to bluish-yellow; pedicel ca 6-12 mm long, filiform. Calyx ca 2-3 mm long, pubescent, teeth lanceolate. Corolla bluish turning yellow; vexillum ca 4.5-5 mm long. Style ca 1 mm long, glabrous. Fruit 1-3.5 x 0.3 - 0.5 cm, linear-oblong, flat, thinly appressed, pubescent, hairs appressed, 6-10-seeded. Seeds suborbicular, smooth, minutely mottled, light brown.

Field note: Common in waste lands, along roadsides in open lands, fields etc. Often missed due to grass-like leaves and pods with green or straw coloured.

Fl. & Fr.: August - November.

Specimen examined: Badnor 612, Chatarpura 1878 (MLVGCB Herb).

Distribution: India, Pakistan, Burma, Laccadive Islands.

Tephrosia uniflora Pers. subsp. ***petrosa*** (Blatt. & Hallb.) Gillett & Ali in Kew Bull. 13: 114. 1958; Raizada, Suppl. Duthie, Fl. Gangetic Plain 59. 1976; Bhandari, Fl. Indian Desert 121. 1978; Singh, in Shetty & Singh Fl. Rajasthan 1: 266. 1987; Sanjappa, Legumes India 259. 1992. *T. pauciflora* Graham ex Baker, in Hook. *f.*, Fl. Brit. Ind. 2: 114. 1876. *T. petrosa* Blatt. & Hallberg, in Journ. Bomb. Nat. Hist. Soc. 26 (1): 239. 1918. (Plate 106)

Vern.: *Baker biyani.*

Semi-erect perennial, up to 1 m tall. Stem pubescent, hairs silky. Leaf imparipinnate, petiole ca 3-8 mm long; rachis ca 2 cm long; leaflets 5-9, 3.0-5.5 x 0.8-1.3 cm, obovate, oblanceolate or elliptic, acute or obtuse, glabrous or pubescent above, appressed pilose below; stipules up to 9 mm long. Flowers 2 - 3 in the leaf-axils, pinkish-red; pedicel up to 4 mm long. Calyx pubescent, tube ca 1.5 mm long, teeth up to 6 mm long. Corolla pink; vexillum ca 10 mm long, pubescent externally. Ovary pubescent; style glabrous; stigma penicillate. Fruit ca 2.8 - 4.8 cm x 3 - 4.5 mm, appressed pubescent, 7 - 8-seeded. Seeds greenish-yellow, roundish.

Field note: Common on rocky slopes.

Fl. & Fr.: August - November.

Specimen examined: Mogar 186, Badnor 1887 (MLVGCB Herb).

Distribution: India, Pakistan and Arabia.

Uses: Seeds boiled in water and taken as food, it is a good remedy for syphilis.

Notes: *Tephrosia petrosa* Blatt. & Hallberg described from Rajasthan has been regarded as conspecific with the African plant which has been named *T. uniflora* Pers. But there are certain differences in the number of leaflets, length of pods, number of seeds per pod and nature of indumentum. According to Gillett (1958) the plants occurring in N.W. Trop. Africa and the Red Sea Coast of Arabia combine the characters of the Indian and Arabian plants. The epithet uniflora is, of course, misleading, since like all other species, it has at least two flowers in each axil (Bhandari, 1978).

Tephrosia villosa (L.) Pers., Syn. Pl. 2: 329. 1807; Baker, in Hook. *f.*, Fl. Brit. Ind. 2: 113. 1876; Duthie, Fl. Gangetic Plain 1: 245. 1903; Cooke, Fl. Bomb. Pres. (reprint. ed.) 1: 347. 1958; Bhandari, Fl. Indian Desert 133. 1978; Singh, in Shetty & Singh Fl. Rajasthan 1: 266.1987; Sanjappa, Legumes India 260. 1992. *Cracca villosa* L. Sp. Pl. 2: 752. 1753. *Galega villosa* (L.) L. Syst. Nat. ed. 10. Buch.-Ham. in Trans. Soc. London 13: 385. 1822. *Tephrosia incana* Sweet, Hort. Brit. ed. 2. 142.1830. *Galega incana* Roxb., Fl. Ind. 3: 385.1832. *Tephrosia villosa* (L.) Pers. var. *incana* (Roxb.) Baker, in Hook. *f. l.c.* 2: 113. 1876.

Annual or perennial, bushy herb; branches appressed pubescent. Leaf imparipinnate, subsessile, ca 5 - 7.5 cm long; leaflets 11 - 19, 1 - 2 x 0.5 - 1.2 cm, narrowly oblanceolate, obtuse, mucronate, subglabrous to sparsely villous above; densely villous below; stipules up to 8 mm long. Inflorescence an elongated raceme, lower flowers in fascicles, the lowest often in the axil of the leaf. Flowers pinkish-white; bract narrowly triangular; pedicel ca 2 - 4 mm long. Calyx densely tomentose, tube ca 2 mm long, teeth ca 7-8 mm long. Corolla red or purple; vexillum up to 11-15 mm long, externally pubescent. Fruit 2.5-4.5 x 0.5-0.8 cm, flat, falcate upwards, velvety tomentose, hairs dense, spreading, 6-8-seeded. Seeds ovate, glabrous, light-brown.

Field note: Common in waste lands, along roadsides, open lands, fields etc.

Fl. & Fr.: November - March.

Specimen examined: Rampura 424, Pander 1672 (MLVGCB Herb).

Distribution: India, Pakistan, Sri Lanka, Egypt, Tropical Africa, Angola, Uganda, Mozambique, Madagascar, Rhodesia, S. Africa.

TRIGONELLA L.

KEY TO THE SPECIES

1a. Fruit cylindrical, 4-5 cm long...... *T. monantha*

1b. Fruit not cylindrical, ca 2 cm long................. 2

2a. Racemes exceeding the leaves, Fruit sickle-shaped.............................. *T. corniculata*

2b. Racemes equal or shorter than leaves, Fruit not sickle-shaped..................... *T. occulta*

Trigonella corniculata (L.) L., Syst. Nat. ed. 10. 1180. 1759; Roxb., Fl. Ind. 3: 389. 1832; Wight. Icon. t. 384. 1840; Baker, in Hook. *f.*, Fl. Brit. Ind. 2: 88. 1876; Duthie, Fl. Gangetic Plain 1: 210. 1903; Bhandari, Fl. Indian Desert 123. 1978; Singh, in Shetty & Singh Fl. Rajasthan 1: 268. 1987; Sanjappa, Legumes India 266. 1992. *Trifolium corniculatum* L. Sp. Pl. 766. 1753.

Vern.: *Ark-methi.*

Erect to procumbent, glabrous or sparingly pubescent. Leaflets 1 - 4 x 0.8 - 1.4 cm, obovate to oblong-cuneate, obtuse to emarginate, glabrous above, sparsely pilose along the nerves below; stipules lanceolate, dentate or incised. Peduncle 1.5 - 6 cm long, 8 - 20-flowered, pedicel ca 3 mm long. Calyx 3 - 4 mm long, teeth subequal, shorter than or as long as the tube. Corolla 6 - 7 mm long, yellow; wings shorter than the keel. Fruit 10 - 20 x 2 - 3 mm, linear, acuminate, curved, glabrous, with transverse anastomosing veins, 4 - 8-seeded.

Field note: Common weed of cultivated fields during winter season.

Fl. & Fr.: January - March.

Specimen examined: Bhilwara 572, Shahpura 1874 (MLVGCB Herb).

Distribution: India, Pakistan, Afghanistan, Europe, Western Mediterranean area.

Trigonella monantha C.A. Mayer subsp. ***incisa*** (Benth.) Ali in Nasir & Ali, Fl. West. Pak. 100: 294. 1977; Bhandari, Fl. Indian Desert 124. 1978; Singh in Shetty & Singh Fl. Rajasthan 1: 269. 1987; Sanjappa, Legumes India 265. 1992. *T. incisa* Benth. in Royle, Illus. Bot. Himal. Mount. 197. 1835. *T. polycerata* auct. non L. 1753; Baker in Hook. *f.*, Fl. Brit. Ind. 2: 87. 1876; Duthie, Fl. Gangetic Plain 1: 209. 1903. (Fig. 51)

Vern.: *Ark- methi.*

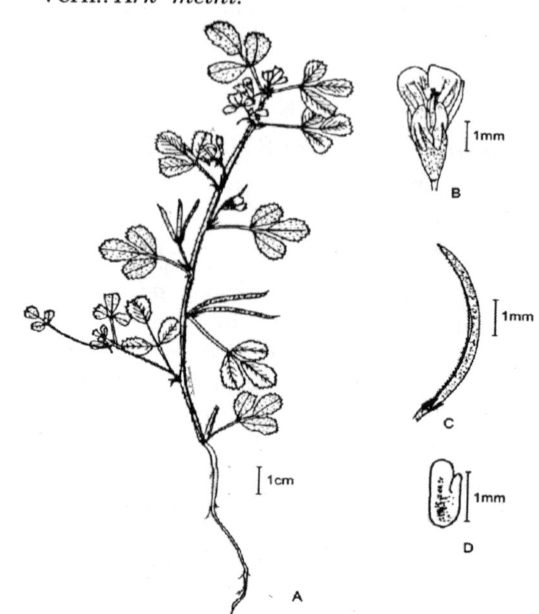

Fig. 51. *Trigonellia monantha* C.A. Mayer subsp. *incisa* (Benth) Ali : A. Habit, B. Flower, C. pod, D. Seed

Prostrate to ascending annual. Leaflets 4-12 mm x 3-8 mm, oblong to cordate-cuneate, dentate or pinnatifid; stipules semisagittate, denticulate. Inflorescence pedunculate, 1-3-flowered, in leaf axils. Calyx campanulate, 5 - 6 mm long, teeth about as long as the cup. Corolla yellow, 6 - 8 mm long. Fruit 4 - 10 cm x 1 - 2 mm, reticulately netted, many-seeded.

Field note: Common weed of cultivated fields.

Fl. & Fr.: January - April.

Specimen examined: Tikar 1832, Ladpura 1885 (MLVGCB Herb).

Distribution: India, Pakistan and Afghanistan.

Trigonella occulta Delile ex DC. Prodr. 2: 185. 1825; Baker, in Hook. *f.,* Fl. Brit. Ind. 2: 87. 1876; Duthie, Fl. Gangetic Plain 1: 209. 1903; Bhandari, Fl. Indian Desert 125. 1978; Singh, in Shetty & Singh Fl. Rajasthan 1: 269. 1987; Sanjappa, Legumes India 265. 1992.

Densely branched annual. Stem glabrous to subglabrous. Leaf pinnately trifoliolate, petiole 6 - 9 mm long, silky when young; leaflets 1.4 - 2.2 x 0.6 - 1 cm, obovate or oblanceolate-cuneate, toothed, subglabrous above, sparsely silky-pubescent below; stipules 6-6.5 mm long. Flowers in 2 - 4-flowered sessile axillary clusters, yellow. Calyx 3 - 3.5 mm long, pubescent, teeth longer than the tube. Corolla slightly longer than the calyx. Fruit ca 3.5 - 4.0 mm long slightly longer than the calyx, oblong-ellipsoid, sparsely pubescent, hairs spreading, white, 1 - 2-seeded. Seeds 1 - 2, oblong, yellowish-brown.

Field note: Common in wet and shady habitats, on the margins of drying ponds in association with *Cyperus pygmaeus* Rottb., *Gnaphalium pulvinatum* Del., *Polygonum plebeium* R. Br. etc.

Fl. & Fr.: January - April.

Specimen examined: Kareda 159, Guda 1865 (MLVGCB Herb).

Distribution: Tropical Africa, Nubia, India (Rajasthan, Punjab, Uttar Pradesh), Pakistan and Egypt.

VICIA L.

KEY TO THE SPECIES

1a. Leaf paripinnately compound. Fruit 6 - 11 mm long, Fruit pilose. .... *V. hirsuta*

1b. Leaf pinnately compound. Fruit 2.3 - 6.5 cm long, glabrous when mature .... *V. sativa*

Vicia hirsuta (L.) S. F. Gray, Nat. Arr. Brit. Pl. 2 : 614. 1821; Baker, in Hook. *f.,* Fl. Brit. Ind. 2: 177. 1876; Duthie, Fl. Gangetic Plain 1: 258. 1903; Singh, in Shetty & Singh Fl. Rajasthan 1: 270. 1987; Kamble in J. Econ. Taxon. Bot. 16: 557. 1992; Sanjappa, Legumes India 269. 1992. *Ervum hirsutum* L., Sp. Pl. 738. 1753.

Vern.: *Vatri.*

Trailing or climbing annual herb, stem pubescent to glabrescent. Leaf paripinnately compound, leaflets 5-20, 4-15 x 1.5-3 mm, linear, obtuse, truncate or marginate, mucronate, sparsely pubescent when young; stipules semisagittate to lanceolate, the lower toothed or laciniate, tendril mostly branched. Inflorescence a 2-7-flowered, axillary, peduncled raceme, peduncle up to 4 cm long. Calyx 1.5 - 3.0 mm long, pubescent, teeth equal to or longer than the tube. Corolla rose or pale-blue; vexillum 3 - 5 mm long. Fruit 6 - 11 x 2.5 - 4 mm, pubescent to subglabrous, 1 - 2-seeded.

Field note: A common weed of wheat fields.

Fl. & Fr.: January - March.

Specimen examined: Sangawa 1190, Manpura 1831 (MLVGCB Herb).

Distribution: India, Pakistan, Nepal, Russia, Middle East, Far East, N. Africa, Europe, Congo, Ethiopia, Angola.

Vicia sativa L. Sp. P1. 736. 1753; Baker, in Hook. *f.,* Fl. Brit. Ind. 2: 178. 1876; Duthie, Fl. Gangetic Plain 1: 258. 1903; Shinde & S. M. Almeida, in J. Bombay Not. Hist. Soc. 84: 466. f. 2. (1987) 1988; Singh, in Shetty & Singh Fl. Rajasthan 1: 271. 1987; Sanjappa, Legumes India 270. 1992. (Fig. 52 & Plate 107) var. ***sativa***

Vern.: *Vatri.*

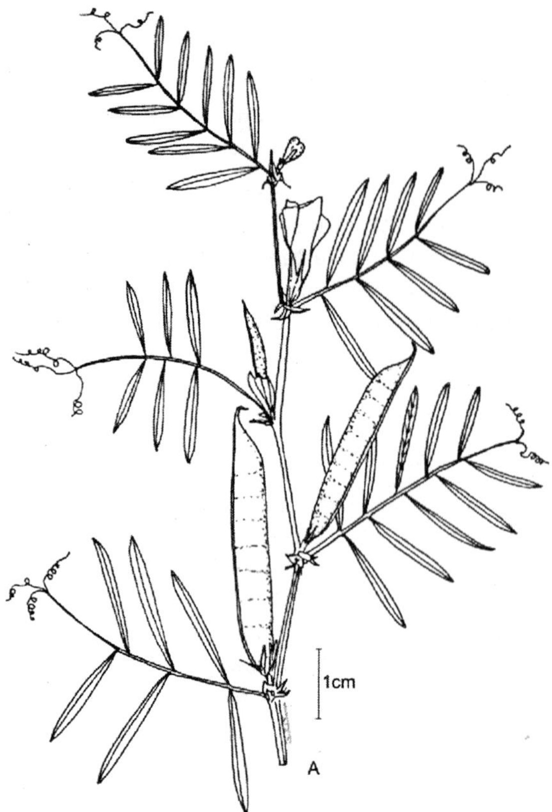

Fig. 52. *Vicia sativa* L.
A. Flowering & Fruiting branch

Annual herb, pubescent to subglabrous, decumbent, erect or climbing. Leaf pinnately compound, petiole less than 1 cm long; leaflets 4 - 18, 1-4 cm x 2 - 15 mm, linear to lanceolate to oblong or obovate, acute, obtuse or emarginate, thinly pubescent or glabrescent; stipules 3 - 8 mm long, semisagittate, dentate: tendril generally branched. Flowers 1-2 rarely 3, axillary, subsessile or shortly pedicellate. Calyx 7 - 20 mm, pubescent, teeth 3-11 mm long, subequal. Corolla pale pink, crimson, purplish violet, rarely white; vexillum 7 - 25 mm long. Fruit 2.3 - 6.5 cm x 4 - 8.5 mm, narrowly oblong, pubescent becoming glabrous when mature, 6 - 12-seeded.

Field note: A common weed of wheat fields.

Fl. & Fr.: January - March.

Specimen examined: Mandal 426, Nayanagar 1525 (MLVGCB Herb).

Distribution: Pakistan, India, Europe and Russia.

Notes: This cosmopolitan taxon is one of the most variable species. A number of infraspecific taxa are sometimes recognised.

Uses: It is good fodder.

VIGNA Savi

KEY TO THE SPECIES

1a. Leaflets distinctly lobed; leaflets ovate. Fruits cylindric. Seeds rounded at the ends .... *V. trilobata*

1b. Leaflets entire or obscurely lobed; leaflets ovate-elliptic to rhomboid. Fruits sub-compressed. Seeds truncate at the ends .... *V. umbellata*

Vigna trilobata (L.) Verdc., in Taxon. 17: 172. 1968 & in Kew Bull. 24: 560. 1970; Bhandari, Fl. Indian Desert 133. 1978; Singh, in Shetty & Singh Fl. Rajasthan 1: 274. 1987; Sanjappa, Legumes India 276. 1992. *Dolichos trilobatus* L., Mant. 101. 1767. *Phaseolus trilobus* auct. plur. non (L.) Aoit 1789; Baker, in Hook. *f.*, Fl. Brit. Ind. 2: 201. 1876; Duthie, Fl. Gangetic Plain 1: 224. 1903.

Vern.: *Mar-mung.*

Annual, prostrate or trailing. Branches glabrous or pubescent. Leaf trifoliolate, petiole 3.7-7.5 cm long; leaflet 2 - 3.5 x 1.5 - 3 cm, ovate, usually as broad, deeply 3-lobed, the middle lobe bigger, oblong or rhomboid, broadly spathulate, obtuse, the lateral lobes oblong, broadly spathulate, obtuse or subacute, glabrous to subglabrous; petiole 1.5 - 2.5 cm long; stipules 4 - 15 mm long. Inflorescence a few flowered raceme. Flowers yellow; peduncle ca 8 - 22.5 cm long. Bracts deciduous; bracteoles ca 3 mm long, below the calyx. Pedicel 2.5 mm long. Calyx 2.5 - 4 mm long, glabrous, teeth minute. Corolla yellow, 5 - 6.5 mm long. Fruit 2.5 - 6.0 cm x 3 mm, glabrous or sparselly pubescent, 6 - 12-seeded. Seeds up to 3 mm long, rounded on both ends, brown.

Field note: Common in grasslands, on boundaries of cultivated fields.

Fl. & Fr.: August - November.

Specimen examined: Rajgarh 1316 (MLVGCB Herb).

Distribution: India, Pakistan, Sri Lanka, Burma, Afghanistan and Malay Isles.

Vigna umbellata (Thunb.) Ohwi & Ohashi, in Journ. Jap. Bot. 44: 31. 1968; Verdc., in Kew Bull. 24: 560. 1970; Singh, in Shetty & Singh Fl. Rajasthan 1: 274. 1987; Sanjappa, Legumes India 276. 1992. *Dolichos umbellata* Thunb. in Trans. Linn. Soc. London 2: 339. 1794. *Phaseolus calcaratus* Roxb., Fl. Ind. 3: 289. 1832; Baker, in Hook. *f.*, Fl. Brit. Ind. 2: 203. 1876.

Vern.: *Mar-mung*.

Plant suberect to prostrate or climbing herb, densely hairy, hairs stiff, deciduous, deflexed, stipules 8-12 mm, biramous, pilose. Leaf pinnately trifoliolate; petiole ca 2.0-15.0 cm, pilose; stipel ca 7 - 8 mm long; petiolule ca 3-4 mm long, rough; leaflets 5-9 x 4-7 cm, ovate-elliptic or ovate rhomboid, 1 - 3-lobed or oblique, acuminate, puberulous on both sides; stipules 5 mm or more. Inflorescence an axillary peduncled raceme, peduncle 5-15 cm long, pubescent; bracts ca 2.5 - 3.0 mm long, minutely pubescent, caducous; pedicel 3 - 4 mm long, pubescent, 2 bracteoles at the base of the calyx, 3 - 3.5 mm long, glabrescent. Calyx ca 4 mm long, pubescent, teeth ca 2 mm. Corolla yellow; vexillum 9 - 12 mm, curved. Fruit 5.0 - 8.0 cm x ca 4.0 mm, glabrous, 8 - 12-seeded. Seeds dark green.

Field note: Common in grasslands, on boundaries of cultivated fields.

Fl. & Fr.: August - November.

Specimen examined: Antali 1328, Lamba 1840 (MLVGCB Herb).

Distribution: India, Pakistan, China and Malaya.

ZORNIA J. F. Gmelin

Zornia gibbosa Span., in Linnaea. 15: 192. 1841; Mohlenbrock in Webbia 16 (1): 112. t. 44 & 76. 1961; Raizada in Ind. For. 92: 326. 1966; Mahesh., Illus. Fl. Delhi f. 64. 1966; Bhandari, Fl. Indian Desert 127. 1978; Singh, in Shetty & Singh Fl. Rajasthan 1: 275. 1987; Sanjappa, Legumes India 280. 1992. *Z. diphylla* auct. plur. non Pers. 1807; Baker, in Hook. *f.*, Fl. Brit. Ind. 2: 147. 1876; Duthie, Fl. Gangetic Plain 1: 268. 1903; Bhandari, in Proc. Raj. Acad. Sci. 10 (1): 42. 1963. *Z. angustifolia* Sm. Rees, Cycl. 39: 1. 1819; nom. illeg. *Z. graminea* Span, in Linnaea 15: 192. 1841. (Fig. 53)

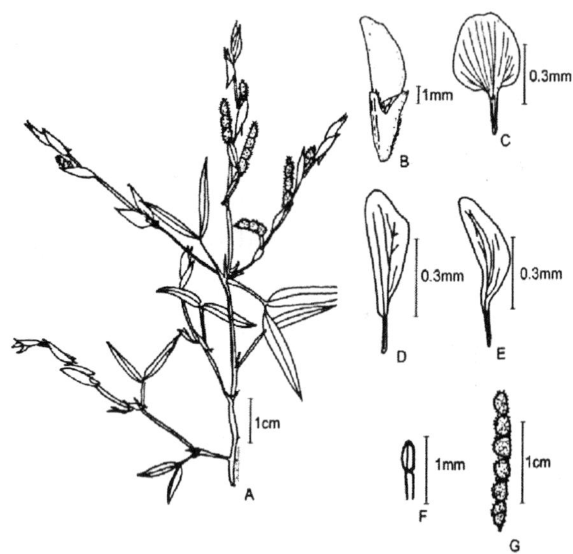

Fig. 53. *Zornia gibbosa* Spam.
A. Habit, B. Flower, C. Vexillum, D. Wing, E. Keel, F. Stamen, G. Fruit

Annual, branches prostrate or ascending, glabrous or pubescent. Leaf bifoliolate, petiole 6.0 - 20 mm long, leaflets 1-2.8 x 0.4 - 0.8 cm, sessile, lanceolate, linear-lanceolate or linear, glandular punctate below, glabrous. Inflorescence a 3 - 12-flowered raceme, 2.5 - 10 cm long. Flowers yellow; bracts foliaceous, up to 1.3 cm long, black dotted, completely enclosing the flower. Calyx 3.5 - 4 mm long, 2 upper teeth obtuse, united, ciliate above, 2 lateral teeth smaller, the lowest longest, acute, ciliate. Corolla 7-8 mm long. Fruit 1-6 jointed, joints ca 2.5 mm long, indehiscent, echinate. Seeds shining brown.

Field note: A slender wiry annual with characteristic leaves and Fruits. Common in grasslands and forests.

Fl. & Fr.: August-September.

Specimen examined: Kishanawaton ki kheri 110 (MLVGCB Herb).

Distribution: India (Punjab, Kumaon, Madras, Bombay, M.P. and Rajasthan), Pakistan, Nepal, Burma, Australia, China, Malaya, New Guinea, Philippines and Thailand.

Notes: This taxon has generally been referred to as *Zornia diphylla* (L.) Pers. (Baker, in Hook. *f.,* Fl. Brit. Ind. 2: 147. 1876) but as shown by Mohlenbrock (l.c.) *Z. diphylla* (L.) Pers. is confined to Sri Lanka and S. India. Therefore, present specimens should correctly be known as *Z. gibbosa* Span.

CAESALPINIACEAE

KEY TO THE GENERA

1a. Leaves simple or unipinnate (paripinnate).....2

2a. Leaves paripinnate......................................3

3a. Corolla with 5 perfect petals, fertile stamens more than 3, free..............................4

4a. Filaments of three abaxial antesepalous stamens sigmoidly curved, bracteoles 1 or 2, pods elongate, cylindric, pulpy or pithy within. seeds 1 or 2 seriate..............................*Cassia*

4b. Filaments of all abaxial stamens straight or simply incurved, bracteoles absent or 2, pods various, seeds variable..................................5

5a. Bracteoles 2. Anther cells ciliolate along sutures. Pods elastically dehiscent.................................... *Chamaecrista*

5b. Bracteoles absent. Anther cells naked along the sutures. Pods indehiscent or inertly dehiscent ....*Senna*

3b. Corolla with 3 perfect petals. fertile stamens 3, monadelphous...............*Tamarindus*

2b. Leaves simple.................................... *Bauhinia*

1b. Leaves bi-pinnate..6

6a. Climbing shrubs.............................*Caesalpinia*

6b. Trees 7

7a. Leaf rachis very short and ending into a sharp spine-like point, pinnae flat and leaf like ....*Parkinsonia*

7b. Leaf rachis not ending into a spine-like sharp point...8

8a. Calyx lobes or teeth imbricate......*Peltophorum*

8b. Calyx lobes valvate ....*Delonix*

BAUHINIA L.

KEY TO THE SPECIES

1a. Flowers creamy white in terminal or leaf opposed racemes. Fertile stamens 10. .... *B. racemosa*

1b. Flowers purple in corymbose racemes. Fertile stamens 5.................................... *B. variegata*

Bauhinia racemosa Lamk. Encycl. 1: 390. 1785; Baker, in Hook. *f.,* Fl. Brit. Ind. 2: 276. 1878; Duthie, Fl. Gangetic Plain 1: 298. 1903; Bhandari, Fl. Indian Desert 127. 1978; Parmar, in Shetty & Singh Fl. Rajasthan 1: 279. 1987; Sanjappa, Legumes India 5. 1992. *B. parviflora* Vahl, Symb. Bot. 3: 55. 1754. *Paliostigma racemosa* (Lam.) Benth, in Mig. Pl. Jungh. 262. 1852.

Vern.: *Kanchnar*.

A small deciduous tree with a short unbranched trunk, and spreading crown; bark dark brown, rough with vertical cracks; young twigs hairy. Leaves 2 - 5 x 2.5 - 7.5 cm, divided half way down into two lobes, glabrous above, hairy below, base usually cordate, 7 - 9 nerved, petiole 7.0 - 18 mm long. Inflorescence terminal or leaf opposed raceme 7.5 - 12.5 cm long. Flowers pedicellate, creamy white, pedicel 5 - 10 mm long, hairy, jointed near the middle; bracts short, linear, acute. Calyx ca 6.0 - 8.0 mm long, pubescent, spathaceous, reflexed. Petals narrow, oblanceolate, acute, white. Stamens 10, all fertile, filaments hairy at the base. Ovary hairy, stigma sessile. Fruits 12.5 - 25 x 1.7 - 2.5 cm; stipe ca 1.2 - 1.5 cm long, sub-turgid, rigid, falcate, blunt at the apex. Seeds 12 - 20, oblong, compressed, black 7 - 8 mm long.

Field note: Common in deciduous forests, also planted along road sides.

Fl. : May - June; *Fr.*: August - November.

Specimen examined: Hamirgarh 1318 (MLVGCB Herb).

Distribution: India (Bengal, Bihar, M.P., Maharastara, Orissa, Punjab, U.P.), W. Pakistan, Sri Lanka, Burma and China.

Uses: The inner bark gives a fibre used for making ropes. Green leaves serve as fodder. Gum and leaves are used medicinally.

Bauhinia variegata L., Sp. Pl. 375. 1753; Baker, in Hook. *f.*, Fl. Brit. Ind. 2: 284. 1878; Duthie, Fl. Gangetic Plain 1: 300. 1903; Parmar, in Shetty & Singh Fl. Rajasthan 1: 290. 1987; Sanjappa, Legumes India 6. 1992. *B. candida* Ait. Hort. Kew. ed. 1. 2: 49. 1789.

Vern.: *Kachnar.*

Medium sized tree upto 6 m high. Bark dark brown, nearly smooth; young shoots pubescent. Leaves upto 10 x 12 cm, petiolate, petiole 2 - 3.8 cm long; lamina ca 4.5 - 15 cm long, as broad as or rather broader than long, with a medium cleft reaching from 1 /4 to 1/3 the way down, lobes obtuse, the base is deeply heart shaped, 9 - 15 nerved, pubescent beneath when young. Inflorescence few flowered corymbose raceme. Flowers purple; pedicel short or absent; bracts and bracteole minute, deltoid. Calyx 2.0 - 2.7 cm long, tomentose, spathaceous, 5 toothed at the apex. Petals 5 - 6.3 cm long, obovate, with long rather broad claw, all white or 4 petals pale purple and fifth darker with purple veins. Stamens 5, all fertile. Ovary hairy; style long, stigma capitate. Fruits 15.0 - 30 x 1 - 2.5 cm, hard, flat, dehiscent 10 -15 seeded. Seeds ovate-orbicular, compressed.

Field note: Generally cultivated as an ornamental.

Fl. & Fr.: February-April.

Specimen examined: Bhilwara 1621, Mandal 1792 (MLVGCB Herb).

Distribution: Burma, China, India (Punjab, U.P., Bengal, Assam, Central India, Madras; Sikkim), Nepal, Pakistan, widely cultivated in Tropics.

Notes: The specimen in the Linnaean Herbarium, London, sheet no. 525/2 bears in Linnaeus's handwriting *'5 virgata'*. But as the specimen is devoid of flowers or buds, which Linnaeus has described, de Wit (l.c.) has rejected it and selected a neotype (e-media).

Uses : Leaves are used as fodder, flowers are used as pot herb and in pickles, wood is used in buildings and for making agricultural implem-ents. The plant yields gum, the bark is used for dyeing and tanning. Plant is reported to have medicinal properties also, the root is tonic and carminative, flowers laxative and the bark astringent, various parts of the plant are reported to have healing properties also.

CAESALPINIA L.

Caesalpinia bonduc (L.) Roxb., Fl. Ind. 2 : 362. 1832; Baker, in Hook. *f.*, Fl. Brit. Ind. 2: 255. 1878; Brenan in Milne-Redhead & Polhill, Fl. E. Trop. Afr. 37. 1967; Parmar in Shetty & Singh, Fl. Rajasthan 1: 279. 1987; Sanjappa, Legumes India 9. 1992. *Guilandinia bonduc* L. Sp. Pl. 381. 1753. *G. bonducella* L. *l.c.* 2. 545. 1762. *Caesalpinia bonducella* (L.) Fleming in As. Res. 11: 159. 1810; Baker, in Hook. *f.*, *l.c.* 2: 254. 1878; Duthie, Fl. Gangetic Plain 1: 302. 1903.

Vern.: *Kanthar.*

Scundent or scrambling shrub, branches hairy, armed with straight prickles. Stipules large, foilaceous and lobed. Leaves 11 - 20 cm long, rachis with 1 - 2 recurved stipellate spines at the base of each pinna and scattered straight or recurved prickles between the pinnae, pinnae 6 - 8 pairs, opposite, 5 - 15 cm long. Leaflets 6 - 10 pairs on each pinna, opposite, 1 - 4.2 x 0.8 - 2.4 cm, elliptic-oblong, obtuse, mucronate, more or less hairy, subsessile. Inflorescence axillary and terminal pedunculate raceme, yellow. Pedicel 5 mm long, bracts and pedicels hairy with brown tomentose, bracts exceeding the calyx. Calyx 5 - 7.5 mm long. Petals 1 - 1.5 cm long, oblanceolate, the upper smaller and sometimes spotted with red. Filaments flattened and hairy in the lower half. Fruits 5 - 7.5 x 3 - 5 cm, shortly stalked, ovoid-oblong, flattened, densely covered with prickles, 1 - 2 seeded. Seeds smooth, lead coloured, shining.

Field note: Common among hedges.

Fl. & Fr.: April-September.

Specimen examined: Triveni 1691, Bijoliya 1835 (MLVGCB Herb).

Distribution: India, Pakistan, widely distributed in the Tropics of the old and the New World.

Uses: Seeds and leaves are used in pain, fever, hydrocele, diarrhoea and rheuma-tism. (Bor & Raizada, Some Beaut. Ind. Cl. Shrubs, 58. 1954). Leaves are astringent, febr-ifuge, anthelmintic, emmenagogue deobstruent. Seeds are used in asthma, chronic fever. Root is anthelmintic, febrifuge and astringent, used in leucorrhoea and blennorrhagia.

CASSIA L.

KEY TO THE SPECIES

1a. Flowers arranged in drooping racemes, Petals yellow to golden. Pods 40 - 60 x 1.5-2 cm .... *C. fistula*

1b. Flowers in pedunculate axillary raceme. Petals pink or red. Pods 20 - 30 x 2 cm.................................... *C. roxburghii*

Cassia fistula L. Sp. Pl. 377. 1753; Baker, in Hook. *f.,* Fl. Brit. Ind. 2: 261. 1878; Duthie, Fl. Gangetic Plain 1: 291. 1903; Parmar, in Shetty & Singh Fl. Rajasthan 1: 283. 1987; Sanjappa, Legumes India 15. 1992; Singh, in Monogr. Indian Cassiinae 25. 2001. *C. rhombifolia* Roxb., Fl. Ind. 3: 334. 1832.

Vern.: *Amaltas*.

Tree, up to 20 m tall. Bark in young tree smooth and ash coloured becoming rough and dark brown in old. Stipules deltoid, acute, 1 - 2 mm long, caducous. Rachis 12 - 25 cm long, terete, glabrous. Leaves compound 22.5 - 40 cm long, with 3 - 8 pairs of opposite leaflets, 6 - 10 cm long, smooth above, hairy below. Flowers arranged in drooping racemes, each raceme ca 10-45 cm long; peduncle 2 - 10 cm long. Pedicel 3.7 - 5.8 cm long, slender, slightly hairy or quite smooth, bracts 8 - 10 mm long, ovate, acute, hairy. Calyx 5, green, folded backward on the stalk, hairy, ovate, 9 mm long. Petals 5, obovate, blunt, distinctly veined. Stamens 10, 3 longest stamens much curled and bear large oblong anthers, 4 smaller medium ones are quite straight, 3 remaining stamens are quite short, erect and sterile. Ovary slender, thinly appressed hairy, style sturdy, stigma punctiform. Pods terete, glabrous, indehiscent, 40 - 60 x 1.5-2 cm, black glossy brown, 40 - 100 seeded.

Field note: Commonly planted in the gardens and along roads for ornamental purposes. The tree become leaflss and flowering occurs on bare branches during spring or early summers. Fruits take about one year to ripe and remain hanging on the tree.

Fl. & Fr.: March - June.

Specimen examined: Bhilwara 1251, Mandal 1550 (MLVGCB Herb).

Distribution: Native of S.E. Asia, India, Nepal, Bhutan, W. Pakistan, Burma and Sri Lanka.

Uses: An ornamental tree, the bark is used as tanning material and wood ash is used as mordant in dyeing. Pulp of pods is a safe purgative and usually given to the children and pregnant ladies. Fruit pulp is given to newly born children as *'Janam Ghutti'* and to cure constipation. Decoction of pods is beneficial in stomachache. Durable wood is used for various purposes. Tribals protect their sleeping children from getting frightened by keeping the dry fruits under their pillows.

Cassia roxburghii DC. Prodr. 2: 489. 1825; Sant. & Wagh, in Bull. Bot. Syrv. India 5: 106. 1963; Singh, In Bull. Bot. Surv. India 18: 97. 1979; Parmar, in Shetty & Singh Fl. Rajasthan 1: 291. 1987; Sanjappa, Legumes India 20. 1992; Singh, in Monogr. Indian Cassiinae 47.2001. *Cathartocarpus roxbur-*

ghii (DC.) Loud. Hort. Brit. 167. 1832. *Cassia marginata* Roxb., Fl. Ind. 2: 338. 1832, non Willd. 1809; Benth., in Trans. Linn. Soc. London 27: 517. 1871; Baker, in Hook. *f.*, Fl. Brit. Ind. 2: 262.1878. *Cathartocarpus marginatus* (Roxb.) G. Don. Gen. Hist. Dichl. Pl. 2: 453. 1832, *nom. illegit.*

Tree, 5 - 10 m tall, with slender downward curving branches. Stipules falcate-reniform. Leaf 10 - 35 cm long, leaflets 8 - 40 pairs, oblong, ovate-elliptical, unequal sided, 1.5 - 4.5 x 1 - 2 cm. Inflorescence pedunculate axillary raceme, peduncle 1-2 cm long, axis 5-9 cm long. Pedicel 1.5-2.5 cm long. Calyx hairy; sepals ovate, 4 - 7 mm long. Petals 10 - 15 mm long, oblong-obovate, hairy externally. Stamens 10, 4 long, 3 medium sized and 3 small, not swollen in the middle. Pod 20 - 30 x 2 cm, cylindrical, glabrous.

Field note: Cultivated as an ornamental. The tree can withstand extreme drought.

Fl. & Fr.: May - June.

Specimen examined: Bhilwara 234 (MLVGCB Herb).

Distribution: A native of Sri Lanka; India, Pakistan, Sri Lanka, Indonesia and Malayasia.

Uses: Wood is used for making naves of wheels and handles of tools. Seeds are good source of industrial gum.

CHAMAECRISTA Moench

KEY TO THE SPECIES

1a. Leaflets 40 - 60 pairs. Fertile stamens 10 .... *Ch. mimosoides*

1b. Leaflets 2 - 20, Fertile stamens 5.................. 2

2a. Erect herb. Leaflets 2 pairs; foliar gland sessile, flattened and ligulate......... .*Ch. absus.*

2b. Prostrate or rarely erect. Leaflets 10 - 20 pairs; foliar gland stipulate, peltate... *Ch. pumila*

Chamaecrista absus (L.) Irwin & Barneby in Mem. New York Bot. Gard. 35 (2): 664. 1982; Singh in Monogr. Indian Cassiinae 55. 2001. *Cassia absus* L. Sp. Pl. 376. 1753; Benth. in Trans. Linn. Soc. London 27: 1871; Baker, in Hook. *f.*, Fl. Brit. Ind. 2: 265. 1878; Duthie, Fl. Gangetic Plain 1: 294. 1903; De Wit in Webbia 11: 279. 1955; Parmar, in Shetty & Singh Fl. Rajasthan 1: 282. 1987; Sanjappa, Legumes India 14. 1992. *C. thonningii* DC. Prodr. 2: 500. 1825. *Senna absus* (L.) Roxb., Fl. Ind. 2: 340. 1832. *S. exigua* Roxb., Fl. Ind. 2: 339. 1832. *Cassia acacalis* Royle in J. Asiat. Soc. Bengal 1: 462. 1835. *C. viscida* Zollinger in Nat. Gen. Arch. 3 : 80. 1846. (Plate 108)

Vern.: *Chaksi.*

Herb or undershrub, 10 - 60 cm tall, erect, widely branching, clothed with viscous glandular hairs. Stipules 4 mm long, linear, persistent, viscous hairy; petiole 1.9 - 4 cm long, rachis 5 - 13 mm long. Leaflets in two pairs, broadly elliptic to obovate, unequal sided, 2 - 2.8 cm x 10 - 20 mm, very puberulous or pubescent on both the surfaces, top rounded to blunt, base rounded or cuneate, petiolules 1 mm long, gland present between the pairs, 0.5 mm long, ligulate-acute, viscid-hairy. Racemes terminal or leaf opposed, ca 4 - 12 cm long, many-flowered, sessile. Flowers yellow; bracts ovate, ca 2 mm long, ovate, viscous hairy. Bracteole 1, about the middle of each pedicel, ca 1 mm long, viscid hairy. Pedicel ca 3 - 4 mm long, rough hairy. Calyx 5, 4 mm long, viscous hairy, imbricate. Petals broadly obovate, 5-7 mm long; claw 1-3 mm long, glabrous. Stamens 5, fertile, staminodes absent. Disc straight, with 1-2 mm long bristles. Ovary densely strigose, sessile, ca 7-ovulate, style glabrous, 2 mm long, stigma, incurved. Pods 2 - 4.5 x 0.5 - 0.6 cm, flat, coarsely hairy, thin valved, dehiscent, 4 cm x 8 mm, short beaked. Seeds black, glossy, obovate.

Field note: Occurring as an undergrowth in the forests and waste lands specially in wet and shady habitats.

Fl. & Fr.: July - December.

Specimen examined: Bhilwara 1165, Mandalgarh 1637 (MLVGCB Herb).

Distribution: India, and everywhere in tropics of the world.

Uses: *'Chaksi'*, extracted from this plant, is reported to have effect on central and peripheral nervous system. Seeds are astringent and strengthen the sight when used as collyrium. Powdered seeds are used in ophthalmia and conjunctivitis. It is also used as cathartic, and in cases of skin infection.

Chamaecrista mimosoides (L.) Green, in Pittonia division 4: 27. 1899; Lock, in Kew Bull. 43 (2) : 337. 1888; Larsen & Hou in Hou *et al.*, in Fl. Malesiana 12 (2): 567. 1996; Singh, in Monogr. Indian Cassiinae 67. 2001. *Cassia mimosoides* L. Sp. Pl. 379. 1753; Benth., in Trans. Linn. Soc. London 27: 579. 1871; Baker, in Hook. *f.*, Fl. Brit. Ind. 2: 266. 1878; Duthie, Fl. Gangetic Plain 1: 296. 1903; De Wit, in Webbia 11: 283. 1955; Parmar, in Shetty & Singh Fl. Rajasthan 1: 284. 1987; Sanjappa, Legumes India 17. 1992. *Senna tenella* Roxb., Fl. Ind. 2: 354.1832. *Chamaecrista mimosoides* (L.) var. *glabriscula* Ghesque in Bull. Jard. Bot. Brux. 9 : 158. t. 1. f. 24. 26 & 28. 1932. *C. mimosoides* L. *forma perennis* De Wit in Webbia 11: 285. 1955. (Plate 109)

Vern.: *Ikrari.*

Annual or perennial herb, sometimes prostrate but more commonly growing as an erect subshrub up to 1.2 m, appressed pubescent or glabrous with age. Leaf-rachis (including petiole) 0.8 - 11 cm long, crenate-crested along upper side; Leaflets 7 - 80 pairs, 0.8 - 1.0 x 0.5 - 0.2 mm, sessile, linear-oblong, acute. Leaves with petioles 1 - 3 mm long, the gland flat, discoid, 0.3 - 1 mm in diam; stipules 4 - 10 mm long, ovate-lanceolate, acuminate-aristate, with strong longitudinal nerves. Inflorescence usually supra-axillary, racemose, 1 to 3- flowered. Flowers yellow; pedicels 4 - 20 mm long, pubescent. Bracts stipule-like; bracteols linear. Sepals acute or acuminate, membranous, pubescent outside. Petals 4 - 10 mm long, obovate or acuminate yellow, claw very short. Stamens 10 all fertile or 1-3 reduced to staminodes, unequal; filaments 1-2 mm long. Ovary sessile, style somewhat 4-gonous, glabrous, stigma truncate or punctiform, ciliate. Pod 15-25 seeded, flat, pubescent. Seeds obliquely obovoid, brown, shining.

Field note: Occurring as an udergrowth in the forests and waste lands specially in wet and shady habitats.

Fl. & Fr.: August - November.

Specimen examined: Tilsawa 1566, Jahajpur 1895 (MLVGCB Herb).

Distribution: Widespread in the tropics of Asia, Africa and S Africa

Chamaecrista pumila (Lam.) Singh in J. Econ. Taxon. Bot. 16 (3): 600. 1992; Singh, in Monogr. Indian Cassiinae 84. 2001. *Cassia pumila* Lamk. Encycl. Meth. 1 (2): 651. 1785; Benth. in Trans. Linn. Soc. London 27: 570. 1871; Baker, in Hook. *f.*, Fl. Brit. Ind. 2: 266. 1878; Duthie, Fl. Gangetic Plain 1: 295. 1903; Bhandari, Fl. Indian Desert 130. 1978; Parmar in Shetty & Singh, Fl. Rajasthan 1: 285. 1987; Sanjappa, Legumes India 19. 1992. *C. prostrata* Roxb. Hort. Beng. 32. 1814, nom. nud. *Senna prostrata* Roxb., Fl. Ind. 2: 352. 1832. *Cassia longipes* Domin in Biblioth. Bot. 89: 242. 1926. *Chamaecrista pumila* (Lam.) Larsen in Nordic J. Bot. 13: 404. 1993; Larsen & Hou in Hou *et al.*, Fl. Malesiana 12 (2): 570. 1996. (Plate 110)

Vern.: *Fori Chuemue, Karola.*

Prostrate or sometimes erect herb or undershrub, 20 - 70 cm tall, short lived, slightly hairy. Stipules 5 - 13 mm long, ciliate, persistent; petiole 2 - 5 mm long, with a slender stiped gland in upper third; rachis 3 - 8 cm long, pubescent with spreading hairs. Leaflets 6 - 25 pairs, linear to linear-oblong, unequal-sided, the larger 7 - 13 x 1.5 - 3 mm, glabrous, or with some hairs scattered near or on the midrib, with a rough often purple margin, top obliquely rounded, base very obliquely truncate, petiolules absent. Flowers 1-3, peduncle and pedicel ca 2.5-5 mm long, pubescent. Bracts ca 4 mm long, linear, caducous, acute; bracteoles 2 - 3 mm long, acute. Sepals lanceolate, acute, 5 - 6 mm long. Petals very unequal, narrowly oblong to broadly orbicular and retuse, 3-5 mm long, narrowed towards the base, nearly clawed. Stamens 5, nearly equally developed. Ovary appressed tomentose, style glabrous, recurved, short, stigma truncate, flat, peltate. Pods flat, thinly pilose, black, not glossy, dehiscent, 2 - 4.5 cm x ca 3.5 - 5 mm, strap-shaped, 6 - 16 seeded. Seeds light brown, glossy, smooth, flattened, rhomboid.

Field note: Occurring as an undergrowth in the forests and waste lands as well as in wet and shady habitats.

Fl. & Fr.: August - November.

Specimen examined: Amet 3918, Bhilwara 1864 (MLVGCB Herb).

Distribution: Tropical Africa, India, W. Pakistan, Sri Lanka, Malaysia, Indonesia, China and Australia.

Uses: Seeds have purgative properties. It is used as a soil binder. It is also planted between the rows of other plants to prevent soil erosion in some parts of the world.

DELONIX Ref.

KEY TO THE SPECIES

1a. Flowers white, turning to yellow, leaf rachis with 4-8 secondary rachis, bearing 10-20 pairs of linear-oblong leaflets .... *D. elata*

1b. Flowers orange-vermilion to deep scarlet, leaf rachis with 11-18 secondary rachis, bearing 20-30 pairs of small oblong leaflets .... *D. regia*

Delonix elata (L.) Gamble, Fl. Pers. Madras. 1: 396. 1919 [1: 280. 1957 (Repr.)]; Blatter, Mill.& Stearn, Some Beaut. Ind. Trees, 2nd. ed. 57. 1954; Bhandari, Fl. Indian Desert 131. 1978; Parmar, in Shetty & Singh Fl. Rajasthan 1: 287. 1987; Sanjappa, Legumes India 27. 1992. *Poinciana elata* L. Cent. Pl. 2: 16. 1756; Baker, in Hook. *f.,* Fl. Brit. Ind. 2: 260. 1878; Duthie, Fl. Gangetic Plain 1: 303. 1903.

Tree, 6-10 m tall, bark ash-coloured. Leaves 6 - 20 cm long, bipinnately compound, pinnae 7 - 20 pairs, 10 - 20 cm long, leaflets 10 - 20 pairs, opposite, sessile or subsessile, linear-oblong, obtuse, sometimes slightly mucronate, narrowed towards the base, 7 - 15 x 2 - 3 mm. Inflorescence terminal raceme, the lower flower with the longest pedicel, pedicel hairy, up to 2.5 cm long. Calyx 2 - 3 cm long, fused at the base, leathery, silky pubescent, oblong, narrow and sharp pointed. Corolla white in the beginning, turning yellow, the upper petal is smaller and darker in colour; sub-orbicular, curled at the margins, ca 2.5 cm long. Stamens hairy, dark coloured, 5 - 10 cm long, thickened at the base, the filaments are orange. Pods 12.5 - 20 cm x ca 6 - 25 mm, smooth, narrow at both the ends. Seeds 4 - 8.

Field note: Locally cultivated as an ornamental, petals at first are white and finally turning pale orange.

Fl. & Fr.: May - July.

Specimen examined: Bhilwara 2168, Jahajpur 1752 (MLVGCB Herb).

Distribution: Found wild in Abyssinia, Arabia, Nubia, India, W.Pakistan.

Uses: Leaves and twigs are employed as manure in some parts of India; wood is well suited for cabinet work. It is the popular belief that the touch of the root removes the pain of a scorpion sting.

Delonix regia (Bojer ex Hook.) Raf., Fl. Tell. 2: 92. 1837; Blatter, Millard & Stearn, Some Beaut. Ind. Trees, ed. 2: 52. 1954; Parmar, in Shetty & Singh Fl. Rajasthan 1: 292. 1987; Sanjappa, Legumes India 27. 1992. *Poinciana regia* Bojerex Hook. in Bot. Mag. 56: 5. t. 2884. 1829. Baker, in Hook. *f.,* Fl. Brit. Ind. 2: 260. 1878; Duthie, Fl. Gangetic Plain 1: 303. 1903.

Vern.: *Gul Mohar*.

Large deciduous tree, ca 10 - 17 m high. Leaves bipinnate, up to 60 cm long, pinnae 10 - 20 pairs, leaflets 5 - 30 pairs on each pinna, oblong, 7.5 - 10 x 3.5 - 4 mm. Stipules lateral, pinnately compound. Inflorescence lax terminal or axillary raceme. Flowers 10 cm across, bright red in colour. Sepals 5, crimson on inside. Petals 5, 4 ca 3.7 cm wide, clawed, 1 petal (standard) ca 5 cm wide, yellow or white, streaked with red. Stamens 10, free, much exserted. Pods 30-50 cm long, 5 cm broad, compressed, firm, rather thick, woody. Seeds 20 - 40, oblong, transverse, mottled.

Field note: Locally cultivated as an ornamental. It is one of the most beautiful avenue trees.

Fl. & Fr.: May - July.

Specimen examined: Bhilwara 1182 (MLVGCB Herb).

Distribution: A native of Madagascar, now cultivated throughout the tropics, common throughout plains of W. Pakistan.

PARKINSONIA L.

Parkinsonia aculeata L. Sp. Pl. 375. 1753; Baker, in Hook. *f.,* Fl. Brit. Ind. 2 : 260. 1878; Duthie, Fl. Gangetic Plain 1: 303. 1903; Bhandari, Fl. Indian Desert 131. 1978; Parmar in Shetty & Singh , Fl. Rajasthan 1: 288. 1987.

Small tree, up to 4 m high, armed, glabrous. Leaf bipinnate; rachis short and ending into a spine, often with a pair of stipular spines at its base, pinnae 2 - 6, ca 15 - 35 cm long, often with much flattened rachis; leaflets 25 - 30 pairs, ca 4 - 5 mm long, obovate-oblong, petiolule 1 - 2 mm. Inflorescence axillary racemes, racemes shorter than the leaf. Flowers pedicellate; pedicel ca 7 - 15 mm. Calyx ca 7 - 8 mm long, segments distinct almost to the base, oblong, obtuse. Petals 5, yellow, 15 - 18 mm long, with a short claw. Stamens 10, free, flattened, villous at the base. Ovary silky hairy, ovules numerous. Pod linear, constricted between the seeds, 7.5 - 10 cm x 6 - 10 mm, glabrous, 1 - 6 seeded. Seed oblong, smooth, mottled.

Field note: Often cultivated for its showy flowers.

Fl. & Fr.: March-June.

Specimen examined: Ajmer 177, Bhilwara 1620 (MLVGCB Herb).

Distribution: Native of tropical America, widely introduced in tropics, India and Pakistan.

PELTOPHORUM (Vogel) Walp. nom. cons.

Peltophorum pterocarpum (DC.) Backer ex K. Heyne, Nutt. Pl. Ned. Ind. ed. 2: 755. 1927; Santapau, Rec. Bot. Surv. Ind. 16 (1). ed. 3: 77. 1967; Parmar, in Shetty & Singh Fl. Rajasthan 1: 292. 1987. *Inga pterocarpa* DC. Prodr. 2: 441, 1825. *Caesalpina inermis* Roxb., Fl. Ind. 2: 367. 1832. *Peltophorum ferrugineum* Benth. Fl. Austr. 2: 279, 1864; Baker, in Hook. *f.,* Fl. Brit. Ind. 2: 257. 1878.

Tree, 12 - 24 m tall; bark smooth grey. Leaves abruptly bipinnate. Main rachis 15 - 30 cm long; pinnae 6 - 20 paired, each pinna with 20 - 30 pairs of leaflets. Leaflets oblong, notched at the apex and unequal sided. Calyx coppery red. Petals wavy, yellow, inversely oval, hairy at the base and much wrinkled at the margins. Stamens 10, densely hairy at the base, anthers golden yellow. Style long and thread like. Pod copper-red, oblong, flat, thin, hard, narrowed at both the ends, 5 - 10 x 2.5 cm. Seeds brown.

Field note: It is a beautiful ornamental tree, widely planted along roadsides.

Fl. & Fr.: March- June.

Specimen examined: Mandalgarh 1817 (MLVGCB Herb).

Distribution: Native of Sri Lanka, India, Malaysia, N. Australia and W. Pakistan.

Uses: The timber is good for cabinet work.

SENNA Mill.

KEY TO THE SPECIES

1a. Foliar glands absent on the petiole and rachis 2

2a. Fertile stamens 10, staminodes absent .... *S. siamea*

2b. Fertile stamens 7, staminodes 3 3

3a. Leaflets obovate or obovate-oblong, rounded or truncate at apex, pods with a ridge of longitudinal crest-like plaite .... *S. italica*

3b. Leaflets elliptic-lanceolate, acute, pods withought plait .... *S. alexandrina*

1b. Foliar glands present on the petiole and rachis 4

4a. Glands on the leaf rachis solitary, inserted beneath the leaflets .... *S. occidentalis*

4b. Glands inserted between one or more pairs of leaflets 5

5a. Leaflets 3 pairs .... *S. tora*

5b. Leaflets more than 3 pairs 6

6a. Leaflets 4 (-5) pairs; stipules auriculate. Pods 10-20 seeded .... *S. auriculata*

6b. Leaflets 8 - 20 pairs; stipules not auriculate, pods 4-10 seeded .... *S. uniflora*

Senna alexandrina Gars. ex Miller, Gard. Dict. Ed. 8. No. 1. 1768; Irwin & Barneby in Mem. New York Bot. Gard. 35 (2): 481. 1982; Singh, in Monogr. Indian Cassiinae 104. 2001. *Cassia senna* L. Sp. Pl. 377. 1753; Sanjappa, Legumes India 20. 1992. exl. var. *C. media* Forssk. Fl. Aegypt.-Arab. 271. 1775, nom. nud., non Vell. 1825. *Cassia angustifolia* Vahl, Symb. Bot. 1: 29. 1790; Benth. in Trans. Linn. Soc. London 27: 553. 1871; Baker, in Hook. *f.* Fl. Brit. Ind. 2: 264. 1878; De Wit in Webbia 11: 289. 1955. *C. lanceolata* auct. plur., non Forssk. 1775; Collad. Hist. Casas. 93. t. 15c. 1816. *C. orientalis* Pers. Syn. Pl. 1 : 457. 1805, pro parte. *C. acutifolia* Del. Mem. Eg. 3: 316. 1808 & Fl. Egypte Expl. Pl. 75. t. 27. f. 1. 1826; Benth. in Trans. Linn. Soc. London 27: 553. 1871. *C. decipiens* Desv. in J. Bot. 3 : 73. 1814. *C. ligustrinoides* Schrank, in Denkschr. Akad. Muench. 6: 179. 1816 - 17. *Senna officinalis sensu* Roxb., Fl. Ind. 2: 346. 1832, non Gaertn. 1791. *Cassia ehrenbergi* Bisch. in Bot. Zeit. 51. 1844. *C. medicinalis* Besch. in Bot. Zeit. 887. 1850. *C. lemtiva* Bisch. in Bot. Zeit. 885. 1850. *Senna angustifolia* (Vahl) Batka, Bot. Zeit. 7 (11): 193. t. 2. f. 2. 1849 & in Monogr. Senna 30, 44. t. 2. 1866. *S. acutifolia* (Del.) Batka, Bot. Zeit. 7 (11): 193. 1849 & in Monogr. Senna 27. 1866. (Plate 111)

Vern.: *Sonna - mukhi.*

Suffruticose perennial, up to 90 cm tall. Branches sparsely appressed pubescent. Leaf rachis pubescent, eglandular; glands like finger-shaped bodies present between the pairs of leaflets. Leaflets 4 - 8 pairs, petiolule ca 1 mm long, lamina 1.2 - 4 x 3.5 - 10 mm, elliptic-lanceolate, acute, glabrous to sparingly hairy on both sides, tip acute. Inflorescence termical or axillary raceme, up to 15 cm long. Young flowers covered with ca 7 - 8 mm long cup-shaped bracts. Pedicel 3 - 4 cm long. Sepals 5, subequal, 7 - 13 x 6 - 8 mm, spoon shaped or cup shaped, light yellow in colour. Petals 5, subequal, 14 - 17 x 7 - 10 mm, obovate, shortly clawed, deep yellow, veins becoming prominent after drying. Stamens 10, upper 3 reduced to staminodes, rest perfect, 2 lower largest. Ovary densely hairy, stipitate. Fruit ca 4 - 5 cm long, ca 16 - 22 mm broad, sparsely hairy, turning black at maturity, generally 4 - 10 seeded; stipe 2 - 3 mm.

Field note: Common in dry waste lands and open forests on sandy and gravelly soils.

Fl. & Fr.: Almost throughout the year.

Specimen examined: Bigod 233, Rampura 1213 (MLVGCB Herb).

Distribution: Native of Tropical Africa, extendiong to Sri Lanka, India, W. Pakistan, U.S.S.R., Egypt, Arabia, Central Sahara, Sudan, Ethiopia, Somaliya, Kenya, Algeria, Libya.

Uses: Dried leaves are valued in medicines for cathartic properties and used in biliousness, gout, rheumatism, skin diseases and against intestinal worms. Pods are cathartic.

Senna auriculata (L.) Roxb., Fl. Ind. 2: 349. 1832. Singh, in Monogr. Indian Cassiinae 119. 2001. *Cassia auriculata* L. Sp. Pl. 379. 1753; Benth. in Trans. Linn. Soc. London 27: 547. 1871; Baker, in Hook. *f.* Fl. Brit. Ind. 2: 263. 1878; Duthie, Fl. Gangetic Plain 1: 294. 1903; De Wit in Webbia 11 : 234. 1955; Bhandari, Fl. Indian Desert 128. 1978; Parmar in Shetty & Singh, Fl. Rajasthan 1: 282. 1987; Sanjappa, Legumes India 14. 1992. *C. densistipulata* Taub. in P. O. A. 200. 1895. (Plate 112)

Vern.: *Anwal.*

Shrub, 1 - 3 m high; Branchlets pubescent or softly hairy. Leaf-rachis ribbed, hairy, Leaf 5 -10 cm long; stipules foliaceous, rotund to reniform, with long apiculae on either side, persistent. Leaflets 5 - 13 pairs, 1.5 - 2.5 cm x 5 - 12 mm, with an erect linear gland between each pair, sometimes exept between the uppermost pair, subulate or fusiform; fingershaped glands -like bodies scattered throughout the groove; identical glands and gland-like bodies are present in the axils of stipules and bracts. Stipules ca 2.5 x 1 cm, reniform, one basal lobe produced into a serrature, ciliate, persistent. Flowers in terminal and axillary corymbose racemes. Pedicel 1.5 -3.5

cm long, puberulous. Bracts ovate-lanceolate to suborbicular, 5-8 mm long, acuminate, pubescent; bracteoles absent. Sepals 9 - 12 mm long, 2 outer much smaller than the other three, broadly ovate, obtuse. Petals 1.2 - 2.5 cm long, broadly elliptic to obliquely rhomboid or orbicular; claw 3 - 7 mm long, bright yellow, veined with orange. Stamens 10, fertile 7; three anthers larger, 3 - 7 mm long, shortly curved; all anthers dehiscing by apical pores and through longitudinal rim; staminodes 3, ovate-oblong or spathulate. Ovary stipulate, appressed hairy; stigma terminal, punctiform. Pod 7.5 - 12.5 x 1.2 - 1.5 cm, straped-shaped, 10 - 20 seeded, undulate between the seeds; valves papery, dark brown, flat. pilose. Seeds 5 - 14 x 5 - 4 mm, obovate-oblong, areole ca 3 x 0.75 mm, linear-oblong or ovate-lanceolate on both faces.

Field note: Common in dry waste lands and open forests on sandy and gravelly soils.

Fl. & Fr.: Almost throughout the year.

Specimen examined: Gulabpura 1404, Sathana 3649 (MLVGCB Herb).

Distribution: Africa, Australia, Burma, Sri Lanka, India, Indonesia, Malaysia, Myanmar and Pakistan.

Uses: Bark of the plant is used for tanning leather, particularly in South India.The bark is astringent and tonic. Leaves and fruits are anthelmintic. Root is used in the treatment of skin diseases.

Senna italica Mill. Gard. Dict. ed. 8. no. 2. 1768; Brenan in Kew Bull. 13 (2): 239. 1958; Irwin & Barneby in Mem. New York Bot. Gard. 35 (2): 482. 1982; Sanjappa, Legumes India 16. 1992; Singh in Monogr. Indian Cassiinae 147. 2001. *C. senna* (L.) Burm. f. Fl. Ind. 96: t. 33. f. 2. 1768, non L. 1753. *Cassia italica* Mill. Gard. Dict. ed. 8. no. 2. 1768. *C. obtusa* Roxb. ex Wight & Arn. Prodr. 288. 1834; Duthie, Fl. Gangetic Plain 1: 294. 1903. *C. obovata* Collad. Hist. Cass. 92: t. 15A. 1816, *nom suprfl.*; Baker, in Hook. *f.,* Fl. Brit. Ind. 2: 264. 1878, *pro parte*. *C. italica* (Mill.) Lam. ex Andrew, Fl. Pl. Anglo-Egypt. Sudan 2: 117. 1952; Bhandari, Fl. Indian Desert 142. 1978; Parmar, in Shetty & Singh Fl. Rajasthan 1: 283. 1987. (Fig. 54 & Plate 113)

Vern.: *Sona-mukhi.*

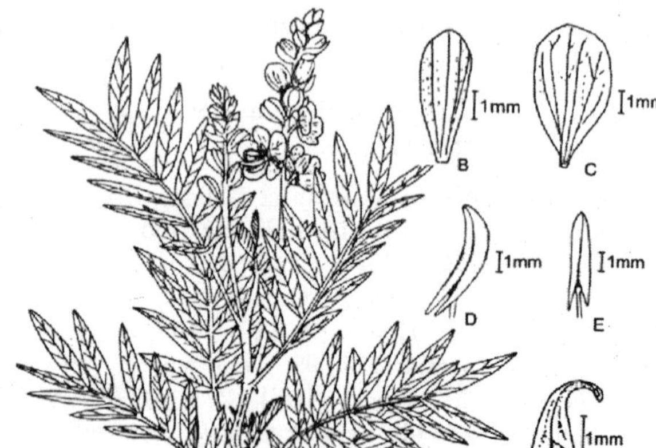

Fig. 54. *Senna italica* Mill.
A. Flowering branch, B. Sepal, C. Petal, D. Larger anther, E. Smaller anther, F. Saminode, G. Pistil, H. Pod, I. Seed

Perenial, erect, herb, 30 - 90 cm tall, stem glabrous, pale green, branches glabrous except the young parts. Leaves 5 - 12 cm long, rachis glabrous, eglandular, petiole 1 - 3.5 cm long, stipules 6 - 7 mm long; leaflet 3 - 7 pairs, 1.8 - 3.5 x 0.7 - 2 cm; broadly oblong or obovate, obtuse, mucronate, glabrous on both sides or sometimes puberulous beneath. Axillary peduncled raceme ca 6-23 cm, pedicel ca 2 mm long. Calyx 8 - 13 mm long, divided to the base, segments oblong, obtuse, membranous, veined. Petals 9 - 20 mm long, obovate to oblong, short clawed, yellow, reticulate with darker veins. Perfect stamens 7, unequal, 2 or 3 lowest larger, ca 8 - 14 mm long, Staminodes minute. Ovary densely hairy. Pods 2.5 - 4.5 cm x 1.2 - 1.8 cm, flat, thin, papery, glabrous, rounded at both the ends, much recurved with a line of prominent longitudinal crest in the middle of the pod. Seeds 6 - 12, wedge shaped, ca 6 x 3 mm, finely reticulate, rugose, with a transverse ridge across the middle of each face, dark brown.

Field note: Common in sandy habitats.

Fl. & Fr.: July- November.

Specimen examined: Bigod 1896, Shahpura 1375 (MLVGCB Herb).

Distribution: Native of Tropical and warm temperate Africa, India, Pakistan Persia, Arabia, Israel, Middle East, Ethiopia, Sudan, Nigeria, West Tropical Africa, N. Africa (Brenan, *l.c.*).

Uses: Leaves are used as a substitute of *Senna alexandrina* Gars. ex Mill.

Senna occidentalis (L.) Link, Handbuch 2: 140. 1831; Irwin & Barneby in Mem. New York Bot. Gard. 35 (1): 436. 1982; Randell in J. Adelaide Bot. Gard. 11 (1): 41. 1988; Larsen & Hou *et al.* Fl. Malesiana 12 (2): 681. 1966; Singh in Monogr. Indian Cassiinae 170. 2001. *Cassia occidentalis* L. Sp. Pl. 377. 1753; Benth. in Trans. Linn. Soc. London 27: 532. 1871; Baker, in Hook. *f.*, Fl. Brit. Ind. 2: 262. 1878; Duthie, Fl. Gangetic Plain 1: 292. 1903; De Wit in Webbia 11: 256. 1955; Bhandari, Fl. Indian Desert 130. 1978; Parmar in Shetty & Singh , Fl. Rajasthan 1: 285. 1987; Sanjappa, Legumes India 19. 1992. *C. falcata* L. Sp. Pl. 377. 1753. *C. occidentalis* L. var. *glabra* DC. Prodr. 2: 497. 1825. *Senna occidentalis* (L.) Roxb., Fl. Ind. 2: 343. 1832. *Cassia occidentalis* L. var. *aristata* Collad. ex Hassk. Pl. Jav. Rar. 405. 1848. *Ditremexa occidentalis* (L.) Britton & Rose in Sci. Surv. Porto Rico 5: 372. 1925 & in N. Amer. Fl. 23 (4): 256. 1930. (Fig. 55)

Vern.: *Kasondi*

An undershrub up to 2 m tall, sometimes an annual. Leaves stipulate, stipules 4 - 6.5 mm, acuminate, caducous. Leaves 12 - 20 cm long, a solitary stalked gland present near the base of the petiole, leaflets 3-5 pairs, opposite, petiolule ca 1 mm, lamina 2.5 - 9 x 1.2 - 3.3 cm, lanceolate or ovate-lanceolate or narrowly to broadly elliptic, apex acute, rounded at the base, glabrous on both the sides, pilose when young. Inflorescence pedunculate, axillary, corymbose raceme forming terminal panicles, peduncle 0.3 - 0.8 mm, enlarging in infructescence. Flowers yellow, 1.2 cm across, pedicel ca 5 - 10 mm long, bracteate, bracts 0.7 - 1 cm long, ovate, oblique, acuminate, glabrous, caducous, white, tinged with pink. Calyx ca 1 cm long, divided up to base, glabrous, oblong, obtuse, white, tinged with pink. Corolla 1.2 cm long, oblong-obovate, more or less unequal, obtuse, orange veined. Stamens 10, 3 lower longer than 4 lateral stamens, 3 staminodes. Pods 9 - 12.5 cm x 7 - 8 mm, slightly falcate, compressed, with transverse septation, sutures thick. Seeds 20 - 30, ovoid, compressed, 6 - 6.5 x 5 mm, rounded at the tip, subacute at the base, smooth, hard, shining, pale-brown.

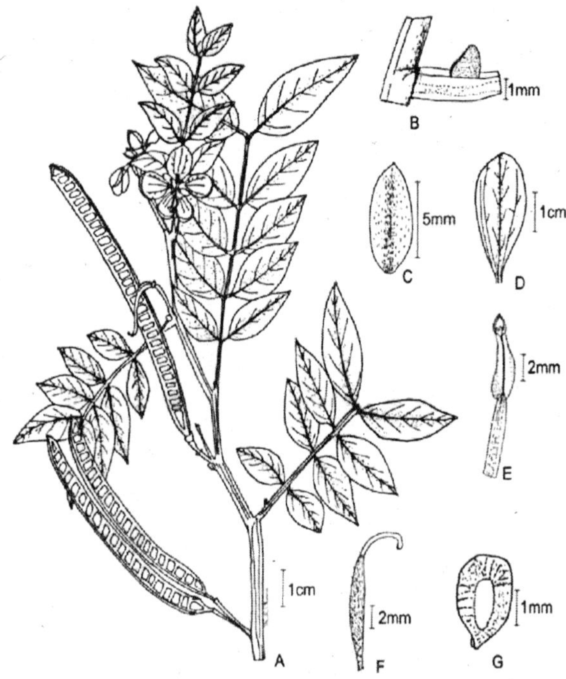

Fig. 55. *Senna occidentalis* (L.) Link.
A. Flowering & fruiting twig, B. Gland, C. Sepal, D. Petal, E. Larger stamen, F. Pistil, G. seed

Field note: Common in waste lands, along roadsides, neglected corners of gardens etc.

Fl. & Fr.: Almost throughout the year, more profuse during August - November.

Specimen examined: Bhilwara 570, Bijoliya 1375 (MLVGCB Herb).

Distribution: Native of S. America; naturalized in Australia, China, Sri Lanka, India, Indonesia, Malaysia, Myanmar, Japan, Arabia, Africa, Taiwan and Pakistan.

Uses: This species is used as green manure in many parts of the world. Seeds are roasted and used as a substitute for coffee in French, Africa and Argentina. Leaves, roots and seeds are considered to be medicinally important i.e. as a febrifuge, purgative, diuretic and tonic. Leaves and seeds are externally applied as antiperiodic and in skin diseases. Root is said to be effective against snake bite also. The tribals of Rajasthan put few drops of leaf decoction into ears to cure earache.

Senna siamea (Lam.) Irwin & Barneby in Mem. New York Bot. Gard. 35 (1): 98. 1982; Larsen & Hou *et al*. Fl. Malesiana 12 (2): 685. 1966; Singh, in Monograph Indian subtribe Cassiinae (Caesalpiniaceae) 193. 2001. *Cassia siamea* Lamk. Encycl. Meth. 1: 684. 1785; Benth. in Trans. Linn. Soc. London 27: 549. 1871; Baker, in Hook. *f.* Fl. Brit. Ind. 2: 264. 1878; De Wit in Webbia 11 : 256. 1955; Raizada. Suppl. Duthie, Fl. Gangetic Plain 63. 1976; Parmar, in Shetty & Singh Fl. Rajasthan 1: 285. 1987; Sanjappa, Legumes India 20. 1992. *C. florida* Vahl. Symb. Bot. 3: 57. 1794. *C. sumatrana* Roxb. ex DC. Prodr. 2: 506. 1825. *C. siamea* Lam. var. *puberula* Kurz in J. As. Soc. Bengal 45 (2): 284. 1876. *Sciacassia siamea* (Lam.) Britton & Rose, N. Amer. Fl. 23 (4): 3252. 1930.

Vern.: *Kasod.*

Evergreen tree, 5-25 m tall. Branches spreading, ribbed, pubescent. Leaf 10.0 x 28.5 cm long, with tranvers bars in the channel between the petiolules, pubescent, eglandular, Leaflets 4 - 15 pairs, oblong to ovate-oblong, ca 2 - 7.5 cm x 10 - 26 mm, lower surface pilose, upper glabrous to hairy. Stipules 3.5 - 5 mm long, linear-subulate or acicular, pubescent, caducous. Flowers in long panicle of corymbs, up to 10 cm long. Pedicel 2.5 - 3.5 cm long. Sepals 5 - 6 mm long, elliptic, ovate, hairy at margins. Petals 1 - 2.0 cm long, ovate, broadly elliptic or orbicular, upper petal clawed, yellow. Stamens 10, all fertile, 2 lateral abaxial largest, 4 lateral and 1 central abaxial medium-sized. 3 adaxial smallest. Ovary somewhat 4-gonous, pubescent; style stout, glabrous; stigma thickened, punctiform. Pods 10 - 30 x 1.2 - 1.6 cm, strap shaped, 20 - 30 seeded. Seeds 7 - 10 x 4 - 6 mm, oblong, flate, truncate at ends, light brown.

Field note: It is a beautiful ornamental tree, widely planted as a roadside tree. plants are resistant to termite.

Fl. & Fr.: Almost throughout the year.

Specimen examined: Bhilwara 1897 (MLVGCB Herb).

Distribution: Native of South-east Tropical Asia. Indigenous in Srilanka, China, India, Indonesia, Malaysia, Taiwan Pakistan etc.

Uses: Leaves are used as fodder and green manure. Pods are, however, reported to be fatal to the cattle, probably due to the presence of some toxic alkaloids. It is one of the hosts of lac insect. In south India it is grown for green manure and to eradicate *Lantana* L. and *Imperata* Cyr. weed species (Singh, 2001).

Senna tora (L.) Roxb., Fl. Ind. 2: 340. 1832; Randell in J. Adelaide Bot. Gard. 11: 45. 1988; Larsen & Hou *et al*. Fl. Malesiana 12 (2): 689. 1966; Singh, in Monogr. Indian Cassiinae 222. 2001. *Cassia tora* L. Sp. Pl. 376. 1753; Benth. in Trans. L. Soc. London 27: 535. 1871, *pro parte*; Baker, in Hook. *f.* Fl. Brit. Ind. 2: 263. 1878, *pro parte*; Cooke, Fl. Bomb. 1: 447. 1903, De Wit in Webbia 11: 276. 1955; Brenan in Kew Bull. 13: 248. 1958; Symon in Trans. Royal Soc. S. Austr. 90: 92. 1966; Parmar, in Shetty & Singh Fl. Rajasthan 1: 286. 1987; Sanjappa, Legumes India 22. 1992. *C. tora* L. var. *barneensis* (Miq.) Miq. Fl. Ind. Bat. 1: 95. 1855. *Emelita tora* (L.) Britton & Rose, Bot. Porto Rico 5: 371. 1925. (Plate 114)

Vern.: *Fumphadia.*

An erect, foetid, usually annual, often profusely branched herb or under-shrub, 30 - 120 cm tall. Stipules linear, hairy, acute, falcate, 8 - 15 x 1 mm, subpersistent. Leaf ca 5 - 10 cm long, leaflets 3 pairs, membranous, mostly obovate, nearly equal sided, ca 0.5 - 5 x 0.5-2.5 cm, both surfaces hairy, glabrescent or glabrous, base oblique, cuneate to rounded, apex delicately acute, petiolules 1.0 - 2 mm long, glands between 2 lower pairs of petiolules. Racemes short, penducle hairy, ca 3 4 mm long, with a single pair of flower, or axillary solitary. Bracts linear-acute, 2 - 4 mm long, pedicel 4 - 10 mm long, in fruit up to 1.5 cm, hairy. Sepals ovate, 5 - 6 mm long. Petals unequal, ovate, top rounded, 8 - 10 mm long. Stamens 7, developed, 3 largest, 4 lateral, 3 upper absent or staminodial. Ovary appressed, densely hairy, ribbed, style glabrous, stigma truncate. Pod glabrescent, subtetragonous, often curved, indehiscent, 10 - 20 cm x 4 - 6 mm. Seeds 20 - 30, rhomboidal, glossy, dark, 5 mm across.

Field note : Common along road sides, as under growth of forests, sparsely distributed in Waste lands.

Fl. & Fr.: August-December.

Specimen examined: Mangrol 135, Kanera 1847 (MLVGCB Herb).

Distribution: India eastwards to Polynesia, W. Pakistan.

Uses: This plant is used as green manure and as fodder in some parts of the world. Leaves are laxative, antiperiodic, asperient and anthelmintic. Seeds are used as tonic. Leaves and seeds are recommended in skin diseases such as ringworms, scabies and leprosy.

Notes: As regard to nativity of this species and its allied species *Senna obtusifolia* (L.) Irwin & Barneby, now seems reliably established that *Senna obtusifolia* (L.) Irwin & Barneby is native to S. America and has been transported as a weed to tropical areas of Africa, Asia and the Pacific Islands. Similarly *S. tora* (L.) Roxb. was native to somewhere in the Asian-pacific region and has been transported to Africa and Australia (Singh, 2001).

Senna uniflora (Mill.) Irwin & Barneby in Mem. New York Bot. Gard. 35 (1): 258. 1982; Singh, in Monograph Indian subtribe Cassiinae (Caesalpiniaceae) 228. 2001; Meena & Yadav, in Nat. Prod. Radiance 8 (5): 525 - 527. 2009. *Cassia uniflora* Mill. Gard. Dict. ed. 8. No. 5. 1768; Raghavan in Bull. Bot. Surv. India 22 (1-4): 225. 1980 (1982); Singh in Cur. Sci. 52 (13): 644. 1983 & in Fl. E. Karnataka 1: 278. 1988; Sanjappa, Legumes India 24. 1992; Samvatsar, Fl. W. Madhya Pradesh.115. 1996. *C. ornithopoides* Lam. Dict. 1: 644. 1785. *C. sericea* Swartz, Prodr. 66. 1788 & Fl. Occ. 2 (1): 724. 1798, nom illegit. ; Benth. in Trans. Linn. Soc. London 27: 536. 1871. *C. sensitiva* Jacq. Coll. 2: 362. 1789 & Ic. Pl. Rar. 3 : t. 459. 1790. *C. ciliata* Hoffm. Verz. Pflan. 208. 1824. *C. monantha* DC. Prodr. 2: 506. 1825, nom. illegit. *Emelista mucronulosa* Pittier in J. Wash. Acad. Sci. 19: 176. 1929. *Serceocassia uniflora* (Mill.) Britton ex Britton & Rose, N. Amer. Fl. 23 (4): 246. 1930. (Fig. 56 & Plate 115)

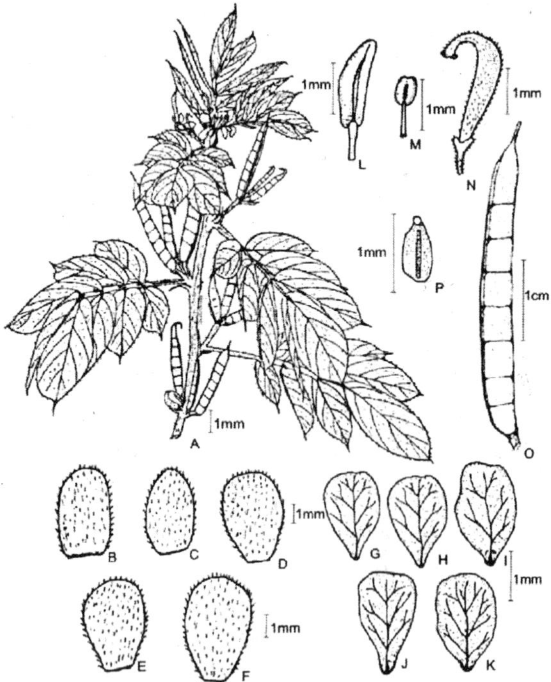

Fig. 56. *Senna uniflora* (Mill.) Irwin & Barneby A. Habit, B-F Sepals, G-K. Petals, L. Fertile stamen, M. Staminode, N. Pistil, O. Fruit, P. Seed

Woody, annual, erect herb, 15 - 20 cm high, young branch appressedly hairy, old glabrescent to glabrous; branches and branchlets 4 - 5-angled; Leaf- rachis 6-12 cm long (including petiole); leaflets 3-5 pairs (usually 4 pairs), ovate-lanceolate to oblanceolate with mucro, 2-5 x 1-2.5 cm, fulvous hairy on lower side, white pubescent to glabrescent on upper side, nerves 4 - 5 pairs, prominent on lower side. Glands ca 3.5 mm long, awl- shaped, slender, stalked, between all or lowest pair of leaflets; petiolules 2 mm long. Stipules 6-10 mm long, linear, bristle-like, persistent. Racemes axillary, 3 to 4-flowered; flowers yellow; peduncles ca 1 cm long; pedicels 2 - 4 mm long, rufous. Bracts linear, hairy. Sepals 4 - 6 mm long, oblong- suborbicular, obtuse, ciliate, outer rufous hairy, inner ones pubescent. Petals yellow, obovate, 6 - 10 mm long, prominently veined. Fertile stamen 7; three large anthers narrowed at the apex like neck of bottle, rest anthers rounded at the top, all anthers dehiscing by apical pores. Staminodes 3. Ovary densely silky hairy; style pubescent; stigma punctiform. Pods 3 - 5 cm long, linear, obscurely 4-angled (2-angles sharply margined), 5 to 10-seeded, appressedly hairy, obscurely pointed at apex. Seeds dark brown 0.4 - 0.7 cm long, rhomboid or subquadrangular, glabrous.; areole like a slit on both faces.

Field note: Occasional along road sides, sparsely distributed in wasteland. Growing intermingled with *Senna tora* (L.) Roxb.

Fl. & Fr.: August-December.

Specimen examined: Hamirgarh 437 (MLVGCB Herb).

Distribution: Brazil, West Indies, Central America and Mexico; Recently introduced in many tropical countries of the world. In India (Rajasthan, Maharastra, M.P. and Karnataka)

Uses: Dried stems are used as firewood.

TAMARINDUS L.

Tamarindus indica L., Sp. Pl. 34. 1753; Baker, in Hook. *f.*, Fl. Brit. Ind. 2: 273. 1878; Duthie, Fl. Gangetic Plain 1: 301. 1903; Bhandari, Fl. Indian Desert 131. 1978; Parmar, in Shetty & Singh Fl. Rajasthan 1: 289. 1987; Sanjappa, Legumes India 36. 1992. (Plate 116)

Vern.: *Imli, Amali.*

A large evergreen tree. Stipules linear, caducous. Leaves 4-10 cm long, paripinnate. Leaflets 6 - 12 pairs, opposite, 1.2 - 1.8 cm x 3.7 - 5 mm, oblong, obtuse. Inflorescence 10 - 15 flowered lax raceme. Pedicel 8 - 10 mm. Calyx ca 1.2 cm long, tube turbinate, teeth lanceolate, the lowest two connate. Petals 3, 1 - 1.5 cm long, pale yellow, with red veins, the two lower reduced to scales. Stamens monadelphous, 3 perfect, others reduced to bristles. Pods 7.5 -20 x 2 - 2.5 cm, slightly compressed, with rough, fibrous, grey epicarp, indehiscent. Seeds smooth, dark brown or black.

Field note : Common all throughout the, but always cultivated near villages along roads etc.

Fl. : November- January; *Fr.*: February - April.

Specimen examined: Pur 111 (MLVGCB Herb).

Distribution: Probably native of Tropical Africa, India, West Pakistan.

Uses: Almost all parts of the plant are used, particularly the pulp of the fruit and young leaves which are regarded as important ingredients of many tasty dishes. Pulp of the fruit, seeds, leaves, flowers and bark are also put to various medicinal uses.

MIMOSACEAE

KEY TO THE GENERA

1a. Stamens indefinite, more than 10 2

1b. Stamens as many as or twice the number of petals 4

2a. Stamens monadelphous 3

2b. Stamens free .... *Acacia*

3a. Pod not circinately coiled .... *Albizia*

3b. Pod circinately coiled .... *Pithecellobium*

4a. Anthers eglandular (pod flat, not pulpy within) 5

4b. Anthers gland crested in early stage 7

5a. Plants and fruits prickly, fruit curved, when ripe, joints separating from the persistent sutures .... *Mimosa*

5b. Plants and fruits not prickly, Fruits straight, dehiscing at the suture 6

6a. Herbs. Leaflets sensitive, heads ca 5 mm in diam. yellow, anthers gland-crested .... *Neptunia*

6b. Small trees. Leaflets sensitive, heads c.1.5 - 2.5 mm in diam. creamy-white, anthers not gland-crested .... *Leucaena*

7a. Flowers dissimilar in colour and sex; pod not septate .... *Dichrostachys*

7b. Flowers all similar in colour and sex. fruit not dehiscent .... *Prosopis*

ACACIA Mill.

KEY TO THE SPECIES

1a. Mature plants with phyllodes, Inflorescence a spike .... *A. auriculiformis*

1b. Mature plants with normal leaves................ 2

2a. Inflorescence a spike...................................... 3

3a. Pinnae 10 pairs or more .... *A. catechu*

3b. Pinnae 3 - 5 pairs or fewer .... *A. senegal*

2b. Inflorescence a head 4

4a. Heads generally terminal, paniculate .... *A. leucophloea*

4b. Heads axillary, pedunculate.......................... 5

5a. Ovary glabrous. 6

6a. Podes rounded. .... *A. farnesiana*

6b. Podes flat. .... *A. jacquemontii*

5a. Ovary pilose 6

6a. Flowers white to creamy. Pod contorted or spirally twisted .... *A. tortilis*

6b. Flowers yellow. Pod not spirally twisted .... *A. nilotica*

Acacia jacquemontii Benth. in Hook., Lond. J. Bot. 1 :499. 1842; Baker in Hook. *f.*, F1. Brit. Ind. 2:293.1878; Duthie, Fl. Gangetic Plain 1: 3143. 1903; Pandey, in Shetty & Singh Fl. Rajasthan 1: 296. 1987

Vern.: *Kikar, Babul.*

A much branched, erect shrub, twigs zigzag, bark greyish brown, young shoots slightly puberulous. Spines stipular, straight, white, connate at the base. Rachis glabrous, usually with a gland between the upper pair of pinnae. Pinnae 1-4 pairs, 5-15 cm long. Leaflets 5-10 pairs. Pod glabrous, rounded at base, reticulately veined, 5 - 6 seeded. Seeds compressed.

Field note: Common in sandy habitats.

Fl. & Fr.: February-May.

Specimen examined: Beawar 4802, Bhudha Pushkar 4934 (MLVGCB Herb).

Distribution: Pakistan, India (Madhya Pradesh, Punjab and Rajasthan).

Acacia auriculiformis A. Cunn. ex Benth. in Hook., Lond. J. Bot. 1:377. 1842; Mahesh., Illus. Fl. Delhi f. 81. 1966; Sen & Naskar, in Bull. Bot. Surv. India 7: 31. 1958; Sanjappa, Legumes India 45. 1992. (Plate 117)

Tree, 8-20 m height, with a straight, single stem. Bark whitish. Mature foliage, phyllodes, straight or falcate, acute or sub-falcate, 10 - 20 x 1.2 - 5 cm, with 4-6 sub-parallel nerves, 3 prominent longitudinal nerves running together towards the lower margin or in the middle near the base, with many fine crowded secondary nerves, a distinct gland at the base of the phyllode. Spikes panicled. Flowers 5-merous, yellow. Pods, slightly woody, glaucous and transversely veined, straight or curved, become very twisted and irregularly coiled on maturity. Seeds encircled by long red, yellow or orange funicle.

Field note: An ornamental tree from Tropical Australia, cultivated in gardens.

Fl. & Fr.: August - April.

Specimen examined: Bhilwara 232, Karoi 1673 (MLVGCB Herb).

Distribution: Endemic to Australia, widely introduced to many tropical countries Asia, Africa and America.

Acacia catechu (L. f.) Willd. Sp. Pl. 4: 1079. 1806; Baker, in Hook. *f.* Fl. Brit. Ind. 2: 295. 1878; Duthie, Fl. Gangetic Plain 1: 316. 1903; Mahesh., Illus. Fl. Delhi f. 83. 1966; Pandey, in Shetty & Singh Fl. Rajasthan 1: 289. 1987; Sanjappa, Legumes India 38. 1992. *Mimosa catechu* L. f., Suppl. Pl. Syst. Veg. 439. 1781. *Acacia polyacantha* Willd. Sp. Pl. 4: 1079. 1806; Brenan, in Kew Bull. 1956: 195. 1957. *A. suma* (Roxb.) Buch.-Ham. ex Voigt, Hort. Subur. Calc. 260. 1845; Baker, in Hook. *f.* op. cit. 294; Cooke, op. cit. (Plate 118)

Vern.: *Khair, Katha.*

Medium sized tree, up to 10 m tall, deciduous; dark grey or greyish brown bark peeling off in long strips or in narrow rectangular plates, brown and red inside. Leaf bipinnate compound; stipules spiny, flattened, hooked and pointed. Pod 5.0 - 14 x 1.0 - 2 cm, flat, thin brown, shining, narrowed at the base into a ca 3 - 7 mm long stalk, triangular beaked at the apex, dehiscent. Seeds 3 - 10 per fruit.

Field note: Common in open forests and on hills.

Fl. & Fr.: May - September.

Specimen examined: Ladpura 112 (MLVGCB Herb).

Distribution: India, Pakistan, Nepal and Burma.

Uses: *Katha* is used with betel leaf and for tanning. Wood is considered to be very durable and is not affected by white ants. It is used for house posts, agricultural implements, wheels etc. It is an excellent firewood and one of the best woods for charcoal preparation.

Gum is used in the form of *"laddu"* specially by tribal males as they are prohibited to use gum of *A. latifolia,* during winter season for 15 days to treat weakness.

Acacia farnesiana (L.) Willd. Sp. Pl. 4: 1083. 1806; Baker, in Hook. *f.* Fl. Brit. Ind. 2: 292. 1878; Duthie, Fl. Gangetic Plain 1: 313. 1903; Pandey, in Shetty & Singh Fl. Rajasthan 1: 296. 1987; Sanjappa, Legumes India 39. 1992. *Mimosa farnesiana* L. Sp. Pl. 521. 1753. (Fig. 57 & Plate 119)

Vern.: *Foro borio, Jari boriyo.*

Shrub or small tree, branches slightly zigzag, marked with grey or pale brown dots, young twigs glabrescent. Spines stipular, in pairs, straight, ca 7 - 18 mm long. Inflorescence globose pedunculate heads in axillary fascicles,. Pod, slightly curved, dark brown, mesocarp pulpy. Seeds numerous, globose, biseriate.

Field note: Rare, in scrub forest of the area.

Fl. & Fr.: November-March.

Fig. 57. *Acacia farnesiana* (L.) Willd.
A. Flowering twig, B. Fruit

Specimen examined: Arnia panth 72, Bangrera 1540 (MLVGCB Herb).

Distribution: Native of America Cosmopolitan in Tropics, India, W. Pakistan.

Plant yields high class gum and the flowers are used in perfumes in South Europe but none of these are used in this area. Bark and seeds are used for tanning.

Acacia leucophloea (Roxb.) Willd. Sp. Pl. 4: 1803. 1806; Baker, in Hook. *f.,* Fl. Brit. Ind. 2 : 294. 1878; Duthie, Fl. Gangetic Plain 1: 315. 1903; Mahesh., Illus. Fl. Delhi f. 80. 1966; Bhandari, Fl. Indian Desert 133. 1978; Pandey in Shetty & Singh Fl. Rajasthan 1: 297. 1987; Sanjappa, Legumes India 41. 1992. *Mimosa leucophloea* Roxb., Pl. Cor. 2: 27, t. 150. 1800. (Fig. 58 & Plate 120)

Vern.: *Khejra*.

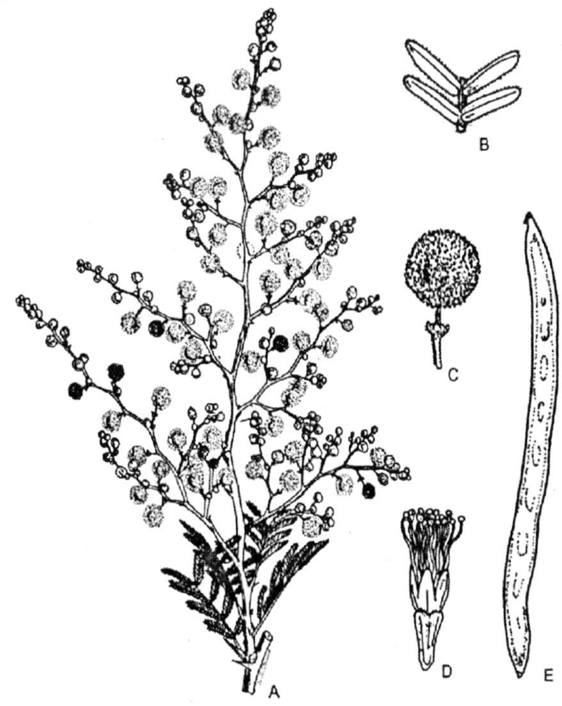

Fig. 58. *Acacia lecuophloea* (Roxb.) Willd. A. Flowering twig, B. Leaflets, C. Inflorescence, D. Flower, E. Pod

Moderate size, deciduous tree with whitish-yellow bark, 3 - 6 m tall.Younger branches velvety pubescenct. Leaves 2-pinnate, subssile, 2.5 - 5 cm long, main rachis pubescent, usually with a cup-shaped gland between each pair of pinnae. Stipular spines upto 3.5 cm long, equal or slightly equal, straight, pinnae 5 - 15 pairs, subsessile. Leaflets 5 - 15 pairs, linear, glabrous, crowded. Flowers in terminal panicales. Heads numerous, globose, 6 - 9 mm in diam, white, turning yellow, fragrant. Pedicels short, downy. Calyx 1.2 mm long. Corolla 2.5 mm long, lobes subacute. Podes 5 - 20 x 0.5 - 0.8 cm, sessile, thin, flat, slightly curved, clothed with pale-brown tomentum. Seeds 10 - 20 pale brown, 4 x 6 mm.

Field note: Common in open forests, easily recognized by galls on stem and branches and whitish bark.

Fl. : September- November; *Fr. :* November - March.

Specimen examined: Hamirgarh 313 (MLVGCB Herb).

Distribution: Native range through South and Southeast Asia is non-contiguous. India, Sri Lanka, Bangladesh, Burma and Thailand.

Uses: Dye obtained from the bark is traditionally used to colour the leather. Heart wood is durable and is used for agriculture equipments. Leaves serve as fodder. Fruits have toxic properties.

Acacia nilotica (L.) Willd. ex Del. Desir. Egypte Hist. Nat. 3: 79. 1813. *Mimosa nilotica* L. Sp. Pl. 521. 1753.

KEY TO THE SUB SPECIES

1a. Plant cupressiform in appearance........ *A.nilotica* ssp. *cupressiformis*

1b. Plant not cupressiform in appearance 2

2a. Pod narrowly and regularly constricted between seeds .... *A. nilotica* ssp. *indica*

2b. Pod not narrowly and regularly constricted between seeds .... *A. nilotica* ssp. *astringens*

subsp. ***astringens*** (Schum. & Thonn.) Roxb. in Candollea. 11: 150. 1948; Ali, in Nasir & Ali, Fl. W. Pak. 36: 12. 1973. *A. adansonii* Guill. & Perr. in Guill. *et al.* Fl. Senegamb. Tent. 249. 1832. *A. arabica* var. *vediana* T. Cooke, Fl. Pres. Bombay 1: 444. 1903 [1: 473. 1958 (Rpr.)] *A. arabica* var. *astringens* (Schum. & Thonn.) Baker, Legum. Trop. Afr. 3: 849. 1930. *A. nilotica* ex Del. ssp. *adansonii* (Guill. & Perr.) Brenan in Kew Bull. 12: 85. 1957. *A. nilotica* var. *adansoniana* (Dubard) Hill, in Bot. Mus. Leafl. Harvard Univ. 8: 99. 1940. *A. nilotica* var. *adansonii* (Guill. & Perr.) O.

Ktze., Rev. Gen. Pl. 1: 156. 1891. *A. nilotica* var. *vediana* (T. Cooke) Vajrav. & Kamble in J. Bombay Nat. Hist. Soc. 79: 708. (1982) 1983. (Plate 121)

Medium sized tree, not hemispherical or cupressiform in shape. Bark deeply cracked. Spines stout, ca 6.2 cm long. Pods flat, shortly stalked, 10 - 12 x ca 2 cm, slightly constricted between seeds.

Field note: Rare, in wastelands and scrub forests.

Fl. & Fr.: August - April.

Specimen examined: Mainal 1619 (MLVGCB Herb).

Distribution: Pakistan, India, Arabia, Sudan, Somalia, Senegal, Ghana, Togoland, N. Nigeria and Guinea.

subsp. ***cupressiformis*** (J. L. Stewart) Ali & Faruqi, in Pak. Journ. Bot. 1: 4. 1969; Bhandari, Fl. Indian Desert 134. 1978; Pandey, in Shetty & Singh Fl. Rajasthan 1: 298. 1987; Sanjappa, Legumes India 42. 1992. *A. arabica* (Lam.) Willd. var. *cupressiformis* Stewart, Punj. Pl. 51. 1869. *A. nilotica* var. *cupressiformis* (J.L. Stewart) Vajrav. & Kamble in J. Bombay Nat. Hist. Soc. 79: 708. (1982) 1983. (Plate 122)

Vern.: *Ramkanta, Ramkati babul.*

Tree, cupressiform in appearance; ascending branches like a poplar. Pods generally narrowly and regularly constricted between the seeds.

Field note : Its narrow erect crown, reminiscent of a Lombardy Poplar, is unique in *Acacia nilotica* (L.) Willd. ex Del.

Fl. & Fr.: August - April.

Specimen examined: Atun 2163 (MLVGCB Herb).

Distribution: India (Rajasthan, Punjab, U. P., Maharashtra, Gujarat) and Pakistan.

subsp. ***indica*** (Benth.) Brenan in Kew Bull. 12: 84. 1957; Bhandari, Fl. Indian Desert 147. 1978; Pandey, in Shetty & Singh Fl. Rajasthan 1: 299. 1987; Sanjappa, Legumes India 42. 1992. *Acacia arabica* (Lam.) Willd. var. *indica* Benth. in Hook. Lond. J. Bot. 1: 500. 1842. *A. arabica auct. non* (Lam.) Willd. 1806; Baker, in Hook. *f.* Fl. Brit. Ind. 2: 293. 1878; Duthie, Fl. Gangetic Plain 1: 314. 1903. *Acacia nilotica* (L.) Willd. ex Del. var. *indica* (Benth) A. F. Hill, Bot. Mus. Leafl. Harvard Univ. 99. 1940. (Plate 123)

Vern.: *Boryo, Kikar.*

Moderate size deciduous tree, up to 12 m high with deep taproot. Young branches slightly pubescent. Bark fissured, brown to black. Stipular spines in pairs, associated with leaves, whitish-brown, glabrous, 50 mm long. Leaves, bipinnate, with glands on the petiole and between the pinnae especially the upper ones; pinnae 4 - 11 pairs; leaflet 10 - 27 pairs, subsessile, linear-oblong, subobtuse, glabrous. Flowers, in axillary, pedunculate, fasciculate, globose heads, 0.8 - 1.5 cm in diam, predominantly yellow, regular, pedicellate. Bracteoles 2, above the middle part of peduncle, ovate, pubescent. Calyx 5, 1 - 1.2 mm long; teeth short. Corolla 5, free, 2.5 - 3 mm long; lobes triangular. Stamens numerous, free of the perianth, free of each other. Anthers dehiscing via longitudinal slits. Ovary monomerous, superior, 1 - celled. Style 1, simple. Podes 3.2 - 22 x 1 - 1.5 cm, moniliform, subcompressed, constricted at sutures between the seeds, beaked, grey pubescent, indehiscent. Seeds 8 - 12 per pod, 6 - 7 mm in diam, suborbicular, brown, smooth.

Field note: Common in wastelands and scrub forests.

Fl. & Fr.: August - April.

Specimen examined: Mainal 615 (MLVGCB Herb).

Distribution: Native to India, but has spread into western Asia, India (Punjab, Uttar Pradesh, W. Bengal, M.P., Madras, Maharashtra) Pakistan, Tanganyika and Africa.

Uses: Trees have been exploited for wood and timber. Branches are commonly lopped for fodder. Pods and leaves, are treated as a nutritious fodder. Pods are used as a supplement to poultry rations in India. Dried pods are particularly sort out by animals on rangelands.

Acacia senegal (L.) Willd. Sp. Pl. 4: 1077. 1806; Baker, in Hook. *f.* Fl. Brit. Ind. 2: 295. 1878; Duthie, Fl. Gangetic Plain 1: 317. 1903;

Bhandari, Fl. Indian Desert 148. 1978; Pandey, in Shetty & Singh Fl. Rajasthan 1: 300. 1987; Sanjappa, Legumes India 44. 1992. *Mimosa senegal* L. Sp. Pl. 521. 1753. *Acacia rupestris* Stock ex Boiss. Fl. Orient. 2: 638. 1853. (Plate 124)

Vern.: *Kumta.*

Medium sized tree, 3-6 m. tall, young shoots pubescent, old branches glaucous-grey, on older stems the bark peels off in thin flakes of a darker colour. Prickles in threes at the base of the petiole, two lateral ones nearly straight or slightly curved upwards, the third recurved ca 5 mm long. Rachis ca 2.5 - 5 cm long, with glands between the lowest and upper most pair of pinnae. Pinnae 3 - 5 pairs, opposite, sometimes alternate, ca 1.2 - 2.5 cm long. Leaflets 8 - 15 pairs, ca 2 - 5 x 1 - 1.5 mm, linear, obtuse, subsessile. Inflorescence a pedunculate spike, peduncle ca 8 - 18 mm long; spike 5 - 10 cm long. Flowers sessile. Calyx 1.5 - 2.5 mm long, broadly campanulate, glabrous. Corolla ca 4 mm long. Stamens indefinite, filaments 6 - 7 mm long. Pod 5 - 7.5 x ca 1.7-2.5 cm, thin, flat, almost straight, shortly stipitate, tip with a slightly curved beak. Seeds 5 - 6, disc like, almost circular, ovate to linear-ovate, 6 - 9 x 5 - 8 mm, with a U-shaped depression on either side, smooth, dark brown to greyish green in colour.

Field note: Commonly found in waste lands and scrub forests.

Fl. & Fr.: July- February.

Specimen examined: Gurala 160 (MLVGCB Herb).

Distribution: India (Rajasthan, South-East Punjab, Delhi), W. Pakistan (Sind, Baluchistan), Arabia, widely distributed in Tropical Africa

Uses: Tree yields a gum of commerce value. Which is used in medicines and most forms of confectionaries. Wood takes beautiful polish and is used for weaver's shuttles, fuelwood and charcoal.

Acacia tortilis (Forsk.) Hayne, subsp. **raddiana** (Savi) Brenan in Kew Bull. 1957: 87. 1957; Sanjappa, Legumes India 44. 1992. *Acacia raddiana* Savi, Acad. Egiz. Mem. 1. f. A-G. 1830; Pandey, in Shetty & Singh Fl. Rajasthan 1: 299. 1987. *Mimosa tortilis* Forsk. Fl. Aegypt. - Arab. 176. 1775. *Acacia tortilis* (Forssk.) Hayne. Darst. Beschr. Arzneyk. Gebrauchl. Gewachse 10. t. 31. 1827. *in part.* (Plate 125)

An umbrella shaped big tree. Branches puberulous when young, glabrous at age, forming flat-topped canopy; bark reddish-brown. Stipular spines up to 1 cm long, straight or some slightly recurved at apex, white. Leaves 4 -5 cm long, puberulous atleast in young age; pinnae 2-5 pairs, leaflets 6-19 pairs, 4-5 x 1.5- 2mm, oblong to oblanceolate, obtuse at apex, slightly cuneate at base, entire. Inflorescence capitate, globose, white to creamy. Pod 7.5-15.0 x 6-9 mm, contorted or spirally twisted. Seeds 6-7 x 4-5 mm, suborbicular, blackish-brown.

Field note: Introduced, but naturalized in the area.

Fl. : July-October; *Fr.* : November-February.

Specimen examined: Puskar 3803, Gurala 1816 (MLVGCB Herb).

Distribution: Algeria to Egypt and Arabia, southwards to S. Africa, introduced in India.

Uses: Gum used in the manufacture of mucilage, pharmaceuticals and medicines.

ALBIZIA Durazz.

Albizia lebbek (L.) Benth. in Hook. Lond. J. Bot. 3: 87. 1844, p.p. ('lebbeck') & in Trans. L. Soc. 30: 562. 1875 p.p.; Baker, in Hook. *f.*, Fl. Brit. Ind. 2: 298. 1878; Duthie, Fl. Gangetic Plain 1: 320. 1903; Bhandari, Fl. Indian Desert 148. 1978; Pandey, in Shetty & Singh Fl. Rajasthan 1: 301. 1987; Sanjappa, Legumes India 56. 1992. *Mimosa lebbeck* L., Sp. Pl. 516. 1753. (Fig. 59)

var. ***lebbek***

Vern.: *Sares.*

An unarned, deciduous 7-12 m tall tree. Bark rough, greyish-brown; inner bark reddish. Branches pubescent when young, glabrous at age. Leaves bipinnate, rachis 20 - 35 cm, with a large gland on petiole near base and other between uppermost pair of pinnae; pinnae 2 - 4 pairs; Leaflets 3 - 11 pairs, oblong to elliptic-oblong, asymmetrical, 15 - 65 x 5 - 35 mm, glabrous, entire, initially bright green and folding at night, maturing to a duller glaucous green and fixed rachis. Fully but briefly deciduous in the dry season. Inflorescence an axillary cluster of 15 - 40 pedicellate flowers, heads in umbellate, greenish to pale yellow, fragrant. Peduncle up to 4 cm, pedicel 1.5 - 5 mm. Calyx campanulate, hairy. Corolla 6-7 mm long, pubescent outside, glabrous within. Filaments numerous, free, connate below in a short tube, 1 - 1.5 cm. Pod flat, oblong 10 - 35 x 2 - 5 cm, stiff-papery when ripe, swollen over seeds, dehiscent. Seeds 3 - 12 per pod, brown, flattened, 7 x 1.5 mm, suborbicular, shinning, glabrous, pale-brown.

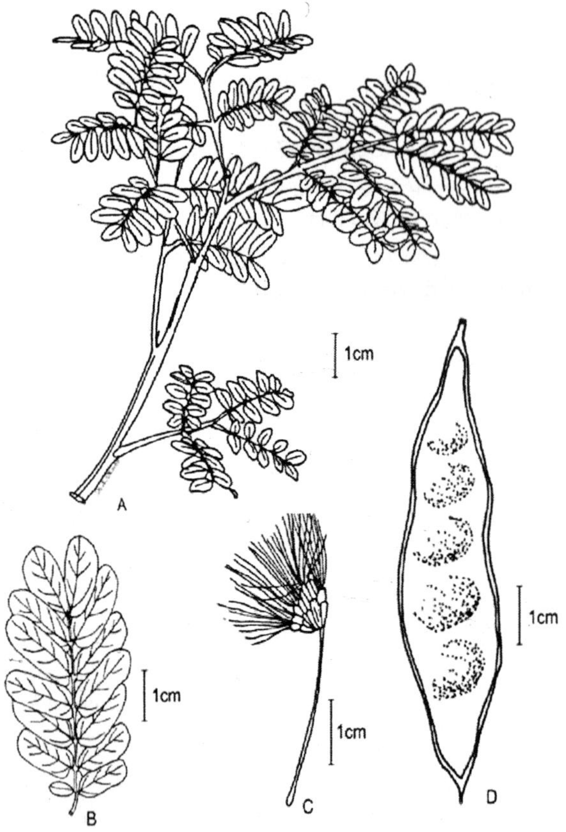

Fig. 59. *Albizia lebbek* (L.) Benth.
A. Twig, B. Leaf, C. Inflorescence, D. Fruit

Field note : Common in waste lands, also planted along road sides as an avenue tree.

Fl. & Fr.: March - November.

Specimen examined: Bhilwara 1358 (MLVGCB Herb).

Distribution: Pantropical.

DICHROSTACHYS (DC.) Wight & Arn. nom. cons.

Dichrostachys cinerea (L.) Wight & Arn. Prodr. 271. 1834; Wight, *l.c.* 2 (1): 4. t. 357. 1840; Baker, in Hook. *f.,* Fl. Brit. Ind. 2: 288. 1878; Duthie, Fl. Gangetic Plain 1: 310. 1903; Bhandari, Fl. Indian Desert 135. 1978; Pandey, in Shetty & Singh Fl. Rajasthan 1: 303. 1987; Sanjappa, Legumes India 65. 1992. *Mimosa cinerea* L., Sp. Pl. 520. 1753. (Fig. 60 & Plate 126)

Vern.: *Goyakhair.*

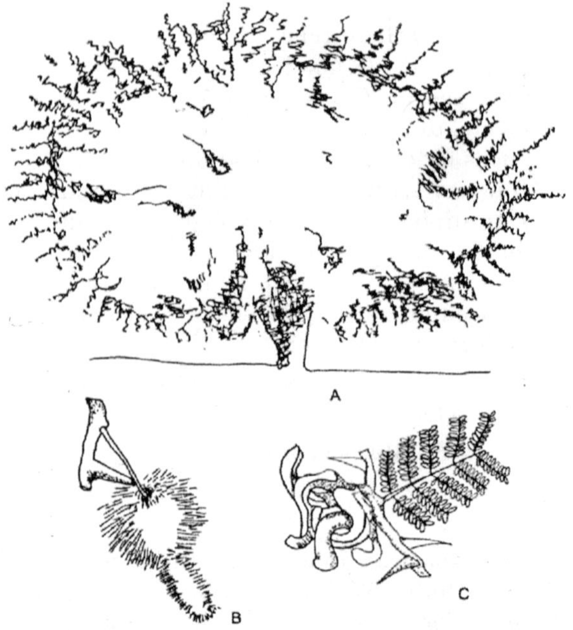

Fig. 60. *Dichrostachys cinerea* (L.) Wight & Arn.
A. Habit, B. Flower, C. Fruit

A much branched shrub or small tree. Branches usually ending in spines, pubescent. Leaves 2-pinnate, 3 - 6 cm long; rachis softly pubescent with a small, erect, stalked glands between each pair of pinnae; pinnae 4 - 14 pairs; leaflets 12 - 24 pairs, upto 3x1mm, linaer to linear-oblong, base oblique, subacute at the apex, ciliate on margins. Flowers small, bisexual and yellow in the upper half of spike, while neuter and white rose or purplish with long filiform staminodia in the lower half. Sepals 5, connate. Petals 5, connate at the base, oblong, valvate. Stamens 10, all fertile in bisexual flowers, distinct, slightly exserted, anthers glandular. Ovary sessile or shortly stipitate. Fruits clustered, linear or narrowly oblong, coriaceous, twisted at maturity, indehiscent or opening irregularly. Seeds smooth, compressed, ovoid to ellipsoid.

Field note: Common in open forests, and can be easily identified by its bi-coloured flowers.

Fl. & Fr.: May- February.

Specimen examined: Ladpura 1144, Kumbalgarh 3301 (MLVGCB Herb).

Distribution: Tropical and sub-ttropical Africa, Arabia, Verde Islands, Australia, Malaysia, Sri Lanka, Indonesia and India.

Uses: The extract of root bark is mixed with extract of stem bark of *Butea monosperma* and *Ziziphus mauritiana* and 1/4 cup juice + 3/4 cup of water is used to cure diarrhoea. Only one dose is sufficient.

LEUCAENA Benth.

Leucaena latisiliqua (L.) Gillis in Taxon 23: 190. 1974; Sanjappa, Legumes India 67. 1992. *Mimosa latisiliqua* L., Sp. Pl. 519. 1753. *M. glauca* L., Sp. Pl. ed. 2. 2: 1540. 1763, *non* L. *1753. Mimosa leucocephala* Lam. Encycl. 1: 12. 1783. *Leucaena glauca* (L.) Benth. in Hook. Lond. J. Bot. 4 : 416. 1842; Baker, in Hook. *f.,* Fl. Brit. Ind. 2: 290. 1878; Duthie, Fl. Gangetic Plain 1: 310. 1903; Bhandari, Fl. Indian Desert 135. 1978; Pandey, in Shetty & Singh Fl. Rajasthan 1: 303. 1987. *L. leucocephala* (Lam.) de Wit in Taxon. 10(2): 54. 1961.

Vern.: *Subabul.*

Large shrub or small, unarmed evergreen tree, ca 9 m tall. Bark greyish brown, slightly fissured, young shoots covered by tomentum. Leaves bipinnate, stipulate, stipule setaceous; rachis 7 - 15 cm long, pubescent, ending in a soft bristle, common rachis 15 - 20 cm long, pinnae 4 - 9 pairs, 5 - 10 cm long, hairy, ending into a bristle, leaflets 10 - 17 pairs, sessile, ca 10 - 13 x 2 - 4 mm, oblique, linear, oblong, acute, glabrous or subglabrous, base oblique. Flowers in globose white heads, often in pairs, peduncle 2.5 - 3.8 cm long, head 1.5 - 2.2 cm in diam. Calyx 2 - 3.5 mm long, tubular-campanulate, teeth short, villous. Corolla 10 - 11 mm long, petals free, valvate, spathulate, pilose outside. Stamens 10, much exserted, anthers eglandular. Ovary stalked, pubescent, style filiform, stigma minute. Pod straight, flat, 12.5 - 20 x 1.5 - 2.0 cm, minutely pilose, acute at the apex. Stipe 1 - 2.5 cm long. Seeds 15 - 25, 5.6 x 4.5 mm, oval, glaucous, dark brown, with a prominent U-shaped mark on either side.

Field note: Commonly planted along the hedges of cultivated fields, near habitations.

Fl. & Fr.: September - March.

Specimen examined: Ladpura 1144, Tikar 17791 (MLVGCB Herb).

Distribution: Probably native of Central America or South America, now pan-tropically cultivated. Sometimes planted in gardens in India and Pakistan.

Uses: Branches, leaves, pods and seeds of the plant are relished by camel, sheep and goats. Leaves are good source of protein, carotenes and can be employed to supplement alfaalfa leaf meal in poultry feed. The wood is used in making charcoal

MIMOSA L.

KEY TO THE SPECIES

1a. Pinnae 1-2 pairs .... *M. pudica*

1b. Pinnae more than 2 pairs 2

2a. Pinnae 5 - 12 pairs. Sutures of pod generally not prickly or with only a few prickles. Pod glabrous .... *M. himalayana*

2b. Pinnae 3-6 pairs. Sutures of pod prickly. Pod velvety .... *M. hamata*

Mimosa hamata Willd., Sp. Pl. 4: 1033. 1806; Baker, in Hook. *f.,* Fl. Brit. Ind. 2: 291. 1878; Duthie, Fl. Gangetic Plain 1: 312. 1903;

Bhandari, Fl. Indian Desert 150. 1978; Pandey, in Shetty & Singh Fl. Rajasthan 1: 304. 1987; Sanjappa, Legumes India 68. 1992. *Mimosa armata* Rottl. ex Spreng. Syst. Veg. ed. 2. 2: 206. 1825. (Plate 127)

Vern.: *Arkar, Arati.*

Medium sized much branched shrub up to 2 m high, branches hairy, very prickly, prickles on internodes also, ca 4 - 5 mm long, hooked. Leaf stipulate, stipules setaceous ca 3 mm long, hairy; bipinnate, rachis 1.2 - 5.0 cm long, sometimes prickly. Pinnae 3 - 6 pairs, ca 7 - 25 mm long, leaflets 4 - 10 pairs, more or less sessile, ca 2 - 3 x 1 - 2 mm, ovate oblong, acute, mucronate, hairy on both the sides, glabrous above, pilose below. Flowers in globose heads, solitary or paired in upper axils, peduncle 1.5-3.5 cm long, hairy; heads ca 10-13 mm in diam., flowers pink, tetramerous, sessile, bracteate, bract linear, spathulate, pilose. Calyx ca 1 mm. Corolla tube ca 3 mm long, lobes ca 1.5 mm long. Stamens 8. Ovary stalked, hairy. Pod 5 - 8 x ca 1.0 cm, flat, falcate, velvety, sutures wavy, prickly, joints 4 - 8. Seed ovate, flattened, reddish brown.

Field note: Common in open sandy plains.

Fl. & Fr.: August - March.

Specimen examined: Nathdwara 521, Kalwara 3384 (MLVGCB Herb).

Distribution: Pantropical.

Uses: It is browsed by cattle. It may be grown as a hedge plant.

Mimosa himalayana Gamble in Kew Bull. 1920: 4. 1920; Bhandari, Fl. Indian Desert 150. 1978; Pandey, in Shetty & Singh Fl. Rajasthan 1: 304. 1987; Sanjappa, Legumes India 68. 1992. *M. rubicaulis* Baker, in Hook. *f.* Fl. Brit. Ind. 2: 291. 1878, *pro parte*, non Lam. 1783; Duthie, Fl. Gangetic Plain 1: 311. 1903. (Plate 128 & 129)

Vern.: *Shiah kanta, Narkanta.*

Large straggling deciduous shrub, branches ribbed, densely hairy; prickles present on nodes or inter nodes, straight or hooked, 4 mm or less long. Leaf bipinnate, rachis 10 - 23 cm long, prickly, prickles mostly hooked, hairy, ribbed; pinnae 5-12 pairs, 2.5 - 6.5 cm long. Leaflets 6-15 pairs, 3-8 x ca 2 mm, oblong, obtuse, mucronate, glabrous above, more or less hairy below, subsessile. Inflorescence globose penduculate head forming terminal panicle; peduncle ca 2.5 - 6.0 cm long, pilose, head ca 1.2 - 1.5 cm across. Flowers pink fading to white, pedicellate, pedicel minute; bracteate, bract small, linear, ciliate. Calyx ca 1 mm, hairy. Corolla ca 3 mm long, tubular, 4 lobed, lobes ca 1 mm long. Stamens 8, long, exserted. Ovary short stalked, glabrous. Fruit 6 - 10 x 1.0 - 1.3 cm, occasionally with prickles on the sutures, strap-shaped, glabrous 4 - 10, 1 seeded joints, which get separated from the sutural frame.

Field note: Common in waste lands and open forests.

Fl. & Fr.: August - January.

Specimen examined: Ladpura 1326 (MLVGCB Herb).

Distribution: India (Punjab, Uttar Pradesh, Bihar, Orissa, W. Bengal, Assam, Madras), W. Pakistan and Afghanistan.

Uses: It is used as for hedge plant. Wood is suitable for tent pegs and for making gunpowder charcoal. Root, leaves and fruits are of minor medicinal importance.

Mimosa pudica L. Sp. Pl. 518. 1753; Baker, in Hook. *f.* Fl. Brit. Ind. 2: 291. 1878; Duthie, Fl. Gangetic Plain 1: 311. 1903; Pandey, in Shetty & Singh Fl. Rajasthan 1: 305. 1987; Sanjappa, Legumes India 69. 1992. (Fig. 61)

Vern.: *Chui-mui, Lajwanti.*

A low spreading shrub, branches hairy and prickly, hairs glandular. Leaf very sensitive; stipulate, stipules linear-lanceolate, ca 7 - 8 mm long; rachis ca 2.5 cm long, grooved and prickly, pinnae 1 - 2 pairs, ca 2.5 - 5 cm long, sessile, situated at the end of the rachis; leaflets 12 - 20 pairs, sessile, 6 - 8 x ca 5 - 6 mm, coriaceous, linear-acute, glabrous above, hairy on the margins and below. Flowers in peduncled more or less globose heads, yellow, peduncle 2 - 2.5 cm long; densely hairy; head ca 6-8 mm in diam. solitary or in axillary pairs along the branches, bracts small, linear-acute. Calyx minute. Corolla pink, ca 2 - 3 mm long, divided into 4 obtuse lobes. Stamens 4, much exserted. Fruit ca 12 - 18 x 2 - 4 mm, flat, recurved, having 3 - 5 or 1 seeded parts, glabrous, light brown; joints separating from the persistent, spinous sutures.

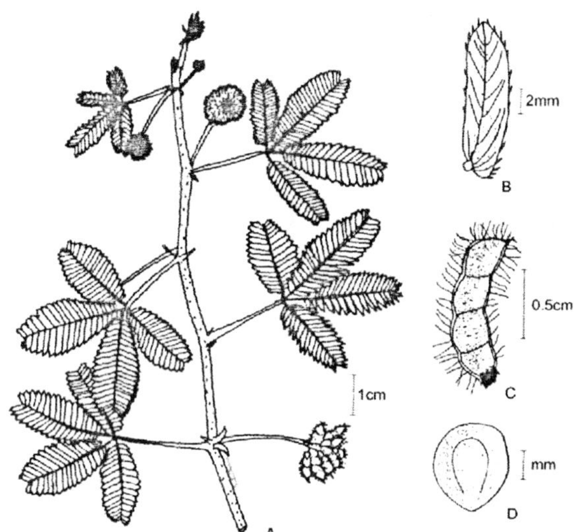

Fig. 61. *Mimosa pudica* L.
A. Flowering & fruiting branch, B. Leaflet, C. Fruit, D. Seed

Field note: Common in waste lands, in the gardens and open forests. Leaves are very sensitive to touch.

Fl. & Fr.: August - January.

Specimen examined: Bhilwara 1126, Mandal 2143 (MLVGCB Herb).

Distribution: Native of Brazil, cultivated in gardens of India, Pakistan.

Uses: The plant is grown as an interesting ornamental plant. The juice of freshly crushed leaves is used internally and externally in piles.

NEPTUNIA Lour.

Neptunia triquetra (Vahl) Benth. in Benth. in Hook. Lond. J. Bot. 4: 355. 1842; Baker, in Hook. *f.*, Fl. Brit. Ind. 2: 286. 1878; Duthie, Fl. Gangetic Plain 1: 307. 1903; Pandey, in Shetty & Singh Fl. Rajasthan 1: 306. 1987; Sanjappa, Legumes India 69.1992. *Mimosa triquetra* Vahl, Symb. Bot. 3 : 102. 1797. *Desmanthus triquetrus* Willd. Sp. Pl. 4. 1045. 1806; Wight, Ic. 3 (1) : 3. t. 756. 1844.

Diffuse or suberect, perennial herb. Leaves 2-pinnate, 2.5 - 5.0 cm long, pinnae 1 - 3 pairs, leaflets 10- 18 pairs, sessile, upto 7.5 x 2.0 mm, linear-oblong, obtuse, ciliate on margins, flowers yellowish; peduncles with 2 large bracts. Pods 1.0-2.5 x 0.6-0.8 cm, elliptic-oblong, rounded, with a small beak at the apex, venulose. Seeds 2 - 8, ca 4.0 x 3.5 mm, oval to suborbicular or obovate-oblong, compressed, polished, blakish-brown.

Field note: Rare, in marshy places, near ponds.

Specimen examined: Jahajpur 2115, Mandal garh 2165 (MLVGCB Herb).

Distribution: Native of India (Rajasthan, Tamil Nadu), Myanmar and Netherlands.

PITHECELLOBIUM Mart. nom cons

Pithecellobium dulce (Roxb.) Benth. in Hook. Lond J. Bot. 3 : 199. 1844; Baker, in Hook. *f.* Fl. Brit. Ind. 2: 302. 1878; Duthie, Fl. Gangetic Plain 1: 322. 1903; Bhandari, Fl. Indian Desert 137. 1978; Pandey, in Shetty & Singh Fl. Rajasthan 1: 306. 1987; Sanjappa, Legumes India 71. 1992. *Mimosa dulsis* Roxb. Pl. Cor. 1: 67. t. 99. 1798. *Igna dulsis* (Roxb.) Willd. Sp. Pl. 4: 1005. 1806; Wight, *l.c.* 1: t. 198. 1839. (Fig. 62 & Plate 130)

Vern.: *Jangal Jalebi, Mar-amli.*

A medium sized, evergreen tree. Leaf stipulate, stipules modified into spines; rachis 1-2.5 cm long, pinnae 1 pair, leaflets two on each pinna, elliptic-oblong, oblique, 2-4.5 cm long. Inflorescence small globose, sessile or short peduncled heads arranged in long panicled raceme. Flowers greenish white. Calyx ca 1 mm long, funnel shaped, pilose. Corolla ca 3-4 mm long, petals united below the middle. Stamens monadelphous, much exserted, not glandular. Style filiform, stigma simple. Pods 10 - 12.5 cm x 2-3 mm, turgid, twisted, sutures indented between the seeds. Seeds 5 - 9, enveloped in pink or white pulpy aril which is edible. Seed ca 2.5 cm long, spiny, black.

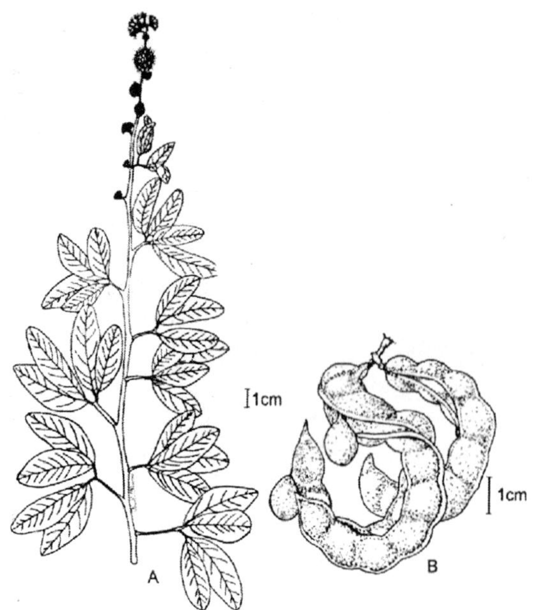

Fig. 62. *Pithecellobium dulce* (Roxb.) Benth.
A. Fowering twig, B. Fruit

Field note: Common along the boundries cultivated fields, often planted as a hedge.

Fl. & Fr.: November - April.

Specimen examined: Pansal 1346 (MLVGCB Herb).

Distribution: It is a native of Mexico, widely cultivated in Tropics.

Uses: The fleshy pulp of the fruit is edible. Leaves are used as fodder. The wood is used as firewood.

PROSOPIS L.

KEY TO THE SPECIES

1a. Tree, Podes cylindrical .... *P. cineraria*

1b. Shrub or small tree. Podes compressed......... 2

2a. Branches zigzag. Leaflets 10 - 40 mm long.. *P. glandulosa*

2b. Branches not as above. Leaflets 3 - 7 mm long.. .... *P. chilensis*

Prosopis cineraria (L.) Druce, in Rep. Bot. Exch. Club. Brit. Isles. 1913. 3: 422. 1914; Burkart, in J. Arnold Arbor 57: 450. 1976; Bhandari, Fl. Indian Desert 151. 1978; Pandey, in Shetty & Singh Fl. Rajasthan 1: 307. 1987; Sanjappa, Legumes India 71. 1992. *Mimosa cineraria* L. Sp. Pl. ed. 2. 1500. 1763 (*'cinerea'*). *Prosopis spicigera* L., Mant. 68. 1767; Baker, in Hook. *f.* Fl. Brit. Ind. 2: 288. 1878; Duthie, Fl. Gangetic Plain 1: 309. 1903. *Adenanthera aculeata* Roxb., Fl. Ind. 2: 371. 1832. (Fig. 63 & Plate 131)

Vern.: *Khejri.*

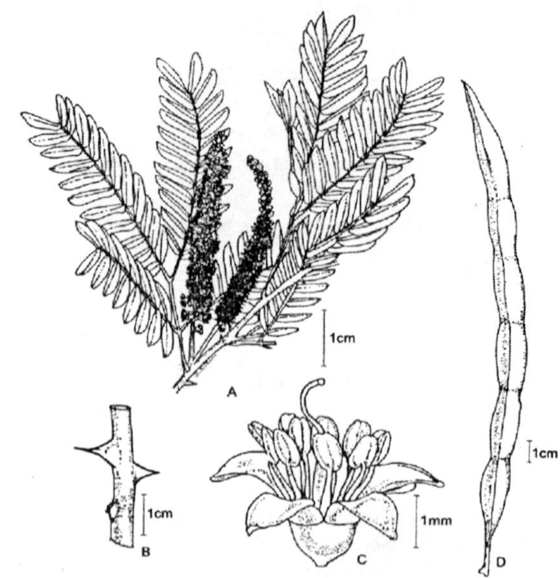

Fig. 63. *Prosopis cineraria* (L.) Druce
A. Flowering branch, B. Young part of stem, C. Flower, D. Fruit

Medium sized tree up to 6 - 15 m tall, branches prickly, prickles curved, compressed. Leaves alternate, bipinnate, rachis 1.2 - 5 cm long; pinnae 1 - 2 pairs, 2.5 - 8.0 cm. long, leaflets 7-12 pairs, more or less sessile, ca 3 - 5 x 2 - 4 mm, oblong, oblique, apex usually mucronate, base rounded, 3 nerved. Flowers creamy white in pedunculate spikes, nearly 5 - 12.5 cm long, peduncle 1.0 - 2.5 cm long. Calyx ca 1 - 1.5 mm long, cup-shaped, truncate or obscurely 5 toothed. Petals 3 - 4 mm long, oblong, tips recurved. Stamens 10, free, shortly exserted, anthers tipped with deciduous glands. Pods 6 - 25 cm x 5 8 mm, slender, pendulous, cylindric, turgid, exocarp coriaceous, mesocarp pulpy, endocarp papery. Seeds 10 - 15, ovoid, compressed.

Field note: Commonly growing throughout the area. Insect galls on stems, branches and petioles have been observed.

Fl. & Fr.: March - June.

Specimen examined: Puskar 3253 (MLVGCB Herb).

Distribution: India (Punjab, Haryana, Rajputana, Bombay, Madras), Pakistan, Afghanistan, Persia and Arabia.

Uses: The foliage is lopped for fodder and the wood is commonly used for fuel. The bark powder is mixed with flour during scarcity of food and famine. Pods are used as fodder, as vegetables or eaten dry (Dry fruits are called as *'Thinj, Zheenj'*). Seeds are used green or dry, raw or cooked.

Prosopis glandulosa Torr. Ann. Lyceum Nat. Hist. New York 2: 192. 1827. Bhandari, Fl. Indian Desert 138. 1993.

Much branched shrub or medium sized tree. Spines axillary, uninodal. Leaves glabrous, uni- or bijugate; petiole (with rachis when extant) 2 - 15 cm long; pinnae 6 - 17 cm long; leaflets 6 to 17 pairs, ca 7 - 18 mm distant on the rachis, linear or oblong, obtuse, glabrous, subcoriaceous, prominently veined below. Racemes in spike. Podes straight, 8–20 cm long x 0.7–1.3 cm broad, rarely subfalcate, compressed to subterete, submoniliform, glabrous. Seeds oblique to longitudinal.

note: Rare.

Fl. & Fr.: Aug. - April.

Specimen examined: Beawar 5802, on the way of Burr 4934 (MLVGCB Herb).

Distribution: SW U.S., India and N Mexico

Prosopis juliflora (Swartz) DC. Prodr. 2 : 447. 1825; Duthie, Fl. Gangetic Plain 1: 309. 1903; Pandey, in Shetty & Singh Fl. Rajasthan 1: 307. 1987; Sanjappa, Legumes India 72. 1992. *Mimosa juliflora* Swartz, Prodr. 85. 1788. *Prosopis chilenisis sensu* Bhandari, Fl. Indian Desert 151. 1978, non (Molina) Stuntz 1914.

Vern.: *Vilayati borio, Angreji borio, Vilayati Babul.*

Large shrub or tree, evergreen, ca 5 m tall, generally armed with stipular spines. Leaves alternate, bipinnate, with 1-3 pairs of pinnae, rachis 1 - 8 cm long, prolonged beyond the last pinnae as a soft bristle. Leaflets 10 - 20 pairs, 5 - 17 x, 2 - 3 mm broad, entire, oblong, obtuse, sometimes mucronate. Stipules spiny, generally 1.0 cm or more, in pair. Inflorescence dense axillary pedunculate spikes 4 - 8.5 cm long, peduncle ca 6 - 12 mm long. Flowers greenish yellow, pedicel 1 mm. Calyx ca 1 mm long, cup-shaped, 5 toothed, teeth small. Petals 5, free, ca 3 mm long, tip and margin hairy. Stamens 10, free, exserted, ca 4 mm long, anthers tipped with deciduous glands. Pod pedicellate, ca 16 - 23 cm x 10 - 12 mm broad, almost straight to semi-circular, light yellow, glabrous, pedicel ca 5 - 7 mm long. Seeds 10 - 18, oblong.

Field note: Commonly growing throughout the area.

Fl. & Fr.: Almost throughout the year.

Specimen examined: Jahajpur 1366 (MLVGCB Herb).

Distribution: Indigenous to West Indies and Mexico, widely naturalized in India and Pakistan.

Uses: It is good firewood and excellent for making charcoal. It is also used to fence posts.

ROSACEAE

POTENTILLA L.

Potentilla supina L. Sp. Pl. 497. 1753; Hook. *f.*, Fl. Brit. Ind. 2: 359. 1878; Duthie, Fl. Gangetic Plain 1: 327. 1903; Bhandari, Fl. Indian Desert 139. 1978; Pandey, in Shetty & Singh Fl. Rajasthan 1: 309. 1987. (Fig. 64)

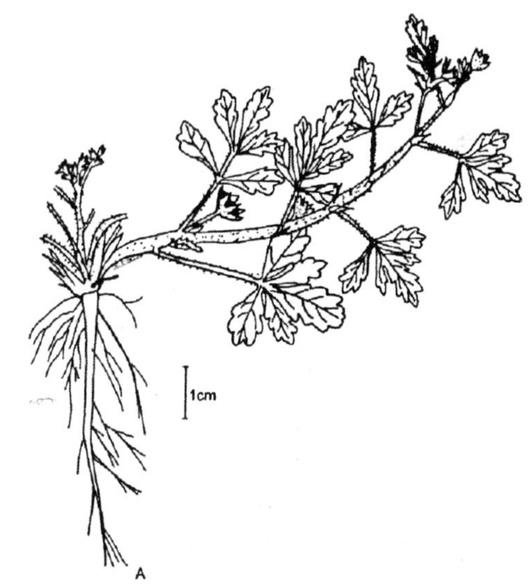

Fig. 64. *Potentilla supina* L. A. Habit

Prostrate or procumbent much branched, annual herb, 10 - 20 cm long. Stem numerous froms the base, slender, spreading, hairy. Leaves variable, basal and lower cauline leaves larger with more leaflets; upper cauline leaves nearly sessile, smaller with fewer leaflets; leaflets 3 - 9, subsessile, 0.5 - 1 x 0.3 - 0.5 cm, obovate, 3-lobulate or incised, coarsely serrate, appressedly hairy, petiolate. Flowers solitary, axillary, yellow, often crowded on the repeatedly branched stems with much reduced leaves. Pedicels 3 - 12 mm long, slender, hairy. Bracteoles 5, oblong-elliptic, obtuse, half as broad as their calyx lobes. Calyx 5 mm long, hairy outside; lobes 5, ovate-oblong, subacate. Petals 5, 3 - 4 mm long, oblong. Fruits globose, smooth. Achenes many, minute, brown, ovoid, receptacles villose, globose.

Field note: Common on moist bed, sandy river banks, field margins, along with *Trigonella occulta* Del. and *Polygonum plebeium* R. Br.

Fl. & Fr.: December - April.

Specimen examined: Patola Mahadev 901, Harni Mahadev 1674 (MLVGCB Herb).

Distribution: Throughout warmer regions of the world.

COMBRETACEAE

KEY TO THE GENERA

1a. Flowers in globose heads, fruits 2-winged ....*Anogeissus*

1b. Flowers in spikes or racemes, fruits smooth or 3 to 5 winged........................ ......*Terminalia*

ANOGEISSUS (DC.) Guill., Perr & A. Rich.

KEY TO THE SPECIES

1a. Leaves 4 - 14 x 2 - 8 cm, glabrous .. *A. latifolia*

1b. Leaves 2 - 6 x 1 - 4 cm, compressed hairy...... 2

2a. Leaf nerves 5 – 7, apex obtuse or acute. Fruit wings narrow.................................. *A. pendula*

2b. Leaf nerves 4 – 5, apex obtuse or subcordate. Fruit wings cordate at the base......... *A. sericia*

Anogeissus latifolia (Roxb. ex DC.) Wall ex Guill. & Perr., Fl. Seneg. Tent. 1: 280. 1832; Clarke, in Hook. *f.*, Fl. Brit. Ind. 2: 450. 1878; Bedd. Fl. Sylv. t. 15. 1869; Duthie, Fl. Gangetic Plain 1: 327. 1903; Scott., in Kew Bull. 33 (4): 560. 1779; Pandey, in Shetty & Singh Fl. Rajasthan 1: 313. 1987. *Conocarpus latifolius* Roxb. ex. DC. Prodr. 3: 17. 1828 & Mem. Combret. 25. 1828; *Anogeissus latifolia* (Roxb. ex DC.) Wall ex Guill. & Perr. var. *glabra* & *villosa* Clarke in Hook. *f.* Fl. Brit. Ind. 2: 450. 1878.

Vern.: *Dhawari.*

Tall deciduous tree, up to 24 m tall with whitish grey smooth bark, younger parts pubescent or glabrous. Leaves alternate or subopposite, broadly elliptic or ovate, coriaceous, pubescent when young. Flowers sessile in dense globose head. Bracteole very minute, nearly rusty pubescent. Calyx pubescent, teeth triangular. Stamens exserted. Fruit almost glabrous, orbicular, 5 (excluding the beak) x 6 mm, laterally winged, long beaked, beak equal to or longer than the main body of the fruit, yellowish brown, crowded in a globose head.

Field note: Common in forests towards southern side.

Fl. & Fr.: March - July.

Specimen examined: Mainal 1639 (MLVGCB Herb).

Distribution: India, Tropical Africa, Sri Lanka and Pakistan.

Uses: Its wood is used as fuel, in agricultural instruments and in buildings. Leaves are used for tanning and a gum extracted from it, is used in calico printing.

Anogeissus pendula Edgew., in Cat. Pl. Banda Dist. 47. 1851. & J. Asiat. Soc. Beng. 21: 171. 1853; Brandis, Fl. 229. 1874; Clarke in Hook. *f.*, Fl. Brit. Ind. 2: 451. 1878; Duthie, Fl. Gangetic Plain 1: 339. 1903; Mahesh, Illus. Fl. Delhi f. 84. 1966; Bhandari, Fl. Indian Desert 140. 1978; Pandey, in Shetty & Singh Fl. Rajasthan 1: 313. 1987.

Vern.: *Dhokra, Dhok.*

Tree, 4 -10 m high. Bark grey. Leaves alternate, young ones elliptic or sub-orbiculate. Petioles upto 3 mm long. Inflorescence solitary head, axillary and terminal yellow, 15 mm in diam. Calyx tube compressed, 2-winged; wings attenuated above the ovary, glabrous, brown, subulate, margins jagged; limb oblong, campanulate, 1.5 mm in diam. Stamens 10, glabrous, exserted, filament brown, anthers yellow. Fruit 4 - 6 x 3 - 4 mm, glabrous exept at the apex winged, wings narrow, beaked.

Field note: Dominant through out the area.

Fl. & Fr.: August - November.

Specimen examined: Bhim 3456, Jahajpur 1689 (MLVGCB Herb).

Distribution: India and Pakistan.

Uses: Its wood is used as fuel, in agricultural instruments and in house building framwork. Leaves are used in dyeing.

Anogeissus sericia Brandis var. ***nummularia*** King ex Duthie, Duthie, Fl. Gangetic Plain 1: 340. 1903; Scott in Kew Bull. 33: 559. 1979. A. coronata Stapf in Bull. Misc. Inf. Kew 1914. A. rotundifolia Blatt. & Hallb. in Journ. Bombay Nat. Hist. Soc. 26: 525. 1919; Pandey, in Shetty & Singh Fl. Rajasthan 1: 314 - 315. 1987.

Vern.: *Indokada, Indrok.*

Large or medium sized tree. Young branches and leaves rusty silky. Leaves alternate, many, elliptic - sub – orbicular, or orbicilar. Flowers in globose, dense heads, solitary or subterminal peduncles. yellow to brownish yellow. Calyx tube compressed, 2-winged, attenuated above the ovary, limb long campanulate, pubescent. Stamens exserted. Fruits glabrous, wings of fruit jagged at the apex only.

Field note: Rare.

Fl. & Fr.: September - June.

Specimen examined: Sopura 5237, Rampura Jahajpur 5288, Jalam pura (Chittorgarh) 5332 (MLVGCB Herb).

Distribution: Endemic to NW India (Rajasthan & Gujarat).

TERMINALIA L. nom. cons.

KEY TO THE SPECIES

1a. Flowers in simple spike or spersely branched, upper flowers of spike male. Fruits not winged............................ *T. bellirica*

1b. Flowers in panical of spike, flowers bisexual, Fruits winged, ... 2

2a. Bark smooth, pale greenish to whitish grey. Fruits with short, hard angles of wings, usually notched near the top, lines on wings oblique and curving up wards.......... .*T. arjuna*

2b. Bark dark grey, rough. Fruits with long, thin, papery wings, usually rounded at the top, lines on wings straight and horizontal....................................... *T. elliptica*

Terminalia arjuna (Roxb. ex DC.) Wight & Arn., Prodr. 314. 1834; Bedd., Fl. Sylv. t. 28. 1869; Clarke in Hook. *f.*, Fl. Brit. Ind. 2: 447. 1878; Pandey, in Shetty & Singh Fl. Rajasthan 1: 316. 1987. *Pentaptera tomentosa* Roxb. ex DC. Prodr. 3: 14. 1828. *cum descript. P. glabra* Roxb., Fl. Ind. 2: 440. 1832, *cum descript. Terminalia glabra* (Roxb.) Wight & Arn. Prodr. 314. 1834; Duthie, Fl. Gangetic Plain 1: 336. 1903. (Fig. 65 & Plate 132)

Vern.: *Arjun.*

Fig. 65. *Terminalia arjuna* (Roxb. ex DC) Wight & Arn. A. Flowering twig, B. Fruit

Deciduous tree, up to 6 - 15 m tall with smooth, pale greenish to whitish grey bark. Leaves usually subopposite, oblong-elliptic or somewhat sub-orbicular, 7 - 18 x 4 - 6 cm, obtuse, rarely subacute with rounded or cordate base, glabrous to subglabrous above, partially pubescent beneath, entire somewhat crenate or serrate in the upper half or throughout; petiole 5 - 10 mm long with 2 rounded glands at the apex. Inflorescence axillary or terminal paniculate spikes, 3 - 6 x 1 cm. Bracteole small, deciduous, linear lanceolate. Flowers yellowish white, sessile. Bracteoles broadly campanulate. Calyx 4 - 5 mm long, teeth triangular c. 15 mm long, glabrous. Stamens much exserted. Ovary glabrous; disc barbate. Fruit ovoid-oblong, 2.5 - 5 cm long, brown with 5 projecting wings, wings striated with 5 ascending veins.

Field note: Common in forest and waste lands particularly in eastern part of the area.

Fl.: March - May; *Fr.*: July - November.

Specimen examined: Mainal 1882 (MLVGCB Herb).

Distribution: India, Sri Lanka and Pakistan.

Uses: Decoction (250 gm dried or 40 gm fresh bark + 750 ml cold water, reduce to about 500 ml after simmering, this makes 3 - 4 doses). 2 tsp of Decoction is taken orally 2 times a day, it cures poor circulation of blood. 1 tsp tincture diluted with 25 ml water of fruit juice is taken 2 - 3 times a day, it is valuable cardiac tonic (Tincture: 200g dried or 300gm fresh bark is powdered and added to 1 liter 35 - 40% alcohol).

Terminalia bellirica (Gaertn.) Roxb., Pl. Corom. 2: 54. t. 198. 1805; Miq., Fl. Ind. Bat. 1. (1): 600. 1855, *'bellerica'*; Clarke in Hook. *f.,* Fl. Brit. Ind. 2: 445. 1878; Duthie, Fl. Gangetic Plain 1: 335. 1903; Pandey, in Shetty & Singh Fl. Rajasthan 1: 316. 1987; Gang. & Chakrab., in J. Econ. Tax. Bot. 21: 338. 1997. *Myrobalanus bellirica* Gaertn., Fruct. 2: 90. t. 97. f. a-d. 1791; *Terminalia bellirica* var. *laurinoides* Clark, in Hook. *f.,* Op.Cit. (Plate 133)

Vern.: *Bahera, Berra.*

Large deciduous tree, 10-50 m tall with ashy bark, often with a bluish ting, young parts profous pubescent. Leaves alternate or fascicled at the end of branches, elliptic or broadly obovate, puberoulous when young, glabrous on maturity, coriaceous, dotted, entire, obtuse to subacute, 8-20 x 7.5-15 cm; petiole 2.15 cm long, eglandular. Spikes axillary and termi-nal, loosely arranged, 5-15 cm long with pubesc-ent floral axis. Flowers greenish yellow, 5-6 mm across, sessile, upper flowers of the spike male, lower flowers bisexual; bracteoles minute caducous. Calyx. 2-2.5 mm long, densly tomentose outside, densly villous with long brown hairs inside, expanded upward in 5 recurved; lobes triangular-ovate, ca 1.5 mm long. Stamens 3-4 mm long; anthers ca 1 mm long inserted on the hypanthium, epigynous. Disc intrastaminal, densly hairy. Ovary ca 2 mm long; style glabrous, fulvous, ca 4 mm long. Fruit ovate or elliptic, 1.5-2.5 cm in diam., covered with minute pale pubescence, stone very thick, indistinctly 5 angled.

Field note: Common in forest and waste lands particularly in eastern part of the area.

Fl.: February - April; *Fr.*: May - August.

Specimen examined: Mainal 836, Bhilwara 1529 (MLVGCB Herb).

Distribution: Sri Lanka, Burma, Indo-china, Siam, Malayan peninsula, India and Pakistan.

Uses: Grows wild in sub-himalayan tracts from Rawalpindi to eastwards. The wood is not very durable and the fruit is used in dyeing and tanning. The fruit powder is taken with milk to cure sexual debility. The powder of seeds mixed in sweets (Ladoos) is taken as health tonic.

Terminalia elliptica Willd., Sp. Pl. 4: 969. 1806; Gang. & Chakrab., in J. Econ. Tax. Bot. 16: 601. 1992 & Ibid., 21: 351. f. 21 A. 1997. T. alata Roth, Nov. Sp. Pl. 379. 1821. *T. crenulata* Roth, Nov. Pl. Sp. 380. 1821; Pandey, in Shetty & Singh Fl. Rajasthan 1: 317. 1987. *T. coriacea* (Roxb.) Wight & Arn. Prodr. 314. 1834 non Mart. 1829; Clarke in Hook. *f.,* Fl. Brit. Ind. 2: 445. 1878. *T. tomentosa* (Roxb.) Wight & Arn. var. *crenulata* (Roth) Clarke in Hook. *f., l.c.* 2: 445. 1878. *T. tomentosa* (Roxb.) Wight & Arn. var. *coriacea* (Roxb.) Clarke in Hook. *f.,* op. cit. (Plate 134)

Deciduous tree with somewhat dark grey rough bark, younger parts more or less pubescent. Upper leaves alternate, lower subopposite, variable; elliptic-lanceolate, elliptic-oblong or ovate, glabrous above, pubescent below, entire or serrulate, 5.0 - 22.5 x 2.5 - 13 cm, petiole 7 - 10 mm long with 2 glands on either side at the apex. Inflorescence terminal or paniculate spike. Flowers all bisexual, dull yellow, with linear lanceolate bracts. Hypanthium ca 2.5 mm long, cup shaped, glabrous or pubescent, teeth triangular, ca 1.5 mm long. Disc velvety hairy. Fruits 2.4 - 5 cm long, obovoid-oblong or globose, dark-brownish, glabrous with 5 broad wings, wings striations horizontal.

Field note: Common in forests and wastelands particularly in eastern part of the area.

Fl.: February- April; *Fr.*: May - August.

Specimen examined: Mainal 811 (MLVGCB Herb).

Distribution: India and Sri Lanka.

Uses: The wood is used in agriculture as well as in household equipments. Leaves are fed to cattle for milk production.

MYRTACEAE

SYZYGIUM Gaertn. nom. cons.

KEY TO THE SPECIES

1a. Large tree. Leaves 3-6 cm broad; petiole 1 - 2 cm long. Calyx turbinate .... *S. cumini*

1b. Small tree. Leaves less than 3 cm broad; petiole 0.5 cm long. Calyx funnel-shaped.. *S. heyneanum*

Syzygium cumini (L.) Skeels, U.S.D. ept Agric. Bur. Pl. Industr. Bull. 248: 25. 1912; Parmar, in Shetty & Singh Fl. Rajasthan 1: 318. 1987. *Myrtus cumini* L. Sp. Pl. 471. 1753. *Eugenia jambolana* Lam. Encycl. 3: 198. 1789; *I.c.* 2 (3): 2. t. 535. 1832; Duthie in Hook. *f.,* Fl. Brit. Ind. 2: 499. 1879 & Fl. Gangetic Plain 1: 342. 1903.

Vern.: *Jamun.*

Tree, 6-20 m tall. Bark thick, rough. Branchlets greyish white when dry, terete. Petiole 1 - 2 cm. Leaves broadly elliptic-oblong or elliptic or ovate, 6 - 18 × 3.0 - 8 cm, leathery, abaxially slightly pale when dry, adaxially brownish green to blackish brown and slightly glossy when dry, both surfaces with small glands, secondary veins numerous, 1 - 2 mm apart, and gradually extending into margin, intramarginal veins ca 1 mm from margin, base broadly cuneate to rarely rounded, apex rounded to obtuse and with a short cusp. Inflorescences axillary on and crowded in heads on ends of laxy, paniculate cymes, to 11 cm. Flowers white, sessile or shortly, pedicelled. Calyx lobes inconspicuous, 0.3 - 0.7 mm, cupular. Petals 4, white or light purple, coherent, ovate and slightly rounded, ca 2.5 mm. Stamens 3-4 mm. Style as long as stamens. Fruit purple or red to black, oblong or ovoid, fleshy, 1.5-2.2 x 1.0 - 1.5 cm, 1- seeded; persistent calyx tube 1 - 1.5 mm. Seeds solitary, with thick, fleshy cotyledons closely appressed on their inner faces

Field note: Common in forests, also cultivated along road sides, in gardens and agriculture fields.

Fl.: Feb - Mar; *Fr.*: April - July.

Specimen examined: Bhilwara 104, Jahajpur 1529 (MLVGCB Herb).

Distribuition: Australia, Bhutan, China, India, Indonesia, Laos, Malaysia, Nepal, Sri Lanka, Thailand and Vietnam.

Syzygium heyneanum (Duthie) Wall. ex Gamble Fl. Madras 482, 1919; Parmar, in Shetty & Singh Fl. Rajasthan 1: 318. 1987. *Eugenia heyneana* Duthie, in Hook. *f.,* Fl. Brit. Ind. 2: 500. 1879 & Fl. Gangetic Plain 1: 343. 1903. *E. jambos* L. Sp. Pl. 470. 1753. *Jambosa vulgaris* DC. Prodr. 3: 286. 1828; Wight, Ic. 2 (2): 2. t. 435. 1841. (Plate 135)

Vern.: *Jamuni.*

Medium tall tree, 3 - 5 m high, Leaves 6 - 14 x 1.5 - 4.5 cm, opposite, oblanceolate or elliptic-lanceolate, subobtuse, glabrous, nerves uniting in an intramarginal nerves. Cyme panical, usually extra-axillary, slender, shorter than leaves. Flowers in small heads at the ends of branches of cymes, sessile or shortly pediceled, white. Calyx obscurely lobed. Petals orbicular, calyptrate. Berries ca 2.5 x 1.6 cm, ovoid-ellipsoid, pale purple, glabrous.

Field note: Rare in forests, particularly in river beds and along streams.

Fl.: Feb - Mar; *Fr.*: April - July.

Specimen examined: Kanera 3704 (MLVGCB Herb).

Distribuition: India.

LYTHRACEAE

KEY TO THE GENERA

1a. Small marshy herbs, flowers small, calyx membranous.. 2

2a. Flowers solitary or sub-solitary, fruits dehiscent, opening by valves................. *Rotala*

2b. Flowers in cymes, fruits indehiscent, opening irregularly........................ *Ammannia*

1b. Shrubs or trees, flowers larger, calyx moderately thick... 3

3a. Flowers tetramerous, stamens 8, seeds angular, pyramidal........................... *Lawsonia*

3b. Flowers trimerous, stamens more than 8, seeds not angular 4

4a. Inflorescence a terminal panical. Capsules loculicidal. Seeds winged........... *Lagerstroemia*

4b. Inflorescence an axillary cyme. Capsules irregularly dehiscent. Seeds obvoid not winged .... *Woodfordia*

AMMANNIA L.

KEY TO THE SPECIES

1a. Petals present, caducous; style distinct.. *A. auriculata*

1b. Petals absent; style absent or minute ....*A. baccifera*

Ammannia auriculata Willd. Enum. P1. Hort. Berol. 1: 7. t. 7. 1803; Parmar, in Shetty & Singh Fl. Rajasthan 1: 321.1987. *A. senegalensis* DC. Prodr. 3: 77. 1828; non Lamk. 1791; Clarke in Hook. *f.,* Fl. Brit. Ind. 2: 570. 1878; Duthie, Fl. Gangetic Plain 1: 350. 1903. (Plate 136 & 137)

Erect or decumbent, 10 - 40 cm tall,. Stem usually unbranched, winged in the upper part. Leaves linear-lanceolate, auriculate, 0.5 - 2.0 x 0.2 - 0.4 cm, acute, cordate or subcordate at base. Cymes 2 - 3 (-13) flowered; peduncle 2 - 5mm long. Pedicel 1 - 4 mm long. Epicalyx minute. Petals obovate-cuneate, pink. Stamens inserted above the middle of the hypanthium. Ovary 1- 2 x 1 - 2 mm; style 1-1.75 mm long. Capsule globose or obtusely 4-gonous, 2 - 3 x 2 - 3 mm, brown. Seeds discoid.

Field note: Common in wet places.

Fl. & Fr.: August-September

Specimen examined: On the way of Bijoliya 105 (MLVGCB Herb).

Distribution: Widely distributed throughout tropical and warm temperate regions.

Uses: The plant is used in rheumatic pains, fevers, etc.

Ammannia baccifera L. Sp. P1. 120. 1753; Clarke in Hook. *f.* Fl. Brit. Ind. 2: 569. 1878; Duthie, Fl. Gangetic Plain 1: 350. 1903; Bhandari, Fl. Indian Desert 157. 1978; Parmar in Shetty & Singh , Fl. Rajasthan 1: 321. 1987. (Fig. 66)

Vern.: *Jal-bhangra.*

An erect, glabrous, 6.5 - 40 cm tall, rarely suffruticose. Stem erect, tetragonous, usually reddish, glabrous, often with opposite axillary branches. Leaves 1 - 3.5 x 0.3 - 1.5 cm, opposite, oblanceolate or oblong-elliptic, acute or obtuse, base usually attenuate, sometimes cuneate or rounded. Flowers in sessile, glomerular cymes. Flowers 3 -many flowered axillary cymes, sessile or subsessile forming whorls in all axils; pedicels 1 - 2 mm long. Bract filiform and shorter than pedicels. Epicalyx obscure. Calyx tube campanulate; lobes 4. Petals absent. Stamens 4, inserted below the calyx tube, filaments glabrous, longer than the red anthers. Style ca 0.25 mm. Fruit a capsule, red, mature not wholly covered by calyx teeth, with numerous bulges throughout. Seeds many hemispheric, minute, excavated on the plane face.

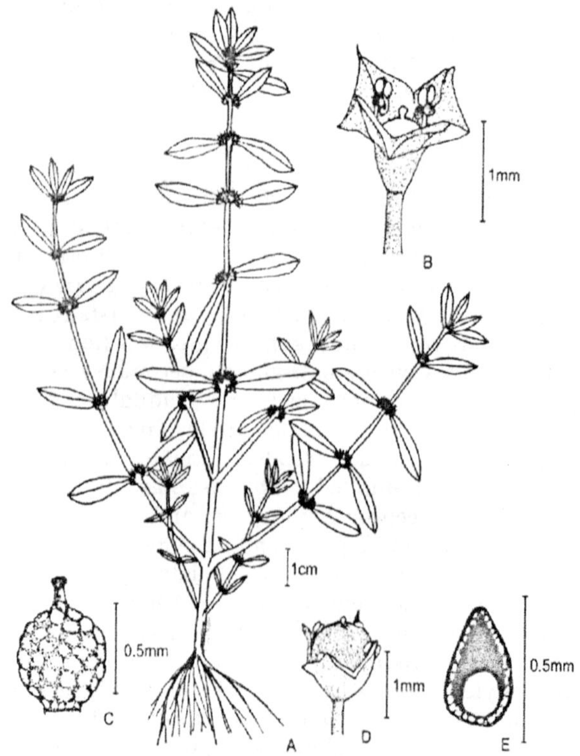

Fig. 66. *Ammannia baccifera* L.
A. Habit, B. Flower, C. Pistil, D. Fruit, E. Seed

Field note: Common along water streams and in marshy places.

Fl. & Fr.: August - September

Specimen examined: Dariba 3473 (MLVGCB Herb).

Distribution: Africa, Afghanistan, Pakistan, India, China, Philippines, New Guinea and Australia.

Uses: Fresh leaves are used in rheumatic pains, fevers, etc. Leaf extract is poisonous and food adulterated with it may cause severe burning pain in the abdomen.

LAGERSTROEMIA L.

Lagerstroemia indica L. Syst. Nat. ed. 10: 1076. 1759; Clarke in Hook. *f.*, Fl. Brit. Ind. 2: 575. 1879.

Vern.: *Chinamehandi, Gulmehandi.*

Shrub or small tree, to 5 m tall. Branchlets slender, 4-angled or subulate, puberulous, glabrescent. Leaves sessile or with petiole to ca 2 mm; leaf blade elliptic, oblong, obovate, or suborbicular, typically at least some suborbicular to obovate and mucronate, 2.0-7 × 1.5-3 cm, papery to slightly leathery, glabrous or with slight indumentum on veins abaxially, lateral veins 3 - 7 pairs, base broadly cuneate to rounded, apex acute, obtuse with small mucro, or retuse. Panicles subpyramidal, 7 - 20 cm, puberulous, densely flowered. Floral tube 6-merous, 7 - 11 mm, smooth walled or obscurely to decidedly 6-ribbed, glabrous; sepals 3.5 - 5.5 mm, adaxially glabrous; annulus present; epicalyx absent. Petals purple, fuchsia, pink, or white, orbicular, 1.2 - 2 cm including claw 6 - 9 mm. Stamens 36 - 42, dimorphic. Ovary glabrous. Capsules ellipsoidal, 1 - 1.3 × 0.7 - 1.2 cm, 4 - 6-valved. Seeds winged at one end, ca 8 mm including wing.

Field note: Commonly cultivated in gardens as an avenue tree.

Fl.: July - November *Fr.*: December - April.

Specimen examined: Bhilwara 544 (MLVGCB Herb).

Distribution: Native of China; widely cultivated throughout the warmer regions of the world.

LAWSONIA L.

Lawsonia inermis L. Sp. Pl. 349. 1753. Cooke, Fl. Pres. Bombay 1: 544. 1958 (Rpr.); Bole. & Almeida, in J. Bombay Nat. Hist. Soc. 79: 312. 1982. *L. alba* Lam. Encycl. 3 : 106. 1780; Clarke in Hook. *f.* Fl. Brit. Ind. 2: 573. 1879; FGS 1: 309. 1978. (Fig. 67)

Vern.: *Henna, Mehndi.*

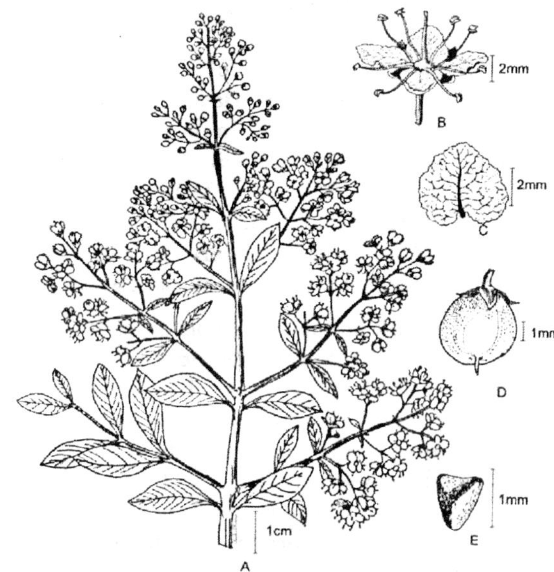

Fig. 67. *Lawsonia inermis* L.
A. Flowering twig, B. Flower, C. Petals, D. Fruit, E. Seed

Glabrous much-branched shrub, up to 2.5 m tall. Branches angular. Leaves sessile, elliptic, ovate or obovate, acute, obtuse or subacuminate, 0.5 - 2.6 x 0.1 - 1 cm. Panicles 3 - 22 cm long. Flowers small or in terminal cymose clusters, perigynous, tetramerous, creamish, fragrant. Pedicels 2 - 3.5 mm long. Calyx 4, ovate. Petals 4, 3 - 4 x 4 - 5 mm, greenish white. Filaments 4 mm long. Ovary 4-locular, placentation axile, ovules many, slyle long, stigma capitate. Capsules exserted, globose, coriaceous, repturing irregularly, veined. Seeds many pyramidal, smooth.

Field note: Cultivated as hedge plant throughout the area and for its orange-red dye obtained from the leaves.

Fl. & Fr.: April - August.

Specimen examined: Bhilwara 2166 (MLVGCB Herb).

Distribution: Africa and Asia.

Uses : The dye obtained from the leaves is used for dyeing clothes, hairs, for staining nails, palms, soles and in medicine.The dried stems are used as firewood as well as in fancing the boundries of agriculture fields.

ROTALA L.

Rotala mexicana Cham. & Schlecht. in Linnaea 5: 567. 1830; Blatte & Hallb. J. Nat. Hist. Soc. 25: 702. 1988; Parmar, in Shetty & Singh Fl. Rajasthan 1: 325. 1987. *Ammania pygmaea* Kurz in Seem. J. Bot. 5: 376. 1867; Clarke in Hook. *f.,* Fl. Brit, Ind. 2: 568. 1878; Duthie, Fl. Gangetic Plain 1: 349. 1903; Babu, in Herb. Fl. Dehradun 186. 1977.

Annual herb, aquatic or amphibious. Stem 3-10 cm, erect or creeping. branched, quadrangular. Leaves 3 - 5(-8)-whorled; aerial leaves narrowly lanceolate to broadly linear, 3 - 15 × 0.5 - 1.5 mm, base narrow, apex truncate and often 2-cleft; submerged leaves linear, to 15 × 0.5 mm. Flowers solitary, sessile, axillary; bracteoles linear, ca as long as floral tube. Floral tube 4-merous, campanulate, translucent, ca 0.5 mm. Calyx 4. Epicalyx absent. Petals absent. Stamens 2 (3 or 4), included. Ovary subglobose; style included, much shorter than ovary. Capsule ca 1 mm in diam, globose, brown. Seeds ca 0.3 mm. ellipsoid, brownish-yellow.

Field note: Rare, in the area near water streams and wet places.

Fl. & Fr.: December - March.

Specimen examined: Ladpura 1345, Mandalgarh 1548 (MLVGCB Herb).

Distribution: Tropics and warm-temperate regions of the world.

WOODFORDIA Salisb.

Woodfordia fruticosa (L.) S. Kurz in J. As. Soc. Bang. 40 (2): 56. 1871; Parmar, in Shetty & Singh Fl. Rajasthan 1: 326. 1987. *Lythrum fruticosum* L., Syst. ed. 10: 1045. 1759. *Woodfordia floribunda* Salib. Parad. Lond. t. 42. 1806; Clarke in Hook. *f.* Fl. Brit. Ind. 2: 572. 1878; Duthie, Fl. Gangetic Plain 1: 351. 1903.

Vern.: *Dhawi.*

Deciduous shrub or under shrub, ca 2.5 m tall. Branches long. Leaves oblong-lanceolate or ovate, rarely falcate, oblique, 1.8 - 11.1 x 1.2 - 2.5 cm, glabrous above and glandular punctate beneath. Flowers axillary or extra-axillary, in fascical cyme, deep-red. Petals 3 - 4.5 x 0.5 - 0.75 mm, brick-red. Stamens 3 - 16.5 mm long. Ovary oblong, 4 - 6 mm long, 1.25 - 2 mm; style 8 - 15 mm long. Capsule 0.8 - 1.0 cm long, ellipsoid, enclosed by persistent calyx tube, oblong, shining brown. Seeds minute, pale brown, smooth, narrowly obconical, trigonous, cuneate at base.

Field note: Rare, in the forests and the foothills.

Fl. & Fr.: January-June.

Specimen examined: Delwara 3807 (MLVGCB Herb).

Distribution: Madagascar, India, Pakistan, Sri Lanka, China, Indonesia.

Uses: Leaves and twigs yield a yellow dye used in printing. Petals yield a red dye. Flowers and leaves are used medicinally as astringent and analgesic.

ONAGRACEAE

LUDWIGIA L.

KEY TO THE SPECIES

1a. Aquatic herbs. Stem with erect clusters of short, spindle-shaped pneumatophores at nodes of floating branches. Flowers 5-merous .... *L. adscendens*

1b. Terristrial herbs. Stem not as above. Flowers 4-merous 2

2a. Capsules narrowly linear.......... *L. hyssopifolia*

2b. Capsules stout, oblong .... *L. perennis*

Ludwigia adscendens (L.) Hara, in J. Jap. Bot. 28: 291. 1953; Raven, in Reinwardtia 6: 387. 1963; Parmar, in Shetty & Singh Fl. Rajasthan 1: 328. 1987. *Jussiaea adscendens* L. Mant. 1: 69. 1767. *J. repens* L. Sp. Pl. 388. 1753; Clarke in Hook. *f.* Fl. Brit. Ind. 2: 587. 1878; Duthie, Fl. Gangetic Plain 1: 355. 1903. (Plate 138)

Perennial herb, with creeping or floating stems, rooting at nodes, with white, erect, short (1 - 3 cm), spindle-shaped pneumatophores in clusters at nodes of floating stems. Floating stems to 1 m, terrestrial stems 20 - 60 cm, much branched, tips ascending, glabrous or densely villous. Petiole 5-20 mm; leaf blade oblong - elliptic, 0.4 - 7.0 × 0.7 - 4.0 cm, glabrous, base narrowly cuneate or attenuate, margin entire, apex obtuse to subacute. Flowers solitary axilary. Sepals 5, deltoid-acuminate, 5-11 mm, glabrous or villous. Petals creamy-white with yellow center, obovate, 9 - 18 × 6 - 10 mm. Stamens 10, epipetalous; filaments white, 2.5 - 4 mm; anthers 0.7 - 1.8 mm; pollen in monads. Style white, 4 - 10 mm, glabrous; stigma discoid. Capsule light brown with dark brown ribs, cylindric, terete, 1.2 - 2.7 cm x 3 - 4 mm., glabrous or villous, thickly walled, tardily and irregularly dehiscent. Seeds numerous, in one row per locule, firmly embedded in coherent cubes of woody endocarp fused to capsule wall, pale brown, quadrate, oblong or elliptic, 1.1 - 1.3 mm, raphe inconspicuous.

Field note: Occurring at wet swampy places, often floating in water at edges of ponds, tanks, ditches etc.

Fl. & Fr.: February - May

Specimen examined: Nagadi 161, Shahpura 1138 (MLVGCB Herb).

Distribution: China, India, Indonesia, Japan, Malaysia, Nepal, Pakistan, Philippines, Sri Lanka, Taiwan, Thailand; widespread in Africa, Asia and Australia.

Uses: Plant is used medicinally for its febrifugal and antiswelling properties.

Ludwigia hyssopifolia (G. Don) Exell, Garcia de Orta. 5: 471. 1957; Parmar, in Shetty & Singh Fl. Rajasthan 1: 328. 1987. Raizada, Suppl. Duthie, Fl. Gangetic Plain 75. 1976. *Jussiaea hyssopifolia* G. Don, Gen. Syst. 2: 693. 1832. *J. linifolia* Vahl. Gamble 3: 1298, 1957 Repr. Ed. (Plate 139)

An erect, annual herb, sometimes woody at base and perennial, Stems 0.5-1.2 m tall, branched, new growth and inflorescence minutely puberulous. Petiole 3-18 mm; leaf blade lanceolate, 1-9 × 0.2-3 cm, subglabrous or puberulous, lateral veins 7-17 per side, submarginal vein inconspicuous, base narrowly cuneate, apex acuminate. Flowers solitary, axillary, 1 cm across, yellow. Sepals 4, lanceolate, 0.2 - 0.4 mm, finely puberulous. Petals yellow, fading to orange-yellow, elliptic, 2 - 3 × 1 - 2 mm. Stamens 2, pale greenish-yellow; anthers 0.4 - 0.6 mm. Style 1 - 1.5 mm; stigma depressed-globose. Disc slightly elevated. Capsule cylindric, subterete, 1.5 - 3 cm, finely puberulous, relatively thinly walled; subsessile. Seeds in inflated upper capsule in 2 or more rows per locule, free, pale brown, ovoid, 0.3 - 0.5 mm, raphe narrow; seeds in lower capsule in one row per locule, embedded in cube of relatively hard endocarp, brown, 0.7 - 0.9 mm, raphe ca 1/3 width of seed.

Field note: Rare, in open wet often disturbed areas, streamsides, roadsides, ditches, clear agricultural land, or in moist forests

Fl. & Fr.: August - November.

Specimen examined: Nagdi 314, Mandal 830 (MLVGCB Herb).

Distribution: Bangladesh, Bhutan, India, Indonesia, Malaysia, Myanmar, Nepal, Philippines, Singapore, Sri Lanka, Taiwan, Thailand, Vietnam; widespread in Africa, SE Asia, Australia, Pacific islands and South America

Ludwigia perennis L. Sp. Pl. 119. 1753; Mahesh. Illus. Fl. Deihi f. 85. 1966; Bhandari, Fl. Indian Desert 143. 1978; Parmar, in Shetty & Singh Fl. Rajasthan 1: 329. 1987. *L. parviflora* Roxb. Hort. Beng. 11. 1814 (nomen); Fl. Ind. 1: 440. 1820; Boiss. Fl. Or. 2: 752. 1872; Clarke in Hook. *f.*, Fl. Brit. Ind. 2: 588. 1879; Duthie, Fl. Gangetic Plain 1: 356. 1903. *Jussiaea perennis* (L.) Brenan, Kew Bull. 8: 163. 1953.

Annual erect herb up to 1 m tall, subglabrous or minutely puberulent on younger parts. Leaves narrowly elliptical to lanceolate, 1 - 11 x 0.3 - 2.7 cm, narrowly cuneate at base, the apex subacute, with petioles 2 - 15 mm long, winged. Flowers solitary or clustered, axillary, minutely pubescent. Sepals 4 (-rarely 5), deltoid, 1.5 - 3.5 x 0.5 - 1.8 mm, glabrous or minutely puberulent. Petals yellow, elliptical, 1 - 3 x 0.7 - 2 mm. Stamens as many as sepals, rarely more. Pollen shed in tetrads. Style 0.7 - 1.5 mm long. Disc slightly elevated Capsules thin-valled, glabrous or puberulent, 0.3 - 1.6 cm long, terete, readily and irregularly loculicidal, sessile or on a pedicel up to 6 mm long, often more or less nodding. Seeds pluriseriate in each locule of the capsule, free, brown with fine brown lines, ellipsoid-rounded, 0.3 - 0.5 x 0.2 - 0.25 mm, with narrow, inconspicuous raphe.

Field note: Common in open wet often disturbed areas, streamsides, roadsides, ditches, clear agricultural land, or in moist forests

Fl. & Fr.: August - January.

Specimen examined: Bijoliya 229, Kachhola 1033 (MLVGCB Herb).

Distribution: Africa, tropical and subtropical Asia, through Malaysia to tropical Australia and New Caledonia.

TRAPACEAE

TRAPA L.

Trapa natans L. var. ***bispinosa*** (Roxb.) Makino in Linuma, Somoku-Dzusetsu ed. 3. 1: 137. 1907; Brenan in Fl. Trop. E. Africa Pt. 6 : 3. 1953; Bhandari, Fl. Indian Desert 144. 1978; Pandey, in Shetty & Singh Fl. Rajasthan 1: 330. 1987. *T. bispinosa* Roxb. Pl. Cor. t. 3: 29. 234. 1815 & Fl. Ind. 1: 449. 1820; Clarke in Hook. *f.*, Fl. Brit. Ind. 2: 590. 1879, *pro parte*; Duthie, Fl. Gangetic Plain 1: 358. 1903. (Fig. 68)

Vern.: *Singhara*

Floating or submerged, annual herb. Stem rooting at nodes. Floating leaves in rosette, 1.5 - 6 x 1.5 - 7 cm, dentate, glabrous above, pubescent to villous below, especially on the nerves; petiole up to 12 cm long, dilated near the leaf base,inflated, villous. Submerged leaves finaly dissected into tooth-like linear segments. Flowers solitary, axillary, pedunculate, white. Calyx ca 7 mm long, ovate, villous, especially in the middle portion. Petals up to 1.5 cm, obovate-oblong, white. Stamens 4. Fruit one seeded, shortly beaked at the apex, 3-5 cm across, 2-horned; horns erectop-atent, usually straight, 1 - 1.8 cm long, sharp at point and reflexed barbed for a little way below it. two of which are shorter.

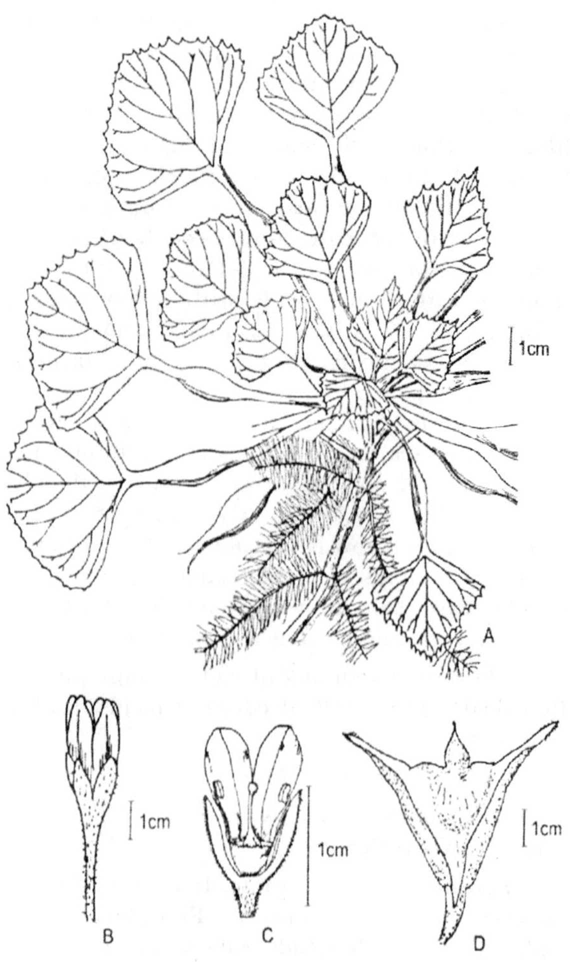

Fig. 68. *Trapa natans* L. var. *bispinosa* (Roxb.) Makino, A. Habit, B. Flower, C. Flower with frontal sepal and petal removed, D. Fruit

Field note: Common in tanks, cultivated for fruits.

Fl.: July-September; *Fr.*: September-November.

Specimen examined: Mainal 106 (MLVGCB Herb).

Distribution: Europe, Asia, and Africa

Uses: This species is distinguished from *Trapa bispinosa* by its smaller leaves, not densely villous and a 4-horned fruit, all the angles of the fruit being spinescent.

CUCURBITACEAE

KEY TO THE GENERA

1a. Petals free 2

2a. Petiole with a pair of glands below the lamina. All flowers solitary. Fruit polymorphic, woody, indehiscent..... *Lagenaria*

2b. Petiole without glands below the lamina. Male flowers racemose, female solitary. Fruit oblong or cylindric, fibrous, dehiscing by a terminal operculum..........*Luffa*

1b. Petals united.. 3

3a. Corolla segments fimbriate, with long delicate filaments............... *Trichosanthes*

3b. Corolla segments not fimbriate...................... 4

4a. Calyx tube of male flowers closed at the bottom by 2-3 incurved scales. Peduncle mostly with large foliaceous bracts .... *Momordica*

4b. Calyx tube of male flowers without scales. Peduncle mostly ebracteate........................... 5

5a. Flowers up to 5 mm long 6

6a. Fruit operculately dehiscent at the base. Disc absent .... *Corallocarpus*

6b. Fruit indehiscent. Disc present...................... 7

7a. Ovules numerous. Seeds not scorbiculate, mostly without a distinct margin..... .*Cucumis*

7b. Ovules few. Seeds scorbiculate, with thick raised margins .... *Melothria*

5b. Flowers more than 5 mm long........................ 8

8a. Leaves pinnately lobed 9

9a. Leaves deeply pinnatipartite. Pollen grains reticulate................................ *Citrullus*

9b. Leaves sparingly pinnatifid. Pollen grains echinate........................... *Praecitrullus*

8b. Leaves entire or palmately lobed................ 10

10a. Leaves densely or roughly hirsute on both sides.. 11

11a. All flowers politary 12

12a. Flowers white, leaves glabrous, punctate on upper surface. ....*Coccinia*

12b. Flowers yellow, leaves densely or roughly hirsute on both sides or somewhat long haired on nerves beneath to scabrescent. ....*Cucurbita*

11b. Male flowers racemose, fasciculate or clustered....................................... *Diplocyclose*

10b. Leaves shortly setose-scarbid, ultimately punctuate with white, tuberculate dots on both sides. Flowers; male solitary; female in fascicles.................................. *Dactyliandra*

CITRULLUS Schrad. ex Eckl. & Zeyh. (nom. cons.)

KEY TO THE SPECIES

1a. Perennial with scabrid stem. Tendrils not robust, simple, rarely 2-fid. Fruit up to 7 cm in diam., fruit pulp bitter in taste .... *C. colocynthis*

1b. Annual with softly hairy stem. Tendrils robust, 2-fid. Fruits more than 15 cm in diam., with sweet pulp .... *C. lanatus*

Citrullus colocynthis (L.) Schrad., in Linnaea. 12: 414. 1838; Clarke in Hook. *f.,* Fl. Brit. Ind. 2: 620. 1876; Duthie, Fl. Gangetic Plain 1: 374. 1903; Bhandari, Fl. Indian Desert 161. 1978; Chakravarty, Fl. India. Fasc. 11: 20. f. 1 - 7. 1882; Pandey, in Shetty & Singh Fl. Rajasthan 1: 339. 1987. *Cucumis colocynthis* L., Sp. Pl. 1011. 1753. *Colocynthis vulgaris* Schrad., Ind. Sem. Hort. Gotting 2: 1833; Chakravarty, in Recd. Bot. Surv. Ind. 17 (1): 113. 1959. (Fig. 69 & Plate 140)

Vern.: *Indryan, Tumba.*

Perennial trailing herb with somewhat woody tuberous root. Stems shortly hairy when young becoming scabrid. Tendrils simple, rarely bifid, slender, glabrous. Leaf blade elongate-ovate in outline, 4 - 10 x 1.5 - 9.0 cm, distinctly scabrid-hairy beneath, smooth except on the nerves above, cordate 3-5-lobed, ultimate lobes pinnately lobulate with central lobe longest, long-ovate in outline; petiolate, petiole rather densely rough-hairy; probracts lanceolate-elliptic, 4 - 5 x l. 5 mm, caducous. Flowers (male and female) axillary, solitary, pale yellow. Male flowers on long pedicels; calyx campanulate, ca 9 mm long;

lobes ca 5 mm broad; corolla ovate-acute, ca 8 x 5 mm; female flowers on longer pedicels than male, receptacle-tube short, lobes lanceolate, ca 5 mm long. Ovary hairy, subglobose or obovate. Fruit on long stalk, ca 5 cm in diam., globose, smooth, longitudinally green striped, 5 - 7 cm in diam., eipcarp thin, filled with a dry spongy very bitter pulp, pulp yellow. Seeds numerous, obovate or ovate-oblong, yellowish brown, not margined, ca 6 x 3 mm.

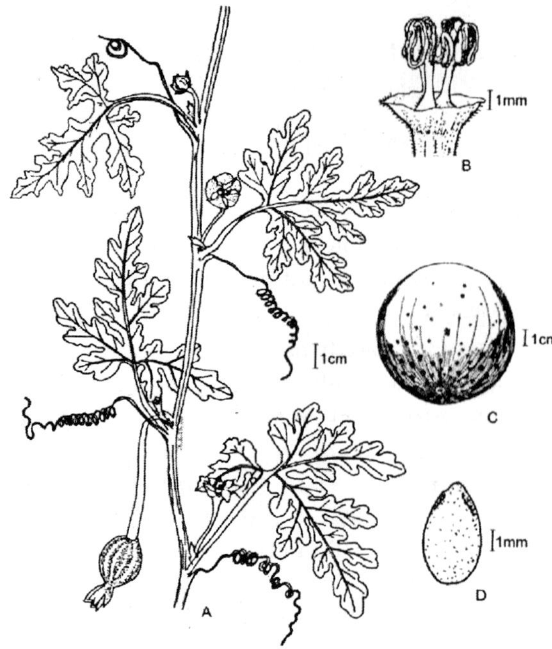

Fig. 69. *Citrullus colocynthis* (L.) Schrad.
A. Flowering & fruiting branch, B. Stamens, C. Fruit, D. Seed

Field note: Commonly found in sandy plains. This is a large, creeping plant with its branches radiating on all sides with a long perennial tap root.

Fl. & Fr.: May - October.

Speciemen examined : Gilund 3169, Asind 819 (MLVGCB Herb).

Distribution: Northern Tropical Africa, Atlantic Islands, North-West India, Pakistan, Spian, Arabia and Australia.

Uses: Fruits are used in various medicines and as a purgative. Fruit pulp is very bitter. Seeds used as an ingridient of sweets.

Citrullus lanatus (Thunb.) Mats.& Nakai, in Cat. Semet. Spor. Hort. Bot. Univ. Imp. Tokyo: 30. 1916; Bhandari, Fl. Indian Desert 163. 1978; Chakravarty, Fl. India. Fasc. 11 : 22. 1882; Pandey, in Shetty & Singh Fl. Rajasthan 1: 334. 1987. *Cucurbita citrullus* L. Sp. Pl. 1010. 1753. *Momordica lanata* Thunb. Prodr. Pl. Cap. 13. 1800. *Cucumis colocyntis* Thunb. Prodr. Pl. Cap. (non Linn. 1753). *Citrullus vulgaris* Schrad ex Eckl. & Zeyh. Enum. Pl. Afr. Austr. 279. 1836; Clarke in Hook. *f.,* Fl. Brit. Ind. 2 : 621. 1876; Duthie, Fl. Gangetic Plain 1: 375. 1903.

var. ***lanatus***

Vern.: *Tarbuz.*

Annual, climber or trailer herb. Stem herbaceous, rather villous. Monoecious. Tendrils robust, pubescent, 2 - 3-fid. Petiole long-villous. Leaves upto 12 x 10 cm, triangular-ovate; lamina cordate at base, hirsute beneath, long-hispid on veins and veinlets; upper surface smooth with transluscent dots or scabrous, margin minutely denticulate, apparently smooth, deeply 3 - 5-lobed, lobes elongated-ovate, obtuse, more or less pinnately sinuate lobulate; young leaves and shoots densely villous; probracts obovate-spathulate, 8 - 10 mm long. Male flowers on ca 20 (-40) mm long pedicel; calyx broadly campanulate, ca 7 mm long, lobes ca 5 (3 - 5) mm long, lanceolate; petals yellow, 12 mm long, obovate-oblong, ca 4 mm broad. Female flowers on ca 6 cm long pedicels, calyx and corolla as in the male, ovary ellipsoid or subglobose, hairy. Fruit large, ellipsoid or subglobose, ca 30 cm or more in diam., hairy, green mottled with paler green and yellowish more or less longitudinal stripes; mesocarp fleshy, red or yellow, rarely white; indehiscent. Seeds ovate in outline, ca 10 x 5 mm, red or yellow, smooth or slightly verrucose.

Field note: Common on sandy soils, often cultivated in the fields and sandy river beds for edible fruits.

Fl & Fr.: January - May.

Specimen examined: Jahajpur 2169 (MLVGCB Herb).

Distribution: Native of the Kalahari region. Cultivated throughout Tropics.

COCCINIA Wight & Arn.

Coccinia grandis (L.) Voigt, Hort. Suburb. Calc. 59. 1845; Bhandari, Fl. Indian Desert 163. 1978; Chakravarty, Fl. India. Fasc. 11: 24. f. 1-9. 1882; Pandey, in Shetty & Singh Fl. Rajasthan 1: 335. 1987. *Bryonia grandis* L. Mant. Pl. 1: 126. 1767. *Coccinia indica* Wight & Arn. Prodr. 347. 1834; Duthie, Fl. Gangetic Plain 1: 376. 1903. *Cephalandra indica* (Wight & Arn.) Naud, in Ann. Sci. Nat. Bot. 5. 5: 16. 1859; Clarke in Hook. *f.,* Fl. Brit. Ind. 2: 621. 1876.

Vern.: *Tindori.*

Perennial, climbing and widely spreading, sometimes prostrate, bright green. Stem much branched, cylindrical, glabrous or white scaly. Petiole slender, striate, 2-5 cm long. Leaves 2.9 - 12. 0 x 2.6 - 11.0 cm, broadly ovate or ovate-triangular, entire or angled or palmately lobed, with few glistening glands on both sides of the midrib towards the base. Flowers large, white, ca 4 cm long, pedicellate, solitary, male and female on different plants. Fruit cylindrical or fusiform, 2.5-5 x 1.2 - 2.5 cm, slightly beaked, subglabrous, many seeded; pulp bright when ripe. Seeds asymmetrically pyriform in outline, 6 - 7 x 2.5 - 4.0 mm, smooth, compressed.

Field note: Most common climber among the hedges of crop fields and gardens.

Fl. & Fr.: Almost throughout the year.

Specimen examined: Bhilwara 1166, Chittorgarh 1882 (MLVGCB Herb).

Distribution: India, Malaya, Pakistan and Tropical Africa

Uses: Fruits are used as vegetable. Various preparations of roots, stems and leaves have been mentioned in indigenous systems of medicine as being efficacious in the treatment of skin diseases, bronchial catarrh, bronchitis and diabetes.

CORALLOCARPUS Welw. ex Hook. f.

Corallocarpus epigaeus (Roettl.) Clarke in Hook. *f.,* Fl. Brit. Ind. 2: 628. 1879; Duthie, Fl. Gangetic Plain 1: 380. 1903; Bhandari, Fl. Indian Desert 165. 1978; Chakravarty, Fl. India. Fasc. 11 : 28. 1882; Pandey, in Shetty & Singh Fl. Rajasthan 1: 336. 1987. *Bryonia epigaea* Rottl., Neue Schrif. Ges. Naturf. Freunde Berlin 4: 212. 1803. *B. glabra* Roxb., Hort. Beng. 104. 1814 (*nom. nud.*); Fl. Ind. 3: 725. 1832. *Aechmandra epigaea* Arn. in Hook. Journ. Bot. 3 : 274. 1841.

Perennial, monoecious, climber up to 4 m long with tuberous root. Stem angular-sulcate, glabrous to finely pubescent, glaucous; tendrils simple, elongated, subfiliform, glabrous. Leaves sinuately 3-lobed, 1 - 2.5 cm long, slightly broader than long, cordate, pale green above, greyish-green below, finely hairy on both surfaces; lobes obovate to oblanceolate, each 1-3-lobulate and minutely apiculate-obtuse. Petiole 10 - 30 mm long, glabrous to shortly hairy. Flowers greenish-yellow. Male in axillary racemes, peduncle 4 - 6 cm long, 5 - 15-flowered; pedicels 1 - 5 mm long, filiform. Calyx lobes lanceolate, ca 1 mm long. Corolla dull or greenish yellow; segments ovate, obtuse, spreading, ca 1 mm long. Female flowers often solitary (sometimes in many-flowered clusters), on contracted axillary branch; (sometimes co-axillary with male flowers); pedicel 1-5 mm long, thickened in fruit; calyx-tube campanulate, ca 2 mm long; petals 1.5-2.5 x 1-1.5 mm, reflexed. Fruit ovoid or ellipsoid, beaked, glabrous, smooth, 10-20 x 5-8 mm, red except for the greenish base and portion of beak which is 2.5 - 5 mm long. Seeds 6-9, asymmetrically pyriform, smooth, ca 4 x 2.5 mm, yellowish, turgid.

Field note: Rare, plant growing among the hedges of crop fields. Perennates by napiform tap roots.

Fl. & Fr.: August- November.

Specimen examined: Bhilwara 1561, Pansal 1744 (MLVGCB Herb).

Distribution: India, Pakistan, Tropical East Africa and Sudan.

CUCUMIS L.

KEY TO THE SPECIES

1a. Fruits softly echinate or tuberculate .... *C. prophetarum*

1b. Fruits glabrous or with few scattered hairs .... *C. melo*

CUCUMIS MELO L.

KEY TO THE VARIETIES

1a. Leaves usually deeply 5(-7)-lobed, scabrid. Corolla up to 10 mm long. Fruit oblong or obscurely trigonus, upto 7 cm long .... *C. melo* var. *agrestis*

1b. Leaves shallowly 5-lobed or angular, softly hairy. Corolla 10-20 mm long. Fruit polymorphic, not trigonous.
.... *C.melo* var. *momordica*

Cucumis melo L. var. ***agrestis*** Naudin, Ann. Sc. Nat. Par. Ser. 11 : 73. 1859; Bhandari, Fl. Indian Desert 151. 1978; Chakravarty, Rec. Bot. Surv. Ind. 17 (1) : 103. 1959; Sharma, Jour. Bomb. Nat. Hist. Soc. 81 (1); 230. 1984.

Vern.: *Kachri*

A much branched prostrate annual or perennial herb. Stem angular, generally covered with scabrous hairs, leaves triangular, ovate, 3 - 5 (-7)-lobed, scabrid with rigid hairs: petiole 1 - 6 cm long. Flowers small, solitary or rarely in pairs or threes, peduncle 5 - 10 mm long, densely hipsid. Calyx ca 1.5 mm long, subulate. Corolla 6 - 8 mm long, yellow, lobes ovate-oblong. Fruit ellipsoid, oval-globose, sometimes obscurely trigonous, smooth and glabrescent, 4 x 2.5 cm, generally with dark green stripes.

Field note: Rare, prostrate or climbing among hedges.

Fl. & Fr.: August - October.

Specimen examined: Nayasamaliya 1694, Atun 1859 (MLVGCB Herb).

Distribution: Throughout the Old-World tropics, adventive in neotropics.

Notes: (i). In many collections from south central Rajasthan ovaries have been observed with rather long adpressed, forward pointing hairs, in this respect being somewhat intermediate with wild subsp. *melo* which occurs in adjacent part to the USSR & S.W. Asia.

(ii). Perennial forms have sometimes been treated as a distinct species, *Cucumis trigonus* Roxb. The status of these taxa needs elucidation by comparative studies in the field.

Cucumis melo L. var. ***momordica*** Duthie & Fuller, Field & Gard. Crops 2: 50. t. 49, 1882; Duthie, Fl. Gangetic Plain 1: 380. 1903; Bhandari, Fl. Indian Desert 169. 1978; Pandey, in Shetty & Singh Fl. Rajasthan 1: 338. 1987. *C. melo* L var *culta* Kurz, in J. Asiat. Soc. Bengal 46 (2): 102. 1877. *Cucumis melo* L. var. *melo sensu* Chakravarty, Fl. India. Fasc. 11: 34. 1882, non L. 1753. (Fig. 70)

Vern.: *Kachro*.

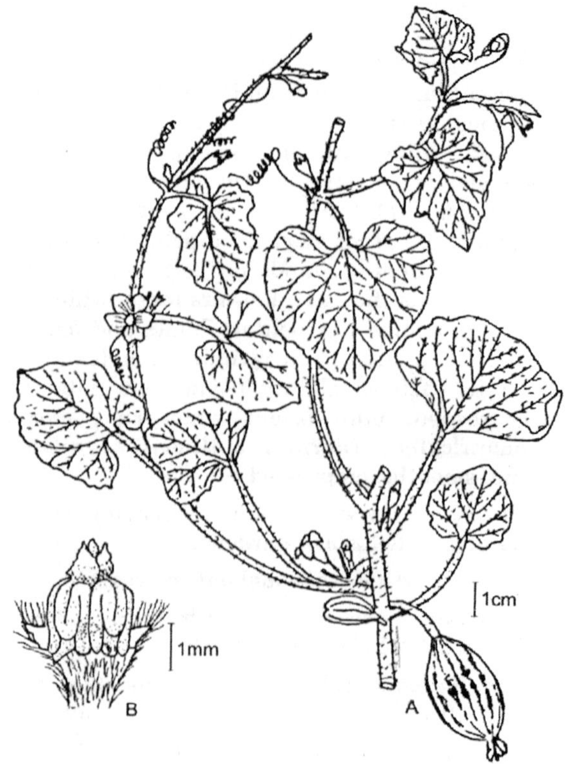

Fig. 70. *Cucumis melo* L. var. *momordica* Duthie & Fuller, A. Flowering & fruiting branch, B. Stamens

Prostrate annual herb, stem and branches angular hispid. Leaves 6 - 10 x 5.5 - 10 cm, suborbicular or reniform, 3-7 lobed; lobes shallowly rounded-cordate at base, more or less dentate, sparsely hairy on both surfaces. Tendril simple, sparsely hispid. Male flowers yellow, 1-3 on axillary peduncles. Calyx tube narrowly campanulate, 4 - 5 mm long, hairy; lobes 5 - 6 mm long, linear-lanceolate, hispid. Corolla lobes upto

1.8 cm long, ovate-oblong, hispid and glandular outside. Staminal filaments glabrous, appendages of the connectives ca 1 mm long. Female flowers axillary, solitary, yellow; peduncles 1-2 cm long, Calyx and corolla same as in male flowers. Staminodes 3. Ovary pyriform, woolly; stigma connivent. Fruits oval or cylindrical smooth, yellow in colour, often spotted with dark green, when ripe bursts spontaneously, varies in size from 30 - 60 x 7 - 15 cm, mealy and insipid flesh. Seeds oblong, obtuse apex, subacute at base, pale yellow.

Field note: Common in waste lands, along boundaries of cultivated fields, grasslands etc.

Fl.& Fr .: July - November.

Specimen examined: Bijoliya 2170 (MLVGCB Herb).

Distribution: Cultivated throughout India and Pakistan.

Uses: Young fruits are used as a substitute for the cucumber or cooked as vegetable. Although it is inferior to the melon, the fruit is edible on ripening.

Cucumis prophetarum L. Cent. 1: 33. 1755; Clarke in Hook. *f.,* Fl. Brit. Ind. 2: 619. 1879; Duthie, Fl. Gangetic Plain 1: 373. 1903; Bhandari, Fl. Indian Desert 169. 1978; Chakravarty, in Rec. Bot. Surv. India 17: 109. 1959 & in Fasc. Fl. India. 11: 35. 1882; Pandey, in Shetty & Singh Fl. Rajasthan 1: 339. 1987. (Plate 141)

Vern.: *Khato-kachro.*

Prostrate or climbing perennial, hispid herb. Stem slender, sulcate, hispid and rough, hairs whitish. Tendrils simple and short, filiform, scabrid. Leaves suborbicular, 2 - 4 cm long. cordate-truncate at base, 3 (-5)-lobed, lobes obtuse or subacute, margin dentate or lobulate, hispid-scabrid on both surfaces; petiole 1 - 4 cm long, densely scabrid with white hairs. Flowers yellow. Male flowers in fascicles of 2 - 3 flowers, rarely solitary; peduncle 4 - 10 mm long, hispid. Calyx-tube campanulate, 3 - 5 mm long, densely, hispid; lobes linear, spreading, ca 1.5 mm long. Corolla sparsely villous, 3 - 5 mm long, lobes ovate-oblong, mucronate. Staminal filament filiform; anthers oblong 2 - 5 mm long, hispid. Female flowers axillary, solitary, yellow; peduncles 2 - 5 cm long, enlarging in the fruit. Calyx and corolla as in male. Ovary muricate. Fruit broadly ovoid or subglobose, puberulous, softly echinate, longitudinally striped green and white, yellow when ripe, 3 - 4 x 2 - 3.5 cm. Seeds pale ashy, ca 4 x 2 mm, oblong or ellipsoid.

Field note: Rare, found in sandy plains.

Fl. & Fr.: August - January.

Specimen examined: Ajmer 3711, Makarwali 3772 (MLVGCB Herb).

Distribution: Paleotropic, throughout North Africa, Arabia to India and Pakistan.

CUCURBITA L.

Cucurbita maxima Duch. ex Lam. Encycl. Meth. Bot. 2: 151. 1786; Clarke in Hook. *f.,* Fl. Brit. Ind. 2 : 622. 1879; Duthie, Fl. Gangetic Plain 1: 377. 1903; Jeffrey, in Kew Bull. 34: 799. 1980; Chakravarty, Fl. India. Fasc. 11: 40. 1882; Pandey, in Shetty & Singh Fl. Rajasthan 1: 340. 1987.

Vern.: *Kaddu, Koro.*

Prostrate or twining, annual herb. Leaves orbicular or reniform in outline shallowly 5-lobed or subentire, coarsely hairy on both surfaces with hispidulous soft hairs, margin densely and minutely dentate. Petiole hairy, not pungent. Flowers axillary, solitary, reddish-yelow, faintly scented. Calyx lobes linear. Fruiting peduncle circular, striated, corky or spongy, not expanded at the attachment. Fruit generally oval or rounded, bluntly ribbed, large, upto ca 1 m in diam.; flesh yellow or reddish when ripe. Seeds white, brown or bronze, strongly marginate.

Field note: Cultivated throughout the area, also found wild near habitations, protected gardens etc.

Fl. & Fr.: August- November.

Specimen examined: Bhilwara 939 (MLVGCB Herb).

Distribution: Probably South American in origin. Cultivated throughout India and Pakistan and elsewhere.

Uses: Fruits are used as vegitables.

DACTYLIANDRA (Hook. f.) Hook. f.

Dactyliandra welwitschii Hook. *f.* in Oliver, Fl. Trop. Afr. 2 : 557. 1871; Bhandari & Singh in Kew Bull. 19 : 133. 1964; Bhandari, Fl. Indian Desert 170. 1978; Chakravarty, Fl. India. Fasc. 11: 46. 1882; Pandey, in Shetty & Singh Fl. Rajasthan 1: 340. 1987. (Plate 142)

Vern.: *Moti-ankhphutanti.*

Annual climber, monoecious with slender, much branched, scabrid. angular stem, furrowed when dry. Leaves 5 - 8 x 3 - 5 cm, digitately 3-5 lobed, deeply cordate at base; lobes oblong to obovate, obtuse to rounded and mucronate or acute at apex, coarsely dentate or serrate along margins. Petiole angled, as long as the lamina, scabrid. Tendril simple, slender, sulcate when dry. Bracts subsessile, ca 1.5 cm in diam., stipuliform, suborbicular or reniform, shortly setose. Male flowers white, bracteates, solitary, axillary. Female flowers axillary, in fascicles of 3; pedicles 1-1.5 mm long. Bracts minute, deciduous. Calyx slightly shorter than corolla, a little longer than wide; teeth 1 - 1.5 mm long. Corolla-lobes ovate-oblong. Stamens 3, attached at the base of hypanthium, connectives not produced. Ovary ellipsoid, glabrous; stigmatic lobes 2, fleshy. Fruits ca 1.3 cm in diam., ellipsoid-globose, smooth, with green and white blotches. Seeds 8 - 15, angled and asymmetrical, discoid, compressed, obliquely narrowed at both ends, truncate at apex, brown.

Field note: Commonly growing among the hedges and shurbs, near habitations.

Fl. & Fr.: August - November.

Specimen examined: Mandalgarh 446, Kishanawato ki kheri 1858 (MLVGCB Herb).

Distribution: South - West Africa, Angola and India.

DIPLOCYCLOS (Endl.) Post & Kuntze

Diplocyclos palmatus (L.) Jeffrey, in Kew Bull. 15 (3): 352.1962; Chakravarty, in Jain et al. Fl. India. Fasc. 11: 48.1882; Pandey, in Shetty & Singh Fl. Rajasthan 1: 341.1987. *Bryonia palmata* L. Sp. Pl. 1012. 1753, *excl. syn.*

Vern.: *Ankh-phutani.*

Perennial glabrous herb. Stem glabrous, becoming thickened and white dotted on the ridges when older. Leaves palmetely 3-lobed, broadly ovate, 3.5 - 14 x 4 - 14.5 cm, lobes linear-lanceolate to elliptic, glabrous; petiole 1.5 - 4.0 cm long; tendrils bifid. Flowers small, white or yellowish, male in sessile clusters of 2 - 8, axillary fascicles, white, along with 5 female flowers in the same axil. Calyx 5-lobed, tube 3 - 4 mm long in male, 1.5-2.5 mm long in female, lobes smaller than tube. Corolla of male larger than female, white. Stamens 3. Ovary ellipsoid, ca 5 x 2.5 mm, with longitudinal white strips; staminoides 3; stigma 3, bilobed. Fruit solitary, or in clusters of 2 - 5, ovoid-subglobose, 1.5 - 2.5 cm. Seeds 5-7 x 2.5 - 4 mm, ovate, attenuate, belted with a partition wall slightly raised all round the belt and the central portion surrounded by the belt.

Field note: Commonly growing among the hedges and shurbs on the boundries of crop fields. The morphology of seeds resembles to the "*Shiv ling*" so people called it locally as "*Shiv lingi*".

Fl. & Fr.: August-October.

Specimen examined: Bari sadari 3212 (MLVGCB Herb).

Distribution: Tropical and North Africa, Tropical Asia, Malaysia, Philippines and Tropical Australia.

LAGENARIA Ser

Lagenaria siceraria (Molina) Standl. in Publ. Field Mus. Nat. Hist. Chicago, Bot. Ser. 3: 435. 1930; Jeffrey, in Kew Bull. 34: 790. 1980; Chakravarty, Fl. India. Fasc. 11: 65. 1882. *Cucurbita siceraria* Molina, Sagg. Chile, 133: 335, 1782. *Lagenaria leucantha* Rushy, in Mam. Torr. Bot. Cl. 6: 43. 1896 nom. illeg. superfl. *L. vulgaris* Ser. Mem. Soc. Phys. Hist. Nat. Geneve, 3 (1): 25. t. 2, 1825; Clarke in Hook. *f.,* Fl. Brit. Ind. 2: 613, 1879; FUGP 1: 366, 1903; Cook, 1: 581, 1958.

Vern.: *Al, Toomba, Lauki.*

Annual, climber or trailer. Tendril 2-fid. Stem densely hairy. Leaves palmate, broadly ovate, reniform or sub-orbicular, cordate, obscurely 3 - 5-lobed, lobes rounded, 10 - 40 x 10 - 40 cm; petiole 2.7 - 13 cm long with a pair of small apical lateral conical glands. Both male and female flowers solitary; male penduncle as long as petiole, female somewhat shorter. Corolla white, petals of male flower obovate, apiculate, 2.7 - 4.5 x 1.8 - 3.7 cm. Staminal filaments 3 - 4 mm long; anthers oblong, coherent, included; thecae triplicate. Petals of female flowers ca 30 x 24 mm. Ovary ovoid, villous, 10-17 x 6-8 mm. Fruit large, subglobose, ellipsoid, lageniform, sometimes biventricose, green or greenish-yellow, hairy, indehiscent. Seeds white, oblong, compressed, slightly tapered, slightly 2-horned on the shoulders, 6 - 24 mm long.

Field note: Commonly cultivated, but wild are also found in Waste lands, near cultivated fields and gardens.

Fl. & Fr.: Almost throughout the year, specially August - November.

Specimen examined: Bhilwara 2173 (MLVGCB Herb).

Distribution: A pantropical species of Asian and African origin. Cultivated throughout India.

Uses: The bottle gourd is extensively cultivated in area all the year round for its young and tender fruits which are used as popular domestic vegetable called *Lauki* or *Al*. The shell of the old fruits are often used as water bottles and for making ladles, pipes, blowing horns, snuff boxes etc. The dry shells are also used in making musical instruments like Sitar and Bia. Pulp has cooling and antibilious effect. Seed oil is applied externally in headache.

LUFFA Mill.

1a. Flowers white. Fruit 2.5 - 5.0 cm long, densely covered with bristles .... *L.echinata*

1b. Flowers pale to deep yellow. Fruit more than 5 cm long, without bristles.................... .2

2a. Stamens 3. Fruit clavate, acutely 10-angled. Seeds beaked.............................. *L.acutangula*

2b. Stamens 5. Fruit cylindrical, not angled. Seeds without beak..................... *L. cylindrica*

LUFFA ACUTANGULA (L.) Roxb.

KEY TO THE VARIETIES

1a. Petals distincly emarginate. Fruit 15 - 30 cm long......... *L.acutangula* var. *acutangula*

1b. Petals round or scarcely emarginate. Fruit 5-8 cm long......... *Luffa acutangula* var. *amara*

Luffa acutangula (L.) Roxb., Fl. Ind. 3: 713. 1832; Clarke in Hook. *f.,* Fl. Brit. Ind. 2: 615. 1879; Duthie, Fl. Gangetic Plain 1: 367. 1903; Chakravarty, Fl. India. Fasc. 11: 67. f. 1-7. 1882; Pandey, in Shetty & Singh Fl. Rajasthan 1: 342. 1987. *Cucumis acutangulata* L. Sp. Pl. 1011. 1753.

var. ***acutangula***

Vern.: *Turai, Kali tori.*

Annual, climber with mostly 3-fid subscabrous tendrils. Stem 5-angled, scabrous along ribs. Leaves pale green, roundish, palmately 5 - 7 - lobed, 15 - 20 cm across, glabrous; petiole 8 - 12 cm long. Probracts fleshy with 3 - 10, glistening glands on upper surface, 3 - 7 x 2 - 4 mm. Flowers pale yellow, pedicellate, ca 4.5 cm across. Male raceme 10 - 15 cm long, erect, axillary 10 - 20-flowered, female flowers solitary in same axil as males. Calyx tube pentagonal, slightly hairy, lobes equal to slightly longer than tube. Petals obcordate, ca 2 x 2 - 2.5 cm, emarginate or not. Stamens 3, filaments ca 4 mm long. Ovary 10-angled. Fruit 10 - 30 x 6 - 10 cm, acutely 10-angled. Seeds black, 10 - 12 x 6 - 8 mm, rugose, emarginate.

Field note: Commonly found on the boundaries of cultivated fields, gardens, among hedges. Also cultivated for its fruits as vegetables.

Fl. & Fr.: Almost throughout the year

Specimen examined: Kotri 906 (MLVGCB Herb).

Distribution: Widely cultivated, e.g. in China, India, Australia, Russia, Pakistan, Tropical Africa, Mascarene Isles and Tropical America

Uses: The tender fruits are used as vegetable.

Luffa acutangula (L.) Roxb. var. ***amara*** (Roxb.) Clarke in Hook. *f.,* Fl. Brit. Ind. 2: 615. 1879; Duthie, Fl. Gangetic Plain 1: 367. 1903; Bhandari, Fl. Indian Desert 154. 1978; Chakravarty, Fl. India. Fasc. 11: 70. 1882; Pandey, in Shetty & Singh Fl. Rajasthan 1: 342. 1987. *Luffa amara* Roxb. Fl. Ind. 3 : 715. 1832.

Vern.: *Mar turai, Ram torai.*

Extensive, scarbrous, monoceious, climbing annual. Stem angular scabrid at angles only. Leaves 7 - 12 x 6 - 10 cm, palmately 5-angled, scarbid on both surfaces, margins denticulate, cordate at base; petioles 1 - 3 cm long, pubescent. Male flowers in 10 - 20 flowered axillary racemes, in the axil of a small, fleshy probract with a few glistening glands on the upper surface. Peduncles upto 15 cm long. Calyx pubescent, lobes obscurely, longitudinal keeled. Corolla obcordate, rounded or scarcely emarginated, spreading. Stamens 3; filaments bearded at base. Female flowers on 5 - 10 cm long peduncles, solitary in the same axils as male peduncles. Ovary strongly 10-ribbed. Fruit obovoid-oblong, both ends obtuse, conical, 5 - 8 x 3 - 4 cm, smaller 10-ribbed not prominently winged. Seeds ovate, 8 - 9 x 5 mm, black, beaked, narrowly winged in the beak region only, verrucose, compressed, small.

Field note: Commonly found on the boundaries of cultivated fields, gardens, among hedges. The whole plant is exceedingly bitter.

Fl. & Fr.: August - October.

Specimen examined: Bari sadari 524 (MLVGCB Herb).

Distribution: India, Pakistan and Yemen.

Luffa cylindrica (L.) M. J. Roem. Syn. Monogr. 2: 63. 1846; Chakravarty, in Rec. Bot. Surv. India 17: 75. 1958 & in Fasc. Fl. India. 11 : 70. 1882; Pandey, in Shetty & Singh Fl. Rajasthan 1: 342. 1987. *Momordica cylindrica* L. Sp. Pl. 1009. 1753. *Luffa aegyptiaca* Mill. Dict. ed. 4. 500. 1785; Clarke in Hook. *f.*, Fl. Brit. Ind. 2: 614. 1879; Duthie, Fl. Gangetic Plain 1: 367. 1903

Vern.: *Ghia torai.*

An extensive annual climber or trailer. Tendrils slightly pubescent, 3 - 6-fid. Stem 5 - angled, finely hairy to glabrous. Leaves palmately 5-lobed, dark green, orbicular-cordate, 7 - 25 cm across, lobes triangular, lanceolate, acute-apiculate, entire or sinuate, scabrous. Petiole 5 - 15 cm long. Flowers bright yellow, pedicellate, 5 - 6 cm across; male racemose, racemes axillary, 12 - 25 cm long, 15 - 20-flowered; female flowers in the same axil as males. Probract fleshy, ovate, with 3 - 7 glistening glands on the upper surface. Calyx tube short, broadly campanulate, slightly pubescent; lobes triangular-lanceolate, longer than tube. Petals obovate-cuneiform, 2.5 - 3.5 x 1 - 2.5 cm, obtuse. Stamens 3 - 5; filaments 6 - 8 mm long. Ovary cylindrical, finely appressed hairy. Fruit cylindrical and fusiform, 20 - 50 x 6 - 10 cm, smooth. Seeds black, smooth, ovoid, ca 10 - 12 x 6 - 8 mm, with ca 1 mm wide margin.

Field note: Widely cultivated for its fruits which are used as vegetable.

Fl. & Fr.: April-October.

Specimen examined: Ladpura 126 (MLVGCB Herb).

Distribution: Distributed in warmer countries of Asia and Africa; cultivated or subspontaneous; introduced in tropical America

Uses: The mature fruit becomes bitter and unpalatable due to hardening of fibrovascular bundles forming loofah sponge which is used for scrubbing and cleaning the utencils. Loofah sponges are also used for cleaning motor cars, glassware and for stuffing pillows, mattresses, saddles and shoulder pads. The tender fruit is considered diuretic and its juice is purgative. Mature seeds are bitter, emetic and catthartic.

Luffa echinata (L.) Roxb., Fl. Ind. 3: 716. 1832; Clarke in Hook. *f.*, Fl. Brit. Ind. 2: 615. 1879; Duthie, Fl. Gangetic Plain 1: 367. 1903; Chakravarty, in Jain *et al*. Fl. India. Fasc. 11: 71. 1882; Pandey, in Shetty & Singh Fl. Rajasthan 1: 343. 1987.

Vern.: *Duttar torai.*

Dioecious, annual, climber. Stem sulcate, slightly hairy-glabrous. Tendrils bifid. puberulous or glabrous. Leaves 5 - 7 x 5 - 8 cm, reniform-suborbicular, obscurely or deeply 5-lobed, apex round or rarely acute, scabrous on both surfaces, margins minutely denticulate. Petiole stout, puberulous, up to 12 cm long. Flowers white, pedicellate, ca 2.5 cm across; male flowers in 5 - 12-flowered, up to 15 cm long raceme; calyx tube 6 mm long, hairy, lobes lanceolate, acute. Petals ovate, 1 - 1.2 cm long, obtuse, hairy at the base. Stamens 3, filaments united, ca 3 - 9 mm long, obtuse, hairy at the base, anthers entire slightly bifid. Female peduncles 1 - 3 cm long. Ovary ovoid. Fruit ashy, oblong, ovoid, 2 - 5 cm long, densely covered with ca 4 - 7 mm long bristles. Seeds ovate, black, slightly verrucose, emarginate, ca 4 - 5 x 3 - 5 mm and ca 2 mm thick.

Field note: Common among hedges and on fringes of forests among shrubs and bushes.

Fl. & Fr.: August - September.

Specimen examined: Hamirgarh 2177 (MLVGCB Herb).

Distribution: India, Pakistan, Bangladesh and Northern Tropical Africa

Uses: The fruit is intensely bitter and used medicinally.

MELOTHRIA L.

Melothria maderaspatana (L.) Cogn., in DC. Monog. Phan. 3: 623. 1881; Duthie, Fl. Gangetic Plain 1: 379. 1903; Chakravarty, Fl. India. Fasc. 11: 83. 1882; Pandey, in Shetty & Singh Fl. Rajasthan 1: 345. 1987. *Cucumis maderaspatanus* L. Sp. Pl. 1012. 1757. *Mukia scabrella* Arn. in Hook. J. Bot. 3 : 276. 1841; Clarke in Hook. *f.*, Fl. Brit. Ind. 2: 615. 1879, *excl. syn. Bryonia scabrella* L. f. 1781. *Mukia maderaspatana* (L.) M. Roem. Syn. Pepon. 47. 1846; Bhandari, Fl. Indian Desert 175. 1978. (Plate 143)

Vern.: *Kachri*

Annual, monoecious, climbing or trailing up to 3 m; stem bristly-hairy. Tendrils simple, filiform. Leaves sagittate, hastate, subtriangular to ovate in outline, usually cordate, up to 10 x 8 cm, and about as broad, 5-lobed to almost unlobed, sinuate-dentate to irregularly lobulate-dentate, scabrid; middle lobe longest, triangular; lateral lobes ovate-triangular; petiole 3 - 80 mm long. Male flowers on 1-3 mm long pedicels, yellow; calyx tube 1 - 2 mm long; lobes 1.5 mm long. Petals 1 - 2 x 1 - 1.2 mm; lobes ovate, oblong, rounded at apex. Anthers oblong. Female flowers axillary, solitary or sub-fasciculate. Fruits 1 - 7, subsessile, axillary, clustered, subglobose, scarlet, smooth, glabrous, bristly when young, 5 - 11 mm in diam.. Seeds, ca 4 x 2.5 mm, ovoid-oblong, suborbicular.

Field note: Common among hedges and on fringes of forests among shrubs and bushes.

Fl. & Fr.: August - September.

Specimen examined: Nikumbh 3199, Pur 2146 (MLVGCB Herb).

Distribution: India, Pakistan, China, Malaya, Africa and Australia.

Uses : The fresh green fruits are consumed by shepherds.

MOMORDICA L.

KEY TO THE SPECIES

1a. Flowers dioecious. Fruits densely prickly ....*M. dioica*

1b. Flowers monoecious. Fruits not prickly 2

2a. Bracts of the male flowers at the apex of the pedicel, ovate-cordate, dentate. Fruit 2 - 7 cm long ....*M. balsamina*

2b. Bracts of the male flowers about the middle or below the middle of the pedicel, reniform or orbicular-cordate, mucronate, entire. Fruit 7 - 25 cm long ....*M. charantia*

Momordica balsamina L. Sp. Pl. 1009. 1753; Clarke in Hook. *f.*, Fl. Brit. Ind. 2: 617. 1879; Duthie, Fl. Gangetic Plain 1: 370. 1903; Chakravarty, in Jain *et al*. Fl. India. Fasc. 11: 88. 1882; Pandey, in Shetty & Singh Fl. Rajasthan 1: 346. 1987. (Plate 144)

Vern.: *Mar-karelo.*

Annual or perennial, climber with unbranched glabrous tendrils, monoecious. Stem pubescent to glabrescent. Leaves orbicular, 1.5 - 5.0 cm across, cordate, glabrous or sparsely hairy, 3-5-lobed, middle lobes broadly ovate or rhombic-ovate, sinuate-dentate or acutely lobulate. Petiole 1 - 3 cm long, pubescent. Flowers yellow, 2.5 - 3.0 cm across, pedicellate. Male flowers solitary, on 1.5 cm long peduncle, bearing near the apex a sessile, broadly ovate-cordate, subglabrous, dentate bract; female flowers on 5-15 mm long basally bracteate or ebracteate peduncles. Calyx tube 5-6 mm long, pubescent, lobes ovate, acuminate, 3-3.5 mm broad. Corolla slightly zygomorphic, brown at base, lobes yellow, obovate, 10 - 15 x 8 - 10 mm. Ovary fusiform, beaked, puberulous, longitudinally tuberculate. Fruit broadly ovoid, narrowed at ends, 2 - 7 x 1 - 2.5 cm, orange-red, tuberculate. Seeds elliptic-ovate or ovate, 1 - 1.2 cm x 6 - 7 mm, compressed, slightly verrucose, grey with red aril.

Field note: Fairly common in the area, clumping on bushes and hedges of boundaries of crop fields.

Fl. & Fr.: August-November.

Specimen examined: Pansal 445 (MLVGCB Herb).

Distribution: Widespread in the drier parts of South Africa and Tropical Africa, Arabia, Tropical Asia and Australia.

Uses: Tender fruits are used as vegetables. Leaves and stem are used as fodder. The fruits contain momordicin which appears to be identical with elaterin.

Momordica charantia L. Sp. Pl. 1009. 1753; Clarke in Hook. *f.*, Fl. Brit. Ind. 2: 616. 1879; Duthie, Fl. Gangetic Plain 1: 369. 1903; Bhandari, Fl. Indian Desert 173. 1978; Chakravarty, Fl. India. Fasc. 11: 89. 1882; Pandey, in Shetty & Singh Fl. Rajasthan 1: 346. 1987.

Vern.: *Karela, Karelo.*

Annual, monoecious, climber or trailer with unbranched tendrils. Stem glabrous or hairy. Leaves suborbicular to orbicular, 4 - 8.5 x 2.5 - 6 cm, ovate, mucronate-dentate, deeply 5 - 7-lobed, lobes ovate-oblong or ovate-elliptic, glabrous or pubescent, acute, base constricted. Petiole 1.5 - 3.5 cm long, villous. Flowers yellow, ca 3 cm across, solitary, male peduncles ca 7.0 cm long, bearing the bracts about the middle or below the middle. Calyx pubescent. Corolla slightly zygomorphic, with obovate, obtuse lobes. Female flowers solitary, often bracteate. Ovary fusiform, muricate. Fruit oblong-fusiform or oval, 7 - 25 cm long, dehiscent. Seeds oblong, 1 - 1.5 cm x 6 - 9 mm, embedded in red pulp, sculptured.

Field note: The plant is extensively cultivated during the warm season for its fruits.

Fl. & Fr.: July - November.

Specimen examined: Pansal 1568 (MLVGCB Herb).

Distribution: Tropical and South Africa, South East and Far East Asia to Australia; naturalized in tropical and South America including W. Indies.

Uses : Fruits are bitter and are used as a vegetable; they are considered tonic, stomachic and carminative, fruits are used in rheumatism, gout, diseases of liver and spleen. The immature fruits are very commonly used during epidemics of small-pox as it is supposed to have a preventive effect.

Momordica dioica Roxb. ex Willd. Sp. Pl. 4 : 605. 1805; Roxb., Fl. Ind. 3 : 709. 1832; Clarke in Hook. *f.*, Fl. Brit. Ind. 2 : 617. 1879; Cooke, *l.c.* 563; Duthie, Fl. Gangetic Plain 1: 370. 1903; Bhandari, Fl. Indian Desert 156. 1978; Chakravarty, Fl. India. Fasc. 11 : 92. 1882; Pandey, in Shetty & Singh Fl. Rajasthan 1: 347. 1987. (Fig. 71 & Plate 145)

Vern.: *Kinkora.*

Perennial climber with unbranched, glabrous tendrils, dioecious. Root tuberous. Leaves broadly ovate, membranous, 4 - 8.5 x 2 - 6.0 cm, glabrous on both sides, deeply 3 - 5-lobed, lobes trianglular ovate or oblong, acute, with undulate or minutely remotely denticulate margin petiole 1 - 3 cm long, puberulous. Male flowers 5 - 6 cm across, yellow, on 4 - 6 cm long peduncles; bracts large, sessile, entire, glabrous, often ciliated or villous on both surfaces, cucullate, suborbicular or reniform, partly enclosing the flower. Petals 2 - 3 x 1.5 - 2.5 cm, obtuse. Female flowers yellow, on ca 2.5 cm long, ebracteate or minutely bracteate peduncles. Calyx lobes linear-lanceolate, 5 - 6 mm long, villous. Corolla similar to male flowers. Ovary ovoid, covered with long and soft papillae. Fruit ovoid or ellipsoid, 3 - 5 cm long, 2 - 3.5 cm across, yellow when mature, densely covered with soft spines, shortly beaked at apex. Seeds somewhat compressed, 6-7 x 5-6 mm, irregularly corrugated.

Field note: Commmon among hedges of cultivated fields, gardens and outskrits of forests.

Fl. & Fr.: July-October.

Specimen examined: Jahajpur 64, Atun 542 (MLVGCB Herb).

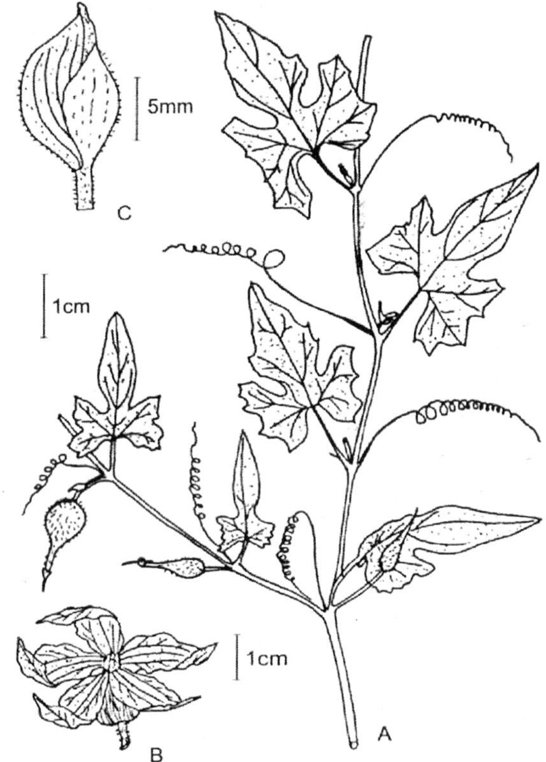

Fig. 71. *Momordica dioica* Roxb. ex Willd.
A. Fruiting twig, B. Flower, C. Bract

Distribution: India, Sri Lanka, Malaysia, Pakistan, Nepal, Bangladesh, Burma and China.

Uses : Fruits are often used as a vegetable. The seed kernels are used for the extraction of a semi-drying oil. The tuberous root is used in medicine. Roots of female plants are larger than those of the male and are preferred for medicinal uses. They are applied in bleeding piles and urinary complaints. The root paste is applied over the body as a sedative in fever.

PRAECITRULLUS Pangalo

Praecitrullus fistulosus (Stocks) Pangalo, Bot. Zhurn. 29: 203. 1944. Stewart, Ann. Cat. Vasc. Pl. W. Pak. 706. 1972; Jeffrey, in Kew Bull. 34: 801.1980. *Citrullus lanatus* (Thunb.) Matsumuras & Nakai, var. *fistulosus* (Stocks) Chakravarty, Fl. India. Fasc. 11: 23. 1882; Pandey, in Shetty & Singh Fl. Rajasthan 1: 335. 1987. *C. fistulosus* Stocks in Hook. Kew J. Bot. 3. : 74. t. 3. 1851; Bhandari, Fl. Indian Desert 162. 1978. *Colocynthis citrullus* (L.) O. Kuntze var. *fistulosus* (Stocks) Chakravarty in Rec. Bot. Surv. India 17 (1): 114. 1959.

Vern.: *Tinda, Tengsi.*

Annual, climber or trailer, monoecious. Stem robust, villous to hispid. Tendrils slender, 2 - 3-fid. Petiole hirsute. Leaves sparingly pinnatifid, lamina sparsely hispid all over, densely hispid on veins and veinlets of under surface; margins minutely denticulate, apparently entire; young leaves villous to densely hispid. Probract spathulate, ca 0.8 cm long. Flowers solitary, yellow, small. Male pedicel c. 1 mm long, calyx campanulate, hairy, lobes obconic, ca 8 mm long. Stamens 3, one free, others connate; female flower: calyx broadly campanulate, lobes ca 5 mm long, lanceolate. Ovary villous or softly hairy. Fruit almost spherical, ca 6 cm in diam.. Seeds ovate-oblong, ca 8 mm long, yellowish-white, smooth.

Field note : Cultivated in most parts of India and Pakistan for its tender fruits which are used as vegetable. There are two forms of this species, one with light green fruits and the other with dark green fruits. The former is generally preferred. Roasted seeds are also edible.

Fl. & Fr.: March-September

Specimen examined: Bhilwara 2179 (MLVGCB Herb).

Distribution: India and Pakistan.

Notes: This genus is very similar to *Citrullus* in its morphological characters, but differs in the stratification of pollen grains, haploid chromosome number and to some extent in leaf morphology. In *Citrullus* the pollen grains are reticulate, haploid chromosome number is 11 and the leaves are deeply pinnatipartite while in *Praecitrullus* pollen grains are echinate, haploid chromosome number is 12 and the leaves are sparingly pinnatifid. The seeds are also devoid of the enzyme urease which is commonly found in the genus *Citrullus* (Jeffrey in Kew Bulletin 20(3): 42. 1966) and the seed coat is 3-layered, not 4-layered as in *Citrullus* (C. Jeffrey, in Kew Bulletin 30 (3): 476-478. 1975).

TRICHOSANTHES L.

Trichosanthes cucumerina L. Sp. Pl. 1008. 1753; Clarke in Hook. *f.,* Fl. Brit. Ind. 2: 609. 1879; Duthie, Fl. Gangetic Plain 1: 364. 1903; Mahesh., Illus. Fl. Delhi f. 88.1966; Chakravarty, Fl. India. Fasc. 11: 112. 1882; Pandey, in Shetty & Singh Fl. Rajasthan 1: 349. 1987.

Vern.: *Mar-chichinda.*

Annual climbing herb. Stem slender, puberulous. Leaves cordate-subreniform, distantly denticulate, petiolate, densely pubescent to sparsely hairy, 10 - 15 x 12 - 18 cm , deeply or shallowly 5-lobed. Tendrils sulcate, puberulous, 2 - 3-fid. Male flowers: peduncle puberulous, 10 - 25 cm long, 8 - 15-flowered. Pedicels up to 2 cm long, filiform. Calyx-tube 1.5 - 3 cm long, lobes reflexed, ca 2 mm petals oblong 3-nerved. Female flowers axillary, solitary; peduncles 8 - 10 mm long Ovary oblong,; style ca 1.6 cm long, filiform. Fruit ovoid-conical, twisted, longitudinally striped white, yellow when ripe. Seeds 8-10, elliptic, compressed, rugulose, 15-17 mm long, with undulate margins.

Field note: Common among hedges on the boundries of crop fields and in forests.

Fl. & Fr.: August - October.

Specimen examined: Pansal 227, Karoi 1530 (MLVGCB Herb).

Distribution: Cultivated throughout India and Pakistan.

Uses: Fruits are used as vegetable.

CACTACEAE

OPUNTIA Mill.

Opuntia elatior Mill. Gard. Dict. ed. 8. 4. 1768; Raizada, Suppl. Duthie, Fl. Gangetic Plain 83. 1976; Bhandari, Fl. Indian Desert 158. 1978; Parmar in Shetty & Singh, Fl. Rajasthan 1: 353. 1987. *O. dillenii auct. plur.* non Haw. 1819; Grah. Cat. Bombay Pl. 546. 1839; Clarke in Hook. *f.,* Fl. Brit. Ind. 2 : 657. 1879, *pro parte*; Duthie, Fl. Gangetic Plain 1: 384. 1903. (Plate 146)

Vern.: *Nagphani.*

Shrub sprawling or erect, 1 - 2 m tall. Stem succulent, jointed, terminal joints green to grey-green, obovate-oblong 10 - 35 × 7.5 - 20 cm. Areoles 2 - 9 mm in diam. Spines 3 - 8 per areole on most areoles, spreading, yellow, ± brown banded or mottled, subulate, straight, 1.2 - 4 cm, basally flattened; glochids yellow. Leaves subulate, 4.5 - 6 mm, deciduous. Flowers solitary from the areoles, 4 - 10 cm in diam., soon turning rose pink. Sepaloids greenish with yellow margins, broadly deltoid-obovate to obovate, 10 - 25 × 6 - 12 mm, margins entire or slightly crisped, apex mucronate. Petaloids spreading, bright yellow, obovate or cuneate-obovate, 25 - 30 × 12 - 20 mm, margins entire or slightly undulate, apex rounded, truncate, or emarginate. Filaments yellow, ca 12 mm; anthers yellow, apiculate, ca 1.5 mm. Style yellowish. Fruit red when ripe, marked with areoles, but bristles and spines deciduous. Seeds light tan, irregularly orbicular, 2-3.5 × 3-4.5 mm.

Field note: Found on thickets, rocks, sandy soils, also cultivated for fencing.

Fl. & Fr.: January - June.

Specimen examined: Pur 107, Hamirgarh 1174 (MLVGCB Herb).

Distribution: Native to South Africa, Introduced in India.

AIZOACEAE (FICOIDEAE)

TRIANTHEMA L.

Trianthema portulacastrum L. Sp. Pl. 223. 1753; Mahesh., Illus. Fl. Delhi f. 93. 1966; Bhandari, Fl. Indian Desert 164. 1978; Parmar in Shetty & Singh , Fl. Rajasthan 1: 355. 1987. *T. monogyna* L. Mant. 69. 1767; Clarke in Hook. *f.,* Fl. Brit. Ind. 2: 609. 1879; Duthie, Fl. Gangetic Plain 1: 385. 1903. *T.obcordata* Roxb. Hort. Beng. 34. 1814 (nom. nud.) & Fl. Ind. 2: 445. 1832; Wight, Ic. t. 228. 1840.

Vern: *Santio, santi.*

Procumbent or prostrate annual herb, up to 35 cm long. Stem branched, glabrous to sparsely pubescent. Leaves 1.5 - 3.5 x 0.8 - 2 cm, opposite, one of the pair smaller than the other, petiolate, stipulate, obovate to rounded or spathulate, obtuse or retuse at apex, sometimes slightly apiculate, obtuse or cuneate at base, glabrous on both surfaces; petiole 4 - 11 mm long, papillose to glabrous, base dilated, clasping, calyculate, with 2 lateral stipular appendages. Flowers axillary solitary, sessile, pink. Calyx tube closely sheathed by the leaf base; sepals 5, oblong, 2.5 - 3.0 mm long, shortly aristate. Stamens 10 - 15, unequal, filaments 1.5-2.2 mm long. Ovary ca 2 mm long, more or less conical; style 1, linear, 1.5 - 2.0 mm long, persistent. Pyxidium (Capsule) ca 4 mm long, circumsessile; amphora 4 - 5-seeded, lid 2.5 - 3.0 mm broad. Seeds 5 - 10, ca 2 mm broad, reniform, lenticular, black, rugose.

Field note: Common weed of Waste lands. In late season the whole plant becomes reddish.

Fl. & Fr.: August - December.

Specimen examined: Bhilwara 138, Piplund 1378 (MLVGCB Herb).

Distribution: Tropical America, Africa, W. Asia, Sri Lanka, India and W. Pakistan.

Uses : It is used as fodder.

MOLLUGINACEAE

KEY TO THE GENERA

1a. Leaves alternate to subopposite. Flowers magenta or pinked; petals many; staminodes petaloid .... *Corbichonia*

1b. Leaves whorled or opposite. Flowers green or greenish yellow; petals as many as sepals; staminodes absent 2

2a. Leaves stellately hairy. Flowers in axillary fascicles, seeds strophiolate.................. *Glinus*

2b. Leaves glabrous. Flowers in terminal cymes, seeds without strophiolate...... *Mollugo*

CORBICHONIA Scopoli

Corbichonia decumbens (Forsk.) Exell. in Journ. Bot. 73: 80. 1935; Raizada, Suppl. Duthie, Fl. Gangetic Plain 84. 1976; Bhandari, Fl. Indian Desert 159. 1978; Parmar, in Shetty & Singh Fl. Rajasthan 1: 357. 1987. *Orygia decumbens* Forsk. Fl. Aegypt. - Arab. 103. 1775; Clarke in Hook. *f.*, Fl. Brit. Ind. 2: 661. 1879. (Fig. 72 & Plate 147)

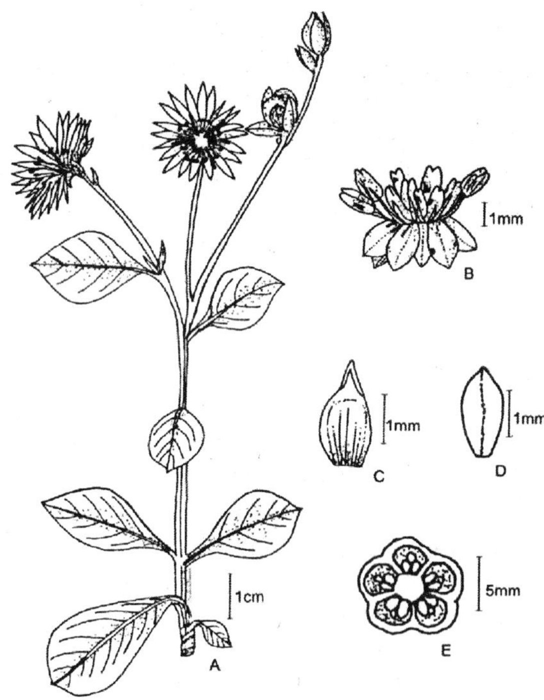

Fig. 72. *Corbichonia decumbens* (Forssk.) Exell. A. Flowering & fruiting branch, B. Flower, C. Perianth lobe, D. Staminode; E. T.S. of ovary

A decumbent or prostrate, glabrous, more or less annual rarely perennial, from woody base. Branches rigid, often reddish; internodes with norrow ridges, decurrent from the petiole-bases. Leaves attenuate, elliptic ovate to obovate, 1.5 - 4.0 x 0.7 - 2.6 cm, mucronate, fleshy; petiole up to 10 mm long, with membranous margins. Inflorescence terminal or leaf opposed, few to many flowered, lax, racemiform. Bracts lanceolate, acute, 2 - 3 mm long, membranous. Pedicels 3 - 8 mm long. Flowers ca 5 mm broad, magenta or pinked. Perianth segments ovate, 4 mm long, margin white; inner narrower, all acute, mucronate. Stamens 20 - 25, erect; anthers buff-coloured; filaments 1 - 2 mm long; petaloid staminodes about 20, ca 2.5 mm long, oblong ovate, adnate. Ovary ca 1.5 mm broad, globose;

styles 5, filiform, spreading as long as or longer than the ovary, linear, deciduous. Capsule papery, 5 - 6 mm broad, dehiscing by 5 valves; valves ovate, more or less included in the perianth, yellow green, shining. Seeds ca 1 mm long, compressed, rugose, shiny black, strophiolate.

Field note: Commonly found in rocky or sandy places, in dry hot areas. The basal leaves turns pink at withering.

Fl. & Fr.: August - October.

Specimen examined: Badnor 108, Theekrana (Beawar) 3124 (MLVGCB Herb).

Distribution: Africa, Asia, India and Pakistan.

GLINUS L.

1a. Plants stellately hairy, leaves broadly obovate or suborbicular .... *G. lotoides*

1b. Plants glabrous, leaves oblanceolate or linear-lanceolate.................... .*G. oppositifolius*

Glinus lotoides L. Sp. Pl. 463. 1753; Sant, in Bull. Bot. Surv. India 3: 17. 1961; Mahesh., Illus. Fl. Delhi f. 90. 1966; Bhandari, Fl. Indian Desert 161. 1978; Parmar in Shetty & Singh, Fl. Rajasthan 1: 358. 1987. *G. dictamnoides* L., Mant. Pl. 2: 243. 1771. *Mollugo hirta* Thunb. Prodr. Pl. Cap. 24. 1794; Clarke in Hook. *f.*, Fl. Brit. Ind. 2: 662. 1879; Duthie, Fl. Gangetic Plain 1: 386. 1903. (Fig. 73 & Plate 148)

Prostrate or procumbent annual herb up to 40 cm long, with stellate woolly parts. Leaves opposite or sub-verticillate, 0.6 - 2.0 x 0.5 - 1.8 cm, obovate or sub-orbiculate or more or less cuneate, stellately pubescent on both surfaces, mucronate; petiole 2 - 8 mm long. Flowers in axillary clusters of 5 (-6), sub-sessile, greenish, white; pedicel up to 1.5 mm long. Sepals 6 in whorls 3 + 3, 4 - 4.5 mm long, up to 7 mm in fruit, persistent, ovate to ovate-oblong, apiculate, imbricate. Stamens 10 - 20; fertile stamens 12; filaments 2 - 3.5 mm long, unequal, persistent; anthers less than 1 mm long, bilobed. Ovary ca 2 mm long, ovoid; styles 5 - 6; stigmas 5, linear, ca 1 mm long, persistent. Capsule sub-globose or oblong, ca 6 mm long, membranous, enclosed in the sepals. Seeds many, less than 1 mm long, tuberculate, strophiolate, with a curved linear appendage ca 3/4 the size of the seed.

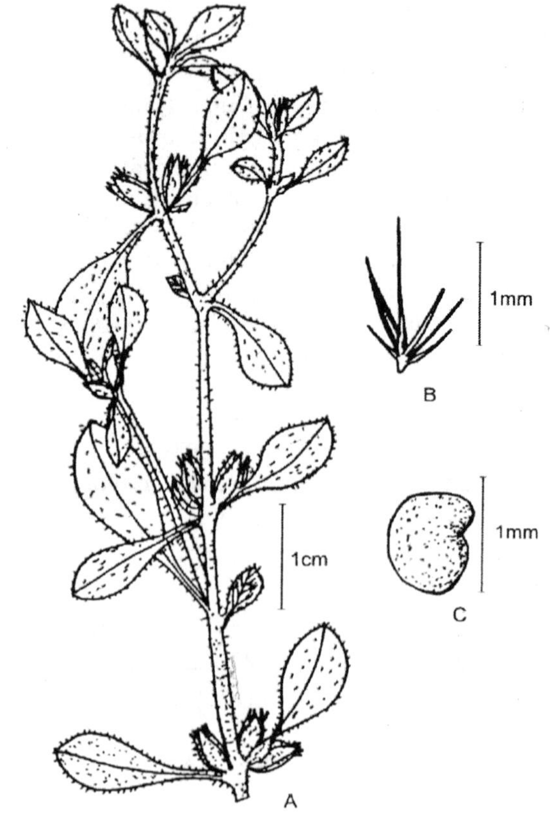

Fig. 73. *Glinus lotoides* L.
A. Flowering plant, B. Stellate hair, C. Seed

Field note: The species is seen on the muddy beds of water resevoirs, on the margins near standing water, in association with *Heliotropium supinum* L. and *Polygonum plebeium* R. Br..

Fl. & Fr.: February - May.

Specimen examined: Bhilwara 448, Pansal 145 (MLVGCB Herb).

Distribution: Most tropical parts of the world; Sri Lanka, Africa, India, W. Pakistan etc.

Uses : Whole plant is boiled in water and is used as a remedy against urinary trobles.

Glinus oppositifolius (L.) DC. in Bull. Herb. Boiss. sér. 2. 1: 559. 1901; Parmar, in Shetty & Singh Fl. Rajasthan 1: 358. 1987. *Mollugo oppositifolia* L. Sp. Pl. 1 : 89. 1753; Duthie,

Fl. Gangetic Plain 1: 387. 1903. *M. spergula* L. Syst. ed. 10. 881. 1759; Clarke in Hook. *f.*, Fl. Brit. Ind. 2 : 662. 1879. (Fig. 74 & Plate 149)

Erect or ascending annual, herb 10 - 40 cm, much branched, pilose or subglabrous. Leaves in pseudowhorls of 3-6 or opposite; leaf blade spatulate-oblanceolate or elliptic, 1 - 2.5 cm × 3-6 mm, base attenuate, margins with sparse teeth, apex obtuse or acute. Pedicels 5 - 14 mm, slender. Flowers usually 2-6 in a cyme, greenish or pinkish-white. Sepals greenish white, yellowish, or milky-colored, oblong, 3-4 mm, margins membranous, 3-veined. Stamens 3-5; free staminodes absent. Ovary 3-locular; styles 3. Capsule ellipsoid, slightly shorter than persistent tepals, 3 - or 4 - valved. Seeds chestnut-brown, subreniform, granulose; strophiole ca 1/5 as long as seed, claviform.

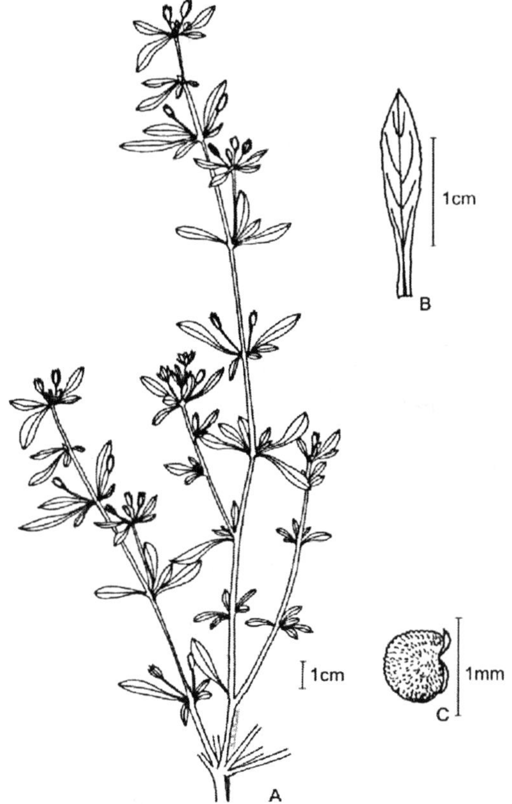

Fig. 74. *Glinus oppositifolius* (L.) DC
A. Habit, B. Leaf, C. Seed

Field note: Rare, on moist sand, specially in riverbeds.

Fl. & Fr.: Almost year-round.

Specimen examined: Bijoliya 139 (MLVGCB Herb).

Distribution: Tropical Africa, Asia, N Australia, Hainan, S Taiwan.

MOLLUGO L.

KEY TO THE SPECIES

1a. leaves cauline and ramal. Inflorescences terminal or seemingly axillary, basal leaves usually soon dry up, seeds not strophiolate .... *M. cerviana*

1b. All leaves radicle, inflorescence arising from a rosette of persistent basal leaves, seeds minutely strophiolate .... *M. nudicaulis*

Mollugo cerviana (L.) Seringe in DC. Prodr. 1: 392. 1824; Clarke in Hook. *f.*, Fl. Brit. Ind. 2: 663. 1879; Duthie, Fl. Gangetic Plain 1: 387. 1903; Mahesh., Illus. Fl. Delhi f. 91. 1966; Bhandari, Fl. Indian Desert 162. 1978; Parmar, in Shetty & Singh Fl. Rajasthan 1: 358. 1987. *Pharnaceum cerviana* L., Sp. Pl. 272. 1753.

Vern.: *Chidion ka khet.*

Erect, annual up to 15 cm tall. Stem many upright or ascending from base, branched, glabrous, wiry. Leaves up to 11 per node, linear, 0.6-2.0 cm x ca 1 mm, glaucous green, basal leaves in rosette, oblanceolate, spathulate; cauline leaves whorled or linear-oblong or subsphathulate, apex apiculate; stipules caducous. Pedicel up to 14 mm long, slender, glabrous. Flowers numerous, in green axillary & terminal cymes. Bracts linear, ca 1 mm long. Sepals free, up to 2 mm long in fruit, slightly inflexed, margins white, obovate, green. Stamens 3, alternating with the sepals; filaments dilated at the base, slightly longer than the sepals. Ovary ca 1.1 mm long, ovoid; stigmas 3, short. Capsule subglobose, ca 2.5 mm broad, slightly longer than the sepals, dehiscing by 3 valves; valves slightly retuse. Seed less than 1 mm, slightly curved, light brown, reticulate and shiny.

Field note: Common on sandy or gravelly soils in Waste lands. A delicate herb bearing a superficial resemblance to the Caryophyllaceous *Spergula* L.

Fl. & Fr.: August-November.

Specimen examined: Atun 1894 (MLVGCB Herb).

Distribution: S. Europe, Greece, Rumania, S. Russia, Siberia, Arabia, Tropical Africa, Galapagos Islands, Australia, Sri Lanka, India and Pakistan.

Mollugo nudicaulis Lamk. Encycl. 4: 234. 1797; Clarke in Hook. *f.,* Fl. Brit. Ind. 2: 664. 1879; Duthie, Fl. Gangetic Plain 1: 388. 1903; Mahesh., Illus. Fl. Delhi f. 92. 1966; Bhandari, Fl. Indian Desert 163. 1978; Parmar, in Shetty & Singh Fl. Rajasthan 1: 360. 1987.

Erect, annual herb, up to 30 cm tall. Leaves all radical, 1.2 - 5.0 x 0.5 - 1.2 cm, spathulate-obovate, entire, glabrous, attenuate. Flowers in dichasial cymes; scapes slender, glabrous; involucral bracts ovate-oblong, 1 - 2 mm long, scarious; pedicels up to 1.5 cm long. Sepals 5, oblong, 2.5 - 3.0 mm long, mucronate, imbricate persistent. Filaments c. 1 mm long, persistent, anthers less than 1 mm long, sub-globose. Ovary sub-globose, 1.5 mm long. Stigmas 3, short, curved. Capsule as long as the sepals, sub-globose, dehiscing by 3 valves. Seeds less than 1 mm, muricate, minutely strophiolate, black and shiny.

Field note: Common weed on roadsides, in open wasteplaces, often found in dry crevices of rocks.

Fl. & Fr.: July-November.

Specimen examined: Bhilwara 1356 (MLVGCB Herb).

Distribution: S. America, tropical Africa, Sri Lanka, India and W. Pakistan.

ALANGIACEAE

ALANGIUM Lam. nom. cons.

Alangium salvifolium (L. f.) Wangerin in Engl. Pflanz. 41: 9. f. 2A-E. 1910; Mukerjee, in Bull. Bot. Surv. India 10: 330. 1968; Parmar, in Shetty & Singh Fl. Rajasthan 1: 366. 1987; Mathew, in Nayer *et al.* Fasc. Fl. India 19: 4. 1988. *Grewia salvifolia* L. f. Suppl. Pl. 409. 1781. *Alangium lamarkii* Thw. Enum. Plant. Zeyl. 133. 1857; Clarke in Hook. *f.,* Fl. Brit. Ind. 2: 741. 1879; Duthie, Fl. Gangetic Plain 1: 402. 1903. (Plate 150)

Vern.: *Ankol.*

Large decidious shrub or small tree, up to 10 m high. Branchlets sometimes spinescent, tomentose to glabrescent. Petiole ca 1 cm; leaf blade oblong to oblong-lanceolate, 5 - 14 × 2.0 - 3 cm, papery to subleathery, glabrescent, base cuneate, margins entire, apex obtuse. Inflorescence sessile, often clusters of 4-8 flowers, usually fewer, sometimes only a solitary flower, densely rusty tomentose. Pedicels 2-8 mm. Flowers fragrant, cream-colored, 1.2-3 cm. Calyx tube ca 2.5 mm; lobes 5 - 10, ca 3 mm, toothed. Petals 4 - 6, basally swollen and apically obtuse, 1.2 - 3 cm, outside tomentose, inside less so. Stamens 10 - 30; filaments 4-12 mm, apically geniculate and bearded. Disk lobed. Ovary glabrous except near base of inner side, 1 - or 2-loculed; ovule 1 per locule, pendulous; style glabrous, 8-24 mm; stigma capitate. Drupe red when ripe, ellipsoid or ovoid, 2.5 - 3.0 cm across.

Field note: Common on hill slopes.

Fl. & Fr.: February - June.

Specimen examined: Gargrar 3299, Jantala Mata 1021 (MLVGCB Herb).

Distribution: Cambodia, China, India, Indonesia, Laos, Malaysia, Nepal, Philippines, Sri Lanka, Thailand, Vietnam and Africa.

RUBIACEAE

KEY TO THE GENERA

1a. Herbs 2

2a. Leaves opposite, stipulate. Seeds with a ventral cavity .... *Borreria*

2b. Leaves whorled, exstipulate. Seeds angular, smooth .... *Hedyotis*

1b. Trees or shrubs 3

3a. Plants armed, fruits fleshy berry....*Ceriscoides*

3b. Plants unarmed, fruits capsular 4

4a. Flowers in corymbiform cymes. Corolla lobes twisted Ovules solitary in each cell........ .*Ixora*

4b. Flowers in globose head. Corolla lobes valvate. Ovules more than one in each cell ....*Mitragyna*

BORRERIA G. F. W. Meyer (nom. cons.)

Borreria articularis (L. f.) F. N. Will. in Bull. Herb. Boiss. ser. 2. 5: 956. 1905; Bhandari, Fl. Indian Desert 187. 1978; Singh, in Shetty & Singh Fl. Rajasthan 1: 369. 1987. *Spermacoce hispida* L. Sp. Pl. 102. 1753; Hook. *f.*, Fl. Br. Ind. 3: 200. 1880; Duthie, Fl. Gangetic Plain 1: 429. 1905. *Spermacoce articularis* L. f. Suppl. Pl.: 119. 1781. *Borreria hispida* (L.) K. Schum. in Engl. & Prantl, Pflanzenfam. 4 (4): 144. 1891, non Spruce ex K. Schum. 1888. (Plate 151)

Annual prostrate, hispid herb. Branches 3-4 at the base, radiating, 4- angled. Leaves opposite, sessile, subacute, entire, elliptic, 2 - 3 x 0.8 - 1 cm, hispid. Inflorescence axillary clustered. Flowers white. Calyx lobes green, narrowly triangular, to 1.5 mm long. Corolla campanulate, tube 1 cm long; outside glabrous except for a few hairs near lobe apex, inside a band of dense moniliform hairs near middle of tube. Stamens inserted at the throat of corolla tube; anthers to 0.7 mm long, oblong, shorter than the filaments. Style equal to or slightly exceeding corolla lobes; stigma bilobed, nectary disc bipartite. Capsule ca 3 mm long, ovoid to ellipsoid, with hairs on upper half, septum usually remaining erect and undamaged as valves dehisce. Seeds dark brown to almost black, obovate or elliptic, to 2.2 - 3 x 1 - 1.2 mm. reticulate, ventrally shallowly grooved, brown.

Field note: Common in open disturbed areas, usually on sand, often at wet places.

Fl. & Fr.: August - November.

Specimen examined: Atun 449, Bhilwara 543 (MLVGCB Herb).

Distribution: Possibly native to Thailand. Naturalized in Africa, parts of Asia (India) and Australia.

CERISCOIDES (Hook. f.) Tirvengadum

Ceriscoides turgida (Roxb.) Tirveng. Bull. Mus. Hist. Nat. (Paris) ser. 3. no. 521, Bot. 35: 1987. *Gardenia turgida* Roxb., Fl. Ind. 2: 557. 1824; ed Carey 1: 711. 1832; Hook., in Benth. & Hook. *f.*, Gen. Pl. 2 :89. 1873; Clarke. J. Linn. Soc. Bot. 17: 310. 1880; Hook. Fl. Br. Ind. 3: 118. 1880; Duthie, Fl. Gangetic Plain 1: 422. 1905. *Gardenia montana* Roxb., Fl. Ind. 2: 556. 1824; ed Carey 1: 709. 1832; Wight, *l. c.* 2 (3): 6. t. 577. 1842. *G. turgida* Roxb. var. *montana* Hook. *f.*, *l. c.* 3: 118. 1880; Singh, in Shetty & Singh Fl. Rajasthan 1: 373. 1987; Duthie, *l. c.* 1: 422. 1905.

Armed deciduous tree, upto 8 m high. Bark bluish-grey, smooth. Branches tomentose usually with erect spines, spines 2 - 3 cm long, strong, leaf-bearing; Leaves, decussate, dimorphic, lower branches ovate, 1.0 - 12.5 x 0.4 - 5.6 cm, upper ones 1 - 9 x 1 - 2.5 cm; stipules interpetiolar, connate into a sheath. Flower 5-merous, dimorphic, solitary in lux cymes, male ones in fascicles; female in solitary, white turing yellow. Calyx lobes 5, included. Corolla hypocrateriform, tube 1.5 cm, lobes elliptic obovate, 1 - 1.5 cm long. Stamens 6 mm, subsessile, included. Ovary turgid, 1-locular with 5 parietal placentation, styles 0.08 mm, branched. Berry globose, turgid 4 - 5 x 4 - 4.6 cm. Seeds many, embedded in fleshy pulp; testa smooth, rough.

Field note: Occasional in the interior of forests near water streams.

Fl. & Fr.: April - July.

Specimen examined: Thago ka khera 62 (MLVGCB Herb).

Disribution: India, Sri Lanka, SE Asia and Maynmar.

Uses: Fruits and roots are used in treating dog bite and in epilepsy.

Notes: Often the plant is unarmed but the variations have been observed in the specimens of this area.

HEDYOTIS L.

Hedyotis corymbosa (L.) Lam. Encycl. 1: 272. 1792; Dutta & Deb, in Tax. Revis. Hedyotis (Rubiaceae) Ind. Subcont. 147 - 149. t. 48. 2004. *Oldenlandia corymbosa* L. Sp. Pl. 119. 1753; Hook. *f.,* Fl. Brit. Ind. 3: 25. 1880, *pro parte*; Duthie, Fl. Gangetic Plain 1: 413. 1905, excl. syn.; Singh, in Shetty & Singh Fl. Rajasthan 1: 379. 1987.

Sub-erect annual herb, up to 20 cm tall. Leaves elliptic-lanceolate or linear, 1.5 - 3.0 x 0.1 - 0.2 cm, linear or elliptic-lanceolate, revolute-margined, glabrous or with appressed hairs on the lower surface; petioles ca 1 mm long, stipules ca 2.5 mm long, connate, membranous, scarious. Flowers white, axillary corymbose cymes, peduncles ca 8 - 30 mm long, glabrous. Calyx 4-toothed, teeth ca 1 mm long, ovate, margins scabridulous. Corolla lobes oblong, acute. Anthers sessile, inserted in the mouth of corolla tube. Style glabrous. Capsules globose, glabrous, brown with persistant calyx. Seeds minute, angled, black.

Field note: Common in wetlands near streams, ponds particularly in wet habitats.

Fl. & Fr.: September - November.

Specimen examined: Mezadam 162 (MLVGCB Herb).

Distribution: Pantropical.

IXORA L.

Ixora coccinea L. Sp. pl. 1 : 110. 1753. Hook. *f.,* Fl. Brit. Ind. 3 : 145. 1880: Dalz. & Gibs. 112. 1861; Wight, Ic. t. 153. 1844. *Ixora bandhuca* Roxb. Hort. Beng. 10. 1814.

Shrub upto 2 m high, Leaves sessile to short-petiolate, 5 - 10 x 3.0 - 5.7 cm; blades elliptic, oblong, or obovate, usually leathery, base cordate to rounded (rarely wedge-shaped), apex rounded, mucronate, or shortly tapering; stipules basally sheathing, lobes triangular and strongly awn-tipped. Inflorescence terminal, usually a 3-branched panicle or corymbose cymes. Flowers sessile, red. Calyx lobes short, triangular, persistent. Corolla tube usually 1 - 1.5 inches long, lobes lanceolate to ovate, less than 0.25 inches long, acute or sometimes obtuse. Fruit thinly fleshy, reddish black, globose, 1 - 2 seeded.

Field note: Prefers a rich acidic soil that is well watered yet has good drainage.

Fl. & Fr.: Almost throughout the year.

Specimen examined: Bhilwara 2180 (MLVGCB Herb).

Distribution: Sri Lanka, China and India.

MITRAGYNA Korth. nom. cons.

Mitragyna parvifolia (Roxb.) Korth. Obs. Naucl. Ind. 19. 1839; Duthie, Fl. Gangetic Plain 1: 408. 1905; Mahesh., Illus. Fl. Delhi f. 97. 1966; Singh, in Shetty & Singh Fl. Rajasthan 1: 376. 1987. *Nauclea parvifolia* Roxb. Pl. Cor. 1: 40. t. 52. 1796. *Stephegyne parvifolia* (Roxb.) Korth., in Verh. Nat. Ges. Bot. 161. 1840; Hook. *f.,* Fl. Brit. Ind. 3: 25. 1880. (Plate 152)

Vern.: *Kadamb*

Deciduous tree, up to 12 m tall. Bark grey, smooth and thin irregularly scaly when mature. Young branchlets angular to subterete, glabrous. Leaves 4 - 15 x 2.5 - 10 cm, simple, opposite, decussate; stipules foliaecous with keeled back, interpetiolar, caducous and leaving scar; lamina ovate, elliptic-obovate to sub-orbiculate, apex abruptly acuminate with blunt tip, rarely acute, base acute to attenuate to subcordate, margin entire, coriaceous, glabrous; petioles 1 - 4 cm long, canaliculate in cross section, glabrous; Inflorescence terminal head; flowers sessile, cream-white or greenish yellow. Calyx 5 - toothed; lobes ciliate along margins. Corolla-tube funnel shaped, glabrous-outside, ciliate along midrib. Capsules, oblong-ellipsoid, 10-ribbed. Seeds many, winged.

Field note: Common tree in forests throughout area.

Fl. & Fr.: July- October.

Specimen examined: Ladpura 109, Bijoliya 1641 (MLVGCB Herb).

Distribution: Indomalayasia and India.

ASTERACEAE (COMPOSITAE)

KEY TO THE GENERA

1a. Involucral bracts spinescent or aristate 2

2a. Heads 1- flowerd, crowded in dense globose balls (compound heads).................... .*Echinops*

2b. Heads many flowered, distinct, simple 3

3a. Heads homogamous...................... .*Carthamus*

3b. Heads heterogamous 4

4a. White wooly herbs. Flowers white or purple ....*Dicoma*

4b. Glabrous or scabrous herbs. Flowers purple ....*Amberboa*

1b. Involucral bracts neither spinescent nor aristate 5

5a. Heads homogamous, either male or female or bisexual 6

6a. Heads with ligulate florets only.................... .7

7a. Pappus of extremely short, 2 to 3-seriate scales. Flowers blue........... .*Cichorium*

7b. Pappus hairs united at the base into a ring. Flowers yellow.. .8

8a. Stem fistular, latex milky white. Leaves mostly cauline, involucre campanulate. Achenes with a narrow base and truncate apex ....*Sonchus*

8b. Stem not fistular, latex yellow. Leaves mostly rosulate, involucre cylindric. Achenes truncate at both ends........................ .*Launaea*

6b. Heads with tubular florets only.................... .9

9a. Heads compound, 1- flowered........... .*Caesulia*

9b. Heads simple.. .10

10a. Leaves opposite, at least in the lower region.. .*Ageratum*

10b. Leaves alternate or radical .... .11

11a. Involucral - bracts 1-seriate......... .*Xanthium*

11b. Involucral -bracts 2 to many seriate.....12

12b. Flowers purple ....*Vernonia*

12a. Flowers yellow or yellowish white. 13

13a. Anther-base sagittate, tailed.......... .*Pulicaria*

13b. Anthers bases obtuse, not tailed. ....*Chrysanthellum*

5b. Heads heterogamous, outer flowers usually female, inner ones bisexual or male .... .13

13a. Stem winged. Heads collected into globular balls ....*Sphaeranthus*

13b. Stem not as above. Heads simple, distinct .... .14

14a. Leaves opposite, at least in the lower region .... .15

15a. Leaves 3-foliate or deeply pinnatifid . ... *Bidens*

15b. Leaves simple, entire or serrate dentate .. .16

16a. Palea concave, embracing the florets or achenes... .17

17a. Ray florets 2 to many seriate, florets white ....*Eclipta*

17b. Ray florets 1 -seriate. florets yellow ....*Acanthospermum*

16b. Palea flat, not embracing the florets or achenes... .18

18a. Heads axillary or at forks of dichotomy, subsessile. Achenes with pointed spine .... *Synedrella*

18b. Heads terminal, stalked. Achenes withought spines .... .19

19a. Erect. Heads white. Pappus bristles 2 - 4 , smooth................................ .*Blainvillea*

19b. Prostrate or decumbent. Heads yellow. Pappus bristles numerous, plumose..... *Tridax*

14b. Leaves alternate or radical........................ .20

20a. Ray florets ligulate.................................... .21

21a. An erect herb. Leaves pinnately lobed Florets white................................ .*Parthenium*

21b. Prostrate or decumbent herb. Leaves 1 to 2-pinnate or pinnatisect. Florets yellow ....*Glossocardia*

20b. Ray florets narrow, tubular, not ligulate .. .22

22a. Anther bases tailed .... .23

23a. Achenes glandular. Involucral-bracts dry, scarious............................... .*Gnaphalium*

23b. Achenes not glandular. Involucral-bracts herbaceous................................ *.Blumea*

22b. Anther bases obtuse, not tailed................. .24

24a. Capitula homogamous. Pappus 1....... *.Cotula*

24b. Capitula heterogamous. Pappus more than one.. *.Grangea*

ACANTHOSPERMUM Schrank

Acanthospermum hispidum DC. in Prodr. 5 : 522. 1836; Raizada, suppl. Duthie, Fl. Gangetic Plain 99. 1976; Bhandari, Fl. Indian Desert 171. 1978; Singh, in Shetty & Singh Fl. Rajasthan 1: 390. 1987; Chowdhey, in Hajra *et al*. Fl. India 13: 361. f. 93. 1995. *A. australe sensu* Raizada & Sharma in Ind. For. 88 (5): 364. 1962, non O. Kuntze, 1891. (Fig. 75 & Plate 153)

Vern.: *Dokanta*.

An erect,dichotomously branched, hispid, herb, 10 - 50 cm. Stem villous, hirsute, specially upwards with rather long, multicellular hairs. Leaves sessile or short petioled, 2 - 7 x 1 - 4.5 cm, obovate-spathulate or oblanceolate, cuneate at the base, obtuse or acute, serrate dentate, hirsute with more or less appressed to loosely spreading hairs. Heads on deflexed peduncles in the leaf-axils or forks of branches. essentially sessile, 2 - 3 cm in diam., yellow. Involucral bracts in 2 - series; outer invulucral bracts 3-5 mm long, lanceolate, villous ciliate margins; inner ones 3 - 5 mm, enveloping the ovary of ray florets, softly prickly with 2 long, straight apical awns. Ray florets 5 - 8, female; corolla yellowish-white, 3-lobed. Styles puberulose, undivided. Disc florets 5 - 6 or 7, bisexual; corolla yellow, tube 5-lobed. Stamens 5; anthers linear with sagittate base. Style shortly bifid. Achenes of ray florets black, triangular, straight or hooked, 3 - 4 mm long, apical, wide spreading spines; achenes of disc florets, glandular.

Field note: Common along forest path and in Waste lands.

Fl. & Fr.: August - October.

Specimen examined: Thago ka khera 160, Kotri 2144 (MLVGCB Herb).

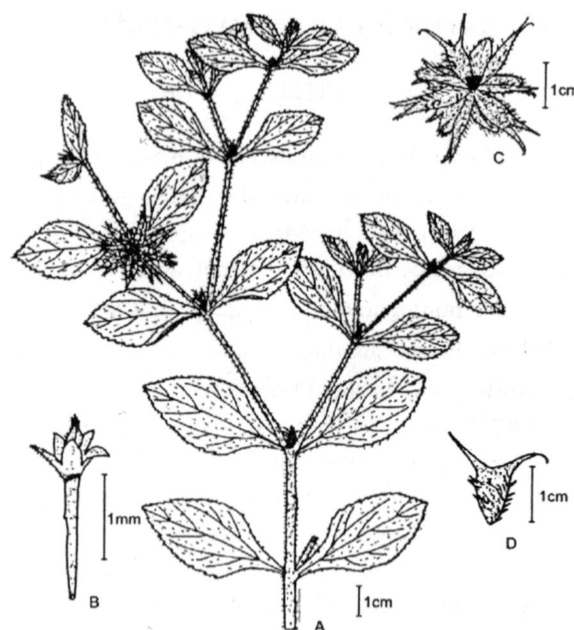

Fig. 75. *Acanthospermum hispidum* DC
A. Flowering & fruting twig, B. Flower, C. Achenes, D. Achene

Distribution: Native of South America; also introduced in Mexico, West Indies, Central America, Europe, Asia, Africa and Pacific Islands (Hawaii).

Uses: Fresh leaves are used as fodder to increase milk production in goats.

AGERATUM L.

Ageratum conyzoides L. Sp. Pl. 839. 1753; Hook.f., Fl. Brit. Ind. 2: 243 .1881; Duthie, Fl. Gangetic Plain 1: 443. 1905; Bhandari, Fl. Indian Desert 172. 1978; Singh, in Shetty & Singh Fl. Rajasthan 1: 391. 1987; Uniyal, in Hajra *et al*. Fl. India 12 : 248. 1995. (Fig. 76)

Simple to much branched, annual herb. Stems erect, 30 - 60 cm tall, hispid, branched, somewhat aromatic. Leaves opposite, petiolate, ovate or rhomboid-ovate, 4 - 9 x 2.5 - 6.5 cm, apex acute, base truncate to rounded, rarely cordate, margins serrate. Heads to 6 mm across, bluish purple, in dense small terminal corymbs of 8-15 heads. Involucreal, bracts 2-many, ca 3.5 seriate, linear, subequal, acute, sparsely pilose outside.

Flowers many (60-75), pale blue or white; corolla tubular, 1-1.5 mm long, limb 5-cleft; tube glandular. Achenes linear-oblong, up to 1.5 mm long, ribbed, short bristtle on the ribs, black, 5-angled; pappus of 5 short scales, scales often serrate basally, ending in a long awn.

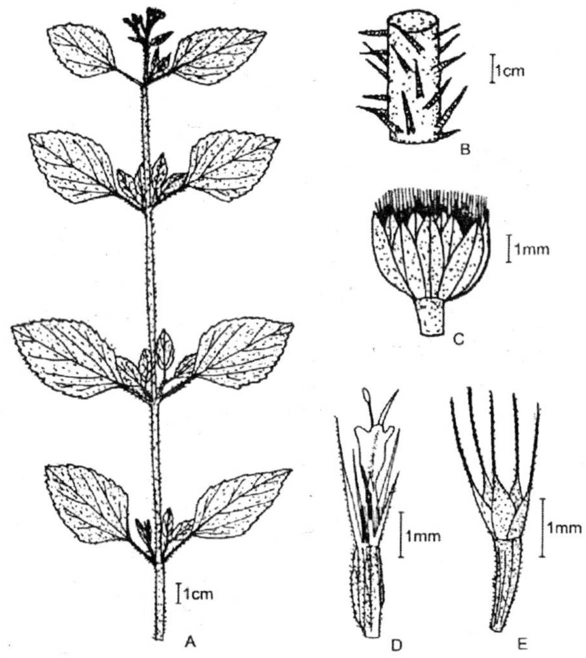

Fig. 76. *Ageratum conyzoides* L.
A. Flowering twig, B. Stem, C. Capitulum, D. Flower, E. Fruit

Field note: Common in wet and shaded habitats.

Fl. & Fr.: August - October.

Specimen examined: Mandal 315, Rupaheli 1568 (MLVGCB Herb).

Distribution: Pantropical.

AMBERBOA (Pers.) Less.

Amberboa ramosa (Roxb.) Jafri in Scientist 3: 29. 1958; Hajra, in Hajra et al. Fl. India 13: 153. f. 44. 1995. *Oligochaeta ramosa* (Roxb.) Wagenitz in Verroffent. Inst. 37: 323. 1962; Bhandari, Fl. Indian Desert 185. 1978; Singh, in Shetty & Singh Fl. Rajasthan 1: 425. 1987; *Centaurea divarcata* Wall. Cat. 2984. 1831. *Carduus ramosus* Roxb., Fl. Ind. 3: 407. 1832. *Volutarella divaricata* Benth., in Benth & Hook. *f.* Gen. Pl. 2 : 476. 1873, *pro parte;* Hook. *f.*, Fl. Brit. India 3: 383. 1881; Duthie, Fl. Gangetic Plain 1: 486. 1905. (Plate 154)

Procumbent annual herb, dichomously branched, up to 40 cm high. Leaves variable, 0.7 - 1 x 0.2 - 3.0 cm; lower leaves oblong or obovate or lyrate, upper ones sinuately pinnatifid, entire or toothed with mucronate. Heads 1.5 - 2.3 x 1 - 1.5 cm, ovoid-oblong, arachnously pubescent; peduncles grooved, glabrous or pubescent, scabrous, with a few, small, foliaceous bracts. Involucral-bracts 1.3 - 1.7 cm long, spinescent, elliptic-oblong; outer bracts smaller, recurved; inner bracts longer, erect. Ray florets tubular, neuter. Corolla pale purple, 1.2 - 1.5 cm long; lobes 3 mm long, linear, spinescent acute. Achenes 4 - 5 x 1.7 mm, acutely angled, often pitted between the angles, narrowed at base. Pappus hairs numerous, unequally, silvery-brown, upto 10 mm long; inner 3 or 4 flattened, upto 15 mm long.

Field note: Common on dry rockey as well as sandy soils throughout the area.

Fl. & Fr.: October-May.

Specimen examined: Bhilwara 111, Shakkargarh 1806 (MLVGCB Herb).

Distribution: India and Pakistan.

BIDENS L.

Bidens biternata (Lour.) Merr. & Sherff ex Shreff in Bot. Gaz. 88: 293. 1929; Bhandari, Fl. Indian Desert 172. 1978; Singh, in Shetty & Singh Fl. Rajasthan 1: 394. 1987; Chowdhey, in Hajra *et al.* Fl. India 13: 367. 1995. *Coreopsis biternata* Lour. Fl. Cochinch. ed. 1. 508. 1790. *Bidens pilosa auct. plur.* non L. 1753; Hook. *f.*, Fl. Brit. India 3: 309. 1881; Duthie, Fl. Gangetic Plain 1: 472. 1905.

Erect annual herb. Stem 30 - 90 cm long, subtetragonous, loosely crisped-pilose, branched. Leaves usually consisting of two sub-opposite pairs; median cauline leaves opposite, to 9 - 19 cm long (including petiole 5 cm long), densely pubescent on both surfaces, once (or twice) pinnatisect, terminal segments often ovate, apex short acuminate, margins distinctly dentate, lateral segments ovate, acute to obtuse, lower segments sometimes pinnatifid. Heads 7 - 10 mm long, peduncles 1.5 - 11 cm long. Involucral bracts 8 - 10, linear, 2 - 6.5 mm long, acute; outer palea oblong, 5 - 6 mm long, pilose; inner ones narrowly oblong to linear, glabrous. Ray florets, 1 - 5, ligulate sterile or female, corolla yellow, 5.5 mm long. Disc florets fertile (bisexual), corolla tubular, pubescent, 4 - 5.5 mm long. Achenes linear, 9 - 30 × 1 - 2 mm, tetragonous, short-strigose, awns 2 - 5 at apex. Pappus of 2 - 4, rigid, retrorsely hispid, awns 3 - 4 mm long.

Field note:Very common in forests, also observed in wet and shady places.

Fl. & Fr.: August- December.

Specimen examined: Mandalgarh 1586 (MLVGCB Herb).

Distribution: Asia, Malaysia, Africa, Fiji, India (Throughout) and Australia.

BLAINVILLEA Cass.

Blainvillea acmella (L.) Philipson, in Blumea 6: 350. 1950; Singh, in Shetty & Singh Fl. Rajasthan 1: 394. 1987; Choudhery, in Hajra *et al.* Fl. India 12: 377. 1995. *Verbesina acmella* L., Sp. Pl. 901. 1753. *Eclipta latifolia* L. f. Suppl. 378. 1781. *Blainvillea rhomboidea* Cass., in Dict. Sci Nat. 29: 494. 1823; Duthie, Fl. Gangetic Plain 1: 469. 1905 *B. latifolia* (L. f.) DC. ex Wight, Contr. Bot. Ind. 71. 1834; Hook. *f.*, Fl. Brit. India 3: 305. 1881 (Plate 155)

Erect, annual herb, up to 120 cm high. Stem clothed with white hairs. Leaves alternate to subopposite, ovate-lanceolate or rhomboid-ovate with cuneate or rounded base, apex acuminate, dentate serrate, 4 - 15 x 2 - 9 cm, hispid hairy on both surfaces. Heads white, soliatry, radiate, terminal or axillary or at the fork of the branches; peduncles 0.4 - 4 cm long. Involucral bracts foliaceous. Ray florets few, white, pistillate. Disc florets tubular, bisexual. Stamnes 5; anthers linear, base obtuse. Achenes triquetrous, transversely rugose; disc achenes laterally compressed, truncate at apex. Pappus setae 2 - 5, plumose.

Field note: Common in Waste lands, forest edges, gardens etc. throughout the area.

Fl. & Fr.: August - November.

Specimen examined: Mandal 1315, Hamirgarh 1876 (MLVGCB Herb).

Distribution: Africa, America, China, India and Sri Lanka.

BLUMEA DC.

KEY TO THE SPECIES

1a. Leaves sessile or subsessile, laciniate. Corolla yellow or rarely pinkish .... *.B. lacera*

1b. Leaves distinctly petiolate. Corolla lobes of disc florets purplish .... *.B. mollis*

Blumea lacera (Burm. f.) DC. in Wight, Contrib. Bot. Ind. 14. 1834; Hook. *f.*, Fl. Brit. India 3: 309. 1881; Duthie, Fl. Gangetic Plain 1: 472. 1905; Mahesh., Illus. Fl. Delhi f. 106. 1905; Randeria, in Blumea 10: 264. 1960; Rao *et al.*, in Fl. India Enum. Aster. 16. 1988; Bhandari, Fl. Indian Desert 173. 1978; Singh, in Shetty & Singh Fl. Rajasthan 1: 396. 1987; Kumar, in Hajra *et al.* Fl. India 13: 128. 1995. *Conyza lacera* Burm. f. Fl. Ind. 180. t. 59. f. 1. 1768. *Blumea subcapitata* DC. Prodr., 5: 439. 1836. *B. lacera* DC. var. *cinerascens* Hook. *f.*, *l.c.* 3: 263. 1881. *B. lacera* DC. var. *glandulosa* Hook. *f.*, *l.c.* 3: 263. 1881

Erect, aromatic, annual herb, 70 - 100 cm tall. Stems simple or more often branched, terete, densely villous with silky hairs or glandular trichomes. Leaves ovate, oblong or elliptic to oblong, sessile or petiolate, 2.0 - 6.0 × 1 - 2.5 cm, apex obtuse, base attenuate, margins doubly serrate or sometimes slightly variously lyrately lobed, tomentose on upper surface, woolly on lower surface, veins 5 - 8 pairs. Heads 5 - 8 x 2.5 - 3.5 mm across, in axillary and terminal dense to lax panicles. Involucral bracts in 2 or 3 series, outer ones shorter, linear-lanceolate, 1.5 - 2.5 × 0.2 - 0.3 mm, densely villous with multicellular and glandular hairs; inner ones linear, 4 - 5 x 0.2 - 0.3 mm, margins ciliate. Receptacle 2 - 3 mm in diam., convex, alveolate, glabrous. Central florets yellowish, 6 - 7 mm long, with 5 triangular lobes,

lobes with sessile glands and multicellular hairs. Outer florets 5 mm long, 2 - 5 lobed, limb glandular. Achenes oblong, sparsely hirsute. Pappus white, 4 mm long.

Field note: Common in Waste lands, forest edges, gardens etc. throughout the area.

Fl. & Fr.: January - May.

Specimen examined: Bhilwara 541, Kotri 1808 (MLVGCB Herb).

Distribution: Southeastern Africa, South-Eastern Asia and Northern Australia. Taiwan, in Waste lands at lower elevation.

Notes: The species is very closely related with *B. mollis* in many characters.

Blumea mollis (D. Don) Merr. Philipp. J. Sci. (Bot.) 5: 395. 1910; Singh, in Shetty & Singh Fl. Rajasthan 1: 398. 1987; Kumar, in Hajra *et al.* Fl. India 13: 135 - 137. f. 41. 1995. *Erigeron molle* D. Don., Prodr. Fl. Nepal. 192. 1825. *Blumea wightiana* DC. in Wight, Contrib. Bot. Ind. 14. 1834; Hook. *f.*, Fl. Brit. India 3: 261. 1881; Duthie, Fl. Gangetic Plain 1: 453. 1905. *B. neilgherrensis* Hook. *f.*, *l.c.* 3: 261. 1881. (Plate 156)

Annual herb, erect, 5 - 90 cm tall. Stems terete, soft-pilose with white spreading hairs and glands, viscid-tomentose particularly on younger parts and on inflorescence axis. Leaves ovate-oblong, 1.0 - 11 × 0.5 - 5.0 cm, apex acute to apiculate, base attenuate, margins closely doubly serrate, densely silky-villous on both surfaces with soft multicellular hairs and glands, veins 5 - 7 pairs. Heads in terminal, narrow, compact, spiciform panicle, subsessile to pedunculate. Peduncle up to 7 mm, densely villous. Involucral bracts in 3 or 4 series, outer ones shorter, linear-lanceolate, 3 - 4 × 0.6 - 0.8 mm, densely soft-villous and with multicellular hairs and glands; inner ones linear, 5 - 6 × 0.2 - 0.3 mm, margins scarious, ciliate. Central florets pinkish-purple apically, whitish or yellowish at base, 3.5 - 4.5 mm long, with 5 triangular lobes, lobes pubescent and with sessile glands. Outer florets 3.5 - 4.5 mm long, 2- or 3-lobed, glandular. Receptacle convex, alveolate, glabrous, 1.8 - 2 mm across. Achenes brown, oblong, 0.8 × 0.2 - 0.3 mm, shining, pubescent. Pappus white, 2 - 4 mm long.

Field note: Common in Waste lands throughout the area.

Fl. & Fr.: February-May.

Specimen examined: Bhilwara 1310, Asind 1643 (MLVGCB Herb).

Distribution: Africa, southern and south-eastern Asia, and China. Taiwan, particularly Waste lands at low elevations in southern Taiwan.

CAESULIA Roxb.

Caesulia axillaris Roxb. Pl. Cor. 1 : 64. t. 93. 1798; Hook. *f.*, Fl. Brit. India 3: 291. 1881; Duthie, Fl. Gangetic Plain 1: 462. 1905; Mahesh., Illus. Fl. Delhi f. 106. 1905; Bhandari, Fl. Indian Desert 174. 1978; Singh, in Shetty & Singh Fl. Rajasthan 1: 400. 1987; Pant, in Hajra *et al.* Fl. India 13: 2. f. 1. 1995. (Plate 157)

Erect or decumbent, succulent, annual herb, up to 60 cm high, often rooting at the lower nodes. Stem stout, branches 1.5 - 4 cm long, glabrous, closely striate and minutely, longitudinally streaked. Leaves 2 - 16 x 0.5 - 2.5 cm, alternate, sessile, lanceolate, distantly serrulate, glabrous, auricled at base and ultimately dilated, scarious; midrib raised below, acute. Heads many, connate in a compound sessile head in the axils of dilated leaf-sheaths, blue or white. Flowers sessile, on a convex, common receptacle, irregularly mixed with linear, acute bracteoles; bracts 2, orbicular or broadly ovate, crenulate or dentate with a dorsal apiculus near the apex. Corolla tubular, blue or white, ca 2 mm long, 5-lobed. Anthers base sagittate. Pappus scales 2, ovate, keeled. Achenes 1.5 x 1 mm, obovate, notched at the tip, winged on the margins, hairy on flat faces, 1-nerved in the middle. Pappus-scales ovate-lanceolate, 2-ribbed, apiculate.

Field note:Very common in shallow ditches as well as damp places, throughout the area.

Fl. & Fr.: August - November.

Specimen examined: Bhilwara 1247, Mezadam 1819 (MLVGCB Herb).

Distribution: India, Pakistan, Nepal, Bangladesh, Myanmar and East Indies.

CARTHAMUS L.

Carthamus oxycantha Bieb. Fl. Taur. Cauc. 2: 283. 1808; Hook. *f.,* Fl. Brit. India 3: 386. 1881; Mahesh., Illus. Fl. Delhi f. 112. 1905; Raizada, Suppl. Duthie, Fl. Gangetic Plain 106. 1966; Bhandari, Fl. Indian Desert 175. 1978; Singh, in Shetty & Singh Fl. Rajasthan 1: 400. 1987; Hajra, in Hajra *et al.* Fl. India 13: 162. 1995.

Spiny prenicious annual, herb. Branches 30 - 40 cm high, corymbose, glabrous, white. Leaves 1.5-4.5 cm long, oblong or oblong-lanceolate, semi-amplexicaul, spinulose-toothed, very spinous. Heads 1.5-2.5 cm long, orange-yellow, homogamous. Outer involucral bracts exceeding the heads, white below the contracted portion, green above, with yellow spines, many-seriate, foliaceous; inner involucral bracts dry; receptacles flat densely bristly; Staminal filaments usually hairy at the middle; anther-bases sagittate, auricles connate; tails short, fimbriate. Achenes obovoid, 4-angled, compressed, smooth, polished, truncate at the apex. Pappus absent.

Field note: Rare, weed of crop fields.

Fl. & Fr.: January - May.

Specimen examined: Hamirgarh 1234, Ladpura 1854 (MLVGCB Herb).

Distribution: India and Pakistan.

Uses: Oil extracted from seeds is used for dressing ulcers and as a remedy of itching.

CHRYSANTHELLUM Rich.

Chrysanthellum americanum (L.) Vatke in Abli, Nat. Ver. Brem. 9: 122. 1887; Singh, in Shetty & Singh Fl. Rajasthan 1: 402. 1987; Chowdhery, in Hajra *et al.* Fl. India 13: 379 - 381. f. 102. 1995. *Artemisia americana* L. Sp. Pl. 895. 1753. *Chrysanthellum indicum* DC. Prodr. 5: 631. 1836; Hook. *f.,* Fl. Brit. India 3: 310. 1881; Duthie, Fl. Gangetic Plain 1: 473. 1905.

Erect or decumbent, annual herb, upto 25 cm high. Stem rigid, grooved, 4-angular. Lower leaves petioled, upper ones sessile, 2 - 7 cm long, 1 to 2-pinnatifid; segments 2.5-4.0 x 1.2 - 1.5 mm, lanceolate-oblong, succulent, mucronate, glabrous. Heads solitary, axillary and terminal, on long peduncles, 5 - 8 mm in diam., enlarging in fruit. Involucral bracts 1 - 3-seriate, 2 - 5 x 1 - 1.5 mm, few, lanceolate, obtuse. Ray florets few, female; corolla yellowish-white; style 2-fid, ligule 3-dentate. Disc floret many, bisexual; corolla tubular, Achenes 2.5 x 1-1.5 mm, oblong-obovate, dorsally compressed, with thickened and minute, crown-shaped pappus.

Field note: Common in shallow ditches and damp places, throughout the area.

Fl. & Fr.: August - October.

Specimen examined: Tikar 1193, Banera 1766 (MLVGCB Herb).

Distribution: India, Medagaskar and Tropical Africa

CICHORIUM L.

Cichorium intybus L., Sp. Pl. 813. 1753; Hook. *f.,* Fl. Brit. India 3: 391. 1881; Raizada, Suppl. Duthie, Fl. Gangetic Plain 107. 1976; Singh, in Shetty & Singh Fl. Rajasthan 1: 403. 1987; Mamgain & Rao, in Hajra *et al.* Fl. India 13: 248. f. 60. 1995.

Vern.: *Kasni, Papro.*

Perennial herb with a long fleshy deep taproot. Stem erect, grooved, 30 - 110 cm tall, with stiff spreading-ascending branches, hirsute or nearly glabrous. Lower leaves variable, oblong, oblong-lanceolate or oblanceolate, petiolate, toothed or more often pinnatifid, 5 - 25 × 1 - 5.6 cm, becoming reduced, sessile, entire or toothed upwards, glabrous or sparsely hairy. Heads 1 - 1.5 x 0.5-0.8 cm, terminal on somewhat thickened stalk and in subsessile clusters of 2 or 3 in axils of upper leaves. Flowers large, bright blue. Involucral bracts ovate, ovate-lanceolate, green; outer bracts ca 8; inner bracts linear-lanceolate, 12 - 14 x 1 - 2mm. Achenes 2 - 3 mm long, angled, many ribbed, smooth, broadly rounded on back, base contracted. Pappus of many minute, 2 to 3-seriate, fimbriate scales, pale.

Field note: Common weed of fodder crop (*Medicago sativa* L.) during winter and early summers.

Fl. & Fr.: February- January.

Specimen examined: Karoi 2159 (MLVGCB Herb).

Distribution: Native of N. Europe, widely naturalized in North America and Taiwan. Introduced concomitantly with seeds of *Trifolium alexandrinum* L. and naturalized in pastures at low elevations (Peng, 1978).

COTULA L.

Cotula anthemoides L., Sp. Pl. 891. 1753; Hook. *f.,* Fl. Brit. India 3: 316. 1881; Duthie, Fl. Gangetic Plain 1: 476. 1905; Singh, in Shetty & Singh Fl. Rajasthan 1: 406. 1987; Naithani, in Hajra *et al.* Fl. India 13: 52 - 53. 1995.

Prostrate or decumbent, ascending, much branched, glabrous, upto 20 cm long, annual herb. Leaves 1.0-5.0 cm long, 2-pinnatisect or pinnatifid; segments linear, lanceolate, sharply acute, entire, glabrous. Petiole half amplexicaul Heads solitary on 0.7-1.0 cm long peduncles, 3-4 mm in diam, yellow. Involucral bracts 2-seriate, narrowly-oblong, obtuse, with a glandular brown midrib. Receptacle naked, tubercled. Ray florets winged, fertile; corolla minute. Disk florets not winged, hermaphrodite, fertile, tubular, 2 x 1 mm; corolla, broadly campunulate, ovate with thick narrow wings, that of ray florets broad, flat. Achenes of ray-florets 1.0 - 1.2 mm long,

Field note: Occassinally in fields, and in wet places.

Fl. & Fr.: November - May.

Specimen examined: Dabla 226, Dunwala 1809 (MLVGCB Herb).

Distribution: Cosmopolitan.

DICOMA Cass.

Dicoma tomentosa Cass. in Bull. Sci. Soc. Philom. Paris 12. 1818; Hook. *f.,* Fl. Brit. India 3: 387. 1881; Duthie, Fl. Gangetic Plain 1: 487. 1905; Bhandari, Fl. Indian Desert 176. 1978; Singh, in Shetty & Singh Fl. Rajasthan 1: 408. 1987; Rao, in Hajra *et al.* Fl. India 13: 171. f. 53. 1995. *D. lanuginose* DC. in Wight, Contrib. Bot. Ind. 26. 1834. (Plate 158)

Vern.: *Vajaradanti.*

Erect or decumbent, annual herb, 10 - 40 cm high, much branched, clothed with white wooly pubescence. Stems terete, covered in whitish felt-like hairs, older branches often tinged purple. Leaves alternate, 2 - 7 x 0.5 - 1.0 cm, linear-lanceolate or linear-obovate, entire, denticulate, clothed with thick, cottony wool below, less so above. Head sub-sessile, axillary, with long, involucral bracts much-seriate, bristly, outer up to 1 cm, spreading; inner 14 - 15 mm long, erect, all lanceolate, appearing arachnoid outside, glabrous inside, tapering into yellow, apical, scabrous spines. Corolla white, up to 6.5 mm long; tube ca 3.5 mm long, more or less spreading, shortly barbellate at apex outside. Anther bases sagittate, tail longer than stalk, barbellate. Achenes 3 x 2 mm, ribbed, densly silky truncate at top. pappus 6 - 7 mm long, duplex; outer bristle slender, barbellate, tawny; inner pappus slightly longer than outer of scarious aristate paleae, strongly nerved, with a rigid midrib.

Field note: Common on gravel or on sandy soils.

Fl. & Fr.: August- March.

Specimen examined: Atun 1173, Bhilwara 1764 (MLVGCB Herb).

Distribution: Tropical Africa, India and Pakistan.

Uses: Fresh roots are kept in mouth, to cure toothache.

ECHINOPS L.

Echinops echinatus Roxb., Fl. Ind. 3 : 447. 1832; Hook. *f.,* Fl. Brit. India 3: 358. 1881; Duthie, Fl. Gangetic Plain 1: 480. 1905; Bhandari, Fl. Indian Desert 176. 1978; Singh, in Shetty & Singh Fl. Rajasthan 1: 408. 1987; Hajra, in Hajra *et al.* Fl. India 13: 178. 1995. (Plate 159)

Erect 30 - 70 cm tall, erect, much branched herb. Leaves 7 - 15 x 2 - 6 cm deeply pinnatifid, amplexicaul at base, midrib raised below; lobes sinuate and spinescent, spines pale, 2 - 3 cm long, glabrous. Heads homogenous, white or violet, campact globose. Involucre with white bristles; middle bract with a sharp spine exceeding the compound head, inner connate for more than half their length into a tube which surrounds the achene. Flowers bisexual. Corolla upto 13 mm long, tube white, narrow; limb as long as tube, pale pink, linear, acute. Anthers ca 1 mm long, ashy-black, anther tails fimbriate. Pappus short, cylindric, forming a 3 mm long, antrorsely hispid, yellowish, cylindrical brush above the achene. Achene obconic, ca 4 mm long, densely villous.

Field note: Common in Waste lands and open forests specially in dry habitats.

Fl. & Fr.: October - June.

Specimen examined: Ajmer 3184, Shahpura 1179 (MLVGCB Herb).

Distribution: India, Afghanistan and Pakistan.

Uses: Pieces of 2 inches size of its fresh root kept at the back of head touching scalpel or coil the fresh root and kept in the naval part before parturition time or during delivery pains for easy delivery in human beings. The Root extract about 1 tsp is given orally twice a day with lukewarm water for a week in whooping cough.

ECLIPTA L. nom. cons.

Eclipta prostrata (L.) L. Mant. Pl. 2: 286. 1771; Chaudhery, *et al.*, in Fl. India 12: 381. f. 103. 1995. *Verbesina prostrata* L. Sp. Pl. 902. 1753. *V. alba* L. Sp. Pl. 902. 1753. *Eclipta erecta* L. Mant. Pl. 2: 286. 1771; Duthie, Fl. Gangetic Plain 1: 468. 1905. *E. alba* (L.) Hassk. Pl. Jav. Rar. 528. 1848; Hook. *f.,* Fl. Brit. India 3: 304. 1881; Singh, in Shetty & Singh Fl. Rajasthan 1: 409. 1987 .

Vern.: *Bhringraj.*

Erect or prostrate, hirsute herb. Stems erect, ascending or prostrate, to 60 cm tall, appressed hairy, often rooting at the lower nodes. Leaves simple, opposite, subsessile, 1.5 - 5 × 0.5 - 2.5 cm, ovate, elliptic-oblong or lanceolate, entire or undulate, apical mucronate, cuneate at base, serrate. Heads solitary, on solitary or paired, unequal, axillary or terminal peduncles, 0.5-1.0 cm in diam., white. Involucral bracts in 2-series, ovate-lanceolate, acute or short acuminate. Ray in 2-3 series with dentate, florets female or sterile. Disc florets tubular bisexual; corolla campanulate, tube short, 4-lobed; lobes ciliate at margins. Stamens 4 - 5, sagittate. Style branches acute, pubescent. Achenes dark brown, oblong-tuberculate, dorsiventrally compressed and sharply angled, the apical ridge with few small hairs. Pappus a ring of thick, ciliate, partially or completely united scales forming cone on the top of the achene.

Field note: Common weed of marshy habitats throughout the area.

Fl. & Fr.: Almost throughout the year.

Specimen examined: Bhilwara 110, Shahpura 1179 (MLVGCB Herb).

Distribution: Warmer parts of America, Africa, Asia, Australia and India.

Uses: Leaves are used as a hair tonic.

GLOSSOCARDIA Cass.

Glossocardia bosvallea (L. f.) DC., in Wight Contrib. Bot. Ind. 19. 1834; Singh, in Shetty & Singh Fl. Rajasthan 1: 413. 1987; Choudhery, in Hajra *et al.* Fl. India 12: 391. f. 107. 1995. *Verbesina bosvallea* L. f. Suppl. 379. 1781. *Glossocardia linearifolia* Cass. in Dict. Sci. Nat. 19 : 62. 1821; Hook. *f.,* Fl. Brit. India 3: 308. 1881; Duthie, Fl. Gangetic Plain 1: 471. 1905. *G. setosa* Blatt. & Hallb. in J. Bombay. Nat. Hist. Soc. 26 (2): 536. 1919; Santapau in J. Bomb. Nat. Hist. Soc. 56(2): 280. 1959; Bhandari, Fl. Indian Desert 178. 1978. (Plate 160)

Vern.: *Chidiyon ka chugga.*

Diffuse, up to 15 cm, usually prostrate, diffused, glabrous, annual herb. Stem teret, more or less sulcate or obscurely grooved, glabrous. Leaves 1 - 5 cm long, alternate, once or twice pinnatisect; segments linear, apiculate, glabrous; petioles 0.5 - 3 cm long, very variable in length, base dilated with a few hairs on inner surface of dilations. Heads radiate, pedunculate, solitary, axillary, ca 10 x 5 mm in diam., yellow. Involucral-bracts 2 to 4-seriate; the outer oblong-lanceolate, 3 - 3.5 x 1 - 1.5 mm, acute, with scarious margins; inner many, oblong, obtuse, longer. Ray-florets ligulate, female, corolla bilobed or trilobed, 3 - 3.5 mm; style slightly bifid, 2 mm long, stigma obtuse. Disk-florets tubular, bisexual; corolla, 2 - 2.5 mm long; lobes ovate, acute; stamens 4; anthers oblong; style 2 - 2.5 mm; stigma puberulous, acute. Achenes 5 - 8 mm long, oblong, flattened, hairy, dark-brown. Pappus of 2, stiff awns, 3 - 4 mm long.

Field note: Common in open dry, grasslands, on hills as well as rocky or gravel habitats.

Fl. & Fr.: July- October.

Specimen examined: Chittorgarh 1736, Dhanop 1753 (MLVGCB Herb).

Distribution: India.

Notes: Bhandari (1978) recorded this species as an endemic to N.W. Rajasthan, but this species is common in other parts of Rajasthan as well as also reported from Madhya Pradesh (Samvatsar, 1996).

GNAPHALIUM L.

KEY TO THE SPECIES

1a. Heads in leafless corymbose clusters ....*G. luteo-album*

1b. Heads in leafy spikes or clusters.................. .2

2a. Prostrate, diffusely branched herb. Heads in axillary or terminal clusters. Stamens striate........................ .*G. pulvinatum*

2b. Erect herb. Heads in leafy spikes. Stamens not as above ....*G. polycaulon*

Gnaphalium luteo-album L. Sp. Pl. 851. 1753; Hook. *f.*, Fl. Brit. India 3: 288. 1881; Duthie, Fl. Gangetic Plain 1: 461. 1905; Bhandari, Fl. Indian Desert 180. 1978; Singh, in Shetty & Singh Fl. Rajasthan 1: 415. 1987; Pant, in Hajra *et al.* Fl. India 13: 87. 1995. *G. luteo-album* L. var. *pallidum* Hook. *f. l.c.* 3 : 288. 1881. *G. luteo-album* L. subsp. *pallidum* (Lam.) Mahesh., in J. Bombay Nat. Hist. Soc. 57. (2): 377. 1960.

subsp. ***luteo-album.***

Erect, ascending or decumbent, up to 20 cm high, softly wooly herb. Leaves 1.5 - 6.5 x 0.2- 1.7 cm, sessile, linear, oblong, oblanceolate, apiculate, woolly on both surfaces, 1-nerved. Flowers white to yellow in dense leafless corymbose heads, 3 - 4 mm across; peduncles ca 2 mm long, densely woolly-tomentose. Involucral-bracts 2 to 3-seriate, ovate-oblong, obtuse, shining, pale-brownish. Ray florets outer, female, ca 2 mm long, many seriate, filiform. Disc floret inner, bisexual, ca 2 mm long, 5-tothed. Receptacles pitted. Achenes oblong-ellipsoid, papillose, brown. Pappus-hairs shining, white, shorter than involucre.

Field note: Common in moist places, throughout the area.

Fl. & Fr.: March - April.

Specimen examined: Karoi 1236 (MLVGCB Herb).

Distribution: Pantropical.

Gnaphalium polycaulon Pers. Syn. 2 : 421. 1807; Bhandari, Fl. Indian Desert 180. 1978; Singh, in Shetty & Singh Fl. Rajasthan 1: 416. 1987; Pant, in Hajra *et al.* Fl. India 13: 91. 1995. *G. indicum auct. plur. non* L. 1753; Hook. *f.*, Fl. Brit. India 3: 289. 1881; Duthie, Fl. Gangetic Plain 1: 461. 1905. *G. strictum* Roxb., Fl. Ind. 3: 424. 1832. *G. multicaule* Roxb., Fl. Ind. 3: 425. 1832.

Prostrate or sub-erect, spreading annual herb with numerous, lateral ascending branches from near the base. Leaves 2.0 - 5.5 x 0.3 - 0.8 cm, linear-oblanceolate to obovate or spathulate, sub-acute at the apex, apiculate, upper ones sessile; lower usually shortly, petioled, woolly. Heads 3 mm in diam., sessile, passing into terminal, leafy spikes, yellowish white. Involucral bracts 2 to 3-seriate, 2.0 - 2.5 mm long, erect, linear-oblong or outer ovate-lanceolate, acute, usually green below, and shining yellow-brown above, outer wooly; inner nearly glabrous. Ray florets female, fertile, with filiform corolla, ca 1 mm long, obscurely toothed. Disc florets female, with corolla ca 1 mm long, 5 - toothed. Receptacles pitted. Achenes ca 0.5 mm long, oblong, hispidulous. Pappus hairs white, 1-seriate, 1.5 mm long, free.

Field note: Common in damp, sandy soils and rocky grounds throughout the area.

Fl. & Fr.: December-April.

Specimen examined: Karoi 1176, Mandalgarh 1860 (MLVGCB Herb).

Distribution: Africa, Australia, Nepal, Myanmar, Senagal, Brazil, China, Burma, Japan, India and Pakistan.

Gnaphalium pulvinatum Del., Fl. Aegypt. 122. t. 44. f. 1812; Hook. *f.,* Fl. Brit. India 3: 289. 1881; Duthie, Fl. Gangetic Plain 1: 462. 1905; Bhandari, Fl. Indian Desert 180. f. 67 (a-f). 1978; Singh, in Shetty & Singh Fl. Rajasthan 1: 416. 1987; Pant, in Hajra *et al*. Fl. India 13: 92. 1995. (Plate 161)

Diffuse, prostrate woolly, 2 - 3 cm long, annual herb. Leaves dull green, 0.4 - 1.5 x 0.1 - 1.0 cm, spathulate, narrowly attenuate at base, mucronate, hairy above and densely white wooly beneath. Heads crowed in subglobose, leafy or involucrate, terminal clusters, yellow, emerged in white wool; peduncle very short. Involucral bracts greenish tinged, narrowly elliptic-linear, membranous, tips acute. Ray florets female, fertile; corolla tube filiform, narrow; style bifid. Disc florets bisexual, few, fertile or sterile; corolla ca 1 mm long. Achens 0.5 mm long, oblong-obovate, minutely papilose. Pappus hairs 2 mm long, white.

Field note: Common in drying beds of ponds throughout the area.

Fl. & Fr.: December - April.

Specimen examined: Karoi 540, Jahajpur 1145 (MLVGCB Herb).

Distribution: India, China, Egypt and Pakistan.

GRANGEA Adans.

Grangea maderaspatana (L.) Poir. in Lam. Encycl. Suppl. 2 : 825. 1812; Hook. *f.,* Fl. Brit. India 3: 247. 1881; Duthie, Fl. Gangetic Plain 1: 446. 1905; Bhandari, Fl. Indian Desert 181. 1978; Singh, in Shetty & Singh Fl. Rajasthan 1: 417. 1987; Hajra, in Hajra *et al.* Fl. India 12: 127. f. 37. 1995. *Artemisia madraspatana* L. Sp. Pl. 849. 1753. (Plate 162)

Prostrate or procumbent annual herb, glandular pubescent, 11 - 65 cm high. Branches spreading Leaves alternate, many, sessile sinuately pinnatifid, 2 - 8 x 0.5 - 1.5 cm, coarsely serrate-dentate, sparsely hairy on both surfaces. Heads globose, solitary or paired on leaf-opposed or terminal peduncles, 5 - 24 mm in diam., yellow. Involucral-bracts 1.5 mm long, oblong, rigid, obtuse hairy. Outer florets pistillate, one to many seriate; corolla tube of ray florets filiform, 2 to 4-fid at the apex. Disc florets bisexual; corolla tube slender; limb 4 - 5 fid at apex. Receptacles convex or conical, naked. Achenes 2 mm long, turbinate, compressed. Pappus form a short tube with fimbriate mouth.

Field note: Common in drying beds of ponds throughout the area.

Fl. & Fr.: Almost throughout the year.

Specimen examined: Pansal 1189, Shahpura 11730 (MLVGCB Herb).

Distribution: Tropical and sub-Tropical Asia, Africa, Sri Lanka, India and Pakistan.

LAUNAEA Cass.

Launaea procumbens (Roxb.) Ramayya & Rajagopal in Kew Bull. 23 : 465. 1969; Bhandari, Fl. Indian Desert 182. 1978; Singh, in Shetty & Singh Fl. Rajasthan 1: 423. 1987; Mamgain & Rao, in Hajra *et al*. Fl. India 12: 303. 1995. *Chondrilla nudicaulis* L. Mant. 273. 1767. *Prenanthes procumbens* Roxb., Fl. Ind. 3: 404. 1832. *Launaea nudicaulis* (L.) Hook. *f.,* Fl. Brit. India 3: 416. 1881 non Less. 1832; Duthie, Fl. Gangetic Plain 1: 494. 1905. Mahesh., Illus. Fl. Delhi f. 99. 1905. (Plate 163)

Annual or biennial, glabrous, herb, up to 30 cm high. Stem prostrate, with creeping stolons. Leaves cauline, in rosette, upto 18 x 4 cm, obovate, oblong, pinnatifid, sinuate-lobed or lyrato-pinnatifid, with round or very obtuse segments, spinnulose or denticulate on margins; denticulations cartilaginous, white on drying; cauline leaves distant, few, narrowly-oblong. Heads 5 - 8, yellow, shortly pedicelled or sessile, in terminal and sub-racemose clusters or solitary. Involucral-bracts 3-seriate, outer ovate, inner ones lanceolate, glabrous. Pappus hairs subequal, soft, white, copious. Stigma longer than the corolla tube, 2-fid, bluish tinged. Achenes 2.5 - 3.5 mm long, polymorphous, inner as if composed of 4, thick ribs, outer slightly curved and flattened with a thick ventral and several thick dorsal ribs, all smooth or obscurely uneven. Pappus up to 6 mm long, simple, white, deciduous.

Field note: Common in Waste lands throughout the area.

Fl. & Fr.: September - April.

Specimen examined: Bhilwara 1167, Sarsiya 1818 (MLVGCB Herb).

Distribution: Afghanistan, India and Pakistan.

PARTHENIUM L.

Parthenium hysterophorus L., Sp. Pl. 988. 1753; Rao, in J. Bombay Nat. Hist. Soc. 54: 218. 1956; Raizada, Suppl. Duthie, Fl. Gangetic Plain 127. 1976; Singh, in Shetty & Singh Fl. Rajasthan 1: 425. 1987; Chowdhery, in Hajra *et al*. Fl. India 12: 403. f.113. 1995.

Vern.: *Gajar ghas.*

Herbaceous aromatic, annual, erect, up to 1 m high. Stem 30 - 100 (-150) cm tall, erect, unbranched at base, much branched apically, puberulous, glandular. Leaves alternate, cauline, initially forming a basal rosette, but this wanting at anthesis, lanceolate; smaller upward, deeply pinnately dissected or bipinnatifid; larger toward base, up to 18 × 10 cm, apex acute, puberulous and glandular on both surfaces, petiolate. Inflorescences corymbose cymes, terminal and axillary, peduncle up to 1 cm long, densely yellowish puberulous. Heads radiate, 3 - 5 mm across, numerous, white. Involucral bracts 2 - seriate; outer involucral bracts 5, narrowly ovate, acute, 2.0 - 2.5 × 1.2 - 1.4 mm, puberulous; inner bracts 5, obovate, thin, transparent, 1.5 - 1.8 × 2 - 2.2 mm wide, margins glandular, subtending a ray floret with 2 disc florets on either side. Ray florets 5, with 2 - seriate disc florets on either side; corolla white to light yellow, ovate-orbicular, bilobed, obtuse. Style 1 mm long with stout, obtuse stigma. Disc florets many; corolla light yellow, 5-lobed, 1.2 - 2 mm long, ovate-acute. Stamens 5; anthers linear. Style 1-5 mm, puberulous, entire. Achenes 2 mm long, flattened. Pappus of 2 broad, strongly reflected awns, 0.3 - 0.5 mm long, puberulous along the sides.

Field note: Abundant in the Waste lands throughout the area.

Fl. & Fr.: July-November.

Specimen examined: Bhilwara 225, Ajmer 1410 (MLVGCB Herb).

Distribution: Native of America; Widely naturalized in India.

PULICARIA Gaerrtn.

KEY TO THE SPECIES

1a. Stem not cottony-tomentose, leaves not crisped, achenes hairy. Pappus hairs not thickened at apex .... *P. angustifolia*

1b. Stem cottony-tomentose, leaves crisped, achenes glabrous. Pappus hairs thickened at apex .... *P. crispa*

Pulicaria angustifolia DC. Prodr. 5: 479. 1836; Hook. *f.*, Fl. Brit. India 3: 299. 1881; Duthie, Fl. Gangetic Plain 1: 465. 1905; Bhandari, Fl. Indian Desert 187. 1978; Singh, in Shetty & Singh Fl. Rajasthan 1: 429. 1987; Kumar, in Hajra *et al*. Fl. India 13: 33. 1995. (Plate 164)

Erect, annual, herb, upto 60 cm high, with woody root stock. Stem pubescent, branched from the base, ascending, pubescent divergent. Leaves 1 - 5 x 0.3 - 1. 5 cm. sessile, linear, oblong or obovate, base narrowed, entire or serrulate, glandular-pubescent on both surfaces. Heads solitary, terminal, peduncled, 8 - 25 mm in diam., yellow. Outer Involucral-bracts 2 mm long, linear-oblong, glandular, subobtuse, pubescent, green; inner ones 3 mm long, scarious, acute, with few hairs near apex. Ray florets ligulate; ligules 7 mm long. Pappus 2-seriate; outer row a setulose-laciniate cup; inner row of white, brabellate hairs, the latter twice as long as achenes. Achenes 1.5 mm long, teret or scarcely quadrangular, with a few minute, scattered hairs. Outer row of pappus forming a setulose-laciniate cup; inner row of white barbellate hairs, up to 3 mm long.

Field note: Common in Waste lands, on hillocks etc. throughout the area.

Fl. & Fr.: October - February.

Specimen examined: Atun 316 (MLVGCB Herb).

Distribution: India and Pakistan.

Pulicaria crispa (Forssk.) Benth & Hook. *f.* Gen. Pl. 2 : 336. 1867 & Hook. *f.,* Fl. Brit. India 3: 299. 1881; Duthie, Fl. Gangetic Plain 1: 466. 1905. Mahesh., Illus. Fl. Delhi f. 103. 1905; Bhandari, Fl. Indian Desert 187. 1978; Singh, in Shetty & Singh Fl. Rajasthan 1: 430. 1987; Kumar, in Hajra *et al.* Fl. India 13: 33. f. 11. 1995. *Aster crispus* Forsskal, in Fl. Ageypt.-Arab. 150. 1775. (Plate 165)

Erect perennial herb or undershrubs whitish-tomentose, much branched. Stem 30 - 70 cm high, leafy throughout, whitish tomentose. Leaves alternate; upper leaves small, sessile, linear-lanceolate, auriculate, margins recurved, obtuse, 1 - 4 x 0.2 - 0.9 cm, undulate-crisped or denticulate. Heads hemispherical, 5 - 16 mm in diam., many-flowered, solitary at the end of paniculate branches, radiate, yellow. Involucral-bracts multiseriate, 2 - 3 mm long, linear, acute, thinly wooly, glandular, recurved, scarious-margined, pubescent outside. Disc florets 4 - 5 dentate. Receptacle punctuate, 3 - 4 mm in diam., Achenes glabrous. Pappus sub-uniseriate, with a few, shorter setae, subplumosely scabrid above.

Field note: Common in sandy, gravel habitats specially in western part of the area.

Fl. & Fr.: September - March.

Specimen examined: Ajmer 1410, Asind 1232 (MLVGCB Herb).

Distribution: Trop. Africa, Arabia, India, Pakistan, Algeria and Europe.

SONCHUS L.

KEY TO THE SPECIES

1a. Leaves with obtuse auricles. Mature achenes not transversely rugulose..... .*S. asper*

1b. Leaves with acute auricles. Mature achenes transversely rugulose..... .*S. oleraceus*

Sonchus asper (L.) Hill, Herbs. Brit. 1: 47. 1769; Hook. *f.,* Fl. Brit. India 3: 414. 1881; Duthie, Fl. Gangetic Plain 1: 493. 1905; Bhandari, Fl. Indian Desert 189. 1978; Singh, in Shetty & Singh Fl. Rajasthan 1: 435. 1987; Mamgain & Rao, in Hajra *et al.* Fl. India 12 : 318. f. 82. 1995. *S. oleraceus* L. var. *asper* L. Sp. Pl. 794. 1753. (Plate 166)

Annual or overwintering, erect, glabrous, herb. Stem stout, erect, glabrous, 20 - 120 cm tall, hollow except at nodes, branched apically. Leaves basal and cauline, basal leaves with more or less winged petiole, cauline leaves sessile, blades lanceolate or ovate-oblanceolate, 4 - 25 x 8 cm, sometimes lyrate-pinnatifid, or pinnately parted with triangular sharply toothed, acute lobes, base auriculate, amplexicaul, with rounded mucronate -toothed auricles. Inflorescence loose corymbs. Florets yellow, outermost purple-tinged at base. Heads 1 - 2 cm; peduncle flat, glabrous with glandular hairs. Involucral bracts in many series; outermost lanceolate, 5 - 7 x 1.5 - 2 mm; innermost lanceolate or oblong-lanceolate, 9 - 12 x 2 - 2.5 mm. Ligules yellow. Achenes ca 3 - 4 mm long, oblanceolate, compressed, never winged, initially yellow then brown, longitudinally 3-ribbed on each face, transversely rugose. Pappus few seriate, creamy white, 7 - 9 mm long.

Field note: Common weed of gardens as well as cultivated fields.

Fl. & Fr.: February- March.

Specimen examined: Bijoliya 821, Karoi 1676 (MLVGCB Herb).

Distribution: Widespread weed of European origin. Naturalized in India throughout.

Sonchus oleraceus L., Sp. Pl. 794. 1753; Hook. *f.,* Fl. Brit. India 3: 414. 1881; Duthie, Fl. Gangetic Plain 1: 492. 1905; Singh, in Shetty & Singh Fl. Rajasthan 1: 436. 1987; Mamgain & Rao, in Hajra *et al.* Fl. India 12: 321. f. 83. 1995.

An annual or biennial; stem stout, 15 - 150 cm tall, often branched, hollow, nearly glabrous. Leaves variable; lower cauline leaves thin, oblong, 8 - 30 × 4 - 9 cm, apex obtuse or acute, base narrowed, winged-petiolate, pinnately parted to divided, segments oblong or deltoid, 3 or 4 pairs, irregularly spine toothed, glabrous on both surfaces; median leaves oblong, 8 - 22 × 4.5 - 10 cm, apex obtuse or acuminate, base auriculate clasping, auricles acute or partite, segments of 2 or 3 pairs, irregularly spiny-dentate; upper leaves oblong-lanceolate to lanceolate, few. Heads ca 2 cm across, in umbels, yellow; peduncle 1.5-5.5 cm long, glandular hairy; peduncle small, flat; involucre many seriate, imbricate; bracts 3 - or 4 - seriate, glandular pilose on midrib outside, oblong-lanceolate, outer bracts 3.4 - 6 mm long, median bracts 6 - 9 mm long, inner bracts 9 mm long, with transparent margins, oblong or oblanceolate. Florets 11 - 12 mm long, many, bright yellow. Achenes oblanceolate, obovoid, 2.5 - 4 × 1 mm, 3 - 5 ribbed, distinctly muricate between the ribs; ribs smooth or transversely rugose. Pappus bristles slender, ca 6 mm long, white.

Field note: Common in cultivated fields, gardens etc.

Fl. & Fr.: January - March.

Specimen examined: Mahuakalan 1209, Mandal 1249 (MLVGCB Herb).

Distribution: Afghanistan, Australia, Bhutan, China, Eurasia, India, Iran, Iraq, N. & S. America, Nepal, Pakistan and Taiwan.

SPHAERANTHUS L.

Sphaeranthus indicus L. Sp. Pl. 927. 1753; Hook. *f.*, Fl. Brit. India 3: 275. 1881; Duthie, Fl. Gangetic Plain 1: 459. 1905; Singh, in Shetty & Singh Fl. Rajasthan 1: 436. 1987; Kumar, in Hajra *et al*. Fl. India 13: 160. f. 49. 1995. *S. hirtus* Willd. Sp. Pl. 2395. 1802. *S. senegalensis* DC. Prodr. 5. 370. 1836; Sharma Bull. Bot. Soc. Beng. 35: 127. 1981; Bhandari, Fl. Indian Desert 189. 1978. (Plate 167)

Much branched prostrate, hoary glandular herb. Branches terete striate at length shining; upper alate; spinulose-denticulate. Leaves 1.5 - 6 x 0.5 - 2.5 cm, glandular hairy, obovate-oblong or lanceolate, irregular toothed or lobed, acute or subacute, mucronate, somewhat narrowed towards the sessile, clasping base which is decurrent into wings. Heads compound, peduncled, 1.0-1.5 cm long, ovoid-oblong or globose, pinkish-purple, terminal. Involucral-bracts multiseriate, linear-lanceolate, ciliate, pubescent. Florets all tubular; 6 - 9 female, 3-toothed, filiform; 2 - 3 disc florets hermaphrodite, with 5-toothed corolla, the later much swollen in lower half and glandular outside. Corolla 1.2 mm across, ovate, acute. Anther base sagittate, not tailed. Style-arms shortly bifid or entire. Achenes not all fertile, 1 mm long, minutely puberulous on ribs; fertile ones pubescent, up to 1.5 mm long.

Field note: Common in Waste lands, drying ponds, neglected corners of fields etc.

Fl. & Fr.: October-April.

Specimen examined: Bhilwara 140, Hamirgarh 1531 (MLVGCB Herb).

Distribution: Africa, Australia, Bangladesh, India, Malaysia, Myanmar, Singapore and Sri Lanka.

Uses: Flower buds are used as a tonic and depurative.

SYNEDRELLA Gaertn. nom. cons.

Synedrella nodiflora (L.) Gaert., Fruct. 2: 456, t. 171. f. 7. 1791; Hook. *f.*, Fl. Brit. India 3: 308. 1881; Singh, in Shetty & Singh Fl. Rajasthan 1: 438. 1987; Chowdhery, in Hajra *et al*. Fl. India 12: 413. f. 117. 1995. *Verbesina nodiflora* L., in Juslenius Cent. Pl. 28. 1755. (Fig. 77 & Plate 168)

Annual herb, up to 1.3 m tall; stem erect, terete, puberulent with ascending white hairs, dichotomously branched. Median cauline leaves ovate, 3 - 12 × 1.0 - 4.0 cm, apex acute rounded or at base, margins short serrulate or subentire, strigose-pubescent on both surfaces, 3-nerved; upper leaves smaller, 3.5 - 4 cm long. Inflorescences globose, in leaf axils. Heads sessile or short pedunculate, arranged in groups 1 - 4 at the forking of the stem or leaf axils enclosed in foliaceous bracts, 4 - 6 mm across. involucral bracts in 2 - 3 series; outermost foliaceous; inner membranous. Raceptacle convex. Ray florets 5 - 8; corolla yellow, bilobed, tube 2 - 3 mm long. Ovary dorsiventrally flattened, oblong, margins erose, with two strigose awns; style branches flattened, slender. Disc florets several; corolla yellow; tube narrow; lobes short. Anthers dark. Ovary slightly compressed; style branches flattened, marginnaly pilose. Achenes dimorphic; ray floret achenes 4 - 5 x 1.5 - 2 mm; achenes of disc florets up to 4 mm long.

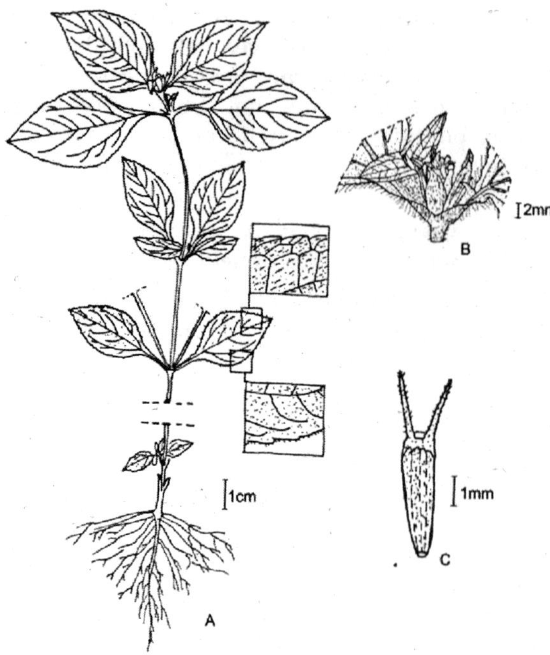

Fig. 77. *Synedrella nodiflora* (L.) Gaert
A. Habit, B. Capitulum, C. Achen

Field note: Rare, on moist sandy soils.

Fl. & Fr.: Almost throughout the year.

Specimen examined: Bijoliya 1214 (MLVGCB Herb).

Distribution: Common weed of the lowland Tropics,Taiwan and India.

TRIDAX L.

Tridax procumbens L., Sp. Pl. 900. 1753; Hook. *f.,* Fl. Brit. India 3: 311. 1881; Duthie, Fl. Gangetic Plain 1: 475. 1905; Bhandari, Fl. Indian Desert 190. 1978; Singh, in Shetty & Singh Fl. Rajasthan 1: 439. 1987; Chodhery, in Hajra *et al*. Fl. India 12 : 418. f. 119. 1995. (Plate 169)

An annual or perennial herb, caulescent, decumbent. Stem procumbent, branched at base, branches slender, spreading or ascending, 20 - 60 cm long, hirsute. Leaves short petiolate, ovate to ovate-lanceolate, few, 0.8 - 9.0 x 0.4 - 0.5 cm, apex acute or acuminate, margins deeply irregularly serrate, pinnatisect, segments few, narrow, base cuneate; petiole 0.4 - 1.6 cm long. Heads solitary, 0.5 - 1.8 cm across on straight, erect peduncles; peduncle 10 - 3 0 cm long. Involucral bracts 2 - 3 seriate, hispid; outer bracts densely glandular hairy, elliptic, ca 5 mm long; inner bracts whitish-green, oval, apiculate. Ray florets pale yellow to white, ligulate, 5 - 8, 2 - 3 lobed; style branches recurved. Disc florets yellow to purple tinged; limb narrower, tubular with 5 lanceolate lobes; style branches revolute, flattened. Achenes narrowly obconical, 1.5 - 2.5 mm long. Pappus 15 - 20 unequal, 2 - 6 mm long, of shiny feathery bristles.

Field note: Common in Waste lands, open forests and dry habitats.

Fl. & Fr.: Most part of the year.

Specimen examined: Kasya 163, Bijoliya 1695 (MLVGCB Herb).

Distribution: Central America, India.

Uses: The paste of the leaves is applied on cut and wound. The fresh leaves are chewed to cure urinary tract inflammation. Decoction of roots is taken to cure loss of consciousness and also in vertigo.

VERNONIA Schrebr. nom. cons.

KEY TO THE SPECIES

1a. Climbers. ....*V. elliptica*

1b. Erect herbs 2

2a. Achenes neither angled nor ribbed, but terete .... *V. cinerea*

2a. Achenes either angled or ribbed, not terete.................................. *V. cinerascens*

Vernonia cinerascens Sch.- Bip. in Schweinf. Beitr. Aethiop. 162. 1867; Hook. *f.,* Fl. Brit. India 3: 237. 1881; Duthie, Fl. Gangetic Plain 1: 441. 1905; Bhandari, Fl. Indian Desert 190. 1978; Singh, in Shetty & Singh Fl. Rajasthan 1: 441. 1987; Uniyal, in Hajra *et al*. Fl. India 13: 364. f. 115. 1995.

Undershurb, 45 - 60 cm high, clothed with ashy pubescence. Leaves sessile, 2.0 - 5.5 x 0.6 - 1.5 cm, sphathulate, alternate, narrow at the base, grey-pubescent, entire or few dentate in the upper half. Heads in corymbose panicle, solitary on divaricating branches, about 10 - 12 flowered, 5 - 6 mm in diam., campanulate, purple, arranged in diffuse. Pedicels 3-10 mm long, with or withought minute bracts. Involucral-bracts 3 - 4 seriate, linear, oblong, pubescent; outer most ca 1 mm long; inner oblong-oblanceolate, ca 3.5 mm long. Corolla 4.5 - 5.5 mm long, purple, gland-dotted. Anther bases obtuse. Achenes 2 - 2.6 mm long, obscurely 5 - 10-ridged, glandular, densely silky, slightly tapering at base. Pappus 4.5 - 5 mm long, biseriate; outer short, rigid; inner long, barbellate, more or equal to corolla, white.

Field note: Common in Waste lands, particularly in wet and shaded habitats.

Fl. & Fr.: August- December.

Specimen examined: Pansal 1126, Jahajpur 1820 (MLVGCB Herb).

Distribution: Tropical Africa, Abyssinia, Baluc-histan, India and Pakistan.

Vernonia cinerea (L.) Less. in Linnaea 4 : 291. 1829; Hook. *f.*, Fl. Brit. India 3: 233. 1881; Duthie, Fl. Gangetic Plain 1: 441. 1905. Mahesh., Illus. Fl. Delhi f. 101. 1905; Bhandari, Fl. Indian Desert 191. 1978; Singh, in Shetty & Singh Fl. Rajasthan 1: 441. 1987; Uniyal, in Hajra et al. Fl. India 13: 367. 1995.

Perennial or an annual, erect, herb, 5 - 80 cm high, Stem ribbed, terete, greyish. Branches occassionally fulvous pubescent, glandular. Leaves 0.5 - 8.5 x 0.3 - 3.5 cm, variable in shape from ovate to obovate, cuneate at base, acute, obtuse, glabrous, entire to dentate, gland-dotted beneath, petioled or subsessile. Heads in terminal corymbose panicle, 4 - 10 mm long, violet or pink, 10 - 21 floweres beneath each head of the florets and with small bracts in the fork of peduncle. Involucral-bracts campanulate, 4-seriate, shorter than heads; bracts lanceolate, keeled, very pointed at the top, adpressedly pilose, mixed with a few shining glands. Corolla purple, 4 - 6 mm long, narrowly infundibuliform, 5-lobed, slightly glandular; lobes narrow, lanceolate, pilose at the apex. Anthers sagittate at base. Pappus biseriate, white, setaceous; setae of the inner row 4-5 mm long, silky, ciliate; of outer row minute, ciliate flat. Achenes 1.0 - 2 × 0.7 mm long, densely short hispid, glandular dotted. Pappus hairs white or fulvous, biseriate; of the outer short.

Field note:Very common in wastelnds, road sides, fields, throughout the area.

Fl. & Fr.: Most part of the year

Specimen examined: Bhilwara 1344, Jahajpur 1820 (MLVGCB Herb).

Distribution: Tropics of the old worlds.

Uses: Plant is useful in the treatment of piles.

Notes: It is a variable species.

Vernona elliptica DC., in Wight, Contrib. Bot. Ind. 5. 1834; Uniyal, in Hajra et al. Fl. India 13: 370. 1995. *Vernona elaegnifolia* DC., Prodr. 5: 22. 1836; Hook. *f.*, Fl. Brit. India. 3: 237. 1881.

Climbing shrub; Branches often hanging over forming dense masses. Leaves oblong or elliptic-lanceolate, 2.5 - 4 x 1.5 cm, glabrous and dotted above, silky beneath. Heads cream, subses-sile, 5-fid aggregated in axillary or corymbs; involucral bracts scarious to herbaceceous; outer ovate-obovate, ca 1 mm long; inner ca 3 mm long, obtuse. Achenes ca 2 mm long, glabrous, 5 - 8 ribbed. Pappus 1-seriate.

Field note: Commonly planted as *curtain bel.*

Fl. & Fr.: February-April.

Specimen examined: Bhilwara 451, Mandalgarh 2004 (MLVGCB Herb).

Distribution: Burma, Myanmar, Thailand and India.

XANTHIUM L.

Xanthium indicum Koen. ex Roxb., Fl. Ind. 3. 610. 1832; Chowdhery in Hajra *et al.* Fl. India 12: 427. f. 123. 1995. *Xanthium strumarium* L. Sp. Pl. 987. 1753, *pro parte*; Hook. *f.*, Fl. Brit. India 3: 309. 1881; Duthie, Fl. Gangetic Plain 1: 472. 1905; Bhandari, Fl. Indian Desert 214. 1978; Singh, in Shetty & Singh Fl. Rajasthan 1: 443. 1987. (Plate 170)

Vern.: *Adhasisi*

An unearned, erect, gregarious annual. Stem 20-100 cm long, stout, pubescent, branched. Leaves alternate, broadly ovate to cordate, 3 - 5 lobed, acute or acuminate, cordate at base, irregularly serrate 3 - 13 x 1.5 - 10 cm, hispid on both surfaces; petiole 2 - 10 cm long, not winged. Male heads globose. Involucral bracts few, 1 - 2 seriate, pubescent. Florets numerous, crowded, each florets enclosed in a translucent scale; corolla tubular, 5 - toothed. Anther free. Ovary and style absent. Female heads ovoid. 9 - 16 mm long; outer involucral bracts few, short; inner many, in one series, narrow, united, covered by hooked bristles and terminating into 2 strong, hooked divergent beaks. Florets 2; corolla absent; style thread-like, producing between beaks, 2 - 5 mm long. Achenes greyish, narrowly ovoid or elliptic, 1 - 2 x 2 - 3.6 mm, smooth, tapering to a fine point at apex.

Field note: Common in Waste lands near habitations, often gregarious in the forests. Very troublesome during plant collection, very efficient dispersal mechanisms with hookes on the achines sticking to cattles tail as well as clothes of human beings

Fl. & Fr.: September - April.

Specimen examined: Bhilwara 112, Bijoiya 1215 (MLVGCB Herb).

Distribution: Eurasia and North America

Uses: Dry fruits of *Xanthium indicum* are burnt and the smoke is inhailed with the help of dried stem of *Calotropis procera,* 3 times a day for 3 days for the treatment of "*Adhasisi*".

PLUMBAGINACEAE

KEY TO THE GENERA

1a. Leaves not perfoliated, flowers white, calyx glandular, 5-nerved wing absent .... *Plumbago*

1b. Leaves perfoliated, flowers golden yellow, calyx eglandular, 5-nerved, winged between the nerve .... *Dyerophytum*

DYEROPHYTUM O. Ktze.

Dyerophytum indicum (Gibs. ex Wt.) O. Ktze. Rev. Gen. Pl. 3: 394, 189; Bhandari, Fl. Indian Desert 192. 1995; Singh, in Shetty & Singh Fl. Rajasthan 2: 201. 1991. *Vogelia indica* Gibs. ex Wight, Ic. 3(4): 5. t. 1075. 1846 & in Cal. Journ. Nat. Hist. 7: 17. 1847; Clarke, in Hook. *f.* Fl. Brit. Ind. 3: 481. 1882; Duthie, Fl. Gangetic Plain 2: 2. 1911. (Plate 171)

An erect, glaucous perennial shrub or undershrub, 1 - 1.5 m high, with terete branches. Leaves perfoliated, alternate, coriaceous, lower 2.5 - 8.0 x 2.0 - 5.5 cm, ovate, elliptic, obtuse or orbicular, covered with scurty scales, those at upper nodes orbiculate, margin undulate, apex acute, obtuse or emarginated, often apiculate, midvien prominent. Flowers in dense paniculate spikes, golden-yellow. Calyx tubular, 8 - 9 mm long, winged between 5 strong ribs; teeth 5, lanceolate, acute, about half of the length of tube. Corolla tube cylindric, lobes ovate-oblong, emarginated. Stamens somewhate exserted. Style filiform, hairy at base, stigmatic branches 5, stigmatose throughout. Capsules 0.6 - 0.8 cm long, glabrous, 5-valved, included in the calyx. Seeds 0.4 cm long, smooth, pyriform, brown.

Field note: Common in clumps among shrubs of hilly regions, forests and on dry hill - slopes.

Fl. & Fr.: October - January.

Specimen examined: Khamnor 1410, Kaliwas 2125 (MLVGCB Herb).

Distribution: India, Western Arabia.

PLUMBAGO L.

Plumbago zeylanica L., Sp. Pl. 151. 1753; Clarke, in Hook. *f.* Fl. Brit. Ind. 3: 480. 1882; Duthie, Fl. Gangetic Plain 2: 2. 1911; Mahesh. Illus. Fl. Delhi f. 166. 1966; Bhandari, Fl. Indian Desert 192. 1995; Singh, in Shetty & Singh Fl. Rajasthan 2: 201. 1991. (Fig. 78 & Plate 172)

Vern.: *Chitrak*.

Erect or straggling, perennial undershrub, ca 1.5 m tall; branches woody, angular-striated.

Leaves ovate, 2.5 - 10.0 x 1.5 - 6.0 cm, ovate-lanceolate or oblong, acute, narrowing suddenly into the petiole; petiole 1 - 2 cm long with a dilated amplexicaul base, glabrous or puberulous, white punctate beneath. Flowers rather densely arranged in 10 - 35 cm long, often branched spikes, white; rachis of the spike often pubescent or glandular. Bracts ovate, acute, 5 - 6 mm long, 3 times shorter than the calyx. Calyx tubular, 1 - 1.5 cm long, covered with stalked glands; limb 5-fid; teeth very short. Corolla gamopetalous; tube 2 times longer than the calyx; limb white, 5-Iobed; lobes cuneate-retuse, shortly mucronate. Stamens free. Style slender glabrous. Fruit 5 - 12 mm long, oblong, longitudinally furrowed, enclosed within persistant calyx. Seeds oblong, smooth.

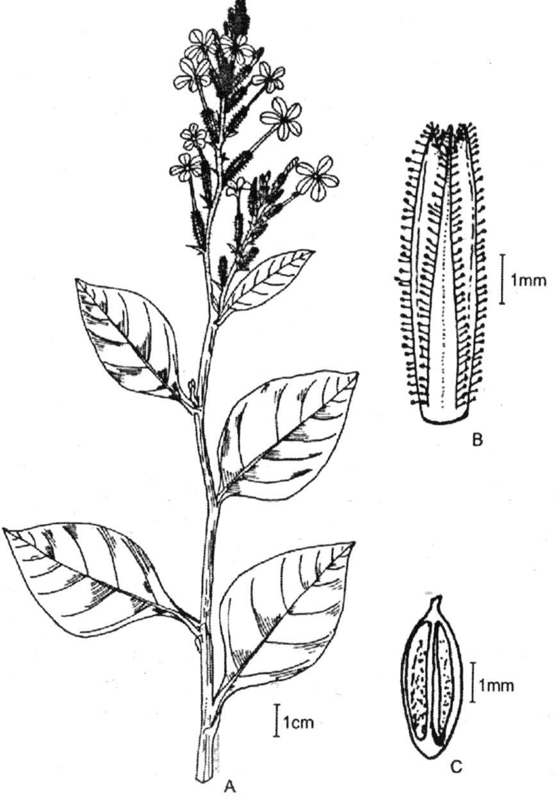

Fig. 78. *Plumbago zeylanica* L.
A. Flowering & fruiting twig, B. Fruit, C. Seed

Field note: Common in Waste lands and open forests.

Fl.& Fr. : July - September.

Specimen examined: Hamirgarh 1588 (MLVGCB Herb).

Distribution: Throughout tropics of the old world.

PRIMULACEAE

ANAGALLIS L.

KEY TO THE SPECIES

1a. Leaves sessile, opposite, 3-nerved, corolla blue longer than the calyx, filaments hairy .... *A. arvensis*

1b. Leaves subsessile, alternate, 1-nerved, corolla white shorter than the calyx, filaments glabrous .... *A. pumila*

Anagallis arvensis L., Sp. Pl. 1: 211. 1753; Hook. *f.* Fl. Brit. Ind. 3: 506. 1882; Duthie, Fl. Gangetic Plain 2: 6. 1911; Mahesh. Illus. Fl. Delhi f. 117. 1966; Singh, in Shetty & Singh Fl. Rajasthan 2: 457. 1991. (Fig. 79 & Plate 173)

Vern.: *Leelo.*

An erect, annual herb, 10-30 cm tall. Stem quadrangular, short winged on ridges, usually branched from base. Leaves opposite, sessile, ovate-elliptic, 1 - 2 x 0.3 - 1 cm, amplexicaul, minutely glandular punctate mainly abaxially, base subrounded, margin entire, apex acute. Flowers axillary, solitary. Pedicel recurved in fruit, 2 - 3 cm. Calyx lobes linear-lanceolate, 3.5 - 6 mm, hyaline margined, apex long acuminate, costate. Corolla blue or red, rotate, 4 - 6 mm, parted nearly to base; lobes obovate-elliptic, 2.7 - 3 mm wide, margin entire to denticulate. Stamens 1/3 - 1/2 as long as corolla; filaments pubescent. Style ca. 1.5 mm, persistent. Capsule *ca.* 0.3 cm in diam. with persistent calyx, dehiscing transversely in the middle. Seeds brownish black, trigonous, tuberculate.

Field note: Common in cultivated fields as a weed throughout the area.

Fl. & Fr.: January - April.

Speciemen examined: Mandal 1898, Janhajpur 1919 (MLVGCB Herb).

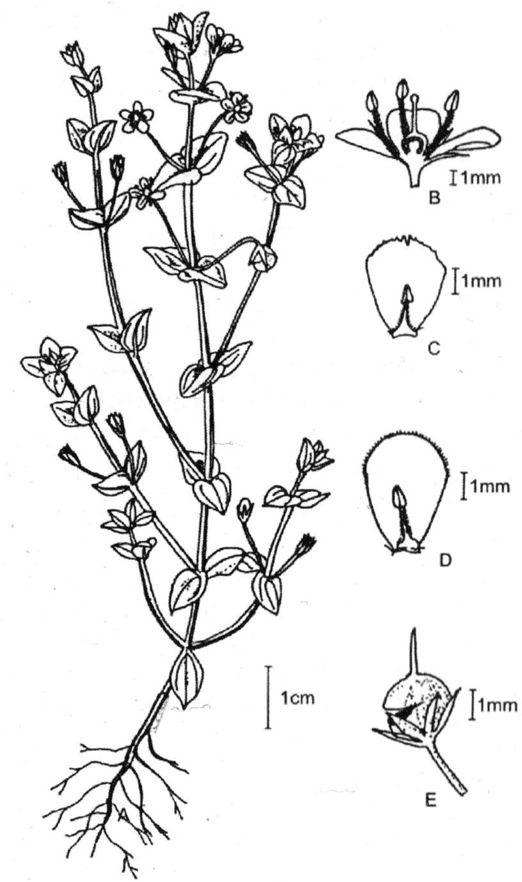

Fig. 79. *Anagallis arvensis* L.
A. Habit, B. Flower, C&D. Petals with stamens, E. Fruit

Distribution: Bhutan, India, China, Japan, Nepal, Pakistan, Russia, N.W. Africa, W Asia, Australia, Europe, North America and South America.

Uses : Used as a fodder.

Anagallis pumila Swartz. Prodr. Veg. Occ. 1 : 40. 1788; Taylor, in Kew Bull. 1955: 342. 1955; Singh, in Shetty & Singh Fl. Rajasthan 2: 458. 1991. *Centrunculus pentandrus* Duby, in DC Prodr. 8: 72. 1844; Hook. *f.* Fl. Brit. Ind. 3: 506. 1882; Duthie, Fl. Gangetic Plain 2: 6. 1911. *C. tenellus* Duby in DC. Prodr. 8: 72. 1844; Wight. Ic. 6 : 14. t. 2000. 1853; Hook. *f.* Fl. Brit. Ind. 3: 506. 1882.

An erect, annual herb, 4 - 15 cm tall. Stem slightly margined. Leaves 0.4 - 1 x 0.2 - 0.3 cm, alternate, rarely opposite, ovate or lanceolate, entire, acute, cuneate at base, glabrous. Flowers axillary, solitary. Calyx divided upto base; lobes 2-3 mm long, lanceolate - oblong, acute or mucronate, longer than the capsule. Corolla - lobes ovate - lanceolate, apiculate, eglandular. Filaments glabrous. Capsule membranous, shorter than calyx, subglobose, dehiscing transversely about the middle. Seeds trigonous, solid.

Field note: Rare weed of cultivated fields, along streams and shady places.

Fl. & Fr.: October- February.

Specimen examined: Karoi 539 (MLVGCB Herb).

Distribution: Native to South Florida. Distributed in Africa, India, Timor, Australia, West Indies, Mexico, Central America and South America.

SAPOTACEAE

KEY TO THE GENERA

1a. Calyx 4-lobed, lobes biseriate. .... *Madhuca*
1b. Calyx 5 or more lobed, lobes uniseriate 2
2a. Calyx segment 6. Fertile stamens 6. Staminodes glabrous .... *Manilkara*
2b. Calyx segment 8. Fertile stamens 8. Staminodes hairy .... *Mimusops*

MADHUCA Buch. - Ham. ex. J. F. Gmelin

Madhuca indica J. F. Gmelin, Syst. Nat. 2(1): 799. 1791; Singh, in Shetty & Singh Fl. Rajasthan 2: 460. 1991. *Bassia latifolia* Roxb. Pl. Cor. 1: 20. t. 19. 1795; Bedd. Fl. Sylv. 1: 41. Pl. 41. 1870; Clarke, in Hook. *f.* Fl. Brit. Ind. 3: 544. 1882; Duthie, Fl. Gangetic Plain 2: 11. 1911. *Madhuca longifolia* (Koen.) Macbr. var. *latifolia* (Roxb.) Chevalier, in Rev. Int. Bot. Appl. 23: 149. 1943. (Fig. 80 & Plate 174 & 175)

Vern.: *Mahua, Morda.*

Tree up to 16 m high, stem glabrous. Bark brown, young growth tomentose. Branches spreading. Leaves scattered or clustered at the end of the branches, obtuse or subacuminate, varying in size, 5-25 x 2-12 cm, glabrate, younger tomentose, broadly elliptic-lanceolate, varying in shape, 12-15-nerved. Stipules caducous. Pedicel tomentose, 2.5-5 cm long. Calyx tomentose, 4-lobed, ovate, acuminate, pubescent. Corolla white, tube succulent, lobes ovate-acuminate. Stamens 20 - 30, roughly in 2 - 3 series. Style exserted. Furit a berry, ovoid, fleshy, 1 - 4 seeded, brown tomentose. Seeds ellipsoid, almost terete, shining brown.

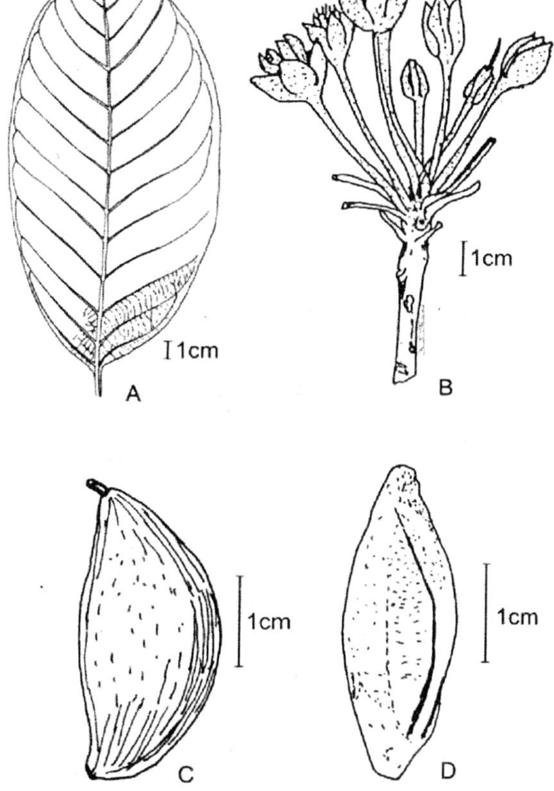

Fig. 80. *Madhuca indica* J.F. Gmelin
A. Leaf, B. Inflorescence, C. Fruit, D. Seed

Field note: Very common in forests and Waste lands particularly in Southern part of the study area.

Fl. : March - April; *Fr.*: May - July

Specimen examined: Hamirgarh 1547, Gadola 3525 (MLVGCB Herb).

Distribution: Sri Lanka, India, Burma and Pakistan.

Notes: Royen recognizes 2 varieties, var. *latifolia* (Roxb.) Chevalier is based on *Bassia latifolia* Roxb., Pl. Cor. 20.t. 19.1795. The present specimen falls in the var. *latifolia* (Roxb.) Chevalier group.

Uses: Fragrant flowers are edible and largely used in the preparation of distilled liquor and vinegar. Flowers are regarded as cooling, tonic and demulcent. They are used cough, cold and bronchitis. The honey from flowers is edible and used medicinally in eye diseases. A low quality of oil is extracted form the seeds which is edible and used in soap making. The wood is very hard and is used for furniture. Bark is used medicinally in leprosy.

MANILKARA Adans. nom. cons.

KEY TO THE SPECIES

1a. Leaves crowded at the end of branches, Staminodes present, berries dull-brown.................................... *M. hexandra*

1b. Leaves spreading over branches, Staminodes absent, berries reddish-yellow .... *M. sapota*

Manilkara hexandra (Roxb.) Dub., in Ann. Mus. Colon. Marseille ser 3. 23: 9. 1915; Royen, in Blumea 7: 408. 1953; Singh, in Shetty & Singh Fl. Rajasthan 2: 460. 1991. *Mimusops hexandra* Roxb. Pl. Cor. 1: 16. t. 15. 1795; Clarke, in Hook. *f.* Fl. Brit. Ind. 3: 549. 1882; Duthie, Fl. Gangetic Plain 2: 13. 1911. *Mimusops indica* DC., in DC. Prodr. 8. 205. 1844; Wight, *l.c.* 4 (4): 13. t. 1587. 1850. (Plate 176)

Vern.: *Ranniya, Khirni.*

Tree, evergreen, 10 - 20 m high; bark with deeply furrowed blakish grey. Leaves obovate-broadly obovate, oblong or elliptic 4.5 - 12 x 2.5 - 6 cm, rounded or emarginate at apex, subacute at base, coriaceous, glabrous, midrib prominent beneath, lateral nerves not prominent; petiole 1 cm or more. Flowers white, axillary, solitary or fascicled of 2 - 6. Calyx 6-lobed, pubescent. Corolla 16 or 24-lobed, in 2 or 3 series,tube short; outer ones linear; inner ones oblanceolate. Stamens 6 or 8, with an equal number of staminodes. Ovary 12-celled, hairy. Fruit berry, ovoid, redish-yellow, 1-seeded, ca 1.5 x 1 cm. Seeds ovoid, reddish, shining.

Field note: It is cultivated for its very sweet edible fruits.

Fl. & Fr.: October - April.

Specimen examined: Hamirgarh 1644, Bhilwara 2128 (MLVGCB Herb).

Distribution: Sri Lanka and India.

Manilkara sapota (L.) P. van Royen, in Blumea. 7: 410. 1953; Moore & Stearn, in Taxon 16(5): 382 - 395. 1967. *Achrus zapota* L., Sp. Pl. 2: 1190. 1753, *quoad typum synonymis sloqnei & Plukenetii exclusis. A. sapota* L., Sp. Pl. 2: 470. 1762; Clarke, in Hook. *f.* Fl. Brit. Ind. 3: 534. 1882. *Manikara achras* (Mill.) Fosb, in Taxon 13: 255. 1964.

Vern.: *Cheeku.*

Small evergreen tree. Leaves crowded at the end of branches, elliptic to elliptic-lanceolate, 7 - 12.0 x 2.5 - 3.5 cm, glabrous, subcoriaceous, midrib prominent. Flowers solitary, axillary, white, 1 - 1.5 cm across, long pedicelled. Calyx in 2 whorls of 3 sepals; sepals brown hairy outside, unequal, 3 larger than the others. Petals 6. Stamens 6, opposite the petals; staminodes as many as the petals, petaloid. Fruit usually a globose, fleshy berry, 5 - 10 cm in diam. epicarp thin, rough rusty brown. Seeds 5 - 12, shining black, ellipsoid, ca 2.0 cm long.

Field note: Commonly cultivated in orchard for its edible fruits.

Fl. & Fr.: Throughout the year.

Specimen examined: Bhilwara 1532 (MLVGCB Herb).

Distribution: A native of W. Indies and tropical America; India, Pakistan.

Uses: Besides fruits, the coagulated resinous latex is derived from the bark and is used commercially for making chewing gum.

MIMUSOPS L.

Mimusops elengi L., Sp. Pl. 349. 1753. Roxb., Pl. Cor. 1. lb t. 14. 1795; Clarke, in Hook. *f.* Fl. Brit. Ind. 3: 548. 1882; Cooke, *l.c.* 95; Talbot, *l.c.* 161; Singh, in Shetty & Singh Fl. Rajasthan 2: 461. 1991. (Fig. 81 & Plate 177)

Vern.: *Morsali, Molchri*

A small evergreen tree. Stem cylindrical, bark dark grey, smooth, not deeply furrowed. Leaves alternate, broadly ovate, 7 - 10.5 x 3.5 - 4.5 cm, base cuneate, apex acuminate, entire or slightly wavy, glabrous, petiolate, petiole 2 - 3.5 cm long. Flowers axillary solitary or fascicled, creamish white, fragrant, ca 15 mm in diam., pedicel ca 2 cm long, hairy. Calyx 6-lobed, lobes free, lanceolate, tomentose. Corolla tube very short, lobes 24 in 2 series. Stamens 8, anthers subsessile, staminodes hairy. Ovary pubescent. Fruit ovoid, 2.5 - 4 cm long, orange, pubescent when young, with persistent calyx.

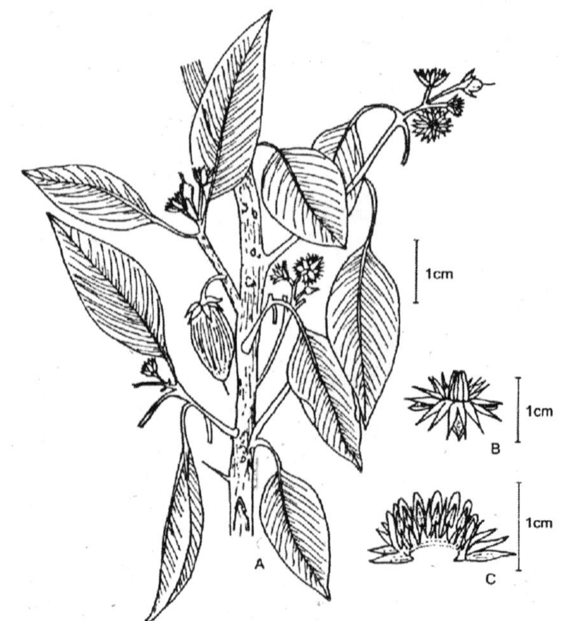

Fig. 81. *Mimusops elengi* L.
A. Flowering & fruting twig, B. Flower, C. Flower showing stamens.

Field note: Commonly cultivated in orchard for its edible fruits and planted as an ornamental.

Fl. & Fr.: September-November.

Specimen examined: Bhilwara 1532 (MLVGCB Herb).

Distribution: India, Sri Lanka, Malay peninsula; often cultivated elsewhere.

Uses: Wood is used for building purposes, bridges, agricultural implements and oil mills. Fruits are edible, it is sometimes used for making pickles. The caducous corolla is used for making fragrant garlands. Bark is astringent and tonic. It is used in diarrhoea and dysentery. A decoction of bark is used for gargles.

EBENACEAE

DIOSPYROS L.

1a. Anthers awnless, Staminodes 8 - 16. .... *D. melanoxylon*

1b. Anthers awened, Staminodes 4 .....*D. montana*

Diospyros melanoxylon Roxb. Pl. Cor. 1: 36. t. 46. 1795; Clarke, in Hook. *f.* Fl. Brit. Ind. 3: 564. 1882; Singh, in Shetty & Singh Fl. Rajasthan 2: 462. 1991. *D. tupru* Buch., in Trans. Linn. Soc. 15: 111. 1827; Clarke, in Hook. *f.* Fl. Brit. Ind. 3: 563. 1882. *D. tomentosa sensu* Duthie, Fl. Gangetic Plain 2: 15. 1911, non Roxb. 1832. (Plate 178)

Vern.: *Tendu Patta, Timroo, Bidi Patta.*

A middle sized decidouos tree, 4 - 12 m high. Bark exfolding in 4 - angular scales. Leaves 4 - 20 x 2.5 - 5 cm, ovate-elliptic or oblong, coriaceous, softly tomentose on both surfaces when young, pubescent when mature. Male flowers 3 - 12 together in panical, drooping cymes. Stamens 4 - 16, linear - lanceolate, not awned. Female flowers solitary. Sepals of fruiting calyx fulvous tomentose with recurved and wrinkled margins. Ovaries 4 - 8 - celled, hairy. Fruits 2 - 4 cm in diam., ovoid or globose, glabrous, yellow when ripe. Seeds obovate - oblong, blackish.

Field note: Common in forests and Waste lands, particularly in South eastern part of the study area.

Fl.: April - May; *Fr.*: June - November.

Specimen examined: Mainal 3275, Bassi 3601 (MLVGCB Herb).

Distribution: India (M.P., Chhatisgarh, Orissa, A.P., Jharkhand, Gujarat and Maharastra).

Uses: Leaves are popularly known as "*Biri patta*" and are used in making *bidis*, popular smoke especially among poor natives. The mature fruits are edible and are sold in market by the name of "*timroo.*"

Diospyros montana Roxb. Pl. Cor. 1: 37. t. 48. 1795; Wight, Ic. 4 (1): 10. t. 1225. 1848; Clarke, in Hook. *f.* Fl. Brit. Ind. 3: 555. 1882, *pro parte*; Duthie, Fl. Gangetic Plain 2: 15. 1911; Singh, in Shetty & Singh Fl. Rajasthan 2: 462. 1991 & Monogr. Indian Diospyros 166. 7. 38. 2005. (Plate 179)

Vern.: *Timroo.*

A shrub or small decidouos tree, 1 - 5 m high. Bark blakish to grey. Leaves 3 - 10 x 2 - 5 cm, ovate - elliptic, cuneate or cordate at base, pubescent beneath, chartaceous. Male flowers in short few-flowered panicles or cymes, white. Stamens 15. Female flowers axillary, solitary. Calyx - lobes 4, flat in fruiting stage. Corolla - lobes 5, ovate, acute, pubescent. Staminodes present in female flowers. Pistillode conical. Fruits 2.0 - 3.5 cm in diam., globose, glabrous, yellow when ripe. Seeds 2 - 8, obovate - oblong, blackish.

Field note: Common in forests and Waste lands throughout the area.

Fl.: April - May; *Fr.*: June - November.

Specimen examined: Pur 1587, Gurla 2157 (MLVGCB Herb).

Distribution: India, Sri Lanka, Malaya and Tropical Africa.

OLEACEAE

KEY TO THE GENERA

1a. Corolla tube orange-red lobes whitish .... *Nyctanthes*

1b. Corolla tube not as above. .... *Jasminum*

JASMINUM L.

KEY TO THE SPECIES

1a. Leaves simple.......................... *J. multiflorum*

1b. Leaves compound, trifoliolate or pinnate....... 2

2a. Leaves alternate. Inflorescence dense capitate .... *J. humile*

2b. Leaves opposite. Inflorescence subumbellate .... *J. officinale*

Jasminum humile L., Sp. Pl. 7. 1753; Clarke, in Hook. *f.* Fl. Brit. Ind. 3: 602. 1882; Duthie, Fl. Gangetic Plain 2: 24. 1911; Singh, in Shetty & Singh Fl. Rajasthan 2: 264. 1991.

Vern.: *Chameli.*

An erect shrub, up to 1 m tall, deciduous or evergreen, glabrous. Branches green, angular. Leaves alternate, very variable in size, 2 - 10 cm long, sometimes revolute; leaflets 3 - 5 (-7), coriaceous, dark green above, paler beneath, variable in shape, elliptic, ovate, or lanceolate, acute or obtuse, terminal sometimes larger than lateral. Flowers in terminal corymbose cymes; pedicels 5 - 15 mm long. Bracts linear, minute. Calyx tube ca 3 mm long, teeth very short. Corolla yellow, tube 1 - 2.5 cm long, lobes 5, broadly ovate-obtuse or round, reflexed when the flower is opened. Berry simple or didymous, globular-ellipsoid, 4 - 6 mm long, black when ripe, full of crimson juice.

Field note: Commonly planted as an ornamental in the gardens.

Fl. & Fr.: August - October.

Specimen examined: Bhilwara 2181 (MLVGCB Herb).

Distribution: Afghanistan, Baluchistan, China, India, Pakistan and South Waziristan,

Jasminum multiflorum (Burm. f) Andrews, Bot. Repos. 8, t. 496. 1807. (Plate 180)

Vern.: *Gulcharia.*

Scandent shrub. Branchlets, pedicels and calyx densely fulvous-villous. Leaves 3.5 - 6.6 x 1.5 - 3.5 cm, opposite, simple, ovate, acuminate, rounded or cordate at base, upper surface pubescent, becoming glabrous, ventral tomentose or pubescent, especially on the nerves; petiole stout, tomentose. Flowers sessile or subsessile, in dense capitate cymes, at the tips of dwarf lateral branchletes or axillary, white, fragrant. Calyx teeth 6 - 9, linear, ca 1 cm long. Corolla white, tube slender, exceeding the calyx, lobes 6 - 9, oblong-lanceolate, acute, 1 - 2 cm long. Berry simple or didymous, ellipsoid, ca 0.6 - 1 cm across, black when ripe, surrounded by the long hairy calyx teeth.

Field note: Cultivated as an ornamentaly.

Fl. & Fr.: December-April and also during the rainy season.

Specimen examined: Bhilwara 1372 (MLVGCB Herb).

Distribution: Pakistan, India, Sri Lanka, Tenasserim, Burma, China.

Jasminum officinale L., Sp. Pl. 7. 1753. Boiss., Fl. Or. 4: 43. 1879; Clarke, in Hook. *f.* Fl. Brit. Ind. 3: 603. 1882; Parker, l.c 316; E. Murray in Rech. f. *l.c.* 9.

Vern. : *Chameli.*

Shrubs, sometimes twining. Branches long, weak, sparsely hairy when young. Leaves 5 - 10 cm long, opposite, imparipinnate, petiole and midrib narrowly margined; leaflets 3 - 7, upper surface slightly pubescent, especially on midrib and margins, lateral leaflets acute or apiculate, sessile or subsessile, the upper pair sometimes with broad connate bases, terminal much larger, ovate or lanceolate, acuminate. Flowers fragrant, 1 - 10 in subumbellate terminal, often leafy cymes; pedicels up to 2 cm. Bracts linear, ca 5 mm long. Calyx teeth linear, 5 - 10 mm long, puberulous. Corolla white, tube 1 - 2 cm long, lobes 4 - 5, oblong, shorter than the tube, more or less involute at the margins. Berry black when ripe, elliptic or globose, 8 - 10 mm long, full of crimson juice.

Field note: Planted as an ornamental in the gardens.

Fl.: May - July; *Fr.*: September - November.

Specimen examined: Bhilwara 223 (MLVGCB Herb).

Distribution: Mediterranean, Persia, Afghanistan, India, Pakistan and China.

NYCTANTHES L.

Nyctanthes arbor-tristis L. Sp. P1. 6. 1753; Beddome, Fl. Sylv. 2: 240. t. 1872; Clarke, in Hook. *f.* Fl. Brit. Ind. 3: 603. 1882; Duthie, Fl. Gangetic Plain 2: 24. 1911; Kundu & Anima, in Bull. Bot. Surv. India 10: 397. 1969; Singh, in Shetty & Singh Fl. Rajasthan 2: 2465. 1991. (Plate 181)

Vern.: *Harsingar, Tomati.*

Large shurb, up to 4 m high. Branches quadrangular; bark grey or greyish-green. Leaves opposite 4-11 x 2-5 cm, ovate or ovate-lanceolate, coriaceous, distinctly serrate, acute or acuminate, cuneate to rounded at base, covered with stiff white hairs. Petiole up to 1 cm long. Flowers fragrant, in 3 - 5 flowered together on axillary solitary or fascicled penduncles or in terminal cymes, white with orange tube, sessile. Bracts ovate. Calyx-lobes 5, turncate. Corolla tube ca 1 cm long, orange red; lobes 5 - 7 mm long, spreading emarginate, whitish. Stamens 2, subsessile, inserted near the mouth of corolla-tube. Ovary 2-locular; style filiform; stigma shortly 2-fid. Capsules orbicular, compressed, 1.3 - 2.0 x 1.2 - 2.0 cm, 2- celled, separating into 2, 1-seeded pyrenes.

Field note: Cultivated since ancient times, naturalised at several places, especially in lower foothills of the area.

Fl. & Fr.: Almost throughout the year.

Specimen examined: Pur 113 (MLVGCB Herb).

Distribution: Indo-Malaysian, India and Pakistan.

Uses: Grown for its fragrant flowers. Leaves are used in medicine as a febrifuge. An orange dye is being extracted from corolla.

SALVADORACEAE

SALVADORA L.

KEY TO THE SPECIES

1a. Leaves elliptic-lanceolate or linear-lanceolate. Panicles compact, axillary, shorter than the leaves, flowers sessile, ripe fruits yellow... *S. oleoides*

1b. Leaves elliptic or ovate. Panicles lax, axillary and terminal, often nodding, longer than the leaves, flowers pedicelled, ripe fruits red.. *S. persica*

Salvadora oleoides Decne., in Jacq. Voy. Bot. 140. t. 144. 1844; Clarke, in Hook. *f.* Fl. Brit. Ind. 3: 620. 1882; Duthie, Fl. Gangetic Plain 2: 29. 1911; Bhandari, Fl. Indian Desert 215. 1995; Singh, in Shetty & Singh Fl. Rajasthan 2: 467. 1991; Rao & Chakraborti, in Hajra & Sanjappa, Fl. India Fasc. 22: 5. t. 5. f. 2. 1996. *S. stocksii* Wight, Ill. Ind. Bot. 2. 229. 1850 & Ic. Pl. Ind. Orient. 4: 18. t. 1621B. 1850. (Fig. 82 & Plate 182)

Vern.: *Kharo jhal, Pilu*

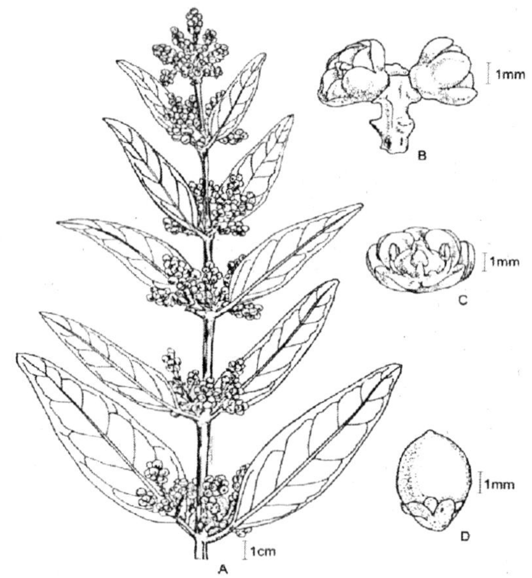

Fig. 82. *Salvadora oleoides* Decne.
A. Fruiting twig, B. Flowers, C. Flower, D. Fruit

A much branched, evergreen medium size tree, 2 - 3 m tall, with short, bent or twisted trunck. Branches drooping, bark grey. Branches stiff, rough, whitish. Leaves, opposite, 5 - 10 x 1 - 1.5 cm, linear-lanceolate or elliptic-lanceolate, acute or mucronate, cuneate at base, coriaceous, whitish-green, glabrous. Petiole 0.2 - 1.2 cm long. Flowers axillary panicles or branched spikes, 2.5 - 4 cm long. Flowers greenish-white, 2 - 3 mm across; pedicel ca 1 mm long or absent. Bracts ovate, acute. Calyx 1.5 - 2 mm long; lobes round, wavy margin, obtuse, divided nearly half way down, glabrous. Corolla ca 2.5 mm long, ovate or oblong; lobes sub-acute or obtuse, recurved. Stamens 4, epipatalous, inserted at the base of the corolla tube. Style absent, stigma peltate. Fruit a drupe, 5 mm in diam., 1-seeded, globose, yellow on maturation.

Field note: Commonly found in the Western part of the study area.

Fl. & Fr.: March - June.

Specimen examined: Masuda 3114, Bhinay 3117 (MLVGCB Herb).

Distribution: Arabia, India (Gujarat and Rajasthan) Pakistan and Aden.

Uses: Wood is used as fuel. Branches and leaves serve as fodder. Sweet fruits are edible and seeds yield a green oil, which is said to be medicinal.

Salvadora persica L., Sp. Pl. 122. 1753; Clarke, in Hook. *f.* Fl. Brit. Ind. 3: 619. 1882 ; Duthie, Fl. Gangetic Plain 2: 28. 1911; Bhandari, Fl. Indian Desert 216. 1995; Singh, in Shetty & Singh Fl. Rajasthan 2: 467. 1991; Rao & Chakraborti, in Hajra & Sanjappa, Fl. India Fasc. 22: 5. t. 1 & 4. f. 3. 1996. *Galena asiatica* Burm. f., Fl. Ind. 88. t. 31. f. 1. 1768. *Salvadora indica* Wight, III. Ind. Bot. 2: 229. t. 181. 1850. (Fig. 83 & Plate 183)

Vern.: *Meetha jal, Pilu.*

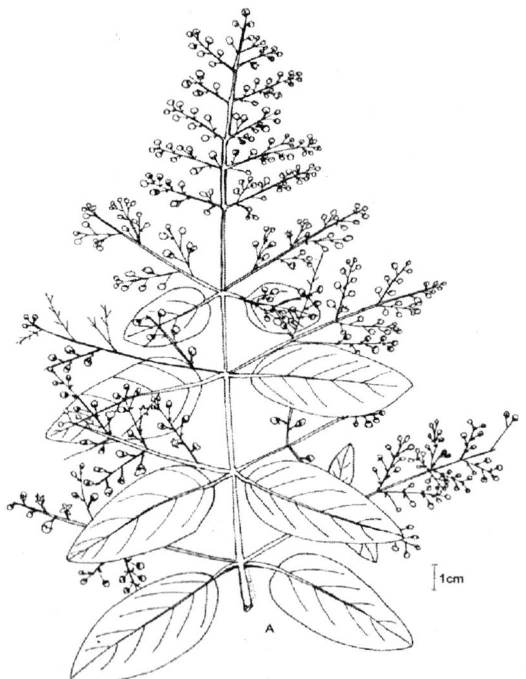

Fig. 83. *Salvadora persica* L.
A. Flowering & fruiting twig

Much branched evergreen, glabrous, small tree, up to 2 - 6 m tall. Branches drooping. Bark whitish yellow to almost white but on the stem somewhat rugose and grey. Leaves opposite, sub-fleshy; petiole 1 - 2.1 cm long; lamina 1 - 5.6 x 0.8 - 4 cm, elliptic ovate, acute or mucronate sometimes obtuse; acute or rarely rounded at base; lateral nerves 5 - 6 pairs. Inflorescence drooping, panicled, 5 - 31 cm long. Flowers greenish yellow, 3 - 4 mm across; pedicel 1 - 3 mm long. Bracts ovate. Calyx ca 1 mm long, glabrous; lobes rounded. Corolla campanulate, 4-lobed, ca 3 mm long, acute to mucronate at apex, reflexed. Stamens 4, epipetalous, inserted at the base of corolla tube, shorter than the corolla but exserted due to reflexed corolla lobes. Ovary 1-locular, glabrous; style absent; stigma more or less peltate. Fruit a drupe, 2.5 - 5 mm in diam., globose, smooth, red on maturation, 1-seeded. Seeds subglobose, brown.

Field note: Common in dry waste lands in association with *Capparis decidua* (Forssk.) Edgew.

Fl. & Fr.: Almost throughout the year.

Specimen examined: Gulabpura 1317, Beawar 65 (MLVGCB Herb).

Distribution: India, Pakistan, Arabia, Palestine, Syria & Africa.

Uses: The branches are commonly used as a toothbrush. Commercially available "Miswak" paste is being prepared from this species. The young branches and the leaves are favourate fodder of camels. Fruits are is edible when ripe and used medicinally.

APOCYNACEAE

KEY TO THE GENERA

1a. Leaves linear. Flowers yellow............ *Thevetia*

1b. Leaves not linear.
Corolla white or dull white 2

2a. Plants armed with spines.
Carpels completly united.
Fruits an indehiscent berry .... *Carissa*

2b. Plants unarmed. Carpels free.
Fruits dehiscent... 3

3a. Erect herbs. Seeds without comas .... *Catharanthus*

3b. Trees or shrubs 4

4a. Leaves in whorls 5

5a. Shrubs .... *Nerium*

5b. Small trees .... *plumeria*

4b. Leaves not in verticles 6

6a. Trees .... *Wrightia*

6b. Shrubs .... *Ervatamia*

CARISSA L. *nom. cons.*

KEY TO THE SPECIES

1a. Spines paired. Corolla tube densely hairy at the base. Berries about 1.5 cm in diam. *C. congest*

1b. Spines solitary. Corolla tube glabrous at the base. Berries about 7 mm in diam. .... *C. spinarum*

Carissa congesta Wight, Ic. 4 (2) : 1. t. 1289. 1848; Singh, in Shetty & Singh Fl. Rajasthan 2: 468. 1991. *C. carandus auct. plur.* non L., 1767; Hook. *f.* Fl. Brit. Ind. 3: 630. 1882, *pro parte*; Duthie, Fl. Gangetic Plain 2: 31. 1911. (Fig. 84 & Plate 184)

Vern.: *Karonda.*

Erect or scandent shrub. Branches dichotomous, spreading, glabrous; spines 2 - 2.5 cm long, stout, mostly straight or slightly curved, sometimes bifurcate. Leaves elliptic-oblong to ovate, 2 - 6 x 1.5 - 4.0 cm, sub-coriaceous, young leaves with pink midribs, base rounded or retuse, acute, mucronate, dark green and shining above. Flowers in corymbose cyme, white, often tinged with pink, ca 2 cm across. Bracts linear, hairy; peduncle pink, ca 1.2 - 2 cm long. Calyx divided more than half, ca 5 mm long, pubescent, lobes lanceolate, acute. Corolla tube cylindrical, glabrous, swollen, pubescent near the top, 1.5 - 2 cm long, lobes white tinged with pink, overlapping to the right in bud, lanceolate, acute, pubescent, ciliolate. Stamens 5, inserted near the top of corolla-tube; anthers oblong, apiculate. Ovary glabrous , 2 locular, with 4 ovules in each cell; style filiform; stigma fusiform, bifid. Fruit 1.25 - 2.5 cm long, globose, ovoid or ellipsoid, white with red shade, purplish black when ripe, with milky juice, 4 - 8 seeded.

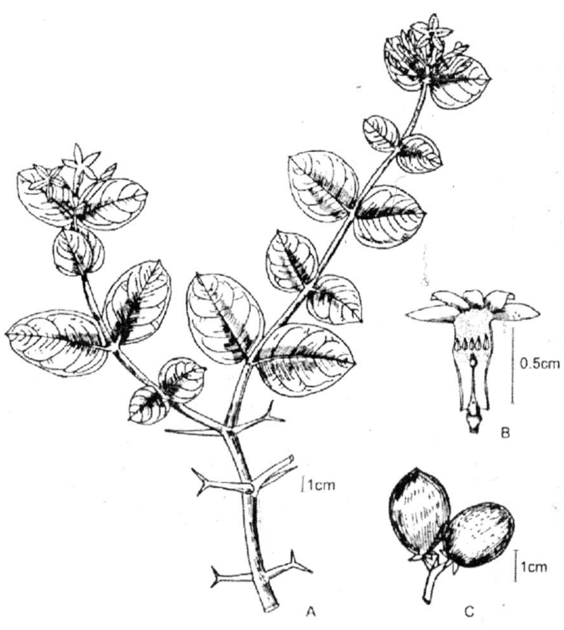

Fig. 84. *Carissa congesta* Wight.
A. Flowering twig, B. Opened corolla showing stamens & pistil, C. Fruit.

Field note: Cultivated throughout the area.

Fl.: January - April; *Fr.*: May - June.

Specimen examined: Gurla 1370, Karoi 1678 (MLVGCB Herb).

Distribution: India, Sri Lanka, Burma, Malacca, Pakistan. Possibly native of S.E. Asia.

Uses: Fruits are edible and used in vegetables and pickels. Extensively grown as a hedge plant.

Carissa spinarum L., Mant. Pl. 2: 559. 1771; Hook. *f.*, Fl. Brit. Ind. 3: 631. 1882; Duthie, Fl. Gangetic Plain 2: 32. 1911; Singh, in Shetty & Singh Fl. Rajasthan 2: 469. 1991.

Vern.: *Karonda.*

Large thorny evergreen shrub, 0.2 - 3 m, occasionally scrambling up to 20 m high; bark grey, slightly rough or smooth; spines simple, rarely forked, 0.5 - 6.2 cm long. Leaves petiolate 1.3 - 3.0 x 1.0 - 2.5 cm, elliptic or broadly ovate - suborbicular, acute or mucronate, entire, often revolute-margined; petiole 0.5 - 6 mm long. Flowers fragrant, white inside, pink to red outside; sepals (narrowly) ovate, 1.5 - 4.5 mm long, 0.4 - 1.7 mm wide, acuminate, ciliate; corolla tube 5 - 25 mm long, tubular, widest at the throat, pubescent at the mouth and lower lobes; lobes overlapping to the right, obliquely oblong or ovate, 4 - 15 x 1 -7 mm, acute at the apex; stamens inserted near the apex of the tube. Fruit red to black, globose or ellipsoid, 5 - 25 mm long, 3 - 20 mm in diam..

Field note: Rare, in waste lands and among hedges on the boundaries of crop fields.

Fl.: January - April; *Fr.*: May - August.

Specimen examined: Bhilwara 317, Bijoliya 1546 (MLVGCB Herb).

Distribution: India, widespread in tropical Africa, tropical Asia and Australia

Uses: Fruits are edible, also extensively grown as a hedge plant.

CATHARANTHUS G. Don.

KEY TO THE SPECIES

1a. Annual herb .... *C. pusillus*

1b. Perennial undershrub .... *C. roseus*

Catharanthus pusillus (Murr.) G. Don, Gen. Hist. 4: 95. 1837; Mahesh. Illus. Fl. Delhi f. 120. 1966; Bhandari, Fl. Indian Desert 194. 1995; Singh, in Shetty & Singh Fl. Rajasthan 2: 468. 1991. *Vinca pusilla* Murr. Novi. Comm. Soc. Reg. Sci. Gotting. 3: 66. t. 2. f. 1. 1773; Hook. f, Fl. Br. Ind. 3: 640. 1882; Trimen, Handb. Fl. Sri Lanka 3: 130. 1895. *Lochnera pusilla* (Murr.) K. Schum. in Eng. & Prantl. Pflanzenfam. 4 (2): 145. 1895; Duthie, Fl. Gangetic Plain 2: 34. 1911. (Fig. 85 & Plate 185)

An erect, glabrous, 10 - 30 cm tall, annual herb. Branches quadrangular. Petiole 0.15 - 0.5 cm long. Leaves 1.5- 6 x 0.3 - 1.5 cm, elliptic-lanceolate, acute, entire, glabrous, base tapering; petioles 3 - 6 mm long; axillary glands subulate. Flowers solitaty or in pairs, axillary, sessile, white. Pedicels shorter than the calyx. Calyx-lobes 3 - 5 mm long, lanceolate-subulate. Corolla white, glabrous; tube ca 8 mm long; the lobes shorter than one-half of the tube, abruptly apiculate at the apex. Filaments short; anthers ex-appendiculate. Follicles 2.5 - 5 cm long, very slender, straight, pointed, membranous, glabrous. Seeds 2.5 - 3 mm long, oblong-cylindric, rounded at both ends, with many mucriculate ribs, black when ripe.

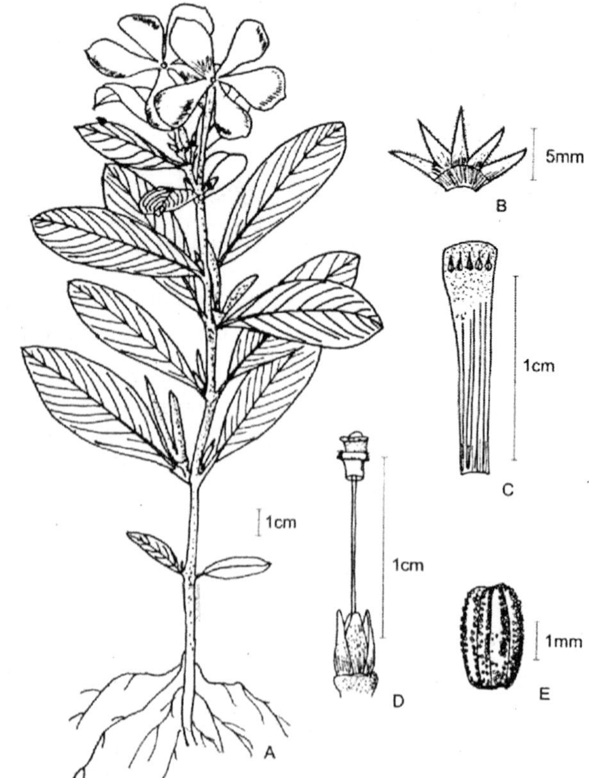

Fig. 85. *Catharanthus pusillus* (Murr.) G. Don. A. Habit, B. Calyx, C. Corolla tube with stamnes, D. Pistil, E. Seed

Field note: A weed of waste lands and cultivated fields in the dry zone.

Fl. & Fr.: July - October.

Specimen examined: Bhilwara 142, Atun 1899 (MLVGCB Herb).

Distribution: India and Sri Lanka.

Catharanthus roseus (L.) G. Don, Syst. 4: 95. 1837; Sant, in Bull. bot. Surv. India 3: 15. 1961. *Vinca rosea* L., Syst. ed. 10. 944. 1759; Clarke, in Hook. *f.* Fl. Brit. India 3: 640. 1882.

Vern.: *Sada bahar.*

Perennial underbshrub, woody at base, 30 - 90 cm tall, young branches pubescent. Leaves oblong 2.5 - 6 x 1.5 - 2.5 cm, entire, obtuse or apiculate at apex, base acute, glabrous or sparingly hairy specially on nerves beneath; petiole short, 3 - 10 mm long, with many glands at the axil. Inflorescence of axillary 1 - 4 flowered cymes. Flowers pink or white; pedicel 1-3 mm long. Calyx tube short, lobes linear, acute, 5-7 mm long, hairy. Corolla tube upto 1 cm long, sparsely pubescent above, throat densely hairy within below the stamens, lobes oblong-rounded, or obovate, 15 - 25 mm long spreading; Disc scales higher than ovary, ca 2-3 mm long. Ovary pubescent; style ca 2.5 cm long; stigma cup shaped. Follicles 2, slender, cylindric, striated. 15 - 25 mm long.

Field note: Cultivated as an ornamental.

Fl. & Fr.: Almost throughout the year

Specimen examined: Bhilwara 165 (MLVGCB Herb).

Distribution: A native of Madagascar; widely cultivated and naturalized in tropics and subtropics of both hemisphere.

Uses: An extract from the root is used to cure the blood cancer. Flowers are also used as a remedy for diabeties.

ERVATAMIA DC.

Ervatamia coronaria (Jacq.) Stapf, Fl. Trop. Africa. 4(1):127. 1904; Duthie, Fl. Gangetic Plain 2: 36. 1911. *Tabernaemontana coronaria* Willd. Enum. Hort. Berol. 275; Hook. f, Fl. Br. Ind. 3 : 646. 1882.

Vern.: *Chandni.*

Shrub upto 1 m high, Branches divaricate, milky latex present. Young shoots green, lenticellate. Leaves opposite, 7 - 15 x 2.5 - 6 cm, elliptic, acuminate or caudate, cuneate at the base, 6 - 8 pairs of lateral nerves; petiole 5 - 7 mm long. Inflorescence 1-few flowered cymes, axillary or terminal, peduncle ca 5 cm long. Flowers 2.5 - 5 cm across, white; pedicel 1 - 1.5 cm long, bract minute or absent. Calyx lobes broad, ovate, acute ca 4 mm long. Corolla tube ca 2 - 2.5 cm, dialated in the middle, lobes obliquely ovate 2.5 - 3.75 cm long, double in cultivated form. Ovary glabrous; ovules numerous; style filiform; stigma papillose. Follicles 2.5 - 4 x 1 cm, sessile, glabrous, 1-3 ribbed, orange or bright red inside, curved to form a beak. Seeds 3 - 6, oblong or irregular in shape, striated.

Field note: Cultivated as ornamental for its jasmine-like flowers.

Fl. & Fr.: May-October.

Specimen examined: Bijoliya 452 (MLVGCB Herb).

Distribution: Native of India; wide spread in the Tropics and warmer regions of the world.

Uses: A decoction of roots is used to treat diarrhoea.

NERIUM L.

Nerium oleander L., Sp. Pl. 209. 1753. *Nerium indicum* Mill. Gard. Dict. n. 2. 1768; Santapau in Rec. Bot. Surv. India 16 (1) : 170. 1953. *Nerium odorum* Soland., in Ait. Hort. Kew 1: 297, 1789; Hook. *f.* Fl. Brit. Ind. 3: 655. 1882; Duthie, Fl. Gangetic Plain 2: 40. 1911. (Fig. 86)

Vern.: *Kaner.*

An erect evergreen shrub, Branches glabrous with milky sap, young branches green. Leaves 10 - 20 x 1 - 2 cm, linear-lanceolate, tapering at both ends, acute to acuminate, coriaceous, midrib prominent, nerves numerous, petiole 5 - 7.5 mm long. Flowers in terminal cymes, white, pink or dark red, single or double in cultivated forms, fragrant 3 - 4 cm across, peduncle and pedicel hairy, bracts small, 5 - 7.5 mm long. Calyx ca 6.25 mm long, divided into 5, linear, acute lobes, hairy with gland at the base inside. Corolla tube 1.8 cm long, hairy within, throat narrow, ending in five twisted petals, tips rounded, corona of 5 scales near the throat of the corolla, cleft into 4 - 7 linear segments. Stamens included, filament short; anthers connivent and adherent to stigma, connectives hairy, produced upward into long thread-like hairy appendages. Ovary with two distinct carpels, style filiform, thickened upward; stigma two lobed. Fruit 12 - 20 cm x 7.5 mm long, cylindrical, straight, rigid. Seeds oblong.

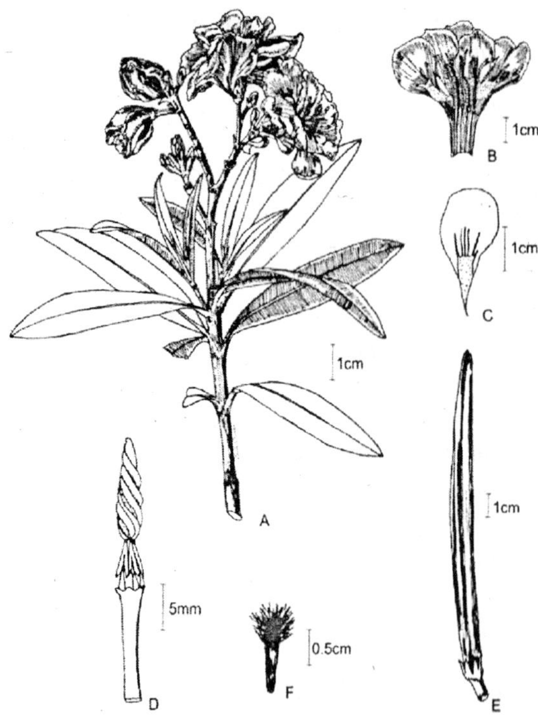

Fig. 86. *Nerium oleander* L.

A. Flowering twig. B. Flower, C. Corolla lobe with coronary appendages, D. Stamens with connectives, E. Fruit, F. Seed.

Field note: Commonly planted as an ornamental for its showy and fragrant flowers.

Fl. & Fr.: April-October.

Specimen examined: Bhilwara 318 (MLVGCB Herb).

Distribution: From the Mediterranean to Persia, China, Japan, India and Pakistan

Uses: All parts of the plant are highly poisonous. Leaves are used in cutaneous eruptions, decoction of leaves is used to kill maggots infesting wounds. Plant is also used as rat poison in southern Europe.

Notes: *Nerium oleander* L. and *Nerium indicum* Mill. are treated by various authors as separate species on the basis of leaf shape, length of protrusion of connectives and the number of teeth of corona appendages. However these characters have not been found to be useful in separating the two taxa. Therefore, in present study Grant, Fosberg & Smith (Smithsonia Contrib. Bot. 17.53.1974) has been followed in treating the two as conspecies.

PLUMERIA L. non. cons

Plumeria rubra L., Sp. Pl. 209. 1753. *Plumeria acuminata* Roxb., Fl. Ind. 2 : 20. 1824. *P. acutifolia* Ait., Hort. Kew. 2. 270. 1789; Hook f.. 3 : 1655. 1882.

Vern.: *Champa.*

Shrub or under tree, bark rough, branches thick, cylindrical, glabrous, swollen and leafy at the tips, leaf scars present. Leaves spirally aranged, obovate-oblong, ca 15 - 30 x 5 - 10 cm, acute at both ends, entire, glabrous above, somewhat pubescent beneath, thickly coriaceous, midrib prominent and thickened, lateral nerves numerous; petiole 2.5 - 3.75 cm long, with gland at the apex. Inflorescence a terminal cyme. Flower white with a yellow centre, ca 5 cm across, externally tinged with pink, fragrant, peduncle ca 12.5 cm long. Calyx small, eglandular; lobes short, rounded. Corolla salvershaped. Fruit a follicle ca 12 cm long, cylindrical.

Field note: Commonly planted in gardens.

Fl.& Fr.: Almost throughout the year.

Specimen examined: Bhilwara 415 (MLVGCB Herb).

Distribution: Probably a native of central America, Mexico and Guetemala.

THEVETIA L.

Thevetia peruviana (Pers.) Merrill in Phil. J. Sci. Bot. 9. 130. 1914; Singh, in Shetty & Singh Fl. Rajasthan 2: 473. 1991. *Cerbera peruviana* Pers. Syn. 1 : 267. 1805; *Thevetia nerifolia* Juss. ex Steud. Nom. ed. 2. 2 : 680, 1841; Duthie, Fl. Gange-tic Plain 2: 42. 1911. (Fig. 87 & Plate 186)

Vern: *Pili kaner.*

Large shrub or small tree upto 6 - 7 m high. Stem and branches glabrous, more or less corky, latiferous. Leaves spiral or alternate, crowded, linear-lanceolate, acute or shortly acuminate, attenuate at base, 8 - 15 x 0.6 - 1 cm, bright green and shining above, sub-coriaceous, margins slightly recurved, midrib prominent, subsessile, with glands in the axils. Inflorescence of few flowered cymes, generally terminal, peduncle very short, bracts variable, caducous. Flowers yellow or orange, pedicel ca 7.5 cm long, bracteoles deltoid-elongated, ca 2 mm long, with glands at the axil. Calyx 6 - 7.5 mm long, glabrous, persistent, divided nearly upto the base; lobes long acuminate, spreading. Corolla ca 5 cm long, tubular at the base, campanulate at the throat, hairy; corona pubescent, opposite, above the stamen, tube 12 - 15 mm long; lobes 3 - 3.5 x 1.7 cm, obovate. Anthers lanceolate, with branched appendages at the tip. Disc thick, cup shaped; ovules 2 in each carpel, stigma broad angular. Fruit somewhat rhomboid, 2 - 4 cm across, glabrous, fleshy, bluntly 4 - angled. Seeds 2, ca 1 cm long.

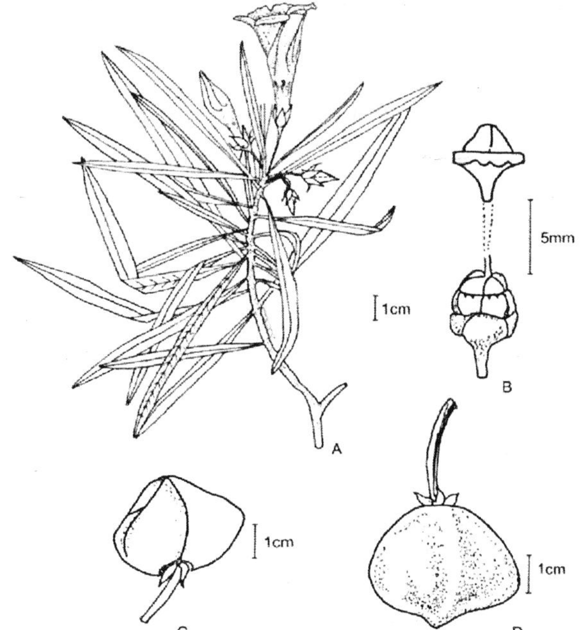

Fig. 87. *Thevetia peruviana* (Pers.) Merr. A. Flowering twig. B. Ovary & stigma, C & D. Fruits

Field note: Common as hedge plant, also planted on roads sides, throughout the area

Fl. & Fr.: Almost throughout the year

Specimen examined: Bhilwara 538 (MLVGCB Herb).

Distribution: A native of Tropical America, commonly cultivated in India and Pakistan.

WRIGHTIA R. Br.

1a. Leaves glabrous beneath. Corona white, fimbriate. Follicles smooth, connate at the base and the apex only. Two fruits produced by each flower .... *W. tinctoria*

1b. Leaves tomentose on both sides. Corona coloured, entire or lobed. Follicles rough with white tubercles, connate throughout. One fruit produced by each flower................. .*W. arborea*

Wrightia arborea (Dennst.) Mabberly, in Taxon 26: 533. 1977; Singh, in Shetty & Singh Fl. Rajasthan 2: 471. 1991. *Periploca arborea* Dennst., Schlussel Hort. Malab. 13. 23 & 25. 1818 & in Fort. Allg. Teut. Gort. - Mag. 3: 32, 41 & 83. 1818. *Wrightia tomentosa* Roem. & Schult., Syst. 4 : 414. 1819; Wight, Ic. 2 (2) : 2. t. 443. 1841; Hook. *f.* Fl. Brit. Ind. 3: 653. 1882; Duthie, Fl. Gangetic Plain 2: 39. 1911. *Nerium tomentosum* Roxb. Hort. Beng. 19, 1814 (nom. nod.) et Fl. Ind. 2: 4. 1832. (Plate 187)

Vern.: *Kalakunda.*

A small deciduous tree, branches tomentose when young, glabrous-puberulous when mature with milky latex. Leaves 3 - 12 x 1.5 - 6.0 cm, elliptic ovate-oblong, acuminate, base rounded, tomentose on both surrfaces, margins thickly undulate, 8 - 14 pairs of lateral nerves, petiole 2.5 - 8 mm long, with glands in the axil. Flowers in terminal corymbose, white turning yellow, c. 2.5 cm across, with unpleasant odour; pedicel tomentose, ca 6 mm long; bracts ovate-lanceolate, acute, tomentose, ca 2.5 mm long. Calyx tomentose; lobes short rounded, tomentose ca 2.5 mm long, glandular within. Corolla-tube about twice as long as the calyx, cylindrical, lobes linear or obovate-oblong, rounded or obtuse, thick and brittle; corona scales variable 5 - 10, orange in colour, laciniate, united to form a cup at the base. Anthers white. Fruit 15 - 20 x 1 - 2 cm, cylindric, stout, laterally compressed, connate throughout, pendulous, grooved at the union of the carpels, beaked, striate with white dots all over. Seeds linear-oblong, brown, with long silky coma.

Field note: Rare in deciduous forests, one locality only.

Fl. & Fr.: April -June.

Specimen examined: Karoi 2129 (MLVGCB Herb).

Distribution: India, China, Sri Lanka, Burma and Penang.

Wrightia tinctoria (Roxb.) R. Br., in Mem. Wern. Soc. 1: 74. 1811; Wight, Ic. 2 (2): 2. t. 444. 1841; Hook. *f.* Fl. Brit. Ind. 3: 653. 1882; Duthie, Fl. Gangetic Plain 2: 38. 1911; Bhandari, Fl. Indian Desert 194. 1995; Singh, in Shetty & Singh Fl. Rajasthan 2: 472. 1991. *Nerium tinctorium* Roxb., Orient. Repert. 1: 39. 1791, cum. tab. (Plate 188)

A small deciduous tree, upto 6 m high, branchlets glabrous or puberulous, with milky latex. Leaves 4.5 - 15.0 x 2.5 - 7.0 cm, oblong-lanceolate, acute or acuminate, base rounded, glabrous or puberlous beneath, with 8 - 14 pairs of lateral nerves; petiole 2.5 - 5 cm long. Inflorescence a lax terminal cyme. Flower white or yellowish, fragrant, 2.5 cm across; pedicel 2 - 1.2 cm long; bracts ca 2.5 mm long, ovate. Calyx ca 2.5 mm long deeply divided, hairy or glabrous, lobes oblong, obtuse, glandular inside, margins scarious. Corolla tube ca 3 mm long, lobes upto 7 mm long, narrow, oblong, obtuse; corona of many linear scales, inserted on the filaments and corolla lobes. Anthers white or yellowish, forming a cone. Follicles 25 - 50 x 0.6 - 0.8 cm, drooping, united at the apex when young, cylindrical, glabrous, pendulous, striate. Seeds 1.25 - 2 cm long, linear-oblong pale brown; coma twice as long as seed.

Field note: Common in dry and moist deciduous forsts.

Fl.: March-June; *Fr.*: August-April

Specimen examined: Karoi 115, Beawar (MLVGCB Herb).

Distribution: Native of central India to Burma and Timor.

Notes: This plant is very similar to *Holarrhena antdysenterica* (Roth.) A. DC., in which the anthers are included in the corolla tube which is without corona and the seeds have the coma at the apex. In *Wrightia tinctoria* R. Br., the anthers are exserted, coronary scales are present and the seeds bear the coma at the base.

ASCLEPIADACEAE

KEY TO THE GENERA

1a. Corona double, corolline and staminal........... 2

2b. Laves reduced. Stem jointed, cylindric .... *Sarcostemma*

2b. Leaves well developed. Stem not as above 3

3a. Roots bulbous. Corolla purple, tubular, swollen at base, petals free above but tips united... *Ceropegia*

3b. Roots not bulbous. Corolla not purple, rotate or campanulate, petals totally free above 4

4a. Latex milky. Flowers dull white with pink tinge. Anthers with membrounous appendages; pollinia stalked. Follicles softly echinate.................. *Pergularia*

4b. Latex in form of sap. Flowers greenish white or yellow. Anthers withought membrounous appendages; pollinia sessile. Follicles glabrous........................... *Leptadenia*

1b. Corona single, coralline, processes in the sinuses of the lobes.. 5

5a. Plants shrub.................................. *Calotropis*

5b. Plants climbing or scrabling........................... 6

6a. All leaves deeply cordate .... *Wattakaka*

6b. Leaves not cordate or very shallowly cordate 7

7a. Flowers in lax, dichasially branched or in dense umbellate cymes, but in latter case paired at the nodes. corolla rotate. Pollinia erect or horizontal .... *Tylophora*

7b. Flowers in umbellate or racemose cymes, cymes solitary at the nodes, sub-sessile. Corolla divided to the base. Pollinia pendulous......................... *Pentatropis*

CALOTROPIS R. Br.

KEY TO THE SPECIES

1a. Leaves distinctly petioled .... *C. acia*

1b. Leaves sessile or nearly so 2

2a. Corolla lobes white. Corona lobes shorter than the staminal column, with two obtuse auricles just below the apex .... *C. gigantea*

2b. Corolla lobes white with purple tip. Corona lobes equaling or exceeding the staminal column, withought auricles below the apex .... *C. procera*

Calotropis acia Buch.-Ham., Trans. Linn. Soc. 14: 247. 1824; Duthie, Fl. Gangetic Plain 2: 49. 1911. *Asclepias herbacea* Roxb., Fl. Ind. ed. 2. 2 : 50. 1832. *Calotropis herbacea* (Roxb.) Wight, Contr. B. Ind. 54. 1834; Icon. Pl. Ind. Or. 2 : t. 492. 1843.

Vern.: *Akro.*

An erect shrub. Stem herbaceous. Leaves petioled, 7-10 x 15 - 22 cm, obovate - oblanceolate or oblong, acute or abruptly acuminate, tapering towards the base, clothed beneath when young with deciduous white floccose tomentose; midrib very stout; main lateral nerves about 6-pairs, prominent beneath, arcuate. Petiole 1 - 2.5 cm. long. Flowers about 2.5 cm. across, arranged in compound umbels. Calyx about half as long as the corolla; lobes lanceolate, acuminate, green tinged with pink, floccose-tomentose outside. Corolla greenish at the base, purplish above; lobes erect, 1 cm. long, tapering to an acuminate apex, margins undulate. Corona-lobes glabrous, pink, shorter than the staminal column, with obtuse auricles below the bifid apex, basal spur abruptly incurved. Follicles subglobose, upto 10 cm long.

Field note: Very rare, the species may be readily distinguished by its stalked leaves.

Fl. & Fr.: October - July.

Specimen examined: Singoli 453 (MLVGCB Herb).

Distribution: Tropical Himalaya (Nepal, Sikkim and Assam).

Calotropis gigantea (L.) R. Br. in, W.T. Aiton f. Hort. Kew. ed. 2. 2 : 78. 1811; Hook. *f.* Fl. Brit. Ind. 4: 18. 1883; Cooke, Fl. Pres. Bomb. 2: 214. 1904; Duthie, Fl. Gangetic Plain 2: 48. 1911; Bhandari, Fl. Indian Desert 195. 1995; Parmar, in Shetty & Singh Fl. Rajasthan 2: 475. 1991. *Asclepias gigantea* L., Sp. Pl. 214. 1753. (Fig. 88 & Plate 189).

Vern.: *Dhoro akro, Safed akro.*

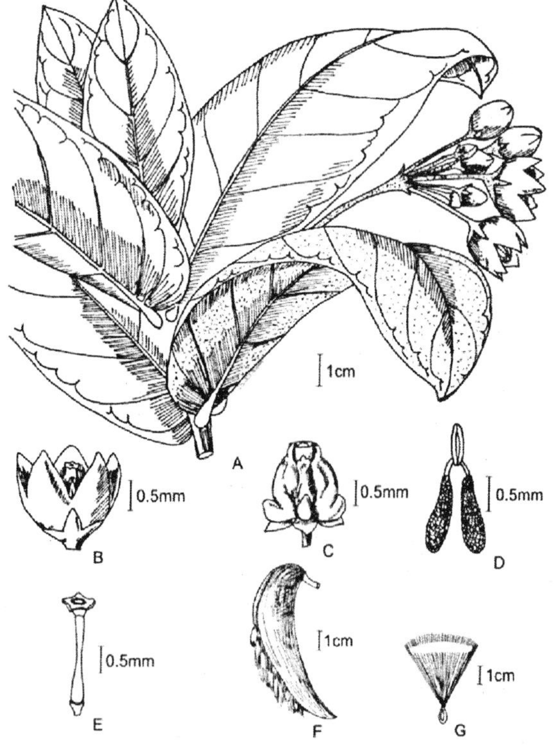

Fig. 88: *Calotropis gigantea* (L.) R. Br.
A. Flowering twig. B. Flower, C. Column & corona, D. Pollinia, E. Pistil, F. Fruit, G. Seed

Undershrub or small tree, 1 - 3 m tall, young branches covered with white cottony tomentum. Leaves 6 - 20 x 3 - 4 cm, elliptic-ovate to obovate, base cordate, acute, glabrous or appressely white floccose; petiole almost absent to 2 cm, young leaves covered with cottony tomentum becoming subglabrous. Flowers in umbelliform cymes, 3 - 4.5 cm in diam., white. Peduncles up to 18 cm long, pedicels 5-8.8 cm long. Calyx 3 - 4 mm long, acute. Corolla 8 - 22 mm long; lobes deltoid, revolute, subulate; corona lobes ca 1.2 cm long. Fruit 7-10 x 2.5-4.0 cm, recurved. Seeds 5-6 mm long, broadly ovoid, flat, minutely tomentose; coma, white, silky.

Field note: Common in open waste lands along roadsides, also cultivated as a sacred plant in houses, gardens etc.

Fl. & Fr.: November- June.

Specimen examined: Bhilwara 1324 (MLVGCB Herb).

Distribution: India, Sri Lanka, Nepal, Maldive Islands, South China and Malaysia.

Calotropis procera (Ait.) Ait. f. subsp. ***hamiltonii*** (Wight) Ali, in Notes Roy. Bot. Gard. Edinberg 38 (2) : 287 - 290. f. 1 C-D. 1980; Parmar, in Shetty & Singh Fl. Rajasthan 2: 475. 1991. *C. hamiltonii* Wight, Contrib. Bot. Ind. 53. 1834. *C. procera auct.* non (Ait.) Ait. f. 1811; Wight, *l.c.* 4 (1): 17. t. 1278. 1848; Hook. *f.* Fl. Brit. Ind. 4 : 18. 1882; Duthie, Fl. Gangetic Plain 2: 48. 1911; Jagtap & Singh, Fl. India Fasc. 24: 11. 1999. (Fig. 89 & Plate 190)

Vern.: *Aak, Madar.*

An erect shrub or small tree up to 3 m tall, much branched from the base, with milky latex; young branches covered with white cottony tomentum. Bark soft, corky. Leaves opposite, 5 - 15 x 1.8 - 10 cm, ovate to obovate, ovate-oblong, entire, base cordate, apex acute, subsessile, young leaves covered with white cottony tomentum, becoming subglabrous. Flowers in axillary terminal corymbose cyme, ca 2.5 cm across, white outside, purplish within, tips darker. Peduncles upto 10 cm long, pedicels 1.5 - 4.2 cm long. Calyx upto 7 x 4 mm; lobes ovate, acute, cottony outside, glabrous inside. Corolla 1 - 1.3 cm across, divided about 2/3 the way down; lobes, acute, glabrous; corona lobes upto 7 x 5 mm, slightly exceeding the staminal column, compressed, 2-fid, purplish above. Fruit 6.5-9.5 x 3-5.1 cm, ellipsoid or ovoid, recurved, many seeded. Seeds 6-8 mm long, flat, broadly ovate, minutely tomentose; coma silky, 3.1 - 3.2 cm long.

Field note: Common species throughout the area.

Fl. & Fr.: Round the year.

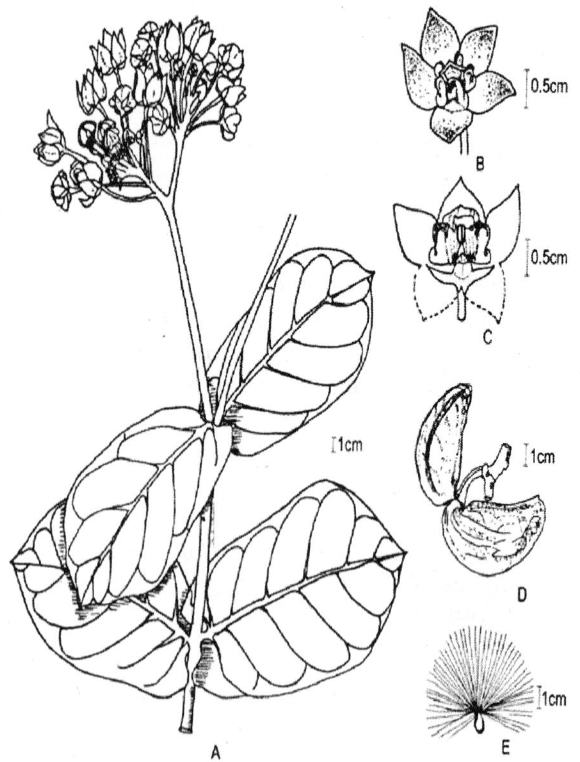

Fig. 89. *Catotropis procera* (Ait.) Ait. f. subsp. *hamiltonii* (Wight) Ali.
A Flowering twig. B. Flower, C. Corona & gynostegium, D. Fruits, E. seed

Specimen examined: Nasirabad 4213, Tikar 1891 (MLVGCB Herb).

Distribution: Tropical Africa, India, Afghanistan, Pakistan, Persia.

Notes: The type subspecies which differs from subsp. *hamiltonii* in having bigger (10.8-14.5 cm x 9.7-11.7 cm) almost globose fruits with the tip invaginated at maturity, is not represented in the study area.

Uses: A medicinal plant. The latex is irritant to the skin and mucous membrane and said to cause blindness. It is also used as a purgative and said to be specific for Guinea worms. The seed floss is used for stuffing mattresses, pillows etc. It is sometimes used to adulterate Indian Kapok but it is inferior to it in resilience and water repellent properties. Being short stapled it cannot be spun by itself.

CEROPEGIA L.

KEY TO THE VARIETIES

1a. Leaves orbicular or ovate, long-petioled.............. *C. bulbosa* var. *bulbosa*

1b. Leaves linear to lanceolate, subsessile........................ *C. bulbosa* var. *lushii*

Ceropegia bulbosa Roxb., Pl. Corom. 1: 11. t. 7. 1795; Hook. *f.* Fl. Brit. Ind. 4: 67. 1883; Duthie, Fl. Gangetic Plain 2: 66. 1911; Cooke Fl. Pres. Bomb. reprint. ed 2: 24.0. 1958; Bhandari, Fl. Indian Desert 197. 1995; Ansari, Fl. India Fasc. 16 : 10. 1984; Parmar, in Shetty & Singh Fl. Rajasthan 2: 477. 1991; Jagtap & Singh, Fl. India Fasc. 24: 217. 1999. *C. acuminata* Roxb. Pl. Cor. 1 : 12. t. 8. 1795; Hook. *f., l. c.* 4: 70. 1883. *C. tuberosa* Dalz. & Gibson. Bomb. Fl. 154. 1861 (non Roxb.), *C. esculenta* Edgew. in J. Linn. Soc. 6: 204. 1862. *C. bulbosa* Roxb. var. *esculenta* (Edgew.) Hook. *f., l. c.* 4: 67. 1883. *C. bulbosa* Roxb. var. *lushii* (Grab.) Hook. *f.* Fl. Brit. Ind. 4: 67. 1883. (Fig. 90 & Plate 191)

var.: ***bulbosa***

Vern.: *Khadula.*

Twinning perennial herb, with underground tuberous stem.Tuber globose. Stem trailing or twining, usually glabrous. Leaves orbicular, or elliptic to obovate, 2.5 - 4.5 x 2.0 - 3.5 cm, slightly cordate, fleshy, glabrous. Flowers in umbelliform pedunculate cymes; peduncle 1.2 - 2.5 cm across. Pedicels 3 - 7 mm long. Sepals 1 - 1.5 cm long, lanceolate, acute. Corolla 1.2 - 3 cm across, purple striate outside, reddish brown inside, 2.5 cm long; lobes narrow, densely pilose. Outer corona lobes rounded to shortly acute, inner lobes linear, ca 2 mm long. Follicles 10 cm long, taper towards apex, glabrous. Seeds 8 - 9 mm long, flattened, broadly-ovate, margins broad, membranous; coma 2.5 - 3.8 cm long, silky.

Field note: Common in grasslands, hillocks etc.

Fl. & Fr.: July - August.

Specimen examined: Gupteshwar Mahadev 1918, Basi 2137 (MLVGCB Herb).

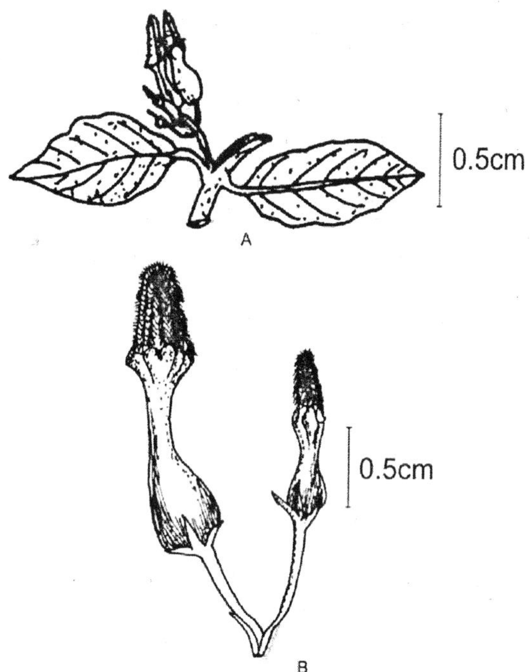

Fig. 90. *Ceropegia bulbosa* Roxb. var. *bulbosa* A. Flowering twig. B. Flowers.

Distribution: India (Punjab, Uttar Pradesh, W. Peninsula) and Pakistan.

Uses: The tubers and leaves are eaten by tribals and shepherds. An alkaloid, ceropegine, is present in tubers which makes them bitter, somewhat like raw turnips.

Ceropegia bulbosa Roxb. var. ***lushii*** (Grah.) Hook. *f.* Fl. Brit. Ind. 4: 68. 1883; Cook, Fl. Pers. Bombay Reprinted 2: 176. 1904; Ansari, Fasc. Fl. India 16: 11. 1984; Parmar, in Shetty & Singh Fl. Rajasthan 2: 477. 1991. *C. lushii* Grah. in Bot. Mag. t. 3300. 1834. *C. esculenta* Edgew. in J. L. Soc. 5: 204. 1862. *C. bulbosa* Roxb. var. *esculenta* (Edgew) Hook. *f. l.c.* 4 : 68. 1883. (Fig. 91)

Vern.: *Suvakhadul.*

Very similar to the *Ceropegia bulbosa* Roxb., but can be distinguished by its linear to lanceolate, sub-sessile leaves.

Field note: Common in grasslands, hillocks etc.

Fl. & Fr.: July-September.

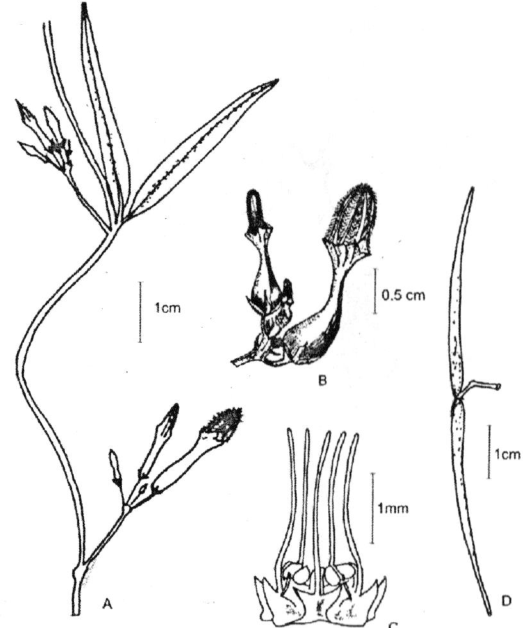

Fig. 91. *Ceropegia bulbosa* Roxb. var. *lushii* (Grah.) Hook.
A. Flowering twig. B. Flowers,
C. Corona & gynostegium, D. Fruits.

Specimen examined: Bijoliya 2183 (MLVGCB Herb).

Distribution: India.

LEPTADENIA R. Br.

KEY TO THE SPECIES

1a. Erect, bushy shrubs, leafless or nearly so. Follicles tapering into straight beak .... *L. pyrotechnica*

1b. Twinins shrubs, leaves well developed. Follicles tapering into curved beak .... *L. reticulata*

Leptadenia pyrotechnica (Forssk.) Decne. in Ann. Sci. Nat. Ser. 2. 9: 269. 1839; Mahesh. Illus. Fl. Delhi f. 128. 1966; Bhandari, Fl. Indian Desert 199. 1995; Parmar, in Shetty & Singh Fl. Rajasthan 2: 480. 1991; Jagtap & Singh, Fl. India Fasc. 24: 259. 1999. *Cynanchum pyrotechnicum* Forssk. Fl. Aegypt.-Arab. 53. 1775. *Leptadenia spartium* Wight & Arn., in Wight. Contrib. India. Bot. 48. 1834; Hook. *f.* Fl. Brit. Ind. 4: 64. 1883; Duthie, Fl. Gangetic Plain 2: 63. 1911. (Fig. 92 & Plate 192)

Vern.: *Khip, Khimp.*

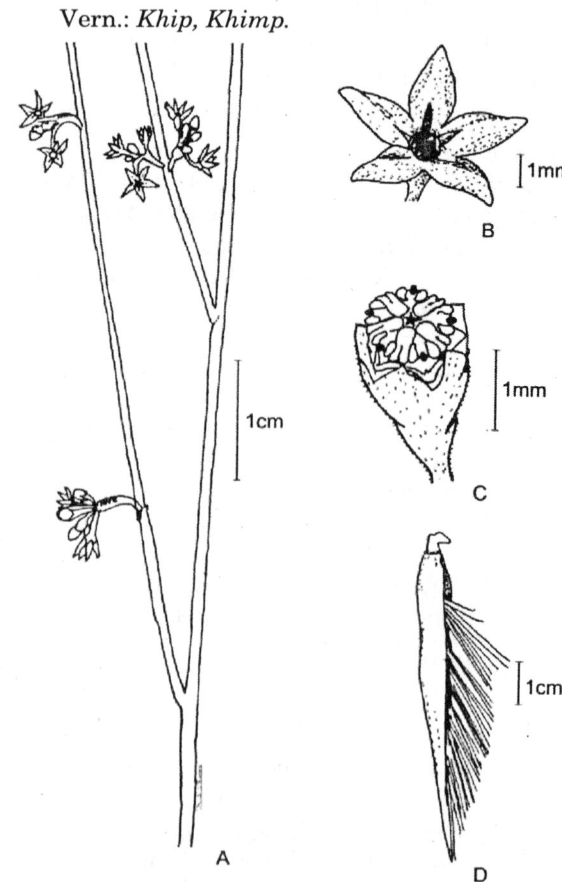

Fig. 92. *Leptadenia pyrotechnica* (Forsk.) Decne.
A. Flowering branch, B. Flower, C. Calyx with gynostegium & corona, D. Fruit.

A much branched erect, generally leafless shrub, 0.6 - 1.5 m tall. Branches erect, slender, green, erect; young parts puberulous. Leaves (when present), sesile, 2.5 - 4 x 0.3 cm, linear to linear-lanceolate, glabrous, acute. Flowers in lateral umbellate cymes. Peduncle 2.5 - 24 mm long, pubescent; pedicel ca 2.5 mm long, pubescent. Calyx ca 1.5 mm long, lobes as long as the tube, acute, pubescent. Corolla 2 - 3.5 mm long, deeply divided near the base, funnel-shaped, glabrous below, lobes pubescent within and without; corona-lobes pubescent at apex. Ovary glabrous. Follicles 7.5 - 11.5 x 0.8 cm, lanceolate, terete, turgid, glabrous. Seeds c. 8 mm long, ovate-lanceolate, flat; coma 2.5 - 4 cm long.

Field note: Common throughout the area in dry habitats.

Fl. & Fr.: August-January.

Specimen examined: Dhanopmata 1355, Ajmer 3689 (MLVGCB Herb).

Distribution: Pakistan, India, Iran, Arabia, Egypt, Sudan, Somalia, Chad, Libya and Algeria.

Uses: This species is medicinally important and is also used for its fibres. The stems are used in makig huts and brooms.

Leptadenia reticulata (Retz.) Wight & Arn. in Wight, Contrib. India. Bot. 47. 1834; Wight, Ic. 2(1): 4. t. 350. 1840; Hook. *f.* Fl. Brit. Ind. 4: 63. 1883; Duthie, Fl. Gangetic Plain 2: 63. 1911; Parmar, in Shetty & Singh Fl. Rajasthan 2: 481. 1991; Jagtap & Singh, Fl. India Fasc. 24: 261. 1999. *Cynanchum reticulatum* Retz. Obs. Bot. 2: 15. 1781. *Daemia reticulata* (Retz.) Moon, Cat. 20. 1824. (Plate 193)

Vern.: *Jivanti.*

Straggling shrub, upto 8 m long, much branched; branches hoary pubescent. Stem with watery sap, corkey deeply cracked bark. Leaves 3.0 - 10.5 x 1.5 - 7.0 cm, broadly ovate to ovate - lanceolate, acute at apex, rounded to slightly cordate at base, sub-coriaceous; petioles 1.2 - 1.8 cm long. Flowers in lateral or subaxillary, many flowered umbellate cymes, greenish. Peduncles 1 - 7 cm long, puberulous. Calyx lobes ovate-oblong, pubescent outside. Corolla - lobes valvate, pale green or white. Outer corona of 5, fleshy quadrate - truncate lobes; Staminal corona angular. Anthers without terminal appendages, pollinia sub-erect; corona double, outer corolline massive, inner staminal, of 5-shallow angular lobes around staminal column. Follicles paired cylindric, 5-9 x 2-3 cm, oblong or broadly lanceolate, with a thick curved beak. Seeds ca 1 cm long, narrowly ovoid or oblong, flate; coma 3 - 4 cm long.

Field note: Common in forests, foot hills and on the hedges of fields.

Fl. & Fr.: Almost throughout the year.

Specimen examined: Iras 1024, Borda 1889 (MLVGCB Herb).

Distribution: Tropical Asia and Africa.

Ueses: The paste of leaves is taken orally with water to cure gangrene.

PENTATROPIS Wight & Arnott

Pentatropis nivalis (J. F. Gmel.) Field & Wood, In Kew Bull. 38: 215. 1983; Jagtap & Singh, Fl. India Fasc. 24: 39. 1999. *Asclepias nivalis* J. F. Gmel., in L., Syst. Nat ed. 13. 2 (1 : 444. 1791. *Pentatropis spiralis* (Forssk.) Decne. in Ann. Sci. Nat. Ser. 2. 9: 327. t. 11. 1838; Hook. *f.* Fl. Brit. Ind. 4: 19. 1883; Bhandari, Fl. Indian Desert 201. 1995; Parmar, in Shetty & Singh Fl. Rajasthan 2: 483. 1991. *Asclepias spiralis* Forssk. Fl. Aegypt. - Arab. 49. 1775. *Pentatropis cynanchoides* R. Br., in Salt. Voy. Abyss. App. 64. 1814; Duthie, Fl. Gangetic Plain 2: 52. 1911. (Fig. 93)

A slender twining shrub. Stem slender, terete. Roots tuberous. Old branches with rough corky bark. Young branches, glabrous or puberulent, green. Leaves opposite, 1 - 5 x 0.6 - 2 cm, ovate or elliptic, acute or obtuse, mucronate, fleshy. Petiole 2.5 - 8 mm long. Flowers in lateral, umbellate cymes. Peduncle up to 5 mm long. Pedicel 5-12.5 mm long. Bracts minute, subulate, pubescent. Calyx lobes ca 1.5 - 2 x 0.5 - 1 mm, ovate-lanceolate, pubescent. Corolla lobes 8 - 13 mm long, narrowly linear from base, glabrous outside, minutely puberulous inside. Corona lobes c. 1.5 mm long. Follicles 5 - 7.5 x 0.8 - 1.5 cm, lanceolate, tapering at apex, glabrous. Seeds ca 6 mm long, ovate, flattened, minutely irregularly crenate at the lower end; coma 1.8 - 2.5 cm long.

Field note: Common in hedges and waste-lands.

Fl. & Fr.: January- August.

Specimen examined: Bhilwara 221, Asind 1560 (MLVGCB Herb).

Distribution: Pakistan, India, Afghanistan, Iran, Arabia and Tropical Africa.

Uses: Flowers are used medicinally and the tubers are reported to be edible.

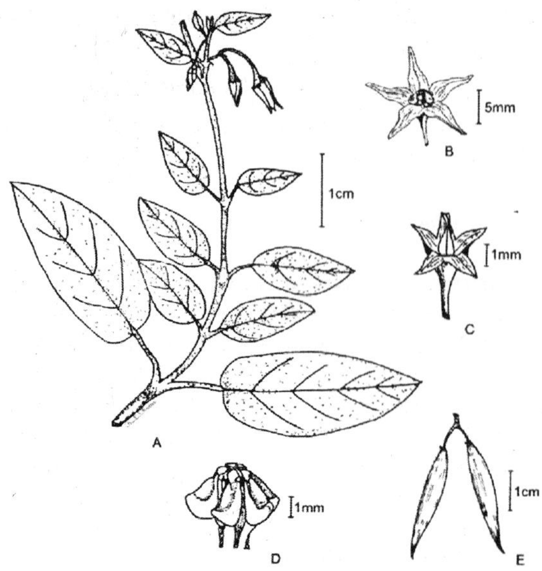

Fig. 93. *Pentatropis nivalis* (J.F. Gmel.) Field & Wood. A. Flowering twig B. Flower, C. Ovary & calyx, D. Corona & gynostegium, E. Fruits

PERGULARIA L.

Pergularia daemia (Forssk.) Chiov. in Result. Sc. Miss. Stefanini Paoli Somal. Ital. 1: 115. 1916; Rech. f. Fl. Iran. 73: 20. 1970; Parmar, in Shetty & Singh Fl. Rajasthan 2: 484. 1991; Bhandari, Fl. Indian Desert 201. 1995; Jagtap & Singh, Fl. India Fasc. 24: 41. 1999. *Asclepias daemia* Forssk. Fl. Aegypt. - Arab. 51. 1775. *Cynanchum extensum* Jacq., Misc. Aust. Bot. 2: 353. 1781. *Daemia extensa* (Jacq.) R. Br. in Mem. Wern. Soc. 1 : 50. 1811; Hook. *f.* Fl. Brit. Ind. 4: 20. 1883; Duthie, Fl. Gangetic Plain 2: 52. 1911. (Fig. 94 & Plate 194)

Vern.: *Bhainsa singha.*

Perennial, hispid, twining undershrub. Branches with spreading hairs. Leaves 3 - 12 x 2.8 - 9 cm, broadly ovate, base cordate, acute to acuminate, glabrous or shortly pubescent above, usually velvety pubescent below. Petiole 1.8 - 6 cm long, pubescent. Flowers drooping, corymbose or racemose cymes. Pedicel 1.8 - 2.5 cm long, pubescent. Bracts minute, lanceolate, acute, pubescent. Sepals ca 2 - 3 mm long, ovate-lanceolate, acute, generally hairy. Corolla greenish-yellow or dull white, campanulate; tube ca 3.5 mm long; lobes ca 5 - 6 x 2.5 - 3.5 cm long, spreading, ovate-oblong, acute, pubescent. Staminal corona biseriate; outer corona membranous, 5-lobed, truncate; inner corona lobes 5 - 8 mm long, including the horns, horns curved over staminal column. Follicles 5 - 8 x 1 - 3 cm, lanceolate, beaked, soft echinate, puberulous. Seeds c. 8 x 4 mm, ovate, tip truncate, velvety pubescent; coma 2.5 - 3.0 cm long, silky white.

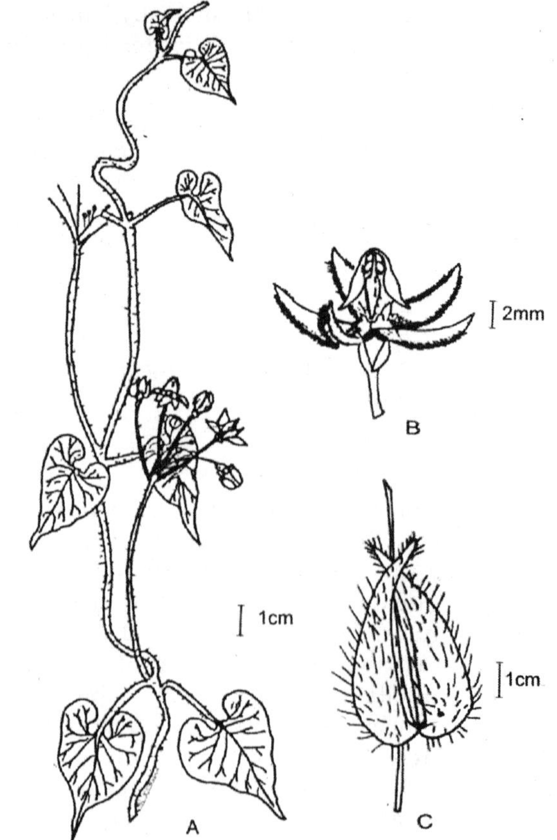

Fig. 94 *Pergularia daemia* (Forsk.) Chiov. A. Flowering branch, B. Flower, C. Fruit.

Field note: Very common among hedges, near villages and on the fencing of fields throughout the area.

Fl. & Fr.: March- December.

Specimen examined: Bhilwara 933, Asind 1821 (MLVGCB Herb).

Distribution: W. Pakistan, India, Sri Lanka, Burma, Malaya, Afghanistan through Arabia to South and Tropical Africa.

Uses: Handful of leaves mixed with salt and made into paste, taken orally to relieve stomachache.

SARCOSTEMMA R. Br.

Sarcostemma acidum (Roxb.) Voigt, Hort. Sub. Calc. 542. 1845; Bhandari, Fl. Indian Desert 202. 1995; Raizada, Suppl. Duthie, Fl. Gangetic Plain 154. 1976. *Asclepias acida* Roxb., Fl. Ind. 2: 31 1832. *Sarcostemma brevistigma* Wt. & Arn. in Wight, Contrib. 59. 1834; Wight Ic. 2 (3): 7 t. 595. 1842; Hook. *f.* Fl. Brit. Ind. 4: 26. 1883. *Sarcostemma viminale* (L.) R. Br. in Mem. Wern. Soc. 1: 51. 1809, printed in 1811 et Prodr. Fl. Nov. Holl. 464. 1810 subsp. *viminale*; Ali, Fl. Pakistan 150: 30. 1983; Parmar, in Shetty & Singh Fl. Rajasthan 2: 485. 1991; Jagtap & Singh, Fl. India Fasc. 24: 50. 1999. *Cynanchum viminale* (L.) L. Mant. Pl. Alt. 2: 392. 1771. (Fig. 95 & Plate 195)

Vern.: *Somlata.*

Straggling shrub. Stem cylindrical, leafless, much branched, 3 - 7 mm in diam., green, glabrous, internodes 10 - 20 cm long. Flowers in sessile many-flowered terminal umbels, sweet-scented. Pedicels 6 - 12 mm long, pubescent. Bracts minute, linear-lanceolate. Sepals 5, 1 - 2 mm long, subacute, margins membranous, glandular inside near the base. Corolla campanulate, pale greenish white, divided almost to the base; lobes ca 5 mm long, ovate-oblong, subacute. Outer corona crenately 10-lobed, lobes of the inner corona thick, fleshy obtuse, longer than the staminal column and concealing the anther. Staminal column very short. Pollen-masses compressed, clavate. Ovaries glabrous, many ovuled; style apex shortly conical. Follicles paired, 4.5 - 12.5 cm long, 3 - 5 mm across, thick, lanceolate, tapering at both ends, tip blunt, slightly diverging when 2 together. Seeds 5 x 3 mm, flattened ovate. Coma 1.8 - 2 cm long, silky white.

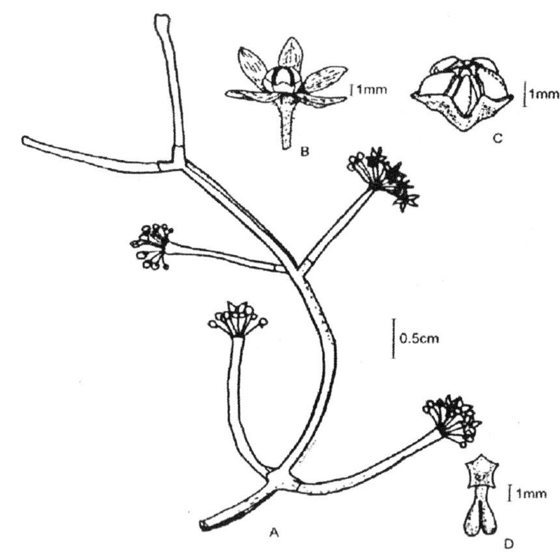

Fig. 95. *Sarcostemma acidum* (Roxb.) Voigt. A. Flowering twig, B. Flower, C. Corona & gynostegium, D. Pollinia.

Field note: Rare in rocky habitats forming an association with *Euphorbia caducifolia.*

Fl. & Fr.: June-November.

Specimen examined: Asind 1228, Badnor 1688 (MLVGCB Herb).

Distribution: Tropical Africa, India (South India, Bengal), and Burma.

Uses: The milky juice of the plant is reported to be drunk by travellers. Stems of this plant are used to extripate white ants from sugar cane fields

TYLOPHORA R. Br.

Tylophora hirsuta (Wall.) Wight & Arn. Contrib. Bot. Ind. 49. 1834; Hook. *f.* Fl. Brit. Ind. 4: 43. 1883; Duthie, Fl. Gangetic Plain 2: 61. 1973; Parmar, in Shetty & Singh Fl. Rajasthan 2 : 487. 1991; Jagtap & Singh, Fl. India Fasc. 24: 157. 1999. *Gymnema hirsutum* Wall. Tent. Fl. Nep. 50. 1824; Wight, Ic. 4(1) : 16. t. 1272. 1848. *Tylophora asthmatica* Wt. & Arn. in Wight, Contrib. 591. 1834; Hook. *f.* Fl. Brit. Ind. 4: 44. 1883. *T. jacquemontii* Decne., in DC. Prodr. 8: 607. 1844. (Fig. 96 & Plate 196)

Vern.: *Antamul.*

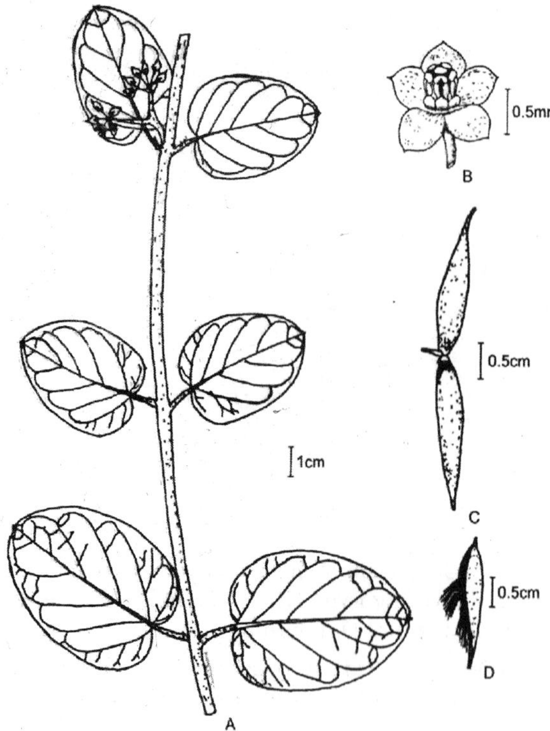

Fig. 96. *Tylophora hirsuta* (Wall.) Wt. & Arn. A. Flowering twig, B. Flower, C. Fruits, D. Dehisced fruit

Twining shrub, provided with dense spreading hairs. Leaves opposite, decussate, 5 - 10 x 4 - 5 cm, ovate-oblong or ovate-lanceolate, base rounded to slightly cordate, tip acute to acuminate, hairy; petioles terete, 0.5 - 3 cm long. Flowers many in extra-axillary umbels, clusters at the angles of the zig-zag peduncle. Peduncle simple or branched at nodes, variable in length, villous. Pedicel upto 1.3 cm, filiform, glabrous. Bracts ca 2.5 mm long, linear. Calyx lobes ca 1.5 - 2 mm long, lanceolate, hirsute. Corolla ca 1 - 1.5 - 4.0 cm across; lobes 5, ca 2.5 x 1.5 mm, lanceolate, apex acute, puberulous within. Coronal processes subglobose, adnate to the staminal-column, not exceeding it; lobes ca 0.5 mm across, ovate. Stamens 5, ca. 1.5 mm long; pollinia 5, pollen masses, solitary in each anther cell, yellow, waxy, horizontal, minute. Gynostegium ca 2 mm long. Follicle 5 - 6.3 cm x 5 - 7.5 mm, solitary or 2, divaricate, tapering to one end, smooth. Seed ca 6.5 x 3.5 mm, dark brown, compressed, truncate, coma ca 2.5 cm long.

Field note: Rare, in the forests, also cultivated for its medicinal properties.

Fl. & Fr.: May- November.

Specimen examined: Bhilwara 219, Mandalgarh 1763 (MLVGCB Herb).

Distribution: India, Sri Lanka, Malaya and Pakistan.

Uses: Fresh roots are used as an expectorant in chronic bronchitis, whooping cough and rheumatism. Root paste is applied externally to relieve gout pain. Leaves are effective in the treatment of allergic disease, particularly bronchial asthma. Leaves are also used as diaphoretics and emetics. Powdered leaves are beneficial for treating diarrhoea and dysentery.

WATTAKAKA Hassk.

Wattakaka volubilis (L. f.) Stapf in Curtis, Bot. Mag. sub. t. 8976. 1923 var. ***volubilis***; Kanjilal *et al.* Fl. Assam 3: 293. 1939; Jagtap & Singh, Fl. India Fasc. 24: 175. 1999. *Asclepias volubilis* L. f. Suppl. Pl. 170. 1781; Roxb., Fl. Ind. 2: 38. 1832. *Dregea volubilis* (L. f.) Benth. ex Hook. *f.* Fl. Brit. Ind. 4: 46. 1883. *Marsdenia volubilis* (L. f.) Cooke, Fl. Pres. Bombay 2: 166. 1905 & 2: 230. 1967 (repr. ed.); Duthie, Fl. Gangetic Plain 2: 56. 1911. *Wattakaka viridiflora* (R. Br.) Hassk. Flora 40: 99. 1857, *nom. illeg. Hoya viridiflora* R. Br. in Mem. Wern. Soc. 1: 27. 1811, *nom. illeg.*; Wight, Contrib. 39. 1834 et Ic. t. 586. 1840-43. (Plate 197)

Vern.: *Padal bel.*

Climbing shrub. Stem woody, much branched; branches terete, with nodes and internodes; nodes swollen, whitish lenticels very common. Leaves 5 - 15.5 x 3 - 8 cm, ovate, base rounded to cordate, tip acute to acuminate, sparsely pubescent on the veins; petiole 1.2 - 5.5 cm long. Flower in many lateral umbellate cymes, drooping. Peduncle 2.5 - 6.0 cm long. Pedicels 1.2 - 3.8 cm long, puberulous. Bracts 7.5 mm long, lanceolate. Calyx divided nearly to the base, lobes 1.8,- 2.5 mm long, glabrous to pubescent. Corolla

6 - 6.5 mm long, cup-shaped, lobes ca 5 mm long, broadly ovate, obtuse. Corona fleshy lobes, rounded on back, obliquely turncate at apex. Follicles 5 - 10 x 1.7 - 4.5 cm, gradually tapering towards the blunt tip, longitudinally ribbed, brown, minutely hairy. Seeds 1.2 - 1.3 cm long, broadly ovate, compressed margins broad; coma ca 3.7 - 5 cm long.

Field note: Common among hedges in forests

Fl. & Fr.: June-September.

Specimen examined: Hamirgarh 932 (MLVGCB Herb).

Distribution: India, Pakistan, Nepal, Cylon and Malaysia.

Notes: The genus *Dregea*, if accepted in a broader sense, includes *Wattakaka* and most of the authors have placed this species in *Dregea*, except Cooke (l.c.) who has included *Dregea* in *Marsdenia*. However, we have followed Huber (l.c.) and Stevens (l.c.) in accepting *Wattakaka* as a genus, separate from *Dregea* and *Marsdenia*.

PERIPLOCACEAE

KEY TO THE GENERA

1a. Leaves elliptic or oblong. Corolla lobes as long as or longer than tube. Follicles ovoid-oblong .... *Cryptostegia*

1b. Leaves ovate, oblong or elliptic. Corolla lobes shorter than tube. Follicles lanceolate...................................... *Cryptolepis*

CRYPTOLEPIS R. Br.

Cryptolepis buchananii Roem. & Schult. Syst. Veg. 4 : 409. 1819 ("buchananii"); Hook. *f.* Fl. Brit. Ind. 4: 5. 1883; Duthie, Fl. Gangetic Plain 2: 45. 1911; Parmar, in Shetty & Singh Fl. Rajasthan 2: 488. 1991; Jagtap & Singh, Fl. India Fasc. 24: 290. 1999. *Nerium reticulatum* Roxb. (Hort. Beng. 19. 1914, *nom. nud.*) Fl. Ind. 2: 8. 1832. (Fig. 97 & Plate 198)

Large shrub, much-branched climbing; branches white, terete, with milky juice. Bark when young purplish red, peeling off in papery flakes; when older brown, rough. Leaves 5 - 16 x 2.5 - 6.0 cm, ovate-elliptic, acute, coriaceous, glabrous and shining above white beneath; secondary nerves numerous, slender, nearly at right angles to midrib and united within the margin; petioles 0.6 - 1.2 cm long. Flowers greenish or yellowish-white, in lax, slender, short-peduncled, axillary, many-flowered cymes. Bracts ovate or lanceolate, with scarious margin. Calyx short, ovate, acute, with 5 scales within. Corolla 6 - 12 cm long, tube short; lobes linear or linear-lanceolate, imbricate in bud, greenish yellow; corona 5, fleshy, clavate, included within the tube. Stamens free; filaments slightly connate by their broad base; anthers short sagittate, with a tuft of hairs at the back, adhering to the broadly conical stigma; pollen-masses granular, in pairs in each cell. Follicles 5 - 12 x 1.2 - 2.0 cm in. diam., stout, straight, rigid, somewhat compressed, in divaricate pairs. Seeds compressed, ca 6 mm long; coma ca 3 cm long.

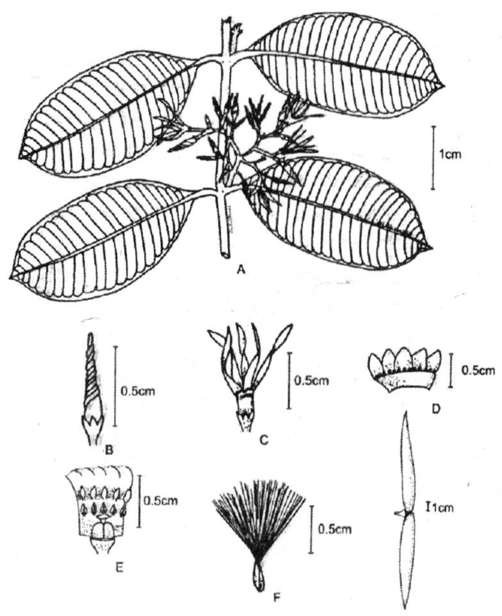

Fig. 97. *Cryptolepis buchananii* R. Br. ex Roem & Schult. A. Flowering branch, B. Flower bud, C. Flower, D. Calyx with glands, E. Opened corolla, F. Fruits, G. Seed

Field note: Very common climber in forests, generally found at edges, in waste places and among shrubly growth.

Fl. : July - October; *Fr.*: November - January.

Specimen examined: Hamirgarh 937, Bijoliya 1871 (MLVGCB Herb).

Distribution: India, Nepal and Pakistan.

Uses: The bark yields a very fine and strong fibre.

CRYPTOSTEGIA R. Br.

Cryptostegia grandiflora R. Br. in Bot. Reg. 5. t. 435. 1820; Wight, Ic. 3(2): 3. t. 832. 1844; Hook. *f.* Fl. Brit. Ind. 4: 6. 1883; Raizada, Suppl. Duthie, Fl. Gangetic Plain 152. 1976; Parmar, in Shetty & Singh Fl. Rajasthan 2: 489. 1991; Jagtap & Singh, Fl. India Fasc. 24: 295. 1999. *Nerium grandiflorum* Roxb. (Hort. Beng. 209. 1814, *nom. nud.*) Fl. Ind. 2: 10. 1832. (Plate 199)

Vern.: *Chabuk-chari.*

A large climbing shrub, 1.5 - 3 m high. Bark grey, sap milky. Branches terete, somewhat swoolen at nodes. Leaves 4 - 11 x 3 - 5 cm, exstipulate, coriaeous, elliptic or oblong, acuminate at the apex, rounded or truncate at base, entire. Flowers in terminal dichotomous cymes, large, 6 - 7 cm long, white or whitish, tinged with light purple, showy, funnel-shaped. Peduncles upto 4 cm long, stout; pedicels 1 - 1.5 cm long, stout, minutely pubescent, terete. Bracts 6 - 12 x 2 - 3 mm, linear, acute, puberulent. Calyx 14 - 30 x 6 - 8 mm, divided nearly to base, ovate, acute at apex. Corolla-lobes contorted, overlapping to the right; limb 4 - 6 cm across. Coronal-scales corolline, cleft to the base into 2 long filiform segments. Filaments short; anthers adnate to the stigma, acute; pollen-masses granular, in pairs in each cell. Follicles thick, straight, divaricate, 8 - 12 x 2.5 cm, ovoid-oblong. Seeds 6 x 2 mm long, ovate, glabrous; coma 2.5 cm long, silky-white.

Field note: Frequently found in gardens or compounds and climbing over trees along roadsides

Fl. & Fr.: April - September.

Specimen examined: Shahpura 119, Bhilwara 1023 (MLVGCB Herb).

Distribution: Native of Medagascar, India.

GENTIANACEAE

KEY TO THE GENERA

1a. Flowers regular, actinomorphic. All stamens perfect 2

2a. Ovary 2-locular. Capsule ovoid, globose or subglobose .... *Exacum*

2b. Ovary 1-locular. Capsule linear, oblong or ellipsoid 3

3a. Flowers pedicillate, in dichotomous cymes. Corolla pink or purple .... *Centaurium*

3b. Flowers sessile, in clusters. Corolla white .... *Enicostema*

1b. Flowers irregular, zygomorphic. Only one perfect stamen 4

4a. Calyx campanulate. Corolla greenish yellow or white .... *Hoppea*

4b. Calyx tubular. Corolla pink or white .... *Canscora*

CANSCORA Lam.

Canscora diffusa (Vahl) R. Br. Roem. ex Schult., Syst. Veg. 3: 301. 1818; Clarke, in Hook. *f.* Fl. Brit. Ind. 4: 103. 1883; Duthie, Fl. Gangetic Plain 2: 76. 1911; Parmar, in Shetty & Singh Fl. Rajasthan 2: 201. 1991. *Gentiana diffusa* Vahl, Symb. Bot. 3 : 47. 1794. *Canscora tenella* Wight, Ic. 4 (2) : 7. t. 1327. f. 3. 1848.

Annual herb, much branched, 12 - 40 cm tall, erect, glabrous throughout. Stem quadrangular, often obscurely winged. Leaves sessile; basal leaves ovate to broadly lanceolate, 1.5 - 5.0 × 0.8 - 2.5 cm, base truncate to cuneate, apex acute, cuneate at base, veins 3 - 5, slender. Flowers in lax cymes, pinkish violet spreading; bracts 2 × 0.5 mm. Pedicel up to 2 cm, narrowly winged. Calyx upto 5 mm; tube narrowly cylindric, upto 3.5 mm; lobes linear-lanceolate, to 1.5 mm, apex acuminate. Corolla pale purple to pink, salverform, tube as long as the calyx; lobes oblong, imbricate, 2.5 - 5 mm. Fertile stamens 3, shorter than corolla tube; sterile stamens 1, ca. as long as tube. Capsules many seeded, oblong, 4 - 6 x 1.5 mm, membranous, as long as the calyx.

Field note: Common in moist rocky places along banks of streams and rivers.

Fl. & Fr.: October - March.

Specimen examined: Fatehpur 537 (MLVGCB Herb).

Distribution: Bangladesh, Bhutan, China, India, Indonesia, Laos, Malaysia, Nepal, Philippines, Sri Lanka, Thailand, Vietnam, Africa and Australia.

CENTAURIUM Hill

Centaurium meyeri (Bunge) Druce, in Rep. Bot. Exch. Club Brit. Isles 1916: 613. 1919; S. Garg, Indian Gent. 3. 1988. *Erythraea meyeri* Bung, in Ledeb. Fl. Alt. 1: 220. 1829; Clarke, in Hook. *f.* Fl. Brit. Ind. 4: 102. 1883. *E. roxburghii* D. Don, Gen. Hist. 4: 206. 1837-1838; Wight, Ic. 4 (2): 7. t. 1325. 1848; Clarke, in Hook. *f.* Fl. Brit. Ind. 4: 102. 1883 ; Duthie, Fl. Gangetic Plain 2: 74. 1911. *Centaurium centaurioides* (Roxb.) R. S. Rao & Hemadri in J. Bomb. Nat. Hist. Soc. 67 (2): 357. 1970; Parmar, in Shetty & Singh Fl. Rajasthan 2: 494. 1991. *Chironia centaurioides* Roxb., Fl. Ind. 1: 584. 1832.

An erect, annual herb, 10 - 25 cm, glabrous. Stem 4-angled, glabrous, branched, branching from the base. Basal leaves arranged in a rosette, larger than upper ones, 0.5 - 4 x 0.6 - 1 cm, obovate or oblong, obtuse; cauline ones smaller, linear - oblong, glabrous, prominantaly nerved below. Flowers arranged in dichotomous cymes, with long pedicelled flowers in each fork, pink. Calyx - segments linear, 5 - 6 mm long. Corolla - lobes elliptic, acute, glabrous, pink or purple. Capsule with persistent calyx, linear-oblong, glabrous, acute. Seeds many, rounded, brown, faintly reticulate.

Field note: Rare, weed of moist river-beds and grasslands. Not common.

Fl. & Fr.: August- March

Specimen examined: Mainal 1133 (MLVGCB Herb).

Distribution: Throughout the world except tropical South-Africa.

ENICOSTEMA Blume nom. cons.

Enicostema axillare (Lam.) Raynal in Adansonia 9: 75. 1969; Bhandari, Fl. Indian Desert 205. 1990; Parmar, in Shetty & Singh Fl. Rajasthan 2: 495. 1991. *Geniana verticillata* L. Syst. Nat. ed. 12. 200. 1767, *pro parte,* non L. 1759. *G. axillaris* Lam. III. Gen. 1 (2): 487. 1793. *Exacum hyssopifolium* Willd. Sp. Pl. 1: 640.1798. *Enicostema littorale auct.* non Blume 1826; Clarke, in Hook. *f.* Fl. Brit. Ind. 4: 101. 1883; Duthie, Fl. Gangetic Plain 2: 273. 1911. *E. verticillatum* (L.) Engl. Pflan. Ostafr. C. 313. 1895, non Baill. 1891. *E. hyssopifolium* (Willd.) Verdoon in Bothalia 7 : 462. 1961; Bhandari, Fl. Indian Desert 230. 1995. (Plate 200)

Vern.: *Nava, Nami.*

An erect, glabrous, 8-35 cm tall herb, with many subquadrangular branches from the base. Leaves opposite, decussate, 1-7 x 0.5-1.5 cm, linear or elliptic lanceolate, subacute, amplexicaul at base, sessile. Flowers, white, sessile or subsessile, in axillary clusters at each node. Calyx deeply lobed; lobes narrowly oblong, obtuse, keeled. Corolla tubular, 5-7 mm long; lobes 2-3 mm long, lanceolate, acute. Stamens inserted at the middle of the corolla-tube, with small, double-hooked, scaly glands at the base; anthers acute. Ovary 1-celled, glabrous; stigma capitate. Capsule ellipsoid-oblong, 3-4.5 mm long, apiculate, glabrous. Seeds many, minute, subglobose, foveolate, brown.

Field note: Common weed of open grasslands, edges of fields, throughout the area.

Fl. & Fr.: July-December.

Specimen examined: Nimbahera 1974 (MLVGCB Herb).

Distribution: Tropical America, Africa and Medagaskar.

Uses: 1/4th cup extract of whole plant is taken once a day for 3 days to cure fever. 1 tsp plant juice is taken for seven days, to relief from bodyache.

EXACUM L.

Exacum pendunculatum L., Sp. Pl. 112. 1753; Wight, Ic. 2(1): 2. t. 336. 1840, *'pedunculare'*;

Clarke, in Hook. *f.* Fl. Brit. Ind. 4: 97. 1883; Duthie, Fl. Gangetic Plain 2: 72. 1911; Parmar, in Shetty & Singh Fl. Rajasthan 2: 496. 1991. *E. penduncularе* Arn., in Ann. Sci. Nat. ser. 2. 11: 176. 1836.

Annual erect glabrous herb, 15 - 30 cm high. Leaves subsessile, 3 - 5 x 0.3 - 1.5 cm, elliptic or lanceolate, obtuse, tapering at the base, 3-nerved. Flowers in terminal much branched panicle cymes, bluish-purple. Calyx lobes 4, tube cup-shaped, ovate, acute, with wedge-shaped wings extending from apex to the base on the outer surface. Corolla lobes lanceolate. Anthers sagittate at base, not spirally twisted after anthesis. Ovary 2-locular, many ovuled. Capsules globose, glabrous, membranous, surrounded by persistent, winged calyx. Seeds minute, glabrous, pale brown.

Field note: Common at moist places in forests.

Fl. & Fr.: December - February.

Specimen examined: Pur 1032, Borda 1323 (MLVGCB Herb).

Distribution: India, Malaya, China and Africa.

HOPPEA Willd.

Hoppea dichotoma Heyne ex Willd. in Ges. Naturf. Fr. Neue Schr. 3 : 435, 1801; Clarke, in Hook. *f.* Fl. Brit. Ind. 4: 100. 1883; Duthie, Fl. Gangetic Plain 2: 76. 1911; Bhandari, Fl. Indian Desert 206. 1995; Parmar, in Shetty & Singh Fl. Rajasthan 2: 496. 1991.

Vern.: *Ramjatta*.

Annual, glabrous, herb, 5 - 12 cm high. Stem quadrangular. Leaves 3 - 6 x 1 - 4 cm, ovate-elliptic or elliptic-lanceolate, acute, glabrous. Flowers in terminal, dense, dichotomous cymes, pale-yellow. Calyx lobes 4, with green veins prolonged into an apiculate tips, campanulate, acute, lanceolate, 2.5 - 3.5 mm long, membranous. Corolla included, tubular, 2 mm long, greenish-yellow; lobes 1 mm long, ovate-oblong. obtuse. Anthers 1 fertile; 3 sterile. Capsule ca 0.2 cm long, ellipsoid, enclosed in persistent calyx. Seeds numerous, oblong or subglobose, red or brown, muricate.

Field note: Common in wet and shady habitats.

Fl. & Fr.: September - February.

Specimen examined: Pur 218, Bijoliya 2118 (MLVGCB Herb).

Distribution: India (Throughout).

MENYANTHACEAE

NYMPHOIDES Seguir

KEY TO THE SPECIES

1a. Leaves purple beneath, corolla white, margins entire, cristate with 3 longitudinal folds, plancentae 2, seeds tuberculate............................... *N. hydrophylla*

1b. Leaves greenish beneath, corolla white with yellow centre, margins fimbriate, not cristate, plancentae 3, seeds smooth... *N. indica*

Nymphoides hydrophylla (Lour.) O. Kuntze, Rev. Gen. Pl. 2: 429. 1891; Parmar, in Shetty & Singh Fl. Rajasthan 2: 498. 1991. *Menyanthes hydrophylla* Lour. Fl. Cochinch. 1: 129. 1790. *M. cristata* Roxb., Pl. Cor. 263 t. 105. 1798. *Limnanthemum cristatum* (Roxb.) Griseb. Gen. Sp. Gent. 342. 1839; Clarke, in Hook. *f.* Fl. Brit. Ind. 4: 131. 1883; Duthie, Fl. Gangetic Plain 2: 79. 1911. *Nymphoides cristata* (Roxb.) O. Kuntze, Rev. Gen. Pl. 2: 429. 1891. (Fig. 98 & Plate 201)

Vern.: *Kumudani*

An aquatic, floating herb. Rhizomes horizontal, short. Branches cylindric, spongy, unbranched, with a terminal leaf. Floating leaves 4 - 13 x 3 - 11 cm, orbicular, deeply cordate at base, with rounded lobs, faintly crenate, greenish above, purplish beneath. Petiole cylindric, 1 - 3 cm, sheathing. Flowers white with yellow centre, in dense clusters at the base of the nodes. Calyx 3 - 5.5 mm, lobed to near base; lobes lanceolate or oblong, apex obtuse. Corolla white with a yellow center, 4 - 8 mm; lobes ovate with a keeled fold from base to apex, apex obtuse. Filaments linear, 1-2 mm; anthers 0.8 - 1 mm. Style short; stigma lobes triangular. Capsules upto 8 x 3 - 5 mm, broadly ovoid-oblong, indehiscent, few seeded. Seeds light brown, globose, 1.3 - 2.0 mm in diam. Seed coat smooth or scabrous.

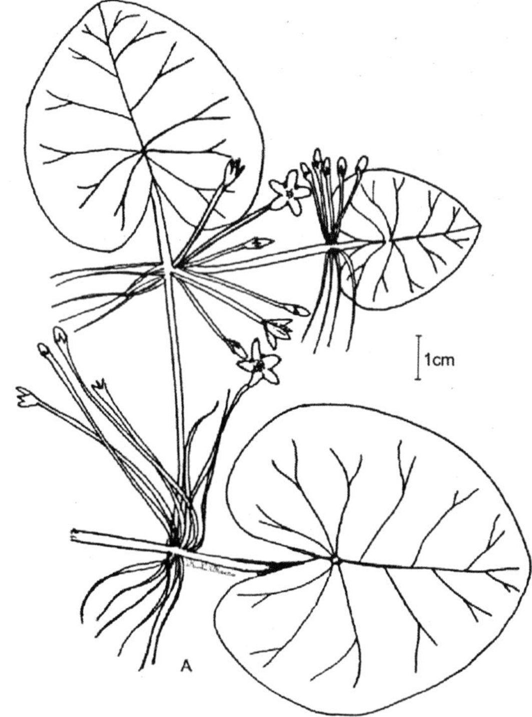

Fig. 98. *Nymphoides hydrophylla* (Lour.) Kuntze A. Flowering branch.

Field note: Common in ponds and tanks in association with *Hydrilla verticillata* (L.) Royle, throughout the area.

Fl. & Fr.: October - December.

Specimen examined: Mainal tank 116, Gambhiridam 4015 (MLVGCB Herb).

Distribution: Fujian, Guangdong, Hainan, Hubei, Hunan, Jiangsu, Sichuan, Taiwan and India.

Nymphoides indica (L.) O. Kuntze, Rev. Gen. Pl. 2: 429. 1891; Parmar, in Shetty & Singh Fl. Rajasthan 2: 499. 1991. *Menyanthes indica* L., Sp. Pl. 1: 145. 1753. *Limnanthemum indicum* (L.) Griseb. Gen. Sp. Gent. 343. 1839; Clarke, in Hook. *f.* Fl. Brit. Ind. 4: 131. 1883; Duthie, Fl. Gangetic Plain 2: 79. 1911. (Plate 202)

Vern.: *Kumudani.*

An aquatic floating herb. Rhizomes horizontal, cylindric, unbranched, with thick scale like leaves. Petiole cylindric, long. Leaf blade broadly ovate to subcordate, 3-12 cm, subcoriaceous, abaxially densely glandular, base cordate, margins serrate-crenate; veins indistinct, palmate. Petiole very short. Flowers many, clustered at nodes, 5 merous, distylous. Pedicels cylindric, 3-5 cm. Calyx 3-6 mm, lobed to near base; lobes lanceolate to narrowly elliptic, apex obtuse. Corolla white with a yellow center, 7-12 mm; lobes ovate-elliptic, outside densely fimbriate -pilose, apex obtuse. Filaments flattened, linear, 1.5 - 1.7 mm; anthers sagittate, 2-2.2 mm. Ovary 1-locular; style cylindric; stigma lobes triangular. Capsules obovoid or oblong-ovoid, up to 8 x 3 mm, few seeded. Seeds brown, globose, 1.2 - 1.5 mm. Seed coat smooth.

Field note: Common in standing water, ponds and tanks in association with *Hydrilla verticillata* (L.) Royle, throughout the area.

Fl. & Fr.: October - December.

Specimen examined: Mainal tank 117, Bari dam 4001 (MLVGCB Herb).

Distribution: Cambodia, China, India, Indonesia, Japan, Korea, Malaysia, Myanmar, Nepal, Sri Lanka, Australia and Pacific Islands.

BORAGINACEAE

KEY TO THE GENERA

1a. Style arising from between the carpels 2

2a. Leaves opposite, amplexicaul or cordate at base. Flowers white changing pale blue to violate. Anthers connivent into a cone .... *Trichodesma*

2b. Leaves alternate, narrowed at base. Flowers yellow. Anthers not connivent into a cone.. *Arnebia*

1b. Style terminal on the ovary, not inserted between the carpels....................................... 3

3a. Leaf margins toothed. Flowers solitary, axillary. Corolla lobes 4. Style 2........ *Coldenia*

3b. Leaf margins entire. Flowers in terminal scorpioid cymes. Corolla lobes 5. Style 1 .... *Heliotropium*

ARNEBIA Forsk.

Arnebia hispidissima (Lehm.) DC. Prodr. 10: 94. 1846; Wight, Icon. Pl. Ind. Or. 4: t. 1393. 1848; Boiss, Fl. Or. 4: 213. 1875; Clarke, in Hook. *f.* Fl. Brit. Ind. 4: 176. 1883; Duthie, Fl. Gangetic Plain 2: 27. 1911; Riedl in Rech. Fl. Iran. 48: 155.1967; Bhandari, Fl. Ind. Desert, 231. 1995; Parmar, in Shetty & Singh Fl. Rajasthan 2: 501. 1991. *Anchusa asperrima* Del., Fl. d'Eg. III 55. 1813. *non. nud. Lithospermum hispidissimum* Sieber ex Lehman. Ic. Nov. Strip. t. 39. 1821. *Arnebia asperrima* (Del.) Hutchison & Dalz. Fl. West. Trop. Afr. ed. 1. 2: 201. 1931. (Fig. 99 & Plate 203)

Vern.: *Ram-buti.*

Annual or perennial branched herb, 20 - 40 cm high. Root exuding a purplish dye. Stem prostrate to decumbent or suberect, diffused from woody base. Branches and leaves hispid, with small (up to 0.6 mm long) weaker hairs. Basal leaves 2 - 8 x 0.2 - 0.8 cm (cauline shorter), linear-lanceolate, entire, acute, attenuate at base, slightly revolute, tapering towards the base. Inflorescence a terminal scorpioid cyme up to 12 cm long in fruit. Bracts foliaceous, 4 - 8 mm long, lanceolate, acute, hispid. Calyx 5-partite, 6 - 9 mm long, hispid, lobes unequal, linear-lanceolate. Corolla pubescent outside, tubular; tube in long styled flowers 8 mm long, infundibuliform, 4 - 5 mm broad; lobes broadly ovate, margin entire to undulate, spreading, throat without hairs. Tube pubescent within just below the anthers (in long styled flowers). Stamens inserted much below the mouth of corolla; anthers 1.5 mm long, oblong. Ovary deeply 4-lobed; style simple, slender; one styles reaching throat of corolla, short styles 4-5 mm long. stigmas 2, flabellate. Nutlets 1.5 - 2.5 mm long, ovoid, broadest at base, brown, minutely tuberculate.

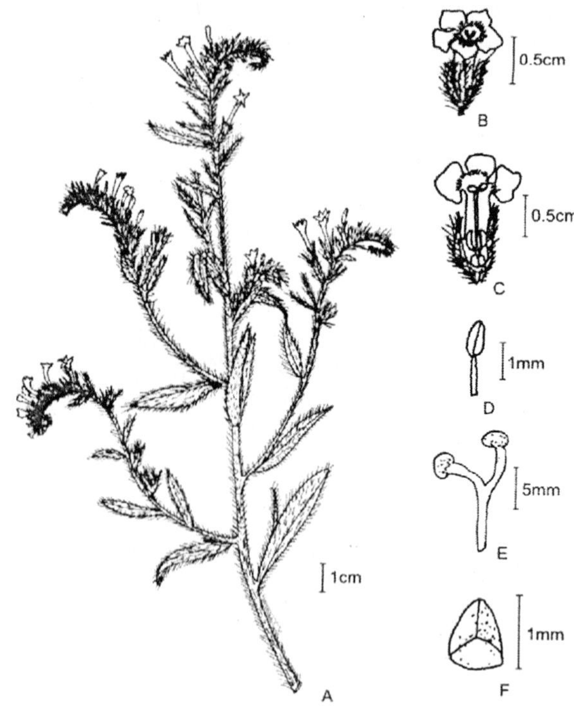

Fig. 99. *Arnebia hispidissima* (Lehm.) DC. A. Habit, B. Flower, C. Flower V.S., D. Anther, E. Stigma, F. Seed

Field note: Common along roads, near streams, more common in dry sandy soils.

Fl. & Fr.: September-April.

Specimen examined: Mangrop 1754 (MLVGCB Herb).

Distribution: N. & E. Africa, India, Arabia and Pakistan.

COLDENIA L.

Coldenia procumbens L., Sp. Pl. 125. 1753; Clarke, in Hook. *f.* Fl. Brit. Ind. 4: 144. 1883; Duthie, Fl. Gangetic Plain 2: 88. 1911; Mahesh. Illus. Fl. Delhi f. 131. 1966; Bhandari, Fl. Indian Desert 208. 1995; Parmar, in Shetty & Singh Fl. Rajasthan 2: 502. 1991.

Prostrate or rarely procumbent branched, annual, herb. Stem rooting at nodes, grooved, with white hairs; young shoots silky-pubescent. Branches up to 35 cm or more long. Leaves 2 - 5 x 0.8 - 3 cm, ovate-oblong, crisped, pinnatifid subobtuse; lobes rounded at apex, oblique or cuneate at base, hairy on lower surface, coarsely serrate or subpinnatifid on margins. Petiole 3-9 mm long, densely hairy. Flowers solitary, axillary, yellow. Calyx ca 2 - 3 mm long; lobes 4, narrow ovate, acute. Corolla 2-2.5 mm long, 4-lobed; lobes obtuse, ca 1.2 mm long. Stamens 4 subsessile; anthers ca 0.4 mm long. Fruit 4-lobed, subpyramidal-shaped, 2-2.5 x 3-3.5 mm; pyrenes beaked, puberulose, 1 -seeded.

Field note: Common weed along banks of ponds, tanks etc.

Fl. & Fr.: September - May.

Specimen examined: Mezadam 167, Janhajpur 1369 (MLVGCB Herb).

Distribution: Tropical India, Sind, Sri-Lanka, Africa, Australia and America.

HELIOTROPIUM L.

KEY TO THE SPECIES

1a. Leaves more than 2.5 cm broad .... *H. indicum*

1b. Leaves less than 2.5 cm broad 2

2a. Calyx shallowly 5-lobed, Fruits completely enclosed in the calyx .... *H. supinum*

2b. Calyx 5-partite, Fruit not enclosed in the calyx.. 3

3a. Plants glabrous. Stem and leaves succulent.............................. .*H. curassavicum*

3b. Plants villous to scabrid. Stem and leaves otherwise 4

4a. Corolla lobes caudate-acuminate................................. *H. subulatum*

4b. Corolla lobes obtuse to ovate or caudate 5

5a. Leaves elliptic oblong to obovate...................................... *H. europaeum*

5b. Leaves linear or linear lanceolate..................................... *H. strigosum*

Heliotropium curassavicum L., Sp. Pl. 130. 1753; DC. Prodr. 9: 538. 1845; Mahesh. & Sharma in Bull. Bot. Surv. India 11: 455. 1972; Raizada, Suppl. Duthie, Fl. Gangetic Plain 158. 1976; Panday *et al.* in Bull. Bot. Surv. India 24: 134. 1982; Bhandari, Fl. Indian Desert 209. 1995; Parmar, in Shetty & Singh Fl. Rajasthan 2: 508. 1991. (Plate 204)

A glabrous, perennial, branched, succulent, herb, up to 45 cm tall. Stem and branches decumbent. Leaves sessile, subopposite, 1.2 - 4.4 x 0.2-1 cm, glabrous, oblanceolate or-linear-spathulate, obtuse, nerves faint. Inflorescence terminal, scorpioid, simple or bifurcate spikes, 3-6 cm long. Flowers white. Calyx 1.5 mm long, persistent, 5-partite into ovate-lanceolate lobes, less subequal. Corolla white, yellow in throat, tube short, ca 1.7 mm long, glabrous; lobes ca 1 mm long, obtuse-undulate, patent. Anthers 0.7-0.8 mm long, sessile, ovoid, acute, broader at the base, attached 0.7 mm from corolla base. Stigma conical, ca 0.5 mm long, sessile, stigmatic ring prominent. Ovary and fruit glabrous. Nutlets 2 mm long, pale brown, back slightly rugose.

Field note: Common in Western Rajasthan but in the South-central Rajasthan it is Rare, particularly in moist sandy saline soils.

Fl. & Fr.: March - December.

Specimen examined: Ajmer 4051 (MLVGCB Herb).

Distribution: Africa, America, Australia, Europe, India and Pakistan.

Notes: The only glabrous species in the study area characterised by its fistular stem and halophytic nature.

Heliotropium europaeum L. var. ***lasiocarpum*** (Fish. & Mey.) Kazmi in Journ. Arn. Arbor. 51 : 176. 1970; Parmar, in Shetty & Singh Fl. Rajasthan 2: 508. 1991. *H. lasiocarpum* Fish. & Mey. Index Sem. Hort. Petrop. 4: 38. 1837. *H. eichwaldii* Steud. var. *lasiocarpum* (Fisch. & Mey.) Clarke, in Hook. f. Fl. Brit. Ind. 4: 150. 1883. *H. eichwaldii auct.* non Steud. ex DC. 1845; Duthie, Fl. Gangetic Plain 2: 92. 1911; *H. ellipticum auct. plur.* non Ledeb. 1831, non R. Br. 1810; Bhandari, Fl. Indian Desert 234. 1978.

An annual, herb, 50 cm or more tall, branched. Branches greyish-green, dense white hairs up to 1 mm long and arising from a bulbous base; Leaves 2 - 6 x 0.8 - 3.5 cm, elliptic-ovate to obovate, subacute at apex, truncate or slightly attenuate at base, hairy on both surfaces; hairs white, appressed, arising from bulbous base, nerves impressed on upper surface. Petiole terete, pubescent, up to 3.5 cm long. Inflorescence terminal spike, 2 - 10 cm long with close set biseriate flowers. Calyx sessile, persistent, 2.5 - 3.0 mm long, lobes ca 2 mm long, ovate-lanceolate, acute, hairy on the outside, glabrous within; throat constricted. Corolla lobes suborbicular to oblong, ca 1.2 mm long, hairy outside, glabrous inside. Anthers ca 1.1 mm long, elongated, branched at base, apex acutish, attached from the corolla tube. Style 2-fid at apex; stigma absent. Nutlets 4, 1.8 - 2 mm long, ovoid, retrorsely dense pilose and minutely verrucose to rugulose on back, brownish-black.

Field note: Weed of moist places, fallow fields and road-sides.

Fl. & Fr.: August-March.

Specimen examined: Atun 319, Bhilwara 1333 (MLVGCB Herb).

Distribution: Trans Jordan, Syria, Iraq, Iran, India, Russia, Afghanistan and Pakistan.

Notes: The taxon is variable throughout its geographical range in leaf size and length of inflorescence. The sinuses between corolla lobes are sometimes appendaged (*Sultanul Abedin* & *A. Hussain* 9634; *Harsukh* 15406). Differs from the type variety in the corolla tube which is constricted at the throat, the stigma which is obscurely 2-fid and densely pilose (not glabrous or subglabrous) nutlets.

Heliotropium indicum L., Sp. Pl. 130. 1753; Clarke, in Hook. *f.* Fl. Brit. Ind. 4: 152. 1883; Duthie, Fl. Gangetic Plain 2: 90. 1911; Parmar, in Shetty & Singh Fl. Rajasthan 2: 509. 1991. *Tiaridium indicum* Lehm. Pl. As. Nucif. 14. 1818; Wight, Illus. t. 171. 1850. (Fig. 100 & Plate 205)

Perennial or annual, broad-leaved, erect, herbs up to 0.5 m high, strigose to woody. Leaves alternate or subopposite, simple, 3 - 9 x 2 - 6 cm, undissected, elliptic or ovate or triangular, base narrowly attenuate or oblique, serrulate, apex acuminate or acute or obtuse, rugose above, pilose beneath. Petiole 0.5 - 4 cm long. Flowers in extraxillary, upto 15 cm long spikes ebracteate, scorpioid at apex; predominantly purple or blue, regular; pedicellate or subsessile or sessile, 1 mm long. Calyx 5, 1.5 - 3 mm long, all sepals joined. Corolla 5, 4 - 7 mm long. Stamens 5, inserted midway down corolla tube, all alternating with the corolla parts; anthers versatile, dehiscing by longitudinal slits. Ovary syncarpous, superior, 4 - celled; ovules 1 per cell; styles 1, simple. Drupes deeply 2-lobed, conical or ovoid. Nutlets compressed, 4-ribbed, beaked, 2-seeded.

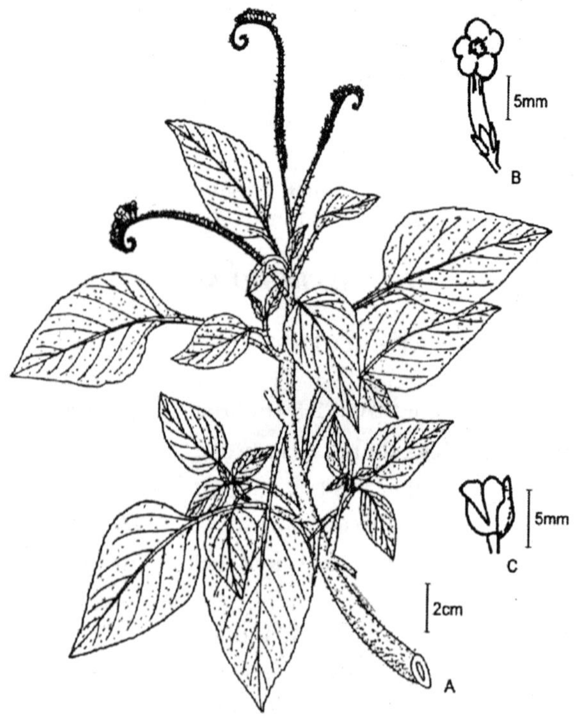

Fig. 100. *Heliotropium indicum* L.
A. Habit, B. Flower, C. Fruit

Field note: Rare, near streams, ponds, dams, river etc.

Fl. & Fr.: Almost throughout the year

Specimen examined: Bhilwara 536, Palri 1589 (MLVGCB Herb).

Distribution: Native of S. America. Also distributed in Asia, Australia, Africa and Caribbean.

Heliotropium strigosum Willd. Sp. Pl. 1: 743. 1798. DC. Prodr. 9: 546. 1845; Boiss., Fl. Or. 4: 143.1875; Clarke, in Hook. *f.* Fl. Brit. Ind. 4: 151. 1883; Duthie, Fl. Gangetic Plain 2: 93.

1911; Bhandari, Fl. Indian Desert 212. 1995; Parmar, in Shetty & Singh Fl. Rajasthan 2: 201. 1991.

Perennial, prostrate to procumbent or suberect, up to 20 cm tall, usually much branched from the base; stem covered with appressed white hairs. Leaves sessile, 0.5 - 2.5 x 0.1 - 0.4 cm, linear-lanceolate or narrowly oblong-lanceolate, acute, appressed, strigosely hairy on both surfaces, with short hairs. Inflorescence in lax bracteate spikes, 2-10 cm long, simple or paired. Lower bracts lanceolate, 2-3 mm long; lower flowers pedicellate, up to 2 mm long; upper ones sessile. Calyx 2.5 - 3 mm long, 5-partite; lobes lanceolate, spreading, hairy outside. Corolla tubular or infundibuliform, strigose outside. Anthers 0.7 - 0.8 mm long, somewhat adnate when young, attached ca. 0.6 mm from base of corolla tube. Stigma conical, 0.6 - 0.7 mm long, minutely pilose towards tip; stigmatic ring prominent. Style slightly shorter, glabrous. Fruit subglobose, depressed or not, 4-lobed, dense pilose. Nutlets 4, dark brown, ca 1.5 mm long, glabrous or appressed hairy.

Field note: Common in waste lands and open forests.

Fl. & Fr.: August - December.

Specimen examined: Palri 2185 (MLVGCB Herb).

Distribution: India, Burma, Pakistan, China, Australia and Malaysia.

Heliotropium subulatum (Hochst. ex DC.) Vatke, Oesterr. Bot. Zeitschr. 2 : 166. 1875; Duthie, Fl. Gangetic Plain 2: 91. 1911; Bhandari, Fl. Indian Desert 212. 1995; Parmar, in Shetty & Singh Fl. Rajasthan 2: 512. 1991. *Tournefortia subulata* Hochst. ex DC. Prodr. 9: 528. 1845. *Heliotropium zeylanicum auct.* non Lam.1789; Wight, Ic. 3(2): 8. t. 892. 1843-45; Clarke, in Hook. *f.* Fl. Brit. Ind. 4: 148. 1883.

A branched annual or perennial, herb, up to 50 cm tall. Stem and branches covered with white to brownish spreading hairs up to 1.8 mm long with a bulbous base. Leaves subsessile or shortly petiolate, 2 - 4 x 0.5 - 1.2 cm, lanceolate, acute, margins entire to undulate, hairy on both surfaces; shorter hairs appressed; longer ones spreading and with a bulbous base; nerves impressed on upper surface Inflorescence terminal or axillary, 4 - 20 cm long, simple or forked, with close set of uniseriate flowers. Calyx subcupular, 2/3rd partite; lobes lanceolate, appressed hairy. Corolla yellowish, tube 3.5 - 4 mm long, appressed hairy; lobes 1 - 1.5 mm long, caudate-acuminate, the sinuses minutely dentate. Anthers 1 mm long oblong, attached 2 mm above corolla base. Stigma 1 mm long, conical, appressed hairy at apex. Style longer than stigma, glabrous. Fruit globose-compressed. Nutlets 4, ca 2.2 mm long, ovoid, brown, back rugose.

Field note: Rare, in dry deciduous forests.

Fl. & Fr.: Almost throughout the year

Specimen examined: Atun 1194, Bhilwara 1371 (MLVGCB Herb).

Distribution: N.E. and C. Africa, India and Pakistan.

Heliotropium supinum L., Sp. Pl. 130. 1753; DC. Prodr. 9: 533.1845; Boiss. Fl. Or. 4: 127. 1875; Clarke, in Hook. *f.* Fl. Brit. Ind. 4: 149.1883; Cooke, Fl. Bomb. Pres. Reprint ed. 2(2): 275. 1908; Duthie, Fl. Gangetic Plain 2: 91. 1911; Bhandari, Fl. Ind. Desert. 237.1995; Parmar, in Shetty & Singh Fl. Rajasthan 2: 513. 1991. *H. malabarica* Retz. Obs. Bot. 4 : 24. 1786. *H. supinum* L. var. *malabarica* (Retz.) Clarke in Hook. *f. l. c.* 4: 149.1883. (Plate 206)

Prostrate branched annual. Branches brown-tomentose with dense appressed hairs up to 1 mm long. Leaves 1 - 2 x 0.5 - 1 cm, ovate to obovate or elliptic, cuneate to slightly oblique, entire-undulate, obtuse, nerves impressed on upper surface, greyish white with dense hairs. petiole up to 15 mm long. Inflorescence lateral and terminal, 2-7 cm long. Flowers subsessile, uniseriate. Calyx deciduous, ca 2.5 mm long, up to 5 mm in fruit, ¼th partite into slightly obtuse lobes, densely hairy. Corolla tube 34 mm long, retroresely hairy to the outside, glabrous within; lobes oblong, shallow, erect. Anther ca 1 mm long, narrowed towards the apex, attached ca 1.5 mm from the corolla base; ovary glabrous; stigma ca 0.6 mm long, conical; stigmatic ring prominent;

style about stigma length, with few retrorse hairs. Fruit globose, enclosed in the persistent calyx. Nutlets ca 2.5 mm long, ovoid in outline, glabrous, dark brown, back rugulose and minutely tubercled.

Field note: Common on drying banks of ponds and tanks, low lying cultivated fields etc.

Fl. & Fr.: December-January.

Specimen examined: Karthana 3206 (MLVGCB Herb).

Distribution: S. Europe, N. Africa, Arabia, Syria, Iraq, Pakistan and India.

TRICHODESMA R. Br. *nom. cons.*

Trichodesma indica (L.) R. Br. Prodr. Fl. Nov. Holl. 496. 1810; DC. Prodr. 10: 172. 1846; Boiss., Fl. Or. 4: 280.1875; Clarke, in Hook. *f.* Fl. Brit. Ind. 4:153.1883; Duthie, Fl. Gangetic Plain 2: 94. 1911, *pro parte;* Parmar, in Shetty & Singh Fl. Rajasthan 2: 515. 1991. *Borago indica* L., Sp. Pl. 137. 1753; *Trichodesma indicum* (L.) R. Br. var. *amplexicaule* (Roth) T. Cooke FPB. 2: 215. 1904; Bhandari, Fl. Ind. Desert. 214.1995.

Vern.: *Chhota kulpha.*

An annual herb with spreading and densely hairs, much branched; hairs dimorphic; shorter ones thin, appressed, up to 0.3 mm long; longer ones stiff, up to 0.6 mm long with smaller bases. Leaves linear-lanceolate, 3.0 - 7.5 x 0.6 - 1.5 cm, subcordate or amplexicaul or not; middle cauline ones the largest, hairy on both surfaces. Inflorescence terminal and leaf opposed, lax, few-flowered. Bracts leafy. Pedicel 0.5 - 1 cm long. Flowers purplish white to pinkish blue or lilac. Calyx 5-partite, hairy, 8 - 12 mm long, elongated in fruit, base subauricled; lobes lanceolate, narrowed towards apex. Corolla exceeding calyx length, infundibuliform, sparsely hairy on outside; lobes ovate-deltoid, subacuminate. Stamens forming a pointed cone, hairy on lateral sides; anthers hairy, aristate, connectives later twisting together. Fruits pyramidal, 4-ribbed. Nutlets 4-5 mm long, oblong with, smooth, shining with back and rugose.

Field note: Common weed of grasslands, open waste lands and forests.

Fl. & Fr.: July - November.

Specimen examined: Atun 2184 (MLVGCB Herb).

Distribution: Afghanistan, India, Pakistan, Philippines and Mauritius.

Notes: Variable in the degree of pubescence. Plants with amplexicaul leaves and a glabrous undersurface of leaves designated as var. *amplexicaul*, occasionally exhibit dense pubescence.

EHRETIACEAE

KEY TO THE GENERA

1a. Trees. Calyx accrescent in fruits. Style 4 - fid. Fruits pulpy.................................... *Cordia*

1b. Shrubs or small tree. Calyx not accrescent in fruits. Style 2 - fid. Fruits dry .... *Ehretia*

CORDIA L.

KEY TO THE SPECIES

1a. Leaves 1-nerved .... *C. gharaf*

1b. Leaves distinctly 3-5-nerved from the base 2

2a. Leaves deeply dentate towards the apex, elliptic, fruits ca 8 mm .... *C. crenata*

2b. Leaves entire throughout, elliptic to ovate, obovate or suborbicular. Fruits 1 - 2.5 cm long ....*C.dichotoma*

Cordia crenata Delile, Fl. Egypt. 195. t. 20. 1813; Duthie, Fl. Gangetic Plain 2: 83. 1911; Parmar, in Shetty & Singh Fl. Rajasthan 2: 518. 1991. *C. lowriana* Brandis, Indian Trees 479. 1906. (Plate 207)

Vern.: *Gundi.*

Small or large shrub, tree, glabrous except minutely hairy inflorescence. Branches teret, suberect. Leaves alternate to subopposite 2.5 - 7.5 cm long, elliptic or elliptic - lanceolate, rounded, sub-acute or shortly acuminate at the apex, obliquely rounded or cuneate at the base, margins crenate - dentate or deeply dentate at least towards the apex, bright green. Flowers in small, compact, pedunculate cymes from the axils of the uppermost leaves. Calyx tube (in flower) 6 mm long, cylindrical; fruiting calyx campanulate 10 mm across, glabrous outside, densely hairy within, margins crenate. Fruit a 1-seeded (by abortion) drupe, ca 8 mm, ovoid, acute. Seeds exalbuminous.

Field note: Very rare, on rockey habitats.

Fl. & Fr.: January-April.

Specimen examined: Taragarh 4053 (MLVGCB Herb).

Distribution: Rajasthan, Endemic.

Notes: According to Duthie (*l.c.*) this species is indigenous to Rajputana. Outside India, i.e. Egypt, it is known under cultivation. Sharma (1958) reported this species from Ajmer without any voucher speciemen, probably on the authority of Duthie (l. c.). The only authentic speciemen of this species is Duthie's 4754 A (DD) collection of January 1886. Probably this taxon is on the verge of extinction from its native home and needs conservation (Shetty & Singh, 1991)

Cordia dichotoma Forster. f. Prodr. 18. 1. 1876; Johnston in J. Arb. 32: 8. 1951, Sant., in Rec. Bot. Surv. Ind. 16: 163. 1967 (3rd Rev. ed.); Kazmi, in J. Arnold. Arbor 51: 141. 1970; Parmar, in Shetty & Singh Fl. Rajasthan 2: 519. 1991. *C. obliqua* Willd. Phytogr. 4. t. 4. 1794; Clarke, in Hook. *f.* Fl. Brit. India 4: 137. 1883. *C. myxa auct. plur.* non L. 1753; Clarke, in Hook. *f. l. c.* 4: 137. 1883; Duthie, Fl. Gangetic Plain 2: 82. 1911.

Vern.: *Lasora, Gonda.*

Medium sized tree, up to 10 m tall. Bark rough. Branchlets pubescent. Leaves variable, 5 - 12 x 4-10 cm, broadly ovate, pubescent, especially on the under-surface, margins sometimes undulate, base cuneate to oblique or rounded. Petiole up to 3.0 cm long. Flowers in dichotomous cymes, ebracteate. Male flowers: calyx ca 2.5 mm long, shallowly 5-lobed, hairy to the inside. Corolla campanulate, ca 6 mm long, tube about equalling the limb, dense hairy within. Lobes oblanceolate or broadly so, recurved. Filaments ca 4.5 mm, lower half hairy and adnate to tube; anthers 2.5-3 mm long. Bisexual flowers: similar but larger than male flowers. Calyx 5 - 6 mm long, campanulate, accrescent and up to 10 mm in fruit. Filaments ca 2 mm long. Style branched. Berries 0.6 - 2.5 cm long, ovoid, pink, glossy, supported by accrescent calyx, pulp muscilaginous.

Field note: Common in dry deciduous forests. Often cultivated for its edible fruits.

Fl. & Fr.: February - June.

Specimen examined: Ajmer 1968, Ladpura 2003 (MLVGCB Herb).

Distribution: China, India, Indo-China, N.E. Australia, New Caledonia, Pakistan and Taiwan.

Uses: Fruits are edible and one of ingredients of pickels.

Cordia gharaf (Forsk.) Ehren. ex Asch., Sitzungber. Ges. Naturf. Freunde Berlin 1879 (4): 1879; Johnston, in J. Arn. Arb. 37: 292. 1956; Sant. & Wag, Bull. Bot. Surv. Ind. 5: 106. (1963) 1964; Kazmi, in J. Arnold. Arbor 51: 140. 1970; Parmar, in Shetty & Singh Fl. Rajasthan 2: 519. 1991. *Cornus gharaf* Forssk. Fl. Aegypt. - Arab. 94. 1775. *Cordia rothii* Roem. & Schult. Syst. Veg. 4: 798. 1819; Wight, Ic. 4 (2): 15. t. 1379. 1848; Clarke, in Hook. *f.* Fl. Brit. Ind. 4: 138. 1883; Duthie, Fl. Gangetic Plain 2: 85. 1911. (Plate 208)

Vern.: *Gondi, Gundi.*

Medium sized tree or shrub, up to 8 m tall, bark deeply furrowed. Leaves 5 - 8 x 2 - 4 cm, oblanceolate to oblong, entire to subundulate, coriaceous to scabrid, cuneate, subacute, nerves impressed above, strigose to pubescent, especially in the axils of the nerves on the under surface. Flowers white, 0.6 - 0.8 cm wide, in terminal or axillary cymes. Calyx 3.5 mm long, tubular-campanulate, pubescent, accrescent, shallowly lobed; lobes obtuse. Corolla ca 5 mm long, white; lobes ligulate-obtuse, reflexed. Filaments ca 1.8 mm long, glabrous. Anthers ca 1 mm long. Drupe 0.6 - 1.2 cm long, ovoid, mucronate, yellow to reddish-brown, pellucid pulpy, 1-seeded.

Field note: Common in forests and waste lands.

Fl. & Fr.: April - December.

Specimen examined: Pansal 1988, Kachhola 2007 (MLVGCB Herb).

Distribution: N. Africa, Arabia, Pakistan, India and Sri-Lanka.

Uses: The edible fruit pulp is consumed by shepherds.

EHRETIA P. Brown

Ehretia laevis Roxb. Fl. Cor. 1: 42. t. 55. 1795; Fl. Ind. ed. (Car. & Wall.) 2 (2): 341. 1824; Clarke, in Hook. *f.* Fl. Brit. Ind. 4: 141. 1883; includ. var. *floribunda* (Benth.) Brandis; Duthie, Fl. Gangetic Plain 2: 86. 1911; Parmar, in Shetty & Singh Fl. Rajasthan 2: 523. 1991. (Plate 209)

Small to medium sized tree up to 8 m tall. Young shoots and branches puberulous to glabrescent. Leaves 4-10 x 4-1.6 cm, suborbicular or ovate-elliptic or obovate, obtuse to acute, minutely pubescent especially on undersurface, but becoming glabrescent, base rounded to cuneate or oblique. Petiole long. Flowers white, in axillary and terminal cymes, sessile or subsessile. Calyx 0.1 - 0.2 cm long, 5-partite, minutely rusty-tomentose; lobes ca 1 mm long, ovate, acute. Corolla 0.2 - 0.4 cm long; lobes ovate-lanceolate, ca 3 mm long, spreading. Filaments ca length of corolla, glabrous, lower half adnate; anthers 1.5 mm long. Ovary 2 mm long, ovoid; styles 2 - 3 mm long; stigma capitate. Drupe 0.3 - 0.4 cm across, 2-lobed, orange when mature, black and wrinkled when dry.

Field note: Common in dry deciduous forests.

Fl. & Fr.: March - May.

Specimen examined: Mangliawas 4055 (MLVGCB Herb).

Distribution: Iran, Pakistan, India, Burma, Hainan, Polynesia, China and Australia.

CONVOLVULACEAE

KEY TO THE GENERA

1a. Pollen echinate.. 2

2a. Stigma usually capitate or entire. Fruits dehiscent.. *Ipomoea*

2b. Stigma linear-lanceolate. Fruits indehiscent.. *Rivea*

1b. Pollen not echinate.. 3

3a. Styles 2, free or more or less united at the base.. 4

4a. Flowers axillary, solitary. Corolla - lobes imbricate. Stamens exserted. Stigma 2.. *Cressa*

4b. Flowers in cymes. Corolla - lobes in duplicate - contorted. Stamens not exserted. Stigma 4.. *Evolvulus*

3b. Styles 1, entire or slightly 2 - lobed at the apex.. 5

5a. Stem and leaves with stellate trichomes or glabrous. Stigma elliptic oblong... *Convolvulus*

5b. Stem and leaves glabrous or with simple trichomes. Stigmas globose or capitate.. *Merremia*

CONVOLVULUS L.

KEY TO THE SPECIES

1a. Leaves long petioled. Corolla 4 - 5 tims as long as the calyx.. *C. arvensis*

1b. Leaves sessile to subsessile. Corolla smallar, equal or at most twice as long as calyx.. *C. prostratus*

Convolvulus arvensis L., Sp. Pl. 1: 153. 1753; Clarke, in Hook. *f.* Fl. Brit. Ind. 4: 219. 1883; Duthie, Fl. Gangetic Plain 2: 106. 1911; Mahesh. Illus. Fl. Delhi f. 146. 1966; Bhandari, Fl. Indian Desert 217. 1995; Singh, in Shetty & Singh Fl. Rajasthan 2: 528. 1991.

Vern.: *Hiranpagi.*

Perennial herb, with woody rhizomes. Stem prostrate or twining, glabrous or sparsely pubescent. leaves ovat or oblong, 1.5 - 5.0 x 0.5 - 3 cm, glabrous or pubescent, base hastate, sagittate, or cordate, apex obtuse, mucronulate; prominently 3-veined basally, upper parts pinnately veined; Petiole 0.3-2 cm long. Inflorescence axillary, cymose, 1-3-flowered; peduncle 3-8 cm; bracts 2, linear, ca 3 mm. Pedicel as long as calyx. Sepals unequal, 3.5 - 5 mm; outer 2 broadly oblong to obovate, shorter, abaxially sparsely pubescent or glabrous, margins somewhat ciliate, apex retuse; inner ones ovate-circular, margins membranous, apex obtuse or retuse. Corolla white or pink, broadly funnelform, 1.5-2.6 cm, midpetaline bands pubescent outside distally; limb shallowly 5-lobed. Stamens included, unequal; filaments minutely scaly basally. Disc cupular. Ovary ovoid, glabrous or sparsely pubescent; stigmas cylindric. Capsule globose, 0.6 - 0.8 cm across, glabrous. Seeds 4 or fewer, dark brown or black, subtrigonous, 3-4 mm, tuberculate.

Field note: Common in waste lands, roadsides, grasslands etc.

Fl. & Fr.: December - April.

Specimen examined: Karoi 1967 (MLVGCB Herb).

Distribution: Asia, Europe, North America and South America.

CONVOLVULUS PROSTRATUS Forssk.

KEY TO THE VARIETIES

2a. Flowers condensed at the end of axillary peduncles .... *C. prostratus* var. *deserti*

2b. Flowers clustered directly in leaf axils .... *C. prostratus* var. *prostratus*

Convolvulus prostratus Forsk. var. ***deserti*** (Hochst. & Steud. ex Baker & Rendle) Parmar, in J. Econ. Taxon. Bot. 18: 251. 1994. *C. deserti* Hoschst. & Steud. ex Baker & Rendle in Dyer, Fl. Trop. Afr. 4 (2): 164. 1905; Singh, in Shetty & Singh Fl. Rajasthan 2: 530. 1991; Bhandari, Fl. Indian Desert 220. 1995. *C. microphyllus sensu* Choisy in DC. Prodr. 9: 1845, non Sieb. ex Spreng. 1825. (Plate 210)

Vern.: *Santari*.

Suffruticose, with slender stems, upto 35 cm high, sparsely clothed with spreading hairs. Leaves subsessile, linear-oblong to linear-spathulate, usually 3 - 4 cm long, few lower ones upto 7 cm long, midrib prominant beneath. Flowers 1-3 on slender axillary peduncles, the latter upto 3 cm long, hairy. Pedicels 2 - 3 cm long, terete, hairy. Bracts narrow, linear, 10 mm long, adpressed with spreading, ferruginous hairs. Sepals upto 7.5 mm long, subequal, ovate, acuminate, green herbaceous tips. Corolla funnel shaped, 11 mm long, midpetaline area darker, pubescent. Ovary ovoid, glabrous, on a disc; style 2.5 mm long. Fruit 3 mm long, pubescent. Seeds 2 - 4 mm long, dark brown, minutely pubescent.

Field note: Rare, in dry sandy habitats.

Fl. & Fr.: August - September.

Specimen examined: Bhilwara 1987, Asind 2148 (MLVGCB Herb).

Distribution: Arabia, Egypt and India.

Convolvulus prostratus Forsk. Fl. Aegypt. - Arab. 203. 1775; Singh, in Shetty & Singh Fl. Rajasthan 2: 530. 1991. *C. microphyllus* Sieb. ex Spreng. Syst. Veg. 1 : 611. 1825; Wight, Ic. 4 (2) : 13. t. 1367. 1848; Clarke, in Hook. *f.* Fl. Brit. Ind. 4: 218. 1883, incl. var. *boissieri*. *C. pluriocaulis* Choisy in Mem. Soc. Phys. Hist. Nat. Geneve 6 : 477. 1834; Clarke in Hook.f. *l. c.* 4 : 218. 1883, incl. var. *macra*; Duthie, Fl. Gangetic Plain 2: 105. 1911. *C. scindicus* Boiss. Diagn. Pl. ser. 2 (3). 123. 1856, non Stocks 1852. (Plate 211)

var. ***prostratus***

Vern.: *Santari*.

Prostrate or procumbent, perennial, herb, 10 - 50 cm high, hairy or ferruginous pilose. Leaves sessile, 1.5 - 7.0 x 0.3 - 0.8 cm, linear to oblong or oblanceolate to lanceolate, obtuse to mucronate at apex, villous; lower leaves usually larger, spathulate, mid-vien prominent beneath. Flowers in 3 - 5 flowered axillary heads; peduncles upto 1 cm long, decreasing in length upwards, finally becoming subsessile. Bracts linear - oblanceolate, acuminate, outer longer than inner ones. Calyx 4 - 7 mm long, villous, ovate-lanceolate. Corolla 10-12 mm long, funnel- shaped, pinkish or white. Stigma 2-3 times longer than style, filiform. Capsules ca 3 mm in diam., sub-globose, somewhat scarrious, glabrous. Seeds 2 - 4, brown or black, pubescent, 2 mm long.

Field note: Common in sandy habitats.

Fl. & Fr.: August - January.

Specimen examined: Shambhupura 4005, Atun 1966 (MLVGCB Herb).

Distribution: Arabia, Egypt, India and Pakistan.

CRESSA L.

Cressa cretica L., Sp. Pl. 223. 1753; Clarke, in Hook. *f.* Fl. Brit. Ind. 4: 225. 1883 ; Duthie, Fl. Gangetic Plain 2: 103. 1911; Singh, in Shetty & Singh Fl. Rajasthan 2: 201. 1991; Bhandari, Fl. Indian Desert 223. 1995. *C. indica* Retz., Obs. Bot. 4: 242. 1786. (Plate 212)

Vern.: *Rudenti*.

Perennial, erect or suberect, hairy, herb, upto 25 cm high. Stem with many branches, densely leafy. Leaves alternate, apparently crowded, 3 - 7 x 1.5 - 3.5 mm, ovate-rounded or narrowed towards base, subacute at apex, densely silky hairy, strigose at age. Flowers whitish-pink, subsessile, solitary or in small clusters in the axils of upper leaves. Bracts 2, linear, hairy, appressed to the calyx. Calyx-lobes 2 - 2.5 mm long, elliptic, obtuse, concave, densely silky. Corolla ca 4 - 6 mm long, campanulate, divided up to middle; lobes oblong, deflexed, hairy outside near th apex, white or purple white. Stamens 5, slightly exserted. Style exserted. Ovary 2 - celled, 4-ovuled. Capsules 3 - 5 mm long, ovoid, crowned by style-base, pubescent at apex, usually 1-seeded. Seeds ca 2 mm long, chestnut-brown, glabrous.

Field note: Common in waste lands.

Fl. & Fr.: November-March.

Specimen examined: Asind 121 (MLVGCB Herb).

Distribution: Throughout the warmer regions of the world.

EVOLVULUS L.

Evolvulus alsinoides (L.) L., Sp. Pl. ed. 2. 392. 1762; Clarke, in Hook. *f.* Fl. Brit. Ind. 4: 220. 1883; Duthie, Fl. Gangetic Plain 2: 104. 1911; Mahesh. Illus. Fl. Delhi f. 136. 1966; Bhandari, Fl. Indian Desert 223. 1995; Singh, in Shetty & Singh Fl. Rajasthan 2: 533. 1991. *Covolvulus alsinoides* L., Sp. Pl. 157. 1753. (Plate 213)

Perennial herb, prostrate or ascending. Stems several to numerous, slender, with appressed and spreading hairs. Leaves petiolate or subsessile, 0.5-3.5 x 0.3-1.0 cm, elliptic oblong, obtuse, emarginate or acute to acuminate adpressedly pilose on both surfaces, somtimes glabrous above. Flowers axillary, pedunculate in 1 to several flowered dichasia or solitary. Peduncles filiform, shorter or exceeding the leaves. Bracteoles linear-lanceolate, 1.5-4 mm. Sepals lanceolate, 2-4 mm long, villous, acute or acuminate. Corolla rotate, 7-11 mm in diam., pale-blue or white. Stamens included; filaments filiform, adnate to base of corolla tube. Ovary glabrous; styles 2, free. Capsule 4 mm long, ovoid, 4-valved. Seeds 4 or fewer, black, smooth.

Field note: Common in waste lands in dry sandy and rocky habitats.

Fl. & Fr.: Almost throughout the year

Specimen examined: Bhinay 4055, Bijoliya 2002 (MLVGCB Herb).

Distribution: Bangladesh, Cambodia, China, India, Indonesia, Japan (Ryukyu Islands), Laos, Malaysia, Myanmar, Nepal, Pakistan, Philippines, Thailand, Vietnam; Africa, Australia, North America, Pacific Islands, South America.

Notes: Highly polymorphic species. Van Oosttstroom (Med. Bot. Mus. Herb. Utrech 14 : 26 - 58. 1934) has recognised 17 varieties under this species, of which five vars. *alsinoides*, *hirsutus* (Lam.) Ooststrt., *decumbens* (R. Br.) Ooststr., *linifolius* (L.) Bak. and *wallichii* Ooststr. are recorded from Rajasthan. The specmens from Rajasthan cited under these species merge together so much that the author could not adopt the present status of these taxa. A detailed study is required to comment upon (Shetty & Singh, 1991).

IPOMOEA L.

KEY TO THE SPECIES

1a. Ovary 4- celled............................ *I. mauritiana*

1b. Ovary 2 - 3 celled... 2

2a. Ovary 3-celled 3

3a. Peduncles usually 1 - 3 flowered. Bracts linear. Capsule globose .... *I. nil*

3b. Peduncles usually 5 - many flowered. Bracts setose. Capsule ovoid .... *I. indica*

2b. Ovary 2-celled 4

4a. Flowers sessile or sub-sessile, in heads or clusters 5

5a. Leaves rotundate in outline, deeply palmately 5 - 9 lobed................. *I. pes - tigridis*

5b. Leaves ovate - cordate or oblong - hastate, entire 6

6a. Capsules glabrous. Seeds villous ...... *I. sindica*

6b. Capsules pubescent. Seeds glabrous .... *I. eriocarpa*

4b. Flowers pedicelled, arranged in variously but not in heads 7

7a. Creeping aquatic or marshland herbs .... *I. aquatica*

7b. Erect or twinning, terrestrial herbs or shrubs or creeper 8

8a. Leaves stipulate, deeply digitately or pedately 5 - 9 partite or foliate.......... *I. cairica*

8b. Leaves exstipulate, entire, angular or somewhat lobed but never digitate 9

9a. Stem muricated or prickly............. *I. turbinata*

9b. Stem glabrous not muricate.............. *I. carnea*

Ipomoea aquatica Forssk., Fl. Aegypt.-Arab. 44. 1775; Clarke, in Hook. *f.* Fl. Brit. Ind. 4: 210. 1883; Mahesh. Illus. Fl. Delhi f. 139. 1966; Bhandari, Fl. Indian Desert 225. 1995; Singh, in Shetty & Singh Fl. Rajasthan 2: 538. 1991. *Convolvulus repens* Vahl, Symb. Bot. 1: 17. 1790, non L. 1753. *Ipomoea reptans* Poir. in Lam. Encycl. Meth. Bot. Suppl. 3: 460. 1814; Duthie, Fl. Gangetic Plain 2: 114. 1911. *Ipomoea repens* (Vahl) Roth, Nov. Pl. Sp. 110. 1821, *non illig.* (Plate 214)

Vern.: *Besharmi.*

Perennial herb, marshy or floating, cree-ping, aquatic; arial parts glabrous. Stems terete, thick, hollow, rooting at nodes. Leaves variable, ovate, ovate-lanceolate, oblong or lanceolate, 3 - 8 x 0.8 - 5.5 cm, glabrous or rarely pilose, base cord-ate, sagittate or hastate, occasionally truncate, margin entire or undulate, apex acute or acuminate. Petiole 3-14 cm, glabrous. Inflorescences in axillary 1-few-flowered, pedunculate cyme. Peduncle 2-6 cm long. Bracts minute narrow, acute. Pedicel 1.5-5 cm. Sepals subequal, glabrous; outer 2 ovate-oblong or oblong-lanceolate, 7-8 mm, margins whitish, thin, apex obtuse, mucronulate; inner 3 ovate-elliptic, *ca.* 8 mm. Corolla white, pink, or lilac, with a darker center, funnelform, 3.5-5 cm, glabrous. Stamens unequal, included; filaments hairy at base. Pistil included; ovary conical, glabrous; stigma 2-lobed. Capsule ovoid to globose, ca 1 cm in diam., 2-4 seeded woody. Seeds densely greyish pubescent, sometimes glabrous.

Field note: Common in marshy and aquatic habitats near ditches, ponds and waste lands.

Fl. & Fr.: Almost throughout the year.

Specimen examined: Mezadam 217 (MLVGCB Herb).

Distribution: Bangladesh, Cambodia, China, India, Indonesia, Laos, Malaysia, Myanmar, Nepal, New Guinea, Pakistan, Philippines, Sri Lanka, Thailand, Vietnam; Africa, Australia, Pacific Islands and South America.

Uses: Plants have minor medicinal uses.

Ipomoea cairica (L.) Sweet Hort. Brit. ed. 1. 287. 1826; Raizada, Suppl. Duthie, Fl. Gangetic Plain 162, 1976; Bhandari, Fl. Indian Desert 225. 1995; Singh, in Shetty & Singh Fl. Rajasthan 2: 538. 1991. *Convolvulus cairicus* L. Syst. Nat. ed. 10. 922. 1759. *Ipomoea palmata* Forsk. Fl. Aegypt.-Arab. 43. 1775; Clarke, in Hook. *f.* Fl. Brit. Ind. 4: 214. 1883, excl. some syns. *I. pulchella* Wight, Ic. 1 (8): 8. t. 156. 1839, non Roth 1821; Duthie, Fl. Gangetic Plain 2: 117. 1911.

var. ***cairica***

Vern.: *Railway creeper.*

Perennial herb, twining. Stems thinly angular, slightly tuberculate or smooth. leaves blade palmately 5 - 7-parted to base; lobes entire or minutely undulate, apex acute or obtuse, mucronulate, basal pair usually again lobed or parted; middle lobe larger, ovate, ovate-lanceolate, or elliptic, 2.5 - 5 x 0.5 - 2.5 cm. Petiole 2-8 cm, base with leafy pseudostipules. Inflorescence in lax 1 to several flowered cyme; peduncle 2-8 cm. Pedicel 0.5 - 2 cm, sometimes verruculose. Bracts and bracteoles early deciduous, small. Sepals unequal, ovate; outer 2 sepals 4 - 6.5 mm; inner ones 5 - 9 mm, glabrous, scarious. Corolla purple or blue purple, with a darker center, funnelform, 2.5-7 cm. Stamens included, unequal. Ovary glabrous; stigma 2-lobed. Capsule subglobose, ca 1-1.5 cm across. Seeds subglobose to ovoid, black, *ca.* 5 mm, densely tomentose, margins with longer hairs.

Field note: Common among bushes, shrubs and hedges of gardens, wasteplaces. Also cultivated ornamentally.

Fl. & Fr.: Almost throughout the year.

Specimen examined: Bhilwara 1590, Mandal 130 (MLVGCB Herb).

Distribution: Pantropical.

Ipomoea carnea Jacq. subsp. ***fistulosa*** (Martius ex Choisy) Austin, in Taxon 26: 237. f. 2. 1977; Singh, in Shetty & Singh Fl. Rajasthan 2: 539. 1991. *I. fistulosa* Mart. ex Choisy in DC. Prodr. 9: 349. 1845; Raizada, Suppl. Duthie, Fl. Gangetic Plain 163, 1976; *Batatas crassicaulis* Benth. Voy. Sulph. 134. 1845. *I. crassicaulis* (Benth.) B.L. Robinson in Porc. Amer. Acad. Sci. 51: 530. 1916. (Plate 215)

Vern: *Besharumi*

Shrub, 1 - 3 m high; axial parts puberulent, later glabrescent; sap milky. Branches terete or angular, stout, becoming sparsely lenticellate with age. Leaves ovate-lanceolate or ovate-oblong, 6 - 15 x 4 - 9 cm, densely puberulent or adaxially subglabrous, base cordate or truncate, margin entire, apex acuminate, mucronulate; midvein 2-glandular abaxially at base, lateral veins 7 - 9 pairs. Petiole 2 - 15 cm. Inflorescence axillary or terminal few to several flowered paniculate cymes; peduncle stout, 5 cm long; bracts early deciduous, ovate. Sepals ovate-elliptic, obtuse to retuse, abaxially puberulent, equal or inner ones longer, 5-6 mm. Corolla lilac or pink, darker inside, funnel shaped, 7 - 10 cm. Stamens included; filaments unequal; anthers linear, base sagittate. Pistil included; ovary puberulent. Style base puberulent; stigma 2-lobed. Capsule pale brown, ovoid, 1.5 - 2 cm, 4-valved, apiculate. Seeds black, ca. 1 cm, brown sericeous - pubescent.

Field note: Common near water sources, also cultivated as hedge plant along agriculture fields as well as grasslands. It forms dense community.

Fl. & Fr.: Almost throughout the year.

Specimen examined: Mandalgarh 2186 (MLVGCB Herb).

Distribution: Cambodia, China, India, Indonesia, Japan, Myanmar, Nepal, New Guinea, Pakistan, Sri Lanka, Thailand; North America, Pacific Islands and South America.

Notes: Verma and Srivastava[J. Econ. Tax. Bot. 4 (3): 903 - 907. 1983] have made an attempt to prove that *I. carnea* Jacq. and *I. fistulosa* Mart. ex Choisy are conspecific ignoring differences pointed out in the habit, leaf - shape character, wood anatomy and pollen morphology by earlier workers. The authors do not agree with them. However, they consider *I. fistulosa* Mart. ex Choisy as a subsp. under *I. carnea* Jacq. due to inadequate differentiating morphological characters which can be observed and assessed easily in the fields as well as laboratory (Shetty & Singh, 1991).

Ipomoea eriocarpa R. Br., Prodr. 484. 1810; Clarke, in Hook. *f.* Fl. Brit. Ind. 4: 204. 1883; Bhandari, Fl. Indian Desert 227. 1995; Singh, in Shetty & Singh Fl. Rajasthan 2: 541. 1991. *Convolvulus hispidus* Vahl, Symb. Bot. 3 : 29. 1794. *Ipomoea hispida* (Vahl) Roem. & Schult. Syst. Veg. 4 : 238. 1819, non Zucc. 1809; Duthie, Fl. Gangetic Plain 2: 113. 1911.

Vern.: *Nakhari.*

An annual herb, twining or prostrate, with slightly retrorsely or spreading hispid axial parts. Stems 1-2 m. Leaves mostly lanceolate to oblong or linear-lanceolate, 2.5 - 6 x 0.5 - 1.5 cm, base cordate, margins entire, apex long attenuate, sometimes acute, appressed hispid; lateral veins 6 or 7 pairs. Petiole 1.5 - 6.5 cm. Flowers axillary sessile or short pedunculate cymes or heads, usually 1 - 7-flowered. Bracts linear-lanceolate, 3 - 8 mm, abaxially pilose. Pedicel very short or absent. Sepals ovate, subequal, 7 - 9 mm, abaxially densely hirsute, apex linear-acuminate; inner sepals slightly narrower. Corolla pink or purplish, campanulate, 7 - 9 mm, midpetaline bands densely hirsute outside; limb triangular-lobed. Stamens included. Ovary hirsute, 2-loculed, 4-ovuled; stigma capitate, 2-lobed. Capsule subglobose, 5 - 8 mm in diam., spreading hirsute, apiculate. Seeds grey-brown, 2.5 - 4 mm, glabrous, minutely reticulate.

Field note: Common weed of fields, thickets, grassy slopes, open places. Twining during rainy season.

Fl. & Fr.: August - December.

Specimen examined: Kishnawaton ki kheri 868 (MLVGCB Herb).

Distribution: Cambodia, China, India, Indonesia, Kashmir, Laos, Malaysia, Myanmar, Nepal, New Guinea, Pakistan, Philippines, Sri Lanka, Thailand, Vietnam; Africa and N Australia.

Ipomoea indica (Burm. f.) Merrill, Interpr. Rumph. Herb. Amboin. 445. 1917; Bhandari, Fl. Indian Desert 228. 1995; Singh, in Shetty & Singh Fl. Rajasthan 2: 542. 1991. *Convolvulus indicus* J. Burm. f. in Rumph. Herb. Amb. 7: 6. 1755. *Ipomoea congesta* R. Br. Prodr. 485. 1810; Prain in J. Asi. Soc. Beng. 63: 104. 1894; van Ooststr. Fl. Males ser. 1 (4) : 465. f. 39 A. 1963; Verdcourrt in Taxon 6: 232. 1957; Santapau & Patel, in Trans. Bose Resch. Inst. Cal. 22: 33. 1958. *I. learii* Paxton, in Paxton, Bot. Mag. 6 : t. 267. 1839; Duthie, Fl. Gangetic Plain 2: 117. 1911.

Vern.: *Morning glory.*

Herb, twining up to 6 m hgh, with densely retrorse pilose axial parts. Stems 3-6 m. Leaves ovate or circular, 4 - 15 x 3 - cm, abaxially densely short, soft, pubescent, adaxially ± sparsely pubescent, base cordate, margin entire or 3-lobed, apex acuminate or abruptly acuminate. Petiole 2 - 18 cm. Inflorescence dense umbellate cymes, several flowered; peduncle 4 - 20 cm; bracts linear, sometimes lanceolate. Pedicel 2 - 5 mm. Sepals subequal, 1.4 - 2.2 cm, gradually linear-acuminate apically, glabrous to appressed pilose; outer 3 lanceolate to broadly lanceolate; inner 2 narrowly lanceolate. Corolla bright blue or bluish purple, aging reddish purple or red, with a paler center, funnelform, 5 - 8 cm, glabrous. Stamens included. Pistil included; ovary glabrous. Stigma 3-lobed. Capsule globose, 1 - 1.3 cm in diam. Seeds ca. 5 mm.

Field note: Common in waste lands. Flower colour changes to pink or red in the afternoon at the time of closing.

Fl. & Fr.: August-March.

Specimen examined: Atun 1721, Pur 1991 (MLVGCB Herb).

Distribution: China, India, Indonesia, Japan, Malaysia, Myanmar, New Guinea, Pakistan, Philippines, Sri Lanka, Pacific Islands and South America.

Notes: Fosberg (Bot. Not. 129: 35-38. 1976) has sorted out the complicated nomenclature for this pantropical species and established that *Ipomoea indica* is the correct name for it.

Ipomoea mauritiana Jacq., Collect. 4 : 216. 1791 & in Hort. Schoenbr. 2 : 39. t. 200. 1797; Shah, Fl. Gujrat 1 : 471. 1978; Singh, in Shetty & Singh Fl. Rajasthan 2: 543. 1991. *Convolvulus paniculatum* L., Sp. Pl. 156. 1753. *Ipomoea digitata* auct. non L. 1759; Clarke, in Hook. *f.* Fl. Brit. Ind. 4: 202. 1883. *I. paniculata* (L.) R. Br. Prodr. 486. 1810, non Burm. f. 1768. (Plate 216)

Perennial herb, twining, with glabrous or minutely muricate axial parts. Roots tuberous. Stems to 10 m, thinly angular. Leaves 5 - 14 x 6 - 16 cm, usually palmately divided to or beyond middle, base cordate or truncate; segments 3 - 7 (-9), lanceolate or elliptic, apex mucronate, glabrous or sparsely pubescent along midvein, entire or irregularly undulate, apex acuminate or acute, mucronulate. Petiole 3-11 cm. Inflorescence axillary, few to many flowered; peduncle 2.5 - 20 cm, terete; bracts early deciduous. Pedicel 0.9 - 2.5 cm. Sepals sub-circular, oblong to broadly elliptic, concave, equal or outer 2 shorter, 6 - 12 mm, glabrous, apex obtuse. Corolla pink or reddish purple, with a darker center, funnel shaped, 5 - 6 cm; limb 5 - 7 cm in diam., undulate. Stamens included. Pistil included; ovary glabrous. Stigma 2-lobed. Capsule ovoid, 1.2 - 1.5 cm long.

Seeds 4, dark brown, ca. 6 mm, woolly-sericeous with long, easily detached hairs.

Field note: Rare in Waste places and forests.

Fl. & Fr.: July - November.

Specimen examined: Bhilwara 1768, Kishnawato ki kheri 2147 (MLVGCB Herb).

Distribution: Cambodia, China, Indonesia, Japan, Laos, Malaysia, Myanmar, New Guinea, Philippines, Sri Lanka, Thailand, Vietnam; Pacific Islands.

Uses: The leaves and roots are used to treat tuberculosis and for the treatment of breast infections.

Notes: The origin of *Ipomoea mauritiana* is unknown, but it may be in tropical America, where the nearest relatives occur. *Ipomoea mauritiana* has often been misidentified as *I. digitata* (L.) L., a West Indian endemic that does not occur in Asia.

Ipomoea nil (L.) Roth, Cat. Bot. 1: 36. 1797; Mahesh. Illus. Fl. Delhi f. 145. 1966; Bhandari, Fl. Indian Desert 229. 1995; Singh, in Shetty & Singh Fl. Rajasthan 2: 543. 1991. *Convolvulus nil* L., Sp. Pl. ed. 2. 219. 1762; *C. hederaceus auct. plur.* non (L.) Jacq. 1760; Clarke, in Hook. *f.* Fl. Brit. Ind. 4: 199. 1883; Duthie, Fl. Gangetic Plain 2: 116. 1911. *Ipomoea hederacea* (L.) Jacq. var. *integrifolia* Clarke in Hook. *f. l.c.* 4 : 200. 1883. (Plate 217)

var.: ***nil***

Vern.: *Nilikalmi.*

An annual, herb twining, with retrorsely hirsute arial parts. Stem 2-5 m. Leaves broadly ovate or nearly circular, 3 - 15 x 2.5 - 12.0 cm, hirtellous, base cordate, margin entire or 3 - lobed, apex acuminate. Petiole 2 - 15 cm. Inflorescences axillary, 1 to 5 flowered umbellate cyme; peduncle 1.5 - 8 cm; bracts linear or filiform, 5 - 6 mm. Pedicel 2 - 7 mm. Sepals lanceolate, linear-lanceolate, 1 - 2.5 cm, abaxially spreading hirsute, subglabrous apically, acuminate, hairs swollen based. Corolla pale to bright blue with whitish tube, fading to pinkish in age, funnel shaped, 5 - 6 cm, glabrous. Stamens included, unequal. Pistil included; ovary glabrous, 3-loculed. Stigma 3-lobed. Capsule straw colored, ovoid to ovoid or globose, 5 - 10 mm in diam., glabrous. Seeds black, ovoid-trigonous, 5 - 6 mm, gray puberulent.

Field note: Common in waste lands, forests, boundaries of agricultural fields etc.

Fl. & Fr.: August- November.

Specimen examined: Bhilwara 215 (MLVGCB Herb).

Distribution: Native of South America, now nearly circumtropical.

Uses: Seeds have a number of local medicinal uses.

Ipomoea pes-tigridis L., Sp. Pl. 1: 162. 1753; Clarke, in Hook. *f.* Fl. Brit. Ind. 4: 204. 1883; Duthie, Fl. Gangetic Plain 2: 116. 1911; Mahesh. Illus. Fl. Delhi f. 143. 1966; Bhandari, Fl. Indian Desert 230. 1995; Singh, in Shetty & Singh Fl. Rajasthan 2: 545. 1991. *I. hepaticifolia* L., Sp. Pl. 161. 1753. *I. capitellata* Choisy in Mem. Soc. Phys. Geneve 6: 446. 1884 & in Dc. Prodr. 9: 365. 1845. *I. pes-tigridis* L. *var. hepaticaefolia* (L.) Clarke in Hook.f. *l.c.* 4: 204. 1883. *I. capitellata* (Choisy) Clarke in Hook.f. *l.c.* 4 : 204. 1883. (Plate 218)

An annual, herb, twining, with spreading hispid axial parts. Stem 0.5 - 3 m long, hirsute. Leaves circular or transversely elliptic in outline, 2 - 10 x 3 - 13 cm, palmately deeply divided; segments 5 - 9, elliptic or oblong, tapered at both ends, densely pubescent, apex mucronate. Petiole 2 - 8 cm. Inflorescences capitate, 3 - 5 -flowered; peduncle 4 - 10 cm long; bracts hirsute; outer involucral bracts oblong to linear-oblong, 2 - 2.5 cm; inner bracts smaller. Sepals lanceolate, slightly unequal, 1 - 1.4 cm, hirsute on both sides. Corolla white, funnel shaped, 3 - 4 cm long, midpetaline bands sparsely pubescent. Stamens included; filaments glabrous. Pistil included; ovary 2-loculed, glabrous; stigma 2-lobed. Capsule ovoid, ca 6 - 8 mm long, 4-valved. Seeds triangular, ca 4 mm, grey tomentellous.

Field note: Common weed of cultivated fields, thickets, roadsides and open forests, Flowers open in afternoon.

Fl. & Fr.: August- December.

Specimen examined: Atun 1406, Badnor 1892 (MLVGCB Herb).

Distribution: Cambodia, China, Indonesia, India, Malaysia, Myanmar, Nepal, New Guinea, Pakistan, Philippines, Sri Lanka, Thailand, Vietnam; Africa, Australia and Pacific Islands.

Ipomoea sindica Staf. in Kew Bull. 93 : 346. 1894; Duthie, Fl. Gangetic Plain 2: 113. 1911; Mahesh. Illus. Fl. Delhi f. 140. 1966; Bhandari, Fl. Indian Desert 231. 1995; Singh, in Shetty & Singh Fl. Rajasthan 2: 2547. 1991. (Plate 219)

Trailing or creeping, hirsute herb. Stems many from base, with long internodes. Leaves, 2.5 x 3-4 cm, oblong, hastate at base, acute or acuminate at apex, appressed hairy on both surfaces, ciliate on the margins, entire. Petioles, hairy. Flowers in axillary, 1-3 flowered cymes; pedicels 3-4 mm long, terete, hairy. Bracts linear, hairy, acuminate. Calyx hispid, ciliate with bulbous-based hairs. Corolla, globose to ovoid, glabrous. Seeds 4, ca 3 x 2 mm, minutely grey-velvety.

Field note: Common in waste lands, forests, boundaries of agriculture fields etc.

Fl. & Fr.: August- November.

Specimen examined: Mandal 1591 (MLVGCB Herb).

Distribution: Circumtropical, India (throughout).

Ipomoea turbinata Lag. Gen. Pl. 10. 1816; Raizada, Suppl. Duthie, Fl. Gangetic Plain 164. 1976; Bhandari, Fl. Indian Desert 232. 1995; Singh, in Shetty & Singh Fl. Rajasthan 2: 549. 1991. *Convolvulus muricatus* L. Mant. Pl. 44. 1767. *Ipomoea muricata* (L.) Jacq. Pl. Hort. Schoenbr. 3 : 40. t. 323. 1798 (1883). *Convolvulus muricatum* (L.) G. Don, Gen. Hist. 4 : 264. 1838; Duthie, Fl. Gangetic Plain 2: 188. 1911. (Fig. 101)

Vern.: *Pherwana*.

Twining or trailing, hispid, annual herb. Stem muricate, sap milky. Leaves oblong-ovate, 2 - 7 x 1 - 2.5 cm, base cordate, margin entire, apex acute to acuminate. Petiole 1 - 3 cm long. Inflorescences 1 to few flowered axillary cymes up to 2.5 mm long. Flowers nocturnal. peduncle up to 2.5 mm long; bracts oblong, ca 4.8 mm long, scarious. Pedicel thicker apically, much thickened in fruit. Sepals lanceolate, fleshy, glabrous, distinctly enlarged in fruit and eventually reflexed; outer 2 sepals 6 - 8 mm, apex attenuate into a thick, suberect awn ca. 4 mm; inner 3 sepals 7 - 8 mm, apex obtuse or emarginate, awn shorter. Corolla pale purple, infundibuliform, 5 - 7.5 cm; tube 3 - 6 cm, shallowly 5-lobed. Stamens slightly exserted or not; filaments inserted in apical part of corolla tube, base sparsely short pubescent; anthers large, base cordate. Pistil slightly exserted or not; ovary glabrous; stigma 2-lobed. Capsule ovoid, 0.5 - 0.6 cm, mucronate. Seeds black, trigonous, 3.5 mm, glabrous.

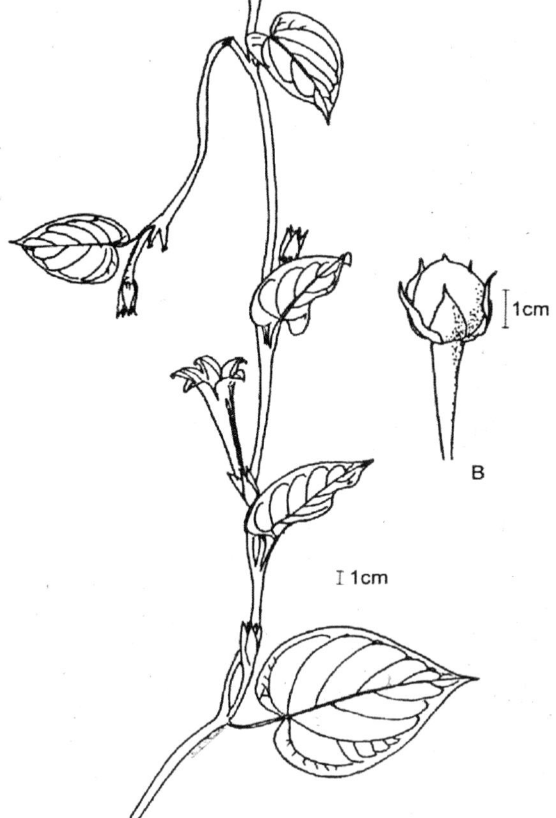

Fig. 101. *Ipomoea turbinata* Lag.
A. Flowering branch, B. Fruit with sepals

Field note: Common in waste lands, forests, boundaries of agriculture fields etc.

Fl. & Fr.: August- November.

Specimen examined: Bhilwara 146, Hamirgarh 1988 (MLVGCB Herb).

Distribution: India, Indonesia, Japan, Myanmar, Nepal, Pakistan, Philippines, Sri Lanka, Vietnam; Africa, North America and South America.

MERREMIA Dennst. ex Endl. nom. cons.

KEY TO THE SPECIES

1a. Prostrate. Leaves simple, crenate. Flowers axillary, yellow .... *M. emarginata*

1b. Twinner. Leaves palmately 5-foliate, Flowers in racemose cymes, white. ....*M. aegyptia*

Merremia aegyptia (L.) Urban Symb. Antill 4: 505. 1910; Ooststroom, in Blumea 3: 327. 1939; Bhandari, Fl. Indian Desert 233. 1995; Singh, in Shetty & Singh Fl. Rajasthan 2: 552. 1991. *Ipomoea aegyptica* L., Sp. Pl. 162. 1753. *Convolvulus pentaphylla* L., Sp. Pl. ed. 2. 233. 1762. *Ipomoea pentaphylla* (L.) Jacq. Collect. 2 : 297. (1788) 1789; Clarke, in Hook. *f.* Fl. Brit. Ind. 4: 202. 1883. *Batatas pentaphylla* (L.) Wight, Ic. 3 (2): 3. t. 834. 1944 - 45. *Merremia pentaphylla* (L.) Hall. f. in Engl. Bot. Jahrb. Syst. 16 : 552. 1893; Duthie, Fl. Gangetic Plain 2: 110. 1911. (Plate 220)

Vern.: *Rotibel*

Twinning annual herb. Stem herbaceous, hirsute with yellow-brown patent hairs. Leaves 3-11 x 1-4 cm, palmately compound; leaflets 5, sessile, elliptic or elliptic oblong, acute at the base, acuminate or acute at apex; entire to dentate, adpressed-pilose on both surfaces. Petiole slender, as long as or longer than blade, patently hirsute. Flowers in few - flowered, lax, fulvous - hairy racemos cymes. Bracts linear. Sepals oblong, enlarged in fruit, 3 outer densely fulvous hairy, 2 inner ones glabrous or with few scattered hairs. Corolla 2-3 cm long, infundibuliform, white. Fruit a capsule, subglobose, 1 - 2 cm across, glabrous, enveloped by persistent densely hairy sepals. Seeds 7-8 mm long, glabrous, shining.

Field note: Common in waste lands among bushes and shurbs, on hedges of cultivated fields as well as gardens.

Fl. & Fr.: August- December.

Specimen examined: Bhilwara 1239 (MLVGCB Herb).

Distribution: Throughout the world.

Merremia emarginata (Burm. f.) Hall. f. in Engl. Bot. Jahrb. Syst. 16: 552. 1893; Duthie, Fl. Gangetic Plain 2: 109. 1911; Bhandari, Fl. Indian Desert 234. 1995; Singh, in Shetty & Singh Fl. Rajasthan 2: 553. 1991. *Evolvulus emarginatus* Burm. f. Fl. Ind. 77. t. 30. f. 1. 1768. *Convolvulus reniformis* Roxb., Fl. Ind. (Carey & Wallich ed.) 2: 67. 1824 & Fl. Ind. 1: 481. 1832. *Ipomoea reniformis* (Roxb.) Choisy. in Mem. Soc. Phys. Geneve 6: 446. 1834; Clarke, in Hook. *f.* Fl. Brit. Ind. 4: 206. 1883. (Plate 221)

Vern.: *Hiran pagi.*

Perennial, herb, prostrate; arial parts sparsely hirtellous or glabrescent. Stems rooting at lower nodes, becoming tuberculate. Leaves reniform to broadly ovate, 0.5 - 3.5 X 0.6 - 3.5 mm, glabrous or sparsely appressed pilose, base cordate, margins entire or coarsely crenate, apex obtuse to broadly rounded or slightly emarginate. Petiole 0.2-3.7 cm. Inflorescence axillary solitary or in 2 - 3 - flowered cymes; peduncles very short. Pedicel 2-4 mm. Sepals ovate to orbicular, pubescent abaxially, margin long ciliate; outer 2 sepals 2.5 - 3 mm, apex obtuse, hoodlike and distinctly mucronate; inner 3 sepals 3 - 6 mm, deeply emarginate. Corolla yellow, tubular-campanulate, 5-9 mm, midpetaline bands distinctly 5-veined, purplish tinged outside, inside pubescent basally; limb slightly 5-lobed. Filaments pubescent basally. Ovary glabrous. Capsule enclosed by persistent calyx, brown-black, subglobular, 5-6 mm, longitudinally grooved, glabrous, apiculate. Seeds greyish brown, ca 2.5 mm, glabrous.

Field note: Common weed of cultivated fields, roadsides, grasslands, clay to sandy soils, forest floors

Fl. & Fr.: October - February.

Specimen examined: Mangrol 4007 (MLVGCB Herb).

Distribution: China, India, Indonesia, Malaysia, Myanmar, Nepal, Philippines, Sri Lanka, Thailand and Africa.

RIVEA **Choisy**

Rivea hypocrateriformis (Desr.) Choisy in Mem. Soc. Phys. Hist. Nat. Geneve 6 : 408. 1834; Clarke, in Hook. *f.* Fl. Brit. Ind. 4: 184. 1883; Duthie, Fl. Gangetic Plain 2: 119. 1911; Mahesh. Illus. Fl. Delhi f. 137. 1966; Bhandari, Fl. Indian Desert 235. 1995; Singh, in Shetty & Singh Fl. Rajasthan 2: 557. 1991. *Convolvulus hypocrateriformis* Desr. in Lamk. Encyl. 3 : 561. 1799. (Plate 222)

Vern.: *Gwariakakari.*

Large woody creeper or climber. Stem terete, woody; Branches silky-pubescent. Leaves 5 - 11 x 6 - 11 cm, broadly ovate to orbicular, cordate - based, shortly acuminate with a pair of purplish glands at the base of lamina. Flowers axillary, 2 - 3-flowered cymes; pedicels 8 - 15 mm long. Bracts 2, spathulate, caducous. Calyx 1 cm long, lobes ovate-elliptic, obtuse, silky hairy. Corolla upto 8 cm long, the tube 4 - 5 cm in dia., white, glabrous. Stamens 5, equal; filaments short, wooly at base. Ovary on a yellow disc; stigma bi-lobed; lobes glandular. Fruits 2 x 1.5 cm in diam., globose-ovoid, mucronate by the style base, woody, chestnut-brown, glabrous, polished subtended by persistant woody deflexed sepals. Seeds 4, ca 8 x 4 mm, brown, glabrous.

Field note: A large climber perticularly on *Euphorbia caducifolia* bushes. The flowers open at about sunset, and remain so until about sunrise, then the corolla wither, become cremy-brown and fall off.

Fl. & Fr.: August- December.

Specimen examined: Jawaice 4059 (MLVGCB Herb).

Distribution: India.

CUSCUTACEAE

CUSCUTA L.

KEY TO THE SPECIES

1a. Corolla white or cream-coloured ...... *C. reflexa*

1b. Corolla yellow 2

2a. Corolla lobes ovate-lanceolate, acute or acuminate, scales absent................ *C. hyalina*

2b. Corolla lobes ovate, obtuse or shortly acute, scales present ....*C. chinensis*

Cuscuta chinensis Lam. Encycl. Meth. Bot. 2 : 229. 1786; Choisy in Mem. Soc. Phys. Hist. Nat. Geneve 1: 279. 1841; Clarke, in Hook. *f.* Fl. Brit. Ind. 4: 226. 1882; Raizada, Suppl. Duthie, Fl. Gangetic Plain 160. 1976; Bhandari, Fl. Indian Desert 238. 1995; Singh, in Shetty & Singh Fl. Rajasthan 2: 560. 1991. (Fig. 102 & Plate 223)

Vern.: *Amarbel.*

Leafless, rootless total stem parasite with much branched stem. Stem slender, 0.6-0.8 mm across, pale yellow, glabrous, smooth. Flower in dense glomerules sessile or shortly pedicellate; pedicel as long as or slightly longer than calyx; bracts narrowly triangular, keeled, entire nearly transparent. Calyx lobes ovate, thickened and fleshy towards the tip to form a carina, also more or less, thickened below the sinuses, slightly overlapping obtuse or acute, 2-2.5 mm long. Corolla slightly longer than calyx, slightly globular, becoming more so as the fruit develops; lobes triangular-ovate or oblong-ovate, spreading, acute or obtuse fleshy with or without a short carina towards the tip. Stamens nearly half the length of corolla lobes; filament flat, pale, longer than the anther; anther ovoid, yellow, scale large; reaching the base of the filaments, fringed with long or medium length processes. Ovary depressed-globose; style 2 stout or slender, thin, divergent, as long or slightly longer than the ovary; stigma capitate. Fruit a capsule, globose, ca 3 mm across, membranous, depressed, dehiscing at the base by an uneven, transverse suture, surmounted by corolla and styles. Seeds 2 - 4, in each eapsule, 1.25 mm long, pale brown, with narrow oblong hilum.

Field note: Uncommon parasite on herbs (*S. tora* (L.) Roxb.) as well as on the shrubs (*i.e. C. multiflorum* (Brum. f.) Kuntze).

Fl. & Fr.: October- January.

Specimen examined: Atun 1893 (MLVGCB Herb).

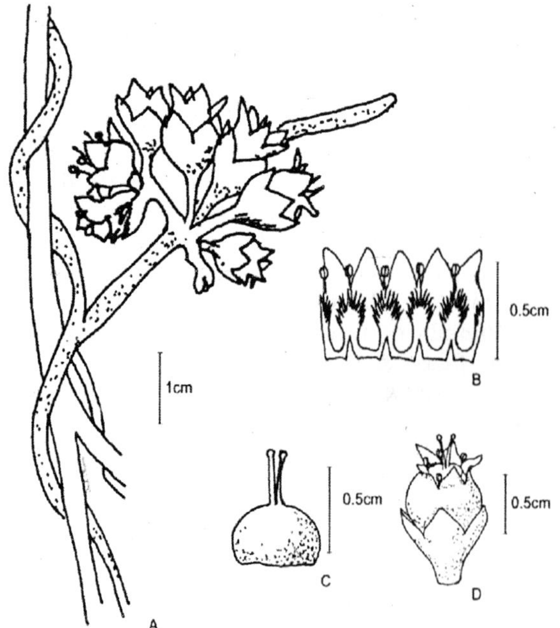

Fig. 102. *Cuscuta chinensis* Lam. var. *chinensis*. A. Inflorescence, B. Corolla with scales & stamens, C. Pistil, D. Fruit with calyx, corolla & stamens

Distribution: Ethiopea, India, Sokotra, Afghanistan, Sri Lanka, Australia and China.

Cuscuta hyalina Heyne ex Roth, Nov. Pl. Sp. 100. 1821; Choisy, in Mem. Soc. Phys. Hist. Nat. Geneve 9: 286. 1841; Engelm. in Trans. Acad. Sci. St. Louis 1: 493. 1859; Clarke, in Hook. *f.* Fl. Brit. Ind. 4: 226. 1883; Raizada, Suppl. Duthie, Fl. Gangetic Plain 161. 1976; Bhandari, Fl. Indian Desert 238. 1995; Singh, in Shetty & Singh Fl. Rajasthan 2: 561. 1991. *C. arabica sensu* Wight, Ic. 4(2): 14. t. 1371. 1848.

Vern.: *Amarbel.*

Stem thin, intermingled, twining, very slender, much branched, leafless, stem-parasite, yellow to brown. Flowers fascicles or shortly pedunculate cyme or racemes; pedicel ca 0.5 mm long. Bracts ca 3 mm long, lanceolate, acute. Calyx lobes 5, scarious, ovate, acute or acuminate, deeply divided, often glandular, 1.5 - 3.0 x 0.5 - 1.5 mm. Corolla lobes yellowish-white, mostly 5, ovate-lanceolate, mostly glandular, acute or acuminate, 1.7 - 2.2 x 0.7 - 1.0 mm; tube 1.7 - 2.0 mm long, scales completely obsolete. Stamens 5, filaments linear, ca 0.5 mm long, attached at the sinus, fused part of the filaments prominent; anther oblong-elliptical, ca 0.5 mm long. Ovary ovoid to globular, 1 - 1.5 mm long; styles 2, linear, unequal, 0.5 - 1.5 mm long; stigma capitate, dark brown to reddish-brown, c. 0.2 mm in diam.; ovules 4, elliptical, 0.5 - 0.7 x ca 0.5 mm. Capsule subglobose, depressed, ca 2.5 mm across, with a intrastylar opening, not definitely circumscissile, the lower part of calyx, surmounted by petals, stamens and styles persisting. Seeds 4, ovoid or suborbicular, dark brown, ca 1.0 mm broad, dull, slightly smooth.

Field note: Common parasite throughout the area usually on *Cassia obtusifolia* L., *Dipteracanthus patulus* (Jacq.) Nees and species of *Amaranthus* L. etc.

Fl. & Fr.: July - October.

Specimen examined: Karoi 168 (MLVGCB Herb).

Distribution: Tropical India to Abyssinia, Kordofan and South Africa.

Cuscuta reflexa Roxb. Pl. Cor. 2: 3. t. 104. 1798; Choisy, in Mom. Soc. Phys. Hist. Nat. Geneve 9 : 273. 1841; Engelm., *l.c.* 1:518. 1859; Clarke, in Hook. *f.* Fl. Brit. Ind. 4: 225. 1883; Duthie, Fl. Gangetic Plain 2: 100. 1911; Bhandari, Fl. Indian Desert 238. 1995; Singh, in Shetty & Singh Fl. Rajasthan 2: 561. 1991. *C. santapaui* Benerji & Das, in J. Arnold Arb. 46: 87. 1965. (Fig. 103 & Plate 224)

Vern.: *Amarbel.*

Stem up to 2 mm, thick, branched, glabrous, or slightly pubescent, succulent, yellowish green. Flowers arranged in paniculate cymes, white, pedicellate, 5.5 - 7.5 mm long; pedicel 0.5 - 3.0 mm long, or obsolete; bracts leaf-like. Calyx lobes 5, fleshy, margin scarious, rounded ovate-obtuse, deeply divided overlapping at the base, 0.7 - 3.0 x 1.5 - 2.5 mm, tube ca 1 mm long. Corolla lobes 5, white, narrowly obtuse, ovate, spreading or reflexed, 6 - 10 mm long; corolla tube 2 - 3 mm long; scales scarious, reaching the middle of the corolla tub, fringed, ca 1 mm long. Stamens 5, filaments linear, 2-3 mm long, fused filaments prominent like a vein in the petal; anther oblong, basifixed, 1 - 1.5 mm long. Ovary conical, fleshy, 1 - 2 mm long; style 1, thick, ca 0.5 mm long, or obsolete; stigmas 2, unequal, conical to elongated, dark brown; ovules 4, oblong-elliptical, ca 0.3 x 0.2 mm. Capsule 1 - 3 x 2 - 3 mm, globose-conical, brown, not depressed, without intrastylar opening, not circumscissile with a definite line of cleavage, i.e. dehiscing irregularly with remains of sepal lobes below, style and stigma at the tip. Seeds 3 - 3.5 mm long, black, ovoid.

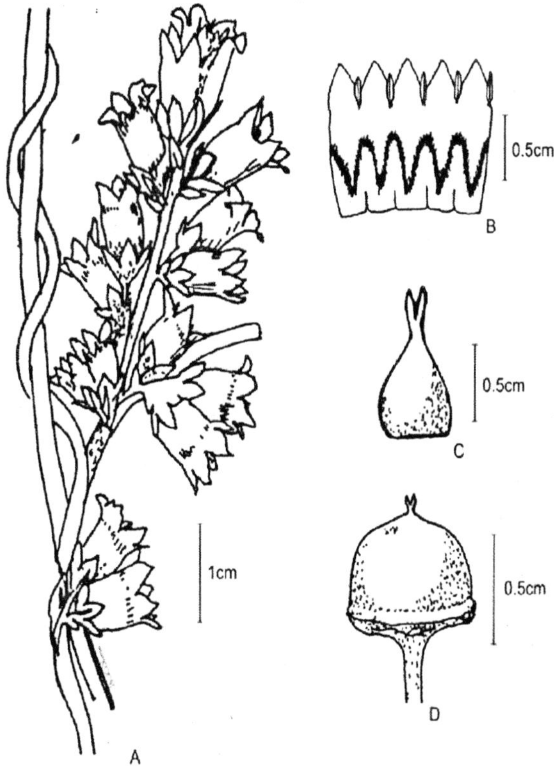

Fig. 103. *Cuscuta reflexa* Roxb.
A. Inflorescence, B. Corolla with scales & stamens, C. Pistil, D. Fruit with calyx

Field note: Common, on *Adhatoda zelanica* Medic., *Acacia nilotica* (L.) Del., *Ziziphus mauritiana* Lam. and *Z. nummularia* (Brum. f.) Wt. & Arn. etc.

Fl. & Fr.: September- February.

Specimen examined: Bhilwara 455, Pansal 1965 (MLVGCB Herb).

Distribution: Afghanistan, throughout northern India to China, Java and Sri Lanka.

Uses: The species has religious importance as it is an essential element of the material offered to the God at the time of inauguration of houses, buldings etc.

SOLANACEAE

KEY TO THE GENERA

1a. Anther dehiscence by apical pore.......*Solanum*

1b. Anther dehiscence longitudinally by slits.......2

2a. Fruit capsular.. 3

3a. Flowers in cymose panicles or lax racemes .... *Nicotiana*

3b. Flowers solitary.................................. .*Datura*

2b. Fruit a berry.. 4

4a. Plants shrubby, armed........................ *Lycium*

4b. Plants herbaceous, not armed 5

5a. Flowers fascicled .... *Withania*

5b. Flowers solitary................................. *Physalis*

DATURA L.

KEY TO THE SPECIES

1a. Plants clothed with erect, glandular hairs .... *D. innoxia*

1b. Plants glabrous or with eglandular hairs 2

2a. Branches with purple tinged. Capsules clothed with numerous short blunt tubercles, irregularly breakig up .... *D. fastuosa*

2b. Branches green. Capsules clothed with numerous long spines, 4 - valved................... 3

3a. Leaves 1.5 to 2.5 times as long as broad. Corolla more than 7 cm long. Spines on the fruits numerous (100 or more). Seeds black... *D. stramonium*

3a. Leaves 1 to 1.4 times as long as broad. Corolla upto 6 cm long. Spines on the fruits numerous (40 - 60). Seeds pale brown .... *D. ferox*

Datura fastuosa L., Syst. Nat. ed. 10. 2: 932. 1759; Wight, Ic. 4(2): 18. t. 1396; Cooke, Fl. Presd. Bomb. Reprint ed. 2: 343.1967; Clarke, in Hook. *f.* Fl. Brit. Ind. 4: 242. 1883; Duthie, Fl. Gangetic Plain 2: 132. 1911; Singh, in Shetty & Singh Fl. Rajasthan 2: 563. 1991. *D. metel* L., Sp. Pl. 179. 1753; Bhandari, Fl. Indian Desert 240. 1995, nom. conf. *D.alba* Nees in Trans. Linn. Soc. 17: 73. 1834; Duthie, *l. c.* 2 : 132. 1911. *D.fastuosa* L. var. *alba* (Nees) Clarke, in Hook. *f. l. c.* 4: 243. 1883. (Plate 225)

Vern.: *Kaladhatura.*

Undershrub, up to 1 m or more, branched, glabrescent, clothed with fine purple dust. Leaves 7 - 12 x 5 - 10 cm, triangular ovate, toothed, acute, oblique at base, glabrescent. Petiole up to 9 cm long, slender. Flowers solitary; peduncles 1 - 2 cm long, pubescent. Calyx 5 - 8 mm long, tubular, half as long as corolla or less, minutely pubescent; lobes 6 - 11 mm long, ovate to triangular, acute or acuminate. Corolla 11 - 17 cm long, deltoid, purple outside, white inside; lobes 5 - 10, ovate, acuminate or cuspidate, minutely pubescent especially on the nerves. Anthers ca 1 cm long; filaments hairy at base, purple. Capsule 3 - 4 cm in diam., globose to subglobose, nodding, tuberculate or spines not exceeding 3 mm in length. Seeds 3 - 5.5 mm in diam., subreniform, compressed, minutely reticulate, pale brown.

Field note: Commonly cultivated near temple for religeous importance, also occur wild in waste lands and near habitations.

Fl. & Fr.: Almost throughout the year.

Specimen examined: Bhilwara 320, Jahajpur 1969 (MLVGCB Herb).

Distribution: Tropical and subtropical regions of Asia. The Mediterranean, Africa, Egypt and America.

Notes: As noted and discussed by A.G. Avery *et al.* (1959), the taxonomy of *D. metel* is very confusing, and it is reasonable to follow Schoenbeck-Temesy in using the name *Datura fastuosa* and considering *D. metel* as *nomen confusum*. The species can be distinguished from *Datura innoxia* mainly by the glabrous nature, the short spines or tubercles of the capsule and corolla which is often purplish suffused and double.

Datura ferox L., Amoen. Acad. 3: 403. 1756; Haegi, in Aust. J. Bot. 24: 427. 1976; Bhandari, Fl. Indian Desert 239. 1995; Singh, in Shetty & Singh Fl. Rajasthan 2: 564. 1991. (Fig. 104 & Plate 226)

Vern.: *Datura.*

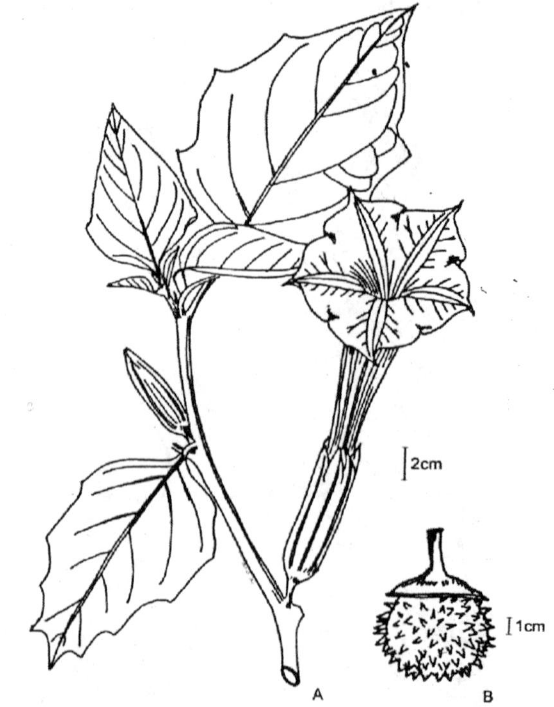

Fig. 104. *Datura ferox.* L.
A. Flowering twig, B. Fruit

Stout, bushy, annual herb. Leaves 10 - 13 x 7 - 10 cm, ovate to rounded - triangular, coarsely lobed, oblique at the base; often slightly decurrent on the petiole, acute or acuminate at th apex. Peduncles 7 - 14 mm long. Calyx 2.5 - 3. 0 cm long, angular, narrowed towards the summit, 5-ribbed, 5 - lobed; lobes tomentose inside. Corolla 4.5 - 6.0 cm long, white, 5 - lobed. Stamens 5, adnate to the corolla - tube. Stigma included in the corolla throat. Capsules 3.0 - 3.5 x 2 - 3 cm, ellipsoid, clothed with sharp conical spines, 4 - valved. Seeds - D-shaped, minutely reticulate - foveolate.

Field note: Rare in waste lands.

Fl. & Fr.: September - December.

Specimen examined: Bhilwara 534, Phuliya kalan 4012 (MLVGCB Herb).

Distribution: Native of China; introduced in Australia and India.

Datura innoxia Mill. Gard. Dict. ed. 8. no. 5. 1768; Safford, in Ann. Rep. Board. Reg. Smith. Inst. 1920: 549. t. 3. f. 4. 1920 & in J. Wash. Acad. Sci. 11 (8) 179. f. 1. 1921; Sant., in Journ. Bombay Nat. Hist. Soc. 47 (4): 659. 1948; Raizada, Suppl. Duthie, Fl. Gangetic Plain 168. 1976; Bhandari, Fl. Indian Desert 240. 1995; Singh, in Shetty & Singh Fl. Rajasthan 2: 564. 1991. *D. metel auct. non* L. 1753; Sim., in Cirtis, Bot. Mag. t. 1440. 1812; Clarke in Hook. *f.* Fl. Brit. Ind. 4: 243. 1883; Duthie, Fl. Gangetic Plain 2: 131. 1911. (Fig. 105)

Vern.: *Datura*.

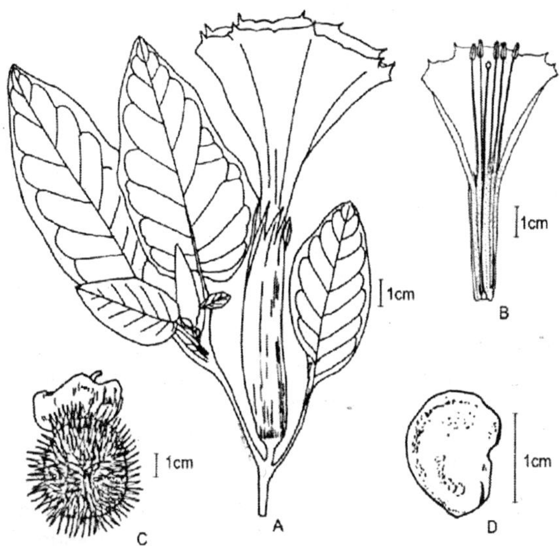

Fig. 105. *Datura inoxia* Mill.
A. Flowering twig. B. Opened flower, C. Fruit, D. Seed

Undershrub, up to 100 cm tall, branched, dense pubescent-villous. Leaves 7 - 10 x 4 - 7 cm, broadly ovate, acute, repand to sinuate-dentate, base oblique to cuneate, pubescent on both surface. Petiole upto 12 cm long, pubescent. Flowers solitary, axillary; peduncle upto 1.5 cm long. Calyx 7.5 - 10.5 cm long, 5-lobed, slightly inflated towards the base; basal portion persistent and strongly reflexed in fruit; lobes unequal, triangular or ovate-lanceolate, acuminate, up to 1 - 2 cm long. Corolla tube 9 - 10 cm long; teeth 10, linear. Stamens 5 - 6, more or less reaching the lenght of corolla tube, white. Ovary softly prickly; stigma 2-lobed. Capsule globose, 4 - 5 cm, dense spiny, nodding; spines acicular, 1 - 1.5 cm long. Seeds 3 - 5 mm long, reniform, compressed, brown, minutely reticulate, rugose-lineate on the back.

Field note: A common plant of roadsides and weedy places.

Fl. & Fr.: Throughout the the year.

Specimen examined: Bhilwara 1646, Mandal 1989 (MLVGCB Herb).

Distribution: America, W. Indies, S. Western USA. Naturalised in the Mediterranean area, Afghanistan, Pakistan, India and Malaysia.

Notes: Native of tropical S. America. *Datura innoxia* is related to *Datura meteloides* DC., but differs in the erect habit, smaller flowers, the larger stigmas and shorter styles. The plant parts as in the other *Datura* species are poisonous. Poisonous seeds have a property of intoxication and are also used to cure hydrophobia (Bhandari, Ind. Desert 269. 1995).

Datura stramonium L., Sp. Pl. 179. 1753; Clarke, in Hook. *f.* Fl. Brit. Ind. 4: 242. 1883; Bhandari, Fl. Indian Desert 241. 1995; Singh, in Shetty & Singh Fl. Rajasthan 2: 565. 1991. *D. stramonium* L. var, *tatula* Clarke, in Hook. *f. l.c.* 2: 242. 1883. (Fig. 106)

Vern.: *Datura*.

An erect annual or perennial herb or under shrub, 60 - 120 cm or more, branched, pubescent; the branches often purplish. Leaves 6 - 17 x 4 - 12 cm, ovate-oblong, acuminate, narrowed and oblique or rounded at base, sinuately dentate, minutely puberulose. Petiole 2-5 cm long. Flowers axillary, solitary. Calyx 3 - 5 cm long, tubular, 5-angled, puberulous, persistent, lobes 6 - 9 mm long, strongly reflexed in fruit, apiculate. Corolla 7-10 cm long, white or purplish suffused; limb up to 8 cm broad. triangular-acuminate. Anthers usually white. Capsule erect, 3 - 7 cm long, ovoid, spiny and densely pubescent, splitting by 4 valves; spines up to 5 mm long. Seeds 3 mm long, reniform, reticulate-foveolate, black.

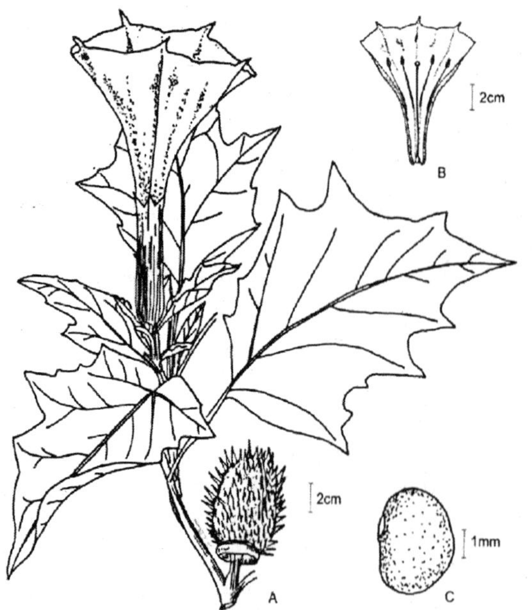

Fig. 106. *Datura stramonium* L.
A. Flowering & fruitng twig. B. Opened corolla, C. Seed

Field note: Common in waste lands and road sides.

Fl. & Fr.: December - May.

Specimen examined: Bhilwara 1592 (MLVGCB Herb).

Distribution: In most temperate and subtropical regions of both the hemispheres.

Notes: Plant contains alkaloid hyoscyamine, which has a powerful narcotic effect. Plant parts are being used in fever, worms, skin diseases, boils and indigestion (Dymock *et al.* Pharmcog. Ind. Reprint. edit. 2: 584. 1891).

LYCIUM L.

Lycium barbarum L., Sp. Pl. 191. 1753; Clarke, in Hook. *f.* Fl. Brit. Ind. 4: 241. 1883; Mahesh. Illus. Fl. Delhi f. 152. 1966; Bhandari, Fl. Indian Desert 241. 1995; Singh, in Shetty & Singh Fl. Rajasthan 2: 565. 1991. (Fig. 107)

Vern.: *Morali.*

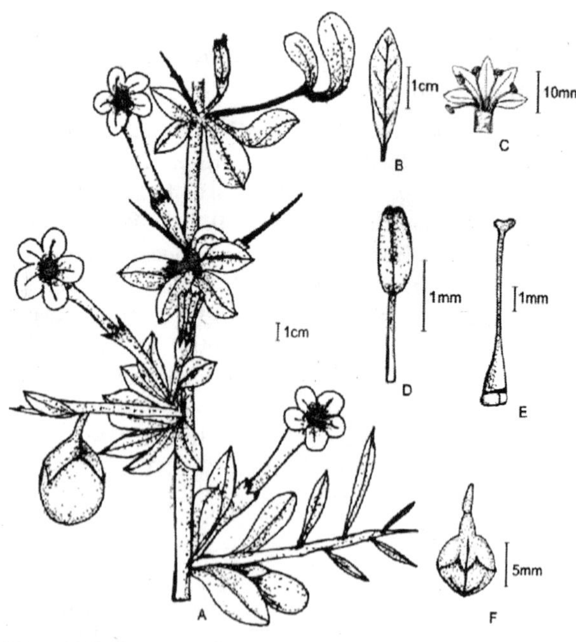

Fig. 107. *Lycium barbarum* L. :
A. Flowering & fruiting branch, B. Leaf, C. Opened corolla with stamens, D. Stamen, E. Pistil, F. Fruit

A much branched, spinous shrub, upto 2.5 m high. Stem and branches white or grey, slender, armed with sharp, conical spines, which often elongate and bear leaves and flowers. Leaves solitary or fascicled, 1 - 3 x 0.6 - 1 cm, oblong-lanceolate or oblong-spathulate or linear, broadest towards the middle, obtuse, attenuated at base, glabrous or nearly so. Flowers solitary or in fascicles of 2 - 5; petioles filiform, 6 - 8 mm long, glabrous. Calyx 3 - 4 mm long, cup-shaped, glabrous, somewhat rugose outside, often becoming irregularly lobed and rarely 2-lipped in fruit; teeth oblong, obtuse, with membranous margins. Corolla 1.2 - 1.5 cm long, funnel-shaped, glabrous; lobes 1.2 - 3 x 1.5 cm long, spreading, oblong, obtuse; tube 5 mm long, gradually expanded. Stamens slightly exserted; filaments flat, densely hairy at the base; anthers orbicular, 1.7 mm long. Ovary ovoid-oblong, glabrous, seated on a large, cup-shaped disc; style 8-9 mm long, glabrous, exserted; stigma mushroom-head-like, with a slightly enlarged, persistent calyx. Berries 4 - 5.5 mm in diam., globose, bright red. Seeds 2 - 2.5 mm in diam., discoid or subreniform, minutely pitted.

Field note: A common species with variable size and shape of the leaves.

Fl. & Fr.: October- January.

Specimen examined: Pansal 124 (MLVGCB Herb).

Distribution: Afghanistan, Persia, Baluchistan, India and Pakistan.

Uses: Powder of leaves is mixed with butter and applied on abscesses. Powder of stem bark is poured into nostrils to cure bronchitis in cattle.

NICOTIANA L.

KEY TO THE SPECIES

1a. Flowers in lax racemes. Stamens inserted near the top of corolla tube .... *N. plumbaginifolia*

1b. Flowers in cymose. Stamens inserted near the base of corolla tube .... *N. tabacum*

Nicotiana plumbaginifolia Viv. Elench. Pl. Hort. Bot. 26. t. 5. 1802; Clarke, in Hook. *f.* Fl. Brit. Ind. 4: 246. 1883; Raizada, Suppl. Duthie, Fl. Gangetic Plain 170. 1976; Deb, in J. Econ. Taxon. Bot. 1: 42. 1980; Bhandari, Fl. Indian Desert 242. 1995; Singh, in Shetty & Singh Fl. Rajasthan 2: 568. 1991. (Fig. 108 & Plate 227)

An erect viscid-pubescent herb, up to 1 m tall, branched. Leaves radical and cauline, glandular white dotted beneath, sessile, variable in size, 6 - 28 x 3.5 - 9.5 cm, elliptic-oblong or oblanceolate, obtuse or acute at apex, entire or wavy, cuneate to decurrent. Panicles up to 15 cm long, in terminal lax racemes. Pedicel up to 10 mm long, glandular-pubescent. Calyx 7 - 12 mm long, glandular hairy, 10-ribbed; lobes linear-lanceolate. Corolla greenish white or pale pink, hypocrateriform; tube 2.5 - 4.0 cm long, limb 10 mm broad; lobes 5, acute. Anthers shorter than filaments, slightly oblong; filaments ca 20 mm long. Ovary ca 2.5 mm long. Capsule 8 - 10 mm long, ovoid, somewhate included in the persistent calyx. Seeds less than 1 mm long, subglobose to angular, minutely rugose-reticulate, brown.

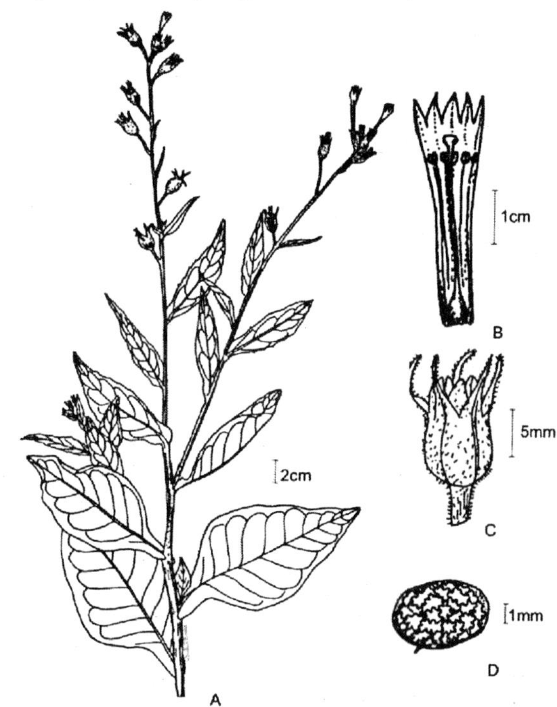

Fig. 108. *Nicotiana plumbaginifolia* Viv. A. Flowering & fruiting twig. B. Opened corolla, C. Capsule partially enclosed in calyx, D. Seed

Field note: Often found in waste places near water.

Fl. & Fr.: Most part of the year.

Specimen examined: Bhilwara 1753, Kekri 4081 (MLVGCB Herb).

Distribution: Native to Mexico and the W. Indies, Tropical America; widely naturalised elsewhere.

Nicotiana tabacum L., Sp. Pl. 180. 1753; Clarke, in Hook. *f.* Fl. Brit. Ind. 4: 245. 1883; Deb, in J. Econ. Taxon. Bot. 1: 43. 1980.

Vern.: *Tambaku, Tambakhu.*

Herb, up to 1 m tall, viscid-pubescent. Leaves cauline, 14 - 27 x 4 - 12 cm, oblong to broadly elliptic-ovate, cuneate or subauricled. Flowers pink, in axillary and terminal compact corymboid panicles. Calyx 10 - 30 mm long, oblong, glandular, pubescent, persistent in fruit; lobes unequal. Corolla tube 3.5 - 4.6 cm long, glandular-hairy, greenish white; limb up to 15 cm broad, lobes acute to apiculate. Anthers 3 mm long, oblong; filaments ca 3.5 cm long. Capsule 2 - 2.3 cm long, ovoid-globose. Seeds less than 1 mm long, angled, minutely ruminate, brown.

Field note: Commnly cultivated in village pur (Bhilwara) for its commercial value. Sometimes found as an escape.

Fl. & Fr.: April-May.

Specimen examined: Pur 533 (MLVGCB Herb).

Distribution: Native to Mexico and northern S. America.

Uses: Tobacco is both smoked and chewed; it contains an alkaloid nicotine, a stimulant, which is harmful to the nervous and respiratory system. Also used for insecticidal purposes.

PHYSALIS L.

KEY TO THE SPECIES

1a. Anthers yellow 2

2a. Leaves entire or subentire. Corolla 4 - 5 mm long. Fruiting calyx not 5 - angled .... *P. micrantha*

2b. Leaves sinuate-dentate. Corolla 8-10 mm long. Fruiting calyx 5-angled .... *.P. minima*

1b. Anthers violet or blue-purple 3

3a. Stem, petiole, lamina and pedicels nearly glabrous. Leaves coarsely incised-serrate. Corolla with 5, small, obscure brown spots inside .... *P. angulata*

3b. Stem, petiole, lamina and pedicels densely glandular pubescent. Leaves subentire to sinuate-dentate. Corolla with 5, large, brown spots inside .... *P. peruviana*

Physalis angulata L., Sp. Pl. 183. 1753; Waterfall in Rhodora 60: 162. 1958; Raizada, Suppl. Duthie, Fl. Gangetic Plain 170. 1976; Singh, in Shetty & Singh Fl. Rajasthan 2: 569. 1991. *P. longifolia auct.* non Nutt. 1837; Sant. *et al.*, in Journ. Bombay Nat. Hist. Soc. 58: 551. 1961. (Fig. 109)

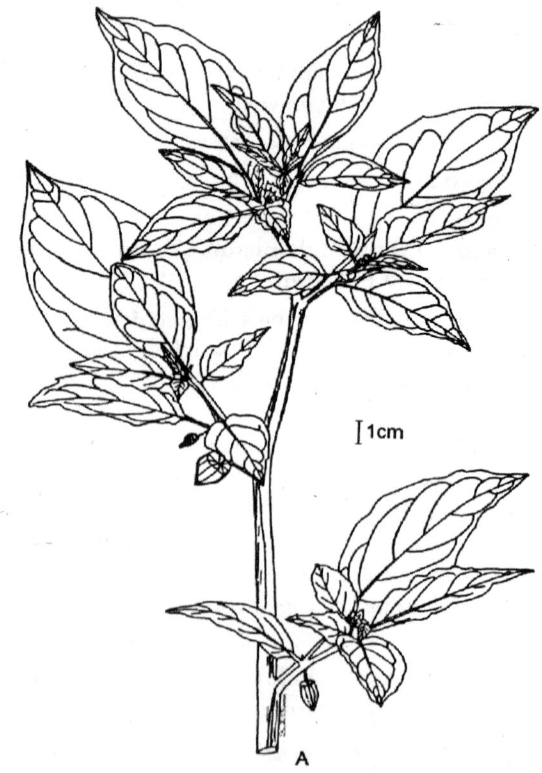

Fig. 109. *Physalis angulata* L.
A. Flowering & fruiting twig

Erect or decumbent, much branched, annual herb, 20-100 cm tall, rarely rooting at the base. Stem angular - ribbed, glabrous, except crispy short hairs when young. Leaves 3 - 10 x 3 - 6 cm, ovate-elliptic, long acuminate, irregularly sinuate-dentate, cuneate or rounded at the base. Flowers solitary, terminal, seemingly axillary, nodding. Calyx 4 - 5 mm long, divided half way down; segments ovate - triangular, acuminate or acute, ciliate. Corolla yellow, with 5, small, brown spots inside. Filaments not dilated upwards; anthers violet or pale blue. Berries 1.0 - 1.5 cm in diam., globose, greenish-yellow, enclosed in ca 4 cm long, 5-angled, bladder - like inflated calyx.

Field note: Common in waste lands, cultivated fields, gardens etc.

Fl. & Fr.: August- November.

Specimen examined: Bhilwara 1723, Bijoliya 1801 (MLVGCB Herb).

Distribution: Native of America, India.

Physalis micrantha Link, Enum. Pl. Hort. Berol. 1 : 181. 1821; Heine in Hutch & Dalz. Fl. W. Trop. Africa ed. 2. 2 : 329. 1963; Raizada, Suppl. Duthie, Fl. Gangetic Plain 172. 1976; Parmar in Bull. Bot. Surv. India 26 : 168. 1986; Singh, in Shetty & Singh Fl. Rajasthan 2: 569. 1991.

Erect or prostrate, branched, glabrous, annual herb, 15 - 50 cm high. Stem angular, ribbed, nearly glabrous. Leaves 1.5 - 6.0 x 1.5 - 4.0 cm, ovate, acute or acuminate, entire or sinuate - dentate, glabrous, rounded or cuneate at the base; petioles 0.5 - 5. 0 cm long. Flowers solitary, terminal, seemingly axillary, nodding; pedicels 5 - 10 mm long, puberulous. Calyx ca 4 mm long, hairy; lobes triangular, acute. Corolla 4-5 mm long, pale-yellow. Filaments hairy at the base; anthers yellow. Berries globose, enclosed in an inflated, bladder-like, 1.5-3.5 x 1.2-2.5 cm calyx.

Field note: Common in fields and waste lands.

Fl. & Fr.: August- November.

Specimen examined: Bhilwara 212, Amarpura 2120 (MLVGCB Herb).

Distribution: India.

Physalis minima L., Sp. Pl. 183. 1753; Clarke, in Hook. *f.* Fl. Brit. Ind. 4: 238. 1883; Duthie, Fl. Gangetic Plain 2: 127. 1911; Sant., in J. Bombay Nat. Hist. Soc. 47: 657. 1948; Deb, in J. Econ. Tax. Bot. 1: 43. 1980; Bhandari, Fl. Indian Desert 243. 1995; Singh, in Shetty & Singh Fl. Rajasthan 2: 570. 1991. *P. indica* Lam. Encycl. Meth. Bot. 2 : 102. 1786. *P. minima* L. var. *indica* (Lam.) Clarke in Hook.f. *l.c.* 4 : 238. 1883; Duthie, *l.c.* 2 : 127. 1911; Bhandari, *l.c.* 243. 1995. (Plate 228)

Vern.: *Papotan*.

Eract, upto 50 cm high, annual herb. Stem angular - ribbed, often purple tinged hairy. Leaves 2 - 6.5 x 1.0 - 5.0 cm, ovate, elliptic or elliptic oblong, sinuate, entire, acute to obtuse, cuneate to subcordate. Petiole 1-5 cm long, slender. Flowers solitary axillary. Pedicel less than 10 mm long, somewhat elongating in fruit. Calyx ca 2.5 mm long, campanulate. inflated, membranous and up to 25 mm in fruit, pubescent. Corolla 5 -10 mm long, shortly tubular, yellow; lobes acute, pubescent. Stamens subincluded. Anthers ca 1 mm long; filaments ca 2 mm long. Ovary 1.1 mm long. ovoid. Style linear, stigma subcapitate. Berry globose, 10 mm in diam., orange or yellow when ripe. Seeds ca 3 mm long, subreniform, compressed; minutely reticulate-undulate. brownish-yellow.

Field note: A fairly common weed of fields in the monsoon season.

Fl. & Fr.: August-October.

Specimen examined: Bhilwara 457, Umari 1686 (MLVGCB Herb).

Distribution: Afghanistan, India, Pakistan and eastward to Nepal.

Notes: var. *indica* apart from the 5-angled calyx, does not differ to any substantial degree from the type variety.

Physalis peruviana L., Sp. Pl. ed. 2 : 1670. 1763; Roxb., Fl. Ind. (Car. ed.) 2. 1 : 562. 1832; Clarke, in Hook. *f.* Fl. Brit. Ind. 4: 238. 1883; Duthie, Fl. Upp. Gang. Pl. 2 : 128. 1911; Raizada, Suppl. Duthie, Fl. Gangetic Plain 172. 1976; Deb, in J. Econ. Tax. Bot. 1: 44. 1980; Bhandari, Fl. Indian Desert 243. 1995; Singh, in Shetty & Singh Fl. Rajasthan 2: 570. 1991.

Vern.: *Rasbharry, Phori papotan.*

Plant 30 - 60 cm tall, pubescent-villous, often densely so. Leaves 4.5 - 14 x 3.5 - 10.5 cm, ovate to ovate-cordate, sinuate to repand or toothed, pubescent-villous. Flowers solitary axillary. Calyx 7 - 9 mm long, campanulate, 1/2-cleft, 35 - 40 mm long and inflated in fruit; lobes triangular-acuminate. Corolla yellow, purple blotched in side, sparse pubescent; lobes shortly ciliate. Anthers 2.3 - 3 mm long, narrow oblong; filaments 3-4 mm long, glabrous. Berry ca 1 mm in diam., globose, orange, glabrous. Seeds 2.4 mm long, subreniform, minutely reticulate, brown.

Field note: Common in waste lands, cultivated fields, gardens etc.

Fl. & Fr.: August- November.

Specimen examined: Bhilwara 1593, Hamirgarh 1790 (MLVGCB Herb).

Distribution: Tropical America, cultivated or naturalised elsewhere.

Uses: Yellow berries are edible and are used for making jams etc.

SOLANUM L.

KEY TO THE SPECIES

1a. Plants unarmed 2

2a. Stellately tomentose undershrubs. Berries 1 - 1.5 cm in diam. .... *S. torvum*

2b. Glabrous or pubescent herbs. Berries 5 - 7 cm in diam.................. *S. nigrum*

1b. Plants armed with prickles 3

3a. Berris less than 1 cm in diam. 4

4a. Calyx lobes acute .... *S. anguivi*

4b. Calyx lobes acuminate to caudate acuminate.............................. *S. elaegnifolium*

3b. Berris more than 1 cm in diam. 5

5a. Prickles recurved. Peduncles paired.. *S. incanum*

5b. Prickles straight. Peduncles solitary............ 6

6a. Erect. Calyx tube prominently inflated in fruit, fruits bright red when ripe .... *S. sisymbriifolium*

6b. Prostrate. Calyx slightly inflated or reflexed in fruit, fruits yellow when ripe................................ *S. virginianum*

Solanum anguivi Lam. Encycl. Meth. Bot. 2 : 23. 1974; Hepper in Hawkes in J. Linn. Soc. 76 (4): 290. 1978; Singh, in Shetty & Singh Fl. Rajasthan 2: 572. 1991. S. indicum auct. non L. 1753; Clarke, in Hook. *f.* Fl. Brit. Ind. 4: 234. 1883; Duthie, Fl. Gangetic Plain 2: 126. 1911; Sant., in J. Bombay Nat. Hist. Soc. 47. 653. 1948; Deb, in J. Econ. Tax. Bot. 1: 47. 1980.

Vern.: *Ringani.*

Erect, prickly, undershrub, upto 1.5 m tall, with dense stellate-tomentose parts. Prickles up to 10 mm long, erect to slightly recurved. Leaves 3 - 10 x 1.5 - 6 cm, ovate or broadly elliptic or elliptic-oblong, acute, both surfaces stellately hairy and prickly on the nerves. Petiole up to 2.5 cm long. Flowers 4 - 15 in number, bluish-purple, in extra-axillary cymes. Calyx ca 4 mm long, campanulate, stellately hairy. Corolla limb 18-20 mm broad; lobes acute. Anthers 6 - 7 mm long; filaments glabrous. Ovary and style pilose-pubescent. Berry globose, 8 - 10 mm in diam., glabrescent, yellow. Seeds ca 3 mm long, subreniform, minutely reticulate, foveolate.

Field note: Rare in the south central Rajasthan.

Fl. & Fr.: Mostly throughout the year.

Specimen examined: Mandal 2188 (MLVGCB Herb).

Distribution: S. & S.E. Asia, Arabia and tropical Africa.

Solanum elaegnifolium Cav. Icon. 3 : 22. t. 243. 1794. Raizada, Suppl. Duthie, Fl. Gangetic Plain 172. 1976. (Fig. 110)

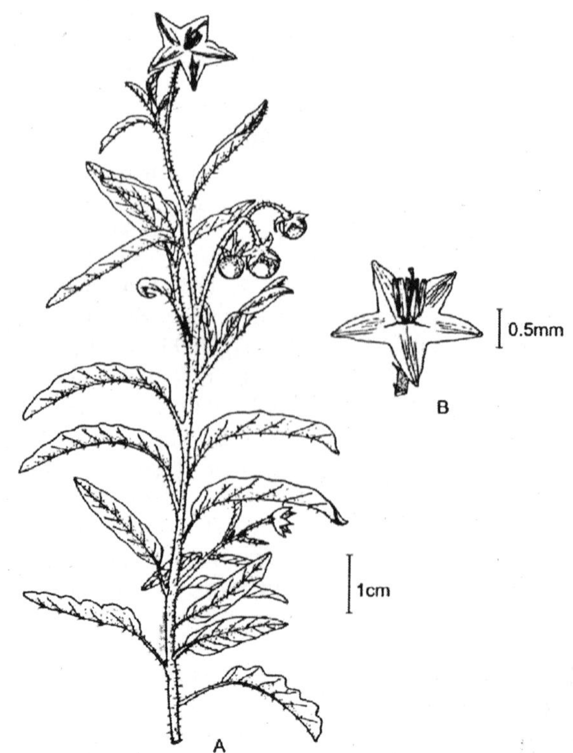

Fig. 110. *Solanum elaegnifolium* Cav. : A. Flowering & fruiting twig, B. Flower.

Perennial, herb, up to 1 m tall. Stem and branches dense stellate-tomentose, prickly. Prickles sparse, acicular, 2 - 5 mm long, yellow. Leaves alternate, 1.5 - 12 x 3.0 cm, oblong-liner, repand-sinuate, base cuneate to oblique, subacute or obtuse at apex. Flowers 1 - 5 in number, violet-blue to purplish, in 2 - 6 flowered corymbs. Pedicel up to 20 mm long. Calyx ca 5 mm long deeply 5-lobed, enlarging in fruit. Corolla 20 - 25 mm broad. Anthers yellow; filaments 3 mm long. Berry globose, 11 - 12 mm in diam., yellowish. Seeds discoid, ca 3 mm long, brown, smooth.

Field note: Rare, among hedges and associated with *Euphorbia caducifolia.*

Fl. & Fr.: July-August.

Specimen examined: Asind 1648 (MLVGCB Herb).

Distribution: Native to temperate America; naturalised else-where in South Europe, India, Baluchistan, Pakistan and Australia.

Solanum incanum L., Sp. Pl. 188. 1753; Duthie, Fl. Upp. Gang. Pl. 126. 1911; Bhandari, Fl. Indian Desert 245. 1995; Singh, in Shetty & Singh Fl. Rajasthan 2: 573. 1991. *S. coagulans* Forsk. Fl. Aegypt.-Arab. 47. 1775; Clarke, in Hook. *f.* Fl. Brit. Ind. 4: 236. 1883. *S. melongena* Clarke, in Hook. *f.* Fl. Brit. Ind. 4: 235. 1883 (*pro parte.*, non. Linn.) (Fig. 111 & Plate 229)

Vern.: *Ringani.*

Shrub up to 120 cm tall. Stem and branches yellowish-green with dense stellate tomentum and prickly; prickles conical, up to 8 mm long, straight or curved, yellow. Leaves 3.5 - 8 x 2.5 - 6.5 cm, ovate-elliptic, sinuate, 4-lobd, greyish-green dense stellately hairy, veins with a few prickles. Peduncles 10-20 mm long, often paired; one bearing bisexual flowers and the other a short raceme of 3 - 5 male flowers. Flowers purple-blue. Pedicel and calyx stellately tomentose and sparsely prickled. Calyx 6 - 7 mm long, often thickened in fruit. Corolla limb 3 - 3.5 cm broad; lobes triangular, 9 - 11 mm long, dense stellate-tomentose, sparsely so within. Anthers 5 - 6 mm long, elongated. Style and ovary glabrescent. Berry globose, 25 - 30 mm in diam., yellow. Seeds ca 2.5 mm broad, subreniform, minutely reticulate-rugose, pale- brown.

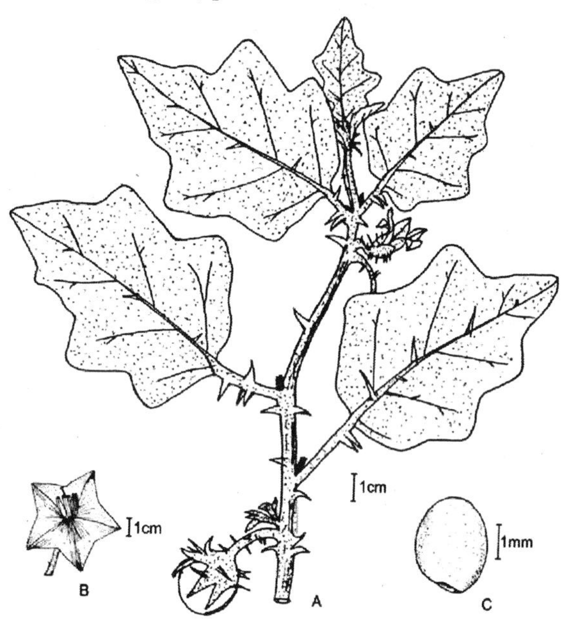

Fig. 111. *Solanum incanum* L. : A. Flowering & fruiting twig, B. Flower, C. Seed

Field note: Rare, in waste lands and open forests.

Fl. & Fr.: August - March.

Specimen examined: Mandal 1545, Mainal 1900 (MLVGCB Herb).

Distribution: Iran, Syria, Egypt, Ethiopia, Sudan, Arabia, Pakistan and India.

Uses: The plant parts are said to be medicinal, being used as a remedy for toothache.

SOLANUM NIGRUM L.

KEY TO THE VARIETIES

3a. Mature berries black. Peduncles longer than pedicels .... *S. nigrum* var. *nigrum*

3b. Berries orange to orange-red. Peduncles shorter than or equalling the pedicels....................... *S. nigrum* var. *villosum*

Solanum nigrum L., Sp. Pl. 186. 1753; Clarke, in Hook. *f.* Fl. Brit. Ind. 4: 229. 1883; Duthie,

Fl. Gangetic Plain 2: 124. 1911; Maheshwari, Fl. Delhi, 245.1963; Bhandari, Fl. Indian Desert 245. 1995; Singh, in Shetty & Singh Fl. Rajasthan 2: 573. 1991. (Fig. 112)

var. ***nigrum***

Vern.: *Makoy*.

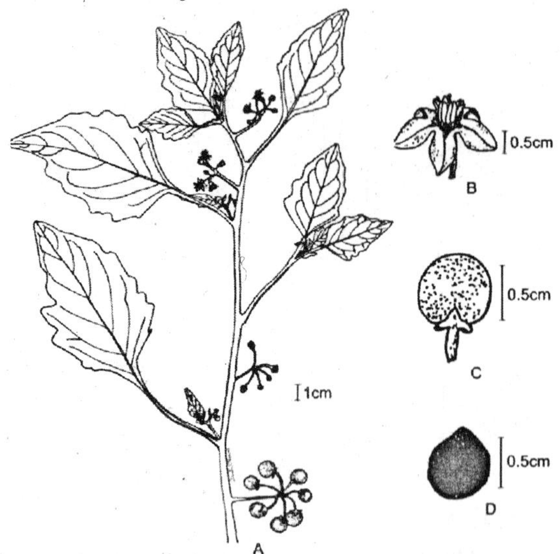

Fig. 112. *Solanum nigrum* L. var. *nigrum* : A. Flowering & fruiting twig, B. Flower, C. Fruit, D. Seed

Annual herb, up to 15 - 60 cm tall. Branches and shoots subglabrous or pubescent to glandular-villous; hairs appressed or patent. Leaves 2.5 - 10 x 2 - 5.5 cm, sinuate to irregularly dentate, glabrous to puberulous-pubescent or glandular. Lamina sub-decurrent. Petiole 8-40 mm long, appressed pubescent or glandular. Peduncle 10 - 25 mm long, exceeding or shorter than the pedicel length. Calyx lobes 1 - 1.5 mm long, glabrous to pubescent. Corolla much exceeding the calyx in length; lobes triangular-acute, 4 - 6 mm long; anthers yellow, dehiscing by 2 apical pores. Filaments as long or shorter than the anthers, pilose. Style included, pubescent at the base. Stigma capitate. Ovary glabrous. Berry globose to subovoid, 5 - 8 mm in diam., black or orange-red. Seeds discoid, minutely reticulate-foveolate.

Field note: Common weed of cultivated fields and gardens etc.

Fl. & Fr.: September - December.

Specimen examined: Bhilwara 125, (MLVGCB Herb).

Distribution: Cosmopolitan. absent from the arctic and subarctic regions.

Notes: Much taxonomic confusion has centred on the *Solanum nigrum* L. complex, especially the treatment of the black and orange-red berries forms. Most authors (Hawkes & Edmonds, 1972; Schoenbeck-Temesy 1972; Baytop, 1995: Hara *et al.* 1982. Enum. of the Flow. Pl. Nep. 3:111.), regard the black berried plants as distinct. Workers dealing with the Indo-Pakistan continent have treated both different colored berried forms under *Solanum nigrum* s.l. Some recent work by Bhaduri (1983, Ind. Bot. Soc. 12.56.64). Tandon & Rao (1966, Current Science. 20.324-325) and Rao, Khan & Khan (1971, Mag. Univ. Tokyo. 88:335-338) have shown that variation in *Solanum nigrum* group (at least for the subcontinent) is often of a continuous nature and the characters show intergrading. During present study numerous specimens of *Solanum nigrum* have been observed in the field and have found that there is an important distinguishing character: The black berried plants have peduncles longer than pedicels, as compared to the orange or orange-red berried plants, where the peduncles tend to be shorter or equalling the pedicel.

Uses: The fruit is edible. Young, shoots are used as a pot herb. Plant parts contain the alkaloid solanine. The juice of the plant is diuretic, and used in the treatment of enlarged liver (Dymoc, Reprint edit. 1972, Pharm., Ind. 2.301).

Solanum nigrum L. var. **villosum** L., Sp. Pl. 186. 1753.

Vern.: *Makoy*.

Charcters of this species differ from *Solanum nigrum* L. var. *nigrum* as mentioned in identification keys and in notes.

Field note: Common weed of cultivated fields and gardens etc.

Fl. & Fr.: September-December.

Specimen examined: Bhilwara 3187 (MLVGCB Herb).

Distribution: Cosmopolitan, absent from the Arctic and subarctic regions of both hemisphere.

Notes: The var. *villosum* is more common than the type variety. The orange to orange-red berried plants have been treated by most European authors as good species (i.e. *Solanum luteum, Solanum alatum*). In local populations here (at least in the material studied) there appears to be no satisfactory treatment to segregate the glandular-hairy forms with terete stems (i.e. *Solanum luteum* Mill.) from those with subglabrous to pubescent parts with stems ± rigid or angled (i.e. *Solanum alatum*).

Solanum sisymbriifolium Lam. Tabl. Encycl. 2 : 25. 1794; Qieshu, Fl. China, 17: 324. 1994; Huang, Flora of Taiwan, Vol. iv (2 ed.), 576. 1998; Yadav & Meena, in J. Econ. Taxon. Bot. 32. (3): 749 - 753. 2008. (Fig. 113 & Plate 230)

Shrubby, annual herb, copious armed, pubescent with glandular, multicellular, simple and stellate hairs, prickles yellow or orange-yellow, and awl-shaped 1.5 mm long. Stem erect branched, with yellow or orange-yellow, subulate, prickles 2 - 16 x 1.5 mm, very sharp. Leaves alternate; leaf-blades ovate-oblong, 10 - 15 x 6 - 10cm, deeply pinnatisect or pinnatified with many prickles, sparsely stellate-hairy above, densely so beneath; lobes rounded, sinuate; petioles 1.5 - 5cm, spiny like stem. Inflorescence racemose, 6 - 10 flowered, upper flowers smaller and male. Peduncle 3 - 7cm long, branched or not, glandular and simple pilose, densely spiny. Pedicels slender, glandular, pilose, slightly spinulose, 8 - 15 mm. Calyx cup-shaped, green, membranous, 6 - 9 x 2 - 4 mm, deeply 5-parted; lobes lanceolate, acuminate, with dense prickles in hermaphrodite flowers. Corolla whitish or lilac bluish and pale violet, rotate, 40 - 45 mm in diam., shallowly lobed, lobes broadly triangular, stellate-pilose outside. Stamens equal; filaments slender, glabrous, 2 - 3 mm; Anther lanceolate 7 - 10 mm long. Ovary ovoid, glabrous; style 1 - 1.2 cm. Fruiting pedicels deflexed, densely glandular-pilose, viscid; fruiting calyx enlarged 11 - 12 x 4 - 6 mm accrescent, densely prickly, enveloping most of berry, longer than fruit ultimately flaring wide and exposing it. Berry bright red, subglobose, 5 - 6 mm diam. Seeds reniform, ca 2.5 x 2 mm.

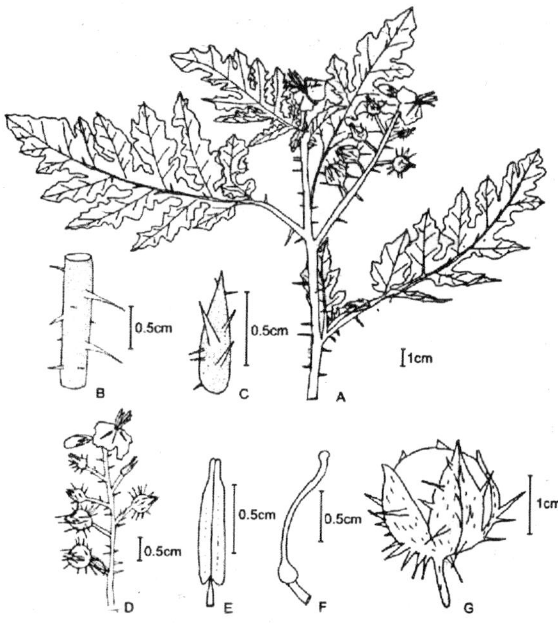

Fig. 113. *Solanum sisymbriifolium* Lam. A. Habit, B. Stem, C. Calyx, D. Flower & fruits, E. Stamen, F. Gynoecium, G. Fruit

Field note: Common in wasteland areas.

Fl. & Fr.: September- November.

Specimen examined: Gurla 142, Borda 207 (MLVGCB Herb).

Distribution: Native of South America; naturalized in North America, Africa, Australia, India (U.P., Bengal, Tamil Nadu and Rajasthan).

Solanum torvum Sw. Prodr. 47. 1788; Clarke, in Hook. *f.* Fl. Brit. Ind. 4: 225. 1883; Raizada, Suppl. Duthie, Fl. Gangetic Plain 176. 1976; Singh, in Shetty & Singh Fl. Rajasthan 2: 574. 1991. (Fig. 114)

Vern.: *Bhurat.*

Much branched, tomentose shrub, upto 2 m high. Stem with dark purple tinged, aculeate with short hooked prickles. Leaves 4 - 15 x 2 - 12 cm, ovate - elliptic to oblong, entire or pinnate-lobed, rounded or cordate at the base, acute or obtuse, sparsely prickly on the midrib. Flowers in many - flowered, sessile or shortly peduncled cymose corymbs. Pedicels glandular hairy with simple and stellate hairs. Calyx glandular hairy; lobes lanceolate, acuminate. Corolla white, stellate pubescent outside; lobes lanceolate, subacute. Ovary glandular hairy; style glabrous. Fruits 1.0 - 1.5 cm across, on a persistent calyx.

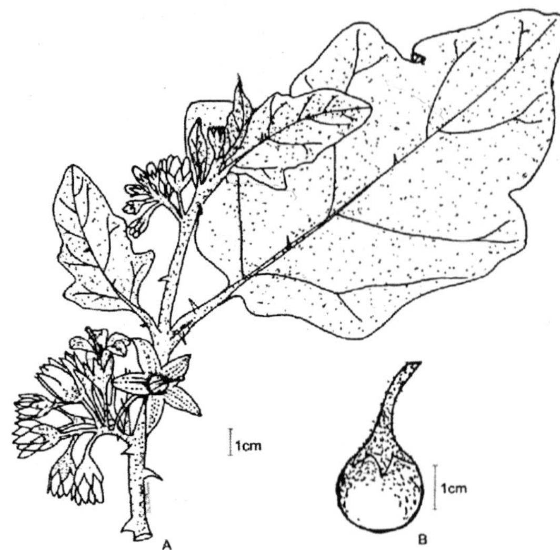

Fig. 114. *Solanum torvum* Sw.
A. Flowering twig. B. Fruit

Field note: Rare in waste lands, along roadsides. Very close to *S. indicum* L.

Fl. & Fr.: September-February.

Specimen examined: Bijoliya 126 (MLVGCB Herb).

Distribution: India.

Solanum virginianum L., Sp. Pl. 187. 1753; Hepper & Jaeger, in Kew Bull. 41: 434. 1986; Singh, in Shetty & Singh Fl. Rajasthan 2: 575. 1991. *S. surattense* Burm. f. Fl. Ind. 57. 1768; Mahesh. Illus. Fl. Delhi f. 150. 1966; Bhandari, Fl. Indian Desert 246. 1995. *S. xanthocarpum* Schrad. & Wendl., in Schrad. Sert. Hannov. 1: 8. t. 2. 1795; Clarke, in Hook. *f.* Fl. Brit. Ind. 4: 236. 1883; Duthie, Fl. Gangetic Plain 2: 125. 1911. (Plate 231)

Vern.: *Bhor-ringni.*

Perennial, prostrate, diffuse prickly herb; prickles up to 1.5 cm long, straight, yellow. Stem and branches zigzag, glabrous to stellate pubescent. Leaves 6 - 15 x 2 - 2.5 cm, ovate-elliptic or lanceolate, sinuate to deeply lobed, dark-green above; lobes unequal, obtuse or acute, often toothed or lobulate. Flowers solitary or in few flowered, extra-axillary cymes. Peduncle 3 - 10 mm long. Calyx lobes 7 mm long, densely hairy, acute, prickly. Corolla limb 2 - 2.8 cm broad; lobes 10 - 12 mm long, ovate-triangular. Anthers 7.5 mm long, elongated, yellow. Berry globose, 1 - 1.8 cm in diam., white with grren veins, yellow when ripe, glabrous. Seeds ca 2 mm in diam., glabrous, brown.

Field note: Common throughout the area, with pretty purple flowers; white flowered forms are uncommon.

Fl. & Fr.: Mostly throughout the year.

Specimen examined: Roopgarh 4063, Shahpura 1800 (MLVGCB Herb).

Distribution: N. Africa, S. & S.E. Asia, Australia and Polynesia.

Uses: Crushed root and bark are used for the preparation of hot poultice & applied locally for the treatment of hernia which is absolutely effective in young patients. Extract of crushed flowers is given orally in the treatment of diarrhoea. The smoke of dried fruit is used to remove worms of teeth. One tsp decoction of fruits is taken in cough & asthma.

WITHANIA Pauquy

KEY TO THE SPECIES

1a. Leaves elliptic-lanceolate or oblanceolate. Berry enclosed tightly within a sub-coriaceous fruiting calyx .... *W. coagulans*

1b. Leaves ovate, oblong or obovate. Berry loosely enclosed in a membranous fruiting calyx .... *W. somnifera*

Withania coagulans (Stocks) Dunal in DC. Prodr. 13 (1): 685. 1852; Clarke, in Hook. *f.* Fl. Brit. Ind. 4: 240. 1883; Raizada, Suppl. Duthie, Fl. Gangetic Plain 178. 1976; Bhandari, Fl. Indian Desert 247. 1995; Singh, in Shetty & Singh Fl. Rajasthan 2: 576. 1991. *Puneeria coagulans* Stocks, in Journ. Asiat. Soc. Bengal 3 : 56. 1849; Wight, Ic. 4 (4): 17. t. 1616. 1850. (Plate 232)

Vern.: *Panirband.*

Branched shrub up to 1 m tall. Shoots rigid, greyish-green, stellate-tomentose. Leaves 2 - 8 x 1.0-3.5 cm, elliptic-ovate to elliptic-lanceolate, obtuse or acute, cuneate, leathery, nerves faint. Petiole 5-10 mm long. Flowers bisexual: calyx cupular-campanulate, stellate-tomentose; lobes 1.5 - 2.5 mm long, acute. Corolla exceeding calyx, campanulate, stellate-tomentose outside, yellowish within. Stamens included. Stigma subexserted. Male flowers: stamens subexserted. Anthers ca 3.5 mm long; filaments 2.0 mm long. Female flowers: stamens included. Anthers smaller than male flowers. Ovary ovoid; style glabrous. Berry globose, 10 - 12 mm in diam., red, included in fruiting calyx. Seeds ca 2.5 mm broad. reniform, brown, minutely rugose-reticulate to ± smooth.

Field note: Rare, found in dry, hot and stony places.

Fl. & Fr.: November - April.

Specimen examined: Near MDS University Ajmer 4017 (MLVGCB Herb).

Distribution: Iran, Afghanistan, Pakistan and India.

Uses: Fruits are emetic diuretic and also has coagulating properties. Used in parts of India for making cheese.

Withania somnifera (L.) Dunal, in DC. Prodr. 13. (1): 453. 1852; Clarke, in Hook. *f.* Fl. Brit. Ind. 4: 239. 1883; Duthie, Fl. Gangetic Plain 2: 128. 1911; Bhandari, Fl. Indian Desert 247 - 248.1995; Singh, in Shetty & Singh Fl. Rajasthan 2: 576-578. 1991. *Physalis somnifera* L. Sp. Pl. 182. 1753. (Fig. 115 & Plate 233)

Vern.: *Asagandh*.

Herb or undershrub, 60 - 90 cm tall. Branches ascending. Shoots stellate-tomentose. Leaves 4 - 10 x 2 - 6 cm, ovate-oblong, entire, subacute, cuneate or oblique, entire, main nerves 5 - 8 pairs. Petiole up to 1.8 cm long, stellate-tomentose. Flowers sessile to subsessile, in axillary clusters of 2 - 5, greenish-yellow,.Calyx 2 - 2.5 x 2.5 - 3 mm, up to 2.0 x 14 mm in fruit, stellately hairy, becoming glabrescent and membranous; teeth 1-1.5 mm long, up to 8 mm in fruit, acute. Corolla lobes 2 - 2.5 mm long, triangular, tomentose outside. Stamens attached near the base of corolla tube; anthers subincluded; filaments 1 - 2.5 mm long. Ovary and style glabrous. Berry globose, 5 - 8 mm in diam., red, enclosed within calyx. Seeds ca 2.5 mm across, yellowish-brown.

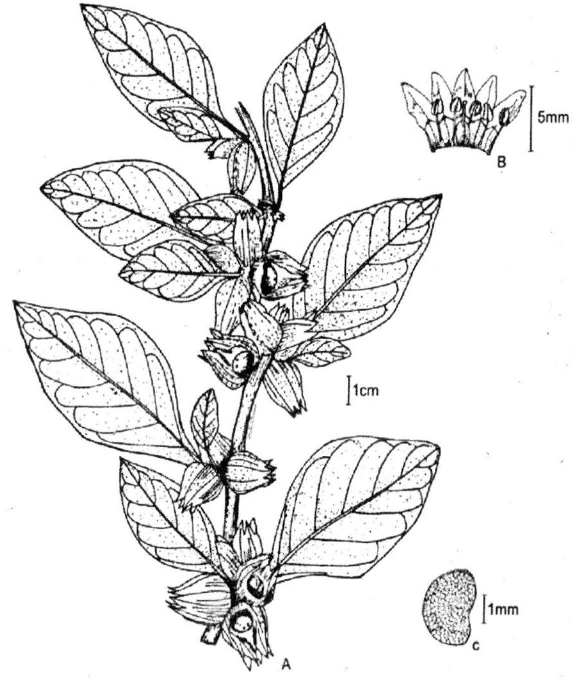

Fig. 115. *Withania somnifera* (L.) Dunal.
A. Fruiting branch, B. Opened flower, C. seed

Field note: Common throughout in waste places.

Fl. & Fr.: Almost throughout the year.

Specimen examined: Nasirabad 4096 (MLVGCB Herb).

Distribution: Canary Islands, Mediterranean, Africa, Iraq, S. Iran, Syria, Turkey, Palestine, Arabia, India and Pakistan.

Uses: Plant parts have alkaloids with sedative properties. Root is used in rheuma-tism and debility. Leaves are used in fever; a decoction of roots is useful for ulcers, boils etc. Fruits are diuretic. Poisonous seeds have a mild coagulating effect.

Notes: The species is variable in the texture and the pubescence of leaves.

SCROPHULARIACEAE

KEY TO THE GENERA

1a. All leaves alternate........................ *Verbascum*

1b. Leaves opposite or whorled, upper ones opposite, whorled or alternate...................... .2

2a. Corolla saccate, gibbose or spurred at the base or throat... .3

3a. Capsules loculicidal .... *Lindenbergia*

3b. Capsules porous or porous and loculicidal both 4

4a. Plants twining. Leaves ovate-lanceolate to triangular, sagittate or hastate at base. Corolla spurred, yellow....................... *Kickxia*

4b. Plants erect. Leaves linear, obtuse or rounded at base. Corolla saccate or gibbose, white... *Antirrhinum*

2b. Corolla rotate, slightly zygomorphic with subequal lobes or distinctly bilabiate.. 5

5a. Fertile stamens 2............................... *Veronica*

5b. Fertile stamens 4... 6

6a. Anther 1-celled, if 2-celled then one cell empty.. *Striga*

6b. Anther 2 -celled, both cells fertile................. 7

7a. Plants with condensed underground stem. Flowers borne on scape.. *Craterostigma*

7b. Plants with aerial stems. Flowers borne on the stem and branches.................................. 8

8a. Corolla 4-lobed................................. *Scoparia*

8b. Corolla 5-lobed... 9

9a. Lower pair of stamens inserted in corolla-throat; lower pair appendiculate..... *Lindernia*

9b. All the stamens inserted in corolla-tube; lower pair not appendiculate................................ 10

10a. Corolla campanulate, obscurely bilabiate.. *Bacopa*

10b. Corolla tubular, distinctly bilabiate....................................... *Limnophila*

ANTIRRINUM L.

Antirrhinum orontium L., Sp. Pl. 617. 1753; Hook. *f.* Fl. Brit. Ind. 4: 253. 1883; Duthie, Fl. Gangetic Plain 2: 140. 1911; Vartak, in J. Bombay Nat. Hist. Soc. 64: 584. 1968; Singh, in Shetty & Singh Fl. Rajasthan 2: 583. 1991. (Plate 234)

Annual herbs, 10 - 40 cm tall, erect, sparingly branched, glabrous to sparsely hirsute in lower parts, glandular - pubescent in upper parts. Leaves 2.5 - 5.0 x 0.2 - 0.3 mm, linear, apex sub acute, sessile or subsessile. Flowers sessile or shortly pedicellate, axillary, passing into spiciform racemes, pedicels very short in flowers, elongating up to 4 mm long in fruit; bracts similar to leaves but much smaller. Calyx 10-17 mm long, the lobes linear, the longest one ca. 1.5 times as long as shortest one, glandular-pubescent. Corolla pink, rarely white, 10-15 mm long, as long as or slightly shorter than calyx. Capsules obliquely ovoid, ca 1 cm long, glandular pubescent. Seeds somewhat flattened, 1 face smooth, keeled and prolonged into a narrow wing, the other face finely tuberculated, with a wide, raised, sinuate, papillose border.

Field note: Rare weed of cultivated fields.

Fl. & Fr.: January - March.

Specimen examined: Autun 209, Bijoliya 1210 (MLVGCB Herb).

Distribution: Native to southwestern and central Europe

BACOPA Aublet *nom. cons*

Bacopa monnieri (L.) Wettst. in Engl. & Prantl. Pflan. 4 (3B) 77. 1891; Pennell, Proc. Acad. Nat. Sci. Philadelphia. 98: 94. 1946; Singh, in Shetty & Singh Fl. Rajasthan 2: 585. 1991; Bhandari, Fl. Indian Desert 250. 1995. *Lysimachia monnieri* L. Cent. Pl. 2: 9. 1756. *Moniera cuneifolia* Michx. Fl. Bor. Amer. 2: 22. 1803; Duthie, Fl. Gangetic Plain 2: 142. 1911. *Herpestis monnieria* (L.) Kunth. Nov. Gen. Sp. Pl. 2: 366. 1818, nom. illeg.: Hook. *f.* Fl. Brit. Ind. 4: 272. 1883. (Plate 235)

Annual herb. Stem suberect or creeping, glandular, rooting at nodes, succulent, glabrous. Leaves sessile, opposite, spathulate, 0.8 - 2 cm x 3 - 6 mm, margins entire or rarely dentate, apex rounded, fleshy. Flowers axillary, solitary. Pedicel 0.5 - 1.5 cm in flower and upto 3.5 cm long in fruits. Bracteoles 2, linear, obtuse, below calyx. Sepals 5, ca 5 mm; lower and upper sepals ovate-lanceolate; lateral 2 sepals lanceolate to linear. Corolla blue, purple or white, 8 - 10 mm, obscurely 2-lipped. Stamens didynamous. Stigma capitate. Capsule ca 5 mm, ovoid, enveloped in persistent calyx, glabrous, apex acute. Seeds yellow-brown, ellipsoid, truncate at one end, longitudinally striate.

Field note: Common in wet places and drying ditches.

Fl. & Fr.: Almost throughout the year.

Specimen examined: Ladpura 532, Nandrai 1559 (MLVGCB Herb).

Distribution: Widespread in tropics and subtropics.

Notes: For a discussion on the nomenclature of the species, see Philcox (Kew Bull. 33: 679 - 680. 1979).

CRATEROSTIGMA Hochst.

Craterostigma plantagineum Hochst. in Fl. 24: 669. 1841; Mahesh. & Singh, in Bull. Bot. Surv. India 13 (1-2): 11. f. 1-9. 1973; Singh, in Shetty & Singh Fl. Rajasthan 2: 587. 1991. *Torenia plantaginea* Benth. in DC., Prodr. 10: 411. 1846. (Fig. 116 & Plate 236)

Densely pilose, acaulescent, small perennial herb, usually less than 10 cm. Leaves in a basal rosette, 3 - 8 in a whorl, 2.5 - 4.0 x 1.5 - 2.5 cm, variable from broadly ovate to lanceolate, spathulate, entire, hairless above, hairy beneath and on the margins; 5-nerved, obtuse, narrowed at the base into a short winged petiole, parallel veins prominent beneath. Scapes 2.5 - 10.0 cm long, unbranched, quadrangular, solitary or many. Flowers usually paired or sometimes more in racemes, violet-purple. Bracts 8 - 15 x 4 - 8 mm, leaf-like, concave, sessile. Pedicels 2 - 4 mm long, sometimes upto 1.5 cm long forming corymbose racemes. Bracts ovate-lanceolate, ciliate, enclosing one or more flowers. Calyx 5-angled or 5-ribbed; teeth 5, 1 - 2 mm long, ciliate, subequal. Corolla 1.2 - 2.0 cm long, 2-lipped; lower lip 3-lobed with the mid-lobe very broad, violet-purple. Stamens 4, didynamous, anterior pair exserted; anthers dimorphic. Capsules 8 - 10 x 3 mm, ovoid or oblong, narrowed at the ends, 5-angled, many seeded.

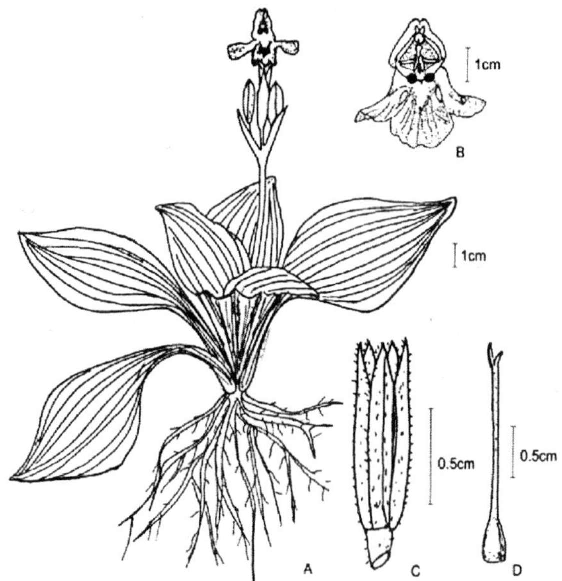

Fig. 116. *Craterostigma plantagineum* Hochst. A. habit, B. Flower, C. Calyx, D. Gynoecium

Field note: Common in gravelly and shallow soils particularly in crevices of rocks. It stays dormant for long periods and spring back to life after some rains.

Fl. & Fr.: August - April.

Specimen examined: Mainal 147 (MLVGCB Herb).

Distribution: Native of Tropical Africa, Naturalised in India.

KICKXIA Dumort.

Kickxia ramosissima (Wall.) Janchen, in Oest. Zeit. 82: 152. 1933; Mahesh. Illus. Fl. Delhi f. 153. 1966; Sant. in Bombay Nat. Hist. Soc. 49 : 7. 1950; Bhandari, Fl. Indian Desert 252. 1995; Singh, in Shetty & Singh Fl. Rajasthan

2: 590. 1991. *Linaria ramosissima* Wall. Pl. Asiat. Rar. 2: 43. t. 153. 1831; Hook. *f.* Fl. Brit. Ind. 3: 175. 1883; Duthie, Fl. Gangetic Plain 2: 262. 1911. (Plate 237)

Twining or trailing, perennial herb, slightly pubescent, 15 - 35 cm high. Stem much branched from root-stock; branches wiry, diffused, 24 - 45 cm long. Leaves variable in shape and size upto 1 - 3 cm long, alternate, lower-ones 5 - 7 lobed, triangular-hastate, upper entire, lanceolate-sagittate, cordate at the base, acute or mucronate, glabrous on both surfaces, ciliate with few hairs on margins. Petioles up to 1.5 cm long. Flowers solitary-axillary. Pedicels 1.0 - 2.5 cm long, glabrous, ebracteate. Calyx 5 mm long, divided almost to base, glabrous, segments linear-lanceolate, acute from an ovate, membranous, margined base. Corolla yellow, 5 - 10 mm long; spur 3 mm long, both pubescent out side; upper lip 2-lobed, the lobes oblong, obtuse, lower lip 3-lobed, the lobes sub-equal, mid-lobe concave, laterals sperading, with two suberect, pubescent, violet palates at the junctions between 3 lobes. Stamens didynamous; filaments with few long hairs at base, more or less papillose above. Ovary subglobose; style glandular-pubescent. Fruits an ovoid or sub-globose capsule, 2 - 3.5 mm in diam., sparsely pubescent at apex only, opening on one side by a valve and other side by a subapical pore. Seeds brown, echinate, minute.

Field note: Common on river banks as well as rock cutings and in ravines as lithophytes, in crevices, walls of old buildings.

Fl. & Fr.: October- January.

Specimen examined: Mainal 1208 (MLVGCB Herb).

Distribution: Temperate rigion of the World. Afghanistan, India (Throughout).

LIMNOPHILA R. Br. *nom. cons.*

KEY TO THE SPECIES

1a. Flowers sessile. Coroll pale-pink. Capsules spherical .... *L. heterophylla*

1b. Flowers distinctly pedicellate. Corolla pale-yellow, Capsules ellipsoid-subglobose .... *.L. indica*

Limnophila heterophylla (Roxb.) Benth. Scroph. Ind. 25. 1835; Hook. *f.* Fl. Brit. Ind. 4: 270. 1884; Philcox, in Kew Bull. 24: 124. 1970; Singh, in Shetty & Singh Fl. Rajasthan 2: 590. 1991. *Columnea heterophylla* Roxb., Fl. Ind. 3 : 97. 1832. *Limnophila reflexa* Benth. Scroph. Ind. 25. 1835. *L. heterophylla* (Roxb.) Benth. var. *reflexa* (Benth.) Hook. *f. l.c.* 4: 270. 1884.

An aquatic, perennial, herb, 10 - 50 cm tall. Leaves sessile, aerial leaves opposite or verticillate, upto 20 x 3.5 mm, oblong lanceolate, subcrenate, semi - amplexicaul, 3-5 parallel nerved, glabrous; submerged leaves multifid with capillary segments, up to 5.5 cm long. Flowers sessile or subsessile, axillary, solitary, forming a lax terminal leafy spike; cleistogamous flowers frequently present on sub-merged stems. Bracteoles absent. Calyx glandu-lar; lobes ovate, shortly acuminate, fruiting ones hemispheric membranous. Corolla pale-pink, glabrous. Capsules spherical, compressed.

Field note: Common in marshy habitats.

Fl. & Fr.: August - December.

Specimen examined: Mainal 1617, Shahpura 1880 (MLVGCB Herb).

Distribution: India.

Limnophila indica (L.) Druce, in Rep. Bot. Soc. Exch. Club. Brit. Isles. 191. 3: 420. 1914; Philox, in Kew Bull. 24: 115. 1970; Singh, in Shetty & Singh Fl. Rajasthan 2: 591. 1991. *Hottonia indica* L. Syst. Nat. ed. 10. 919. 1759. *Limnophila gratioloides* R. Br. Prodr. 442. 1810; Hook. *f.* Fl. Brit. Ind. 4: 271. 1884, *incl.* vars. *elongata* and *intermedia*. *L. racemosa* Benth. Scroph. Ind. 26. 1835; Wight, Ic. 3 (2): 5. t. 861. 1843-45; Hook. *f., l.c.* 4: 271. 1884; Duthie, Fl. Gangetic Plain 2: 146. 1911. (Plate 238)

Perennial, amphibious herb. Submerged stem much branched, glabrous. Aerial stem up to 15 cm, simple or branched, glabrous, with stalked or sessile glands to nearly smooth. Submerged leaves in whorls of 6 - 12, 1.5 - 2.5 cm, pinnatisect; segments flattened to capillary. Aerial leaves usually whorled, pinnately lobed, 0.4 - 2 cm, sometimes few opposite, crenate-serrate lacerate. Flowers axillary from aerial leaves, solitary or in terminal racemes. Pedicel slender, 2-10 mm, usu-ally longer than bract, with sessile or stalked glands. Bracteoles 2, 1.5-3.5 mm, margins entire or sparsely serrate, apex acute. Calyx 3.5-7 mm long, with sessile glands, without raised veins in fruit; lobes 2 - 3 mm, ovate to lanceolate, apex short acuminate. Corolla white, purple or bluish, 1-1.4 cm. Capsule dark brown, compressed, ellipsoid or subglobose, ca 3.5 - 5 mm long. Seeds truncate or rounded.

Field note: Common in pounds, marshy and aquatic habitats.

Fl. & Fr.: August - December.

Specimen examined: Bari sadari 4029, Shahpura 1594 (MLVGCB Herb).

Distribution: Cambodia, China, India, Indonesia, Japan, Laos, Malaysia, Nepal and Pakistan.

LINDENBERGIA Lehm.

Lindenbergia indica (L.) Vatke. Oesterr. Bot. Zeitschr. 25: 10. 1875; Sant. in Bombay Nat. Hist. Soc. 49: 45. 1950; Singh, in Shetty & Singh Fl. Rajasthan 2: 592. 1991. *Dodartia indica* L., Sp. Pl. 883. 1753. *Lindenbergia urticaefolia* Lehm. in Link & Otto, Ind. Sem. Hort. Berol. 5. 1829; Hook. *f.* Fl. Brit. Ind. 4: 262. 1884; Duthie, Fl. Gangetic Plain 2: 160. 1911; *L. polyantha* Royle ex Benth. Scroph. Ind. 22. 1835; Hook.f. *l. c.* 4: 262. 1884; Duthie, *l.c.* 2: 160. 1911. *L. abyssinica* Hochst. ex Benth. in DC. Prodr. 10: 377. 1846.

Erect or decumbent annual herb, 10-30 cm high. Stem glandular hairy, often purple tinged. Leaves opposite, 1.5-5.0 x 1-2 cm, ovate or elliptic, subobtuse or acute, crenate - serrate, glandular hairy, lower leaves petiolate; upper leaves sessile, passing into bracts. Flowers axillary, 1 to many-together, passing into axillary and terminal leafy racemes. Calyx 3 - 6 mm long, campanulate, glandular hairy; lobes 5, obtuse or subacute, ciliate. Corolla 2-lipped, upper lip slightly hooked, reflexed at tip, palate gibbose, glandular hairy. One anther-cell higher than other. Ovary glabrous or pubescent. Capsules ovoid, laterally compressed, beaked, pubescent. Seeds ellipsoid.

Field note: Common on old walls and waste places throughout the area.

Fl. & Fr.: January-March.

Specimen examined: Harni Mahadev 2189 (MLVGCB Herb).

Distribution: Asia and Africa (Warmer parts of the world)

LINDERNIA All.

KEY TO THE SPECIES

1a. Fertile stamens 2, staminodes 2.........*L. ciliata*

1b. Fertile stamens 4, staminodes absent...*L. crustacea*

Lindernia ciliata (Colsm.) Penn., in Brittonia 2: 182; Philcox, in Kew Bull. 22: 51. f. 10/12-15. 1968; Silver. & Mathew, in J. Bombay Nat. Hist. Soc. 80: 133. 1983; Singh, in Shetty & Singh Fl. Rajasthan 2: 595. 1991. *Grantiola ciliata* Colsm., Prodr. Descr. Grat. 14. 1793. *Bonnaya brachiata* Link & Otto, Ic. Pl. Select. 25. t. 11. 1820; Hook. *f.* Fl. Brit. India 4: 284. 1884; Cooke, Fl. Pres. Bombay 2: 369. 1958 (Repr.).(Plate 239)

Erect or decumbent herb, 7 - 15 cm high, diffusely branched. Leaves sessile, 1 - 3.5 x 0.3 - 1.3 cm, oblong, elliptic-oblong or obovate, apex subacute or obtuse. Flowers in lax, terminal racemes, bluish-pink. Calyx deeply divided; segments linear-lanceolate, ciliate, aristate. Corolla white or pink. Fertile stamens 2. Capsules ca 0.6 cm long, narrowly oblong, acute. Seeds truncate at both ends.

Field note: Common in moist places throughout the area.

Fl. & Fr.: August-December.

Specimen examined: Mainal 458 (MLVGCB Herb).

Distribution: India, Australia, China, Malaysia, Myanmar, Philippines and Vietnam.

Lindernia crustacea (L.) F. v. Muell. Syst. Census Austr. Pl. 97. 1882; Mukerjee, in J. Indian Bot. Soc. 24: 130. 1945; Singh, in Shetty & Singh Fl. Rajasthan 2: 596. 1991. *Capraria crustacea* L. Mant. Pl. 87. 1767. *Vandellia crustacea* (L.) Benth Scroph. Ind. 35. 1835; Wight, Ic. 3 (2): 5. t. 863. 1843-45; Hook. *f.* Fl. Brit. Ind. 4: 279. 1884; Duthie, Fl. Gangetic Plain 2: 149. 1911.

Diffuse branched annual herb. Stem 4-angled hairy on the angles, sometimes rooting at nodes. Leaves 0.8 - 2.0 x 0.4 - 1.5 cm, ovate, entire or crenate, rounded or cordate at base, obtuse at apex, pinni-nerved. Petioles 0.2 - 0.8 cm long. Flowers axillary, solitary and subracemose. Pedicels 0.5 - 2.0 cm long. Calyx 5 - ribbed, 5-angled, not divided below the middle before anthesis, afterwards divided near to the base, up to 4 mm long; teeth deltoid or lanceolate, acute. Corolla white, violet or purple, yellow red in throat, 8 - 14 mm long. Stamens 4; anterior filaments appendiculate. Capsules 2 - 8 mm long, oblong or ovoid, apiculate. Seeds ovoid, pale yellow, granulate, with minute tail at one end.

Field note: Common in moist places throughout the area.

Fl. & Fr.: August-December.

Specimen examined: Mainal 456 (MLVGCB Herb).

Distribution: Throughout the tropics and old Worlds.

SCOPARIA L.

Scoparia dulcis L., Sp. Pl. 1: 116. 1753; Hook. *f.* Fl. Brit. Ind. 4: 289. 1884; Duthie, Fl. Gangetic Plain 2: 153. 1911; Singh, in Shetty & Singh Fl. Rajasthan 2: 602. 1991. (Plate 240)

Herb or suffrutescent subshrub, erect, up to 50 cm high. Branches 4 - 6 angld, puberulent at nodes. Leaves in whorls of 3, upper ones opposite, petiolate; leaf blade rhomboid-ovate to rhomboid-lanceolate or elliptic, 1.2-3.5 x 0.5-1.5 cm, glabrous, glandular punctate, base cuneate, margin toothed above middle, sometimes slightly double serrate or subentire, apex obtuse. Flowers usually axillary, 3 - 6 per node. Pedicels 5-12 mm long, glabrous. Calyx lobed to base; lobes 4, ovate-oblong, ca 2 mm, margins ciliate, apex obtuse. Corolla white, ca 4 mm across; tube densely hairy at throat; lobes 2 - 3 mm, margins erose-denticulate, apex obtuse; upper lobe slightly larger than others. Stamens exserted. Style erect; stigma truncate to 2-parted. Capsule 2-3 mm across, columella persistent. Seeds ovoid or obovoid.

Field note: Rare, in waste places.

Fl. & Fr.: Almost throughout the year.

Specimen examined: Bhilwara 1742 (MLVGCB Herb).

Distribution: Throughout tropics and subtropics.

STRIGA L.

KEY TO THE SPECIES

1a. Plants scabridly hairy. Leaves well developed. Calyx 10-15 ribbed .... *S. angustifolia*

1b. Plants glabrous. Leaves reduced to subulate scales. Calyx 5-ribbed............... *S. gesnerioides*

Striga angustifolia (D. Don) Saldhana, in Bull. Bot. Surv. India 5: 70. 1963; Singh, in Shetty & Singh Fl. Rajasthan 2: 604. 1991; Bhandari, Fl. Indian Desert 255. 1995. *Buchnera angustifolia* D. Don Prodr. Fl. Nep. 91. 1825. *B. euphrasioides* Benth. Scroph. Ind. 41. 1835. non Vahl. 1794. *Striaga euphrasioides sensu* Benth., in Hook. Comp. Bot. Mag. 1: 364. 1836, *excl. basionym Buchnera euphrasioides* Vahl; Wight, Ic. 3 (2): 5. t. 863. 1843 - 45; Hook. *f.* Fl. Brit. Ind. 4: 299. 1884; Duthie, Fl. Gangetic Plain 2: 149. 1911. (Plate 241)

Erect, annual, scabrid herb, simple, branched, 15-60 cm high. Leaves mostly alternate, sessile, 1-5 x 0.1-0.4 cm, linear to linear-lanceolate, entire or 1-2-dentate, subacute, hispid. Flowers subsessile, axillary, solitary, passing into terminal spikes. Upper bracts linear to subulate, pubescent; lower bracts foliaceous; bracteoles linear, 3 - 6 mm long. Calyx 6 - 9 mm long, hispid, 15-ribbed, each lobe terminates with 3-ribs. Corolla-tube abruptly incurved at or above the middle; lower lobes outside in bud. Capsules ovoid, ca 7 mm long, 2-valved, apiculate, enclosed in the calyx.

Field note: Common among grasses in wasteplaces.

Fl. & Fr.: August - October.

Specimen examined: Bari sadari 4093, Bhilwara 1733 (MLVGCB Herb).

Distribution: India, Sri Lanka and Java.

Striga gesnerioides (Willd.) Vatke, Oesterr. Bot. Zeits. 25: 11. 1875; Pennel, in Acad. Nat. Sc. Phil. Mon. 5: 97. 1943; Santapau, in J. Bomb. Nat. Hist. Soc. 49: 42. 1950 & in Rec. Bot. Surv. Ind. 16 : 206. 1953; Saldanha, in Bull. Bot. Surv. Ind. 5: 68. 1963; Singh, in Shetty & Singh Fl. Rajasthan 2: 605. 1991; Bhandari, Fl. Indian Desert 256. 1995. *Buchnera gesnerioides* Willd. Sp. Pl. 3: 3389. 1800. *B. orobanchoides* R. Br. ex Endl., in Flora 15: 388. t. 2. 1832; Wight, Ic. 4 (3): 2. t. 1414. 1849; Hook. *f.* Fl. Brit. Ind. 4: 299. 1884; Duthie, Fl. Gangetic Plain 2: 156. 1911.

var.: ***gesnerioides*** (Plate 242)

Vern.: *Gwariya mehndi.*

Erect, rigid, glabrous or pubescent, purple, root-parasitic herb, 10 - 35 cm high. Stem angular, with purple tinge. Leaves opposite or alternate, 5 - 10 x 1.0 - 3.5 mm, scally, ovate-lanceolate, acute, purple, upper ones passing into bracts. Flowers in dense terminal spikes. Bracteoles shorter than calyx. Calyx tubular, 3 mm long, 4 to 5-ribbed, hispid, somewhat scarious, splitting between the teeth. Corolla tube incurved above the middle, hairy outside and in the throat. Capsules ca 5 mm in diam., ovoid-globose. Seeds oblong, minute.

Field note: Common root parasite on *Lepidagathis trinervis* Wall. ex Nees and *Euphorbia caducifolia* Haines.

Fl. & Fr.: August- October.

Specimen examined: Bhilwara 1924, Mandal 1902 (MLVGCB Herb).

Distribution: Afghanistan, China and India.

Uses: The paste prepared from plants is applied for healing wounds during rainy season. For early healing, whole plant is kept in shoes.

VERBASCUM L.

Verbascum chinense (L.) Sant. Fl. Purandhar 90. 1958 & in Rec. Bot. Syrv. India 16: 177. 1967; Bhandari, Fl. Indian Desert 256. 1995; Singh, in Shetty & Singh Fl. Rajasthan 2: 607. 1991. *Scrophularia chinensis* L., Mant. Pl. 2: 250. 1771. *Celsia coromandeliana* Vahl, Symb. Bot. 3: 79. 1794; Wight, Ic. 4 (3): 2. t. 1406. 1849; Hook. *f.* Fl. Brit. Ind. 4: 251. 1884; Duthie, Fl. Gangetic Plain 2: 139. 1911. *Verbascum celsioides* Benth., in DC., Prodr. 10: 229. 1846. *V. coromandelianum* (Vahl) Kuntze, Rev. Gen. Pl. 1: 468.1891. (Plate 243)

An erect, annual or biennial. Stem 20 - 100 cm long, branched apically, sparsely glandular pilose. Radial leaves in rosette; leaf blade 5 - 9 x 3.5 cm, obtuse at apex, dentate, lyrate-pinnatifid; lobes 3 - 5, terminal lobe ovate or oblong, margins serrate distally, usually double serrate or lobed proximally. Other stem leaves short petiolate or sessile; leaf blade ovate, elliptic, or ovate-triangular. Petiole of basal and lower stem leaves 3-8 mm. Raceme to 25 cm, simple or branched. Flowers in lax, simple or branched, 15 - 25 cm long, glandular pubescent racemes. Pedicel to 5 mm, elongating to 1 cm in fruit. Calyx 3-4.5 mm; lobes oblong or ovate, acute, pubescent. Corolla yellow, ca 0.5 cm in diam. Stamens 4, didynamous; filaments woolly; anthers reniform. Capsule sub-globose or ovoid, 5 - 8 mm across, calyx persistent, glandular. Seeds oblong, longitudinally grooved.

Field note: Common in moist places along rivers, ponds etc.

Fl. & Fr.: December - April.

Specimen examined: Bhilwara 324, Karoi 1755 (MLVGCB Herb).

Distribution: Afghanistan, Cambodia, China, India, Laos, Pakistan, Sri Lanka and Thailand.

VERONICA L.

Veronica anagallis-aquatica L., Sp. Pl. 12. 1753; Hook. *f.* Fl. Brit. Ind. 4: 294. 1884; Duthie, Fl. Gangetic Plain 2: 154. 1911; Singh, in Shetty & Singh Fl. Rajasthan 2: 609. 1991.

Perennial herb, simple or branched, 15 - 50 cm high. Stem creeping and rooting at nodes. Leaves 2.5 - 15 x 0.7 - 2.5 cm, oblong-lanceolate or ovate-lanceolate, obtuse at apex, semi-amplexicaul at base, entire, serrate. Flowers axillary, lax, 7.5 - 15 cm long racemes, pinkish. Pedicels filiform, 8 - 10 mm; bracts linear lanceolate. Calyx segments oblong-lanceolate, 2 - 3 mm long, sub-acute. Corolla lobes 4, 4 mm across, white or pink. Capsules, 2 - 3 x 3 - 4 mm, orbicular, ciliate, notched at the apex. Seeds oblong.

Field note: Common in marshy and aquatic habitats.

Fl. & Fr.: December-May.

Specimen examined: Bhilwara 1535, Mandalgarh 1916 (MLVGCB Herb).

Distribution: Europe, N. Africa, China, Japan and India.

LENTIBULARIACEAE

UTRICULARIA L.

1a. Peduncle without floats. Seeds 3 - 4 .. *U. aurea*

1a. Peduncle with a whorl of floats. Seeds 1 - 2 .... *U. stellaris*

Utricularia aurea Lour. Fl. Cochinch. 26. 1790; Singh, in Shetty & Singh Fl. Rajasthan 2: 612. 1991. *U. flexuosa* Vahl, Enum. Pl. 1 : 198. 1804; Clarke in Hook.f. Fl. Brit. Ind. 4: 329. 1884 *pro parte*; Duthie, Fl. Gangetic Plain 2: 166. 1911. *U. fasciculate* Roxb., Fl. Ind. (Carey & Wallich ed.) 1: 143. 1820. (Plate 244)

Slender, aquatic herb, usually submerged, with much-branched, terete stolons. Leaves dichotomously branched into linear segments. Traps suborbicular, lateral or at the point of forking. Scape erect, long with floating vescicles, 5 - 7-fid. Racemes 5 - 7-flowered; scales and bracteoles absent. Bracts orbicular, rounded or subacute at apex, bracteoles absent. Calyx-lobes 2, subequal, ovate, persistant calyx thick, shorter than capsule. Corolla yellow, 10 x 4 mm across; upper lip suborbicular, folded in the middle, entire to undulate; lower lip larger than upper, orbicular. Capsules ca 1 mm across, globose. Seeds prismatic, 5 - angled, narrowly winged along the angles; testa obscurely reticulate.

Field note: Rare, in ponds, tanks and water - logged marshlands. Large, green, bladders turned black with age.

Fl. & Fr.: December - March.

Specimen examined: Mainal 128, Bari Dam 4028 (MLVGCB Herb).

Distribution: India to china and Japan to Australia and Malaysia.

Utricularia stellaris L., f. Suppl. 86. 1781; Wight, Ic. 4 (4): 9. t. 1567. 1850; Clarke, in Hook.f. Fl. Brit. Ind.4: 328. 1884; Duthie, Fl. Gangetic Plain 2: 165. 1911; Subram. Aquatic Ang. 33. f. 23. 1962; Singh, in Shetty & Singh Fl. Rajasthan 2: 614. 1991. *U. inflexa* Forsk. var. *stellaris* (L. f.) Taylor, in Mitt. Bot. Staat. Munch. 4 : 96. 1961 & in Kew Bull. 18 : 189. 1964. (Plate 245)

Free floating, aquatic herb. Stolons up to 1 m × 2.5 mm, terete, filiform, smooth, glabrous. Leaves whorled, auricled at the base, deeply divided into linear - segments. Traps usually numerous, suborbicular or obliquely ovoid; mouth lateral; upper lip with 2 appendages. Scapes erect, slender, 3 - 30 cm. high; flowers 2 - 14, with a whorl of ellisoid, white, spongy floats above the middle; floats with filiform, external pinnules at their apices; pedicels deflxed in fruit; scales and bracteoles absent; bracts basifixed, broadly ovate, 2-3 mm. long. Calyx-lobes subequal, broadly ovate, decurrent. Corolla yellow. Filaments linear, dilated above; anther-thecae subdistinct. Ovary globose; style short but distinct; stigma inferior lip semi-circular, ciliolate, superior very short or obsolete. Capsule globose, circumseissile, more or less 5 mm. in diam. with the style more or less elongated. Seeds numerous, prismatic, 4 - 6 angled, more or less 3 times as wide as high, usually narrowly winged on the angles; testa cells distinct, elongated more or less 0.1 mm. long.

Field note: Rare, in still and slow flowing water, in lakes, rivers and marshes

Fl. & Fr.: Mid April - May.

Specimen examined: Mainal 1238, Vindhiyabhata 1336 (MLVGCB Herb).

Distribution: Widespread in Africa from Mauritania and Egypt to S. Africa (Cape Prov.) India to Vietnam, absent from Malaysia, but in northern Australia.

BIGNONIACEAE

KEY TO THE GENERA

1a. Leaves simple.................................. *Tecomella*

1b. Leaves compound.. 2

2a. Capsule suborbicular-compressed. Flowers purple-mauve .... *Jacaranda*

2b. Capsule elongated, cylindrical, subterete or compressed. Flowers of various colours but not purple-mauve........................ 3

3a. Shrubs .... *Tecoma*

3b. Trees... 4

4b. Leaves 2 - 3-pinnate .... *Millingtonia*

4a. Leaves 1-pinnate..5

5a. Flowers in lax panicles. Fruits woody, gourd-like, hanging on cord-like peduncles..... *Kigelia*

5b. Flowers in terminal, few-flowered racemes Fruits not as above. ....*D. falcata*

DOLICHANDRONE (Fenzl) Seem. nom. cons

Dolichandrone falcata (Wall. ex DC.) Seem. in J. Bot. 8: 381. 1870; Clarke, in Hook. *f.* Fl. Brit. Ind. 4: 380. 1884, excl. syn. *Bignonia atrovirens* Heyne ex Roth; Duthie, Fl. Gangetic Plain 2: 172. 1911; Singh, in Shetty & Singh Fl. Rajasthan 2: 619. 1991. *Spathodea falcata* Wall. ex DC. Prodr. 9: 206. 1845. *Dolichandrone lawii* Seem. *l.c.* 8: 380. 1870; Clarke, in Hook. *f. l.c.* 4: 380. 1884.

Deciduous tree. Leaves 1 - pinnate; leaflets ovate, suborbicular or obovate, abruptly acuminate, entire. Flowers in terminal, few-flowered racemes. Capsules falcate, 2-valved, glabrous. Seeds broadly 4-angular, winged at the both ends.

Field note: Rare.

Fl. & Fr.: Jan. - April.

Specimen examined: Mainal 4536 (MLVGCB Herb).

JACARANDA Juss.

Jacaranda mimosifolia D. Don in Bot. Reg. t. 631. 1822 & in Edin. N. Phil. Jour. 264. 1823; Singh, in Shetty & Singh Fl. Rajasthan 2: 620. 1991. *J. ovalifolia* R. Br. in Curtis, Bot. Mag. t. 2327. 1822. (Plate 246)

Vern.: *Nili gulmohar.*

Tree 6 - 10 m tall. Leaves alternate, 18 - 45 cm long; bipinnate; rachis grooved, pinnae, ca 20 pairs; leaflets 12 - 16 pairs, with an odd one, oblong to rhomboid; terminal one linear-lanceolate, 9 - 12 x 2 - 4 mm, entire, acuminate, sessile. Panicles up to 45 cm long, erect, puberulous. Calyx 5 - toothed, campanulate, truncate, pubescent; teeth linear, ca 0.6 mm long. Corolla tube 1.5 - 2 cm long, abruptly dilated towards base; lobes subequal, 8 - 9 mm long, upper 2 lobes white inside; lower 3 lobes hairy within. Filaments of longer stamens ca 12 mm long, staminodes subexserted, ca 2.6 mm long. Disk fleshy. Ovary oblong, 2 - 2.5 mm long, appressed pubescent; style 18 - 20 mm long, slightly curved and exceeding the stamens; stigma narrow oblong. Capsule ovoid-orbicular, ca 7 mm long, flat, woody. Seeds obovoid, 5-6 x 4 - 4.5 mm, wing 16 - 18 mm broad.

Field note: Frequently planted as a roadside tree. The flowers are very attractive and slightly fragrant.

Fl. & Fr.: May - November.

Specimen examined: Bhilwara 1246 (MLVGCB Herb).

Distribution: A native of Brazil and NW. Argentina. Widely cultivated in warmer parts of the world.

KIGELIA DC.

Kigelia africana (Lam.) Benth. in Hook., t., Nigir. Fl. 463. 1849; Singh, in Shetty & Singh Fl. Rajasthan 2: 620. 1991. *Bignonia africana* Lam. Encycl. 1: 424. 1885. *Crescentia pinnata* Jacq. Collect. 3: 203. t. 18. 1791. *Tecoma africana* G. Don, Gen. Hist. 4: 224. 1837 - 1838. *Kigelia pinnata* (Jacq.) DC. Prodr. 9: 247. 1845. *K. aethiopica* (Fenzl) Decne., in Deless. Icon. sel. Pl. 5. 39. t. 93. 1846.

Vern.: *Jharfanoos.*

Tree, much-branched, up to 10 m tall. Leaves 9 - 50 cm long, rachis ribbed, tomentose or sparsely so; leaflets 5-9, ovate-oblong to elliptic-ovate, 5-11 x 3-5.5 cm, coriaceous, scabrid, entire, acute or mucronate, base often oblique, midrib and veins prominent on undersurface, strigose. Flowers in lax panicles, pendulous, 6 - 7 cm broad. Calyx campanulate, glabrous, greenish. Corolla throat wide dark wine-red, 5 - 10 cm long, basel tube aslong as the calyx. Anther cells parallel, versatile or dorsifixed. Fruit 30-40 x 10 - 13 cm, woody, gourd-like, hanging on cord-like peduncles. Seeds wingless.

Field note: The *African Sausage* (*K. africana*) tree is grown sparingly in gardens and along the roads as an avenue tree. The flowers are nocturnal, foul smelling and bat pollinated.

Fl. & Fr.: May-November.

Specimen examined: Bhilwara 531(MLVGCB Herb).

Distribution: Native of Mozambique. Cultivated elsewhere.

MILLINGTONIA L. f.

Millingtonia hortensis L. *f.* Suppl. 291. 1781; Wight, III. Ind. Bot. t. 161. f. 6. 1850; Clarke in Hook. *f.* Fl. Brit. India 4: 377. 1884; Talbot, For. Fl. Bomb. Pres. & Sind, 2: 318. 1911; Maheshwari, Fl. Delhi, 259. 1963; Singh, in Shetty & Singh Fl. Rajasthan 2: 621. 1991. (Plate 247)

Vern.: *Neem chameli.*

Tree, 8 - 12.5 m high. Bark rough, cracking. Young shoots pubescent. Leaves 40 - 50 cm long; leaflets 5 - 7, ovate-elliptic, 3.5 - 7 x 2 - 4.5 cm, dentate to subentire, acuminate, cuneate-subtruncate, nerves prominent on undersurface. Flowers in widely branched, erect panicles. Calyx cupular, ca 2 mm long, truncate, lobes indistinct, persistent. Corolla tube cylindrical. 2 - 3 cm long, slender, lobes unequal, elliptic-ovate to oval, 6.5 - 12.5 x 5.6 - 6.5 mm. Stamens attached near mouth of corolla; filaments of longer stamens 10.5 mm; fertile anther cell oblong, 4 mm long, abortive cell linear, curved, ca 1.2 mm long, staminodes filiform. Ovary linear-oblong, ca 4.5 mm long; style filiform, exserted; stigma bilobed, lobes subovate, ca 1.2 mm long. Capsule ca 30 x 2 cm.

Field note: Widly cultivated as roadside tree or along avenues. The flowers are fragrant and fully open at night.

Fl. & Fr.: October - March.

Specimen examined: Bhilwara 904 (MLVGCB Herb).

Distribution: Native of Burma & Malaya. Widely cultivated in the tropics and sub-tropics.

TECOMA Juss.

Tecoma stans (L.) Juss. ex Kunth in, Humb., Bonol & Kunth, Nov. Gen. Sp. 3 : 144. 1818; Nicolson, Gard. Dict. 4 : 13.1889; Talbot, For. Fl. Bomb. Presd. & Sind, 2 : 319. 1911; Bailey, Stand. Cyclop. Hort. 6 : 3317. 1917; Parker For. Fl. Punj. 375. 1918; Bor & Raizada, Ind. Climb. & Shrubs, 37.1954; Maheshwari, Fl. Delhi, 258.1963. *Bignonia stans* L. Sp. Pl. ed. 2. 871. 1762.

Shrub up to 2.5 m tall, glabrous. Leaves opposite, imperipinnate, with 3-7 leaflets; leaflets elliptic-ovate or lanceolate, 5-10 x 2-4 cm, acuminate at apex, serrate, glabrous, sessile. Flowers in short terminal, suberect panicles, bright yellow, fragrant. Calyx tubular-campanulate, ca 4.5 mm long; lobes triangular, acuminate, ciliolate. Corolla tube 2 - 2.5 cm long, abruptly contracted at the base, sparse pubescent within and glandulose towards the base; lobes subequal. Stamens included; filaments of longer stamens 20 mm long, curved, base glandulose; anther cells 2, linear, pubescent; staminodes filiform, 5 - 6, 5 mm long. Disc shallow cupular, 5-lobed, fleshy. Ovary ca 3.5 mm long, elongated, minutely papillose. Style 18 - 20 mm long; stigma 2-lobed, lobes oval to ovoid, ca 1.5 mm long. Capsule 10 - 16 x 0.6 - 0.8 cm. Seeds orbicular-compressed, 5 x 6 mm; wing 16 - 18 mm broad.

Field note: Commonly planted in gardens and along roads.

Fl. & Fr.: Most of the year.

Specimen examined: Bhilwara 931, Mandal 1536 (MLVGCB Herb).

Distribution: Native of tropical S.America. Widely cultivated in the tropics and subtropics.

TECOMELLA Seem.

Tecomella undulata (Sm.) Seem. Ann. Mag. Nat. Hist. Ser. 3. 10 : 30. 1862 & in J. Bot. 1: 18. 1863; Duthie, Fl. Gangetic Plain 2: 171. 1911; Mahesh. Illus. Fl. Delhi f. 159. 1966; Singh, in Shetty & Singh Fl. Rajasthan 2: 623. 1991; Bhandari, Fl. Indian Desert 260. 1995. *Bignonia undulata* Sm. Exot. Bot. 1 : 35. t. 19. 1804. *Tecoma undulata* (Roxb.) G. Don. Gen. Hist. 4 : 223. 1838; Clarke, in Hook. *f.* Fl. Brit. Ind. 4: 378. 1884. (Fig. 117 & Plate 248)

Vern.: *Rohira, Ruhis.*

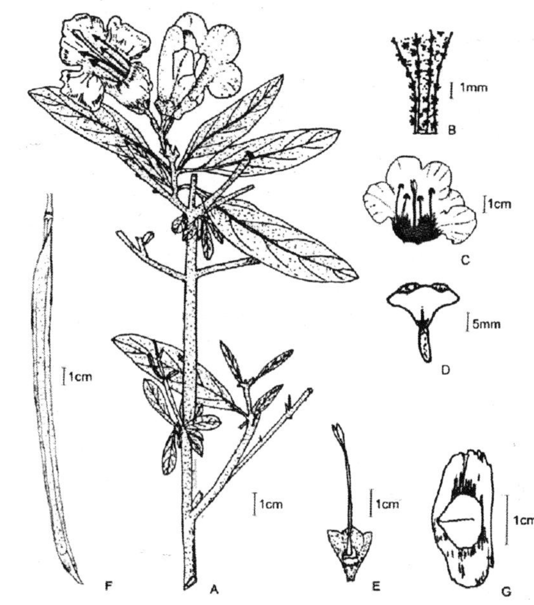

Fig. 117. *Tecomella undulata* (Sm.) Seem. A. Flowering twig. B. Petiole with stellate hairs, C. Flower, D. Calyx, E. Pistil, F. Fruit, G. Seed

Small tree, 2.5-5 m tall, with drooping branches, minutely stellate hairy when young. Leaves subopposite, simple, entire; lamina elliptic-oblong to elliptic-lanceolate or linear-oblong, 4 - 9 x 1 - 2 cm, margins undulate. Petiole 6 - 18 mm long. Racemes condensed, 2 - 7-flowered. Bracts linear, 1.5 - 2 mm long, stellate pubescent; bracteoles 2, similar, ca 1.5 mm long. Pedicel 12-14 mm long. Flowers large, attractive, orange-red. Calyx campanulate, 12 - 13 mm long, unequally 5-lobed; obtuse-apiculate, minutely and sparsely glandular-pubescent. Corolla infundibuliform, dilated at base; lobes sub-orbicular, 0.7-1 x 1-1.5 cm. Stamens unequal-size, exerted; filaments of longer stamens, 1.5-1.8 cm long; staminode 16-17 mm long. Disk cupular-compressed, 5-lobed, yellow. Ovary broad linear, ca 4.5 mm long; style ca 40 mm long, glabrous; stigmas 2, elliptic, compressed, 4.5 - 5 mm long. Capsule 15-35 x 0.8 -1.2 cm, linear-oblong, laterally compressed, slightly curved, smooth, apex beaked with wing, 20 mm broad. Seeds winged at apex, 2-2.5 x 0.9-1cm.

Field note: Rare, in plains perticularly on sandy soils. The colour of flowers is variable. Due to over exploitation for its timber wood it has become threatened in Rajasthan.

Fl. & Fr.: January - April.

Specimen examined: Jawaja 4093, Karoi 1536 (MLVGCB Herb).

Distribution: Arabia, S. Iran, E. Afghanistan, Pakistan & NW. India.

Uses: The wood is said to be fire resistant. Bark is medicinal. Leaves are used as a fodder. Wood is of good quality for furniture and carving. Seeds are used in the treatment of sores and abscesses etc.

PEDALIACEAE

KEY TO THE GENERA

1a. Flowers yellow. Capsules 2-loculed, spinous. Seeds 2, 1 per locule............ *Pedalium*

1b. Flowers white, purplish rosy. Capsules 4-loculed, not spiny. Seeds many per locule.. *Sesamum*

PEDALIUM L.

Pedalium murex L. Syst. Nat. ed. 10 : 1123. 1759; Wight, Ic. 4 (4) : 17. t. 1615. 1850; Clarke, in Hook. *f.* Fl. Brit. Ind. 4: 386. 1884; Singh, in Shetty & Singh Fl. Rajasthan 2: 624. 1991; Bhandari, Fl. Indian Desert 261. 1995. (Plate 249)

Vern.: *Bara gokhru.*

Diffuse annual herb, much branched, spreading, succulent, glandular, up to 60 cm tall. Roots similar to turmeric in colour. Leaves simple, opposite, broadly ovate or orbicular, 2 - 5 x 1.8 - 3.5 cm, truncate or obtuse, irregularly and coarsely crenate-serrate, glabrous above, minutely scaly below; petiole 1 - 4 cm. Flowers yellow, axillary, solitary, 1.5 - 2 cm across; pedicel 1 - 2 mm long, increasing up to 4 mm in fruit. Calyx 5-partite, ca 2 mm long; teeth linear, scaly outside, persistent. Petals connate into a broad tube, 1 - 3 cm long; lobes obtuse. Stamens included, 0.5 - 1 cm long; filaments dilated, glandular hairy at the base; anthers kidney shaped. Capsules pyramidal-ovoid, 4-angled, with a patent spreading spine at each basal corner of the broader part, 1 - 1.8 x 0.5 - 1 cm; spine horizontal, sharp, 2 - 4 mm long. Seeds 2 or 1 per locule, oblong.

Field note: Common in waste lands.

Fl. & Fr.: August - December.

Specimen examined: Mezadam 130, Palri 1650 (MLVGCB Herb).

Distribution: Pakistan, India, Sri Lanka and Tropical Africa.

Uses: Leaves and branches, when briskly stirred in cold water yield a thick mucilage similar to the white of an egg. Leaves and fruits are reported to possess medicinal properties, Leaves are also cooked as vegetable. The powder of mature fruits is taken orally to cure diarhora and dysentery. Pills made by mixing the fruits powder with wheat flour and butter are taken thrice a day for 10 days to cure dropy and debility.

SESAMUM L.

Sesamum indicum L., Sp. Pl. 634. 1753; Clarke, in Hook. *f.* Fl. Brit. Ind. 4: 387. 1884; Duthie, Fl. Gangetic Plain 2: 175. 1911; Mahesh. Illus. Fl. Delhi f. 161. 1966; Singh, in Shetty & Singh Fl. Rajasthan 2: 625. 1991; Bhandari, Fl. Indian Desert 262. 1995. *Sesamum orientale* L. Sp. Pl. 634. 1753. *S. mulayanum* Nair in Bull. Bot. Surv. India 5: 251. 1963. (Plate 250)

Vern.: *Til.*

Erect annual, up to 80 cm tall, galndular - pubescent herb. Stem sulcate, 4-angled. Leaves very variable on the same plant, upper ones lanceolate, entire; lower 2 - 3 segmented near base, serrate, subentire, lobed or 3-partite, rounded or obtuse at base, apex acute, minutely pubescent, petiole 0.5-12 cm long. Flowers axillary, solitary; pedicel up to 5 mm long with 2 glands on either side at the base. Calyx 2-6 mm long, lobes linear to narrowly lanceolate, hairy, persistent. Corolla 2 - 3 cm long, pubescent outside, obtuse, pink, 2-lipped. Filaments glabrous; anthers ca 3 mm long. Ovary 1-2 mm long, pilose, oblong. Capsule oblong, loculicidal, acuminate, 1.5 - 2.5 x 5 - 7 mm, slightly scabrid-pilose. Seeds black, brown or white, 2 - 3 x 1 - 1.5 mm, wingless, smooth.

Field note: Found wild in grasslands, also cultivated as seeds yield edible oil during rainy season.

Fl. & Fr.: August- November.

Specimen examined: Pur 129, Mezadam 2172 (MLVGCB Herb).

Distribution: Original home uncertain, Asia or Tropical Africa, widely cultivated in Pakistan, India, Sri Lanka, Tropical Africa etc.

Uses: Oil is used in confectionery, in the manufacture of "ghee", soap and hair oil. Leaves, seeds and oil are claimed to be medicinally important. Stems are used as fire wood.

MARTYNIACEAE

MARTYNIA L.

Martynia annua L., Sp. Pl. 618. 1753; Mahesh. Illus. Fl. Delhi f. 162. 1966; Singh, in Shetty & Singh Fl. Rajasthan 2: 625. 1991; Bhandari, Fl. Indian Desert 263. 1995. *M. diandra* Glox. Obs. Bot. 14. t. 1. 1785; Clarke, in Hook. *f.* Fl. Brit. Ind. 4: 386. 1884 ; Duthie, Fl. Gangetic Plain 2: 176. 1911. (Fig. 118 & Plate 251). Vern.: *Hemra.*

Erect, branched herb, up to 1.5 m tall. Stem succulent, tomentose-glandular. Leaves opposite, ovate-orbicular, 7 - 15 x 7 - 20 cm, distantly dentate with the veins ending in denticles at the margin, glandular; petiole slightly shorter than or as long as the lamina, glandular. Flowers in lax racemes, borne in the bifurcation of upper branches, drooping, showy, lilac, mottled red, blue or yellow inside; pedicel ca 1.5 cm long, up to 3 cm in fruit, glandular villous; bracteoles 1-3, oblong, unequal, ca 1 cm long, petaloid or lilac, glandular. Calyx 5, free, oblong lanceolate, unequal, 1 - 2 cm long, glandular. Corolla tube ca 2.7 cm long, gibbous, glandular outside, 5 - lobed, the posterior lobe the largest; lobes purple tinged, obtuse. Fertile stamens 2; filaments ca 1.3 cm long, glandular at the base; anthers bilobed; lobes divergent, oblong, ca 7 mm long, adnate when young to the other anther pair, dehiscing longitudinally; staminodes 3, ca 4 mm long, slightly curved. Ovary globose, 2 mm broad, glabrous, situated on the top of the subglobose disc; placentae intruded to form false septa; style ca 2 cm long, slender, persistent; stigmas 2-lamellate, sensitive, flat. Drupes green, black on drying, ca 3 cm long, ovoid, 2 valved, crowned by 2, strong, claw-like, sharp hooks. Seeds oblong.

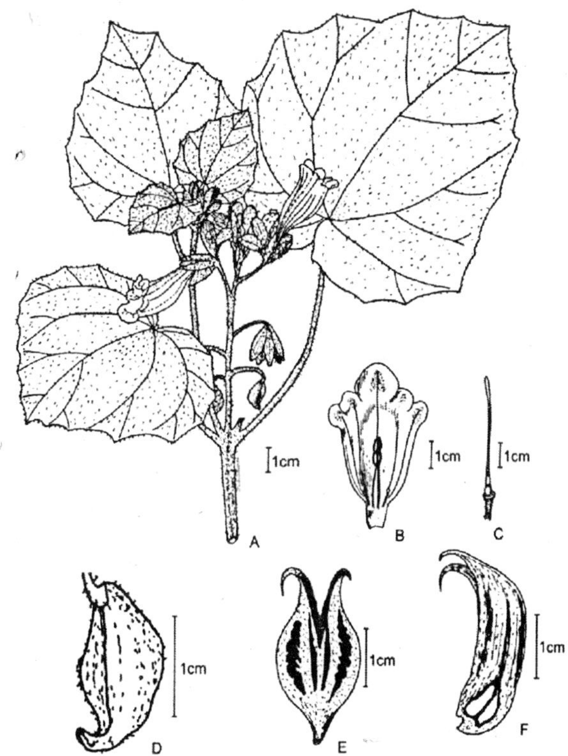

Fig. 118. *Martynia annua* L.
A. Flowering & fruiting twig. B. Opened flower, C. Pistil, D. Young fruit, E. Fruit (Dorsal view), F. Fruit (lateral view)

Field note: Common in waste lands during rainy seasons, flowers are showy.

Fl. & Fr.: August-November.

Specimen examined: Bhilwara 476 (MLVGCB Herb).

Distribution: Tropical and sub-tropical America, Mexico, India, Burma and W. Pakistan.

Uses: The fruits are used for medicinal purposes and locally in the soles of shoes (chapals).

ACANTHACEAE

KEY TO THE GENERA

1a. Leaves radical, borne in rosette on the ground. Seeds not supported on hard retinaculae.. *Elytraria*

1b. Leaves ramal and cauline. Seeds supported on hard upcurved retinaculae 2

2a. Corolla without an upper lip, lower lip large, expanded, 3-lobed.................... *Blepharis*

2b. Corolla distinctly 2-lipped or with 5 subequal lobes 3

3a. Corolla lobes twisted to the left in bud 4

4a. Ovules 2 in each cell i.e. capsules 1 to 4 seeded... *Eranthemum*

4b. Ovules more than 2 in each cell i.e. capsules 6 or more seeded.............................. 5

5a. Corolla distinctly 2 -lipped............ *Hygrophila*

5b. Corolla sub equally 5-lobed............................ 6

6a. Inflorescence a lax, axillary cyme, passing into terminal panicles .... *Ruellia*

6b. Inflorescence a simple spike or capituliform of compound spike or 1-3 flowers in leaf axils.... 7

7a. Inflorescence a capititulum of compound spikes surrounded by 4 large ovate bracts..................................... *Gentelbua*

7b.Inflorescence a simple spike or flowers in leaf axils.. 8

8a. Inflorescence a simple, elongate spike ....*Hemigraphis*

8b. Inflorescence of single axillary flower or in axillary triads or 2-3 flowers superposed.............................. *Dipteracanthus*

3b.Corolla -lobes imbricate in bud....................... 9

9a. Ovules 3 to many in each cell i.e. capsules 6 or more - seeded.......................... 10

10a. Flowers in unilateral racemes.. *Indoneesiella*

10b. Flowers clustered in the midst of axillary cladodes .... *Haplanthodes*

9b. Ovules 1 to 2 in each cell i.e. capsules 2 to 4- seeded 11

11a. Corolla lobes 5, subequal .... *Barleria*

11b. Corolla distinctly 2- lipped......................... 12

12a. Stamens 4................................ *Lepidagathis*

12b. Stamens 2.. 13

13a. Placentae separating elastically from the valves from the base upwards.......... *Dicliptera*

13b. Placentae not separating elastically from the valves.. 14

14a. Anther-cells, at least the lower one with a basal spur-like appendage.................. *Justicia*

14b. Anther- cells not spurred at base.............. 15

15a. Corolla white, with irregular rose - coloured bars in the throat, seeds 2, glabrous .... *Adhatoda*

15b. Corolla rosy. Seeds 4, minutely glandular hairy .... *Peristorphe*

ADHATODA Mill.

Adhatoda zeylanica Medic. Hist. & Commentat. Acad. Elect. Sci. Theod. - Palat. 6: 393. 1790; Matthew, Illus. Fl. Tam. Carnatic pl. 519. 1982; Pandey & Singh, in Shetty & Singh Fl. Rajasthan 2: 630. 1991; Bhandari, Fl. Indian Desert 264. 1995. *Justicia adhatoda* L., Sp. Pl. 15. 1753; Duthie, Fl. Gangetic Plain 2: 207. 1911. *Adhatoda vasica* Nees in Wall. Pl. Asiat. Rar. 3: 103. 1832; Clarke in Hook. *f.* Fl. Brit. India 4: 540. 1885; Santapau, in Univ. Bombay Bot. Mem. 2: 92. 1951.

Vern.: *Adusa.*

Erect shrub, upto 1.5 m hight with many ascending branches, terete, stem glabrous, young parts densely appressed pubescent, with swollen nodes. Leaves Opposite, 5 - 20 x 1.5 - 8.0 cm, elliptic-lanceolate or ovate-lanceolate, acuminate or acute, base of lamina tapering, dark - green abaxially, base cuneate or decurrent, glabrous when mature; Flowers. in dense leafy spikes, white or creamish-white, spikes 4.5 - 12.5 cm long. Bracts leafy, ciliate. 1.8 - 2.5 x 0.6 - 1.5 cm, Ovate - elliptic, 3 - 5 - veined at base, acute. Calyx upto 1.2 cm long, lobes oblong - lanceolate, acute. Corolla 2-lipped, upto 2.5 cm long, with pink or purple stripes in the throat; upper lip ovate - oblong notched; lower lip 3-lobed, lobes oblong - rounded. Staminal filaments hairy at base; lower anther cells apiculate. Capsules 2 x 3 cm, 4 seeded, hard, dark brown. Seeds c. 5 mm across, sub orbicular, rugose, brown.

Field note: Fairly common in scrub forests, along the road sides and in waste lands.

Fl. & Fr. : October - March

Specimen examined: Jahajpur 128, Pansal 2171 (MLVGCB Herb).

Distribution: Sri Lanka, Burma, Malaysia, India (Throughout)

Uses: Plant locally known as "*Adusa*" or "*Vasaka*". or "*Aatroosh*" (Sanskrit) is reported to carry medicinal importance in Ayurveda. Plant parts used as crude drug; fresh or dried leaves, often mixed with stems or other aerial parts. It is used as an expectorant and alterative; flowers and roots are used in asthma, chronic bronchitis and other chest infections. 1-2 gm of drug is prescribed as an expectorant either in the form of liquid extract (1-2 ml) or syrup (2-4 ml). Dried hollow internodes when smoked as cigars by local people (without filling anything inside) relieve congestion in the chest, even in chronic cases.

Powder of whole plant of *Adhatoda zeylanica* and *Tinospora cordifolia* is mixed, boiled with water and sugar at least for an hour. After cooling root powder of *Piper longum* L. is added and two-tea spoonful of this decoction is taken twice a day, for 15-30 days by the tribals to cure chronic bronchitis, tuberculosis and asthma. It acts as an expectorant. Half cup of decoction of leaves along with a pinch of salt is taken orally twice a day for 5 days in cold & cough. The smoke of dried leaves is inhaled to cure asthma. The extract of fresh leaves is used for gargling to cure tonsillitis. Flower extract is used in ophthalmic problems.

BARLERIA L.

KEY TO THE SPECIES

1a. Stem and branches without spines in or near the leaf axils............................ *B. cristata*

1b. Stem and branches with sharp spines in or near the leaf axils. 2

2a. Flowers white, usually more than 6 cm long. Capsule upto 1.2 cm long................................ *B. acanthoides*

2b. Flowers golden yellow, less than 5 cm long. Capsule 1.5-2 cm long .... *B. prionitis*

Barleria acanthoides Vahl, Symb. Bot. 1: 47. 1790; Boiss., Fl. Orient. 4: 524. 1879; Clarke in Hook. *f.* Fl.Brit. India 4: 484. 1884; Cooke, Fl. Pres. Bomb. 2: 458. 1906; Duthie, Fl. Gangetic Plain 2: 200. 1911; Pandey & Singh, in Shetty & Singh Fl. Rajasthan 2: 632. 1991; Bhandari, Fl. Indian Desert 265. 1995. (Plate 252)

Vern.: *Vajardanti.*

A harsh, prickly, small, much branched, up to 60 cm, densely greyish-pubescent shrublet, armed with simple, up to 5-6 mm long spins in leaf axils. Leaves oblong or obovate, 1.5 - 4 cm x 12 - 20 mm, basally attenuate, pubescent on both sides, margins entire, apex obtusely mucronate, dark green,. Petiole 4 - 7 mm long, hairy. Inter-petiolar spines simple, 2 - 4 together or combined. Flowers white, large, 6 - 7.5 cm long, in 1 - 8-flowered, congested, scorpioid racemes; bracts leafy, membranous, elliptic, 14 - 18 mm long, pinnatifid-spinescent or reduced to pinnatifid spines. Calyx deeply lobed; outer 2 lobes broadly ovate, 1.2-1.7 cm x c. 8-10 mm, subequal, sharply mucronate, becoming yellowish-brown; inner lobes lanceolate, 7-8 mm long. Corolla tube cylindrical 5 - 7.5 cm long, slightly widened below the 5-lobed limb, lobes ovate, 6-10 mm long, acute. Capsule ellipsoid, 1-1.2 cm x 4-5 mm, attenuate at both ends, glabrous, 4-seeded, ovoid, 3 - 4 mm in diam., silky hairy. Seeds ovoid, compressed.

Field note: Found on rocks in western part of the study area. A highly variable species in size, mucro of the leaves, outer sepal, spines of bracts, length of the corolla tube and limb lobes.

Fl. & Fr.: September-February.

Specimen examined: Asind 328, Badnor 1372 (MLVGCB Herb).

Distribution: Egypt, Sudan, Ethiopia, Somalia, Djibouti, Pakistan and N.W. India.

Uses: Used as fodder.

BARLERIA CRISTATA L.

KEY TO THE VARIETIES

3a. Flowers purple blue.... *B. cristata* var. *cristata*

3b. Flowers white........... *B. cristata* var. *dicotoma*

Barleria cristata L., Sp. Pl. 636. 1753; Sims in Bot. Mag. 39. t. 1615. 1814; Nees in Wall. Pl. Asiat. Rar. 3: 92. 1832; in DC., l.c. 229; Wight, Icon Ind. Or. t. 453. 1841; Clarke in Hook. *f.* Fl. Brit. India 4: 488. 1884; Duthie, Fl. Gangetic Plain 2: 201. 1911; Santapau in Univ. Bombay Bot. Mem. 2: 59. 1951; Pandey & Singh, in Shetty & Singh Fl. Rajasthan 2: 632. 1991. (Fig. 119 & Plate 253)

var.: ***cristata***

An erect, up to 1.5 m tall, much branched, unarmed shrublet with appressedly hairy twigs, densely so at the nodes, Leaves with 1 - 2 cm long; petiole hairy; lamina elliptic-oblong to lanceolate, 3 - 12.5 x 1 - 3.5 cm, hairy both surfaces, attenuate at the base, entire, apically acute-acuminate. Flowers white, 4.5 - 5 cm long, in short, 1 - 5-flowered, axillary or terminal spikes; bracteoles membranous, acute. Calyx deeply 4-cleft, outer 2 lobes much larger than the inner pair, ovate-lanceolate, 1.5 - 2.5 cm x 6 - 8 mm, acuminate, bristly toothed on margins, hairy or glabrescent, inner 2-lobes linear, 7 - 8 mm long, pointed. Corolla glandular-pubescent outside, tube 3.5 - 4 cm long, infundibuliform, limb with oblong-obovate, up to 2 cm long, unequal lobes. Filaments hairy, anthers oblong, ca 3 mm long. Ovary oblong-conical; style ca 4 cm long. Capsule ellipsoid, 1.5 - 2 cm long, glabrous, pointed at the base and apex, 4-seeded. Seeds orbicular, ca 4 mm across, appressed hairy.

Field note: Rare in the area along rivers and grasslands, often cultivated in gardens for its showy flowers.

Fl. & Fr.: October - February.

Specimen examined: Jogniya Mata 4028 (MLVGCB Herb).

Distribution: S. China, Nepal, Burma, Indo-China, Philippines, Bangladesh, India, Sri Lanka and Pakistan.

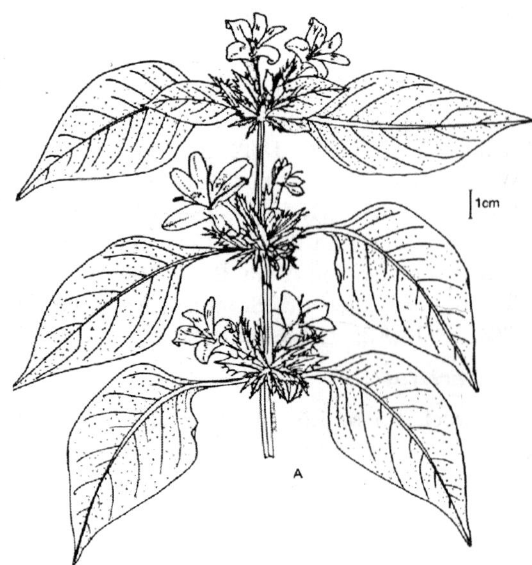

Fig. 119. *Barleria cristata* L. var. *cristata*
A. Flowering twig

Barleria cristata L. var. ***dichotoma*** (Roxb.) Prain, Bengal Pl. 812. 1903; Wight, Ic. 2 (2): 3. t. 453. 1841; Santapau in Univ. Bombay Bot. Mem. 2: 59. 1951; Pandey & Singh, in Shetty & Singh Fl. Rajasthan 2: 633. 1991; Bhandari, Fl. Indian Desert 266. 1995. *B. dichotoma* Roxb. Hort. Beng. 45. 1814 (nom. nud.) et Fl. Ind. 3: 39. 1832. (Plate 254)

A perennial, erect undershrub, upto 1.5 m high. Stem adpressed hairy. Leaves 5 - 15 x 2 - 5 cm, elliptic-oblong, acuminate, tapering at base, pubescent on both sides. Flowers solitary axillary of few in leaf axils forming dense capitate spike, purple blue. Bracts foliaceous or linear or absent. Calyx 4: outer 2 larger, 15 - 25 x 7 - 8 mm, lanceolate, long acuminate, mucronate with subspinous, bristle tipped teeth on the margins, with raised conspicuous nerves and prominent reticulations, pubescent; inner sepals 8 x 2 mm, linear-lanceolate, acute. Corolla tube 4 cm long, glandular-pubescent out side, purple blue, widely infundibuliform; lobes 1.5 cm long, obovate-oblong, unequal. Capsules 1 - 15 cm long, ellipsoid, tapering at both ends, 4-seeded. Seeds orbicular, compressed.

Field note: Not common, some speciemens were collected from Jogniya mata valley.

Fl. & Fr.: October - February.

Specimen examined: Jogniya Mata 4093 (MLVGCB Herb).

Distribution: India.

Barleria prionitis L., Sp. Pl. 636. 1753; Wight, Ic 2(2): 3. t. 452. 1840-1843; Clarke in Hook. *f.* Fl. Brit. India 4: 482. 1884; Duthie, Fl. Gangetic Plain 2: 200. 1911; Santapau in Univ. Bombay Bot. Mem. 2: 57. 1951; Pandey & Singh, in Shetty & Singh Fl. Rajasthan 2: 635, 1991; Bhandari, Fl. Indian Desert 266. 1995. (Fig. 120)

subsp. ***prionitis*** var. ***prionitis.***

Vern.: *Kala bans.*

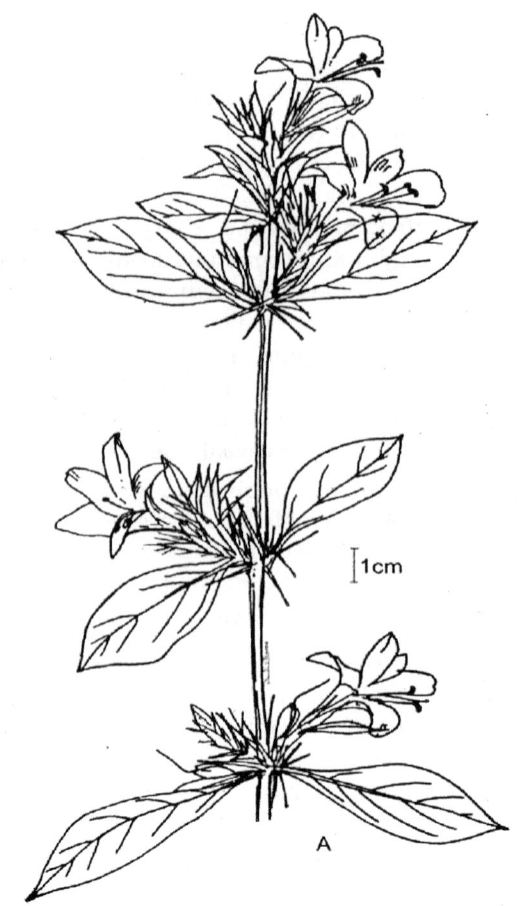

Fig. 120. *Barleria prionitis* L. subsp. *prionitis* var. *prionitis*
A. Flowering & fruiting twig

• Glabrous, much branched, 0.5-1.5 m high, prickly undershrub with whitish bark. Stems 4 - gonous, sparsely pubescent just below the nodes and in between the opposing petioles. Leaves 3.0 - 10.0 x 2.0 - 5.0 cm elliptic, acuminate, bristle-tipped, pubescent, base tapering into a 1.5 - 2.5 cm long petiole; main nerves 5 pairs; inter - petiolar spines 1.5 - 2.0 cm long, rigid, acicular, simple, 2, opposite or often 3 or 4 (1 or 2 of the succeeding pairs, also developed). Flowers sessile often solitary in the lower axils, becoming spicate above; orange yellow to golden-yellow. Bracts foliaceous, 15.0 - 17.0 x 3.0 - 4.0 mm, oblong, lanceolate, acute, keeled, bristle-tipped, glabrous or a few white hairs dorsally on the keel; bracteoles 1.0 cm long, otherwise like bracts. Calyx divided to the base; one of the 2 outer sepals 15.0 - 17.0 mm long, slightly longer than the other; the 2 inner 13 mm long, ovate - lanceolate, mucronate. Corolla 4.0 - 4.5 cm long, orange-yellow, minutely pubescent outside, glabrous inside, obscurely 2-lipped, upper lip of 4 lobes, the lobes unequal in width, oblong, ovate, rounded, the lower lip of one lobe, entire, 2.5 cm long; tube 1.5 cm long. Stamens 2, with filaments exserted and minutely pubescent at base; staminodes 2, with very short filaments, woolly at attachment. Ovary glabrous on an annular disc; style more than 3 cm long; stigma orange. Fruit 1.5 cm long, capsules ovoid-conic, beaked, 2-seeded. Seeds ca 5 mm across, sub-orbicular, appressedly hairy.

Field notes: A common shrub of hill slopes.

Fl. & Fr. : October - February

Specimen examined: Hamirgarh 822, Mainal 2190 (MLVGCB Herb).

Distribution : Tropical Africa & Asia, Sri Lanka, S. Africa and India (Western Peninsula)

Uses: 1 tsp. decoction of leaves is given twice a day for 2 days in the treatment of cough. The roots and leaves are chewed to relieve toothache and bodyache. Fresh roots are kept under painful teeth to relieve pain. The juice of leaves is appllied on feet during the rainy season to prevent the maceration and cracking of the sole. In case of catarrhal infection in children which accompanied with fever and much phlegm, it is generally taken with little honey or sugar and water in the quantity of two-teaspoonful twice a day. The dried bark is given in whooping cough. A paste made of root, is applied on boils and glandular swellings. A medicated oil (Prepared by boiling the leaves and stems with oil of groundnut till all the water has been evaporated) is applied as a cleansing agent to wounds.

BLEPHARIS Juss.

KEY TO THE SPECIES

1.a. Leaves petiolate, acute, membranous, glabrous beneath. Flowers axillary clusters of 2 - 3 .... *B. maderaspatensis*

1 b. Leaves sub-sessile, obtuse thick, hairy beneath. Flowers solitary axillary..... *B.repens*

Blepharis maderaspatensis (L.) Heyne ex Roth, Nov. Pl.Sp 320.1821;Santapau in univ. Bombay Bot. Mem. 2 : 15. 1951; Pandey & Singh, in Shetty & Singh Fl. Rajasthan 2: 637. 1991. *Acanthus maderaspatensis* L., Sp. Pl. 639. 1753. *Blepharis boerhaviaefolia* Pers. Syn. Pl.2: 180. 1806; Wight, Ic. 2 (2): 3. t.458. 1840 - 43; Clarke in Hook. *f.*, Fl. Brit. India 4; 478. 1884; Duthie, Fl. Gangetic Plain 2: 183. 1911.

Diffuse or prostrate, perennial, pubescent herb with long wiry stem, branching divaricately, often rooting at the nodes. Leaves in whorls of 4, 2 smallar and 2 larger at each node, 3 - 8.5 x 1.5 - 3.5 cm, elliptic, ovate - lanceolate, ovate elliptic or oblanceolate, acute, cuneate, subentire, glabrescent. Flowers sessile, axillary, clusters of 2 - 3 at each node, often combined into spikes; white or dirty-white with dark purple streaks. Bracts 8, in 4 opposite pairs, spathulate, strongly 3 - nerved, upper margins with retrorsely hairy bristles, Calyx lobes 4, unequal, oblong, ciliate. Lower lip of corolla with yellow spots within. Larger anther - cells bearded. Capsules ca 7.5 mm long, ellipsoid, brown, 2-seeded. Seeds ca 3.5 mm across, sub orbicular or ovoid, compressed, echinate with white spines.

Field note: Common in rock-crevices and other shaded and moist situations in the hilly areas.

Fl. & Fr. : September - January.

Specimen examined: *Gurla* 908, Ladpura 2131 (MLVGCB Herb).

Distribution: Africa, Asia-temperate, Asia-Tropical (India, Sri Lanka, Myanmar)

Uses: An aquous paste of seeds is prepared and is applied on male genitals in order to get good stamina and retention time. It is applied twice a day. Handful of leaves, mixed with onion bulb, made into a paste and applied externally on cuts and wounds.

Blepharis repens (Vahl) Roth. Nov.Pl. Sp.321. 1821; Pandey & Singh, in Shetty & Singh Fl. Rajasthan 2: 637.1991; Bhandari, Fl. Indian Desert 267. 1995. *Acanthus repens* Vahl, Symb. Bot. 2 : 76. 1791. *Blepharis mollugini-folia* Pers. Syn. Pl 2. 180. 1806; Clarke, in Hook. *f.* Fl. Brit. India 4: 479.1884; Duthie, Fl. Gangetic Plain 2: 183. 1911; Mahesh. Illus. Fl. Delhi f. 164. 1966.

Prostrate, much branched, annual, hairy herb. Stem diffuse, wiry, slender, branches 6-20 cm long. Leaves sub sessile, whorled, usually 4 at each node, upto 0.4 - 3 x 0.5 - 1.0 cm, oblong or obovate, entire, hispid, slightly succulent. Flowers axillary, solitary, sessile, blue or white. Bracte-oles 8 - 10, spinosus, in opposite pairs. Calyx segments linear-lanceolate, ca 8 mm long, acute, membranous-margined, outer one 2-fid at tip; inner linear. Corolla blue or violet; tube ca 3 mm long. Capsules ellipsoid, compressed, glabrous, ca 6 mm long, completely enclosed within the calyx, 2-seeded. Seeds broadly ellipsoid, hairy.

Field note: Common in waste lands, fallow fields and grassy localities.

Fl. & Fr. : October - January.

Specimen examined: Kakroliyaghati 1322, Dhaneswar 1823 (MLVGCB Herb).

Distribution: Sri Lanka, India, extending to tropical and South Africa.

Uses: Plants are given to cattle for milk production. The paste of leaves is applied on cuts and wounds.

DICLIPTERA Juss. *nom. cons.*

Dicliptera verticillata (Forssk.) C. Christen-sen, in Dansk. Bot. Ark. 4 (3): 11. 1922; Wood et al. in Kew Bull. 38: 450. 1983; Pandey & Singh, in Shetty & Singh Fl. Rajasthan 2: 637. 1991; Bhandari, Fl. Indian Desert 268. 1995. *Dianthera verticillata* Forssk. Fl. Egypt.-Arab. 9. 1775. *Dicliptera micranthes* Nees in Wall. Pl. Asiat. Rar. 3: 112. 1832; Clarke, in Hook. *f.* Fl. Brit. India 4: 553. 1885; Duthie, Fl. Gangetic Plain 2: 213. 1911; Santapau in Univ. Bombay Bot. Mem. 2: 79. 1951.

Much branched, decumbent, annual, glabrous herb, with long internodes. Stem glabrous but pubescent on the angles below the nodes with aculeate hairs, 6-angled. Leaves unequal, 1.0 - 6.5 x 0.5 - 3.75 cm, elliptic or ovate, acuminate-acute, entire, main nerves 3 - 5 pairs, base attenuated or rounded; petioles 2.0 - 3.0 cm long. Flowers in dense axillary, globose clusters, extending often to the lower axils; bracts 2, longer of the two 7 mm, smaller 6 mm long, herbaceous, obovate - lanceolate, scabrous on both surfaces, suddenly narrowed into a spinescent mucro, green, lower half margin scarious, ciliate, upper half green; bracteoles 4, upto 4-5 mm long, linear - subulate, pubescent. Calyx 3 - 3.5 mm long, 5-partite to near the base, linear- subulate, hairy, very acute. Corolla 3 - 3.5 mm long, pubescent outside, lower lip slightly longer than the upper, throat with dark purple spots. Capsules ca - 5 mm long, ellipsoid, compressed, acute and pubescent at the apex, stalked brown. Seeds 4, 1 mm in diam., orbicular, echinulate with small tubercles, with stellately radiating tips.

Field note: Found as a forest undergrowth, particularly near shaded moist situations. The plant is very variable in its appearance in the late season, it becomes almost entirely devoid of leaves and gives quite a different look. The corolla are very small and remain enclosed by the bracts and bracteoles. Young plant of this species and that of *Peristrophe paniculata* are apparently very similar.

Fl. & Fr.: August - November.

Specimen examined: Baghera 4041, Pansal 1990 (MLVGCB Herb).

Distribution : India (Rajasthan)

Uses: The paste of fresh leaves is applied externally on mosquitoes as well as ants bites.

DIPTERACANTHUS Nees emend. Bremek.

Dipteracanthus patulus (Jacq.) Nees var. ***alba*** (Saxton) Bhandari in Bull. Bot. Surv. Ind. 6 (2-4): 327. 1964; Pandey & Singh, in Shetty & Singh Fl. Rajasthan 2: 637.1991; Bhandari, Fl. Indian Desert 269. 1995. *Ruellia petula* Jacq. Misc. Bot. 2: 358. 1781; Clarke, in Hook. *f.* Fl. Brit. India 4: 412. 1885; Duthie, Fl. Gangetic Plain 2: 188. 1911. *Ruellia petula* Jacq. var. *alba* Saxton in Rec. Bot. Surv. Ind. 6 : 288. 1918; BH. 26(3): 811. 1919.

Hoary, pubescent undershrub. Stem sub-quadriangular, much branched. Leaves 1.5 - 5.0 x 3.0 - 6.0 cm, ovate or elliptic, obtuse, usually apiculate, densely pubescent on both surfaces, base rounded; petiole 5 - 18 mm long. Flowers axillary, usually solitary, rarely 2- 3 together, white sessile; bracteoles foliaceous, 12 mm long, densely pubescent, elliptic, obtuse, ciliate. Calyx 6 mm long, deeply divided, segments subequal, acute. Corolla white, 3.0 - 5.0 cm long, tube widely infundibuliform; lobes 6 mm long, sub orbicular, rounded at apex. Filaments glabrous. Ovary glabrous; style pubescent. Fruit 12 - 17 mm long capsule, clavate, glabrous. Seeds sub orbicular, margined with hygroscopic, white hairs.

Field note: Common in open rocky habitats, particularly on low hillocks. Flowers are highly caducous.

Fl. & Fr.: July - November.

Specimen examined: Atun 922 (MLVGCB Herb).

Distribution : Tropica Africa, India (Gujarat, Rajasthan).

ELYTRARIA A. Mich. *nom. cons.*

Elytraria acaulis (L. f.) Lindau in Engl. & Prantl. Pflanz. Nacht. 1: 304. 1897; Santapau in Univ. Bombay Bot. Mem. 2: 11. 1951; Bremek. in Reinwardtia 3: 250. 1955; Mahesh. Illus. Fl. Delhi f. 163. 1966; Pandey & Singh, in Shetty & Singh Fl. Rajasthan 2 : 637. 1991. *Justicia acaulis* L. f. Suppl. Pl. 84. 1781. *Elytraria crenata* Vahl, Enum. 1: 106. 1804; Clarke in Hook.f.Fl. Brit. India 4: 394. 1884. *Tubiflora acaulis* (L. f.) Kuntze, Rev. Gen. Pl. 1: 500. 1891; Duthie, Fl. Gangetic Plain 2: 180. 1911.

Vern.: *Pathar-chatta.*

A plant without aerial stem; herb, upto 20 cm high, with fibrous roots. Leaves radical, in a rossete, 6 - 22.0 x 3 - 8.0 cm, oblong-obovate or spathulate, rounded or obtuse at apex, base cuneate, crenate, glabrous; petiole short. Flowers on several procumbent scapes arising in the central portion in between the leaves, scapes upto 22.0 cm long, many. Flowers in simple or branched spikes, spike usually unbranched, white. Bracts ovate, acuminate; bracteoles 2, linear, villous or pubescent. Calyx- lobes unequal, lanceolate with scarious and ciliate margins, acute. Corolla bi-lipped, 5-lobed. Capsules 0.5 - 0.75 x 0.25 - 0.35 cm. ovoid, acute, glabrous, brownish. Seeds many more or less ovoid, minute, flattened, papillose, brown.

Field note: Common in moist and shady places viz. rock - crevices, along the streams and other rocky gravelly habitats.

Fl. & Fr. : April - October

Specimen examined: Mandalgarh 202, Gangrar 4026 (MLVGCB Herb).

Distribution : Southern Africa, Asia and America.

Uses: The paste prepared from leaves is applied on cuts and wounds by local inhabitants. It is also applied locally on sores and skin eruptions whereas rubbed locally in case of urticaria. Decoction of root is mixed with equal amount of local liquor, one cup of this mixture is taken daily for 3 - 4 days in the early morning for easy expulsion of giunea worms. Half tea spoonful of root extract is given to children once a day for two days in asthma. One tea spoonful extract of roots is given orally to children to cure stomachache. Two tea spoonful of leaf powder taken orally with water once a day for 10 days to remove stone from kidney. The roots are also reported to be beneficial in migraine.

ERANTHEMUM L. emend. Radlk.

Eranthemum purpurascens Wight ex Nees, in Wall. Pl. Asiat. Rar. 3: 106. 1832; Santapau in Univ. Bombay Bot. Mem. 2: 35. 1951; Pandey & Singh, in Shetty & Singh Fl. Rajasthan 2: 645. 1991. *E. pulchellum* Roxb., Fl. India 1: 111. 1832, *non* Andr., 1800. *Justicia pulchella* Roxb., Pl. Coromandel 2: 41. t. 177. 1805. *Daedalacanthus purpurascens* (Wight ex Nees) T. Anders. in Journ. Linn. Soc. 9: 488. 1867; Clarke, in Hook. *f.* Fl. Brit. India 4: 420. 1884; Duthie, Fl. Gangetic Plain 2: 194. 1911. *Eranthemum pulchellum* Andr. var *purpurascens* Kuntze, in Kanjilal et al., Fl. Assam 3: 439. 1939. (Plate 255)

Erect herb or undershrub, stout up to 1 m tall. Leaves on 1.5 - 3 cm long petiole; lamina elliptic or ovate-elliptic, 3.5 - 18 x 2.5 - 10.5 cm, oblong-ovate, or ovate - elliptic, acuminate, cuneate at the base. Flowers blue-purple, ca 2.5-3 cm long, in dense 5 - 30 cm long spikes forming a close terminal panicle; bracts leafy, ca 2.5 cm long, ovate-lanceolate or acuminate, ciliolate or prominently nerved. Bracteoles linear-lanceolate, scarious glandular pubescent, ciliate. Calyx white, 6 - 9 mm long, pubescent on both sides; lobes lanceolate. Corolla lilac or bluish purple, 3 - 5 cm long; lobes ovate to obovate, pubescent outside. Filaments glabrous. Capsules ca 1.5 cm long, clavate - oblong, glabrous. Seeds ellipsoid, hairy.

Field note: Rare, in this area recorded only from mainal.

Fl. & Fr.: October- March.

Specimen examined: Mainal 170 (MLVGCB Herb).

Distribution: A native of Indo-Pakistan subcontinent; introduced in gardens elsewhere.

GANTELBUA Bremek.

Gantelbua urens (Heyne ex Roth) Bremek. in Verh. Nederl. Akad. Wet. (1). 40 (I) : 147. 1944; Santapau in Univ. Bombay Bot. Mem. 2 : 28 . 1951; Pandey & Singh, in Shetty & Singh Fl. Rajasthan 2 : 646. 1991. *Ruellia urens* Heyne ex Roth, Nov. Pl. Sp. 302. 1821. *R.dura* Nees in Hook. Comp. Bot. Mag. 2 : 311. 1836. *Hemigraphis dura* (Nees) T. Anders. in J. Linn. Soc. Bot. 9: 461. 1867. *excl. syn. Ruellia crispa* Nees, non L. 1753; Clarke, in Hook .f. Fl. Brit. India 4: 422. 1884; Duthie, Fl. Gangetic Plain 2: 190. 1911. (Plate 256)

Vern.: *Dubrasoos.*

Herb, procumbent or diffuse, upto 70 cm high. Stem somewhat 4-angled, covered with stout white hairs. Leaves 1.5 - 4 x 0.9 - 2 cm, spathulate-oblong or oblong-lanceolate, obtuse at apex, tapering at base, subcrenate. Inflorescence a capitulum of compound spikes surrounded by 4 large ovate bracts. Flowers in involucrate heads, densely clothed with white and yellowish brown hairs, bright rosy-purple. Bracts upto 3.5 cm long, spathulate or ovate-oblong, subacute, densely hairy. Calyx 5 - partite, ca 1.2 cm long, linear, acute, densely hairy and ciliate with jointed hairs. Corolla blue or violet, ca 1.5 cm long, subequally 5 lobed, twisted to the left in bud. Stamens 4. capsules ca 1.5 cm long, linear - oblong, glabrous, 6 to 8 - seeded. Seeds discoid, compressed, yellowish-brown.

Fl. & Fr.: October - March

Field note: Rare, found in scrub forests.

Specimen examined: Bangreda 4038 (MLVGCB Herb).

Distribution: India (Rajasthan)

HAPLANTHODES Ktze.

Haplanthodes neilgherraensis (Wight) Majumdar in Bull. Bot. Soc. Bengal 25: 76. 1971; Pandey & Singh, in Shetty & Singh Fl. Rajasthan 2 : 647. 1991. *Justicia verticillata* Roxb., Fl. India 1: 135. 1832. *Haplanthus verticillaris* Nees, in DC., Prodr. 11: 513.1847; Clarke, in Hook. *f.* Fl. Brit. India 4: 506. 1884. *Haplanthus neilgherryensis* Wight, Ic. 4 (4): 7. t. 1556. 1850; Santapau in Univ. Bombay Bot. Mem. 2: 52. 1951. *H. tentaculatus* (L.)

Nees var. *nilgherrensis* (Wight) Clarke in Hook. f. Fl. Brit. India 4: 507. 1884. *Bremekampia verticillata* (Roxb.) Sreemadh., in Bull. Bot. Surv. India 6: 323. 1965. (Plate 257)

Annual erect, glandular - pubescent herb, upto 1 m high. Stem slender, pubescent. Leaves 5 - 10 x 3 - 5 cm, ovate, acuminate. Cladodes 0.75 - 1.75 cm long, slender, densely hairy, apex with 2 or 3, flattened, villous teeth. Flowers blue - lilac, in terminal spikes, amongst cladodes. Bracteoles shorter than the calyx, subulate. Calyx - lobes hairy, lanceolate, ca 4 mm long, pointed, glandular pubescent. Corolla ca 1.5 cm long; lobes violet; ube white. Capsules ca 1 cm long, oblong, pointed, hairy, green, with 2 - 3, white horns, grooved, glabrous. Seeds compressed, minute, brown.

Field note: Rare, along the rivers, streams.

Fl. & Fr.: August - February.

Specimen examined: Triveni 1368 (MLVGCB Herb).

Distribution: India.

HEMIGRAPHIS Nees emend. T. Anders.

Hemigraphis hirta (Vahl) T. Anders. in Journ. Linn. Soc. Bot. 9: 462. 1867; Clarke in Hook .f. Fl. Brit. India 4: 422. 1884; Duthie, Fl. Gangetic Plain 2: 184. 1911; Pandey & Singh, in Shetty & Singh Fl. Rajasthan 2: 650. 1991. *Ruellia hirta* Vahl, Symb. Bot. 3: 84. t. 67. 1794. (Fig. 121 & Plate 258)

Hairy, small, low-growing, prennial herb with beautiful flowers. Stem much branched, hirsute-villous, creeping. Leaves 0.75 - 2.5 x 0.5 - 1.5 cm, broadly ovate to elliptic, crenate, serrate, acute or obtuse at apex, prominently hairy. Flowers in spikes in leaf axils, condensed in terminal heads, bluish purple, funnel shaped. Bracts upto 1.5 cm long, leafy, elliptic, wooly. Calyx linear or subspathulate in fruit, ca 8 mm long. Corolla glabrous, with two hairy lines in the palate, violat with deep-violet throat, 10 - 15 mm long. Stamens 4. Style hairy. Capsules 0.5 - 1.0 cm long, linear - cylindric, glabrous. Seeds 12, compressed, discoid, brownish.

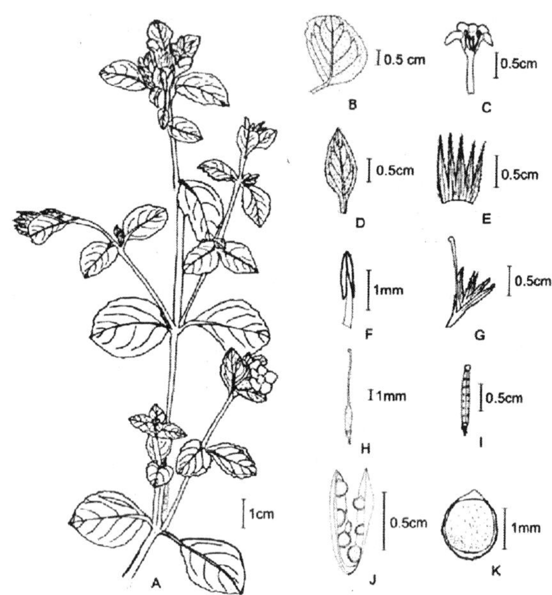

Fig. 121. *Hemigraphis hirta* (Vahl) T. Anders. A. Flowering & Fruiting branch, B. Leaf. C. Flower, D. Bract, E. Calyx, F. Stamen, G. Calyx & pistil, H. Pistil, I. Capsule, J. Dehisced capsule, K. Seed

Field note: Rare in moist places and waste lands.

Fl. & Fr.: April - May.

Specimen examined: Triveni 148, Nagdi river 1903 (MLVGCB Herb).

Distribution: India.

HYGROPHILA R. Br. emend. Heine

Hygrophila auriculata (Schum.) Heine in Kew Bull. 16: 172. 1962; Pandey & Singh, in Shetty & Singh Fl. Rajasthan 2: 646. 1991; Bhandari, Fl. Indian Desert 270. 1995. *Barleria longifolia* L. Cent. Pl. 2:22. 1756, non *Hygrophila longifolia* Nees, 1847. *Barleria auriculata* Schum. in Schum.& Thonn. Beskr. Guin. Pl. 285. 1827. *B. longifolia* L., Cent. Pl. II: 22. 1756. *Asteracantha longifolia* (L.) Nees in Wall Pl. Asiat. Rar. 3:90. 1832; Wight, Ic. 2(2): 3. t. 449. 1841; Duthie, Fl. Gangetic Plain 2: 184. 1911. *Hygrophila spinosa* T. Anders, in Thw. Enum. Pl. Zeyl. 225. 1860; Clarke, in Hook. *f.* Fl. Brit. India

4: 408. 1884. *H. longifolia* (L.) Kurz, in J. Asiat. Soc. Bengal 39: 78. 1870, *non* Nees, 1847. (Fig. 122)

Vern.: *Oont katela.*

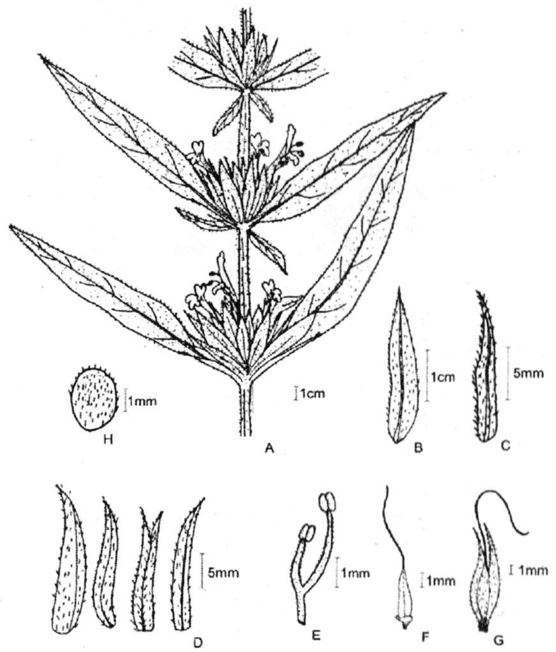

Fig. 122. *Hygrophila auriculata* (Schum.) Heine, A. Flowering & fruiting twig. B. Bract, C. Bracteole, D. Sepals, E. Stamen, F. Pistil, G. Capsule with calyx. H. Seed

Hispid annual herb, 75 cm to 1 m high, with thickened nodes. Stem sub-quadrangular, thickened at nodes, hispid with white, spreading hairs, densely so below the nodes. Leaves in whorls of 6 at a node, 5.0 - 20.0 x 1.0 - 3.0 cm, sessile, oblong or linear-lanceolate, sparingly hispid and pubescent on both surfaces, minutely spinous - serrate on margins, sub-acute at apex, unequal, each leaf with a straight, 2.5 - 4.0 cm long, sharp, yellow spine in its axil. Flowers many in each axillary whorl, purple - blue. Bracts 2-seriate, upto 4 cm long, lanceolate, hairy, ciliate at base; bracteoles linear - lanceolate, hyaline - margined. Calyx 4-partite, lobes linear-lanceolate, hairy the upper lobe upto 20 mm long, broader and longer than the other 3, all linear, lanceolate, cancave, hairy on the back, ciliate. Corolla purple - blue; tube 1.2 cm long, widening abruptly, 2-lipped; lips 2 cm long, sub-equal, the upper bifid with oblong, truncate lobes, the lower with 2, entire, crest-like, longitudinally folded lobes. Filament glabrous, one short and other long, each pair united at base. Style shortly pubescent. Capsules ca 8 cm long, pointed, linear - oblong, glabrous, 4-8 seeded.

Field note: Abundant in swampy habitats, waste lands along ditches and canals in association with grasses and sedges.

Fl. & Fr.: August - December.

Specimen examined: Palri 472, Bhilwara 1869 (MLVGCB Herb).

Distribution : Tropical Africa, Sri Lanka, India (Throughout)

Uses: Handful of leaves mixed with seeds of *Cuminum cyminum* L. and made in the form of juice. 1 glass of juice is taken for three days to relieve cough. It is used in disorders of the genito-urinary system and is helpful in the cases of impotency, spermatorrhea and seminal debilities.

INDONEESIELLA Sreemadhavan

Indoneesiella echioides (L.) Sreemadh., in Phytologia 16: 466. 1968; Pandey & Singh, in Shetty & Singh Fl. Rajasthan 2: 646. 1991; Bhandari, Fl. Indian Desert 271. 1995. *Justicia echiodes* L., Sp. Pl. 16. 1753 *Andrographis echoides* (L.) Nees in wall. Pl. Asiat. Rar. 3: 117. 1832; Wight, Ic 2(4): 4. t. 467. 1841; Clarke, in Hook. *f.* Fl. Brit. India 4: 505. 1884; Duthie, Fl. Gangetic Plain 2: 197. 1911; Santapau, in Univ. Bombay Bot. Mem. 2: 51. 1951. *Neesiella echioides* (L.) Sremadh., in Phytologia 15: 271. 1967, *non. illeg.*

Vern.: *Jodapatta, Pattar.*

An erect, symmetrical, annual herb, upto 50 cm high. Stem quadrangular, grooved, villous with long white hairs. Leaves 2.0 - 10.0 x 1.0 - 4.0 cm, sparsely pubescent on both surfaces, margins usually ciliate, sessile, oblong, obtuse, rounded or rarely cuneate at base. Flowers on divaricate, simple or branched, axillary racemes, only on the upper side of a glandular rachis, pinkish - white, pink or white. Bracts 2 mm long, linear-lanceolate; bracteoles very small, subulate. Calyx 4 - 6 mm long, glandular hairy, divided almost to base; sepals narrowly linear, acute, ciliate, elongating in fruits. Corolla pale- purple spotted, 10 - 15 mm long, 2-lipped, lips slightly shorten than the tube, upper lip obovate- oblong, rounded, lower 3 - lobed, blotched with purple, the middle lobe slightly longer than the laterals. Filaments flat, sparsely hairy; anthers bearded, red. Ovary hairy; style pubescent. Fruit 10 - 11 mm long capsule, half as broad, elliptic-lanceolate, compressed, acute at both ends, hairy. Seeds upto 2.5 x 2.5 mm, oblong, rugosely pitted, glabrous, brownish.

Field note: Commonly found as a forest undergrowth, along the streams and in waste lands, particularly in moist and shady localities. This plant is typical due to its conspicuously symmetrical branching. Flowers are pale-purple, blotched with dark - purple spots; anthers are bearded at base.

Fl. & Fr.: August - November.

Specimen examined: Bhilwara 530, Atun 1762 (MLVGCB Herb).

Distribution : Ceyon, Pakistan, Tropical India (Common in the drier parts).

Uses: Juice of leaves is used as a febrifuge, tonic, cholagogue, anthelmintic and in stomachache.

JUSTICIA L.

KEY TO THE SPECIES

1.a. Calyx 5 - partite .... *J. trinervia*

1.b. Calyx 4 - partite 2

2.a. Bracts linear, obtuse, glabrous ....*J. quinqueangularis*

2.b. Bracts elliptic-lanceolate, acute, hairy, clothed with simple hairs .... *J. procumbens*

Justicia procumbens L., Sp. Pl. 15. 1753; Clarke, in Hook. *f.* Fl. Brit India 4: 539. 1885; Santapau in Univ. Bombay Bot. Mem. 2: 90. 1951; Pandey & Singh, in Shetty & Singh Fl. Rajasthan 2: 646. 1991; Bhandari, Fl. Indian Desert 273. 1995. *Rostellaria procumbens* (L.) Nees in Wall. Pl. Asiat. Rar. 3: 101. 1832. *Rostellularia procumbens* (L.) Nees in DC. Prodr. 11: 371. 1847. Dalz. & Gibs. Bomb. Fl. 193. 1861.

Vern.: *Makhania ghas.*

Sub erect or diffuse herb, stem diffuse, slender, with many divaricate branches often rooting at lower nodes. Leaves up to 15 - 40 x 6 - 15 mm, elliptic or ovate, obtuse at both ends, softly pubescent; petioles upto 1 cm long, with a few white hairs. Flowers in dense, cylindric, terminal spikes, 15 - 60 mm long; bright rosy purple or lilac. Bracts 4.5 x 1.5 mm, linear - lanceolate, acute, hairy with scarious, ciliate margins; bracteoles, linear, lanceolate, acute, hairy with scarious margins. Calyx divided almost to the base, 4.0 - 4.5 mm long, segments linear - subulate, narrower than the bracteoles, with scarious, ciliate margins. Corolla pale - violet - pink, hairy out side, 6 mm long, tube as long as the limb, funnel - shaped; upper lip broadly ovate, concave below, notched above; lower lip shortly 3 - lobed. Stamens 2, with glabrous filaments. Ovary glabrous. Capsule 5 x 2 mm, shortly pointed, oblong, pubescent at apex,. Seeds finely tuberculate, brown.

Field note: commonly found in mixed habitats. This is a highly polymorphic species with regards to habit, shape and size of leaves, capsules, seeds and hairiness.

Fl. & Fr.: Almost throughout the year.

Specimen examined: Bhilwara 1595 (MLVGCB Herb).

Distribution: Australia, Malaysia, Sri Lanka, Pakistan, India (Maharashtra, Gujarat, Rajasthan)

Uses: The plant is used as alterative, anthelmintic, antiplogistic, diaphoretic, diuretic, expectorant, febrifuge and laxative.A decoction or an infusion is used in the treatment of asthma, cough, backache, flatulence and many skin disorders.

Justicia quinqueangularis Koen. ex Roxb., Fl. Ind. 1: 134. 1832; Clarke, in Hook. *f.* Fl. Brit. India 4: 536. 1885; Duthie, Fl. Gangetic Plain 2: 209. 1911; Santapau, in Univ. Bombay Bot. Mem. 2: 89. 1951; Pandey & Singh, in Shetty

& Singh Fl. Rajasthan 2: 646. 1991; Bhandari, Fl. Indian Desert 273. 1995. *Rostellaria quinqueangularis* (Koenig ex Roxb.) Nees, in Wall. Pl. Asiat. Rar. 3: 101. 1832. *Rostellularia quinqueangularis* (Koenig ex Roxb.) Nees, in DC. Prodr. 11: 375. 1847.

var. ***quinqueangularis.***

An erect or ascending, glabrous annual herb, upto 50 cm high. Leaves in distance pairs, 2 - 8.0 x 0.3 - 1.4 cm, elliptic or oblanceolate, base tapering, apex obtuse. Spikes terminating the branches, 1.0 - 4.5 cm long. Flowers pink. Bract and bracteoles linear, glabrous, obtuse, margins scarious. Calyx 4 partite; lobes linear, obtuse, 4 - 5 mm long, equal, glabrous. Corolla pink coloured, or white with purplish blotch, ca 9 mm long. Stamens 2, filaments glabrous. Capsules 0.75 cm long, oblong, pointed, stalked, 4-seeded. Seeds ca 0.5 mm across, compressed, reticulately rugose.

Field note : Common in moist places on sandy soil.

Fl. & Fr. : Aug. - December.

Specimen examined: Mangrol 4037 (MLVGCB Herb).

Distribution: India (almost all over)

Uses: Fresh aerial parts of the plant are used as fodder.

Justicia trinervia Vahl, Enum. 1: 56. 1804; Clarke, in Hook. *f.* Fl. Brit India 4: 526. 1885; Santapau in Univ. Bombay Bot. Mem. 2: 86. 1951; Sharma & Tiagi, in Fl. N-E Rajasthan 319 1979; Pandey & Singh, in Shetty & Singh Fl. Rajasthan 2: 646. 1991. *Adhatoda trinervia* (Vahl) Nees, in Wall. Pl. Asiat. Rar. 3: 103. 1832. Bhandari, Fl. Indian Desert 263. 1995. (Plate 259)

Procumbent herb, with a woody root-stock. Branches many, slender, quadrangular. Leaves upto 4.0 x 1.0 cm, variable; upper sessile, lower ones petiolate, elliptic-oblong or obovate - oblong, Obtuse. Flowers in continuous, somewhat lax 2.5 - 12.5 cm long spikes, pinkish. Bracts with green veins, lanceolate, acute; bracteoles, lanceolate, acute. Calyx 5-partite, pubescent; lobes lanceolate, acute, 3 - nerved corolla pubescent outside. Filaments glabrous above, pubescent at the base. Capsules ca 1.0 cm long, oblong - clavate, shortly pointed, pubescent, 4-seeded. Seeds 0.3 mm across, compressed, densely clothed with acute tubercles.

Field note: Rare, found in rocky habitats in moist situations.

Fl. & Fr.: September - October.

Specimen examined: Bhilwara 1537 (MLVGCB Herb).

Distribution : India (Rajasthan and Maharashtra)

Note: A species of *Synchytrium* infects the plants causing the wart disease, wherein small, transparent, yellowish - white eruptions are seen in clusters on the stem and leaves.

LEPIDAGATHIS Willd.

KEY TO THE SPECIES

1a. Flowers in radical or sub radical globose heads. Calyx 4-partite, Bracts spinous-pointed...................... *L.cristata*

1b. Flowers in axillary, sessile, ovoid spikes; Calyx 5-partite, bracts glabrous, with long recurved spines .... *L. trinervis*

Lepidagathis cristata Willd. Sp. Pl. 3: 400. 1800; Wight, Ic. 2(2): 3. t. 455. 1841; Clarke, in Hook. *f.* Fl. Brit. India 4: 516. 1885; Santapau in Univ. Bombay Bot. Mem. 2: 71. 1951; Bhandari, Fl. Ind. Desert 307. 1978; Pandey & Singh, in Shetty & Singh Fl. Rajasthan 2: 646. 1991; Bhandari, Fl. Indian Desert 275. 1995.

Vern.: *Pattafori, Bangari.*

An annual or perennial, procumbent under shrub, with numerous branches from a highly reduced main stem. Branches spreading on all sides close to the ground, quadrangular, almost winged, glabrous. Leaves opposite, sessile, 6.0 - 8.0 x 0.6 - 1 cm, linear-oblong, hairy above and on nerves beneath, margins spinuous-serrulate. Flowers in sub radial, globose heads, on the lower part of the leafy branches; bracts 8 - 9 mm long, ovate, acuminate, spinous-pointed, hairy;

bracteoles membranous, hairy, spine tipped. Calyx 8 mm long, 4-partite to near the base; outer segments 3 mm broad, elliptic, acute; lower segment 2-fid; inner segment 1.1 mm broad, all ciliate, densely softly hairy on both surfaces and spinous pointed. Corolla 12 mm long, hairy outside, white or pale pink, dotted with brown and or purple spots; tube narrow, constricted below the tip and then suddenly expanded upwards; upper lip 4 mm long, oblong, obtuse, notched at apex; lower lip 6 mm long, 3-lobed; midlobe the broadest. Stamens slightly exserted, filaments glabrous. Ovary glabrous; style slightly pubescent. Capsules 5 mm long, acute, ovoid, glabrous, with scarious back. Seeds 2 - 3 mm long, ovoid-oblong, densely, clothed with long hygroscopic hairs.

Field note: Commonly found in association with grasses, in rocky habitats.

Fl. & Fr.: October -January.

Specimen examined: Kanera 4096 (MLVGCB Herb).

Distribution: India (Maharashtra, Rajasthan)

Uses: Plants are used as fodder.

Lepidagathis trinervis Wall. ex Nees in Wall. Pl. Asiat. Rar. 3: 96. 1832; Clarke, in Hook. *f.* Fl. Brit. India 4: 517. 1885; Duthie, Fl. Gangetic Plain 2: 204. 1911; Santapau in Univ. Bombay Bot. Mem. 2: 72. 1951; Pandey & Singh, in Shetty & Singh Fl. Rajasthan 2: 646. 1991; Bhandari, Fl. Indian Desert 276. 1995.

Vern.: *Patharphor.*

Branched, suffructicose, pubescent perennial with a woody root stock; branches many, slender, quadrangular, 10.0-43.0 cm long. Leaves sessile, 2.5-6.0 x 0.4-1 cm, linear-lanceolate, acute, base tapering, margins scabrid, 3 - nerved at the base, midvein prominent. Flowers pink, pale purple or violet, rarely whitish; spikes axillary, solitary or many, packed together forming ovoid - globose heads in the basal portion of the plant. Bracts ovate - elliptic; bracteoles oblong lanceolate, aristate. Calyx-lobes subequal, spinous-tipped, densely hairy within and glabrous outside. Corolla - throat with transverse streaks and mottled with yellow, brown and purple. Capsules ca 6 mm long, compressed, beaked, glabrous. Seeds ca 3.75 mm long, rounded at the top, grey, covered with long silky white hairs.

Field note: Common on low hillocks in rocky to gravelly habitats.

Fl. & Fr. : Almost throughout the year.

Specimen examined: Puskar 4061, Sahara 1972 (MLVGCB Herb).

Distribution: India (more or less throughout).

Uses: Root powder is taken orally two to three times a day in dysentery. Roots and leaves are soaked in water overnight and infusion is used in problems related to giddiness. Extract of whole plant is taken orally thrice a day for 3 - 5 days to cure typhoid. The leaves are boiled in water and after cooling, the water is used for herbal bath to cure the typhoid. The paste of leaves is applied on gums as an ointment.

PERISTROPHE Nees

Peristrophe paniculata (Forsk.) Brum. in Kew Bull. 38: 451. 1983; Pandey & Singh, in Shetty & Singh Fl. Rajasthan 2: 646. 1991; Bhandari, Fl. Indian Desert 276. 1995. *Dianthera paniculata* Forsk. Fl. Egypt - Arab. 7. 1775. *D. bicalyculata* Retz., in Vet. Acad. Handle. 297. t. 9. 1775. *Justicia bicalyculata* (Retz.) Vahl, Symb. Bot. 2 : 13. 1791. *Peristrophe bicalyculata* (Retz.) Nees, in Wall. Pl. Asiat. Rar. 3: 113. 1832; Clarke, in Hook. *f.* Fl. Brit. India 4: 554. 1885; Duthie, Fl. Gangetic Plain 2: 210. 1911; Santapau in Univ. Bombay Bot. Mem. 2: 93. 1951; Mahesh Illus. Fl. Delhi f. 166. 1966. (Plate 260)

Vern.: *Kanker.*

An erect, much branched herb or undershrub, upto 1.75 m high, hispid. Stem ribbed, 6-angled usually; leaves 4.0-8.0 x 2.0 - 3.5 cm, ovate or ovate lanceolate, acuminate, more or less hairy above, densely hairy on the veins and nerves beneath, rounded at base; petioles 6-12 mm long. Flowers in terminal or axillary, lax, divaricate, leafy paniculate cymes, pink or purple. Bracts 2, opposite, very unequal, the longer of the pair upto 17 mm, linear, acute, mucronate, margins membranous; bracteoles 4, similar to the bracts but shorter, sub-equal. Calyx segments 3 - 4 mm long, divided near to the base, segments lanceolate, margins cililate, acute. Corolla 14 - 15 mm long, throat hairy and lower lip mottled with dark mauve coloured spots. Filaments hairy; anther cell almost one above the other. Ovary pubescent. Capsules ca 1 cm, narrowed into a cylindrical stalk, with glandular and simple hairs throughout except the stalk. Seeds orbicular, minutely glandular hairy, with a white aril at point of attachment.

Field note: Common among the hedges of cultivated fields, in waste lands and grassy localities.

Fl. & Fr. : September - December.

Specimen examined: Peesangan 4032 (MLVGCB Herb).

Distribution: Tropical Africa, Egypt, India (Throughout)

Uses: 1/4 cup of leaf paste mixed with sugar and taken twice a day for 3 days to cure intermittent fever.

RUELLIA L. emend. Bremek.

Ruellia tuberosa L. Sp. Pl 635. 1753; Bremek. in Verh. Kon. Nederi. akad. wet. 45 (1): 11 1948 & in Densk. Botan. Arkiv. 20: 64. 1961; Maheshw., Fl. Delhi 272. 1963; Raizada, Suppl. Duthie, Fl. Gangetic Plain 209. 1976; Pandey & Singh, in Shetty & Singh Fl. Rajasthan 2: 646. 1991; Bhandari, Fl. Indian Desert 276. 1995 .

An erect, branched, glabrescent, annual herb, upto 50 cm high with several tuberous roots. Stems 4-angled, grooved, swollen, purplish at the nodes, tender parts sparsely hairy. Leaves with up to 2 cm long petiole, opposite, upto 3.5 - 11.0 x 1.3 - 5.4 cm, elliptic or elliptic-obovate, base attenuated, lateral veins 9 pairs, margins undulate - crenate, apex acute or obtuse. Flowers, campanulate blue, in axillary cymes; blue-violet, showy, 5 - 5.5 cm long, ca 3.5 cm across, solitary or in 1 - 3-flowered, terminal or axillary cymes; bracts and bracteoles linear-lanceolate, ca 5 mm long. Calyx lobes subequal, linear, 1.3 - 2 cm long, ciliate, acute. Corolla pubescent outside, tube ca 3.5 cm long, abruptly narrowed at base, limb 5-lobed, lobes subequal, ovate, ca 1.5 cm long, patent. Stamens with oblong-sagittate, ca 4 mm long, puberulous anthers. Ovary oblong, ca 4 mm long, glabrous; style 2.3 - 2.5 cm long. Capsule linear fusiform, 2.2 - 3 cm long, pointed, grooved, glabrous or scabrescent, 24 - 28-seeded. Seeds flattened, orbicular, ca 2 mm across, hairy.

Field note: Common in waste lands. Leaves frequently vary from ovate-oblong, narrowly oblanceolate-obovate to broadly obovate within the same individual.

Fl. & Fr.: August - November.

Specimen examined: Bhilwara 1653 (MLVGCB Herb).

Distribution: A native of Tropical America; introduced and naturalized in Africa, Pakistan, India to S.E. Asia.

Uses: Used as an anthelmintic, against joint pains and strained muscles.

VERBENACEAE

KEY TO THE GENERA

1a. Leaves digitately compound (rarely 1-foliolate) .... *Vitex*

1b. Leaves simple, rarely deeply pinnatisect....... 2

2a. Inflorescence indefinite and centripetal, spicate or racemiform, rarely capitate 3

3a. Herbs, trailing and rooting at nodes. Flowers in compact, spicate heads. Pyrenes 1-loculed, 1-seeded, separating or not .... *Phyla*

3b. Shrubs. Flowers in dense heads or compact spikes. Pyrenes 2-loculed, 2-seeded (rarely 1-loculed and 1-seeded by abortion), not separating. Fruits juicy .... *Lantana*

2b. Inflorescence indefinite and centrifugal, cymes mostly congested, panicled, umbellate, thyrsoid or a head, rarely reduced to a single flower when axillary 4

4a. Corolla regular. Stamens 4-6, equal ... *Tectona*

4b. Corolla irregular. Stamens 4, didynamous ... 5

5a. Corolla-tube usually short, cylindrical .... *Clerodendrum*

5b. Corolla-tube large mostly infundibuliform and greatly enlarged above .... *Gmelina*

CLERODENDRUM L.

Clerodendrum multiflorum (Brum. f.) Kuntze, Revis. Gen. Pl. 3: 526. 1891. *Volkameia multiflorum* Brum. f., Fl. Ind. 137. t. 45. f. 1. 1767. *Clerodendrum phlomidis* L. f. Suppl. 292. 1781; Wight, Ic. 4 (3): t. 1473. 1849; Clarke, in Hook. *f.* Fl. Brit. Ind. 4: 590. 1885; Duthie, Fl. Gangetic Plain 2: 262. 1911; Pandey, in Shetty & Singh Fl. Rajasthan 2: 677. 1991; Bhandari, Fl. Indian Desert 278. 1995. (Plate 261)

Vern.: *Irni, Arni.*

Shrub 1.5 - 3 m tall, stem ashy-grey, branches pubescent. Leaves opposite, ovate to ovate-rhomboid, 3 - 6 x 1 - 4 cm, subentire to crenate, serate acute or shortly acuminate at apex, acute or subcordate at base; petiole up to 2.5 cm long, densely hirsute. Flowers in dichotomously branched, axillary, forming a rounded terminal panicuate cymes, creamy-white or pale yellowish, ca 1.5 cm across; pedicels 5 - 10 mm long, densely hirsute; bracts ovate lanceolate, foliaceous. Calyx campanulate, glabrous, pale or somewhat yellowish green, somewhat inflated, 5-lobed; lobes 4 - 10 mm long, ovate-triangulate, Corolla-tube 2 - 2.5 cm long, much narrower than the calyx, pubescent externally; lobes 5, subequal, ovate-elliptic, 7-8 mm long, obtuse. Drupe broadly obovoid, 8 - 10 mm long, black, wrinkled, usually 4 - lobed, enclosed by the persistent calyx. Seeds oblong, white.

Field note: Common among hedges and along boundaries of cultivated fields.

Fl. & Fr.: August - March

Specimen examined: Pansal 1616, Atun 2191 (MLVGCB Herb).

Distribution: Pakistan, India, Sri Lanka and Burma.

Uses: The root is given as a demulcent in gonorrhoea, and decoction of the plant is considered as an alterative. It cures stomach troubles and swellings in cattle. (Watt, Dict. Econ. Prod. Ind. 2: 374. 1889).

GMELINA L.

Gmelina arborea Roxb., Pl. Corom. 3:41. 1819; Brandis, For. Fl. 364. 1874; Wight, Ic. 4 (3): 12. t. 1470. 1849; Clarke, in Hook. *f.* Fl. Brit. Ind. 4: 581. 1884; Duthie, Fl. Gangetic Plain 2: 220. 1911; Pandey, in Shetty & Singh Fl. Rajasthan 2: 679. 1991. *Prema tomentosa* Miq. *apud*. Clarke, in Hook. *f. l.c.* 4: 581. 1885. (Plate 262)

Vern.: *Havanper, Gumhar.*

Unarmed tree, up to 15 m tall, with yellowish-tomentose young shoots. Leaves ovate, long acuminate, 6 - 20 x 5 - 15 cm, entire, tomentose when young, glabrous with age, hairy on the nerves beneath; petiole 5 - 12 cm long, glabrous, glandular near the lamina. Flowers in axillary or terminal panicule arranged in the axils of linear-lanceolate bracts, brownish-yellow. Calyx broadly campanulate, ca 5 mm long, 5-toothed, fulvous outside. Corolla-tube densely pubescent outside, obliquely funnel-shaped; upper lip 2-lobed; lower lip 3-lobed; middle lobe much longer and broader. Drupe obovoid or ovoid-pyriform, 2 - 2.5 cm long, orange-yellow, glabrous.

Field note: A rare plant of the area. It is also cultivated as an ornamental in gardens and avenues.

Fl. & Fr.: March - August.

Specimen examined: Bhilwara 1735, Triveni 1735 (MLVGCB Herb).

Distribution: S.E. Asia, India, Nepal, Bangla Desh, Srilanka, Pakistan and Trop. Africa.

Uses: Roots, bark and seeds sometimes used medicinally, timber is used for ornamental works.

LANTANA L.

KEY TO THE SPECIES

1a. Infloresence, short, cylindrical to spicate. Flowers usually white or pinkish. Bracts ovate-acuminate .... *L. wightiana*

1b. Inflorescence umbellate. Flowers usually scarlet, yellow or mixed coloured. Bracts linear.................................... *L. camara*

Lantana camara L., Sp. Pl. 627. 1753; Clarke, in Hook. *f.* Fl. Brit. Ind. 4: 562. 1882; Duthie, Fl. Gangetic Plain 2: 216. 1911; Pandey, in Shetty & Singh Fl. Rajasthan 2: 201. 1991. *L.camara* L. var. *aculeata* (L.) Moldenke in Torreya 34: 9. 1934. (Fig. 123)

Vern.: *Jharmari.*

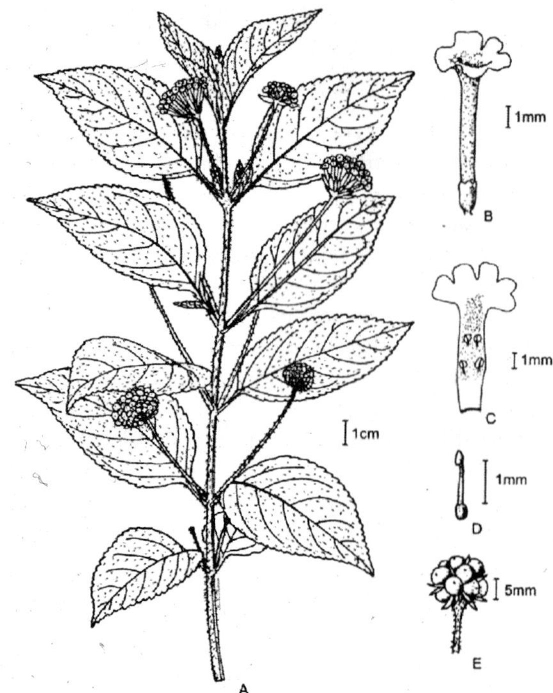

Fig. 123. *Lantana camara* L.
A. Flowering & fruiting twig. B. Flower
C. Opened corolla with stamens, D. Pistil,
E. Fruit

Evergreen shrub with rambling or straggling branches, 1 - 2 m, tall; branches usually minutely or inconspicuously pubescent, unarmed to conspicuously prickly with hooked spines. Leaves opposite, decussate, ovate, 5 - 10 x 2 - 6 cm, rounded or cordate at base, crenate-serrate, acute to shortly acuminate, subrugose, scabrid; petiole 5 - 10 mm long. Flowers in pedunculate, capitate spikes, shorter to exceeding the subtending leaves, 2 - 3 cm across, 5 - 8 mm across, mostly orange or yellow, turning to red or scarlet later. Bracts lanceolate to linear, 5 - 7 mm broad, acute to subulate, rarely a few larger ones also present. Calyx 2-3 mm long, thin, pubescent. Corolla-tube 5 - 11 mm long, pubescent, slightly enlarged and curved above the middle; limb 4 lobed with spreading, subrounded lobes. Drupe 3-5 mm in diam., globose, fleshy, black, shining, 2-seeded.

Field note: A favourite ornamental of gardens with highly variable flower colours, stature, indumentum and prickliness. A number of varieties, of indefinite consatancy in flower colour have been recognized by some authors, some of these, obeserved in the the area, are as follows:

1. var. *camara* : flowers orange-yellow, turning red or scarlet.
1. var. *flava* (Medic.) Moldenke: flowers yellow.
2. var. *rubella* Moldenke: flowers pink.
3. var. *sanguinea* (Medic.) L.H. Bailey: Flowers open saffron yellow but chang to bright red later.
4. var. *aculeata* (Linn.) Moldenke: Plants with conspicuous prickles.
5. var. *alba* Moldenke: Flowers white.

A hybrid, x *Lantana callowiana* Monrov., very similar to this species, has also been recorded from National highway near Bhilwara.

Fl. & Fr.: Throughout the year.

Specimen examined: Jawaja 4082, Janhajpur 1992 (MLVGCB Herb).

Distribution: A native of tropical America, widely introduced and naturalized in many tropical and subtropical regions.

Uses: Decoction is given in tetanus rheumatism and malaria. Essential oil contains camerene, isocamerane and micranene.

Lantana wightiana Wall. ex Gamble, Fl. Madras 2 : 761. 1087. 1924; Matthew, Illus. Fl. Tamil Nadu Carnatic t. 551. 1982; Pandey, in Shetty & Singh Fl. Rajasthan 2: 201. 1991. *L. indica* Roxb. (Hort. Beng. 46. 1814, nomen) Fl. Ind. (Ed. Carey) 3 : 89. 1832; Bhandari, Fl. Indian Desert 279. 1995, var. *albiflora* Wt. ex Fl. Ind. 4:89. 1832; Clarke, in Hook. *f.* Fl. Brit. Ind. 4: 562. 1884; Pandey *et al.*, in Bot. Surv. India 24 : 137. 1982. (Plate 263)

An erect, branched shrub, 1-2.5 m tall, roughly hairy. Leaves ovate or lanceolate, shortly petiolate, or subsessile, 3 - 10 x 2 - 6 cm, crenate-serrate, acute, often rugose, scabrid above, softly and densely pubescent below with minute glands hidden within the pubescence. Flowers in ovoid, spicate heads, sessile, 3-5 mm across, whitish, or pinkish, often with yellowish tube; peduncles 3 - 5 cm long. Bracts 5 - 8 x 4 - 6 mm, ovate or elliptic, acuminate, softly pubescent, outer one involucrate, inner ones smaller. Calyx 6 - 8 mm long, pubescent outside. Corolla ca 8 mm long, pubescent outside, pink or pink-purple to dull purple; limb ca 6 mm wide; tube yellowish. Drupe 3 - 8 mm across, globose, glabrous, purplish with a 2-celled stone, 1 seeded; fruiting calyx thin.

Field note: Common in rocky places and in waste lands.

Fl. & Fr.: September - December.

Specimen examined: Nikumbh 4031 (MLVGCB Herb).

Distribution: India and Pakistan.

Uses: Leaves are used in snakebite.

PHYLA Lour.

Phyla nodiflora (L.) Greene, in Pittonia. 4: 46. 1899; Meikle in Hutch. & Dalz., Fl. W. Trop. Afr. ed. 2. 2: 435. 1963; Pandey, in Shetty & Singh Fl. Rajasthan 2: 681. 1991; Bhandari, Fl. Indian Desert 279. 1995. *Verbena nodiflora* L. Sp. Pl. 20. 1753. *Lippia nodiflora* (L.) A. Rich. in Michaux, Fl. Bor. - Amer. 2: 15. 1803; Wight, Ic. 4 (3): 11. t. 1463. 1849; Clarke, in Hook. *f.* Fl. Brit. Ind. 4: 563. 1885; Duthie, Fl. Gangetic Plain 2: 217. 1911. (Fig. 124 & Plate 264)

Vern.: *Jal-nim.*

Perennial, prostrate herb with somewhat woody rootstock, rooting at nodes, appressedly pubescent to glabrescent. Leaves obovate-spathulate, somewhat fleshy, 1 - 2.5 x 0.5 - 1 cm, serrate above the middle, entire below, glabrous to appressedly pubescent, subsessile to sessile, obtuse, rarely subacute. Flowers in condensed or elongated pedunculate spikes, very small, white, rarely pinkish, ca 3 mm long; bracts ca 2 mm long, broadly ovate, concave, apiculate, mucronate or acuminate, imbricate, sparsely ciliate. Calyx flattened, shorter than bracts, cupular; lobes 2, lanceolate, acute. Corolla slightly exceeding the bracts, 2 - 2.5 mm across; pinkish white; tube ca 1.5 mm long. Drupes globose, ca 1-1.6 mm long, subcompressed, enclosed by the persistent calyx, separating at maturity into two.

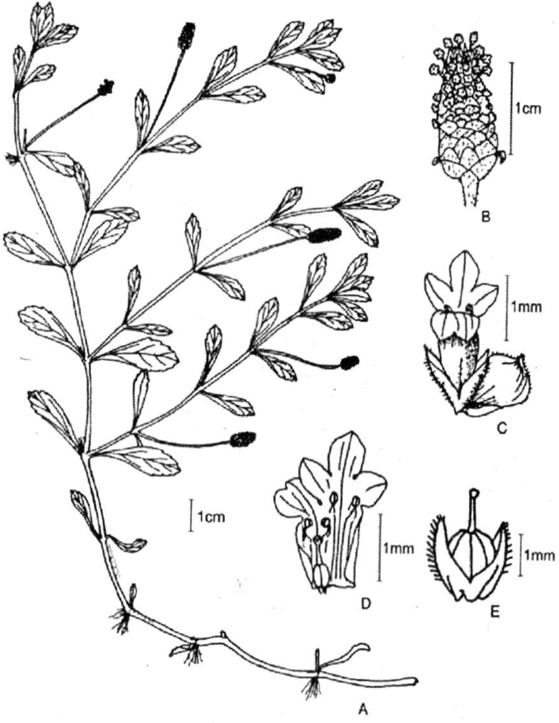

Fig. 124. *Phyla nodiflora* (L.) E. Greene, A. Habit, B. Inflorescence, C. Flower with subtending bracts, D. Flower showing pistil & stamens, E. Capsule

Field note: Common in moist waste lands throughout the area, often in gregarious patches.

Fl. & Fr.: Throughout the year.

Specimen examined: Pansal 1787 (MLVGCB Herb).

Distribution: Throughout tropical and subtropical regions.

Uses: Leaves and young shoots are sometimes used in curing indigestion in children; its decoction is considered as cooling agent and used as a demulcent in cases of venereal diseases.

TECTONA L.

Tectona grandis L. f., Suppl. 151. 1781; Beddome, Fl. Sylv. S. India t. 250. 1872; Clarke, in Hook. *f.* Fl. Brit. Ind. 4: 570. 1884; Duthie, Fl. Gangetic Plain 2: 220. 1911; Pandey, in Shetty & Singh Fl. Rajasthan 2: 682. 1991. (Fig. 125 & Plate 265)

Vern.: *Teak, Sagwan.*

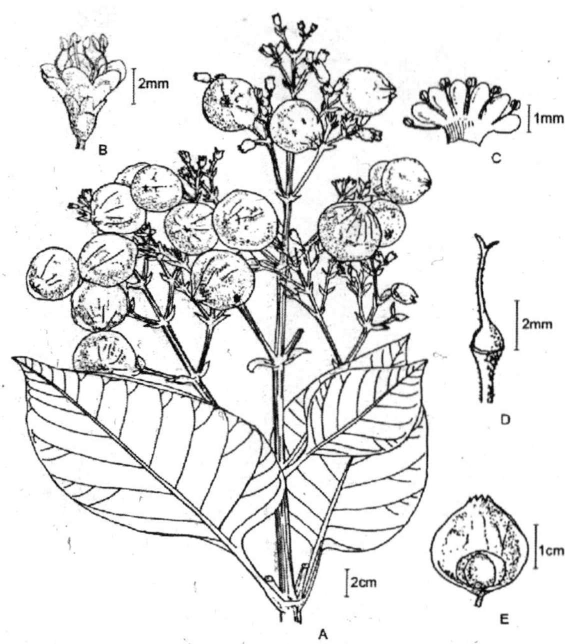

Fig. 125. *Tectona grandis* L.
A. Flowering & fruiting twig. B. Flower, C. Corolla with stamens, D. Pistil, E. Fruit

Deciduous tree, up to 20 m tall, younger parts densely pubescent with stellate hairs and branchlets stout, 4-angled. Leaves opposite, ovate-elliptic, 15 - 50 x 6 - 30 cm, stout, entire to repand-dentate, acute or shortly acuminate, glabrescent, shining-green above, paler and tomentose beneath. Flowers in dichotomously branched, 20 - 40 cm long, tomentose panicles, ca 6 mm across, white, rarely pinkish. Calyx light green, 3 - 4.5 mm long, campanulate, 5 - 7-toothed, bladder-like and enlarged up to 2.5 cm in fruit, brittle and dry. Corolla-tube 1.5 - 3 mm long; lobes 5 - 7, obovate-elliptic, 2.5 - 3 mm long, straight or reflexed; tube broadly cylindrical, Stamens 2.5-4 mm long, exserted. Drupes subglobose to tetragonally flattened, ca 1.5 cm in diam., tomentose, usually slightly 4-lobed at the apex.

Field note: Introduced and cultivated in some parts of the area. It is very sensitive to frost when young.

Fl. & Fr.: August - November.

Specimen examined: Asawarimata 4049, Mandal 197 (MLVGCB Herb).

Distribution: Native of tropical southern Asia and Malaysia; widely cultivated for its highly valuable timber.

VITEX L.

Vitex negundo L., Sp. pl. 638. 1753; Wight, Ic. 4 (3): 1. t. 519. 1842; Clarke, in Hook. *f.* Fl. Brit. Ind. 4: 583. 1885; Duthie, Fl. Gangetic Plain 2: 224. 1911; Pandey, in Shetty & Singh Fl. Rajasthan 2: 684. 1991. (Plate 266)

Vern.: *Nirgud, Nirgundi.*

Shrubs usually 1-2 m tall. Leaves opposite-decussate, 3 - 5-foliolate; leaflets oblanceolate, 5 - 12 x 1.5 - 4 cm, middle one largest, entire to irregularly denticulate, chartaceous, glabrous above, grey pubesecnt beneath, subsessile to petiolulate; petiole 3-6 cm long;. Flowers in and upper axillary, ca 20 cm long panicles, small, 3-5 mm across, usually blue or violet, subsessile to shortly pedicelled (pedicels up to 1 mm long). Calyx ca 2 mm long, increasing up to 3 mm in fruit, white-spotted, ca 3 mm long, persistent, campanulate, 5-toothed. Corolla purple to violet; tube as long as the calyx; limb slightly 2-lipped, with 5, unequal lobes, densely ciliate, up to 2 mm long, largest one obovate-orbicular, undulate or crenulate, the others oblong, smaller. Stamens 4, didynamous, protruding, anther cells divaricating later. Drupe globose, ca 5 mm in diam., usually 4-celled, with 1 seed in each cell, black.

Field note: It is often planted along water channels to check soil erosion, and readily grows from cuttings.

Fl. & Fr.: Almost throughout the year.

Specimen examined: Ladpura 1981, Mezadam 1993 (MLVGCB Herb).

Distribution: India, Pakistan, W. Asia and N. Africa; introduced and widely cultivated as an ornamental elsewhere.

Uses: Its leaves are sometimes used for curing inflammatory swellings of joints, headache etc.

LAMIACEAE

KEY TO THE GENERA

1a. Perfect stamens 2 .... *Salvia*

1a. Perfect stamens 4 2

2a. Leaves deeply divided into linear segments... *Lavandula*

2b. Leaves entire or serrate................................ 3

3a. Posterior sepal widest, broad .... *Ocimum*

3b. Posterior sepal not as above 4

4a. Flowers in compact dense heads...................... Stamens declinate......................... *Anisochilus*

4a. Flowers not as above. Stamens erect, spreading or ascending.................................. 5

5a. Corolla 4-lobed .... *Pogostemon*

5b. Corolla 4-lobed, 2-lipped............................... 6

6a. Flowers blue or violet, upper lip of corolla glabrous or pubescent.................... *Anisomeles*

6b. Flowers white or orange-red, upper lip of corolla densely villous.................................... 7

7a. Flowers white, lowerlip of corolla longer than the upper...................................... *Leucas*

7a. Flowers oronge-red, lowerlip of corolla shorter than the upper....................... *Leonotis*

ANISOCHILUS Wall. ex Benth

Anisochilus carnosus (L. f.) Wall ex Benth. in Wall. Pl. As. Rar. 2 : 18. 1831; Hook. *f.* Fl. Brit. Ind. 4: 627. 1885; Duthie, Fl. Gangetic Plain 2: 240. 1911; Parmar, in Shetty & Singh Fl. Rajasthan 2: 687. 1991. *Lavandula carnosa* L. f. Suppl. Pl. 273. 1781.

var.: ***carnosus***

Vern.: *Panjiri*.

Erect, annual herb. Stem erect, 30 - 60 cm tall, 4-angled, fleshy, dull pinkish-red, branched, villous, subglabrous at apex. Leaves broadly ovate, 2 - 6.5 × 1 - 6 cm, subacute or obtuse at apex, rounded or subcordate at base, crenate-serrate, glabrous above, densely scabrid beneath, rather fleshy; petioles 0.6 - 6.6 cm long, puberulous. Panicles diffusely branched; Spikes 1.5 - 5 cm, long, subglobose to oblong-ovoid; bracts broadly ovate, truncate at base. Calyx tubular; upper lip deltoid-lanceolate, upto 2.5 x 2 mm, acute; lower lip slightly truncate; tube up to 2 mm long. Corolla pinkish-purple, ca 1.5 cm long, densely pubescent outside; tube slender, exserted, recurved at middle, dilated at throat; upper lip 4-lobed; lower lip broadly ovate, up to 3 mm long; lobes shollowly orbicular; tube 4 - 9 mm long. Anterior stamens slightly exserted. Nutlets suborbicular, compressed, smooth, brown.

Field note: Common in the forests and waste lands.

Fl. & Fr.: September - January.

Specimen examined: Bijoliya 1786 (MLVGCB Herb).

Distribution: China, India, Myanmar and Sri Lanka.

ANISOMELES R. Br.

Anisomeles indica (L.) O. Kuntze, Rev. Gen. Pl. 2: 512. 1891; Raizada, Suppl. Duthie, Fl. Gangetic Plain 217. 1976; Parmar, in Shetty & Singh Fl. Rajasthan 2: 688. 1991. *Nepeta indica* L., Sp. Pl. 2: 571. 1753. *Anisomeles ovata* R. Br. in Ait. Hort. Kew. ed. 2. 2: 364. 1811; Wight, Ic. 3 (2): 5. t. 865. 1843-45; Hook. *f.* Fl. Brit. Ind. 4: 672. 1885; Duthie, Fl. Gangetic Plain 2: 245. 1911. *Ballota disticha* L., Mant. Pl. 83. 1767. (Fig. 126 & Plate 267)

Suffruticose herb, upto 1.5 m high, sparingly hairy or densely pubescent. Stem erect, 4-angled, 1 - 1.5 m tall, branched, densely appressed white pubescent. Leaves broadly ovate, 3 - 10 × 2.5 - 8 cm, base broadly truncate-rounded or cordate at base, rarely cuneate slightly thick, margin irregularly dentate, apex acute or short acuminate, softly pubescent. Petiole 1 - 4.5 cm, densely hairy. Verticillasters axillary, interrupted below, dense and spicate above; bracts 3 - 5 mm long, linear, pubescent. Calyx *ca.* 6 mm, hirsute, glandular pubescent, yellow glandular; teeth purple-red, lanceolate, ca 2.7 mm, acute, ciliate. Corolla bluish purple, ca 1.3 cm, glabrous outside; tube funnelform, mouth to 3.5 mm wide; upper lip oblong, 4.5 - 5 × 3 mm, margin entire; lower lip subhorizontally spreading, ca 9 × 5 mm; middle lobe obcordate, ca 3 × 4.5 mm, slightly undulate, bearded; lateral lobes ovate. Ovary glabrous. Nutlets ca 3 mm long, broadly ovoid, with rounded ends, subcompressed, polished, black.

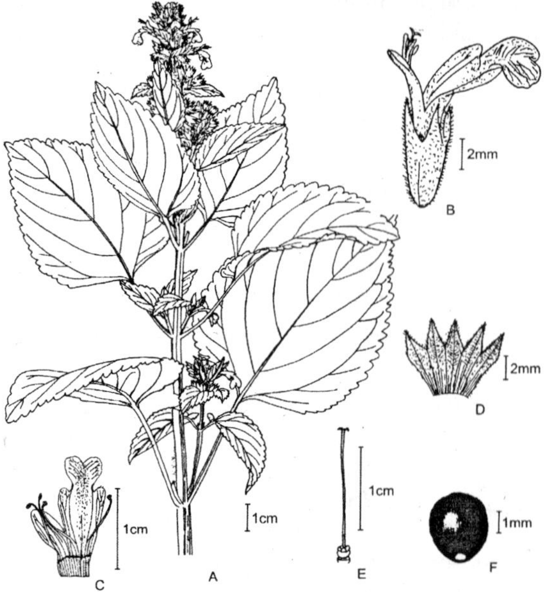

Fig. 126. *Anisomeles indica* (L.) O. Kuntze, A. Flowering twig. B. Flower, C. Opened corolla with stamens, D, Calyx, E. Pistil, F. Nut

Field note: Fairly common in waste lands, open forests and neglected corners of gardens.

Fl. & Fr.: August - December.

Specimen examined: Mezadam 405, Harigarh 1870 (MLVGCB Herb).

Distribution: Cambodia, China, India, Laos, Malaysia, Myanmar, Philippines, Thailand and Vietnam.

Uses: Plants are used medicinally for rheumatism, cold, fever, abdominal pain, skin sores, and snake bites.

LAVANDULA L.

Lavandula bipinnata (Roth) O. Kuntze var. ***rothiana*** O. Kuntze, Rev. Gen. Pl. 2 : 521. 1891; Bhandari, Fl. Indian Desert 280. 1995; Parmar, in Shetty & Singh Fl. Rajasthan 2: 691. 1991. *L. multifida* Burm. F. Fl. Ind. 126. 1768, excl. t. 38. f. 1., non L. 1753. *L. burmanni* Benth. Lab. Gen. *et* Sp. 151. 1833; Wight, Ic. 4 (3): 8. t. 1438. 1849; Hook. *f.* Fl. Brit. Ind. 4: 631. 1885; Duthie, Fl. Gangetic Plain 2: 240. 1911. (Fig. 127 & Plate 268)

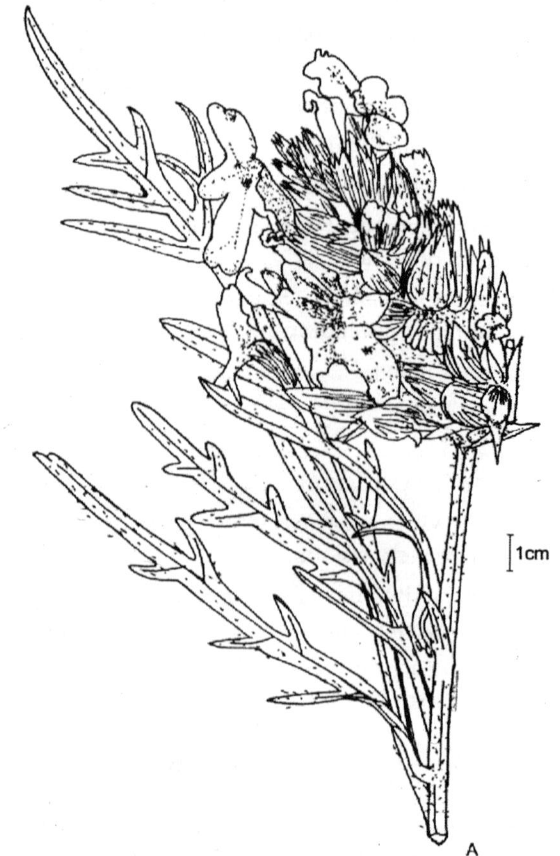

Fig. 127. *Lavandula bipinnata* (Roth) O. Kuntze. A. Flowering & fruiting twig

An erect slender, annual herb, upto 1 m high. Stem simple or branched, quadrangular, pubescent. Leaves sessile or nearly so, 2 - 10 cm long, pinnatisect; segments linear or linear - lanceolate, entire or dentate, pale beneath. Flowers in simple or branched axillary, subumbellate spikes, blue.

Bracts 6 - 8 mm long, longer than calyx, broadly ovate at base, glandular pubescent, produced into a long capillary awn at the apex, persistent. Calyx tubular, 4 - 6 mm long, slightly curved; finely hoary pubescent, glandular; teeth equal, lanceolate, acute with penicillate tips. Corolla bilabiate, ca 12 mm long; tube slender below, slightly dialated above, hairy outside. Stamens declinate; anther cells confluent. Nutlets minute, oblong-ellipsoid, smooth, brown or black with large white areola on the dorsal side.

Field note: Not common, found in hilly slopes and along hedges of crop fields.

Fl. & Fr.: October - February.

Specimen examined: Bansi 4014 (MLVGCB Herb).

Distribution: Chiefly Mediterranean species.

LEONOTIS (Pers.) R. Br.

Leonotis nepetifolia (L.) R. Br. in Ait. Hort. Kew ed. 2 (3) : 409. 1811; *Phlomis nepetifolia* L., Sp. Pl. 586. 1753; Hook. *f.* Fl. Brit. Ind. 4: 631. 1885; Duthie, Fl. Gangetic Plain 2: 253. 1911. (Fig. 128 & Plate 269)

Vern.: *Baraguma.*

Erect herb, up to 2 m high. Stem stout, 4-angled, puberulous. Leaves opposite, upto 20 cm long, decussate, broadly ovate, cuneate at base, margins deeply serrate, apex acute, pubescent, hairs simple; petiole 45 - 80 mm long,. Flowers in axillary, globose, 3 - 7 cm heads; floral leaves deflexed, lanceolate; pedicel 0.5 - 2 mm long; bracts linear, deflexed, spine-tipped. Perianth 2 - whorled. Calyx 2 - 3 cm long; teeth unequal, spine-tipped; upper one ovate, ca 8 mm long, acuminate; lower one deltoid, smaller; mouth oblique; throat glabrous; tube incurved, ribbed, puberulous below, densely hispid above with long hairs. Corolla 2.5 - 3.0 cm long, orange-red, glabrous below, densely villous above as well as on the long upper lip; tube hairy inside, with 3, parallel rings of white hairs. Stamens 4 (in 2 unequal pairs), adnate to the perianth (inserted at mouth of corolla tube), all alternating with the corolla parts, free of each other. Anthers dehiscing via longitudinal slits. Ovary syncarpous, superior, 4 -celled. Ovules 1 per cell. Style 1, simple; stigma 2-lobed. Nutlets linear oblong, ca 4 mm long, obliquely truncate.

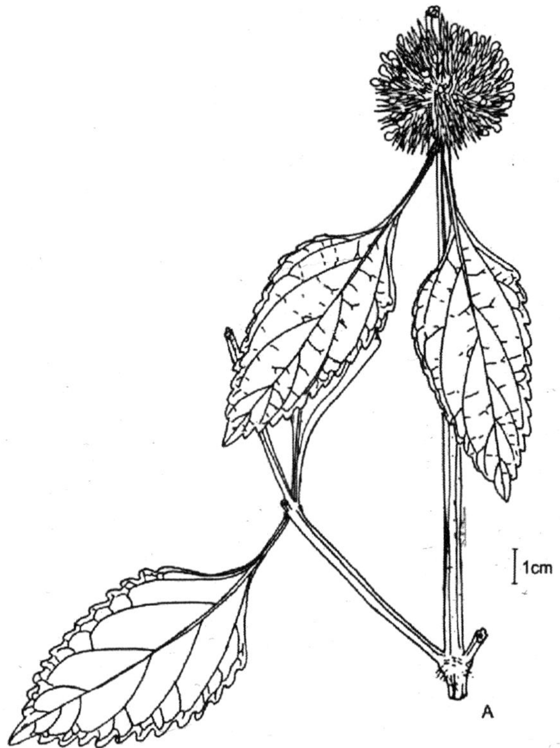

Fig. 128. *Leonotis nepetifolia* (L.) R. Br.
A. Flowering & fruiting branch

Field note: Common in grasslands along road sides.

Fl. & Fr.: September - December.

Specimen examined: Bijoliya 198(MLVGCB Herb).

Distribution: Native of South America, but is a pantropical weed.

LEUCAS R. Br.

KEY TO THE SPECIES

1a. Throat of calyx glabrous or somewhat pilose with very short hairs 2

2a. Stem with short spreading hairs. Bracts ca 1/2 length of calyces .... *L. aspera*

2b. Stem with very short subretrorse hairs. Bracts more than 1/2 length of calyces.. *L. urticifolia*

1b. Throat of calyx prominently hairy with long hairs... 3

3a. Calyces prominently curved, some slightly reflexed. Calyx teeth upto 2.5 mm. Bracts not imbricate .... *L. nutans*

3b. Calyces slightly curved, not reflexed. Calyx teeth upto 1 mm. Bracts imbricate, numerous..................................... *L. cephalotes*

Leucas aspera (Willd.) Link, Enum. Hort. Berol. Ait. 2: 113. 1822; Benth. in DC., Prodr. 12: 532. 1848; Hook. *f.* Fl. Brit. Ind. 4: 690. 1885; Duthie, Fl. Gangetic Plain 2: 250. 1911; Mukerjee in Rec. Bot. Surv. Ind. 14, 1: 166. 1940; Parmar, in Shetty & Singh Fl. Rajasthan 2: 694. 1991; Bhandari, Fl. Indian Desert 280. 1995. *Phlomis aspera* Willd. Enum. Pl. Hort. Berol. 2: 621. 1809. (Fig. 129 & Plate 270)

Vern.: *Gotta.*

Densely hairy, erect, annual, usually with many branches. Stem 15 - 60 cm with spreading short hairs. Leaves linear to linear lanceolate, crenate, obtuse at apex, narrowed at base, up to 1 - 9 x 0.5 - 1.5 cm with adpressed hairs on upper surface and a denser indumentum of short spreading hairs below especially on nerves; petiolate or subsessile Verticillasters terminal or axillary, very dense, ca 16 - 20 flowered, 1 - 4, ca 2.5 cm in diam.. Bracts linear, subspinulose, almost equalling the calyx, long-ciliate at margins. Calyx 7 - 12 mm, elongating in fruit, pale green, curved, usually smooth and glabrous below, ribbed and scabrid or ribbed above; teeth short, deltoid, spinulose, upper ones largest; mouth small, very oblique, produced on the upper side. Corolla 8 - 10 mm, white; upper lip short, densely bearded; lower lip clearly longer than upper and projecting forward. Nutlets trigonous, ca 2.5 x 1.5 mm, smooth, brown.

Field note: A common weed of cultivated areas and open grasslands throughout.

Fl. & Fr.: September - December.

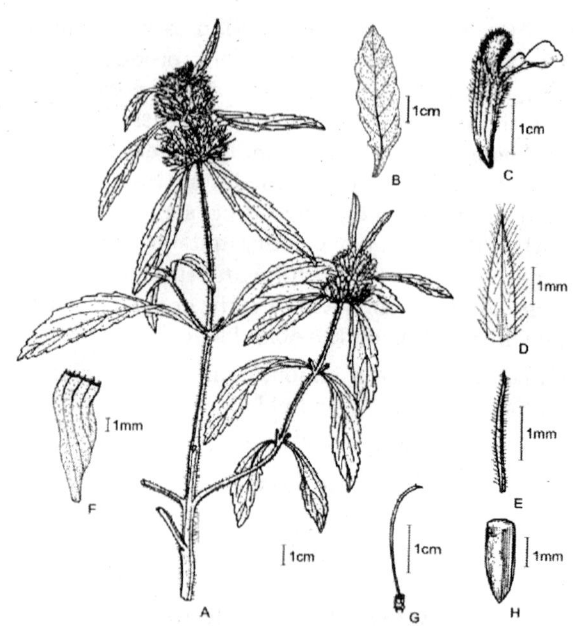

Fig. 129. *Leucas aspera* (Willd.) Link. A. Habit. B. Leaf. C. Flower, D. Bract, E. Bractiole, F. Calyx, G. Pistil, H. Nutlets

Specimen examined: Atun 529, Dungala 4061 (MLVGCB Herb).

Distribution: Indo-China and Malaysia.

Uses: As a culinary herb.

Leucas cephalotes (Koen. ex Roth) Spreng., Syst. Veg. 2 : 743. 1825; Benth. in DC., Prodr. 12: 532. 1848;. Stewart in Journ. Agr. Hort. Soc. India 1, 1: 99. 1867; Hook. *f.,* Fl. Brit. Ind. 4: 689. 1885; Duthie, Fl. Gangetic Plain 2: 251. 1911; Mukerjee in Rec. Bot. Surv. Ind. 14, 1: 168. 1940; Wealth of India, Raw Materials 6: 79. 1962; Parmar, in Shetty & Singh Fl. Rajasthan 2: 695. 1991; Bhandari, Fl. Indian Desert 280. 1995. *Phlomis cephalotes* Koen. ex Roth, Nov. Pl. Sp. 262. 1821. *Leucas capitata* Desf. in Mem Mus. Hist. Nat. Paris 11 : 8. t. 4. 1824.

Vern.: *Gotta*

Erect, annual, stout, herb, unbranched or with few branches, 10 - 60 cm high, leafy, pubescent with spreading and adpressed retrorse hairs. Leaves narrowly ovate or ovate lanceolate, cuneate, crenate-serrate, up to 8 x 2 cm, with long and short eglandular hairs below; shortly petiolate. Verticillasters usually terminal, densely up to 50-flowered. Bracts numerous, imbricate, narrow, lanceolate, acute, covering the floral heads, ciliate on margins. Calyx up to 2 cm long, tubular, slightly curved, clearly nerved, with a slightly oblique mouth, distinctly hairy at throat, with eglandular and sometimes also with very short glandular hairs; teeth 10, subequal, shortly triangular, mucronulate, ca 1 mm. Corolla ca 2.5 mm long, upper lip bearded; lower lip slightly longer than upper. Nutlets narrow oblong, bluntly trigonous, ca 3 x 2 mm, brown. smooth.

Field note: Common weed of cultivated areas and open grasslands throughout.

Fl. & Fr.: July-October.

Specimen examined: Mezadam 1596, Ladpura 1881 (MLVGCB Herb).

Distribution: E. Afghanistan, Pakistan and India.

Leucas nutans (Roth) Spreng., Syst. Veg. 2 : 743. 1825; Benth. in DC., Prodr. 12: 532. 1848; Hook. *f.,* Fl. Brit. Ind. 4: 688. 1885; Duthie, Fl. Gangetic Plain 2: 251. 1911; Mukerjee in Rec. Bot. Surv. Ind. 14, 1: 169. 1940; Parmar, in Shetty & Singh Fl. Rajasthan 2: 698. 1991; Bhandari, Fl. Indian Desert 281. 1995. *Phlomis nutans* Roth, Nov. Pl. Sp. 263. 1821.

Erect, annual herb, 4 - 12 cm high, unbranched or with few lateral branches, often straggling, diffuse, laxy hairs. Leaves oblong-or oblong-lanceolate, closely serrate or subserrate, pubescent on both surfaces, usually with long hairs on nerves beneath, 4 x 1 cm; petiole on lower leaves up to 1 cm, sessile above. Verticillasters terminal or subterminal many flowered. Bracts lanceolate, foliaceous, acuminate, equal to or somewhat shorter than calyx, ciliate on margins. Calyx ca 10 mm, membranous, tubular, prominently curved in the middle, almost reflexed in fruit, hairy in throat, with a somewhat oblique mouth, gland-dotted and with spreading eglandular hairs; teeth 10, unequal, lanceolate, acuminate, longest teeth up to 2.5 mm. Corolla white 10-12 mm; upper lip short bearded; lower lip longer than upper; tube slender annulate near base. Nutlets narrow oblong-ellipsoid, scarcely trigonous, 2.5 x 1 mm, smooth.

Field note: Rare species of cultivated fields. Recognized by the spreading hairs on the stem and the prominently curved or reflexed calyces which are hairy in the throat. Usually a species of the plains.

Fl. & Fr.: August-October.

Specimen examined: Kotri 1221 (MLVGCB Herb).

Distribution: Pakistan, India and Burma.

Leucas urticifolia (Vahl) R. Br., Prodr. 504. 1810; Benth. in DC., Prodr. 12: 524. 1848; Hook. *f.,* Fl. Brit. Ind. 4: 680. 1885; Duthie, Fl. Gangetic Plain 2: 249. 1911; Mukerjee in Rec. Bot. Surv. Ind. 14. 1: 166. 1940; Parmar, in Shetty & Singh Fl. Rajasthan 2: 698-99. 1991; Bhandari, Fl. Indian Desert 282. 1995. *Phlomis urticifolia* Vahl, Symb. Bot. 3: 76. 1794. (Fig. 130)

Erect, annual herb. Stem simple or branched, 10-50 cm high, with a tomentose indumentum of very short usually retrorse egla-ndular hairs. Leaves ovate to ovate-lanceolate, coarsely serrate, cuneate at base, acute above, with short adpressed hairs; lamina 3-5 x 1-3 cm; petiole 5-15 mm long. Verticillasters usually axillary, globose, dense, ca 16-40-flowered. Bracts linear to linear-lanceolate, pale green, ca 5-8 x 1-2 mm, ciliate at margins, softly subulate. Calyx 7-12 mm, membranous, elongating in fruit to ca 11 mm, slightly curved, with a clearly oblique mouth, transversely-reticulately nerved and whitish above, pubescent or villous out side, sparsely hairy inside; teeth 8-10, narrow triangular, spinulose. Corolla 7-9 mm, white; upper lip short, entire, white-bearded, slightly straight; lower lip clearly longer than upper, projecting forward. Nutlets ca 1.5 - 2 x 1 mm, apically finely glandular.

Field note: Common in waste lands and open hilly to moist banks of river, tank and ponds etc.

Fl. & Fr.: July-October.

Specimen examined: Mezadam 1354, Jahazpur 1615 (MLVGCB Herb).

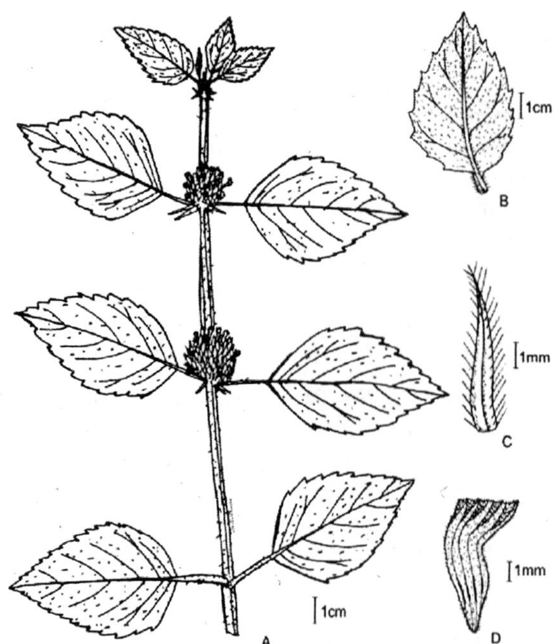

Fig. 130. *Leucas urticifolia* (Vahl) R. Br. A. Habit. B. Leaf. C. Bract. D. Calyx.

Distribution: Pakistan, India, Arabian peninsula and E. Africa.

Notes: According to R. R. Stewart (1972), it is a a weed of the plains. Sebald (l.c.) has recognized three varieties in east Africa: the type variety (to which present specimen belongs) and two other localised endemic varieties.

OCIMUM L.

Ocimum canum Sims. in Curtis, Bot. Mag. 51: t. 2452. 1824; Hook. *f.* Fl. Brit. Ind. 4: 607. 1885; Duthie, Fl. Gangetic Plain 2: 234. 1911; Parmar, in Shetty & Singh Fl. Rajasthan 2: 701. 1991. *Ocimum americanum* L., Sp. Pl. 833. 1753 et in Cent. Pl 1: 15. 1755; Santapau in Rec. Bot. Surv. Ind. 16 (1): 242. 1953.

Vern.: *Mar tulsi*

Annual herb, 20 - 60 cm tall. Stem erect, some what 4-angled, retrorse puberulent, nodes bristly, much branched from stem base. Leaves elliptic-lanceolate 1.7 - 4.2 cm long, narrowed at ends, puberulent, serrulate-subentire or faintly dentate, ciliate, apex acute, lateral veins 4- or 5-paired. Petiole 1 - 2.5 cm long, puberulent and pilose; Verticillasters densely flowered, in 7 - 20 cm long, spiciform racems. Flowers subsessile; pedicels short, erect, often closely appressed to the rachis; bracts awned, cilioate with long white hairs, smaller than the calyx. Calyx campanulate, ca 3 × 1 mm, villous inside, slightly pubescent out side; upper calyx tooth broad and rounded, awned, ciliate; other 4 teeth lanceolate, subulate. Corolla straight, 5 - 6 mm long, limb sparsely puberulent, glabrous inside; tube conspicuously exserted, obconical; lower lip oblong, concave, margin entire. Stamens free, ca 2 × as long as corolla. Nutlets black, narrowly ellipsoid, ca 1.5 × 0.6 mm, punctulate.

Field note: Common weed of open waste lands.

Fl. & Fr.: Almost throughout the year.

Specimen examined: Bhilwara 121 (MLVGCB Herb).

Distribution: India, China, Indonesia, Malaysia, Myanmar, Philippines, Sri Lanka, Africa and SW Asia

Uses: Leaves are made into a paste which is used for skin diseases.

POGOSTEMON Desf.

Pogostemon benghalense (Burm. f.) O. Ktze. Rev. Gen. Pl. 2 : 529. 1891; Parmar, in Shetty & Singh Fl. Rajasthan 2: 694. 1991. *Origanum benghalense* Burm. f. Fl. Ind. 128. t. 38. f. 1768. *Pogostemon pleetranthoides* Desf. in Ann. Mus. Natl. Hist. Nat. Paris 2 : 156. t. 6. 1808; Hook. *f.* Fl. Brit. Ind. 4: 632. 1885; Duthie, Fl. Gangetic Plain 2: 241. 1911.

Herb and undershrub, strongly aromatic, 1 - 2 m high, pilose with simple hairs or glabrous. Branches often dark purple; young parts hoary pubescent villous. Leaves 5 - 15 x 2 - 8 cm, ovate, acute, double serrate, cuneate or acute at base, glabrous or pubescent, somewhat velvety beneath. Verticillasters close, many flowered, in axillary or terminal, stout, villow or hoary, paniculate spikes. Bracts ovate, ca 8 mm long, acute, softly pubescent, imbricating, each bract subtending a fascicle of sessile flowers. Calyx 5-toothed; often with purple tinge in upper part, ca 5 mm long; teeth lanceolate; tube slender, subcylindrical. Corolla bilabiate, rose, purple or white. Stamens 4, exserted. Nutlets smooth, shining brown or black when ripe, ellipsoid, brown.

Field note: Common on hill slopes and as forest undergrowth.

Fl. & Fr.: November - March.

Specimen examined: Mainal 149 (MLVGCB Herb).

Distribution: India.

SALVIA L.

Salvia aegyptiaca L., Sp. Pl. 23. 1753; Hook. *f.* Fl. Brit. Ind. 4: 656. 1885; Duthie, Fl. Gangetic Plain 2: 256. 1911; Mukerjee in Rec. Bot. Surv. Ind. 14 (1) : 107. 1940; Parmar, in Shetty & Singh Fl. Rajasthan 2: 707. 1991; Bhandari, Fl. Indian Desert 283. 1995. (Fig. 131 & Plate 271)

Vern.: *Maur.*

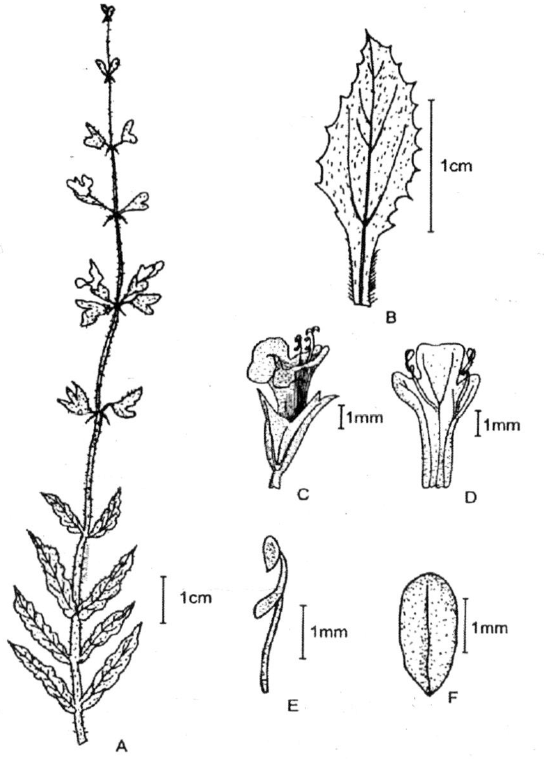

Fig. 131. *Salvia aegyptiaca* L.
A. Flowering & fruiting twig. B. Leaf. C. Flower, D. Opened corolla, E. Stamen, F. Seed

Suffruticose much branched herb, forming small tufts. Stem somewhat leafy, erect-ascending, ca 10 - 25 cm tall, divaricate hairy. Leaves linear-lanceolate, 0.5-3 x 0.2-0.8 cm, crenate, rugulose, hairy on both surfaces. Verticillasters distant, 2 - 6-flowered in erect or ascending racemes. Bracts and bracteoles present. Pedicels ca 2 mm long elongating to ca 3.5 mm in fruit, spreading-ascending. Calyx ovate to tubular campanulate, ca 5 mm in flower and ca 7 mm in fruit, hairy. Corolla violet-blue, pale lavender or white with purple or lilac markings on lips ca 6-8 mm long; upper lip straight or reflexed, much shorter than lower; tube somewhat annulate. Lower thecae fertile. Nutlets smooth, bluish-black, ca 2 x 1 mm, oblong-elipsoid, mucilaginous on wetting.

Field note: Common throughout the area.

Fl. & Fr.: December - March.

Specimen examined: Atun 122, Bhilwara 2149 (MLVGCB Herb).

Distribution: Cosmopolitan.

PLANTAGINACEAE

PLANTAGO L.

KEY TO THE SPECIES

1a. Stem well developed. Leaves cauline, opposite .... *P. exigua*

1b. Stem subcaulescent . Leaves all radical .... *P. ovata*

Plantago exigua Juss. ex Murr., Comm. Soc. Reg. Sci. Gotting. 1: 94. t. 5. f. 1 - 3. 1779; Parmar, in Shetty & Singh Fl. Rajasthan 2: 711. 1991. *P. pumila* L. f. Suppl. 125. 1781; Hook. *f.* Fl. Brit. Ind. 4: 707. 1885; Raizada in Journ. Ind. Bot. Soc. 15: 167. 1736 & Suppl. Duthie, Fl. Gangetic Plain 228. 1976. (Plate 272)

Annual herb with well developed stems. Stems 15 - 45 cm long, many from the base, decumbent to ascendent, usually erect or branched or crowded in the upper parts. Leaves opposite, in whorls of 3 or 4 upwards 2 - 4 cm long, 0.5 - 0.75 mm broad, narrow linear to filiform, entire, near the base sparsely ciliate, hairs long, on the other parts covered with scattered short hairs, somewhat fleshy. Inflorescences many flowered, ovoid, 1 - 4 cm long; peduncles filiform, solitary, terminal or axillary, 2 - 5 cm long. Flowers minute, yellowish white. Bracts in the lower flowers 2 - 2.5 mm long, ovate-membranous, attenuated into 1 cm long narrow, subulate, obtuse apex, upper parts covered with mixed glandular and non glandular long hairs, bracts in the upper flower similar to the lower ones but with smaller produced apex. Sepals 4.5 - 5 mm long, slightly unequal, acuminate with produced keels, sparsely covered with long hairs, posterior concave, broad lanceolate, apices shortly produced, hairs lax. Corolla lobes 1.75 - 2 mm long, ovate-lanceolate, slightly reflexed; tube swollen at base. Capsules ellipsoid or ovoid, 3.5 - 4 mm long, 2-celled. Seeds brown, linear-elliptic, shining, 3 mm long.

Field note: Rare weed of cultivated fields. Usually confused with *Spergula* species in vegetative state.

Fl. & Fr.: January - March.

Specimen examined: Mandalgarh 123 (MLVGCB Herb).

Distribution: From Egypt to south Iran, Afghanistan and Pakistan.

Plantago ovata Forssk., Fl. Egypt - Arab. 31. 1775; Hook. *f.* Fl. Brit. Ind. 4: 707. 1885; Duthie, Fl. Gangetic Plain 2: 262. 1911; Cooke, Fl. Bomb. Pres. 2nd Repr. ed. 2: 562. 1967; Parmar, in Shetty & Singh Fl. Rajasthan 2: 285. 1991; Bhandari, Fl. Indian Desert 285. 1995. *Plantago trichophylla* Nab. in Publ. Fac. Sci. Uni. Masaryk (Brono), 105: 6. 1929.

Vern: *Isabgol*.

Annual subsaculescent herb, up to 5 - 12 cm tall, sparsely or thickly covered with soft hairs. Leaves in rosettes, narrow linear, 2 - 10 x 1 - 1.5 mm, obtuse to acute, entire or distantly denticulate, 3-nerved, densely covered with appressed woolly hairs, attenuated at bases, bases slightly dilated. Scapes ascendent, 1 - 9 cm long, glabrous, or at the upper part pubescent. Flowers in dense capituliform or short cylindrical, 0.5 - 2 cm long thick spikes. Bracts ovate to rotundate, 3 mm long, concave, membranous, upper margins ciliate. Sepals usually glabrous, elliptic obovate, up to 3 mm long. Corolla lobes broad ovate to broad rotundate cordate, glabrous. Capsules ca 8 cm long, ellipsoid, obtuse, membranous, glabrous. Seeds 2, small ovoid-oblong, boat-shaped, smooth, yellowish-brown.

Field note: Introduced and under cultivation for its seeds of medicinal value.

Fl. & Fr.: January - March.

Specimen examined: Karthana 172 (MLVGCB Herb).

Distribution: Mediterranian regions to the desert of Kizil Kum, Afghanistan, Pakistan and India.

Uses: Medicinal in the different pharmacopeias under the name of *Ispaghula*.

NYCTAGINACEAE

KEY TO THE GENERA

1a. Armed climbing shrubs. Bracts conspicuous, large, brightly coloured .... *Bougainvillea*

1b. Herbs. Bracts inconspicuous, small, not bright coloured 2

2a. Flowers more than 2 cm large, showy, perianth tube 2 - 3 cm long. Fruits ovoid .... *Mirabilis*

1b. Flowers small less than 1 cm, in umbels, not showy; perianth tube up to 1 cm long. Fruits various 3

3a. Anthocarp 5-ribbed, glabrous or glandular, without wart-like glands.................. *Boerhavia*

3b. Anthocarp 10-ribbed, with large wart-like glands......................... *Commicarpus*

BOERHAVIA L.

Boerhavia diffusa L., Sp. Pl. 3. 1753; Duthie, Fl. Gangetic Plain 3: 2. 1915; Clarke, in Hook. *f.* Fl. Brit. Ind. 4: 175. 1882; Duthie, Fl. Gangetic Plain 2: 262. 1911; Mahesh. Illus. Fl. Delhi f. 71. 1966; Mukherjee, in J. Econ. Tax. Bot. 5: 582. 1984; Pandey in Shetty &

Singh Fl. Rajasthan 2: 201. 1991; Bhandari, Fl. Indian Desert 97. 1995. *B. repens* L. var *diffusa* (L.) Hook. *f.,* Brit. India 4: 709. 1885. (Fig. 132)

Vern.: *Punarnavwa.*

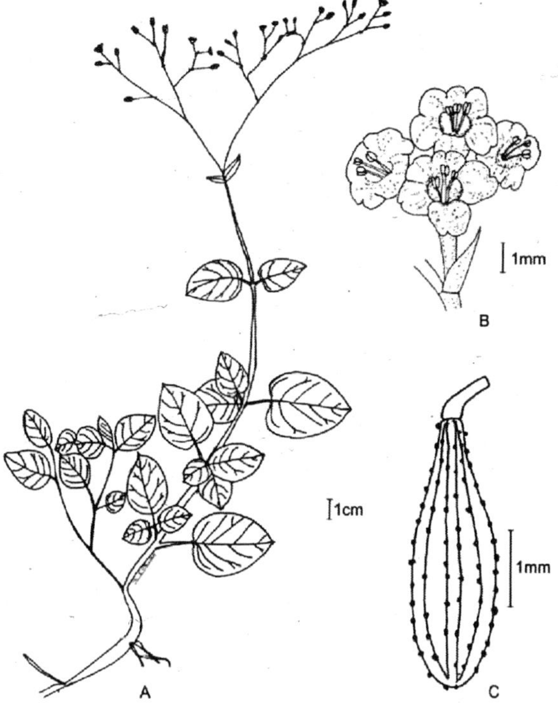

Fig. 132. *Boerhavia diffusa* L.
A. Habit, B. Inflorescence, C. Fruit

Decumbent-ascending or diffuse, straggling herb, perennial. Stem woody trailing, to 1 m; glabrous or sparsely pubescent; branches pinkish. Roots thick, fleshy. leaves ovate-oblong, 1 - 5 × 1 - 4 cm, both surfaces sparsely pubescent, abaxially grey-yellow, wrinkled when dry, base rounded or cuneate, margins undulate, apex obtuse or acute. Petiole 0.4-2 cm. Flowers in terminal, diffuse panicles, much exceeding the foliage, ca 3 mm long, pinkish mauve, sessile; peduncle slender, sparsely pubescent. Bracts small, ovate, acuminate pubescent. Perianth limb bright purple or purple-red, viscid glandular, 1.5 - 3 mm. Stamens 3, slightly exserted or included; filaments pink; anther-lobes yellow; stigma capitate. Anthocarp fusiform, 3 - 3.5 mm long, 5-ribbed, with viscid glands and sparse pubescence, apex rounded.

Field note: Common in waste lands throughout the area.

Fl. & Fr.: Almost throughout the year.

Specimen examined: Bhilwara 332, Mandalgarh 528 (MLVGCB Herb).

Distribution: Cambodia, China, India, Indonesia, Japan, Laos, Malaysia, Myanmar, Nepal, Philippines, Thailand, Vietnam; Africa, America, Australia, Pacific Islands.

Uses: The paste of leaves is used orally 4-5 times daily to stop bleeding after delivery.

BOUGAINVILLEA Comm. ex Juss. *nom. cons.*

KEY TO THE SPECIES

1a. Branches and leaves glabrous. Spines slightly curved, up to 1 cm long. Involucral bracts acute, magenta coloured .... *B. glabra*

1b. Branches and leaves pubescent. Spines curved, up to 5 cm long. Involucral bracts obtuse, rosy-red to crimson coloured...................................... *B. spectabilis*

Bougainvillea glabra Choisy in DC., Prodr. 13. (2): 437. 1849.

Vern.: *Bouganvillea.*

An evergreen climbing shrub. Spiny hooks up to 1 cm long, slightly curved. Leaves 4 - 10 x 1.7 - 5 cm, ovate to cordate, upper ovate acuminate, acute or shortly acuminate, obtuse at base, entire or undulate, sparsely pubescent; petiole 0.3 - 1 cm long. Bracts 3, 3.5 - 4 mm long, ovate, acuminate, nerved, magenta coloured. Flowers in cymes, usually in clusters of 3, ca 2 cm long; sepals 5, acute, ca 2.5 mm long, tip villous, yellow above, lilac beneath, alternating with 5 irregular and adnate yellow lobed appendages. Pedicels subsessile, ca 3.5 mm long, adnate to the bract. Perianth base adnate to bracts; filament 8 - 13 mm long, unequal, inserted. Ovary 2 mm long, more or less elongated. Style 1 mm long, stigma 2.5 mm long, plumose.

Field note: Commonly cultivated as an ornamental.

Fl. & Fr.: Almost throughout the year.

Specimen examined: Bhilwara 1824 (MLVGCB Herb).

Distribution: A native of Brazil; widely cultivated in the tropics and subtropics.

Bougainvillea spectabilis Willd. Sp. Pl. 2: 348. 1799.

Vern.: *Bouganvillea*

Evergreen climber. Branches and leaves densely pubescent; spines recurved, 2 - 3 cm long. Leaves orbicular-ovate, acuminate at apex. Bracts dark red or light purple-red, elliptic-ovate, 2.5 - 6.5 × 1.5 - 4 cm, base rotund to cordate. Perianth tube green, narrowly tubular, rounded, 1.6 - 2.4 cm, densely pubescent, apex 5 - 6-lobed, lobes spreading, yellow, 3.5 - 5 mm, hairs copious, spreading, to 1 mm. Stamens 8. Fruit 1 - 1.5 cm, densely hairy.

Field note: Commonly grown as an ornamental.

Fl. & Fr.: Winter to spring.

Specimen examined: Bhilwara 1408 (MLVGCB Herb).

Distribution: A native of Brazil.

COMMICARPUS Standley

Commicarpus chinensis (L.) Heimerl in Engler & Prantl, Nat. Pflanzenf., ed. 2. (16c): 117. 1934; Pandey, in Shetty & Singh Fl. Rajasthan 2: 715. 1991. *Valeriana chinensis* L., Sp. Pl. 1: 33. 1753. *Boerhavia repanda* Willd. Sp. Pl. 1: 22. 1797; Wight, Ic. 5(2): 1. t. 1766. 1852; Hook. *f.* Fl. Brit. Ind. 4: 709. 1885; Duthie, Fl. Gangetic Plain 3: 3. 1915. *B. chinensis* (L.) Aschers. & Schweinf. Beitr. Fl. Aethiop. 1: 167. 1867. (Fig. 133)

Scanded diffuse glabrous herb. Stem thick, branches spreading, to 1 m, glabrous or sparsely pubescent. leaves ovate or cordate-ovate, 2.5 - 7.25 × 1 - 5.0 cm, chartaceous, glabrous or abaxially strigose, veins 3 - 4 pairs, base truncate or subcordate, margin & undulate, apex acuminate or acute. Petiole 1 - 3 cm, pubescent. Flowers axillary, long pendulate umbels, mostly from axil of the pair only, pink, 7 - 10 mm across. axillary or terminal; peduncle 2 - 4 cm. Pedicel slender 3 - 12 mm. Perianth tube ca 2 mm, with few protuberances; Perianth limb pink, funnel-form, 6-8 mm, deciduous. Stamens 2 - 4, exserted. Ovary elliptic. Stigma peltate. Fruit obconic, 6 - 7 mm, longitudinally 10-ribbed, with minute glands.

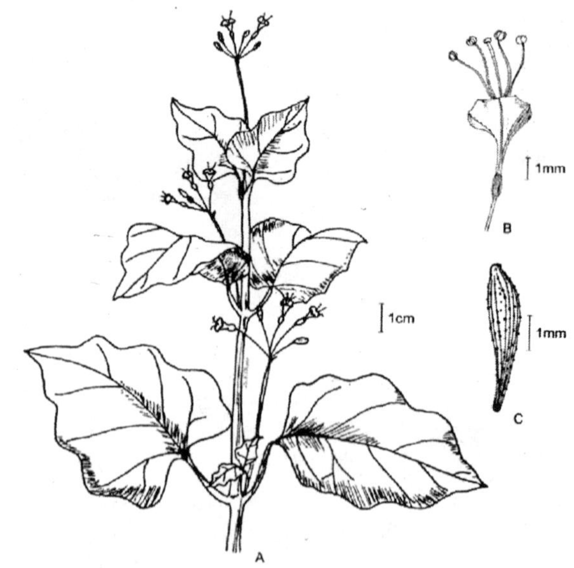

Fig. 133. *Commicarpus chinensis* (L.) Helmet, A. Flowering & fruiting branch, B. Flower, C. Fruit

Field note: Common in gravelly-rockey habitats.

Fl. & Fr.: September - December.

Specimen examined: Bhinayaka (Bari Sadari) 4009, Bhilwara 477 (MLVGCB Herb).

Distribution: Burma, Malaysia, S. India, Sri Lanka and Pakistan.

MIRABILIS L.

Mirabilis jalapa L., Sp. Pl. 177. 1753; Duthie, Fl. Gangetic Plain 3: 3. 1915; Mukherjee in J. Econ. Taxon. Bot. 5: 587. 1984; Pandey, in Shetty & Singh Fl. Rajasthan 2: 715. 1991.

Vern.: *Gul abbas.*

Herb up to 1.5 m tall. Roots tuberous. Stem dichotomous, glabrous to minutely pubescent, often swollen at the nodes, reddish, glaucescent and flaccid. Leaves ovate, 5 - 12 x 2 - 6 cm, acuminate, puberulous, base slightly oblique. Involucral bracts ovate, united at the base, ca 9 mm long, sepaloid, puberulous and nervose. Flowers subsessile, in clusters of 4 - 5; pedicel ca 2 mm long. Perianth tube 2.5 - 3 cm long, puberulous outside, limb plicately 5-lobed; lobes emarginate. Stamens 5, exserted; filaments ca 4 cm long, reddish, lower half adnate to the tube, basally connate and surrounding the ovary. Ovary 1.5 mm long, ovoid; ovule solitary, basal, surrounded by a few linear scales. Style reddish; stigma capitate. Anthocarp 9 mm long, ovoid, 5-ribbed, black, tuberculate. Nut subglobose, ca 7 mm long.

Field note: A garden annual with pretty red flowers. The faintly fragrant flowers open in late afternoon.

Fl. & Fr.: Almost throughout the year.

Specimen examined: Bhilwara 1655 (MLVGCB Herb).

Distribution: A native of S. America; widely cultivated and found as an escape in many tropical areas.

Uses: The leaves and roots are said to be medicinal.

AMARANTHACEAE

KEY TO THE GENERA

1a. Plants with fertile flowers subtended by modified sterile flowers forming of antler-like scales or hooked spines 2

2a. Leaves alternate, modified sterile flowers represented by antler-like scales........... *Digera*

2b. Leaves opposite, modified sterile flowers represented by hooked spines............. *Pupalia*

1b. Plants with fertile flowers never subtended by modified sterile flowers........ 3

3a. Bracteoles with a vertical crest along the dorsal surface of the midrib. Filaments fused almost to the tip .... *Gomphrena*

3b. Bracteoles not dorsally cristate. Filaments fused at the base only 4

4a. Ovary with few to many ovules........... *Celosia*

4b. Ovary uniovuled........ 5

5a. Pseudo-staminodes absent........ 6

6a. Flowers hermaphrodite. Tepals villous. Stamens 1 - 2. Stigma solitary, capitate .... *Nothosaerva*

6b. Flowers unisexual. Tepals not villous. Stamens 3 - 5. Stigmas 2-3, linear........ *Amaranthus*

5b. Pseudo-staminodes present........ 7

7a. Bracteoles strongly spinescent with the long-excurrent midrib. Flowers in pedunculate spikes, the lower flowers increasingly distant in fruit......... *Achyranthes*

7b. Bracteoles mucronate or aristate, but not strongly spinescent. Flowers in dense, axillary, sessile or pedunculate heads or spikes 8

8a. Tepals glabrous or almost so, or with tufts of glochidiate hairs. Stigma capitate........ *Alternanthera*

8b. Tepals somewhat densely lanate. Stigmas 2, distinct, divergent................ *Aerva*

ACHYRANTHES L.

KEY TO THE SPECIES

1a. Plants sub - glabrous. Leaves membranous, nealry glabrous .... *A. aspera*

1b. Plants hairy to densely tomentose. Leaves thick, villous to silky beneath .... *A. bidentata*

Achyranthes aspera L. var. ***argentea*** (Thw.) Hook. *f.* Fl. Br. Ind. 4: 730. 1885; Sharma & Tiagi, Fl. North-East Rajasthan 345. 1979; Parmar, in Shetty & Singh Fl. Rajasthan 2: 719. 1991; Bhandari, Fl. Indian Desert 288. 1995. *Achyranthes argentea* Thw. Enum. Pl. Zeyl. 249. 1864. (Fig. 134 & Plate 273)

Vern.: *Andijhada.*

An erect herb or stragling undershurb. Stem simple or branched angular, ribbed, hairy, nodes more or less shrunken on drying. Leaves 2 - 8 x 1.3 - 6.0 cm, opposite, elliptic to ovate, pointed at tip, acute, cuneate at base. Flowers ca 3 - 5 mm long. greenish-white, arranged in long, terminal or axillary spike. Peduncles 1 - 6 cm long; bracts lanceolate 1.5 - 5mm long, membranous, glabrous; bracteoles 1.5 - 4mm long. Basal wings 1/3 - 1/4th of the length of spines and adnate to it. Tepals whitish, pale green to red or purple, 5, 3 - 6 mm long. Stamens 5; filaments 1.5 - 4 mm long, pseudostaminodes, subquadrate; style 1 - 4 mm long; Utricles 1 - 3 mm long. Seeds cylindrical, smooth, filling the utricle, raddish brown.

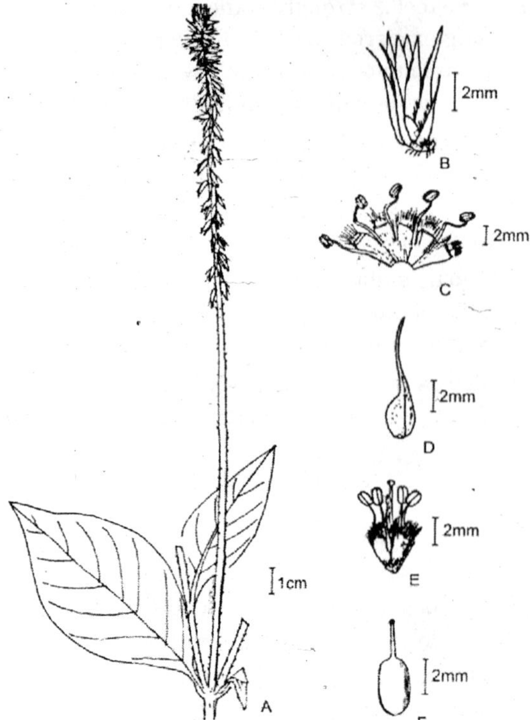

Fig. 134. *Achyranthes aspera* L. var. *argentea* (Thw.) Hook. f. A. Flowering & fruiting twig. B. Flower, C. Flower with stamens & pistil, D. Bracteole, E. Stamens, F. Pistil

Field note: Common in open dry areas.

Fl. & Fr.: Almost throughout the year.

Specimen examined: Bhilwara 1325, Jahazpur 2132 (MLVGCB Herb).

Distribution: Trop. Asia, India, Sri Lanka, Africa, America and Australia.

Achyranthes aspera L. var. ***porphyristachya*** (Wall. ex Moq.) Hook. *f.* Fl. Br. Ind. 4: 730. 1885; Raizada, Suppl. Duthie, Fl. Gangetic Plain 229. 1976; Parmar, in Shetty & Singh Fl. Rajasthan 2: 719. 1991. *Achyranthes porphyristachya* Wall. ex Moq. in DC. Prodr. 13 (2): 316. 1849. (Fig. 135)

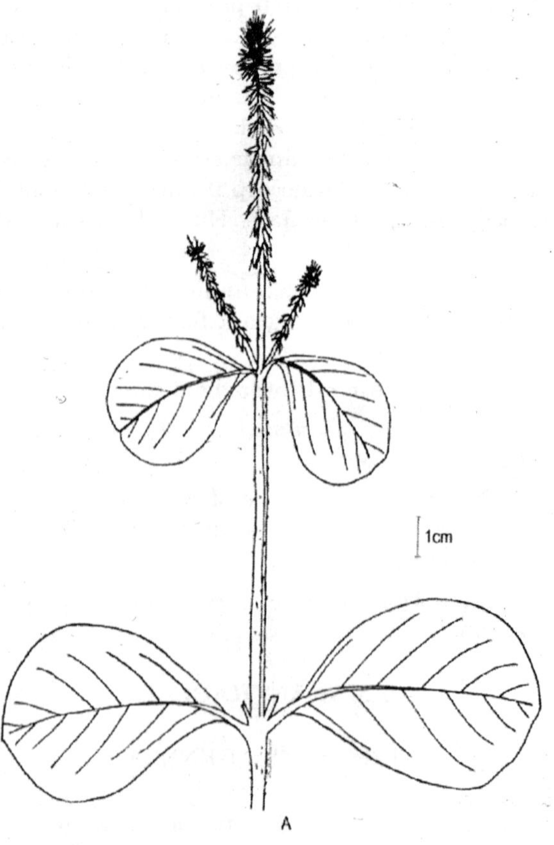

Fig. 135. *Achyranthus aspera* L. var. *porphyristachya* (Wall. ex Moq.) Hook. f. : A. Flowering & fruiting twig

An erect herb or straggling undershurb. Stem angular, ribbed, hairy. Leaves 3.7 - 15.0 x 1.5-8.0 cm, opposite, membranous, broadly elliptic, rounded at apex, cuneate at base. Flowers greenish-white, arranged in long, terminal or axillary spike; perianth pointed, spine-like, hardens in fruit. Seeds reddish brown.

Field note: Common in open dry areas.

Fl. & Fr.: Almost throughout the year.

Specimen examined: Chanderiya 4057 (MLVGCB Herb).

Distribution: India.

AERVA Forssk.

KEY TO THE VARIETIES

1a. Leaves upto 6 mm broad, spike often interrupted with oblong or globose lateral clusters, leafy .... *A. javanica* var. *bovei*

1b. Leaves more than 6 mm broad, spike dense, cylindric, naked .... *A. javanica* var. *javanica*

Aerva javanica (Burm. f.) Juss. ex J.A. Schultes var. ***bovei*** Webb in Hook., Niger Fl. 173. 1849; Boiss, Fl. Or. 4: 993. 1879; Parmar, in Shetty & Singh Fl. Rajasthan 2: 721. 1991. *Aerva bovei* (Webb) Edgew. in Journ Linn. Soc. 6: 206. 1862. *Aerva pseudotomentosa* Blatter & Hallberg, in Journ. Bombay Nat. Hist. Soc. 26: 817. 1919. *Aerva persica* (Burm. f.) Merril. var. *bovei* (Webb) Chiov. Fl. Somalia 286. 1929.

Much branchrd, densely woolly - tomentose, woody undershrub, upto 1 m high. Stem erect, obscurely furrowed, often glabrous, hairy, stellate-tomentose. Leaves alternate, 6-8 x 0.4-6.0 cm, linear-lanceolate, attenuated at base, sub-obtuse or acute at apex, densely tomentose on the dorsal surface, nearly glabrous on upper surface, midrib and laterals prominent below. Spikes more slender, often interrupted with oblong or globose lateral clusters, all spikes except the terminal frequently shortly cylindrical or oblong. Flowers unisexual dioecious. Female flowers bracteolate; bracts more or less equal to the sepals, widely ovate, acute, hyaline, red apex of dorsal surface barbed. Tepals 5.2 mm long. Ovary shorter than the sepals, stigma 2, purple, glandular. Male flowers smallar than female flowers. Perianth broad. Stamens 5, filaments small, anthers oblong, bilocular, dehiscence longitudinally. Pistillode shortly stipiltate, style very short. Stigma minute.

Field note: The species is common on sandy soils some times intermixed with *Aerva javanica* var. *javanica*.

Fl. & Fr.: September - December.

Specimen examined: Badnor 1727, Asind 2134 (MLVGCB Herb).

Distribution: India, Iran, Afghanistan, Egypt, Sahara, Niger, Sudan, Somalia and Arabia.

Notes: var. *bovei* certainly grades into var. *javanica* by intermediates, as demonstrated by a series of speciemens. These have narrow leaves, but all the spikes are elongate. We would have included them in his concept of var. *bovei* (DC. Prodr. 13(2): 300, 1849), as an examination of the IDC microfiche of the De Candolle Herbarium indicates; but in south central Rajasthan the clearest break in the series occurs between "typical" *Aerva javanica* with broad leaves and stout, elongate spikes, and the *"A. pseudotomentosa"* form with narrow leaves and short, interrupted spikes.

Aerva javanica (Burm. f.) Juss. ex J.A. Schultes, Ann. Mus. Hist. Nat. par. 11: 13. 1808 & Syst. Veg. ed 15. 5: 565. 1819; Boiss., Fl. Or. 4: 992. 1879; Hook. *f.* Fl. Brit. Ind. 4: 727. 1885; Parmar, in Shetty & Singh Fl. Rajasthan 2: 721. 1991. *Iresine javanica* Burm. f. Fl. Ind. 212. t. 65. f. 2. 1768. *Iresine persica* Burm. f. Burm. f. l. c. 212. t. 65. f. 1. 1768. *Aerva tomentosa* Forssk. Fl. Aegypt.-Arab. 122. 170. 1775; Duthie, Fl. Gangetic Plain 3: 15. 1915. *Aerva persica* (Burm. f.) Merrill in Philip. J. Sci. 19: 348. 1921; Bhandari, Fl. Indian Desert 288. 1995. (Plate 274)

var. ***javanica.***

Perennial herb, erect clumped, 0.3 - 1.5 m, branched about the base with simple hairs. Stem with long, ascending, branches, terete, striate, whitish or yellowish-tomentose. Leaves alternate, very variable in size and shape, linear-oblong to oblong-spathulate, densely whitish or yellowish tomentose on both surfaces, sessile or with a short and indistinct petiole. Spikes sessile, cylindrical arranged in terminal leafless panicles. Flowers dioecious, dull-white. Bracts 0.5 - 2.25 mm, broadly deltoid-ovate, hyaline, persistent. Bracteoles broadly ovate, acute, white, hyaline, also persistent. Male flowers smaller, perianth lobes 5, elliptic-oblong, subobtuse, woolly on dorsal surface. Outer tepals 1.5 - 2.25 mm, ovate; filaments delicate, anthers about equalling to the perianth. Female flowers with outer 2 tepals 2 - 3 mm, oblong-obovate to obovate-spathulate, lanate, acute to obtuse or apiculate at the tip, inner 3 slightly shorter, elliptic-oblong, ± densely lanate, acute. Anthers absent. Style slender, distinct. Ovary small, style very short, stigma 2, bifid, rudimentary, equalling the ovary. Capsule 1-1. 5 mm, rotund, compressed. Seed 0.9 - 1.25 mm, rounded, slightly compressed, brown or black, shining, smooth or very faintly reticulate.

Field note: Common in sandy soils. Sometimes it forms gregarious patches on sand.

Fl. & Fr.: August - January.

Specimen examined: Mezadam 1598 (MLVGCB Herb).

Distribution: E. & W. Trop. Africa, Cape Verde Island, N. Africa, Arabia, Pakistan, Sri Lanka and India.

Uses: The seeds are used in stuffing beds and pillows.

ALTERNANTHERA Forssk.

Alternanthera pungens Kunth in H.B.K., Nov. Gen. Sp. Pl. 2: 206. 1817; Mahesh. Illus. Fl. Delhi f. 180. 1966; Raizada Suppl. Duthie, Fl. Gangetic Plain 233. 1976; Parmar, in Shetty & Singh Fl. Rajasthan 2: 724. 1991; Bhandari, Fl. Indian Desert 290. 1995. *Achyranthes repens* L. Sp. Pl. 205. 1753. *Alternanthera repens* (L.) Link, Enum. Pl. Hort. Berol. 1: 154. 1821, non Gmel. 1791. (Plate 275)

Prostrate mat-forming perennial herb with a stout vertical rootstock, rooting at the lower nodes, much-branched from the base outwards, mats up to ca 1 m across. Stem and branches terete, striate, villous with long, white hairs when young. Leaves broadly rhomboid-ovate, 1.5 - 4.5 x 0.3 - 1 cm, rounded and mucronate at the apex, narrowed at base, glabrous or thinly appressed-pilose on both surfaces. Flowers sessile, axillary, solitary or 1 - 3 flowered heads, globose to shortly cylindrical, 0.5 - 1.5 cm across; bracts and bracteoles membranous, white or stramineous, 3 - 5 mm across, lanceolate-ovate, glabrous or dorsally pilose. Tepals extremely dissimilar; the 2 outer (abaxial) deltoid-lanceolate, 5 mm, very rigid, 5-nerved below; inner (adaxial) tepal oblong, flat, 3 mm, blunt and strongly dentate at the apex, 3-nerved below; lateral tepals ca 3 mm, sinuate in side view with the two sides of the lamina connivent and denticulate above, sharply mucronate; abaxial and adaxial tepals with small tufts of glochidiate and barbed whitish bristles about the basal angles, the lateral tepals each with a large tuft about the centre of the midrib. Stamens 5, all fertile. Ovary compressed, narrowed below, style very short, wider than long. Utricles lenticular, rounded or retuse above, 2 mm. Seed rounded, ca 1.25 mm across, brown.

Field note: Common weed of gardens, road sides and waste lands.

Fl. & Fr.: Almost throughout the year.

Specimen examined: Rajsamand 4056 (MLVGCB Herb).

Distribution: A native of tropical America; now widespread as a weed of waste ground in the tropics and subtropics of both Old and New Worlds.

Notes: For the nomenclautre of *Alternanthera sessilis* (L.) R.Br. ex DC. and *Alternanthera pungens,* see Melville in Kew Bull. 1958: 171-174. 1958.

AMARANTHUS L.

KEY TO THE SPECIES

1a. Capsules circumscissile 2

2a. Leaf axils with paired spines.......... *A.spinosus*

2b. Leaf axils without paired spines 3

3a. Flowers in terminal leafless spike or in spicate panicles. Perianth segments usually 5...................................... *A. hybridus*

3b. Flowers in axillary cymose clusters, never in terminal laefless spike. Perianth sgments 2 or 3 4

4a. Capsule hyaline and transparent, with 4-5 prominent green longitudinal ribs. Perianth segments 2................................ *A. tenuifolius*

4b. Capsule neither hyaline and transparent nor longitudinally ribbed. Perianth segments 3.. 5

5a. Perianth segments longer than the capsules... *A. tricolor*

5b. Perianth segments shorter than the capsules .... *A. graecizans*

1b. Capsule indehiscent, irregularly rupturing 6

6a. Capsule scarcely exceeding the female perianth. Seeds with shallow, scurfy verrucae on the reticulate pattern of the testa .... *A. viridis*

6b. Capsule distinctly exceeding the female perianth. Seeds without shallow verrucae.. *A. lividus*

Amaranthus graecizans L., Sp. Pl. 990. 1753; Parmar, in Shetty & Singh Fl. Rajasthan 2: 726. 1991. *A. angustifolius* Lam. Encycl. 1: 115, 1783, *nom. illegit. A. blitum* L. var. *graecizans* (L.) Moq. in DC. Prodr. 13 (2): 263. 1849. *A. polygamus* L. var. *angustifoliya* (Lam.) Hook. *f.* Fl. Brit. Ind. 4: 721. 1885.

Annual herb, branched from the base, erect, decumbent or prostrate, mostly up to ca 45 cm. Stem slender to stout, angular, glabrous or few multicellular hairs. Leaves broadly rhomboid-ovate, retuse at tip, narrowed at base. Petioles upto 3.5 cm long. Flowers in axillary sessille, cymose clusters, male and female intermixed, males commonest in the upper whorls. Bracts and bracteoles narrowly lanceolate-oblong, pale-membranous, acuminate, membranous bracteoles subequalling or usually shorter than the perianth. Perianth segments 3, all 1.5 - 2 mm; those of the male flowers lanceolate-oblong, acute or subacute; those of the female flowers lanceolate-oblong to linear-oblong, gradually to abruptly narrowed to a very short to rather long mucro. Stigmas 3, slender, flexuose, ca 0.5 mm. Utricles subglobose to shortly ovoid, 2 - 2.25 mm, usually strongly wrinkled throughout with a very short, smooth neck, slightly exceeding the perianth, circumscissile or sometimes not, even on the same plant. Seeds shining, compressed, black, 1 - 1.5 mm, faintly reticulate.

Field note: Rare, along channels of cultivated fields, wet waste lands and the forests.

Fl. & Fr.: July - December.

Specimen examined: Pansal 2124, Guwardi Dam 2135 (MLVGCB Herb).

Distribution: S.W. Asia (Jordan, Iraq, Arabia, southern Iran etc.); India and Pakistan.

Amaranthus hybridus L. subsp. ***hybridus*** var. ***erythrostachys*** Moq. in DC. Prodr. 13 (2) : 259. 1849; Townsend in Nasir & Ali, Fl. W. Pakistan 71 : 11. 1974 et in Fl. Sri Lanka 1 : 13. 1980; Parmar, in Shetty & Singh Fl. Rajasthan 2: 727. 1991.

Erect annual herb. Stem stout, branched, angular, glabrous or slightly hairy. Leaves long - petioled, 3 - 20 x 1.5 - 10.0 cm, broadly lanceolate to rhomboid or ovate, gradually narrowed to the blunt to subacute, mucronate tip, attenuate or shortly cuneate at base. Inflorescence red, axillary and terminal spikes of cymose clustres, much-branched panicle. Utricles subglobose to ovoid or ovoid - urceolate. Seeds black and shining, compressed, faintly reticulate around the margins.

Field note: Grown as a crop, and as an ornamental.

Fl. & Fr.: Almost throughout the year.

Specimen examined: Mandalgarh 2194 (MLVGCB Herb).

Distribution: Throughout the warmer regions of the world as a spontaneous naturalised weed, frequent in the temperate regions as a casual (e.g. as a "wool adventive" in Europe). Grown in many areas as a crop, and almost cosmopolitan (including Pakistan) as an ornamental,

Notes: The red colouration of this variety vanishes to a greater or lesser extent on drying, and only a limited number of specimens bear colour.

Amaranthus hybridus L. subsp. ***cruentus*** (L.) Thell. var. ***paniculatus*** (L.) Thell. in Asch. & Graebn. Syn. 5 (1): 247. 1914; Parmar, in Shetty & Singh Fl. Rajasthan 2: 727. 1991. *A. paniculatus* L. Sp. Pl. ed. 2. 1406. 1763, non Hook. *f.* Fl. Brit. India 4: 718. 1885; Duthie, Fl. Gangetic Plain 3: 11. 1915.

Very similar to subsp. ***hybridus*** var. ***erythrostachys*** Moq., but can be distiguished by its green inflorescence, longer bracteoles of the female flowers, mostly 1-1.5 times as long as the perianth. Stigma-bases and upper part of lid of fruit scarcely swollen, fruit with no distinct neck.

Field note: Common in waste lands, neglected corner of fields and gardens.

Fl. & Fr.: Almost throughout the year.

Specimen examined: Railmagura 4004 (MLVGCB Herb).

Distribution: Throughout the warmer regions of the world, chiefly in Asia eastwards from Malaya (Indonesia, New Guinea, the Philippines etc.) and in tropical Africa.

Amaranthus lividus L., Sp. Pl. 990. 1753; Parmar, in Shetty & Singh Fl. Rajasthan 2: 728. 1991. *A. blitum* L., Sp. Pl. 990. 1753, *nom. confusum;* Hook. *f.* Fl. Brit. Ind. 4: 721. 1885, excl. var. *sylvestris*; Duthie, Fl. Gangetic Plain 3: 13. 1911. *A. oleraceus* L., Sp. Pl. ed. 2. 1403. 1763. (Plate 276)

Annual herb, erect ascending or prostrate. Stem angular, yellow-green, almost glabrous. Leaves 3-8 x 0.3 - 5.5 cm, elliptic-ovate or rhomboid-ovate, broad and always emarginate-mucronate at apex, cuneate at base. Petioles 2-4 cm long, terete. Flowers greenish-white, in slender to stout, terminal and axillary, cymose clusters or panicles upto 7 cm long. Utricles ca 2 mm long, subrotund to shortly pyriform, compressed, wrinkled on drying, exceeding the perianth. Seeds rounded, compressed, shining, brownish-black, glabrous.

Field note: Common in cultivated fields.

Fl. & Fr.: July - March.

Specimen examined: Hamirgarh 1571 (MLVGCB Herb).

Distribution: Throughout the warmer regions of both Old and New Worlds.

Amaranthus spinosus L., Sp. Pl. 991. 1753; Wight. Ic. 2(2): 8. t. 513. 1840-43; Hook. *f.* Fl. Brit. Ind. 4: 718. 1885; Duthie, Fl. Gangetic Plain 3: 10. 1915; Parmar, in Shetty & Singh Fl. Rajasthan 2: 729. 1991; Bhandari, Fl. Indian Desert 292. 1995.

Annual herb, erect, much-branched, bush, up to 1.5 m in height. Stem stout, sometimes reddish, branched, terete or angular, glabrous or increasingly furnished above (especially in the inflorescence) with long, multicellular, flocculent hairs. Leaves glabrous or thinly pilose on the lower surface of the primary nervation, long-petiolate; lamina ovate to ovate-oblong or lanceolate-, ca 1.5 - 12 x 0.8 - 6 cm, subacute mucronate, cuneate or attenuate at the base; each pair of spines axillary, upto 2.5 cm long, slender to stout. Petioles up to ca 9 cm. Flowers in dense, axillary clusters or in axillary or teminal up to 15 x 1 cm spikes; lower clusters and lower part of spikes bearing only female flowers; upper part of spikes with male flowers. Bracts and bracteoles lanceolate to acicular, pale-membranous; bracteoles shorter than, subequalling or little exceeding the perianth, commonly smaller than the bracts. Perianth segments 5, those of the female flowers ca 1.5 - 2 mm, oblong, obtuse, mucronulate; those of the male flowers broadly lanceolate or lanceolate-oblong, acute or acuminate. Stamens 5, filaments equalling the upper perianth or slightly longer or shorter than bracts. Ovary ovoid, style 3, stigmas 3 or 2, flexuose or reflexed, 1 - 1.5 mm. Capsule ovoid-urceolate with a short inflated neck below the style base, ca 1.5 mm, regularly or irregularly circumscissile, little below the middle up to 2.5 mm long. Seed with a thin margin, 0.75-1 mm, compressed, black, shining, very faintly reticulate.

Field note: Commonly occurs as a weed of cultivation, roadsides, waste places etc.

Fl. & Fr.: Almost throughout the year.

Specimen examined: Rashmi 4054, Badnor 2193 (MLVGCB Herb).

Distribution: Of presumed American origin; now a cosmopolitan weed in the warmer regions of the world and also occurring as a casual in some temperate regions.

Amaranthus tenuifolius Willd. Sp. Pl. ed. 4. 4: 381. 1805; Wight, Ic. 2 (4): 7. t. 718. 1843; Hook. *f.* Fl. Brit. Ind. 4: 722. 1885; Duthie, Fl. Gangetic Plain 3: 14. 1915; Parmar, in Shetty & Singh Fl. Rajasthan 2: 730. 1991.

Prostrate or ascending annual herb, 8 - 35 cm, much-branched from the base. Stem slender, slightly angular or sulcate, yellowish-green to reddish, glabrous or occasionally shortly papillate. Leaves glabrous, oblong, 0.8 - 3.2 x 0.2 - 0.5

cm, attenuate at the base, apex rounded or mucronate. Petiole long. Flowers in axillary cymose clusters, up to ca 8 mm in diam.; male and female flowers intermixed or the males mostly towards the top of the cymes. Bracts and bracteoles oblong-elliptic, hyaline, 0.75 - 1 mm, obtuse. Tepals 2, one frequently smaller than the other, 0.75 - 1 mm, similar to the bracteoles, often concave-subcucullate at the apex; stamens of male flowers 2, free; anthrs 2 - celled. Stigmas 2, erect, very short, highly papillose; ovary ovoid; ovule solitary, erect. Capsule ellipsoid, compressed, ca 1.25 mm, delicate, hyaline, indehiscent, 4 - 5 ribed. Seeds globose, ca 0.8 mm, compressed, smooth, dark brown.

Field note: Common in waste lands.

Fl. & Fr.: October - April.

Specimen examined: Deogarh 4052, Kotri 2123 (MLVGCB Herb).

Distribution: India, Pakistan and Europe.

Notes: It can hardly be confused with any other species of the genus.

Amaranthus tricolor L., Sp. Pl. ed. 1: 989. 1753; Backer in Fl. Malaysiana Ser. 1, 4 (2): 77. 1949; Parmar, in Shetty & Singh Fl. Rajasthan 2: 730. 1991; Bhandari, Fl. Indian Desert 292. 1995. *A. tristis* L. Sp. Pl. 989. 1753. *A. mangostanus* L. Cent. Pl. 1: 32. 1755; Hook. *f.* Fl. Brit. Ind. 4: 720. 1885. *A. polygamus* L. Cent. Pl. 1: 32. 1755, non Hook. *f. l. c.* 4: 721. 1885; Duthie, Fl. Gangetic Plain 3: 12. 1915. *A. gangeticus* L. Syst. Nat. ed. 10. 1268. 1759; Hook. *f. l. c.* 4: 719. 1885; Duthie, *l. c.* 3: 12. 1915, including var. *tristis*.

Vern.: *Chauli*.

Annual herb, ascending or erect, attaining ca 60 cm or more in cultivation. Stem stout, usually much-branched; branches angular, glabrous or furnished in the upper parts sparsely hairy with crisped hairs. Leaves glabrous, very variable in size, long-(up to ca 8 cm) petiolate, the lamina, rhomboid-ovate or broadly elliptic, acute at the apex, at the base shortly cuneate. Petiole 4 - 5 cm long. Flowers green or crimson, in axillary, subglobose clusters, ca 4 - 25 mm in diam. Bracts and bracteoles deltoid-ovate, awned at apex; bracteoles subequalling or shorter than the perianth. Tepals 3, 2.5 - 3 mm long, elliptic or oblong-elliptic, narrowed above, pale-tipped awn; female flowers with the perianth segments slightly accrescent in fruit. Stigmas 3, erect or recurved, ca 2 mm. Capsule ovoid-urceolate with a short neck below the style-base, 2.25 - 2.75 mm, circumscissile, membranous, obscurely wrinkled. Seed ca 1 mm in diam., black or dark brown, shining, very faintly reticulate, lenticular.

Field note: Common weed of cultivation, gardens, roadsides, waste places etc.

Fl. & Fr.: April - June.

Specimen examined: Karoi 943, Bijoliya 1026 (MLVGCB Herb).

Distribution: Asia (India, Pakistan, China, Japan, Indonesia), Africa, West Indies etc.

Amaranthus viridis L., Sp. Pl. ed. 2 : 1405. 1763; Hook. *f.* Fl. Brit. Ind. 4: 720. 1885; Duthie, Fl. Gangetic Plain 3: 13. 1915; Parmar, in Shetty & Singh Fl. Rajasthan 2: 731. 1991; Bhandari, Fl. Indian Desert 292-93. 1995. *A. gracilis* Desf. Tabl. Ecol. Bot. 43. 1804. (Fig. 136)

Vern.: *Mar chouli, Arak chauli.*

Annual herb, erect or more rarely ascending, 10 - 75 cm high. Stem rather slender, sparingly to considerably branched, angular, thinly pubescent when young, glabrous with age. Leaves ovate or rhomboid or oblong, 2 - 8 x 1.5 - 3.5 cm, the margins occasionally obviously sinuate, shortly cuneate to sub-truncate below, obtuse and narrowly to clearly emarginate at the tip, minutely mucronate glabrous or shortly to fairly long-pilose on the lower surface of the primary and most of the venation, long-petiolate; petioles up to ca 10 cm long. Flowers green, lower clusters, axillary, upper ones in terminal, rather dense, continuous or interrupted male or female spikes which are ca 7 mm in diam.; male and female flowers intermixed but the latter more numerous. Bracts and bracteoles ovate minutely mucronate, bracteoles shorter than the perianth (ca 1 mm.). Perianth segments 3, very rarely 4, those of the male flowers oblong-oval, acute, concave, ca 1.5 mm, shortly mucronate; those of the female flowers narrowly oblong to spathulate, 1.25 - 1.75 mm, mucronate or not. Ovary rudimentary. Stigmas 2-3, short, erect or almost so. Capsule subglobose, 1.25-1.5 mm, not or slightly exceeding the perianth, indehiscent or rupturing irregularly, very strongly rugose throughout. Seeds ca 1-1.25 mm, round, only slightly compressed, dark brown to black, shining, reticulate and with shallow scurfy verrucae on the reticulum.

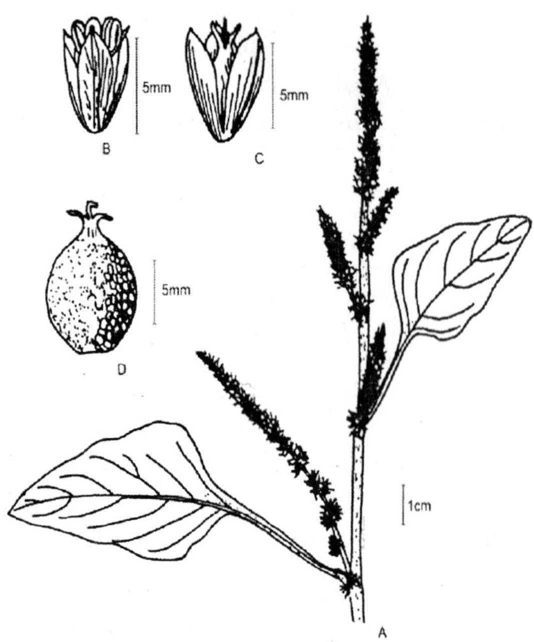

Fig. 136. *Amaranthus viridis* L
A. Flowering & fruiting twig. B. Male flower, C. Female flower, D. Fruit

Field note: A common weed of waste and cultivated fields.

Fl. & Fr.: Almost throughout the year.

Specimen examined: Bhilwara 1135 (MLVGCB Herb).

Distribution: Throughout the tropical and subtropical regions of the world, and penetrating further into the temperate regions than most of the tropical species.

CELOSIA L.

Celosia argentea L., Sp. Pl. ed. 1: 205. 1753; Moq. in DC., Prodr. 13 (2): 242. 1849; Hook. *f.* Fl. Brit. Ind. 4: 714. 1885; Boiss., Fl. Or. 4: 987. 1879; Duthie, Fl. Gangetic Plain 3: 7. 1915; Parmar, in Shetty & Singh Fl. Rajasthan 2: 731. 1991; Bhandari, Fl. Indian Desert 293. 1995. *C. cristata* L., Sp. Pl. 205. 1753; Hook. *f.*, l.c. 4: 715. 1885. *C. coccinea* L., Sp. Pl. ed. 2. 297. 1762. (Fig. 137 & Plate 277)

Vern.: *Safed murgha, Seergai*

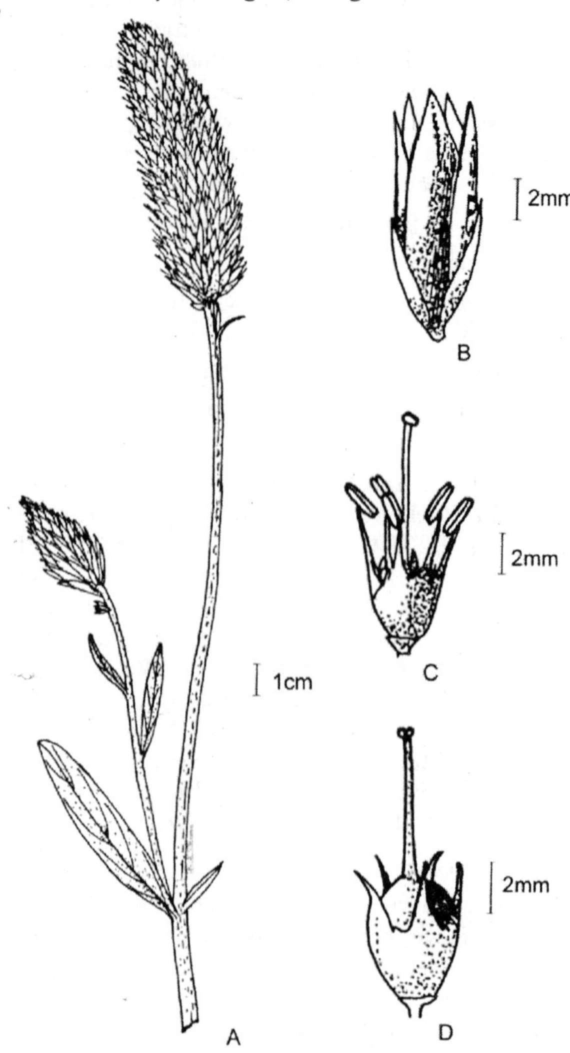

Fig. 137. *Celosia argentea* L.
A. Flowering & Fruiting twig, B. Flower, C. Stamens & Pistil, D. Fruit

Annual herb, erect, 0.4 - 2 m, simple or with many ascending branches. Stem and branches strongly ridged, often sulcate, glabrous. Leaves lanceolate-oblong to narrowly linear, acute to obtuse, shortly mucronate, glabrous, 2 - 15 x 0.1 - 3.2 cm, tapering at the base on slender petiole; Flowers in terminal, dense, many flowered, silvery to pink spike, 2.5 - 20 x 1.5 - 2.2 cm, conical at first but becoming cylindrical in full flower, terminal on the stem and branches; peduncles up to 20 cm long. Bracts and bracteoles lanceolate or towards the base of the spike deltoid, 3 - 5 mm, hyaline, persistent. Tepals segments 6 - 10 mm, narrowly elliptic-oblong, acute or obtuse. Filaments very delicate, free part subequalling or exceeding the staminal sheath, sinuses rounded; anthers and filaments creamy or magenta. Stigmas 2 - 3, very short, the filiform style 5 - 7 mm long; ovary 4 - 8 -ovulate. Capsule 3 - 4 mm, ovoid to globose. Seeds ca 1.25 - 1.5 mm, lenticular, black, shining, testa very finely reticulate.

Field note: Common elsewhere on waste-ground, in cultivated fields, along roadsides etc. on sandy or stony soils.

Fl. & Fr.: August - December.

Specimen examined: Nimbahera 4002, Jahajpur 1825 (MLVGCB Herb).

Distribution: Pantropical.

DIGERA Forssk.

Digera muricata (L.) Mart. Beitr. Amar. 77. n. 2. 1825 *et* Nov. Act. Acad. Phys. - Med. Acad. Caes. Leop. - Carol. Nat. Cur. 13 (1): 285. 1826; Parmar, in Shetty & Singh Fl. Rajasthan 2: 732. 1991; Bhandari, Fl. Indian Desert 293-94. 1995. *Achyranthes muricata* L. Sp. Pl. ed. 2. 295. 1762. *A. alternifolia* L., Mant. Pl. 50. 1767. *Digera arvensis* Forssk. Fl. Aeg. - Arab. 65. 1775; Hook. *f.* Fl. Brit. Ind. 4: 717. 1885; Duthie, Fl. Gangetic Plain 3: 8. 1915. (Plate 278)

Vern.: *Khanjaru, Kondra.*

Annual herb, erect, upto 80 cm high, glabrous, simple or with ascending branches from near the base; stem and branches glabrous or very sparingly pilose, with pale ridges. Leaves broadly ovate, 1.5-8.0 x 0.5-4.0 cm, glabrous, acute or acuminate at apex, narrowed at base; petiole slender, up to ca 3.5 cm. Flowers in axillary, 15-30 cm long racemes, white tinged with pink to carmine or red, usually becoming greenish-white in fruit; peduncles slender, the lower up to ca 14 cm in length, glabrous or sparingly spreading-hairy; bracts persistent, deltoid-lanceolate, acuminate, 1-2.75 mm, glabrous, membranous. Central flower fertile, the 2 membranous, outer perianth segments ca 3 - 4.5 mm long, oval or oblong, 7 - 12-nerved, acute; 2 - 3 inner segments slightly shorter, more delicate, blunt or erose, hyaline; stamens sub-equalling or shorter than the style; style ca 1.5 - 4 mm, the 2 stigmas finally recurved. Lateral flowers appressed, 1 - bracteolate, bracteoles similar in form to the bract; flowers much reduced, increasingly so in the upper part of the spike (sometimes absent there), modified into accrescent, antler-shaped scales, these scales with the lateral lobes narrow and wing-like. Fruit subglobose, slightly compressed, 2 - 2.5 mm, bluntly keeled along each side, surmounted by a thick rim, style persistent.

Field note: Common as a weed of cultivation and waste places.

Fl. & Fr.: August - November.

Specimen examined: Danthal 923, Bagor 1614 (MLVGCB Herb).

Distribution: Widespread in southern Asia from tropical Arabia and Yemen to Afghanistan, India, Sri Lanka, Malaysia and Indonesia. Also in Southern, Central and Eastern tropical Africa and Madagascar.

GOMPHRENA L.

Gomphrena celosioides Mart. Bietr. Amar. 93. 1825 & Nov. Act. Phys. - Med. Acad. Caes. Leop. Carol. Nat. Curios. 13 (1): 301. 1826; Mahesh. Illus. Fl. Delhi f. 179. 1966; Raizada, Suppl. Duthie Fl. Gang. Plain 235. 1976; Panday *et al.*, in Bull. Bot. Surv. Ind. 24: 138. 1982; Parmar, in Shetty & Singh Fl. Rajasthan 2: 732-33. 1991; Bhandari, Fl. Indian Desert 294. 1995.

Annual herb, prostrate and mat-forming to ascending or erect, 7-30 cm, much-branched from the base and also above. Stem and branches striate, often sulcate, when young hairy, glabrescent with age. Leaves narrowly oblong to oblong-elliptic or oblanceolate, 1.5 - 4.5 x 0.5 - 1.3 cm, obtuse to subacute at the apex, mucronate, narrowed to a poorly demarcated petiole at base, with long whitish hairs on the margins and lower surface. Flowers in spikes, spikes sessile above the uppermost pair of leaf, initially subglobose, 1.25 cm in diam., finally elongate and cylindrical, 2 - 7 cm long, sessile; bracts deltoid-ovate, 2.5 - 4 mm, mucronate; bracteoles strongly laterally compressed, navicular, 5 - 6 mm, mucronate. Tepals ca 4 - 5 mm long, lanceolate, 1-nerved, the outer 3 more or less flat, lanate only at the base, nerve thick and greenish below, thinner and excurrent in a short mucro at the tip; inner 2 sigmoid. Staminal tube subequalling the perianth. Style and stigmas together ca 0.75 -1 mm, style very short, stigmas divergent. Utricles ca 1.75 mm long. Seeds compressed-ovoid, ca 1.5 mm, brown, faintly reticulate, shining.

Field note: Common weed of waste lands.

Fl. & Fr.: Almost throughout the year.

Specimen examined: Sanwariyaji 4011 (MLVGCB Herb).

Distribution: Native of S. America (S. Brazil, Paraguay, Uruguay and Argentina). Widely distributed in tropical and subtropical regions.

NOTHOSAERVA Wight

Nothosaerva brachiata (L.) Wight, Ic. Pl. Ind. Orient 6 : 1. 1853; Hook. *f.* Fl. Brit. Ind. 4: 726. 1885; Duthie, Fl. Gangetic Plain 3: 17. 1915; Parmar, in Shetty & Singh Fl. Rajasthan 2: 733. 1991; Bhandari, Fl. Indian Desert 294-95. 1995. *Achyranthes brachiata* L., Mant. Pl. 50. 1767. *Illecebrum brachiatum* (L.) L. Mant. Pl. 2: 213. 1771; *Pseudanthus brachiatus* (L.) Wight, Ic. Pl. Ind. Orient 5 (2): 3. t. 1776 bis. 1852.

Annual herb, upto 40 cm high; branches many, spreading from about the base upwards. Stem and branches subterete, striate, glabrous or thinly hairy. Leaves narrowly to broadly elliptic, elliptic-oblong, entire, thinly hairy to glabrous or almost so, obtuse to subacute at the tip, lamina of the lower main stem-leaves 10 - 40 x 6 - 20 mm, gradually or more abruptly narrowed to a petiole about half the length of the lamina, upper and branch leaves becoming shorter and narrower. Flowers in dense, 3 - 15x 2 - 2.5 mm spikes, which are clustered in the leaf-axils of the stem and branches or on very short axillary shoots; spikes sessile, or the terminal spike on axillary shoots, shortly pedunculate, inflorescence axis thinly to rather densely pilose; bracts hyaline, minutely erose, concave, acute or shortly acuminate, ca 0.5 mm, glabrous or very thinly hairy, nerveless; bracteoles minute, hyaline. Tepals broadly ovate, ca 1 - 1.25 mm, sub-acute to shortly acuminate, villous on the outer surface. Stamens longer than the ovary and style. Capsule included, ca 0.75 mm, falling with persistent perianth. Seed ca 0.4 mm, chestnut-brown, smooth, shining.

Field note: Not common found in sandy depressions or ditches in which water stands during some part of the year

Fl. & Fr.: September - November.

Specimen examined: Hamirgarh 125 (MLVGCB Herb).

Distribution: S. and S.E. Asia from India to Burma & Sri Lanka, also reported from Borneo; tropical Africa from Nigeria and the Sudan south to Angola and Rhodesia, Mauritius.

PUPALIA Juss.

Pupalia lappacea (L.) Juss. Ann. Mus. Hist. Nat. 2: 132. 1803; Boiss., Fl. Or. 4: 995. 1879; Hook. *f.* Fl. Brit. Ind. 4: 724. 1885; Duthie, Fl. Gangetic Plain 3: 19. 1915; Mahesh. Illus. Fl. Delhi f. 184. 1966; Parmar, in Shetty & Singh Fl. Rajasthan 2: 734. 1991; Bhandari, Fl. Indian Desert 295. 1995. *Achyranthes lappacea* L. Sp. Pl. 204. 1753. *A. atropurpurea* Lam. Encycl. 1: 546. 1785. *Pupalia atropurpurea* (Lam.) Moq. in DC. Prodr. 13 (2): 331. 1849; Hook. *f. l.c.* 4: 724. 1885

Annual or biennial herb, erect upto 1m tall. Stem branched and swollen at the nodes, 4-angled to almost terete, thinly pilose to densely tomentose. Leaves variable in shape and size, ovate-elliptic to oblong or orbicular, 2 - 12 x 1 - 6 cm, acuminate at apex, cuneate at the base, petiole 2 - 25 mm long. Flowers in spikes; Spikes terminal on the stem and branches, dense when young, elongating to as much as 50 cm in fruit, axis subglabrous to tomentose, peduncle ca 1 - 10 cm; bracts lanceolate, 1.5 - 2.5 mm, persistent, subdeflexed after the fall of the fruits, subglabrous. or pilose, sharply mucronate. Hermaphrodite flowers mostly in subsessile clusters of 3, upper often solitary; bracteoles of hermaphrodite flowers broadly cordate-ovate, 2.75-5 mm, conspicuous, sharply mucronate. Tepals oblong-ovate to lanceolate-ovate, 3.5-6 mm, glabrous to slightly pilose dorsally, 3-nerved. Spines of modified flowers glabrous except sometimes near the base, yellowish to purple, 3-4 mm; 3-flowered clusters falling together to form a "burr" up to ca 2.5 cm in diam. Filaments 2-3 mm. Style slender, 1.25-2 mm. Capsule ovoid, 2-2.5 mm. Seeds oblong-ovoid with a prominent radicle, 2 mm long, dark brown, shining, testa at first faintly reticulate but finally smooth or punctulate.

Field note: Occurs as a weed of dry sandy or stony waste grounds, cultivation or on broken rocky hill slopes

Fl. & Fr.: Almost throughout the year.

Specimen examined: Bagor 930, Fatehpur 1788 (MLVGCB Herb).

Distribution: Tropics and Sub-Tropics of the Old World. Asia (Arabia, India, Malaya, Indonesia, the Philippines and New Guinea), Egypt, Africa, S. Africa, Madagascar and Australia.

CHENOPODIACEAE

KEY TO THE GENERA

1a. Leaves dentate, often hastate; perianth segments not hardening in fruit.. *Chenopodium*

1b. Leaves entire-sinuately lobed; perianth segments and bracteoles coalescing and hardening in fruit 2

2a. Flowers hermaphrodite; leaves entire, sometimes with a wavy margins.............. *Beta*

2b. Flowers unisexual (plants dioecious); leaves mostly lobed .... *Spinacia*

BETA L.

Beta vulgaris L., Sp. Pl. 222. 1753; Hook. *f.*, Fl. Bit. Ind. 5: 5, 1886; subsp. *maritima* (L.) Archangeli, Comp. Fl. Ital. 593.1882; Stewart, Ann. Cat.Vasc. Pl. W. Pak. Kashm. 220. 1972; Hedge in Rech. f. Fl. Iran. 172: 21. 1997. *Beta maritima* L. Sp. Pl. ed. 2 : 322. 1762.

Vern.: *Chukander*

Annual to perennating stout herb, glabrous to hirsute, up to 60 cm long, decumbent to erect, branched and leafy, green to purplish-violaceous. Roots 5 - 10 cm in diam., napiform, purple-red within, with sugar storage. Leaves usually up to 3 - 20 x 1.5 - 15.0 cm, petiolate; broadly ovate, shortly cuneate; cauline rhombic-oblong to linear-lanceolate. Floral clusters in terminal paniculate spikes, usually with smaller subtending leaves (bracts); bracteoles very small or aborted. Tepals herbaceous, green, 2 - 5 mm long, as long as or exceeding the diam. of the fruit, united and indurated below, usually incurved, keeled. Fruits globose.

Field note: It is usually cultivated as commercial crop. A highly polymorphic taxon, in which great variability is often present in leaf shape, size and in the general facies of perianth.

Fl. & Fr.: March-May.

Specimen examined: Bhilwara 2196 (MLVGCB Herb).

Distribution: Europe, N. Africa, Asia; widely cultivated (beet).

CHENOPODIUM L.

KEY TO THE SPECIES

1a. Plants with simple and glandular hairs or glands, aromatic. ....*C. botrys*

1b. Plants glabrous to vesicular hairy (farinose), not aromatic but sometimes bad smelling... .2

2a. Secondary nerves not very prominent. Terminal panicle much larger than the axillary ones. Seeds smooth, blackish brown, shining .... *C. album*

2b. Secondary nerves very prominent. Terminal panicle not much larger than the axillary ones. Seeds rugos, black ....*C. murale*

Chenopodium album L., Sp. Pl. 219. 1753; Boiss. Fl. Or. 4: 901. 1879; Hook. *f.* Fl. Brit. Ind. 5: 5. 1886; Duthie, Fl. Gangetic Plain 3: 22. 1915; Parmar, in Shetty & Singh Fl. Rajasthan 2: 737. 1991; Bhandari, Fl. Indian Desert 296. 1995. (Fig. 138 & Plate 279)

Vern.: *Bathua, Sil, Chil.*

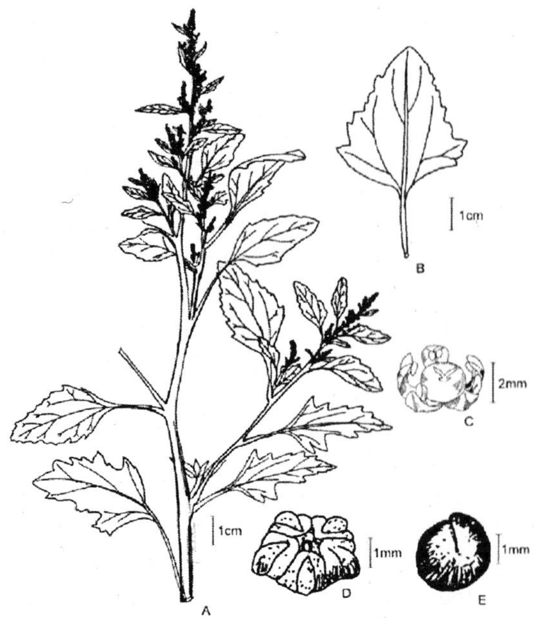

Fig. 138. *Chenopodium album* L.:
A. Flowering & fruiting twig, B. Leaf,
C. Flower, D. Perianth, E. Seed

Annual herb, 10 - 50 cm, erect. Stem yellowish to green, green-striated, sometimes reddish or with red spots at leaf axils. Lower and medium leaves petiolate, ovate-rhomboid to lanceolate, serrate to entire, often somewhat 3-lobed, teeth mostly acute, often unequal in size; uppermost leaves elliptic-oblong, lanceolate, acute or obtuse at apex, acute or rounded at base, less deeply incised or entire, shortly petiolate. Flowers in terminal, leafy paniclulate, mostly terminal, 5-merous. Tapals 5, ovate, 1.5 - 2.2 mm long, obtuse, dorsally keeled. Stigmas 2, short. Utricles depressed globose, finely papillate, enclosed by incurved tepals. Seeds horizontal, black, 1.1 - 1.5 mm in diam., somewhat ovate, margin weakly acute; testa with faint radial striae, otherwise almost smooth.

Field note: Common weed of cultivated fields, gardens and irrigated places.

Fl. & Fr.: November - February.

Specimen examined: Gangapur 195, Delwara 1015 (MLVGCB Herb).

Distribution: Cosmopolitan.

Chenopodium botrys L., Sp. Pl. 219. 1753; Hook. *f.*, Fl. Br. India 5: 4. 1886. *Ambrina botrys* (L.) Moq. Chenop. Enum.: 37. 1840. (Fig. 139 & Plate 280)

Aromatic annual, erect, 5 - 40 cm, densely hairy, most hairs stalked, glandular. Leaves 1 - 7 cm, glaucous, elliptic to ovate, cuneate to subcordate at base, obtuse at apex, pinnatifid with 2-5 broad lobes and several teeth on each side; lower leaves sometimes repand. Flowers in terminal or axillary dense dichasial cyme, main axis to 5 cm wide. Terminal flowers solitary, larger and earlier flowering than lateral ones. Perianth segments 5, ca 1 x 0.5 mm, elliptic to ovate, acuminate, free to base, often not contiguous at base, herbaceous with membranous margins, often whitish in fruit, weakly keeled, with stalked glandular hairs on back. Stamens 1 - 5. Stigmas 2. Pericarp free. Seeds horizontal, black, 0.6 - 0.8 mm, round in outline, margin rounded often truncate in part; testa almost smooth.

Field note: Common weed of cultivated fields and gardens during winter season.

Fl. & Fr.: November - February.

Specimen examined: Bhilwara 482, Ajmer 4120 (MLVGCB Herb).

Distribution: Mediterranean area to SW and C. Asia and NW India; introduced and naturalized in C Europe, S. Africa and N America.

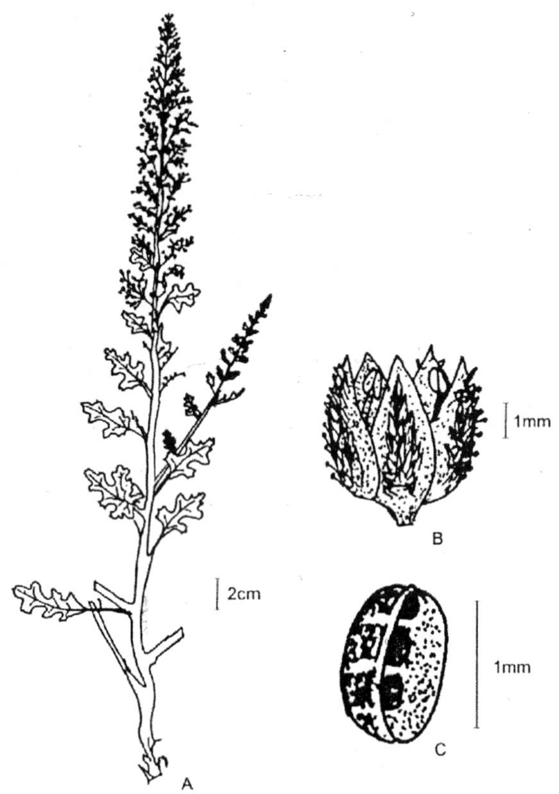

Fig. 139. *Chenopodium botrys* L.
A. Habit, B. Flower, C. Seed

Chenopodium murale L., Sp. Pl. 219. 1753; Boiss., Fl. Or. 4: 902. 1879; Hook. *f.* Fl. Brit. Ind. 5: 4. 1886; Duthie, Fl. Gangetic Plain 3: 23. 1915; Parmar, in Shetty & Singh Fl. Rajasthan 2: 738. 1991; Bhandari, Fl. Indian Desert 296. 1995. (Fig. 140 & Plate 281)

Vern.: *Moti-sil.*

Annual herb, erect or ascending, 15 - 40 cm high, branched. Stem yellowish to green-striped, rarely tinged with red. Leaves deltoid-ovate, somewhat fleshy, 1 - 8 cm long, coarsely dentate, teeth often slightly incurved, apex acute or obtuse, base variously attenuate to truncate; uppermost leaves lanceolate, sometimes entire. Flowers in cymes forming axillary racemes or panicle bisexual and sometimes female. Tepals 5, connate below the middle, oblong, subacute, covering fruit. Stamens 5. Stigmas 2. Fruits depressed-globose, 1.2 - 1.5 mm across, falling with perianth. Seeds orbicular, black, 0.15 cm across, compressed, rugose, margins acute; embryo completely annular.

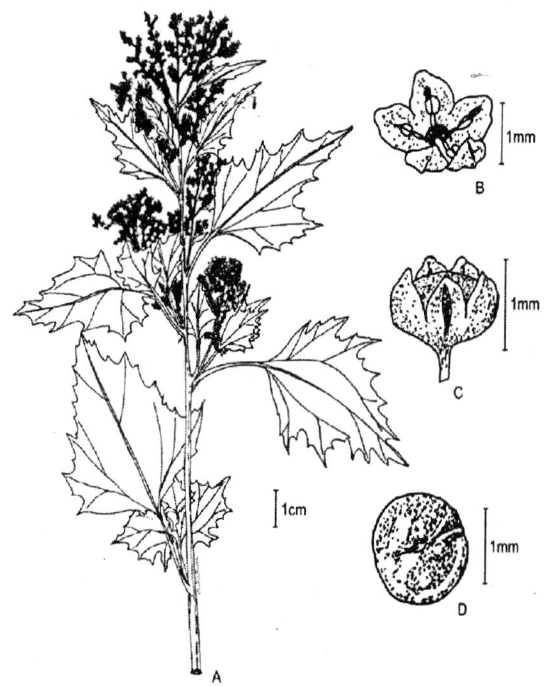

Fig. 140. *Chenopodium murale* L.
A. Flowering & fruiting branch, B. Flower, C. Perianth containing fruit D. Seed

Field note: Most common in cultivated fields and gardens. Usually easily recognized by its rather hidden inflorescence, small, somewhat densely arranged glomerules, yellowish to pure green colour, clearly pitted testa and acute margin of the seed.

Fl. & Fr.: December - April.

Specimen examined: Mandal 150, Dhaneshwar 2174 (MLVGCB Herb).

Distribution: Europe, Africa, Iran, Afghanistan, Turkmenistan, India, Sri Lanka and Arabian peninsula; introduced in subtropical and tropical world.

Uses: 1 tsp leaf paste taken with *"Gur"* (Sugar), + 1 glass of cow milk at morning in an empty stomach for 3 days to cure piles..

SPINACIA L.

Spinacia oleracea L., Sp. Pl. 1027. 1757; Hook. *f.* Fl. Brit. India 5: 6. 1886.

Vern.: *Palak*.

Annual, dioecious herb, 20 - 90 cm tall, erect, light green, glabrous. Leaves radical ones 4 - 15 x 2 - 8 cm, oblong to ovate, acute or obtuse at apex, cuneate and with projecting lobes at base on either sides, entire or broadly pinnatifid. Staminate flowers in interrupted spiciform panicles, with 4 (-5) perianth segments and stamens; pistillate flowers in dense axillary sessile clusters, each separated and falling so in fruit; bracteoles in fruit orbicular-obovate, usually broader than long, free, with or without divergent spines at the apex. Fruits compressed; fruiting perianth with 2 - 3 dorsal spines.

Field note: Widely cultivated and sometimes naturalized or found as an escape from cultivation.

Fl. & Fr.: February - May.

Specimen examined: Bhilwara 2197 (MLVGCB Herb).

Distribution: Probably originated from *S. tetrandra* Stev. in W. Asia.

Uses: Leaves are commonly used as vegetable. It contains a considerable amount of vitamin A, B and C, iron, phosphorus, sufficient proteins and thus recommonded during general weakness and in case of anaemia; it is said to increase secretions of stomach and function of pancreas.

BASELLACEAE

BASELLA L.

Basella rubra L., Sp. Pl. 272. 1753; Wight, Ic. 3 (2) : 9. t. 896. 1845; Hook. *f.* Fl. Brit. Ind. 5: 20. 1886; Duthie, Fl. Gangetic Plain 3: 27. 1915; Parmar, in Shetty & Singh Fl. Rajasthan 2: 743. 1991. *B. alba* L., Sp. Pl. 272. 1753. *B. cordifolia* Lam. Encycl. 1 : 382. 1785. (Plate 282)

Vern.: *Kalimirach*.

Perennial twining herb. Stem stout at the base, 1.5-2.0 cm in diam., upper branches slender, fleshy, climbing. Leaves broadly ovate, 5 - 10 x 2.5 - 8 cm, acute or acuminate, basally cordate, cuneate or truncate; petiole 0.5 - 3 cm long. Flowers sessile, in lax, pedunculate spikes white or red, subsessile, remaining closed at anthesis; rachis stout. Bracts small, apiculate; bracteoles longer than the perianth, oblong, obtuse. Perianth divided half-way down. Sepals 3-5 mm long, united up to the middle, lobes short, 2 - 3 mm broad. Bracts scaly, small; bracteoles similar to calyx, acute. Stamens included; filaments short; anthers cordate. Ovary 1-locular. Fruit black or dark purple enclosed within the persistent fleshy calyx, ca 1.0 cm long. Seed globose, indehiscent.

Field note: Common among the hedges of cultivated fields, gardens and houses. There is dramatic variation in growth form, colour of stem and the shape of inflorescence, according to age.

People confuse with *"Kalimirch"* and commonly grow it.

Fl. & Fr.: September - December.

Specimen examined: Bhilwara 2176 (MLVGCB Herb).

Distribution: A native of Tropical and subtropical Africa; introduced and widely cultivated in S.W. Asia to E. Asia, Fiji, Hawai, West Indies, Brazil and Guiana.

POLYGONACEAE

KEY TO THE GENERA

1a. Perianth-lobes 6, 2 seriate stigma fimbriate .... *Rumex*

1b. Perianth lobes 3-5, 1 seriate stigma capitate 2

2a. Flowers in simple or panicle, terminal, speciform racemes or heads. All stamens of the same size, usually not enlarged at the base. Tepal venation trifid .... *.Persicaria*

2b. Flowers axillary, solitary or in clusters. Stamens usually enlarged and inflated at the base, inner larger. Tepal venation dendricular or anastomosing .... *.Polygonum*

PERSICARIA (L.) Mill.

Persicaria barbata (L.) Hara in Fl. E. Himal. 70. 1966. *Polygonum barbatum* L., Sp. Pl. 362. 1753; Hook. *f.*, Fl. Brit. Ind. 5: 37. 1886;

Duthie, Fl. Gangetic Plain 3: 36. 1915; Parmar, in Shetty & Singh Fl. Rajasthan 2: 746. 1991. *P. hispidum* Buch.-Ham. ex D. Don, Prodr. Fl. Nep. 71. 1825. *P. stagnium* Buch.-Ham. ex Meissn., in Wall. Pl. As. Rar. 3: 56. 1832; Hook. *f. l.c.* 5: 37. 1886. *Persicaria glabra* Gomez, in Ann. Inst. Segunda Enset. Habana 2: 278. 1896. (Plate 283)

Erect to suberect, upto 1 m tall, sometimes decumbent or prostrate simple-branched, annual, herb. Stem stuot, simple or branched, glabrous, reddish below. Leaves 7 - 20 x 1 - 3 cm, elliptic-ovate, tapering at base, margins entire, apex acuminate, glabrous, gland-dotted, dark brown, shining when dry. Flowers in racemes terminal pseudospikes, 4 - 12 cm, usually with several dense spikes aggregated and paniculate, pink. Bracts oblong, truncate, not ciliate, usually each 3 or 4 flowered. Pedicel, 0.75 - 1.5 mm long, apiculate at apex. Perianth white or pinkish or red, 5 -parted. Tepals 5, 1.5 - 2.5 x 0.75 - 1.25 cm, oblong, obtuse, biseriate, unequal, eglandular. Stamens 6-8; filaments long, equal. Ovary broadly ovate-orbicular, trigonous, styles three, long, free till half of the length with capitate stigmas. Nuts biconvex, 1.8 - 3 x 1.2 - 1.5 mm, glabrous, black-brown, shining.

Field note: Common in marshy and aquatic habitats.

Fl. & Fr.: September - March.

Specimen examined: Bhilwara 174 (MLVGCB Herb).

Distribution: Africa, W. Asia, Himalayas, India, Pakistan, China extending to Japan and Australia.

POLYGONUM L.

Polygonum plebeium R. Br. Prodr. 420. 1810, *'plebijum'*; Hook. *f.*, Fl. Brit. Ind. 5: 27. 1886; Duthie, Fl. Gangetic Plain 3: 31. 1915; Parmar, in Shetty & Singh Fl. Rajasthan 2: 750. 1991. (Plate 284)

var. ***plebeium***

Prostrate, diffusely branched, annual herb. Stem stout or slender, purplish green, white hairy, branches numerous. Lower internodes longer than upper. Leaves 0.3 - 1 cm, alternate, lanceolate or oblong linear, entire, obtuse at apex, margins recurved. Ochrea membranous, ovate, tubular, lacerate-fimbriate, 1 - 2 mm long, sessile or shortly petiolate. Flowers in axillary, solitary or in clusters, 3 - 4 -flowered, sunken between ochrea or hardly exserted, 1 - 2 mm across, sessile to subsessile. Ochreolae small, hyaline, lacerate - fimbriate. Perianth pink, ca 2 mm long, divided nearly base. Tepals 5, 1.5 - 3 x 0.5 - 1.0 mm, oblanceolate-lanceolate, unequal, outer tepals slightly longer, acute, inner too obtuse. Stamens 7 - 8, filaments long with broaden base, equal. Ovary small, trigonous with three styles; stigma capitate. Nutlets 1.0 - 1.75 mm long, rhomboid-trigonous, shining, black, glabrous.

Field note: Abundant in dried ponds and ditches or dried river beds throughout the area.

Fl. & Fr.: November - May.

Specimen examined: Matri kundiya dam 3645, Vindiyabhata 1654 (MLVGCB Herb).

Distribution: Afghanistan, Pakistan, India, Bangladesh, Nepal, Tropical East Africa, Madagascar Bhutan extending to Australia through Indonesia and Malaysia.

RUMEX L.

Rumex dentatus L. subsp. ***klotzschianus*** (Meisn.) Rchb. *f.* Bot. Centralb. 49: 19. 1932 & in Candollea 12: 119 1949. *R. klotzschinus* Meisn, in DC., Prodr. 14(1): 57. 1856. , Mant. Pl. 226. 1771. *R. dentatus auct. non* L. 1771; Hook. *f.*, Fl. Brit. Ind. 5: 59. 1886 no L. 1771; Duthie, Fl. Gangetic Plains 3: 41. 1915; Parmar, in Shetty & Singh Fl. Rajasthan 2: 754. 1991; Bhandari, Fl. Indian Desert 301. 1995.

Annual, erect herb, glabrous. Stem fistular, purplish, grooved, glabrous. Leaves oblong, 7.0 - 10 cm long, radial and cauline: lower leaves long petioled, 7 - 10 x 2.0 - 4 cm, linear-lanceolate to oblong or elliptic-oblong. Petiole equal to or shorter than the lamina. Flowers in leafy or leafless whorls, pedicellate, green to greenish white. Tepals 2 -3 mm long; inner tepals broad ovate, much enlarged in fruit, reticulately veined. Nutlets 3 - angled, ca 2 mm long, brown.

Field note: Common weed of waste places and cultivated fields throughout the area.

Fl. & Fr.: February - April.

Specimen examined: Karoi 3654, Vindiyabhata 4334 (MLVGCB Herb).

Distribution: Europe; tropical and subtropical Asia; N. Africa; introduced elsewhere.

ARISTOLOCHIACEAE

ARISTOLOCHIA L.

Aristolochia bracteolata Lamk., Encyl. Bot. 1 : 258. 1783; Parmar, in Shetty & Singh Fl. Rajasthan 2: 756. 1991; Bhandari, Fl. Indian Desert 301. 1995. *A. bracteata* Retz. Obs. Bot. 5: 29. 1788; Hook. *f.*, Fl. Brit. Ind. 5: 75. 1886; Duthie, Fl. Gangetic Plain 3: 44 1915. (Fig. 141 & Plate 285)

Vern: *Kidamar, Hukkabel.*

Perennial prostrate, herb, slender, glabrous, with an unpleasant smell. Stem and branches angular; branches spreading, decumbent, striate, glabrous. Leaves, glabrous, 3 - 8 cm across, reniform or broadly ovate, obtuse to subacute at apex, cordate at the base, margin wavy, glabrous above, glaucous beneath; petiole striate. Flowers solitary, axillary, bracteate. Pedicel 1 - 2 cm long. Bracts large, orbicular to subreniform, 1 - 1.3 cm long, glabrous. Perianth 2.5 - 4 cm long, irregular, tubular with a linear lip, glabrous externally, hairy within, dark purple on the side of the lip, greenish yellow on the other side and in the tubular portion; tube cylindrical, with trumpet-shaped mouth. Stamens 6, sessile, anthers united with a 6-lobed gynostegium, lobes subovate, acute, ca 1 mm long. Ovary inferior; placentation parietal in upper portion, placentae intruded, meeting in the centre, axile below. Capsule 1 - 3 cm long, 12-ribbed, glabrous. Seeds many, 5 - 7 mm long, flattened, deltoid, with cordate base, dark coloured.

Field note: Common among hedges, roadsides and in dry habitats.

Fl. & Fr.: August - December.

Specimen examined: Badnor 483, Matri Kundiya Dam 1822 (MLVGCB Herb).

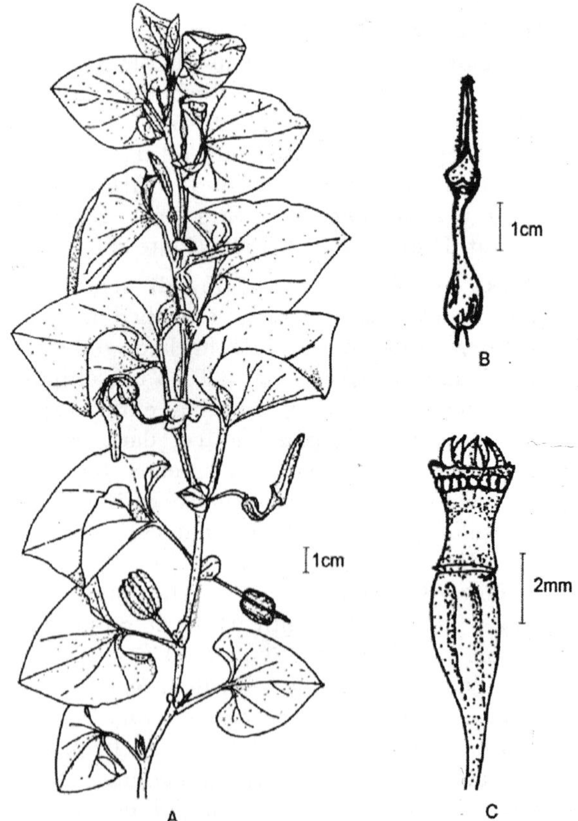

Fig. 141. *Aristolochia bracteolata* Lam.
A. Flowering & fruiting branch, B. Flower, C. Pistil

Distribution: Tropical Africa, Arabia, Sri Lanka, India and Pakistan.

Uses: Said to possess medicinal properties, used by local physicians as a purgative and for its anthelminthic properties. Dried root powder is referred as abortifacient. Leaves pounded with castor oil are used for the treatment of eczema and snakebite.

LORANTHACEAE

DENDROPHTHOE Mart

Dendrophthoe falcata (L.f.) Etting. in Denkschr. Akad. Wiss. Math.- Naturwise. 32: 52 - 53, 58. t. 13. f. 14. 1872; Parmar, in Shetty & Singh Fl. Rajasthan 2: 759. 1991. *Loranthus falcatus* L. f. Suppl. 211. 1781. *L. longiflorus* Desr. in Lamk., Encycl. Meth. 3 : 598. 1972;

Wight, Ic. 1: t. 302. 1840; Hook. *f.*, Fl. Brit. Ind. 5: 214. 1886, incl. var. *falcata* (L. f.) Kurz & var. *amplexifolia* (DC.) Thw.; Duthie, Fl. Gangetic Plain 3: 63 1915. (Plate 286)

Vern: *Parand, Dudhibel.*

A partial parasitic, perennial, dichotomously branched, glabrous shrub with grey bark. Branches woody, pendulous or erect on host. Leaves sub-sessile or shortly petioled, 4 - 15 x 2 - 8 cm, elliptic-oblong, obovate or ovate-lanceolate, mostly opposite, obtuse or acute at apex, entire, leathery, coriaceous, cordate or cuneate at base. Flower buds cylindrical, acute. Flowers 5 - 10 cm long, racemes, often clustered in leaf axils, red, 3.5 - 4 cm long; pedicels 2 - 4 cm long. Bract orbiculate, persistent. Calyx ca 5 cm long, entire or obscuely toothed, persistent, glabrous, splitting into 4-5 truncate lobes. Corolla bright red; tube 2 - 5 cm long, scarlet, 5-lobed; lobes ca 1 cm long, reflexed, linear-oblong, greenish. Stamens 5, filaments 6 mm long, glabrous; anthers 5 mm long. Style 3 - 3.5 cm long; stigma capitate, 1 mm long. Fruit 1 cm long, ovoid-oblong, glabrous, bright red, black when ripe, crowned by persistent calyx base. Seed 8 - 9 mm long, oblong.

Field note: A shrub with pretty red flowers, parasitic on *Boswellia serrata* Roxb. and other trees of the foot-hill zone.

Fl. & Fr.: August - February.

Specimen examined: Aroli 1538, Gurla 1726 (MLVGCB Herb).

Distribution: Tropical Africa, Indo-Malaysia, India, Sumatra, New-Guinea, Pakistan and Sri Lanka.

SANTALACEAE

SANTALUM L.

Santalum album L., Sp. Pl. 349. 1753; Hook. *f.*, Fl. Brit. Ind. 5: 231. 1886; Duthie, Fl. Gangetic Plain 3: 67. 1915; Parmar, in Shetty & Singh Fl. Rajasthan 2: 761. 1991. (Plate 287)

Vern: *Chandan*

A small to medium-sized tree, evergreen, semi-parasitic on root, with slender branches, sometimes reaching up to 10 m. in height, 2.4 m. in girth. The bark is reddish or dark-grey or nearly black, rough, with deep vertical cracks on old trees. Leaves glabrous, thin, elliptic or lanceolate, 1.5 - 8 x 1.6 - 3.2 cm, rounded at base, entire, subcoriaceous. Petiole 1 - 1.5 cm long. Flowers in terminal and axillary trichotomous paniculate cymes, small, 5-merous, reddish-brown, Pedicels as long as the perianth tube, opposite. Perianth purplish brown inside; lobes obovate, minutely ciliate; tube campanulate. Disc of 5, fleshy, ovoid scales. Drupe globose, 1.3 cm across, purple-black, annulate above with hard, ribbed endocarp. Seeds 1, globose or obovoid.

Field note: Common in hedges on the boundaries of crop fields.

Fl. & Fr.: March - October.

Specimen examined: Mandal 193, Mangrol 3822 (MLVGCB Herb).

Distribution: Sri Lanka, India (It is commonly found in comparatively dry regions of peninsular India. It has also been introduced into Rajasthan, Uttar Pradesh, Madhya Pradesh and Orissa, where it has become naturalized at some places).

Uses: A paste is made form wood by handgrinding it over stones; this paste is used for '*tilak*' on forehead in *Hindu* religion. Sandalwood oil has been used traditionally to treat skin diseases, acne, dysentery, gonorrhoea, and a number of other conditions. The oil from the seeds is used in skin troubles. Dried seeds are eaten by shepherds.

EUPHORBIACEAE

KEY TO THE GENERA

1a. Perianth totally absent, flowers arranged in cyathia 2

2a. Cyathia subactinomorphic. Glands of involucre free, bracts of male flowers also free .... *Euphorbia*

2b. Cyathia conspicously zygomorphic. Glands of involucre united in a ring, bracts of male flowers forming a tube round the female flowers .... *Pedilanthus*

1b. Flowers with perianth of sepals atleast present. 3

3a. Perianth two seriate, composed of five sepals and five petals in both types of flowers 4

4a. Trees 5

5a. Leaves palmately lobed. Capsules echinate .... *Ricinus*

5b. Leaves simple. Capsules not echinate ....*Bridelia*

4b. Shrubs or herbs... 6

6a. Shrubs .... *Jatropha*

6b. Herbs or undershrubs... 7

7a. Leaves entire or lobed. Inflorescence axillary. Stamens erect in bud. Seeds eastrophiolate... *Chrozophora*

7b. Leaves serrate. Inflorescence terminal or in bifurcations of the stems. Stamens incurved in bud. Seeds with white spongy cap... *Croton*

3b. Perianth one seriate, composed of sepals only 8

8a. Ovule 2, per chamber... *Phyllanthus*

8b. Ovule 1 per chamber... 9

9a. Female flowers with acrescent leafy bracts. Anther vermiculiform .... *Acalypha*

9b. Female flowers without such leafy bracts. Anther not vermiculiform .... *Baliospermum*

ACALYPHA L.

Acalypha ciliata Forssk. Fl. Aegypt.-Arab. 162. 1775; Hook. *f.,* Fl. Brit. Ind. 5: 417. 1887; Duthie, Fl. Gangetic Plain 3: 108. 1915; Parmar, in Shetty & Singh Fl. Rajasthan 2: 763. 1991; Bhandari, Fl. Indian Desert 304. 1995. (Plate 288)

Erect, annual herb, 25 - 80 cm high. Stem shortly pubescent with recurved hairs, at length glabrous, striate. Leaves petiolate, 3.0 - 7.5 x 1.8 - 4.5 cm, ovate-elliptic or rhomboid-ovate, caudate-acuminate, rounded at base, crenate-serrate, membranous, very slightly scabrous on both surfaces, digitately 5-nerved at base; lateral nerves slightly raised an both surfaces. Petioles 2.5 - 8 cm long. Flowers in axillary 1 - 1.8 cm long spikes, pale green; male flowers few, ebracteate sessile, at the top of spikes. Tepals minute. Stamens 8. Female flowers many, crowded in the lower part of the spike, greenish. Bracts sub-orbicular, 2 - 3 mm across, glabrous with about 20, laciniate teeth; teeth shortly ciliate. Tepals 3, ovate, very small, membranous, ciliate. Ovary trilobed, sparsely pubescent near apex; styles 3, filiform, laciniate. Capsules ca 2 mm long, whitish, enclosed in persistent bracts, glabrous, twisted after dehiscence. Seeds ovoid, ca 1.5 mm across, minutely pitted.

Field note: Common weed of moist-shaddy places, gardens, cultivated fields etc.

Fl. & Fr.: August - December.

Specimen examined: Rashmi 2119, Mandal 1658 (MLVGCB Herb).

Distribution: Tropical Africa, Arabia, Sri Lanka and India

BALIOSPERMUM Blume

Baliospermum montanum (Willd.) Muell-Arg., in DC. Prodr. 15 (2): 1125. 1866; Parmar, in Shetty & Singh Fl. Rajasthan 2: 766. 1991. *Jatropha montana* Willd. Sp. Pl. 4: 563. 1805. *Baliospermum axillare* Blume, Bijdr. 604. 1826; Hook. *f.,* Fl. Brit. Ind. 5: 461. 1887; Duthie, Fl. Gangetic Plain 3: 107. 1915. *B. polyandrum* Wight, I. c. Pl. Ind. Orient 5 (2): 23. t. 1885. 1852. (Plate 289)

Vern: *Jamal gota*

A stout herb or undershrub, perennial, erect, 1 - 1.5 m high; branches herbaceous from the roots, young branches hairy. Leaves alternate, variable in shape and size, ovate-oblong or elliptic-lanceolate, 5 - 22 x 3 - 18 cm, acute or acuminate at apex, rounded, subcordate or cuneate at base, pinninerved, simple, sinuate-toothed, dentate; lower ones large, 20 - 27 cm long, base rounded or acute with 2 glands, sometimes palmately 3-5 lobed. Petiole 5-12 cm long; Stipules glandular, glands 2. Flowers, in axillary or terminal panicles greenish-yellow. Male flowers terminal. Tepals 2 - 5 - 3 mm long, green, glabrous or hairy; lobes rounded, orbicular. Female flowers at base. Disk entire, glands soft, lobulate. Stamens 15 - 25. Ovary hairy. Capsules, 0.8-1.3 x 0.6-1.0 cm, 3-lobed, subglobose, obovoid, pubescent. Seeds ellipsoid, trigonous, smooth, glabrous, creamy-white, mottled with light black.

Field note: Common in waste lands and also as undergrowth in the forests.

Fl. & Fr.: September - March.

Specimen examined: Bangreda 1405, Keli 3984 (MLVGCB Herb).

Distribution: India and Malaya.

BRIDELIA Willd.

Bridelia retusa (L.) Spreng. Syst. Veg. 3: 48. 1826; Hook. *f.,* Fl. Brit. Ind. 5: 268. 1887, excl. vars.; Duthie, Fl. Gangetic Plain 3: 85 1915; Parmar, in Shetty & Singh Fl. Rajasthan 2: 767. 1991. *Cluytia retusa* L., Sp. Pl. 2: 1042. 1753. *Bridelia spinosa* (Roxb.) Willd. Sp. Pl. 4: 979. 1805; Hook. *f. l.c.* 5: 268. 1887; FGS 2: 614. 1978. (Fig. 142 & Plate 290)

Vern: *Khasai.*

Tree up to 7 m tall, spinous when young, bark grey. leaves obovate, elliptic-oblong, 4 - 15 × 2.5 - 6 cm, obtuse at apex, rounded at base, glabrous and bright green above, glacuous and finaly tomentose beneath; secondary nervs 15 - 20 pairs. Petiole 0.7 - 1.2 cm, slightly stout. Flowers in axillary clusters or long, axillay or terminal spikes with male and female flowers intermixed, greenish yellow, sessile or shortly pedicellate. Bracts ovate-triangular, 1.5-3 mm. Male flowers small, 4 - 6 mm in diam.; pedicels ca 1 mm. Tepals deltoid-ovate, ca 2 - 3 mm long, fleshy. Disc cup-shaped, up to 1.5 mm in diam.; staminal column ca 1.2 mm; free filaments ca 1 mm; anthers broadly ovoid; rudimentary ovary terete, apex undivided. Female flowers: tepals spathulate. Disc pulvinate and urceolate, enclosing and as long as ovary. Ovary globose to ovoid, up to 2 mm tall; styles 2, bifid. Capsules ovoid or depressed globose, 7-8 × 5-9 mm, black, supported by persistent calyx. Seeds semiglobose with an adaxial groove, ca 5 mm in diam., reddish brown, smooth.

Field note: Common in dry deciduous to evergreen forests towards the eastern part of the area.

Fl. & Fr.: September - March.

Specimen examined: Mainal 1219 (MLVGCB Herb).

Distribution: Bhutan, Cambodia, China, India, Indonesia (Sumatra), Laos, Myanmar, Nepal, Sri Lanka, Thailand and Vietnam.

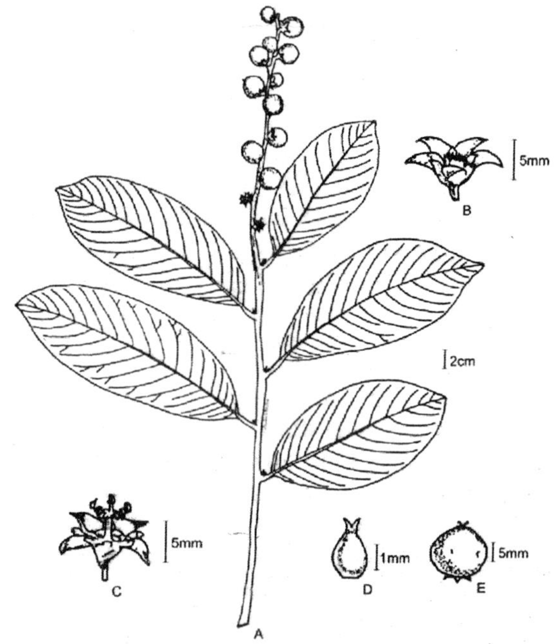

Fig. 142. *Bridelia retusa* (L.) Spreng. A. Flowering & fruiting twig. B. Male flower, C. Female flower, D. Ovary E. Fruit

Uses: The durable wood is used in various ways; bark contains up to 40% tannin.

CHROZOPHORA A. H. L. Juss. *nom. cons.*

Chrozophora rottleri (Geis.) A. Juss. ex Spreng. Syst. Veg. 3 : 850. 1826; Duthie, Fl. Gangetic Plain 3: 105. 1915; Balak, in Bull. Bot. Surv. India 15 4: 1976; Parmar, in Shetty & Singh Fl. Rajasthan 2: 771. 1991; Bhandari, Fl. Indian Desert 305. 1995. *Croton rottleri* Geis., Croton Monogr. 54. 1807. *Chrozophora tinctoria* non A. Juss. 1824; Hook. *f.,* Fl. Brit. Ind. 5: 408. 1887, *'pro parte'*. *Chrozophora plicata* (Vahl) Juss. ex Spreng. var. *rottleri* (Geis.) Müll.-Arg. in DC., Prodr. 15 (2): 747.1866. (Plate 291)

Herb to undershrub, annual, up to 60 cm high, stellate tomentose Stipules 1.3 - 2.7 x ca 0.3 mm. Leaves alternate, broadly ovate or suborbicular, entire or 3-lobed, 2.6 - 8 x 2.6 - 7 cm, truncate cuneate at base, undulate-crenate, stellately hairy, lower surface with 2 glands near the base. Flowers in axillary or terminal racemes. Bracts linear, ca 3 mm long, caducous. Male flowers sessile, 4 - 6 mm in diam., yellow; outer tepals ovate, 2.5 - 3 mm long, stellately hairy; inner tepals yellow, ovate-lonceolate, 2.2 - 5 mm long. Stamens 5 - 15; filaments connate at base, 1 - 3 seriate; anthers 0.9 - 1.3 x 0.7 mm, yellow. Female flowers pedicellate, 3.2-3.3 mm in diam., greenish - yellow; pedicel 1.4 - 2 mm long, elongating in fruit to up to 1.1 cm; outer tepals deltoid, ca 2 mm long, acute, hairy; inner tepals yellow, shorter, setaceous; style 3, 2-fid, 0.5 - 0.8 mm long, red; stigmas erect, up to 2.3 mm long, apically split for up to 1.8 mm, red. Capsules suglobose, ca 7 - 9 mm, stellataly tomentose; style persistent. Seeds globose, ca 3 mm long, ash-coloured.

Field note: Common in drying pounds ditches. stream beds. Prefers clay (mud) soil.

Fl. & Fr.: August - May.

Specimen examined: Nimbahera 114, Potla 2200 (MLVGCB Herb).

Distribution: India, Myanmar, Thailand, Andaman Islands, and Malaysia.

Uses: Fruits provide a blue-colouring dye, though the plant has never been cultivated for this purpose.

CROTON L.

Croton bonplandianum Bail., in Adansonia 4: 339. 1863-1864; Croizat, in J. Bombay Nat. Hist. Soc. 41: 573. 1940, "*bonplandianum*"; Raizada, in Suppl. Duthie, Fl. Gangetic Plain 247. 1976. *C. sparsiflorus* Morong, in Ann. N. Y. Acad. Sci. 7: 221. 1893; Parmar, in Shetty & Singh Fl. Rajasthan 2: 771. 1991. (Plate 292)

Annual to perennial herb, erect, diffusely branched, up to 60 cm high. Leaves crowded towards the top of branches, ovate-lanceolate, 2.5 - 7 x 1 - 3 cm, acute at apex, obtuse-subtruncate at base, serrate. Flowers in 5 - 8 cm long terminal racemes, greenish; male flowers above, female flowers fascicled in axil of minute bracts. Males, outer tepals obovate, minute; inner tepals linear-oblong. Stamens 10-15. Female tepals lanceolate. Ovary densely stellate-hairy; stigmas 2-fid. Capsules trigonous, 5-6 mm long, 3-angled, stellate-hairy. Seeds oblong, ca 4 mm long, shining, with white caruncle.

Field note: Common in waste lands.

Fl. & Fr.: February - May.

Specimen examined: Ajmer 3371, Fuliya kalan 2136 (MLVGCB Herb).

Distribution: Pantropical

EUPHORBIA L.

KEY TO THE SPECIES

1a. Plants dendroid. Armed with stipular spines. 2

2a. Leaves caducious, oblong-ovate. Style bifid.................................*E. caducifolia*

2b. Leaves not caducious, ovate-oblong. Style simple. ....*E. neriifolia*

1b. Plants not dendroid. unarmed........................3

3a. Leaves alternate..4

4a. Involucre with 4 to 5 well developed glands along the upper margin.........*E. dracunculoides*

4b. Involucre with a single large gland on one side ....*E. heterophylla*

3b. All leaves opposite..5

5a. Glands with a petaloid limb............................6

6a. Floral leaves distichously imbricate, concealing the involucre/capsule.......*E. elegans*

6b. Floral leaves neither distichously imbricate nor concealing the involucre/capsule...............................*E. indica*

5b. Glands without petaloid limb.........................7

7a. Erect or ascending herbs. Leaves upto 1.2 - 4 cm long ....*E. hirta*

7b. Prostrate or ascending herbs.
Leaves less than 1.2 cm long 8

8a. Stem hairy. Capsule pubescent
on angles.............................. ......*E. chamaesyce*

8b. Stem glabrous or nearly so.
Capsule glabrous.......................................9

9a. Leaves nearly as long as broad.
Seeds triangular ....*E. heyneana*

9b. Leaves much longer than broad.
Seeds distinctly 4-angled......... ......*E.clarkeana*

Euphorbia caducifolia Haines in Indian For. 40 : 154. 1914 et Bot. Bihar & Orissa 2 : 143. 1921; Fisher in Kew Bull. 341. 1925; Parmar, in Shetty & Singh Fl. Rajasthan 2: 775. 1991; Bhandari, Fl. Indian Desert 306. 1995. *E. neriifolia auct. plur.,* non L. 1753, nec. Roxb. 1832.

Vern: *Thor, Thur, Dandathor.*

Fleshy dendroid shrub, 1.5 - 2 m tall. Branches numerous arising from the very base, cylindrical, 3 - 5 cm in diam., smooth with small distant, slightly raised tubercles arranged in 5 spirals; each tubercle with a black areole bearing a pair of pointed, divergent spines, reddish when young, black at maturity. Leaves 1-9 x 1-4 cm, ovate or ovate-oblong, thick, fleshy, glabrous, apex acute, caducous. Flowers solitary or 2-3 together on short fleshy peduncles; when 3, the central usually male, lateral hermaphrodite. Pedicels stout, 1-5 mm long, arising in the axil of triangular, ovate bracts. Involucral lobes 4 - 5 mm in diam., broadly oblong, cuneate, toothed, scarsely fimbricate; glands, tumid, transversely oblong, glabrous; anthers yellow, subglobose, erectopatent, non-apiculate. Female flowers pedunculate. Perianth lobes 3, minute, rounded at apex. Style 3, connate upto their half length. Capsule sharply 3-angled, ca 8 mm across, plano-convex, glabrous. Seeds greyish-brown, smooth, ca 3 mm across, ovoid,with a marginal line.

Field note: Common hedge plant throughout the area, in western part of the study area it is growing on barren rocks and covers whole hillocks. It provides favourite support to a number of climbers.

Fl. & Fr.: January - April.

Specimen examined: Deogarh 3222, (MLVGCB Herb).

Distribution: India.

Uses: The plants are used to make fencing of the crop fields. Leaves are eaten by goats.

Euphorbia chamaesyce L., Sp. Pl. 455. 1753; Shah & Yogi, in Journ. Bombay Nat. Hist. Soc. 69: 488. 1972; Raizada, Suppl. Duthie, Fl. Gangetic Plain 247. 1976; Parmar, in Shetty & Singh Fl. Rajasthan 2: 776. 1991; Bhandari, Fl. Indian Desert 306. 1995. *E. prostrata* Aiton, Hort. Kew. 2: 139. 1879; Hook. *f.,* Fl. Brit. Ind. 5: 266. 1887. *Chamaesyce prostrata* (Aiton) Small. Fl. S. E. U. S. 713. 1903. (Fig. 143)

Vern: *Dudhi.*

Annual, herbs, 15 - 30 cm tall. Stems many from base, prostrate, usually light red or red, occasionally green or yellow-green, glabrous or sparsely pubescent. Leaves 2 - 7× 1 - 5 cm, ovate-oblong, slightly oblique, rounded at apex, cordate-rounded at base, adaxially green, sometimes with light red or red abaxially, margins entire or irregularly finely serrulate. Cyathia single, axillary or few clustered, peduncle 2 - 3 mm; involucre turbinate, ca. 1 × 1 mm, hairy outside, marginal lobes 5, triangular or rounded; glands 4, red, suborbicular, appendages white. Male flowers many, usually shorter than cup. Female flower pedicellate, exserted from involucre; ovary sparsely pubescent on angles; styles nearly connate at base; stigma 2-lobed. Capsule subglobose, 3-angular, ca. 1.5 × 1.4 mm, smooth, glabrous except for white hairs along angles. Seeds ovoid-tetragonal, ca 1 × 0.5 mm, obtuse at apex, truncate at base, yellow, each side with 6 or 7 transverse furrows; caruncle absent.

Field note: Common on moist sandy soils, along roadsides and cultivated fields throughout the area.

Fl. & Fr.: Almost throughout the year.

Specimen examined: Banera 1327 (MLVGCB Herb).

Distribution: Tropical and subtropical Americas; naturalized in many parts of the Old World.

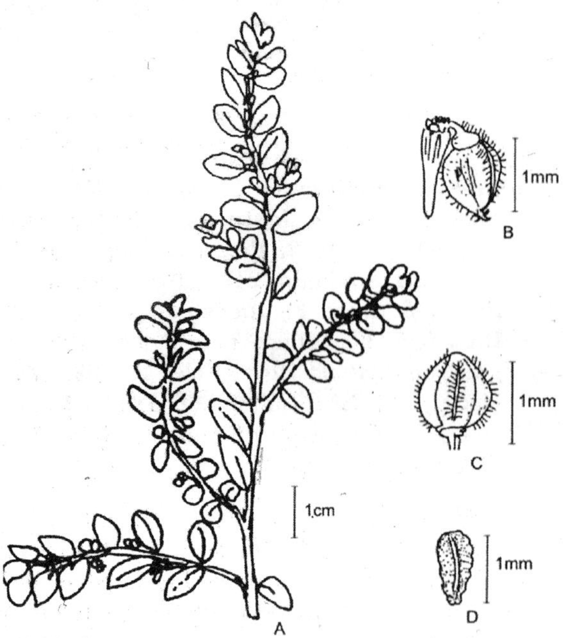

Fig. 143. *Euphorbia chamaesyce* L.
A. Flowering & fruiting branch, B. Cyathium, C. Fruit, D. Seed

Euphorbia clarkeana Hook. *f.*, Fl. Brit. Ind. 5: 253. 1887; Duthie, Fl. Gangetic Plain 3: 82. 1915; Parmar, in Shetty & Singh Fl. Rajasthan 2: 776. 1991; Bhandari, Fl. Indian Desert 307. 1995. *E. heyneana* Bioss., in DC. Prodr. 15 (2): 35. 1862, *pro parte*, non Spreng. 1826.

Vern: *Dudhi*.

Diffuse, prostrate or decumbent herbs with many stems, filiform, glabrous. Leaves 3 - 6 x 1 - 4 mm, opposite, much longer than broad, obliquely linear-oblong, base almost auricled on one side, entire or toothed at the rounded tip, glabrous, often crowded towards the ends. Petiole very small (less than 1 mm long). Stipules rather large, setaceous from a broad toothed base. Cyathia usually solitary in each axil. Involucre stalked, campanulate, glabrous, with 5, lanceolate, toothed lobes; glands minute, without limb. Capsules ca 2 mm across, glabrous, trigonous; cocci keeled. Seeds ca 1.5 mm long, ovoid, acutely 4-angled, reddish, transversly rugose.

Field note: Common in cultivated fields.

Fl. & Fr.: September - December.

Specimen examined: Banera 1613 (MLVGCB Herb).

Distribution: Pakistan and India.

Euphorbia dracunculoides Lam., Dict. 2 : 428. 1786 et Encycl. 2 : 428. 1788; Hook. *f.*, Fl. Brit. Ind. 5: 262. 1887; Duthie, Fl. Gangetic Plain 3: 78. 1915; Mahesh. Illus. Fl. Delhi f. 192. 1966; Parmar, in Shetty & Singh Fl. Rajasthan 2: 777. 1991; Bhandari, Fl. Indian Desert 307. 1995. *E. angustifolia* Buch.-Ham. ex D. Don, Prodr. Fl. Nepal. 62. 1825; Hook. *f.* *l. c.* 5: 265. 1887. (Plate 293)

Vern: *Marbui*.

Annual, glabrous, 10 - 40 cm tall herbs. Stem dichotomously branched. Leaves, 1 - 3 × 0.2 - 0.5 cm, sessile, alternate, linear or linear-lanceolate, acute, entire. Cyathium subsessile, solitary in each axil, hairy inside, involucre campanulate, 2 - 3 × 3 - 5 mm, glabrous outside, hairy inside, subsessile, lobes ovate ciliate; glands 4, pale brown, crescent-shaped, apex 2-horned, horns light green. Male flowers many, not exserted from involucre. Female flower: ovary exserted from cup, smooth, glabrous; styles free, persistent; style arms 2-lobed. Fruiting peduncle ca 3 mm. Capsule, 3-lobed, 3-seeded, subglobose, ca 3.5 × 3.5 mm, smooth or obscurely reticulate, glabrous. Seeds ellipsoid, ca 2 - 2.5 × 1.5 - 2 mm, testa white, leprous, tuberculate; caruncle present, stipitate.

Field note: Common weed cultivated fields during winters. It is regarded as a problematic weed of certain crops in the area.

Fl. & Fr.: February - May.

Specimen examined: Atun 192 (MLVGCB Herb).

Distribution: Tropical Africa, Arabia, Pakistan and India (Throughout plains).

Euphorbia elegans Spreng, Syst. Veg. 3: 794. 1826; Hook. *f.*, Fl. Brit. Ind. 5: 246. 1887; Duthie, Fl. Gangetic Plain 3: 79. 1915; Parmar, in Shetty & Singh Fl. Rajasthan 2: 777. 1991; Bhandari, Fl. Indian Desert 307. 1995. (Plate 294)

Annual erect herb, upto 40 cm, clothed with long spreading hairs. Branches dichotomous, often reddish. Leaves 0.5 - 2.5 x 0.4 - 1.5 cm, subsessile, opposite, distant, obtuse, ovate or orbicular at apex, serrulate. Cyathia 1-3 axillary/ subterminal within distichous imbricate. Floral leaves (bracts) broadly elliptic or obovate, obtuse, hairy, softly inside, much imbricating, forming short spikes at the ends of the branches, subsissile. Involucre turbinate, pubescent on both sides; lobes rosy, obovate, cunnate, sinuate at apex. Glands 4, limb of glands large, sinuate, rosy. Capsules pubescent, subglobose, 3 - 4 mm across. Seeds ca 2 mm long, tetragonous, unequal ovoid, parallel-grooved, glabrous, yellowish-brown.

Field note: Rare, in forest undergrowth.

Fl. & Fr.: September - December.

Specimen examined: On the way of Bijoliya 1245 (MLVGCB Herb).

Distribution: Western India.

Euphorbia heterophylla L., Sp. Pl. 453. 1753; A. R. Smith in Kew Bull. 26: 264. 1972; Tripathi *et al.*, Fl. Elem. Madhya Pradesh 183. 1994; Shukla, in Fl. Madhya Pradesh vol. II: 549. 1997. *E. geniculata* Ort., Nov. Pl. Desr. Dec. 18. 1797. *E. prunifolia* Jacq., Pl. Hort. Schoenbr. 3: 15. 1798; Hook. *f.*, Fl. Brit. Ind. 5: 266. 1887, in adnot. *Poinsettia geniculata* (Ortega) Klotzsch & Garcke, Montasber. Akad. Berlin. 1859: 253. 1859. *P. heterophylla* (L.) Klotzsch & Garcke, Monatsber. Akad. Berlin 1859: 253. 1859; Dressler, in Ann. Mo. Bet. Gard. 48(4): 339. 1962. *P. prunifolia* (Jacq.) Klotzsch & Garcke, Monatsber. Akad. Berlin. 1859: 253. 1859. (Plate 295)

Vern.: *Dudhi.*

An erect subglabrous or sparingly-pubescent annual herb, up to 90 cm tall. Stem fistular, green, branched in upper part. Leaves variable, mostly opposite, all petiolate, the petioles up to 4 cm long. Leaf-blades variable, linear-lanceolate, oblong, ovate or elliptic to obovate, rounded, 1 - 15 x 0.5 - 8 cm, the upper often narrower than the lower, acute, subacute or obtuse at the apex, cuneate or rounded-cuneate at the base, entire, denticulate or dentate, lateral nerves up to 20 pairs, subglabrous and dull green above, sparingly-puberulous and whitish beneath, the upper leaves sometimes faintly purplish-blotched at the base. Stipules sessile, glandular. Inflorescence a pseudo-dichasium or rarely a triradiate pseudo-polychasium, with the cyathia aggregated together in dense terminal clusters. Lobes 5, ovate, fimbriate. Glands 1 - 2 (-3), stipitate, circular, concave. Capsules roundly trilobate, 4 - 4.5 x 5 - 5.5 mm, smooth, glabrous. Styles connate at the base, 1 mm long, bifid. Seeds is quadrangular-ovoid, basally truncate, 2 - 2.5 x 1.5 - 2 mm, coarsely bluntly tuberculate, dark brown, ecarunculate.

Field note: Common in crop fields, farms and gardens

Fl. & Fr.: September - December.

Specimen examined: Nimbahera 4019 (MLVGCB Herb).

Distribution: Throughout Tropical America from S. USA southwards; widely introduced into the Old World Tropics.

Uses: It is a good fodder.

Euphorbia heyneana Spreng., Syst. Veg. 3 : 791. 1826; Parmar, in Shetty & Singh Fl. Rajasthan 2: 779. 1991; Bhandari, Fl. Indian Desert 309. 1995. *E. microphylla* Heyne ex Roth, Nov. Pl. Sp. 229. 1821, non Lam. 1788; Hook. *f.*, Fl. Brit. Ind. 5: 252. 1887; Duthie, Fl. Gangetic Plain 3: 82. 1915. *E. orbiculata* Miq. Fl. Ind. Bat. 1 (2) : 421. 1859, non H.B.K. 1817. *E. bombaiensis* Santapau in Bull. Bot. Soc. Bengal 8 : 17. 1955, *excl. syn.*

Vern: *Dudhi.*

Annual, prostrate herb, 8-17 cm tall. Stems many from base, dichotomously branched, 1-2 mm thick, glaucous or green. Leaves opposite, 2 - 8 × 1 - 5 mm, obliquely oblong, base extremely oblique, margin subentire, entire at base, serrate in upper part, apex rounded, glabrous; stipules membranous, subulate, caducous. Petiole ca 1 mm long. Cyathia single or paired, axillary. Involucre numerous, subsolitary, axillary, broadly campanulate, ca 0.5 × 0.4 mm, glabrous; lobes 5, deltoid, apex acute; glands 4, narrowly elliptic, appendages white, very narrow. Male flowers 3-5, not exserted from involucre or reaching margin. Female flower pedicellate, nearly reaching margin of involucre; ovary glabrous; styles connate at base; stigma 2-lobed. Capsule 3-angular-globose, ca. 1.7 × 1.5 mm, smooth, glabrous. Seeds ca 1 mm long, ovoid, obtusely 4-angled, brown, smooth; caruncle absent.

Field note: Common in rocky crevices and marshy places.

Fl. & Fr.: August - December.

Specimen examined: Mandalgarh 1539 (MLVGCB Herb).

Distribution: India and Java.

Euphorbia hirta L., Sp. Pl. 2: 454. 1753; Duthie, Fl. Gangetic Plain 3: 80. 1915; Mahesh. Illus. Fl. Delhi f. 193. 1966; Parmar, in Shetty & Singh Fl. Rajasthan 2: 780. 1991; Bhandari, Fl. Indian Desert 309. 1995. *E. pilulifera auct. plur.* non L. 1753; Hook. *f.*, Fl. Brit. Ind. 5: 250. 1887. *Chamaesyce hirta* (L.) Millsp., in Field Mus. Bot. 2: 303. 1909. (Fig. 144)

An annual, herbs 20 - 35 cm tall, usually few branched. Root fibrous, 3 - 5 mm. Stem branched from middle or above, ascending to erect, rarely prostrate, ca 3 mm thick, with mixture of long yellow-brown multicellular hairs and much shorter white hairs. Leaves opposite, ovate-oblong to elliptic, 1.5 - 2.5 × 0.5 - 1.5 cm, adaxially green to red, both surfaces pilose, denser abaxially, base slightly oblique, margins entire or few serrulate below middle, finely serrulate above middle, apex acuminate or obtuse. Stipules membranous, triangular, 0.8-1.7 mm, caducous. Petiole 1-3.5 mm. Cyathia in dense, often headlike, pedunculate cymes at upper nodes, peduncle to 25 mm, all parts very hairy; involucre campanulate, ca 1 × 1 mm, pilose; marginal lobes 5, acute, hairy, triangular-ovate; glands 4, red, rounded to transversely elliptic, center slightly sunken, appendages white to reddish, orbicular, to 0.3 × 0.2 mm, margins entire to slightly undulate, truncate at apex. Male flowers 4 or 5; anthers red. Female flower: pedicel short, exserted from involucre; ovary 3-angular, sparsely pilose; styles free; stigma slightly 2-lobed. Capsule 3-angular, depressed globose, 1 - 1.5 × 1 - 1.5 mm, smooth, shortly pilose; fruiting peduncle to 1.5 mm. Seeds globose, 4-angled, 7 - 9 × 4 - 5 mm, reddish, sides transversely furrowed; caruncle absent.

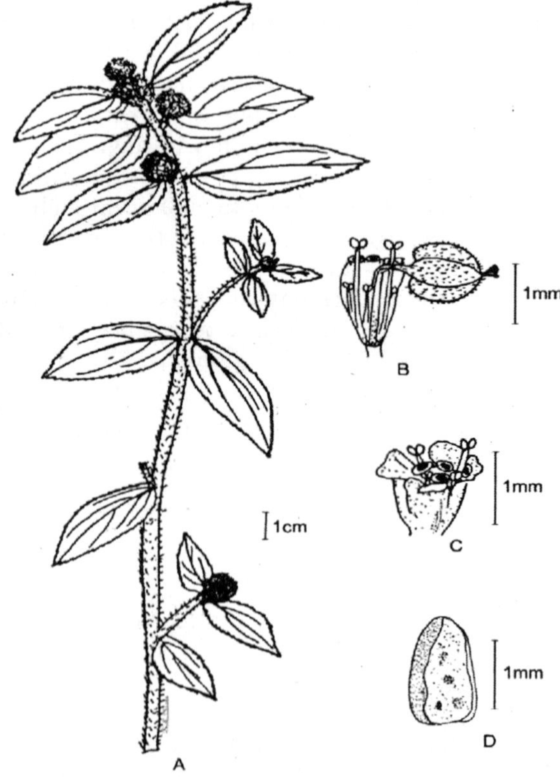

Fig. 144. *Euphorbia hirta* L.
A. Flowering & fruiting branch, B. Involucre with glands & flowers, C. Flower, D. Seed

Field note: Very common in fields, roadsides, scrub and open forests.

Fl. & Fr.: Almost throughout the year.

Specimen examined: Bhim 1569 (MLVGCB Herb).

Distribution: Tropical and subtropical regions in both hemispheres.

Uses: This species is used medicinally. Decoction of whole plant is effective in bronchial infection, cough, asthma and in de-worming. Whole plant water extract, checks diarrhoea & gonorrhoea.The milky juice of plant is useful in promoting hair growth, in the treatment of warts & cracking of skin, it should be applied externally on the affected area. Leaf extract is applied on burns and skin disorders. Leaves are crushed and applied as paste in joints pain.

Euphorbia indica Lam., Encycl. 2 : 423. 1788; Raju & Rao in Indian Juorn. Bot. 2 : 205. 1979; Parmar, in Shetty & Singh Fl. Rajasthan 2: 780. 1991; Bhandari, Fl. Indian Desert 310. 1995. *E. hypericifolia auct.*, non L. 1753; Hook. *f.*, Fl. Brit. Ind. 5: 249. 1887; Duthie, Fl. Gangetic Plain 3: 80. 1915. *E. perviflora auct.*, non L. 1759; Roxb., Fl. Ind. 2 : 472. 1832; Santapau in Rec. Bot. Surv. Ind. 16 : 273. 1959; Bhandari, Fl. Indian Desert 343. 1978. *E. hypericifoliya* non L. var. *parviflora* Hook. *f.* in Prain, Beng. Pl. 2: 924. 1903.

Vern: *Dudhi.*

Annual herb, upto 60 cm tall. Stems spreading or erect, often purplish tinged, sparsely white pubescent. Leaves opposite, ovate, 1.5- 4 x 0.5-2.0 cm, adaxially almost glabrous, abaxially glaucous, base obliquely rounded, margin obscurely toothed, apex rounded; stipules triangular, upto 1.5 mm, laciniate; petiole to 3 mm; Cyathia in axillary or terminal pedunculate capitate cymes, peduncle to 30 mm; involucre cuplike, ca 1 × 1 mm, pubescent, marginal lobes triangular, apex acute; glands 4, green, rounded, appendages white, upto 1 mm in diam., often smaller. Male flowers slightly exserted. Female flower exserted from involucre; ovary pubescent; styles free; stigma deeply 2-lobed. Capsule 3-angular-ovoid, ca 1.5 × 2 mm, smooth, pubescent; fruiting pedicel ca 1.5 mm. Seeds ovoid-tetragonal, ca 1 × 0.75 mm, reddish brown, sides obscurely transversely furrowed; caruncle absent.

Field note: Common in waste lands and as a weed of cultivated fields.

Fl. & Fr.: September - December.

Specimen examined: Bhilwara 182 (MLVGCB Herb).

Distribution: Tropical Africa and Asia.

Notes: Until recently, this species was consistently named as *Euphorbia hypericifolia*, but that New World species is completely glabrous and with smaller capsules and dimpled seeds (see Raju & Rao, Indian J. Bot. 2: 202. 1979).

Euphorbia neriifolia L., Sp. Pl. 1: 451. 1753; Hook. *f.,* Fl. Brit. Ind. 5: 255. 1887; Duthie, Fl. Gangetic Plain 3: 76. 1915, *pro parte*; Mahesh. Illus. Fl. Delhi f. 191. 1966. *E. ligularia* Roxb., Fl. India 2: 465. 1832. (Plate 296)

Vern: *Thor, Thur.*

Large shrub or small tree, 2 - 3 m tall, branched, glabrous. Stems 6 - 15 cm thick, green, with 5 spiral ranks of tubercles, ascending, branching from upper parts. Stipular spines sharp, thick. Leaves alternate, crowded towards the apex of branched, persistent, succulent,; obovate-oblong or spoon-shaped, 15 - 25 × 5 - 8 cm, fleshy, base attenuate, margins entire, apex obtuse, apiculate; veins inconspicuous. Petiole short, 2 - 4 mm Cyathia in subterminal cymes, peduncle ca 3 mm; cyathophylls membranous, caducous; involucre campanulate, ca 4 × 5 - 6 mm, lobes 5, rounded; glands 5, succulent, thick, entire. Male flowers many, bracts linear. Female flowers rarely developed. Capsules 1 mm across, obovoid-trigonous, deep red or pale-brown, smooth, glabrous. Seeds ca 5 mm across, smooth.

Field note: Common in rocky habitats, also planted as hedge plant along the boundaries of cultivated fields.

Fl. & Fr.: March- July.

Specimen examined: Banera 1242, Chhajela khera 1915 (MLVGCB Herb).

Distribution: Native to India; widely cultivated in tropical Asia.

Uses: Stem and leaves are used medicinally.

JATROPHA L.

KEY TO THE SPECIES

1a. Plants eglandular. Leaves palmately lobed. Flowers greenish yellow. Petals connate to central column .... *J. curcas*

1b. Plants glandular hairy. Leaves palmately lobed. Flowers red. Petals free .... *J. gossypifolia*

Jatropha curcas L., Sp. Pl. 2: 1006. 1753; Hook. *f.,* Fl. Brit. Ind. 5: 383. 1887; Duthie, Fl. Gangetic Plain 3: 116. 1915; Parmar, in Shetty & Singh Fl. Rajasthan 2: 784. 1991.

Vern: *Ratan jot.*

Shrub or small tree, 2 - 5 m tall, with watery latex; bark smooth; branches glaucous-gray, glabrous, sparsely lenticellate, pith larger. Leaves broadly ovate, cordiform, 4 - 15 × 2.5 - 13 cm, entire or shallowly 5 - lobed, apex acute; palmate, veins 5 - 7. Stipules small; petioles 6 - 18 cm. Flowers in terminal cymose panicles, small, yellowish; bracts lanceolate, 4 - 8 mm. Male flowers: sepals 5, ca 4 mm, connate at base; petals oblong, green-yellow, ca 6 mm, connate to middle, hairy inside; disc glands 5, nearly terete; stamens 10, 2-seriate; outer 5 filaments free, ca 3 mm long, inner filaments connate in lower part. Female flowers: pedicels elongate; sepals free, ca 6 mm; petals and disc glands as in male; ovary 3-locular, ca 2 mm across, glabrous; styles ca 5 mm across, bifid at apex. Capsules ellipsoidal or globose, 2.5 - 3 cm, yellow when ripe. Seeds ellipsoidal, 1.5 - 2 cm, black.

Field note: Commonly grown as hedge plant along boundaries of fields, grasslands etc. The seeds are poisonous and many times it cause, serious problems in childrens.

Fl. & Fr.: Almost throughout the year.

Specimen examined: Bhilwara 66, Ganeshpura 1572 (MLVGCB Herb).

Distribution: Native to tropical America; widely introduced.

Uses: *Jatropha curcas* seeds are the source of curcas oil, a very powerful purgative, also used for making candles and soap, for lighting, and as a lubricant. It is biofuel crop for areas too dry for normal crops. The bark is used as medicine for lepriasis (leprosy), itchy skin, eczema, chronic ulcers, etc.

Jatropha gossypifolia L., Sp. Pl. 2: 1006. 1753; Hook. *f.*, Fl. Brit. Ind. 5: 383. 1887; Raizada, Suppl. Duthie, Fl. Gangetic Plain 251. 1976; Parmar, in Shetty & Singh Fl. Rajasthan 2: 784. 1991.

Vern: *Ratan jot.*

Shrub to about 1.5 m tall. Branches glandular hairy, deep reddish-purple. Leaves 5 - 13 x 5 - 14, subcordate, palmately 3 - 5-lobed; lobes obovate, 3-nerved, purplish, stick with capitate glandular hairs, slightly serrate, cordate at the base, acute at apex. Petioles about 6-9 cm long. Flowers in axillary or terminal corymbose cymes, small dark red. Bracts ovate, acute. Outer tepals ca 6 mm long, glandular hairy, acuminate; inner tepals ovate, ca 4 mm long, obtuse, ciliate. Stamens 8, 5 + 3; filaments of inner whorls connate at the base, longer than outer ones. Ovary ca 2 mm across; styles ca 1.5 mm long. Capsules ca 2 cm across, oblong, 3-lobed, smooth, glabrous, brown when ripe. Seeds oblong, 3-angled, glabrous, greyish-brown with pale-yellow or brown strophiole.

Field note: Planted in the gardens and on the edges of fields.

Fl. & Fr.: Almost throughout the year.

Specimen examined: Bhilwara 67 (MLVGCB Herb).

Distribution: Tropical America, Africa and Asia.

PEDILANTHUS Neck. ex Poit. *nom. cons.*

Pedilanthus tithymaloides (L.) Poit., Ann. Mus. Natl. Hist. Nat. Pairs 19: 390. t. 19. 1812; Hook. *f.*, Fl. Brit. Ind. 5: 239. 1887; Parmar, in Shetty & Singh Fl. Rajasthan 2: 795. 1991. *E. tithymaloides* L., Sp. Pl. 453. 1753.

Vern: *Nagphanhi.*

Erect subshrub 40-70 cm tall; stem stout, slightly fleshy, strongly flexuous, glabrous or pubescent when young. Leaves distichous ovate or ovate-lanceolate, 3.5 - 8 × 2.5 - 5 cm, fleshy, both surfaces pubescent, glabrescent, truncate-cuneate at base, acute-obtuse at apex; midrib elevated abaxially, Stipules small, with a rounded ca 1 mm in diam. gland. Cyathia in dense terminal or axillary cymes; involucres 2 - 13 mm long, shoe-shaped, deep-red or purple-red, equilateral, ca. 1 cm, glabrous, apex nearly labiate-shaped, 2-fid, lobes small, oblong, 3-serrulate at apex, other lobe boat-shaped, ca. 1 cm, 2-partite at apex. Male flower: pedicels slender, 2.5-4 mm, glabrous, similar to filaments; anthers globose, slightly shorter than filaments. Female flower inserted at center of involucres, oblique, exserted from involucre; pedicels 6-8 mm, glabrous; ovary fusiform; styles usually connate; stigma 3, 2-fid.

Field note: This species is cultivated as an ornamental, often as hedge plant.

Fl. & Fr.: March-July.

Specimen examined: Bhilwara 68, Nimbahera 2934 (MLVGCB Herb).

Distribution: Native to Central America, cultivated throughout the tropics.

Uses: All parts are used as medicine for traumatic injury and fractures.

PHYLLANTHUS L.

KEY TO THE SPECIES

1a. Large trees .... *P. emblica*

1b. Small herbs 2

2a. Stipules peltate. Leaves obovate ....*P. maderaspatensis*

2b. Stipules subulate. Leaves elliptic-oblong obtuse .... *P. fraternus*

Phyllanthus emblica L., Sp. Pl. 2: 982. 1753; Hook. *f.,* Fl. Brit. Ind. 5: 289. 1887; Duthie, Fl. Gangetic Plain 3: 96. 1915; Parmar, in Shetty & Singh Fl. Rajasthan 2: 788. 1991. *Emblica officinalis* Gaertn. Fruct. 2: 122. t. 108. 1791; Wight, Ic. 5 (2): 26. t. 1896. 1852.

Vern: *Anvara, Amla.*

Tree 3-15 m tall, monoecious, deciduous; bark brownish, exfoloating into scales; main stem terete, sparsely lenticellate, with very reduced short shoots producing groups of leafy shoots; leafy shoots angular, tawny pubescent. Leaves distichous, oblong, 8 - 14 × 1.5 - 3 mm, papery to leathery, base shallowly cordate and slightly oblique, by subcordate at base, obtuse at apex, mucronate or retuse at tip; lateral veins 4 - 7 pairs. stipules triangular-ovate, 0.8 - 1.5 mm, brown, margins entire or denticulate, ciliate; petiole 0.3 - 0.7 mm long. Flowers in axillary clusters towards lower portion of branches, greenish yellow. Male flowers: pedicels 1-2.5 mm; sepals 6, membranous, yellow, obovate or spatulate, subequal, 1.2 - 2.5 × 0.5 - 1 mm, apex obtuse or rounded, margins entire or shallowly denticulate; disc glands 6, subtriangular; stamens 3; filaments coherent into column, 0.3 - 0.7 mm; anthers erect, oblong, 0.5 - 0.9 mm, longitudinally dehiscent, apex mucronate. Female flowers: pedicels ca. 0.5 mm; sepals 6, oblong or spatulate, 1.6 - 2.5 × 0.7 - 1.3 mm, apex obtuse or rounded, thicker, margins membranous, ± lobate; ovary ovoid, ca. 1.5 mm, 3-celled; styles 3, (1-)2.5 - 4 mm, connate at base, deeply bifid, lobes divided at tip. Drupe globose, 1 - 5 cm across, exocarp fleshy, obscurely 6 - lobed, pale green or yellowish white, endocarp crustaceous. Seeds 6 mm long, reddish, 3 - angled, 5 - 6 × 2 - 3 mm.

Field note: Commonly cultivated for its commercial fruits, also distributed in dry open sparse forests.

Fl. & Fr.: February - April.

Specimen examined: Kumbhalgarh 3120 (MLVGCB Herb).

Distribution: Bhutan, Cambodia, China, India, Indonesia, Laos, Malaysia, Myanmar, Nepal, Philippines, Sri Lanka, Thailand; South America (cultivated).

Uses: The mature fruits are very sour and contain 1%-1.8% vitamin C. They are used raw or sweetened or preserved. The seeds, roots, and leaves are used as medicine.

Phyllanthus fraternus Webster, Contr. Grey. Herb. 176: 53. 1955; Raizada, Suppl. Duthie, Fl. Gangetic Plain 254. 1976; Parmar, in Shetty & Singh Fl. Rajasthan 2: 788. 1991. *P. niruri auct. plurl.* non L. 1753; Wight, Ic. 5 (2): 25. t. 1894. 1852; Hook. *f.,* Fl. Brit. Ind. 5: 298. 1887; Duthie, Fl. Gangetic Plain 3: 98. 1915. *P. asperulatus auct. plur.* non Hutch. 1920; Sharma & Tiagi. Fl. N.E. Rajasthan 379. 1979.

Vern: *Bhumi amla.*

Annual erect herb, up to 60 cm tall, glabrous. Branched spreading, resembling a compound leaf. Leaves 0.5 - 1.7 x 0.3 - 0.9 cm, oblong, rounded at apex, cuneate at base, glabrous, distichous often overlapping. Petioles ca 0.5 - 0.8 mm long. Stipules 0.8 - 1 x 0.2 - 0.4 mm, lanceolate, truncate at base. Flowers monoecious, yellowish-green, solitary or several in axillary fascicles, often one female flower and several male ones (sometimes absent) per fascicle; male flowers: pedicels ca. 0.5 - 1 mm long. Perianth lobes 6, elliptic-obovate or obtuse, 1.2 - 1.5 x 1 - 1.5 mm, apex obtuse or rounded, yellow-green at middle part, reddish at base, margins membranous; disc-glands 6, obovate, 0.25 - 0.4 mm wide; stamens 3, filaments completely connate at base forming a column, 0.2 - 0.3 mm high, anthers nearly globose, anther-cell longitudinally dehiscent. Female flowers solitary; pedicels 1.5 - 4 mm long; perianth lobes 6, unequal, linear depressed-globose, 1.5 - 2.3 x 1.2 - 1.8 mm, apex obtuse or rounded, green at middle part; disc discoidal; ovaries globose, smooth, 3-celled; styles 3, free, bifid at apex, lobes revolute. Capsules globose, ca 2 - 3 mm across, brown-red, smooth, 3-gonous cocci when maturity, columella and sepals persistent. Seeds 3-gonous, longitudinally ribbed, pale-brown.

Field note: Common in fields, roadsides, waste place and in forest edges.

Fl. & Fr.: August - December.

Specimen examined: Khamnor 3820, Kakroliya 1728 (MLVGCB Herb).

Distribution: China, Taiwan, India, Indochina, Philippines to Tropical America

Uses: Whole plant is used as herbal medicine for hepatitis

Phyllanthus maderaspatensis L., Sp. Pl. 2: 982. 1753; Wight, Ic. 5 (2) : 25. t. 1895. f. 3. 1852; Hook. *f.,* Fl. Brit. Ind. 5: 292. 1887; Duthie, Fl. Gangetic Plain 3: 97. 1915; Parmar, in Shetty & Singh Fl. Rajasthan 2: 789. 1991; Bhandari, Fl. Indian Desert 313. 1995.

Perennial, erect, 45 cm tall herb, monoecious, glabrous throughout; stem usually woody at base; branching mostly basal, all similar, erect or ascending. Leaves spiral, oblong or obovate, 1.5 - 4 × 0.5 - 1 cm, leathery, grey-green, acute at base, mucronate at apex. Stipules ovate, 1.5 - 2mm; petiole very short. Flowers in axillary fascicles, most axils with 1 female flower, distal axils with 1 - 4 male flowers. Male flowers: pedicels ca 1 mm; sepals 6, suborbicular, ca 1 × 0.8 mm, margins entire; disc glands 6, small and smooth; stamens 3; filaments connate at base; anthers longitudinally dehiscent. Female flowers: pedicels 1.5 - 2 mm; sepals 6, suborbicular to subspatulate, ca 2 × 1.5 mm, margin entire; persistent in fruit; disc glands 6; ovary 3-celled, smooth; styles 3, free. Capsules, globose, 1.2 - 3 × ca 3 mm across, olive green, smooth. Seeds 3-sided, ca 1.3 mm, 3-angled, light brown, with rows of minute tubercles on back and sides.

Field note: Common elsewhere often in disturbed areas.

Fl. & Fr.: August - October.

Specimen examined: Bhilwara 129 (MLVGCB Herb).

Distribution: China, India, Indonesia, Pakistan, Sri Lanka; Africa, SW Asia and Australia.

RICINUS L.

Ricinus communis L., Sp. Pl. 1007. 1753; Hook. *f.,* Fl. Brit. Ind. 5: 457. 1887; Duthie, Fl. Gangetic Plain 3: 113. 1915; Parmar, in Shetty & Singh Fl. Rajasthan 2: 791. 1991.

Vern: *Arandi, Endiyo.*

Large shrub to small tree, evergreen, soft wooded, up to 6 m high. Leaves alternately 10 - 40 cm in diam., palmately 5 - 13 lobed; lobes ovate-lanceolate, acuminate at apex, irrregularly serrate, green, red or brown. Stipules triangular, symmetric, parallel-nerved. Petioles 8 - 25 cm long, with one or two glands at the apex. Flowers in terminal pyramidal, leaf-opposed panicles, fragrant; pedicels jointed; tepals 5, ovat-lanceolate, dehiscing often into 3 parts. Male flowers: tepals unequal, ca 5 mm long, yellowish green, pale yellow; stamens cream, white, or light yellow. Female flowers: tepals 8 - 10 mm long, green to red; ovary usually echinate outside with long papillae ending in a translucent stinging hair on top; styles 3, 2-fid. Capsules depressed-globose, of three 2-valved cocci, echinate. Seed oblong, smooth, usually marbled with various shades of brown and white, shiny.

Field note: Cultivated for oil-yielding seeds, also naturalized in waste lands.

Fl. & Fr.: Almost throughout the year.

Specimen examined: Bhilwara 175, Railmagara 934 (MLVGCB Herb).

Distribution: World-wide in tropical and subtropical areas.

Uses: The seeds are used to produce oil (castor oil, wonder oil), but the plants also have a high horticultural value. Plant is taken under improvement schemes carried out to increase its oil content.

Local uses: In case of muscular injury without bleeding, the leaves along with mustard oil are applied on the affected area.The leaf is put on to head to relieve headache. Leaf is boiled with maize grains and used as a rat killer.

Notes: This species is morphologically very variable, presumably due to centuries of cultivation and escapes from cultivation (seeds are known from Egyptian pharaoh graves dating from 4000 B.C.). Therefore, only a qualitative description has been given of the species. Many species and infraspecific taxa have been described (see Müll. Arg., 1866; Pax & K. Hoffm., 1919), but all species names are synonyms of *Ricinus communis* (or synonyms of taxa accidentally confused with *Ricinus*). The infraspecific taxa and also species names probably refer to certain cultivars. No synonyms will be treated here, partly because there are too many and partly because few or none have been described for Thailand.

CULTIVATED SPECIES

Euphorbia milii Des Moulins, Bull. Hist. Nat. Soc. L. Bordeaux. 1: 27-30. Pl. 1. 1826; Parmar, in Shetty & Singh Fl. Rajasthan 2: 795. 1991. *E. sphendens* Bojer ex Hook.*f.* in Curtis, Bot. Mag. 56. t. 2902. 1829.

Shrub, scrambling, much branched, 60 - 90 cm. Stems obscurely 3-5-angled, densely spiny. Leaves alternate, clustered at stem apex; stipules forming spines 1 - 1.3 cm; spines needle like, 1.2 - 2.5 cm long; petiole absent or nearly so. Leaves obovate-oblong, spathulate, 1.5 - 6 × 0.8 - 2.8 cm, base attenuate, margins entire, apex mucronate. Cyathia in subapical, dichasial cymes, peduncle 4 - 7 cm; cyathophylls 2, reniform-rounded, 8 - 10 × 12 - 14 mm, usually bright red (pink, white, or yellow in some cultivars), apex rounded; involucre campanulate, 3 - 4 × 3.5 - 4 mm, lobes 5, lanceolate; glands 5, ovoid or rounded, ca 1 × 2 mm, red. Male flowers many; bracteoles linear, apex pilose. Female flower: ovary smooth and glabrous, usually included with involucre; styles connate below middle; stigma 2-lobed. Capsules 3-lobed, ovoid, ca 3.5 × 4 mm, smooth, glabrous. Seeds ovoid, terete, ca 2.5 × 2 mm, grey-brown, reticulate; caruncle absent.

Field note: *Euphorbia milii* is widely cultivated as an ornamental and frequently grown as a low hedge.

Fl. & Fr.: March - June

Specimen examined: Bhilwara 1218 (MLVGCB Herb).

Distribution: Native to Madagascar; widely cultivated elsewhere.

Euphorbia pulcherrima Willd. ex Klotz., Allg. Gartenzeitung. 2: 27. 1834; Parmar, in Shetty & Singh Fl. Rajasthan 2: 795. 1991. *Poinsettia pulcherrima* R. Graham, in Edinb. New Phil. J. 20: 412. 1836; Hook. *f.* ,Fl. Brit. India 5: 239. 1886.

Vern: *Dudhi*

Herb. Stem glabrous. Leaves alternate, ovate or elliptic-lanceolate, often shallowly lobed, 6 - 15 × 4 - 10 cm, abaxially pubescent, adaxially shortly pubescent or glabrous, cuneate or rounded at base attenuate, apex acuminate or acute. Stipules minute, membranous, caducous; petiole 2-5 cm, glabrous. Cyathia in terminal, dichoto-mous lax cymes. Bracts oblanceolate, 5 - 10 x 2 - 4 cm, acuminate, bright red or occassionally yellow. Peduncle 3-4 mm; involucral lobes 5, fimbriate, triangular, ciliate; glands usually 1, rarely 2, orange, compressed, 2-lipped, 4-5 × ca. 3 mm. Male flowers many, usually exserted from involucre; bracts linear, pilose. Female flower: ovary glabrous, pedicellate, exserted from involucre, smooth; styles connate below middle; stigma deeply 2-lobed. Capsules, 3-lobed-globose, 1.5-2 × ca. 1.5 cm, smooth and glabrous. Seeds ovoid, ca. 10 × 8-9 mm, gray or light gray, subsmooth, without caruncle.

Field note: Common weed of cultivated fields. Also cultivated worldwide on a massive scale for its beautiful leaves.

Fl. & Fr.: Almost throughout the year.

Specimen examined: Bangreda 3134 (MLVGCB Herb).

Distribution: Native to Central America.

ULMACEAE

HOLOPTELEA Planch.

Holoptelea integrifolia (Roxb.) Planch. in Ann. Sci. Nat. Bot. ser. 3. 10: 269. 1848; Wight, Ic. 6: 8. t. 1968. 1853; Hook. *f.,* Fl. Brit. Ind. 5: 481. 1887; Duthie, Fl. Gangetic Plain 3: 120. 1915; Parmar, in Shetty & Singh Fl. Rajasthan 2: 797. 1991. *Ulmus integrifolia* Roxb., Pl. Coromandel. 56. t. 78. 1796-1798 & Fl. India 2: 68. 1832.

Vern: *chil*

Large, deciduous tree, up to 15 m tall, with grey, pustular bark that is smooth when young, exfoliating in corky scales on older trees. Leaves alternate, elliptic-ovate, 8 - 14 x 3.2 - 7.5 cm, glabrous, entire, apex acute or acuminate, base rounded or cordate, main nerves 5 to 7 pairs; stipules lanceolate. Flowers usually male and hermaphrodite mixed, small, greenish-yellow to brownish, pubescent, borne in short racemes or fascicles at the scars of fallen leaves. Tepals often 4, free, pubescent, imbricate. Stamens 7 - 8; anthers hairy. Fruit suborbicular samara, 1.5 - 2.5 cm across, membranous, reticulately veined wings. Seed ovate or ovate-oblong, flate.

Field note: Common in dry deciduous forests.

Fl. & Fr.: January - May.

Specimen examined: Bhim 63 (MLVGCB Herb).

Distribution: Tropical Africa, India & Ceylon.

Uses: 100 ml seed oil of *Holoptelea*, locally known as *Kanjeri* oil is added with 50 gm of Sulphur and it is kept for 5 days, after 5 days it is applied on the affected part of skin. Crushed leaves burn the skin.

MORACEAE

KEY TO THE GENERA

1a. Male and female flowers borne inside a closed, urceolate fleshy receptacle (hypanthodium). Fruit a syconus .... *Ficus*

1b. Male and female flowers borne externally, in a dense to lax catkin, spike. Female flowers in ovoid or cylindrical catkins, with free imbricate sepals. Fruit sorosis .... *Morus*

FICUS L.

KEY TO THE SPECIES

1a. Receptacles sessile or subsessile 2

2a. Leaves not rounded or truncate at base, obtuse at apex .... *F. benghalensis*

2b. Leaves rounded or truncate at base, acuminate at apex 3

3a. Acumen (apical tail) of leaf longer 1/3 to 1/2 of the lenght of lamina .... *F. religiosa*

3b. Acumen (apical tail) of leaf shorter, less than 1/3 of the lenght of lamina...... *F. rumphii*

1b. Receptacles distinctly penduncılate 4

4a. Leaves ovate-oblong. Receptacles in fascicles on old trunk or on entire leafless branches .... *F. racemosa*

4b. Leaves broadly ovate-orbicular. Receptacles solitary, axillary .... *F. palmata*

Ficus benghalensis L., Sp. Pl. 1059. 1753; King in Hook. *f.,* Fl. Brit. Ind. 5: 499.1888; Brandis, Ind. Tr. 600. 1906; Duthie, Fl. Gangetic Plain 3: 147. 1915; Corner, in Gard. Bull. Singapore 17: 381 1960; ibid, 21(1): 14. 1965; Parmar, in Shetty & Singh Fl. Rajasthan 2: 801. 1991; Bhandari, Fl. Indian Desert 314. 1995. *F. cotonaeifolia* Vahl, Enum. Pl. 2: 189. 1806. *F. indica* L., Sp. Pl. 1060. 1953. *emend.*; Lam., Encycl. 2: 494. 1788 *non sensu*; King 1887. *Urostigma benghalense* (L.) Gasp., Nov. Gen. Fic. 7. 1844; Wight, I.c. Pl. Or. t. 1989. 1853. (Fig. 145)

Vern: *Bar, Bargad*

A large, evergreen tree, up to 15 m tall, with wide leafy crown; branches spreading up to 100 m or more with pillar-like prop roots and accessory trunks. Trunk massive, fluted, bark grey, smooth, young softly white puberulous. Leaves with stout, 2 - 6 cm long, dorsiventrally compressed hairy petiole; lamina coriaceous, ovate or obovate to elliptic, 6 - 22 x 4 -15 cm, glabrous above, finely pubescent beneath, base rounded, margins apically obtuse, lateral nerves 4 -7 pairs, stout, 1.5 - 2.5 cm long, acute; cystoliths abundant on side, few or absent below. Receptacles sessile, in axillary pairs on young depressed-globose, 1.5 - 2.0 cm in diam., green, hairy, subtended by 3, reniform ca 3 - 4 mm long, ca 6 - 7 mm wide, minutely hairy basal bracts, flat bracts, internal bristles absent. Male flowers: numerous, near the ostiole of receptacle, shortly pedicellate; sepals 2 - 3; stamen solitary, with shortly mucronate anther. Female flowers: sessile, mixed with gall flowers; sepals small; ovary with an elongated style. Gall flowers numerous, pedicellate; sepal as in female, ovary with a short style. Figs globose to depressed-globose, 1.5 - 2.5 cm in diam., pinkish-red, hairy. Achenes globose-ellipsoid.

Field note: Common in waste lands, walls of old buildings, wells and on other trees. Also planted as a avenue tree.

Fl. & Fr.: Almost throughout the year.

Specimen examined: Kelwa 3875, Mandalgarh 1741 (MLVGCB Herb).

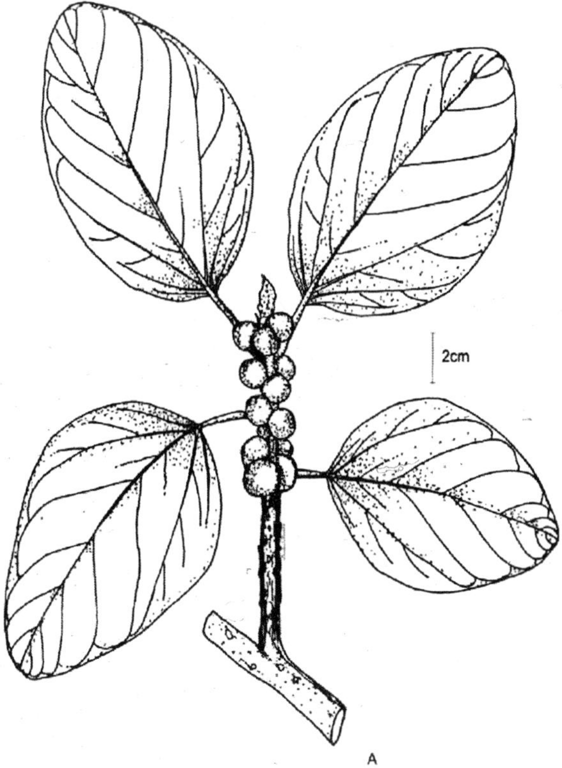

Fig. 145. *Ficus benghalensis* L.
A. Fruiting twig.

Distribution: Pakistan, India, Bangla Desh, introduced and widely cultivated in tropics and elsewhere.

Notes: Linnaeus in his protologue of *Ficus berghalensis* cited Hort. Cliff. Van and Amoen: Acad. 1: 29. No specimens have been found that link with these references except for Van Royen's sheet L 908186144. This specimen consists of a single leaf. No specimen of *Ficus benghalensis* is preserved in the herbaria in London and Stockholm and although Linn. 1240.8 could typify *Ficus indica* L it could not typify *Ficus benghalensis*. The remaining elements in Linnean protologue are Commelin's plate 62 and Rheede's Peralu, Hort. Mal. 1: 49. t. 28. Since Linnaeus obtained his epithet from Commelin and Since Cited t. 62 in all his publications on this taxon, in present study. Commelin's plate 62 has been designated as a suitable lectotype of *Ficus benghalensis* in preference to Rheede's plate and the Van Royen's specimen.

Uses: Ripe fruits are eaten during famine. Birds and monkeys devour the fruits helping in dispersal. The leaves are heated and applied as poultice to abcesses. An infusion of bark and young buds is used in dysentery, diarrhoea and diabetes. Patashas are filled with latex and 1 patasha is given daily for seven days to children to cure weakness.

Ficus palmata Forsk. subsp. ***virgata*** (Roxb.) Browicz in Rechinger, Fl. Iranica 153: 12. 1982; Parmar, in Shetty & Singh Fl. Rajasthan 2: 803. 1991; *F. virgata* Roxb., Fl. Ind. ed. 2 (3): 529. 1832. *F. caricoides* Roxb., *l.c.* 2 (3): 529. 1832; Wight, *l.c.* (2) 4: 1. t. 634. 1843. *F. palmata sensu* King in Hook. *f.*, Fl. Brit. Ind. 5: 530.1888, non Forsk. 1775; Duthie, Fl. Gangetic Plain 3: 158. 1915. (Plate 297)

Vern: *Arak-anjir.*

A large deciduous shrub or small tree, up to 3 m tall. Trunk and branches. without aerial roots, bark smooth, brownish-grey, young twigs densely hairy. Leaves 12 - 24 x 11 - 19 cm, membranous, broadly ovate or orbicular, acute to apiculate at apex, cordate at base, denticulate, mostly undivided or sometimes palmately 3 - 5: partite, tomentose, lateral nerves 3-5 pairs, raised below; stipules ovatc, 6 - 10 mm long, acute. Receptacles axillary solitary, on ca 1 - 2.5 an long, tomentose peduncles, subglobose to pear-shaped, ca 2 cm across, tomentose, subtended by 3, deltoid, acute basal bracts. Male flowers: numerous in the upper half, pedicellate; pedicels hairy; sepals 4 - 5, free, lanceolate, hairy; stamens 3 - 6. Female and gall flowers sessile or pedicelled: basal, numerous; sepals 5, basally united, hairy; ovary ovoid with subterminal, long hairy style. Figs constricted or gradually narrowed at base, 1.5 - 2.5 cm long, yellow or purple, hairy. Achenes 3-angled.

Field note: Common in waste lands, neglected corners of gardens and along with water steams.

Fl. & Fr.: February - July.

Specimen examined: Bhilwara 1656, Ladpura 1912 (MLVGCB Herb).

Distribution: Nepal, India, Pakistan, Afghanistan, Iran, Arabian Peninsula, Somalia, Sudan, Ethiopia and S. Egypt.

Notes: This is a highly variable and common wild fig. Two subspecies are recoginzed. The type subspecies from E. Africa and Saudi Arabia has more elongate, distinctly acute or acuminate leaves with slight pubescence. The present material belongs to ssp. *virgata* (Roxb.) Borwicz (*l.c.* 12) with virgate shoots, leaves longer than broad and densely pubescent beneath. It is sometimes confused with *F. carica*. Its fruits demulcent, laxative and are used in diseases of lungs and the bladder.

Ficus racemosa L., Sp. Pl. 1060. 1753; Corner in Gard. Bull. 21 (1): 34. 1965; Rani & Mathew, in. Mathew, Fl. Tamil.- Carn. 2: 1526; Parmar, in Shetty & Singh Fl. Rajasthan 2: 805. 1991. *Covellia glomerata* Miq. in Hook Lond. J. Bot. 7: 465. 1848. *Ficus glomerata* Roxb., Pl. Corom. 2: 13. f. 123. 1798; Boyle, Ill. Hot. Himal. 337. 1839; Wight, Ic. Pl. Ind. Or 667. 1840; King in Hook. *f.* . Fl. Brit. Ind. 5: 535. 1888; Cooke, Fl. Pres. Bomb. (reprint ed.) 3: 154. 1958. *Ficus goolerea* Roxb., Fl. Ind. 3: 538. 1832. (Fig. 146)

Vern: *Gular, Umer.*

Small to large tree, 10 - 20 m tall, evergreen. Trunk up to 2 m in circumference, with spreading branch with or without aerial roots, bark whitish to pinkish-brawn, smooth, young twigs with fine white pubescence, soon glabrous. Leaves alternate, ovate-oblong, 6-15 x 2.5-6 cm, rounded or acute at base, margins entire, obtuse or acute at apex, glabrous on both sides; lateral nerves 4 - 7 pairs; stipules triangular-ovate, 12 - 15 x 4 - 5 mm, acute-acuminate, brown; cystoliths present only on the lower side. Petiole 2.6 cm long, grooved minutely hairy, brownish-scurfy. Receptacles on 8-40 long peduncles, in large clusters from tubercles on the main trunk and main leafless branches (cauliflorous), pyriform-subglobose, ca 1.5-3 cm, green to red. Male flowers: sessile, in 2-3 ring near the ostiole, united, lobes dentate-lacerate, red; stamens 2, pistillode present. Female flowers: sessile or subsessile. Sepals as in male; ovary substipitate, with lateral, 2.3 long, glabrous style; stigma simple. Gall flowers pedicellate, dispersed among female. Figs depressed subglobose or pyriform, 2.54 cm in diam., red, usually streaked. Achenes lenticular, ca 1 mm long.

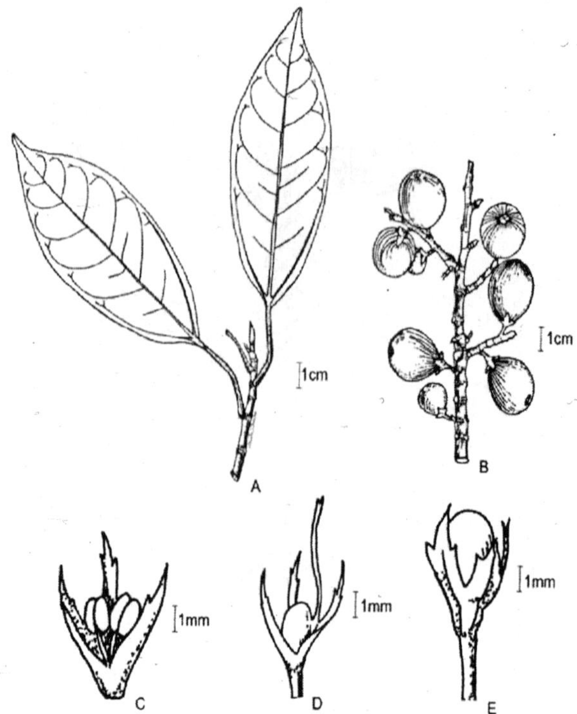

Fig. 146. *Ficus racemosa* L.
A. Vegetative branch, B. Fruiting branch,
C. Male flower, D. Female flower,
E. Gall flower

Field note: Commonly planted as avenue tree.

Fl. & Fr.: Almost throughout the year.

Specimen examined: Armoda 3473, Bijoliya 1936 (MLVGCB Herb).

Distribution: India, Sri Lanka, Bangladesh, China, Burma, Pakistan, Thailand, Malayasia, Indonesia to N. Australia.

Uses: Mature fruits are astringent, stomachic and carminative. They are eaten by locals. A decoction of the bark is used as a wash for wounds.

Ficus religiosa L., Sp. Pl. 1059. 1753; King in Hook. *f.*, Fl. Brit. Ind. 5: 513. 1888; Bamber, Punj. Pl. 16. 1916; Duthie, Fl. Gangetic Plain 3: 150. 1915. *F. peepul* Grifff. Nat. Pl. Asiat 4: 393. 1854. *Urostigma religiosum* (L.) Gasp., Ric. Caprific. 82. t. 7. f. 1-5. 1985; Wight, Ic. Pl. Ind. Or. t. 1967. 1853. (Fig. 147)

Vern: *Pipal, fifri.*

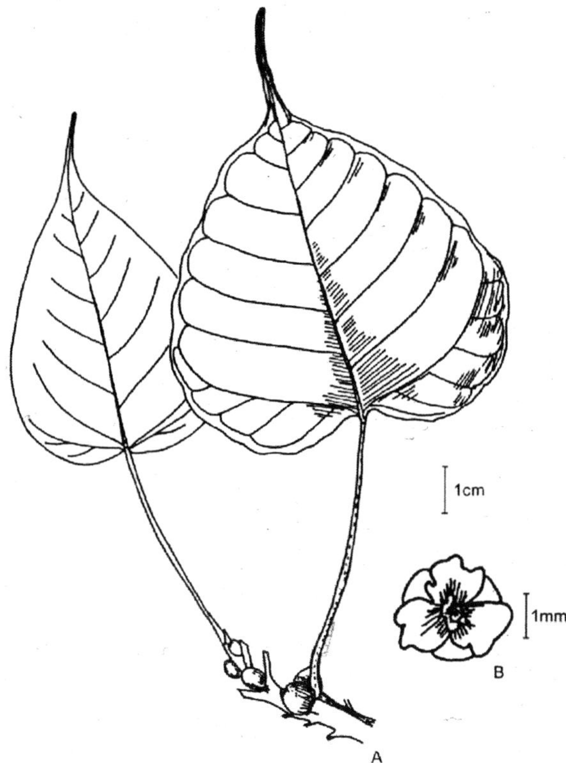

Fig. 147. *Ficus religiosa* L.
A. Fruiting branch, B. Bottom view showing involucral bracts

A large or medium sized, deciduous tree, 6 - 15 m tall. Trunk ca 2 - 3 m in circumference, with spreading branches and usually without aerial roots, bark grey, fissured; young twig pubescent with pink new leaves. Leaves with a pale-green, slender, 4 - 10 cm long, terete petiole; lamina broadly ovate, 6 - 15 x 4 - 13 cm, margins entire or repand, 3 - 5-costate at the truncate or cordate base, apex .abruptly long-acuminate, acumen nearly half as long as lamina, lateral nerves 6 - 9 pairs with several zigzag intercostals; stipules yellowish-brown, deltoid-acuminate, 8 - 12 mm long; cystoliths present only on the lower side. Receptacles sessile, in axillary pairs, ca 5 - 6 mm in diameter, yellowish-green, subtended by 3 silky-puberulent to glabrescent, subglobose, obtuse, 3 - 5 mm long basal bracts, apical orifice closed by 3 apical bracts. Male flowers, few, sessile, near the ostiole; sepals 2 - 3, free, ovate-lanceolate. Female and gall flowers: sessile or pedicellate; sepals 3 - 4, lanceolate. Figs depressed globose, ca 10 - 12.5 mm in diam., dark-purple at maturity. Achenes smooth.

Field note: Planted as an avenue or roadside tree. The tree is believed to sacred commonly planted near temples.

Fl. & Fr.: Almost throughout the year.

Specimen examined: Bhilwara 1573, Jahajpur 1976 (MLVGCB Herb).

Distribution: India, Bangladesh, Ceylon, China, Burma, Pakistan and Thailand; introduced and cultivated in S.E. Asia, Middle East, North Africa (Egypt, Libya), U.S.A. and elsewhere.

Uses: Fruits are commonly eaten by birds and in times of famine by human beings. The leaves and twigs are lopped for cattle. Wood is used for packing cases and in sacrificial fires by Hindus. Leaves and tender shoots are used as purgative and in skin diseases. The fruit is laxative, alterative and cooling.

Ficus rumphii Blume, Bijdr. 437. 1825; King, in Hook. *f.*Fl. Brit. India 5: 512. 1888; Brandis, Ind. Tr. 601. 1906; Duthie, Fl. Gangetic Plain 3: 151. 1915; Bamber, Punj. Pl. 16. 1916; Cooke, Fl. Pres. Bomb. (reprint ed.) 3: 140. 1958; Corner, in Gard. Bull. Singapore 21: 11. 1965. *F. cordifolia* Roxb., Fl. Ind. 3: 548.

1832. (non Blume); Brandis, For. Fl. 416. t. 48. 1874; Wight, Ic. Pl. Ind. Or. t. 640. 1843. *Urostigma cordifolium* (Roxb.) Miq., in Hook., Load. J. But. 6 : 564. 1847.

Vern: *Paras pipal.*

A large, deciduous tree, upto 10 m tall, glabrous. Trunk ca 1.5 m in circumference, with suberect branches without aerial roots, bark grey or pale-grey, flaky, young twigs glabrous. Leaves with 4 - 10, cm. long, grooved or flattened petiole; lamina ovate-oblong or ovate, 9 - 14 x 6 - 9 cm, truncate or subcordate at base, margins entire, apex gradually narrowed into 2 - 3 cm long acuminate, glossy above, glabrous on both sides, lateral nerve 4 - 6 pairs; stipules narrowly ovate-lanceolate, 10 - 25 mm long, acute, brownish. Recepacles sessile, in axillary paired and around old leaf scars, globose, 8-10 m in diam., whitish with dark spots, subtended by 3, rotundate basal bracts; internal bristles absent. Male flowers: few, near the ostiole; sepals 3, spathulate; stamen solitary, filament as long, as anther. Female and gall flowers: sepals 3, united, lanceolate; ovary obovoid with long style, stigma clavate. Figs globose, 12-15 mm in diameter, black smooth. Achenes minutely tuberculate.

Field note: It is often planted near temples, along raodsides etc.

Fl. & Fr.: March- June.

Specimen examined: Bhilwara 1559, Fuliyakalan 2113 (MLVGCB Herb).

Distribution: India.

Uses: Twigs and leaves are lopped for cattle. The juice and fruits are reported to be emetic and anthelmintic.

MORUS L.

Morus alba L., Sp. Pl. 986. 1753; Boiss., Fl. Or. 4: 1153. 1879; Bomber, Pl. Punj. 15. 1916; Yarnsolenko, Fl. U.R.S.S. 5: 377. 1936; Townsend, Pl. Iraq 4(1) : 81. Pl. 16. f. 7-11. 1980. *M. alba* L.var. *purpurea* Bureau in DC., Prodr. 17: 244. 1873. *M. atropurpurea* Roxb., Fl. 3: 595. 1832. *M. indica* L., Sp. Pl. 986. 1753, non. Roxb. Hook. *f.,* Fl. Brit. Ind. 5: 492. 1888. (Fig. 148)

Vern: *Shahtut, hetut.*

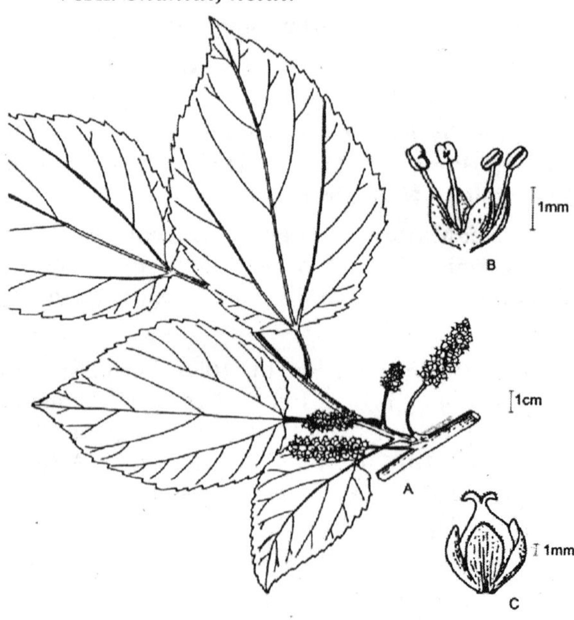

Fig. 148. *Morus alba* L. A. Fruiting twig. B. Male flower, C. Female flower

Dioecious or monoecious, deciduous, 3 - 5 m tall tree with a dense, compact leafy crown. Trunk with dark grey-brown, rough, vertically fissured bark, tender twigs hairy to glabrescent. Petiole 1-3 cm long, crisped hairy, filiform. Leaves 5 - 11 x 3 - 9 cm, narrow to broad, ovate to lanceolate, 3-costate from truncate to shallowly cordate base, upper surface glabrous, midrib and principal wins pubescent, secondary veins and ultimate veinlets glabrous, margins regularly serrate, apex acute; stipules lanceolate, brownish membranous, hairy. Catkins pendulous, green. Male catkins: 10 - 20 mm long including slender, hairy peduncle, ca 5.6 mm broad, with lax flowers. Male flowers: sepals free, broadly ovate, ca 2.5 mm long, obtuse, glabrous to slightly hairy; staminal filaments somewhate equal to sepals, with ovate, exserted anthers. Female catkins ovoid, 3 - 11 mm long, without equally long or slightly longer peduncle. Female flowers: sepals suborbicular, as long as or slightly larger than of male flowers, glabrous or ciliate on margins; ovary with glabrous free styles. Sorosis ovoid, (10-) 15 - 25 mm long, 5 - 8 mm across, purple to red when ripe, sweet.

Field note: Cultivated in orchard, gardens, parks and along roadsides.

Fl. & Fr.: February - May.

Specimen examined: Bhilwara 190, Chittorgarh 3059 (MLVGCB Herb).

Distribution: Native of China and probably in temperate Asia; now widely distributed in Japan, China, Malaya, Burma, India, Pakistan, N. Africa, South & Central Europe; introduced in the New world.

Uses: The white mulberry tree is cultivated and subspontaneous in all parts of south central Rajasthan. It is grown for its edible fruits which are cooling and laxative. The fruit is also used for sore throat, dyspepsia and melancholia. Bark possesses vermifuge and purgative properties. Wood is also suitable for house building, agricultural implements, furniture and fuel.

URTICACEAE

URTICA L.

Urtica dioica L., Sp. Pl. 984. 1753; Hook. *f.,* Fl. Brit. Ind. 5: 548. 1888; Ball, in Tutin et aL. Fl. Europ. 1: 68. 1964; Stewart, Ann. Cat. VAs., Pl. W Pak. & Kash. 197. 1972; Chrtek, in Rech. f. Fl. Iran. 105: 2. t. 2&3. 1974; Parmar, in Shetty & Singh Fl. Rajasthan 2: 812. 1991.

Vern: *Bicchu-booti.*

Dioecious, perennial, 50-150 cm herb with a dense indumentum of stinging hairs. Stem angled, sparingly branched. Leaves with 1.5 - 4 cm long petiole; lamina ovate-lanceolate, oblong or lanceolate, 5 - 6 long, cordate at base, serrate, apex acute-acuminate; stipules free lateral, oblong-lanceolate, 8-12 mm long, ciliate. Flowers in axillary, unisexual, rarely androgynous cymes, pale-greenish or whitish, bracteate; bracts of male flowers smaller than those of female flowers. Sepals pubescent. Achenes up to c. 1.5 mm long, ovoid-ellipsoid, pale green or greenish-brown.

Field note: Common in moist waste lands.

Fl. & Fr.: June - December.

Specimen examined: Bhilwara 71(MLVGCB Herb).

Distribution: Widespread in the temperate regions of both hemispheres.

Uses: It is used to check excessive manustrual flow.

CASUARINACEAE

CASUARINA L.

Casuarina equisetifolia L. Amoen. Acad. 4 : 143. 1759 *'equisetifolia'*; Hook. *f.,* Fl. Brit. Ind. 5: 598. 1888; Duthie, Fl. Gangetic Plain 3: 162. 1915; Bailey, Man. Cult. Pl. 316. f. 43. 1949; Matthew, Illus. Fl. Tam. Carnatic. t. 681. 1982. *C. muricata* Roxb., Fl. Ind. 3: 519. 1832.

Dioecious tree up to 15 m tall. Trunk straight; crown conical; bark scaly, adaxially red-brown on old trees. Ultimate branchlets usually drooping, greyish green. Leaves 3 - 4 cm long, scaly, erect and appressed to branchlets, 6 - 7 per node, ovate-lanceolate or lanceolate, 1-3 mm. Male spikes axillary, ovoid, cone like, becoming woody, shortly pedunculate; flowers, yellow, scented, 1-4 cm. Female flowers arranged in small cone-like clusters at the end of lateral branches, ellipsoid, 1.2-2.5 cm, greyish green or yellowish brown tomentose when young, glabrous at maturity, base and apex truncate to obtuse; apex of bract-eoles slightly obtuse or acute. Fruits (samaras), crowded into a woody cone, 5-8 mm including wings.

Field note: Commonly planted around rest houses, along roads, gardens as a an ornamental tree.

Fl. & Fr.: Almost throughout the year.

Specimen examined: Borda 188 (MLVGCB Herb).

Distribution: India, Indonesia, Malaysia, Myanmar, Papua, New Guinea, Philippines, Thailand, Vietnam, Australia and Pacific Islands.

CERATOPHYLLACEAE

CERATOPHYLLUM L.

Ceratophyllum demersum L., Sp. Pl. 992. 1753; Hook. *f.,* Fl. Brit. Ind. 5: 639. 1888; Duthie,

Fl. Gangetic Plain 3: 168. 1915; van Steenis, Fl. Males. I. 4: 41. f. 1. 1949; Cooke, Fl. Bomb. Pres. (Rep. ed.) 3 (3): 165. 1958; Boiss., Fl. Or. 4: 1202. 1879; Subram., Aq. Angiosp. 52. 1962; Rech. f. Fl. Iran. 28: 1. 1966. (Fig. 149 & Plate 298)

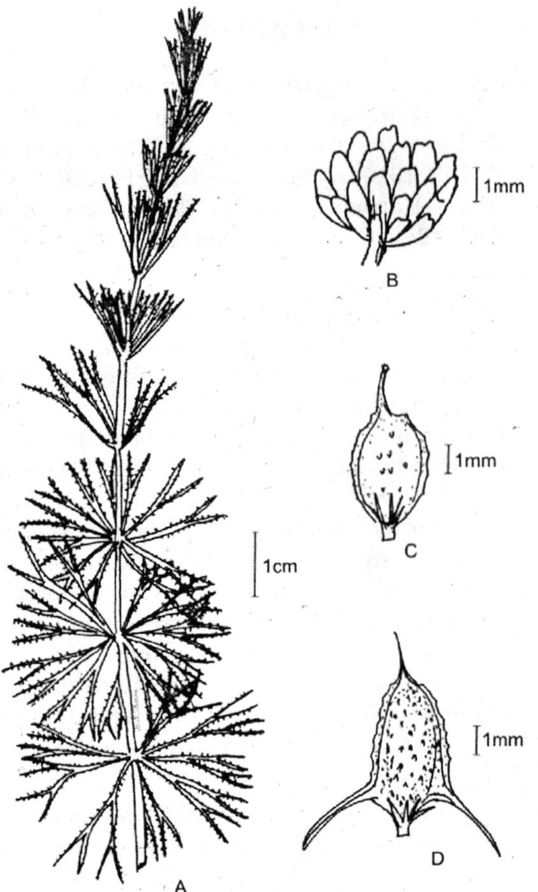

Fig. 149. *Ceratophyllum demersum* L.
A. Habit, B. Male flower, C. Pistil, D. Fruit

Much branched, rootless, submerged, aquatic plant, 15 - 60 cm long. Leaves 4 - 9 in a whorl, 1 - 4 times dichotomously branched, minutely dentate segments, often terminated by 1 or 2 sharp bristles. Flowers solitary, axillary, minute, unisexual, male and female flowers at different nodes. Perianth segments 6 - 15, upto 1 mm long, connate at the base, each segment terminated by 2 bristles. Male flowers, spirally arranged on a convex receptacle, narrowed into the base, subsessile, apex terminated by two spinous incurved processes: tepals 9 - 12, linear, greenish white, ciliate at apex; stamens 8 - 30, anthers erect, 2-lobed, pollen elliptical or round. Female flower: tepals as in male: carpel sessile; ovary 1-celled, 1-ovuled; style simple; stigma oblique, style elongating in fruit forming the apical spine. Nutlets, 3.5 - 4.5 x 2.5 -3 mm (excluding spines), sessile with persistent perianth, ellipsoid or ovoid, compressed, provided with 3 spines; one apical and 2-lateral, near the base, apical spines 5 - 9 mm long; basal spines 5 - 7 mm long, straight or slightly curved.

Field note: Common in ponds, tanks and slow running water.

Fl. & Fr.: September - January.

Specimen examined: Mainal 522, Gambhiri dam 3928 (MLVGCB Herb).

Distribution: Cosmopolitan.

Uses: The plant is used as a cooling agent for biliosness and scorpion sting. The achenes are eaten by wild ducks.

GNETACEAE

EPHEDRA L.

Ephedra ciliata Fisch. & Meyer, ex C. A. Meyer, Monogr. Gatt. Ephedra. 100. 1846; Bobrov, in Komarov, Fl. URSS. 1:198.1934; Pandey, in Shetty & Singh Fl. Rajasthan 2: 815. 1991. *E. foliata* Boiss. & Kotschy ex Bioss. Diagn. ser. 1. 7: 101. 1846; Bhandari, Fl. Indian Desert 439. 1978. *E. penduncularis* Bioss. Fl. Orient. 5: 717. 1884; Hook. f. Fl. Brit. India 5: 640. 1888. *E. foliata* Bioss. & Kotschy ex Bioss var. *ciliata* (Fisch. & Meyer ex Meyer) Stapf, Die Artend. Gatt. Ephedra 49. 1889.

Vern.: *Andho khimp.*

Scandent or trailing plant, much branched. Branches slender, often fascicled, smooth, striated, knoted. Leaves 2-4 at the upper nodes, 0.8 - 3.0 x 0.1-1.5 cm, setaceous, connate at base, linear. Male strobili solitary or 2-3 per node; peduncle 1-2 cm long. Flowers 4-6 pairs, bracteate; bracts connate, obtuse. Staminal column slightly exserted; anthers 3-4, sessile. Female strobili pedunculate, ovoid, 2 - 3-flowered. Bracts connate; outer ca 1 mm long; inner ones 4 - 5 mm long. Berry 6-7 x 5.5-6.5 mm, ovoid-globose, fleshy, white. Seeds 2, 4-6 x 2-2.5 mm, plano-convex, acute at apex, glabrous, red or brownish-black.

Field note: Found in sandy habitats along with *Maytenus emarginata* (Willd.) Ding Hou, *Acacia senegal* (L.) Willd. and *Prosopis cineraria* (L.) Druce. Restricted in Budha Pushkar and Mangliyawas only.

Fl. & Fr.: March - April.

Specimen examined: Budha Puskar 4062 (MLVGCB Herb).

Distribution: C. Asia, Iran, Afghanistan and Pakistan.

Uses: Used as a fodder.

Notes: Only wild gymnosperm in this area.

HYDROCHARITACEAE

KEY TO THE GENERA

1a. Stems present. Leaves cauline........... *Hydrilla*

1b. Stems absent or plants with only stolons. Leaves radical 2

2a. Leaves linear, narrow, sessile. Male peduncles many-flowered...... *Vallisneria*

2b. Leaves linear-lanceolate to broadly ovate or suborbicular, distinctly petiolate. Flowers solitary ....*Ottelia*

HYDRILLA L. C. Rich.

Hydrilla verticillata (L. f.) Royle, Ill. Bot. Himal. Mts. t. 376. 1839; Hook. *f.* . Fl. Brit. Ind. 5: 659. 1888; Duthie, Fl. Gangetic Plain 3: 173. 1920; Hartog in van Steenis, Fl. Males. 1(5): 385. f. 1-3. 1957; Subram. Aquat. Angiosp. 55. f. 37. 1962; Cook & Luond, Aq. Bot. 13: 485-504. 1982; Pandey, in Shetty & Singh, Fl. Rajasthan 2: 816. 1991; Bhandari, Fl. Indian Desert 315. 1995. *H. ovalifolia* Rich., *l.c.* 12(2): 9, 96, t. 2 (upper part). 1814. *H. polysperma* Blatt. in J. Proc. Asiat. Soc. Beng. (n.s.) 26: 356. 1930. *Serpicula verticillata* (L. f.) Roxb., Fl. Ind. 3: 751. 1832. *Vallisneria verticillata* Roxb., Hort. Bengal. 71. 1814. (Fig. 150)

Caulescent, glabrous, slender, profusely branched herb; internodes 5 - 50 mm long; roots fibrous at some nodes. Leaves 3 - 8 in whorl, oblong-linear, 3 - 12 x 1 - 3 mm, green, entire or serrulate, apex acute-apiculate, with a strong midrib, sessile; nodal scales 2, axillary, membranous with orange brown cilia on margins. Male spathe 1.25-1.50 mm long, bursting open to liberate male flowers, beset with subulate appendages. Male flower: sepals ovate, 1.5 - 3 mm long, green; petals oblong, equalling the sepals, patent or reflexed, obtuse; stamens with minute filaments. Female spathe ca 5 mm long, reddish-brown striped, bidentate. Sepals and petals as in males, narrow. Ovary oblong, 3.4 mm long, with 1.5 - 10 cm long slender rostrum. Fruit terete. 5 - 7 mm long. Seeds dark brown, ca 2.5 mm long.

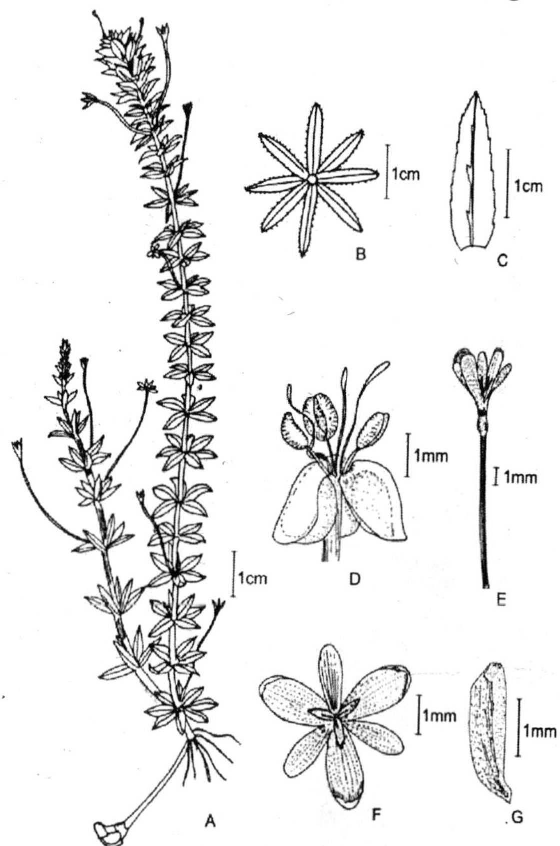

Fig. 150. *Hydrilla verticillata* (L.f.) Royle, A. Habit, B. Whorl of leaves, C. Leaf, D. Male Flower, E. Female flower, F. Female flower upper surface view. G. Seed.

Field note: Gregarious in ditches, pools, lakes, wet ricefields and slow-running water streams.

Fl. & Fr.: September - January.

Specimen examined: Triveni 69 (MLVGCB Herb).

Distribution: S.E. Europe, Africa, Asia and Australia from sea level to 6000 ft; introduced in America and elsewhere.

Uses: The plant is a food of some fresh water fishesh and may be used as manure wherever it occurs in large quantities.

OTTELIA Pers.

Ottelia alismoides (L.) Pers. Syn. Pl. 1: 400. 1805; Hook. *f.*, Fl. Brit. Ind. 5: 662. 1888; Duthie, Fl. Gangetic Plain 3: 177. 1920; Hartog in van Steenis, Fl. Males. 1(5): 398. 1957; Subram. Aquat. Angiosp. 61. f. 43. 1962; Pandey, in Shetty & Singh, Fl. Rajasthan 2: 818. 1991. *Damasonium alismoides* (L.) R. Brown, Prodr. Fl. Nov. Hollandiae. 344. 1810. *Damasonium indicum* Willd., Sp. Pl. 2: 276. 1799. Roxb., Fl. Ind. 2 : 216. 1832. *Stratiotes alismoides* L., Sp. Pl. 535. 1753. (Fig. 151 & Plate 299)

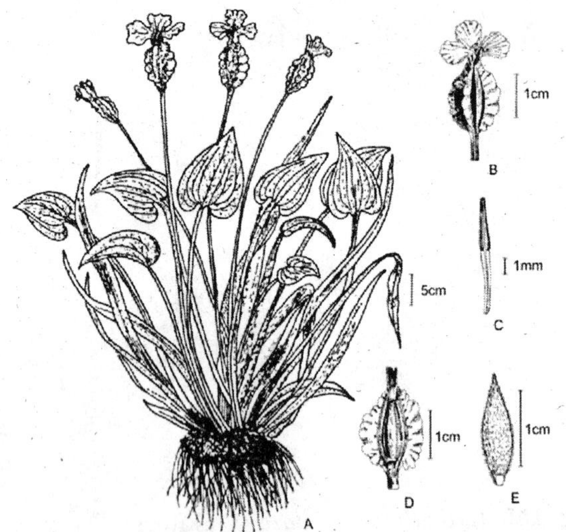

Fig. 151. *Ottelia alismoides* (L.) Pers.
A. Habit. B. Flower, C. Stamen, D. Fruit, E. Seed

Submerged to partly floating and rooted in the mud, aquatic herbs. Root fibrous. Leaves thin, transparent, light green, submerged to partly emergent or floating, lamina variable in shape and size, submerged ones linear-lanceolate, attenuate to short petiole, floating leaves broadly ovate to suborbicular, 6 - 20 x 3 - 16 cm, cordate-truncate at the base, apex acute or acuminate; veins 7 - 10, parallel convergent; petioles 8-50 cm long, triangular, with pale broad sheath at the base. Spathe elliptic-ovate, 2.5 - 3.5 cm long, 2-lobed, lobes acute, glabrous, with undulate margins; peduncle 10 - 30 cm long, 4 - 5-angled. Flowers bisexual, sessile, solitary. Sepals linear-oblong, 1 - 1.5 cm long, obtuse, persistent. Petals obovate-suborbicular, 2 - 3 mm long, 2 - 2.5 cm broad, obtuse, white with yellow spot at the base. Stamens 6, filaments ca 4 mm long, glandular-hairy; anthers basifixed, linear. Styles 6 - 9, each split into 2 unequal, papillate or ± hairy branches. Fruit ellipsoid, 2 - 5 cm long, ca 3 - 4 cm in diam., 6-valved, enclosed in the spathe. Seeds oblong, 1.5 x 1 mm long.

Field note: Common in stagnant or slow running water, in ponds, tanks, streams, ditches, rivers etc.

Fl. & Fr.: Throughout the year.

Specimen examined: Mainal 70, Triveni 2102 (MLVGCB Herb).

Distribution: N.E. Africa eastwards to Malayasia, tropical Australia, Japan and the Solomon Islands; introduced and naturalized in N. Italy in the ricefields.

VALLISNERIA L.

Vallisneria spiralis L. var. ***denseserrulata*** Makino in Bot. Mag. Tokyo 28: 27. 1914; Lowden in Aquat. Bot. 13: 288. 1982; Pandey, in Shetty & Singh, Fl. Rajasthan 2: 818. 1991. *V. spiralis* L. Sp. Pl. 1015. 1753; Wight, Illus. Ind. Bot. t. 33. f. 24. 1840; Boiss., Fl. Orient. 5: 3. 1881; Hook. *f.*, Fl. Brit. Ind. 5: 660. 1888; Duthie, Fl. Gangetic Plain 3: 174. 1920; Subram., Aq. Angiosp. 57. f. 39. 1962; Bhandari, Fl. Ind. Desert. 316. 1995. *Physkium natans* Lour. Fl. Cochinch. 663. 1790.

A caulescent submerged annual herb with coarse, up to 10 cm long, terete stolons rooting in mud. Leaves redical, linear, ribbon-shaped, 5 - 40 x 0.3 - 0.8 cm, sheathing at base, faintly 7-9-costate, costae parallel. Male spathes oblong, ca 6 mm long, peduncles 2-3 cm long. Male flowers pedicellate, trimerous, many enclosed in an ovoid spathe. ca 0.5 mm across; sepals 3, ovate-oblong; petals absent; stamens 1 - 3, solitary by fusion or 2, erect. Female spathe 1-2 cm long, enveloping the ovary to more than half its length, bilobed at apex, peduncle filiform, longer to nearly as long as leaves, spirally coiled in fruit. Female flower in a tubular, 3-toothed spathe; peduncle ca 1 cm long, filiform: solitary; sepals ovate-oblong. 1.5 - 4 x ca 1-2 mm, obtuse, persistent; ovary 1.5 - 2.5 cm long, styles 3, bifid, alternating with the scarious rudimentary petals, stigmas hairy. Fruits 5 - 7 cm long, greensih-yellow, many-seeded. Seeds oblong-fusiform to cylindrical, 1.5 - 3 mm long.

Field note: Common in ponds and ditches also grown in aquaria.

Fl. & Fr.: October - December.

Specimen examined: Mainal 1729, Gambhiri dam 3545 (MLVGCB Herb).

Distribution: South Europe, West Asia to Indo-Pakistan subcontinent.

MUSACEAE

MUSA L.

Musa paradisiaca L., Sp. Pl. 1403. 1753; Cheesman, in Kew Bull. 1948: 145. 1948; Pandey, in Shetty & Singh, Fl. Rajasthan 2: 829. 1991. *M. paradisiaca* L. subsp. *sapientum* (L.) K. Schum., in Engl., Pflanzen. 4(45): 20. 1900. *M. paradisiaca* var. *sapientum* (L.) Kuntze, Revis. Gen. Pl. 2: 692. 1891; *M. sapientum* L. Syst. Nat. ed. 10. 1303. 1759; Baker, in Hook. *f.,* Fl. Brit. Ind. 6: 262.1892.

Vern: *Kela.*

Tree-like, stoloniferous herb, up to 4 m in height. Leaf sheaths tubular, forming a thick trunk. Leaves 1 - 2.5 x 0.4 - 0.7 m, oblong, undulate, usually ragged in appearance, splitting between the transverse parallel veins. Flowers in terminal, solitary, drooping spike ca 1 m. Peduncle thick. Flowers yellow; lower female and upper ones male. Bracts opening in succession, 15 - 25 cm, ovate, concave, dark red, somewhat fleshy, caducous. Outer tepals 1 cm, 5-toothed, fleshy, tinged pink; inner tepals 19 - 20 mm, ovate, acute, concave. Stamens 5. Fruit oblong, 3 - 5 angled, fleshy, 6 - 16 cm long, seedless.

Field note: Commonly cultivated along water streams, gardens and orchards for its fruits.

Fl. & Fr.: Almost throughout the year.

Specimen examined: Bhilwara 2203 (MLVGCB Herb).

Distribution: Cultivated throughout the tropics.

Notes: It is impossible to give a satisfactory account about the nomenclature of the cultivated bananas. Two of the earliest specific epithets *Musa paradisiaca* and *Musa sapientumn* L. are cultivars derived from hybrids. Many authors believe that the only way out of the confusion is to disregard latin names altogether and a species-concept, and use only clonal or cultivar names. However, for the purposes of this floral account we have used the "blanket" name *Musa paradisiaca* to cover the cultivated bananas-even though at least some of the cultivars have probably been derived from other sources. A full account of bananas and plantains is given by Simmonds N.W. *The evolution of Bananas.* London. 1962., *Bananas,* ed. 2. London. 1966.

AMARYLLIDACEAE

CRINUM L.

KEY TO THE SPECIES

1a. Leaves 1.5 - 3 cm broad.
Umbels 15 - 20 flowered.............. *C. asiaticum*

1b. Leaves 10 - 15 cm broad.
Umbels 6 - 12 flowered.................... *C. defixum*

Crinum asiaticum L., Sp. Pl. 293. 1753; Hook. *f.,* Fl. Brit. Ind. 6: 280. 1892; Duthie, Fl. Gangetic Plain 3: 248. 1920.

A stout perennial stoloniferous herbs. Bulbs ovoid, 5 - 7 cm across; neck coverd with old leaf sheaths. Leaves clustered, linear-lanceolate, ca 30 - 100 × 7 - 12 cm, coriaceous, flate, with green sheathing at base, margin undulate, apex acuminate, sharply pointed. Flowers in 15-20-flowered umbel, fragrant at night, pedicels 0.5 - 2.5 cm long; bracts ovate-lanceolate, ca 6 cm long, greenish. Flowers fragrant; pedicel 0.5 - 2.5 cm. Perianth hypocratriform; tube greenish white, straight, 7 - 10 cm × ca. 2 mm; lobes white, linear, 4.5 - 9 × 0.6 - 0.9 cm, apex acuminate. Stamens red; filament 4 - 5 cm; anther linear, more than 1.5 cm, apex acuminate. Ovary fusiform, less than 2 cm. Capsule subglobose, 3 - 5 cm across, 1 - 2 seede. Seed not seen.

Field note: Occasional in moist and shady places, also cultivated ornamentally.

Fl. & Fr.: August - October.

Specimen examined: Bhilwara 2204 (MLVGCB Herb).

Distribution: Tropics and Sub-tropics.

Crinum defixum Ker-Gawl in Journ. Sci. & Arts 3: 105. 1817; Hook. *f.,* Fl. Brit. Ind. 6: 281. 1892; Duthie, Fl. Gangetic Plain 3: 248. 1920; Parmar, in Shetty & Singh, Fl. Rajasthan 2: 830. 1991. (Plate 300)

Stout, perennial herbs, with ovoid bulbs, stoniferous fusiform base, ca 5 cm across, neck ca 18 cm long, underground, clothed with leaf-scars. Leaves clustered at tip of short caudex, 40 - 90 x 3 - 5 cm, linear-oblong, channelled, obtuse or acute. Flowers in 6-12 flowered umbels, 17 - 20 cm long, white. Bracts 2, enclosing flower-bud. Perianth-tube straight, hypocratriform; lobes linear, 7-10 cm long. Stamens scarlet. Fruits subglobose or ovoid, 2.5 cm across. Seeds 1 - 2, rugose.

Field note: Common in marshy and muddy places along rivers and water streams; often cultivated in the gardens.

Fl. & Fr.: August - October.

Specimen examined: Triveni 130, Bijoliya 1886 (MLVGCB Herb).

Distribution: India and Sri Lanka.

AGAVACEAE

AGAVE L.

Agave americana L., Sp. Pl. 323. 1753; Duthie, Fl. Gangetic Plain 3: 242. 1920; Parmar, in Shetty & Singh Fl. Rajasthan 2: 833. 1991. (Plate 301)

Vern: *Narwas.*

Undershrub, perennial. Stem short, stout. Leaves in a lax rosette, stout, oblong-lanceolate, curving outwards, the ends drooping, up to 0.9 - 1.8 m long, 15 - 20 cm broad, widest about the middle, bluesh-green, sometimes glaucous, often with pale yellow borders or otherwise variegated, shallowly channelled only at the apical region; apical spine ca 1 cm, dark reddish brown-black; marginal spines pointing downwards, dark brown. Scape with 5 - 8 m high panicles; primary branches almost horizontal. Fascicles of flowers crowded at the end of subsidiary ascending branches. Tepals amber-coloured, ovate-lanceolate, obtuse. Capsules oblong-obovate, beaked.

Field note: Common along hedges of cultivated fields.

Fl. & Fr.: January - August.

Specimen examined: Sawaipur 1411, Nandrai 1941 (MLVGCB Herb).

Distribution: Mexico, naturalized in the Mediterranean region, India and Pakistan.

Notes: The American aloe is cultivated in gardens throughout the area. It is not naturalized here and grown as hedge plant as well as for its fibres from leaves. This species is monocarpic, flowering after about 10 years of growth.

HYPOXIDACEAE

CURCULIGO Gaertn.

Curculigo orchioides Gaertn. Fruct. Sem. Pl. 1: 63. t. 16. f. 11. 1788; Hook. *f.,* Fl. Brit. Ind. 6: 279. 1892; Duthie, Fl. Gangetic Plain 3: 247. 1920; Pandey, in Shetty & Singh, Fl. Rajasthan 2: 834. 1991. *Curculigo brevifolia* Dryand in Ait., Hort. Kew. ed. 2: 253. 1811; Wight, Ic. 6: t. 2043. f. 2. 1853. *Curculigo malabarica* Wight, Ic. 6: 22, t. 2043, f. 1. 1853. *Hypoxis orchioides* Kurz in Mig. Ann. Mus. Bot. Lugd. Bot. 4: 177. 1869. (Plate 302)

Vern: *Kalimusli.*

Perennial herb, 10 - 35 cm tall, without aerial stem. Root tubers, elongated, thick, covered with fleshy roots and remains of old leaves. Leaves very variable, radical, sessile or subsessile, 10 -

40 x 1 - 3 cm, linear-lanceolate, acute or acuminate, thin but firm in texture, sparsely long ciliate above; sheaths persistent, fibrous. Scape short, hidden among the leaf bases. Flowers appear almost at the ground level, 2 to 3-together on 5 - 10 cm long axillary, peduncle hidden among leaf-sheaths, bright yellow, subses-sile, upper staminate, lower hermaphrodite, bracteate; bracts lanceolate. Perianth 6-lobed; lobes elliptic-oblong, 5-7 mm long, yellow; outer ones hairy outside; inner ones sparsely hairy along nerves. Stamens 6, epiphyllous, about half the length of the perianth segments; anther linear. Ovary pubescent, separated from the perianth by a stipe, lanceolate, locules imperfect; ovules 6-8; stigmas 3. Capsules subulate to ovoid, 1.5 - 1.8 cm long, 1 - 4-seeded, sessile, beaked. Seeds globose, grooved, strophiolate, black.

Field note: Common in grasslands. In vegetative stage it is confused with the seedlings of *Phoenix sylvestris*.

Fl. & Fr.: August - November.

Specimen examined: Bijoliya 187, Mainal 1611 (MLVGCB Herb).

Distribution: India and Pakistan.

Uses: Rhizomes are used medicinally.

DIOSCOREACEAE

DIOSCOREA L.

Dioscorea bulbifera L., Sp. Pl. 1033. 1753; Wight, Ic. 3 (2): 6. t. 878. 1844; Duthie, Fl. Gangetic Plain 3: 254. 1920; Prain & Burkill, in Ann. Roy. Bot. Gard. Calc. 14 (1): 111. 1936; Pandey, in Shetty & Singh, Fl. Rajasthan 2: 836. 1991. *D. sativa auct.* non L. 1753; Hook. *f.,* Fl. Brit. Ind. 6: 295. 1892. (Fig. 152 & Plate 303)

var. ***bulbifera.***

Large, dioecious, glabrous climbing herb. Tubers solitary, globose or pyriform, 10 - 15 cm across, subsessile. Stem glabrous, twining clockwise bearing numerous brown warted bulbils in leaf-axils; bulbils ca. 2.5 cm in diam., axillary, rounded. Leaves alternate, simple, 10 - 30 x 8 - 28 cm, broadly ovate, cordate at the base, long acuminate, 7 - 13-nerved at the base, petiole 2.5-12 cm long. Male spikes slender, drooping, 6.5-10 cm long, clustered in the axils or sometimes panicled; flowers white, pink or brownish white, fragrant. Stamens 6, fertile. Female spikes simple, fascicled in the leaf axils, 10-25 cm long; perianth-lobes linear; staminodes 6; styles conical; stigmas 3, recurved. Capsule 2.0 x 0.7 - 1.3 cm, quadrately oblong, winged, reflexed; wing rounded, 20 - 22 x 7 - 10 mm. Seeds winged at the base.

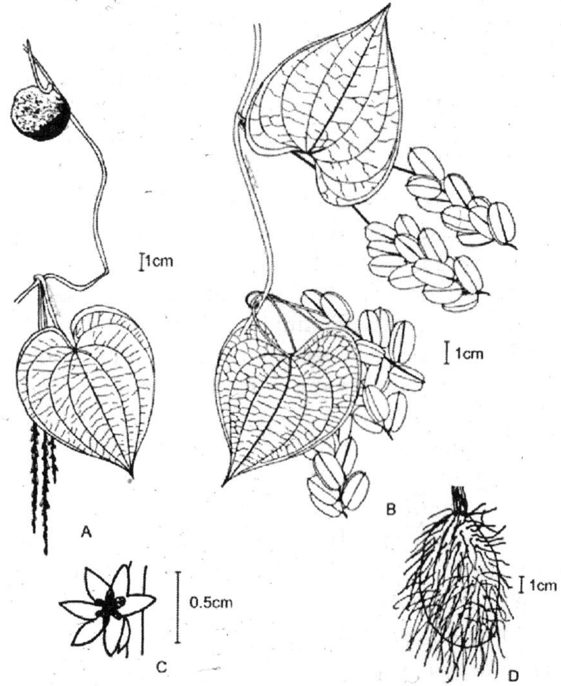

Fig. 152. *Dioscorea bulbifera* L. var. *bulbifera,* A. Male flowering branch. B. Fruiting branch, C. Male flower, D. Tuber

Field note: Common among hedges and in foothills.

Fl. & Fr.: August - December.

Specimen examined: Nikumbh 113, Bari Sadari 3756 (MLVGCB Herb).

Distribution: Tropical and sub-tropical parts of Asia & Africa.

Uses: Tubers and bulbils are edible. It is reported to contain a poisonous glycoside.

LILIACEAE

ASPARAGUS L.

Asparagus racemosus Willd. Sp. Pl. 2: 152. 1799; Wight, Ic. 6 : 26. t. 2056. 1853; Hook. *f.*, Fl. Brit. Ind. 6: 316. 1892; Duthie, Fl. Gangetic Plain 3: 259. 1920; Parmar, in Shetty & Singh Fl. Rajasthan 2: 841. 1991; Bhandari, Fl. Ind. Desert. 317. 1995. (Plate 304)

Vern: *Satawari, dashmool.*

An extensively scandent, much branched, spinous, undershrub hermaphroditic. Roots tuberous, many in large fascicles. Stem climbing, branched, to 2 m; branches usually distinctly striate-ridged, ridges ± cartilaginous denticulate. Spines suberect. Cladodes in clusters of 2 - 6, oblong-linear, 1 - 3 cm × ca 1 mm, triquetrous, falcate. Leaf spur spinescent; spine straight or subrecurved, 1.5-2 cm on main stems, 5-10 mm on branches, woody, sharp. Flowers in axillary, solitary or fascicled, simple or branched racemes, small, white; pedicels jointed in the middle. Bracts ovate or triangular, ca 1 mm. Perianth spreading, 2-3 mm. Tepals white to creamy, linear-oblong, 3 mm long, equal; stamens 6, slightly shorter than the perianth parts; anthers yellow, minute; ovary 3-locular with 4 – 6 ovules in each locule; style 0.75 – 1 mm long, 3-branched; stigma 3, spreading. Berry globose, 4 mm in diameter, red when ripe, 1 – 2-seeded.

Field note: Common climber in the forests. The flowers are fragrant.

Fl. & Fr.: August - March.

Specimen examined: Atun 486 (MLVGCB Herb).

Distribution: Asia, Africa and Australia.

Uses: 1 tsp fresh juice from tuberous roots is given orally in dysentery, acidity and to increase the lactation after delivery. The juice and tuber paste is applied externally to the burnt skin and to heal the wounds. Tuberous root powder is used to cure leucorrhoea and gout. Fresh roots are eaten to beat the heat, for strength as well as to cure bodyache, boils, cough and urinary infection. Medicated oil prepared from tuberous roots is beneficial for nervous and rheumatic complaints.

ASPHODELACEAE

KEY TO THE GENERA

1a. Leaves distinctly succulent, often margined with prickly teeth .... *Aloe*

1b. Leaves not as above, margins not as above 2

2a. Root stock bulbous .... *Iphigenia*

2b. Root stock not as above 3

3a. Leaves tubular, subterete .... *Asphodelus*

3b. Leaves flat .... *Chlorophytum*

ALOE L.

Aloe vera (L.) Burm. f. Fl. Ind. 83. 1768, non Miller, Gard. Dict. Ed. 8. no. 20. 16 April, 1768; Gandhi in Sald. & Nicolson, Fl. Hassan 793. 1976; Boiss. Fl. Or. 5: 329. 1882; Parmar, in Shetty & Singh Fl. Rajasthan 2: 839. 1991; Bhandari, Fl. Ind. Desert. 317. 1995. *A. perfoliata* L. var. *vera* L., Sp. Pl. 320. 1753. *A. barbadensis* Mill. Gard. Dict. no. 2. 1768. (Fig. 153)

Vern: *Guwar patha.*

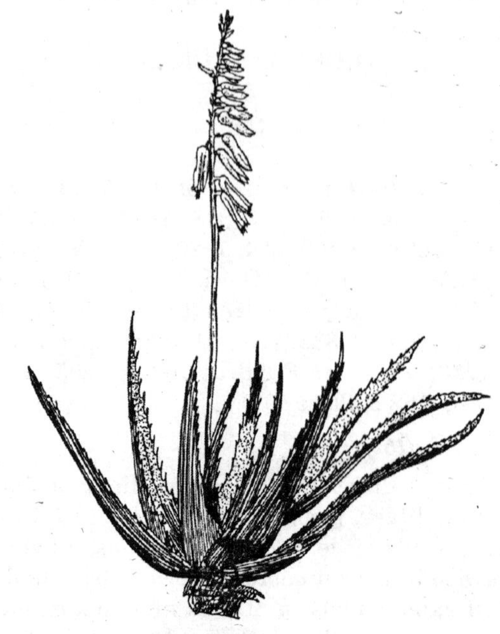

Fig. 153. *Aloe vera* (L.) Burm. f.
A. Habit

Perennial herb, succulent more or less acaulescent or with subterranean stem, producing rosette of leaves from the ground and reaching up to 60 cm high. Stem short, producing suckers at the base. Leaves sessile, thick, erect, lanceolate, 15 - 50 x 4 - 8 cm, glabrous, glaucous, margins sparsely spinecent, 2-3-spine-tipped. Flowers erect 60 - 100 cm long raceme, simple or with 1-2 branches, reflexed, pedicel about half as long as bract; the lower flowers falling off as the racemes elongates. Bracts ovate-lanceolate, 9 - 12 x 5 - 6 mm, persistent. Perianth 2.5 - 3 cm long, tubular, dull reddish, lobes 6, almost equalling the tube. Filaments slender, shortly exserted. Ovary trilocular, placentation axile, many ovules in each locule; style elongated, stigma obscurely 3-lobed. Capsule trigonous, oblong, loculicidally 3-valved, 1.5 cm. Seeds trigonous, winged.

Field note: Common in corners of grazing fields, mostly growing in Graveyards where it is considered as providing peace and calm.

Fl. & Fr.: January - May.

Specimen examined: Pur 72 (MLVGCB Herb).

Distribution: Mediterranean region for a very long time; place of origin uncertain; widely naturalized in Portugal, Turkey, China (S. Yunnan), Pakistan, India, Nepal, West Indies and Central America.

Notes: Plants in South central Rajasthan with reddish flowers are different from the populations of *A. vera* growing in Europe, Turkey and China which are reputed to have yellow to pale yellow flowers. It may be worthwhile to investigate it further.

Uses: It is used as a stomachic, purgative and emmenagogue as well as in the treatment of piles and rectal fissures. The mucilage is cooling and is used to poultice inflammations (Wealth of India 1: 61.1948).

ASPHODELUS L.

Asphodelus tenuifolius Cav., Anal. Cienc. Nat. 3 : 46. t. 27. f. 2. 1801; Boiss., Fl. Or. 5 : 314. 1882; Hook. *f.,* Fl. Brit. Ind. 6: 332. 1892; Cooke, Fl. Pres. Bomb. 3 : 279. 1958 (Reprint. ed.); Wendelbo in Rech. f. Fl. Iran. 151 : 4. 1982; Parmar, in Shetty & Singh Fl. Rajasthan 2: 841. 1991. *A. clavatus* Roxb., Fl. Ind. 2: 148. 1832. *A. fistulosus* L. var. *tenuifolius* (Cav.) Baker in J. L. Soc. Bot. 15 : 272. 1876. *A. fistulosus* L. 1753; Duthie, Fl. Gangetic Plain 3: 265. 1920; Mahesh. Illus. Fl. Delhi f. 201. 1966. *A. parviflorus* Wight, Ic. 6 : 27. t. 2062. 1853.

Vern: *Kanda kori.*

An annual, erect, glacuous, herb, 15 - 50 cm tall. Roots fibrous stout, numerous. Stem scapose, not leafy. Leaves 15 - 40 cm x 0.5 - 3 mm, erect, linear, semiterete, fistular, sessile, acute, sheathing at the base, finely puberulous. Scapes many from the base, much branched, 30-60 cm. Flowers in long branched racemes, white, bracteate; pedicel 4 - 6 mm, jointed below the middle, bract ca 2.5 mm, cymbiform, scarious with a brownish keel. Perianth segment oblong, with a brownish costa. Stamens 3 - 3.5 mm long, anthers 1 - 1.5 mm long. Style *ca.* 2 mm long, stigma subcapitate, minutely 3-lobed. Capsule 2 - 4 mm in diameter, globose, 3-gonous, 3-valved, glabrous. Seeds *ca.* 3 mm, obovoid, rounded and rugose on the back, black.

Field note: Common weed of cultiveted fields during winters.

Fl. & Fr.: December - April.

Specimen examined: Kanera 427 (MLVGCB Herb).

Distribution: North Africa, S. W. Europe, South West Asia, India and Pakistan.

Notes: Richardson & Smythies in Tutin *et al.* Fl. Europ. 5: 17, 1980 have treated *Asphodelus tenuifolius* Cav. as a synonym of *Asphodelus fistulosus*. We have followed Wendelbo (1982) who has accepted *Asphodelus tenuifolius* Cav. as a separate species till more is known about the variation patterns of *Asphodelus fistulosus* L. from the entire range.

Uses: The whole plants are fed to cattle as green fodder. Plant is eaten raw as well as in cooked form, particularly during famine. Seeds are considered to be diuretic.

CHLOROPHYTUM Ker.

Chlorophytum laxum R. Br, Prodr. 277. 1810; Hook. *f.,* Fl. Brit. India 6: 336. 1892; Cooke, Fl. Bombay 2: 773. 1907; Parmar, in Shetty & Singh Fl. Rajasthan 2: 842. 1991. *C.*

perviflorum Dalz., in Hook. Kew J. Bot. Misc. 2: 141. 1850. *Phalangium parviflorum* (Dalz.) Wight, Ic. 6: 21. t. 2039. 1853.

Vern: *Dholi musali.*

Low herb. Roots fibres ending into ellipsoid tubers. Leaves subdistichous, sessile, strongly recurved, grass like or conduplicate or flat, linear-lanceolate, falcate, 8 - 20 cm × 1.5 - 6 mm, wavy or crisped on margins, sessile, glabrous. Flowers in simple or shortly branched racemes; scape usually 2 or 3, erect or arching, 10 - 40 cm long, with small sheath. Flowers solitary or paired; pedicels 2-5 mm, articulate proximally; bracts narrowly deltoid to lanceolate, very small. Perianth greenish white, segments oblong, closely 3-veined. Stamens alternately short and long, shorter than tepals; filaments longer than anthers; anthers subglobose. Capsule obcordate, 3-winged, ca 3 × 5 mm. Seeds angular, 1 in each valve.

Field note: Common in gravelly open grasslands, waste lands etc.

Fl. & Fr.: July - December

Specimen examined: Mainal 1600 (MLVGCB Herb).

Distribution: China, India, Indonesia, Malaysia, Myanmar, Sri Lanka, Thailand; tropical Africa, Australia.

IPHIGENIA Kunth.

Iphigenia indica (L.) A. Gray, in Kunth, Enum. Pl. 4: 213. 1843; Hook. *f.*, Fl. Brit. India 6: 357. 1892; Duthie, Fl. Gangetic Plain 3: 263. 1920; Parmar, in Shetty & Singh Fl. Rajasthan 2: 844. 1991. *Melanthium indicum* L., Mant. Pl. 226. 1771.

Vern: *Santhal-chutia.*

Perennial herb, 5 - 25 cm high. Corm globose, 0.5 - 1.5 cm across; sheaths brown. Stem flexuous, 10 - 25 cm. Leaves few, alternate, sessile, linear, 7 - 15 × 0.3 - 0.9 cm, midvein conspicuous, base sheathing and clasping; upper ones much smaller. Flowers solitary or in corymbs; pedicels 1 - 2 cm long; corymb 2-10-flowered; bracts linear, foliaceous, 1.5-2.5 cm. Perianth dark purple, linear-subulate, 5-7 cm long, spreading and reflexed. Stamens 2-3 mm; filaments flatte-ned; anthers ca 1 mm. Stigma lobes recurved. Capsule oblong-ellipsoid, 3-grooved, obtuse, 3-5 × 3-3.5 mm. Seeds subglobose, angled, brown.

Field note: A troublesome weed in ponds, slow running channels, ditches and marshy areas forming pure formations.

Fl. & Fr.: July - October.

Specimen examined: Armia panth 1540 (MLVGCB Herb).

Distribution: Cambodia, China, India, Indonesia, Myanmar, Nepal, Philippines, Sri Lanka, Thailand, Vietnam; N Australia.

HYACINTHACEAE

URGENIA Steinheil

Urgenia indica (Roxb.) Kunth, Enum. 4: 333. 1843; Wight, Ic. 6: 27. t. 2963. 1853; Hook. *f.*, Fl. Brit. Ind. 6: 347. 1892; Duthie, Fl. Gangetic Plain 3: 264. 1920; Cooke, Fl. Pres. Bomb. 3: 277. 1958 (Reprint. d.); Parmar, in Shetty & Singh Fl. Rajasthan 2: 845. 1991; Bhandari, Fl. Ind. Desert. 318. 1995. *Scilla indica* Roxb., Fl. Ind. 2: 147. 1832. *Drimia indica* (Roxb.) Jessop in Afr. Bot. 43: 272. 1977.

Vern: *Mar kando.*

Perennial herb. Bulb globose, thick, 3 - 5 cm across, pale or white. Leaves appearing after flowers, linear, ensiform, 15-40 x 1-2.5 cm, acute; scape to 60 cm, erect, brittle. Flowers pale brown, racemes erect, 15-30 cm long. Bracts minute, deciduous. Pedicels 2.5-7 cm long, spreading or decurved. Perianth greenish-white, campanulate, linear-oblong, pale brown, white at margins segments. Stamens 6-7 mm, filaments flattened at base, anthers versatile, yellow. Capsule ellipsoid, tapering at both ends 1.5 - 1.8 cm long, 6 - 9 seeded. Seeds elliptic, flattened, black.

Field note: Common in gravelly open grasslands, waste lands etc.

Fl. & Fr.: April - May.

Specimen examined: Bigod 1660 (MLVGCB Herb).

Distribution: Myanmar, India, Pakistan, Uganda, Kenya, Tanzania, Africa West to Mauritania and South to South Africa.

Uses: The plant is medicinally important. It is used as a substitute for true squill, *Drimia maritima* (L.) Stearn.

COLCHICACEAE

GLORIOSA L.

Gloriosa superba L., Sp. Pl. 305. 1753; Wight, Ic. 6: 25. t. 2047. 1853; Hook. *f.*, Fl. Brit. Ind. 6: 348. 1892; Duthie, Fl. Gangetic Plain 3: 262. 1920; Parmar, in Shetty & Singh Fl. Rajasthan 2: 844. 1991. *G. simplex* L. Mant. Pl. 62. 1767. *G. minor* Rendle, J. Bot. 34: 132. 1896; *G. virescens* Lindl., Bot. Mag. 52: t. 2539. 1825; *G. carsonii* Baker, Bull. Misc. Inform. Kew, 74. 1895. (Plate 305)

Vern: *Kalahari.*

Perennial, climbing, sometimes erect herb up to 2 m long; stem annual, glabrous and sparsely branched; tuber perennial, horizontal, abruptly bent in a V or L shape, roots fibrous. Leaves in whorls of 3- 4, opposite or alternate, simple, sessile; blade ovate to lanceolate, 5 - 15 × 1.5 - 4 cm, cordate at base, glabrous, entire, apex of upper leaves with or without 1 - 2 cm long tendril, with prominent mid nerve. Peduncles 5 - 8 cm long, deflexed at the tip. Flowers axillary, solitary, bisexual, regular, 6-merous, 4.5 - 7 cm in diam., showy, pendulous; pedicel 4 - 20 cm long. Perianth segments free, oblong-lanceolate, 5 - 8 × 1 - 1.5 cm, often with undulate margins, strongly reflexed when mature, persistent, with crisply waved margins, at first yellow, finally scarlet to crimson. Stamens with filaments 2-5 cm long, spreading; anthers spreading, dorsiflxed, 7 - 10 mm long, opening by longitudinal slits. Ovary superior, 3-celled, carpels coherent only by their inner margins; style filiform, 2 - 4 cm long, bent at a right angle basally. Capsules loculicidal, linear-oblong, 4 - 6 × 1 - 3 cm, containing up to 20 seeds. Seeds ovoid, 4 - 5 mm in diam., surrounded by a fleshy, red sarcotesta.

Field note: Common among hedges, shrubs and trees in the forests and waste lands.

Fl. & Fr.: August - December.

Specimen examined: Mainal 111 (MLVGCB Herb).

Distribution: Africa, India and southeastern Asia, now a days distributed widely throughout the tropics, and worldwide as a pot plant.

PONTEDERIACEAE

KEY TO THE GENERA

1a. Leaves rosulate or alternate. Flowers zygomorphic. Perianth gamophyllous. Stamens unequal in length; all filaments without appendages .... *Eichhornia*

1b. Leaves solitary or alternate. Flowers actinomorphic. Perianth free. Stamens of sub-equal length; one filament with a lateral appendage and a large anther ...... *Monochoria*

EICHHORNIA Kunth *nom. cons.*

Eichhornia crassipes (Mart.) Solms in DC. Monogr. Phan. 4 : 527. 1883; Subram. Aquat. Angiosp. 70. f. 48. 1962; Raizada, Suppl. Duthie, Fl. Gangetic Plain 279. 1976; Pandey, in Shetty & Singh Fl. Rajasthan 2: 847. 1991; Bhandari, Fl. Ind. Desert. 319. 1995. *Pontederia crassipes* Mart., Nov. Gen. Sp. Pl. 1: 9, t. 4. 1823-1824. *Eichhornia cordifolia* Gandoger, Bull. Soc. Bot. France 66: 294. 1920, non. Richard 1850. (Fig. 154)

Vern: *Jal-kumbhi.*

Floating perennial, aquatic herb, 30-50 cm high, with short stem and many long fibrous roots, floating or sometimes rooting. Stolons axillary, stout, spongy semicylindric. Leaves radical, rosette-forming, glabrous, 15 - 20 x 15 - 20 cm; broadly ovate to rhomboid, entire, obtuse, rounded-subcordate at base, fleshy, shining green. Petiole glabrous, spongy, 6 - 30 cm long, with a fusiform bulbous portion about the middle. Peduncle spongy. Flowers in dense, up to 15 cm long spikes, lilac, tinged blue, withering soon. Perianth-lobes 1.5 - 1.8 cm long, slightly curved, tube green, glandular; segments 2.5 - 3 cm long, obovate to ovate-oblong, posterior segment with a yellow spot. Stamens 3 longer, ca 21 mm, filaments curved, glandular hairy; shorter filaments ca 6 mm long, glabrous; anthers purple-blue. Ovary ovoid, glabrous; style glandular; stigma 3-lobed, glandular. Capsules ovoid-oblong. Seeds minute, ovoid, ca 1 mm long, ribbed.

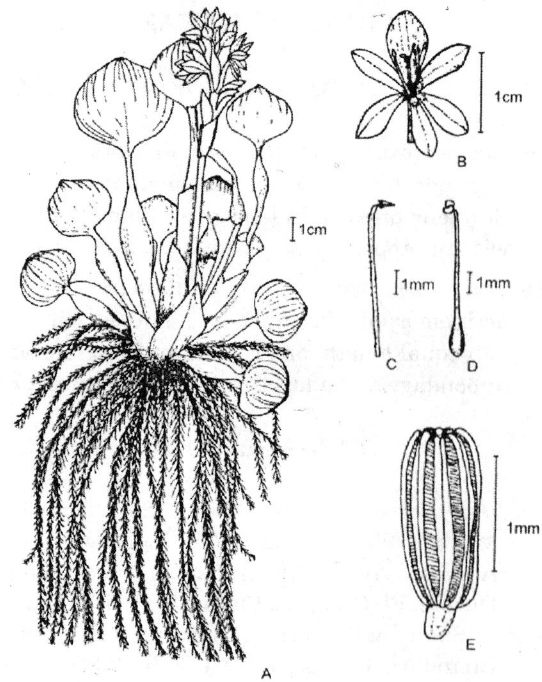

Fig. 154. *Eichhornia crassipes* (Mart.) Solms, A. Habit, B. Flower, C. Stamen, D. Pistil, E. Seed

Field note: A troublesome weed of ponds, slow running channels, ditches and marshy areas forming pure formations. The *'Water Hyacinth'* is very common in the Rajasthan inhabiting vast marshy areas, propagating by stolons and multiplying very fast. It covers vast areas of water surface forming impenetrable growth. The lilac flowers present the beautiful scene during the flowering season. Difficult to eradicate, due to its quick growth, it has become a troublesome weed throughout the state.

Fl. & Fr.: September - December.

Specimen examined: Jahajpur 74 (MLVGCB Herb).

Distribution: Native of Brazil. introduced and naturalised in many tropical countries.

MONOCHORIA K. Presl

Monochoria vaginalis (Burm. f.) K. Presl, Rel. Haenk. 1: 128. 1827; Hook. *f.,* Fl. Brit. Ind. 6: 363. 1892; Duthie, Fl. Gangetic Plain 3: 270. 1920; Subram. Aquat. Angiosp. 69. f. 47. 1962; Pandey, in Shetty & Singh Fl. Rajasthan 2: 848. 1991. *Pontederia vaginalis* Burm. f. Fl. Ind. 80. 1768. *P. plantaginea* Roxb., Fl. Ind. 2: 123. 1832. *Monochoria plantaginea* Knuth, En. 4: 135. 1843. *M. vaginalis* (Burm. f.) Prresl var. *plantaginea* (Roxb.) Solms in DC. Monogr. Phan. 4: 523. 1883; Hook. *f. l. c.* 6 : 363. 1892; Duthie, *l. c.* 3: 270. 1920. (Fig. 155 & Plate 306)

Vern: *Cachiee.*

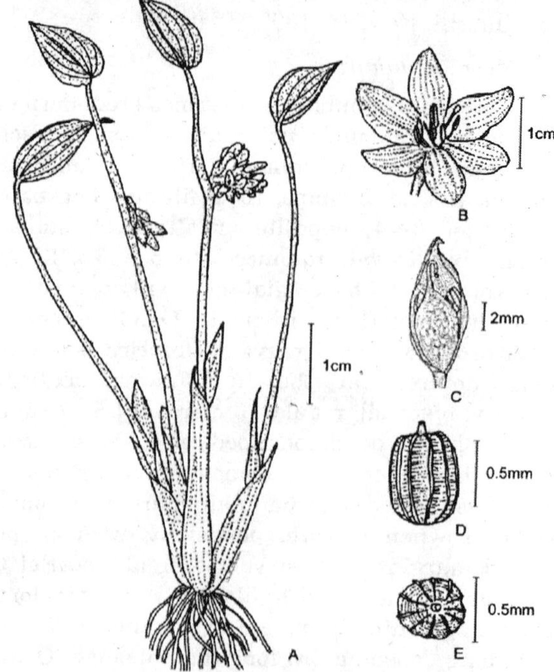

Fig. 155. *Monochoria vaginalis* (Burm. f.) Presl. A. Habit. B. Flower, C. Gynoecium, D. Fruit, E. Seed

Annual or perennial, amphibious, erect, herb, upto 50 cm tall. Root stock small with adventitious roots. Stem suberect to erect. Leaves 5 - 15 x 3 - 11 cm, broadly ovate to ovate-oblong, acute or acuminate, base rounded to subcordate; nerves convergent. Racemes erect, subtendd by leaf sheaths when young, finally deflexed. Pedicel ca 7 mm long. Perianth 6, blue-violet, spotted red; outer oblong, 8 - 10 mm long, subacute; inner obovate, 10 - 12 mm long. Stamens c. half as long as the perianth. Ovary ovoid, 3 mm long; style ca

2 mm long, stigma 3-lobed. Capsule ovoid, ca 8 - 10 mm long, pale with brown ribs. Seeds 1 mm long, ovoid, with many brown ribbed, transversely striated.

Field note: Common in ponds, ditches and in swampy habitats.

Fl. & Fr.: September - December.

Specimen examined: Mainal tank 131, Sareri 1996 (MLVGCB Herb).

Distribution: SE Asia to China, Japan and throughout Malaysia.

Notes: The vegetative parts of the plant vary in size and shape. Flowers are ephemeral (Backer and Heemstede, *l.c.*). Parts of the plant are used as a vegetable; root is chewed to cure toothache; bark is used medicinally for asthma.

COMMELINACEAE

KEY TO THE GENERA

1a. Stamens 6, all perfect 2

2a. Petals free. Stem short and thick, crowded with long, sublanceolate leaves. Ovules solitary .... *Rhoeo*

2b. Petals connate or forming a tube below. Stem often elongated and thin, usually with distant small leaves. Ovules 2 in each locule.. *Cyanotis*

1b. Stamens 3 perfect; staminodes 1 - 3 3

3a. Inflorescence usually lax, panicled; bracts not spathaceous. Petals equal. Filaments bearded .... *Murdannia*

3b. Inflorescence short, condensed, slightly enclosed within spathaceous bract. Petals unequal. Filaments glabrous .... *Commelina*

COMMELINA L.

KEY TO THE SPECIES

1a. Leaves almost as broad as long. Spathes sessile or subsessile, Seeds pitted............................ .*C. benghalensis*

1b. Leaves longer than broad. Spathes distinctly peduncled, Seeds smooth .... *C. forsskalaei*

Commelina benghalensis L., Sp. Pl. 41. 1753; Wight, Ic. 6: 29. t. 2065. 1853; Hook. *f.*, Fl. Brit. Ind. 6: 370. 1892; Duthie, Fl. Gangetic Plain 3: 274. 1920; Parmar, in Shetty & Singh Fl. Rajasthan 2: 853. 1991; Bhandari, Fl. Ind. Desert. 321. 1995. (Fig. 156)

Vern: *Bokano, Mokano.*

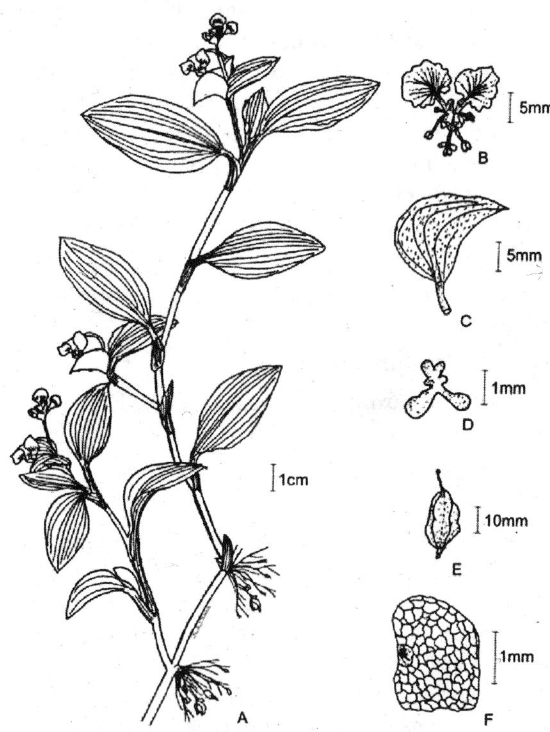

Fig. 156. *Commelina benghalensis* L.
A. Habit, B. Flower, C. Spathe, D. Staminode, E. Capsule, F. Seed

Decumbent-ascending, annual, herb, 20 - 60 cm long, rooting at lower nodes, Branches many, some of them growing underground, bearing reduced leaves, spathes with apetalous closed flowers, often abnormal, 1 - 2 seeded capsules. Leaves ovate, elliptic or suborbicular, subsessile or narrowed into 0.5 - 1 cm long petiole, with sheathing base, 1 - 8 x 1 - 4 cm, entire, acute to sub-rounded at apex; sheaths 1 - 2 cm long, ciliate on the margins. Aerial flowers: spathes 2 - 3 at the end of branches, terminal or leaf opposed, subsessile, recurved, auricled on one side at the base, hairy outside, blue, in short cymes subtended by spathes, maturing earlier than the subterranean flowers. Calyx: posterior sepal broadly boat-shaped, concave with a green inturned tip, anterior sepals obovate, orbicular, joined for a short distance from base. Petals pale-violet or lilac, posterior pair reniform-deltoid, unequal-sided at base, 6 - 7 mm broad claw. Stamens 3, two laterals with slightly curved, filaments blue; anthers dark-blue; median stamen with shorter filament and a large yellow anther with curved and divergent cells. Staminodes 3, filaments slender, colourles, 4 large and 2 small. Ovary 3-lobed. Fruit a capsule broadly ovoid-oblong. Capsules oblong, trigonous, unequally 3-celled. Seeds 5, pitted, wrinkled. Cleistogamous flowers bisexual, solitary, born in pear-shaped spathes.

Field note: Common weed of cultivated fields, gardens during rainy seasons.

Fl. & Fr.: July - October.

Specimen examined: Tikar 176, Kekri 43 (MLVGCB Herb).

Distribution: Tropical and subtropical Asia and Africa.

Uses: A plant of moist and shaded places, used as fodder and sometimes as vegetable; medicinally used to cure skin inflammations, leprosy and as laxative. The young leaves are eaten as vegetable, also in making pakodas. A paste derived from the plant is used to treat burns, and root-juice is used to treat indigestion.

Commelina forsskalaei Vahl, Enum. 2: 172. 1805; Hook. *f.*, Fl. Brit. Ind. 6: 371. 1892; Mahesh. Illus. Fl. Delhi f. 204. 1966; Raizada, Suppl. Duthie, Fl. Gangetic Plain 281. 1976; Parmar, in Shetty & Singh Fl. Rajasthan 2: 854. 1991; Bhandari, Fl. Ind. Desert. 323. 1995. (Fig. 157 & Plate 307)

Vern: *Bokano, mukano.*

Straggling, creeping, annual herb, diffusely branched, rooting at the lower nodes. Leaves 3-6 x 0.5-1.5 cm, linear or narrowly oblong-elliptic, obtuse, undulate margined; sheath 0.7-1.2 cm long, hairy, mouth ciliate. Aerial spathes 1.6 x 0.8 - 0.10 cm, irregularly funnel-shaped, acute, inner surface glossy, hairy outside, in axillary branched cymes; cymes 2 - 3 flowered; inner branch with single male flowers outer ones with 2 - 3 flowers. Cleistogamous flowers solitary in a spathe, bisexual, reduced. Calyx: posterior sepals broadly boat-shaped. Corolla orbicular-cordate, blue. Staminodes 2, filament slender. Stamens 2, leteral with long curved filaments flattened in upper halves; anthers greenish blue. Ovary vestigial. Perfect flowers of much of the same structure but slightly smaller. Capsule orange-brown, oblong, one seeded, laterally compressed. Seeds oblong, rounded at both ends, mottled, brown.

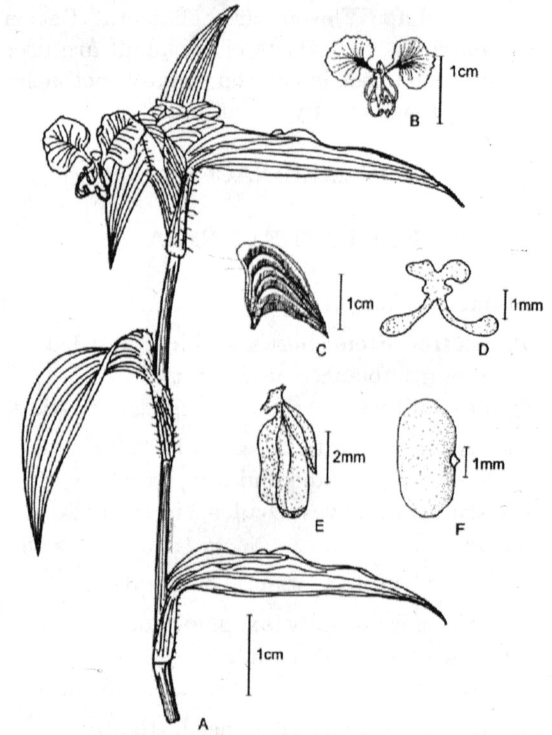

Fig. 157. *Commelina forskalaei* Vahl.
A. Flowering & fruiting branch, B. Flower, C. Spathe, D. Staminode, E. Capsules, F. Seed

Field note: Common weed of cultivated fields, gardens during rainy season.

Fl. & Fr.: August - October.

Specimen examined: Mokhampura 132, Dunwala 2111 (MLVGCB Herb).

Distribution: Tropical Africa, Arabia, India and Pakistan.

CYANOTIS D. Don. *nom. cons.*

Cyanotis cristata (L.) D. Don, Prodr. Fl. Nepal. 46. 1825; Wight, Ic. 6: 32. t. 2082. 1853; Hook. *f.,* Fl. Brit. Ind. 6: 385. 1892; Duthie, Fl. Gangetic Plain 3: 270. 1920; Parmar, in Shetty & Singh Fl. Rajasthan 2: 857. 1991; Bhandari, Fl. Ind. Desert. 324. 1995. *Commelina cristata* L., Sp. Pl. 42. 1753. *Tradescantia cristata* (L.) Jacq., Hort. Vindob. 2: 67. t. 137. 1844. *T. imbricata* Roxb., Fl. Ind. 2: 120. 1824. (Fig. 158 & Plate 308)

Vern: *Bokani.*

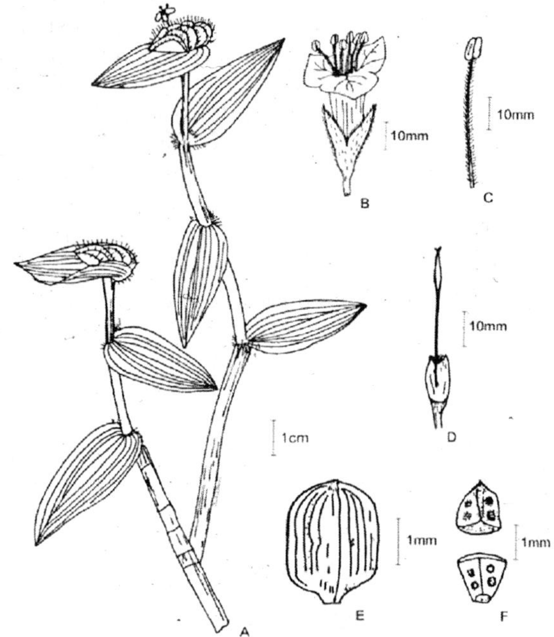

Fig. 158. *Cyanotis cristata* (L.) D. Don.
A. Flowering & fruiting branch, B. Flower, C. Stamen, D. Pistil, E. Fruit, F. Seeds

Prostrate herb with erect or ascending branches, 15 - 30 cm. long, glabrous to slightly pubescent; branches rooting below. Leaves ovate-oblong or oblong-oblanceolate, sessile, distant, acute, entire, 5 - 10 x 1 - 1.5 cm, base cordate sub-rounded to rounded, glabrous or sparsely hairy, margins villously ciliate, sessile; sheaths short hairy. Flowers blue or violet purple, usually in terminal clusters subtended by ovate-falcate, or scorpioid recurved cymes which are longer than the bracteoles, ca 1.5 cm long, linear-ovate, falcate, acute, folded spathes. Petals blue-violet. Capsules 3-4 mm long, oblong, trigonous, membranous, nearly glabrous, usually 6-seeded (sometimes less due to abortion). Seeds ca 1.2 mm long, black striated or piled.

Field note: Common in moist and shady places.

Fl. & Fr.: July - December

Specimen examined: Hamirgarh 133, Bherukhera 2126 (MLVGCB Herb).

Distribution: Most of tropical Asia and Africa, especially on hills.

MURDANNIA Royle *nom. cons.*

Murdannia nudiflora (L.) Brenan in Kew Bull. 7: 189. 1952. in Hutch. & Dalz., Fl. W. Trop. Afr. ed.2. 3(1): 24.1968; Stewart, Ann. Cat. Vas. Pl. W. Pak. & Kash. 37.1972; Parmar, in Shetty & Singh Fl. Rajasthan 2: 858. 1991. *Commelina nudiflora* L. Sp. P1. 41. 1753. *p.p. Aneilema nudiflorum* Wall. ex. Hook. *f.,* Fl. Brit. Ind. 6: 378. 1894. *Murdannia malabarica* (L.) Bruck., in Engl. & Prantl, Nat. Pflanzenfam. ed. 2, 15a: 1930. *Aneilema malabarica* (L.) Merrill, in Philip. J. Sc. Bot. 7: 232. 1912. *Phaeneilema malabarica* (L.) Narayan Swami ex Biswas in Ind. For. Rec. (n.s.) Bot. 3 : 55. 1941. *Tradescantia malabarica* L., Sp. Pl. ed. 2: 412. 1762. (Fig. 159 & Plate 309)

Vern: *Kanshura.*

Procumbent or suberect or diffuse herb, subsimple to branched, 15 - 50 cm long, rooting at base; roots fibrous. Leaves linear-lanceolate, rarely ovate-lanceolate, alternate, narrowed into a basal sheath, glabrous or subglabrous, entire, acute or acuminate, rounded or cordate at base, 3 - 10 cm long, 7 - 11 mm broad; basal sheath 6 - 10 mm long, slightly pubsescent with ciliate margins. Flowers in terminal and axillary lax, cymose panicle, subcorymbose or subglobose at the apex, ca 5 mm across, blue or purple; pedicels 3 - 4 mm long; bracts c. 3 mm long, boat-shaped, caducous. Sepals 1.5 - 2.5 mm long, ovate, pubescent, obtuse. Petals orbicular, 2 - 3 x 1.5 - 2 mm, purple, blue or whitish. Perfect stamens 2, included; staminodes 3-4; filaments bearded; anthers *ca.* 0.5 mm long. Capsules 6-seeded, sub-globose-trigonous, 4 mm long, *ca.* 2 mm broad, membranous, mucronate, 3-loculed with 2 seeds in each locule. Seeds *ca.* 1 mm long, trigonous, tubercled, dark-brown.

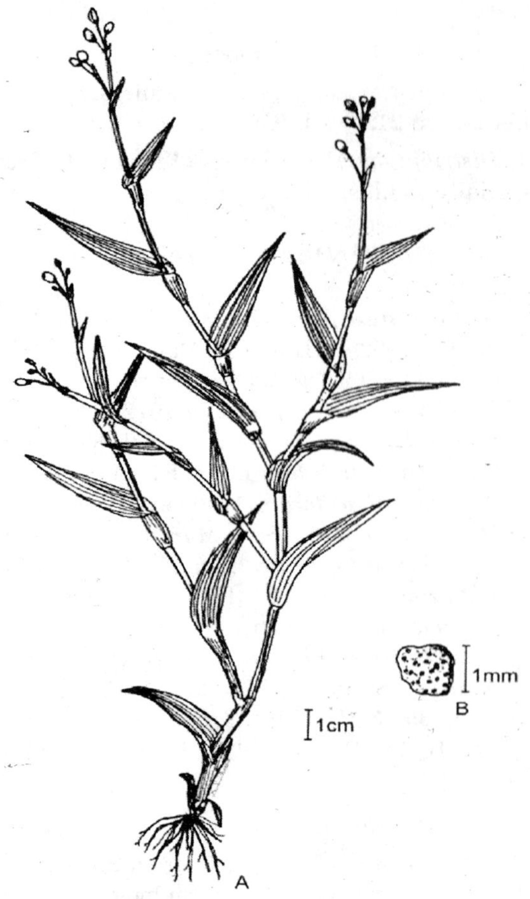

Fig. 159. *Murdannia nudiflora* (L.) Brenan, A. Habit, B. Seed.

Field note: Common in wet and shady habitats.

Fl. & Fr.: August - October.

Specimen examined: Chittorgarh fort 3119 (MLVGCB Herb).

Distribution: Pakistan, India, Bangla Desh, Sri Lanka and Malaysia.

Uses: Paste of entire plant is used for relief in burns, itching, sores etc.

RHOEO Hance

Rhoeo spathacea (Swartz) Stearn in Baileya. 5: 198. 1957; Brenan in Hutch. & Dalz., Fl. W. Trop. Afr. 3(1): 23. 1968. *Tradescantia spathacea* Swartz, Prod. Veg. Ind. Occ. 57. 1788. *Tradescantia discolor* L'Herit., Sert. Angl. 8: t. 12. 1789. *Ephemerum discolor* (L'Herit.) Moench., Suppl. Meth. P1. 78. 1802. *Rhoeo discolor* (L'Herit.) Hance in Walp., Ann. 3: 660. 1853.

Vern: *Rohio*

Perennial, robust herb with short stem arising from the underground rhizome, 30 - 60 cm tall. Leaves crowded on congested nodes, usually greenish above, purplished beneath. Flowers in several dense umbell, 2 cymiform. Sepals 3, ovate-lanceo-late, free, ca 5 mm long. Petals 3, ovate. Stamens 6, all perfect; filaments hairy; anthers ca 1 mm long, sometimes 1 or 2 partly modified into peta-loid structures. Ovary sessile, 3-locular with 1 ovule in each locule. Capsules ovoid, loculicidal, 3-valved.

Field note: Commonly planted, as an ornamental.

Fl. & Fr.: Almost throughout the year.

Specimen examined: Bhilwara 2205 (MLVGCB Herb).

Distribution: A native of C. America and W. Indies, India and Pakistan.

JUNCACEAE

JUNCUS L.

Juncus bufonius L., Sp. Pl. 328. 1753; Boiss., Fl. Or. 5: 361. 1884; Hook. f., Fl. Brit.Ind. 6: 302. 1892; Duthie, Fl. Gangetic Plain 3: 282. 1929; Pandey, in Shetty & Singh, Fl. Rajasthan 2: 860. 1991; Bhandari, Fl. Ind. Desert. 325. 1995.

A slender, annual, erect, ascending to spreading herb, 4 - 25 cm high, fibrous rooted, often gregarious and seemingly tufted or caespitose, usually with several basal leaves and one cauline leaf. Stem dichotomously branched, slender. Leaves usually 1 - 5 cm x 1 - 1.15 mm, setaceous or filiform, with a non-auricled broader sheathing base, deeply channeled. Inflorescence a much branched to subsimple panicle, often occupying greater part of the plant with distantly placed 3 - 5 (or more) flowers on each branch. Flowers sessile, greenish to straw-coloured. Perianth in two rows 3 inner much smallar, than the outer. Stamens 6, up to 2 mm long with subequal anthers and filaments. Capsules 4 - 5 x 2 mm, ovoid to oblong with rounded to obtuse apices, very minutely pointed at the apex, buff, yellowish to brownish. Seeds ovoid-rounded to broadly ellipsoid, 0.3 - 0.5 mm long, ca 1.5 times as long as broad, usually reddish-brown, mucilaginous.

Field note: Widespread in damp habitats.

Fl. & Fr.: November - March.

Specimen examined: Nagdi dam (Jahajpur) 4735 (MLVGCB Herb).

Distribution: Australia, India, North America, India and Pakistan.

ARECACEAE

PHOENIX L.

Phoenix sylvestris (L.) Roxb., Fl. Ind. 3: 787. 1832; Beccari. & Hook., in Hook. *f.*, Fl. Brit. Ind. 6: 363. 1892; Duthie, Fl. Gangetic Plain 3: 287. 1920; Parmar, in Shetty & Singh, Fl. Rajasthan 3: 861. 1993. *Elate sylvestris* L., Sp. Pl. 1189. 1753.

Vern.: *Khajoor*.

A moderate sized monoecious tree, 7.5 - 15 m tall with straight or crooked trunk, without root suckers, stem clothed with remains of petiole bases. Leaves 96 cm - 4 m long, quite glabrous, pinnate, divided into numerous leaflets; leaflets 15 - 60 x 8 - 2.5 cm, alternate, opposite or fascicled, in several planes, apex almost sharply pointed; lower leaf-lets modified into hard spines, up to ca 12 cm long; petiole short, glabrous, spiny. Inflorescence spadix covered by a hard boat-like bracts. Female inflorescence 90 - 120 cm long, main stalk flat. Flowers rounded, green, distant. Calyx lobes 3, united at the base, forming a cup-like structure, upper portion thin, translucent, lower one hard, greenish. Sepals ca 4 x 2 mm. Petals more than two times larger than the sepals, rounded. Male inflorescence much smaller, 12 - 25 cm long. Flowers sessile, white, sweet-scented, much larger than the female flower; stamens 6; anthers erect, filaments short, subulate; pistillodes minute, three, scale-like. Petals 3-lobed. Fruit drupe sweet, edible. Seeds woody, longitudinally grooved on one side.

Field note: Abundant in low-lying areas throughout the district.

Fl.: March - April. *Fr.*: May - August.

Specimen examined: Bijoliya 1602, Ghatarani 1910 (MLVGCB Herb).

Distribution: According to Aitchinson this wild date palm is indigenous of Indus basin.

Uses: Medicinally the decoction of roots is used to cure the loss of consciousness and in vertigo. The meristematic parts of the young shoots as well as fruits are used as food. Trunk and leaves are used in making house hold as well as agricultural equipments.

PANDANACEAE

PANDANUS S. Parkinson

Pandanus odoratissimus L., *f.* Suppl. 64, 1782; St. John, in Taxon 12: 201. 1963; Stone, in Matthew, Fl. Tamiln. Carn. 2: 1680. 1983. *P. fascicularis* Lamk. Encycl. 1: 372. 1785; Hook. *f.*, Fl. Brit. Ind. 6: 363. 1892; Duthie, Fl. Gangetic Plain 3: 270. 1920; Pandey, in Shetty & Singh, Fl. Rajasthan 3: 863. 1993.

Vern.: *Kewra*.

A gregarious shrub or small tree, erect, up to 5 m high with bent stems supported by aerial roots. Stem cylindrical, grey, marked with scars of fallen leaves. Leaves ensiform, caudate-acuminate, 0.75-1.75 m x 4.5 cm, spines on margins and midrib; spines pointing forward, drooping. Inflorescence a spadix, with numerous branches. Male flowers: a cylindric spike of numerous flowers, 4-11 x 2-2.5 cm, enveloped in white fragrant spathes; spathes yellow, caudate-acuminat, prickly. Stamens numeruos; staminal column long cuspidate; anthers longer than filament, basifixed. Female flowers: carpels in groups of 6-16; ovules solitary; stigmas short, reniform. Fruit a large syncarpous of drupes, red or yellow red, spinescent, spines recurved, 15-25 cm long.

Field note: Found along streams, planted in gardens, near temples for its fragrant flowers.

Fl. & Fr.: July - May.

Specimen examined: Bijoliya 186, Bari Sadari 1164 (MLVGCB Herb).

Distribution: India, Burma, Sri Lanka, Pakistan, Andaman, New Hebrides, Tahiti and Australia.

Notes: Female plants seem to be very rare in study area, as only one female plant was observed from Vindiyabhata village only, during this study.

TYPHACEAE

TYPHA L.

Typha angustifolia L., Sp. Pl. 971. 1753. *T. domingensis* Pers. Syn. Pl. 2: 532. 1807. *T. angustata* Bory & Chaub., Exp., Sci. Moree. Bot. 1: 388. 1832; Hook. *f.*, Fl. Brit. Ind. 6: 489. 1892; Subram., Aquatic Angiosp. 76. 1962; Duthie, Fl. Gangetic Plain 3: 293. 1920; Mahesh. Illus. Fl. Delhi f. 206. 1966; Pandey *et al.* in Bull. Bot. Surv. India 24: 139. 1982; Pandey in Shetty & Singh, Fl. Rajasthan 3: 863. 1993; Bhandari, Fl. Ind. Desert. 325. 1995. *Typha australis* Schum. & Thonn., Danske Vidensk. Afhandl 4, Ser. 4: 175. 1829; Riedl in Rech. f. Fl. Iran. 71: 5. 1970. (Plate 310)

Vern.: *Kanro, pitera.*

Robust rhizomatous perennial herb, up to 3.0 m high. Stem somewhat stout, creeping rhizome. Leaves alternate with sheathing base, usually longer than the stem, semicylindric above sheath spongy, leaf sheath usually open at throat, margins free; lamina 5 - 15 mm broad. Spikes cylindrical, upper male and lower female. Staminate spikes covered with pointed or linear hairs, hairs often laciniate at apex; covered with bracts; apiculate, translucent; pollens usually free. Female spikes with ca 1 mm long peduncles; ovary born on gynophore; stigma linear and usually equal to the hairs and bracts. Nutlets minute, ellipsoid, brown.

Field note: Common in pure formations along the margins of ponds and ditches throughout the area.

Fl. & Fr.: Almost throughout the year.

Specimen examined: Karera 2163 135 (MLVGCB Herb).

Distribution: Pantropical.

Uses: The whole plant sometimes used in making huts.

LEMNACEAE

KEY TO THE GENERA

1a. Roots absent. Fronds veinless, subglobular .... *Wolffia*
1b. Roots present. Fronds provided with veins, not globular 2
2a. Root solitary on each frond. Fronds submerged or floating, with 1-5 veins, usually green on both surfaces .... *Lemna*
2b. Roots 2-18, never solitary. Fronds floating, with 3-15 veins, usually green on upper surface and often purplish beneath .... *Spirodela*

LEMNA L.

KEY TO THE SPECIES

1a. Fronds sessile. Root solitary, sheath wingless, root cap obtuse. Fruits symmetrical .... *L. minor*
1b. Fronds tailed. Root solitary, sheath winged, root cap acute. Fruits asymmetrical .... *L. perpusilla*

Lemna minor L., Sp. Pl. 970. 1753; Hook. *f.*, Fl. Brit. Ind. 6: 556. 1894; Hartog & Plas, in Blumea 18: 362. 1970; Raizada, Suppl. Duthie, Fl. Gangetic Plain 287. 1976; Pandey, in Shetty & Singh, Fl. Rajasthan 3: 870. 1993.

Minute, floating, annual herb. Roots solitary with a wingless sheath, root cap obtuse, floral pocket lateral: Fronds free floating, almost symmetrical, rounded-elliptic, elliptical or ovate, 1.5 - 4.5 x 1.5 - 3 mm, flat on both surfaces, nerves 3, never gibbous and more or less rounded apices, usually green above, solitary or several joined together by a hyaline stipe. Papillae absent or if present only at apex and above node with many meshes. Flowering in marginal clafts of fronds, surrounded by 2-lipped spathe. Stamens 2, each representing a male flower. Ovary with solitary ovule; style long. Fruits ellipsoid. Seeds horizontal, smooth, semi antropous, albuminous.

Field note: Common in ponds and ditches.

Fl. & Fr.: July - October.

Specimen examined: Triveni 116 (MLVGCB Herb).

Distribution: Temperate Eurasia, India, Iran, Pakistan Afghanistan, Africa, North America and New Zealand.

Lemna perpusilla Torrey, Fl. New York 2: 245. 1843; Hartog & Plas, in Blumea 18: 363. 1970; Pals, in van Steenis, Fl. Males I. 7: 231. 1971; Pandey, in Shetty & Singh, Fl. Rajasthan 3: 871. 1993. *L. minor auct. non* L., 1753; Hook. *f.*, Fl. Brit. Ind. 6: 556. 1892,*p.p. L. paucicostata* Engelm., in Gray, Manual ed. 5. 681. 1867; Hook. *f.*, *l. c.* 6: 556. 1892; Parker, in Duthie, Fl. Gangetic Plain 3: 305. 1929; Subram. Aquat. Angiosp. 78. 1962. *L. aequinoctialis* Welw. in Ann. Conselho ultram. 55: 578. 1859. (Plate 311)

Minute, aquatic, free floating annual herb. Root solitary and winged, root cap acute; sheath with two lateral wings. Fronds pale green, obovate to oblong, flat on both surfaces, 1.5 - 5 x 0.7 - 3 mm, asymmetrical, thin, obtuse or slightly acute at apex, entire; ventral side flat to convex, obscurely 3-nerved; several individuals lying closely. Flowering pocket coinciding with a margin of frond; spathe open, membranous; staminate flowers 2, stamens 2; pistillate flower 1, ovary 1. Seeds 7 - 30 ribbed.

Field note: Common in ponds and ditches.

Fl. & Fr.: July - October.

Specimen examined: Mandal tank 1170 (MLVGCB Herb).

Distribution: Throughout the warmer regions of both the hemispheres.

SPIRODELA Schleid.

Spirodela polyrhiza (L.) Schleid., in Linnaea. 13: 392. 1839; Parker, in Duthie Fl. Gangetic Plain 3: 305. 1929; Subram. Aquat. Angiosp. 77. 1962; Pandey, in Shetty & Singh, Fl. Rajasthan 3: 871. 1993. *Lemna polyrrhiza* L., Sp. Pl. 970. 1753; Hook. *f.*, Fl. Brit. Ind. 6: 557. 1893.

Free floating. Fronds orbicular-ovate, 1 - 3.5 x 1 - 2.5 mm, asymmetrical, obtuse or rounded at apex, entire, 5 - 11-veined, green above, tinged with pink or purple beneath. Daughter fronds budding from the vicinity of root insertion in a slit in the mother plant. Roots 3-18 per frond. Flowers surrounded by a small, open spathe in a lateral slit-pauch. Spathes open at the top, with two male and one female flowers. Stamen one. Pistillate flowers with 2 ovules. Fruits one seeded, 1 - 1.5 mm, slightly winged on margins. Seeds semi-anatropous, smooth or faintly reticulate.

Field note: Common in stagnant water in ponds and ditches.

Fl. & Fr.: September - December.

Specimen examined: Gupteshwartank (Mandalgarh) 1662 (MLVGCB Herb).

Distribution: Eutrophic, in temperate to tropical regions.

WOLFFIA Horkel ex Scheid. nom. cons.

KEY TO THE SPECIES

1a. Fronds flat above, with 5-6 corners, conical beneath .... *W. microscopica*

1b. Fronds ovate-orbicular, dark green on upper side .... *W. arrhiza*

Wolffia arrhiza (L.) Horkel. ex Wimmer, Fl. Schles. ed. 3 : 140. 1857; Hook. *f.*, Fl. Brit. Ind. 6: 557. 1893, *pro parte*; Parker, in Duthie, Fl. Gangetic Plain 3: 306. 1929; Subram. Aquat. Angiosp. 79. 1962; Pandey, in Shetty & Singh, Fl. Rajasthan 3: 872. 1993; Bhandari, Fl. Ind. Desert. 326. 1995. *Lemna arrhiza* L. Mant. 294. 1771.

A minute, free floating, glandular, root less annual, herb. Fonds 0.8 - 1 x 1 - 1.5 mm, like small dots, orbicular to ovate, sub-globose, dorsal side flat to slightly convex; ventral side inflated, globose. Flower pockets median, infundi-buliform, with circular opening. Inflorescence dorsal, of 1 staminate and 1 pistillate flower. Anther 1-thecous. Ovary with solitary ovule. Fruits globose. Seeds smooth.

Field note: Abundent in pools and stagnant water. It covers the entire water surface, wherever if occurs forming green scum.

Fl. & Fr.: July - October

Specimen examined: Pivnia tank Shahpura 1169 (MLVGCB Herb).

Distribution: Mostly subcosmopolitan

Wolffia microscopica (Griff. ex Voigt) Kurz in J. L. Soc. London (Bot.) 9: 265. 1866; Hook. *f.*, Fl. Brit. Ind. 6: 558. 1894; Parker, in Duthie, Fl. Gangetic Plain 3: 306. 1929; Pandey, in Shetty & Singh, Fl. Rajasthan 3: 872. 1993. *Grantia microscopica* Griff. *ex* Viogt, Hort. Suburb. Calcutt. 692. 1845.

Aquatic, floating herb. Fronds flat above and mostly with 5 - 6 angled, green, 0.5 - 0.5 mm in diameter, transversely appendaged, ventral surface tapering or conical and nearly 1 - 8 mm long. Male and female flowers solitary, inserted on upper surface of the fronds. Anthers 4-thecous. Ovary globose, with solitary ovule. Fruits globose. Seeds almost globose.

Field note: Rare in ponds and ditches.

Fl. & Fr.: July - October.

Specimen examined: Nagdidam 2138 (MLVGCB Herb).

Distribution: India and Pakistan.

ALISMATACEAE

KEY TO THE GENERA

1a. Flowers male and bisexual. Achenes turgid, beaded, neither thin nor winged but with 2 lateral air sacs between endocarp and exocarp .... *Limnophyton*

1b. Flowers mostly unisexual. Achenes laterally compressed, thin, winged and beaked .... *Sagittaria*

LIMNOPHYTON Miq.

Limnophyton obtusifolium (L.) Miq., Fl. Ind. Bat. 3: 243. 1856; Hook. *f.*, Fl. Brit. Ind. 6: 560. 1892; Cooke, Fl. Bomb. Pres. (reprint. ed.) 3 : 344. 1958; Subram., Aquatic Angiosp. 86. f. 52. 1962; Hepper, in Hutch. & Dalz., Fl. W. Trop. Afr. 3: 11. 1968; Raizada, Suppl. Duthie, Fl. Gangetic Plain 288. 1976; Parmar, in Shetty & Singh, Fl. Rajasthan 3: 873. 1993. *Alisma obtusifolium* (L.), Thw. Enum. P1. Zeyl. 332. 1864. *A. sagittifolium* Willd. Sp. Pl. 2: 277. 1799. *Sagittaria obtusifolia* L., Sp. Pl. 993. 1753. (Plate 312)

Perennial, robust, scapigerous, erect, 60 - 90 cm tall, aquatic herb with a stout root-stock. Leaves broadly ovate, sagittate, with spreading, acute triangular lobes, 10 - 15 x 7 - 9.5 cm, floating, pellucid-punctate, with ovate-triangular acute lobes, obtuse or emarginate at the apex; petiole 30 - 80 cm long, triangular, spongy. Peduncle as long as the petioles, triangular, hairy. Flowers pale bluish white, whorls on 15 - 25 cm long scape, ca 1 cm across; pedicels 2.5 - 5 cm long, reflexed and thickened in fruit; bracts triangular-lanceolate, 1.5 - 2.5 cm long, ca 1 cm broad, bracteoles minute. Sepals elliptic-obtuse, 4 - 6 x 2 - 3.5 mm, obtuse, green. Petals delicate, suborbicular, larger than the sepals. Stamens with ca 3 mm long, basally flattened and thickened filaments; anthers oblong. Achenes obovoid, minutely stipitate, 4-7 x 3-4 mm, reticulately ribbed, cuneate at base, shortly beaked.

Field note: Common along margins of streams, ponds ditches etc.

Fl. & Fr.: August - October.

Specimen examined: Mainali stream 1014, Bhagtabara ka pond 1926 (MLVGCB Herb).

Distribution: Far east Asia to Madagascar and topical Africa through Indo-Pakistan subcontinent and Sri Lanka (Ceylon).

SAGITTARIA L.

Sagittaria guayanensis Kunth subsp. ***lappula*** (D. Don) Bogin in Mem. New York Bot. Gard. 9: 192. f. 5. 1955; Subram., Aquatic Angiosp. 87. 1962; Parmar, in Shetty & Singh, Fl. Rajasthan 3: 874. 1993. *Lophiocarpus guayanensis* (Kunth) M. Micheli in DC., Monogr. Phan. 3: 62. 1881. *Lophotocarpus guayanensis* var. *lappula* (D. Don) Buchen. in Engler, Pflanzenreich. 4 (15): 36. 1903. *Sagittaria blumei* Kunth, Enum. Pl. 3: 158. 1841. *S. cordifolia* Roxb., Fl. Ind. 3: 647. 1832. *S. lappula* D. Don, Prodr. Fl. Nepal. 22. 1825. *S. guayanensis auct. pl.* non. Kunth 1816; Hook. *f.*, Fl. Brit. Ind. 6: 561. 1892; Duthie, Fl. Gangetic Plain 3: 310. 1920. *S. obtusissima* Hassk., Cat. Pl. Hort. Bog. 26. 1844. (Plate 313)

Aquatic or semi-aquatic, annual herb with floating leaves. Leaves 2 - 10 x 1.5 - 8 cm, deeply cordate at base, obtuse or rounded at apex, glabrous; petiole up to 34 cm long, sulcate, minutely hairy. Inflorescence an unbranched scapose raceme, up to 45 cm long, minutely hairy towards apex. Flowers in 2 - 6 whorls of 2 - 3 flowers each, lower whorls with bisexual, upper ones with male, ca 1.5 cm across, greenish white; pedicel stout, 1 - 2 cm long, thickened and elongated up to 3.5 cm in fruit; bracts elliptic, 8 - 15 mm long, obtuse with membranous margins. Sepals obovate-suborbicular, 8 - 10 x 6 - 9 mm, acute. Petals obovate-suborbicular, yellow-spotted at the base, larger than the sepals. Stamens with ca 2 - 3 mm long, papillate filaments. Head of achenes ca 1 cm across, achenes flat, ellipsoid, 3 - 4 mm long, with distinctly blunt toothed dorsal wing; wing broad, blunt. Seeds brown.

Field note: Common in ponds and marshy places.

Fl. & Fr.: July - October.

Specimen examined: Mainali stream 1826 (MLVGCB Herb).

Distribution: Tropical Africa and tropical and subtropical Asia.

LIMNOCHARITACEAE

TENAGOCHARIS Hochst.

Tenagocharis latifolia (D. Don.) Buchen. Abh. Naturw. Ver. Bremen 2 : 2, 3, 6. 1868; Subram. Aquat. Angiosp. 88. t. 4. f. 53. 1962; Pandey, in Shetty & Singh, Fl. Rajasthan 3: 875. 1993; *Butomus latifolius* D. Don, Prodr. Fl. Nepal. 22. 1825. *B. lanceolatus* Roxb., Fl. Ind. 2 : 315. 1832. *Butomopsis latifolia* (D. Don) Kunth, Enum. Pl. 3: 165. 1841. *Butomopsis lanceolata* (Roxb.) Kunth, Enum. Pl. 3: 165. 1841; Hook. *f.*, Fl. Brit. Ind. 6: 562. 1892; Duthie, Fl. Gangetic Plain 3: 311. 1920.

An aquatic annual, herb. Root short. Leaves 4 - 15 x 1.5 - 6.0 cm, elliptic to oblong-lanceolate, mucronate, upto 5- nerved. Petioles 10 - 20 cm long. Flowers pale bluish or white, in terminal umbels, sometimes flowers in two whorls. Peduncles 25 - 50 cm long; pedicels unequal, 6 - 14 cm. long, angular; bracts membranous, triangular-lanceolate, bracteoles about 6, membranous, smaller than the bracts. Sepals up to 9 mm long. Petals slightly long, white, very delicate. Stamens with filaments 2 mm long elongating to 4 mm long at maturity; anthers 2 mm. long. Ovary reticulated veined, stigma yellow. Carpels about 5.2 mm., with sessile stigmas. Fruit 9 - 12 mm long, reticulately veined, tapering into short beak. Seeds many, less than 0.5 mm long, elliptic, shiny brown.

Field note: Rare, in ponds and tanks.

Fl. & Fr.: September - December.

Specimen examined: Mainal 513, Kasya tank 1914 (MLVGCB Herb).

Distribution: China, India, Nepal; N Africa, throughout SE Asia, N Australia.

NAJADACEAE

NAJAS L.

KEY TO THE SPECIES

1a. Sheath prolonged with narrowly triangular auricles. Male flowers without spathe; seeds greenish brown; areole quadrangular, rectangular-hexagonal ....*N. graminea*

1b. Sheath with short rounded-truncate auricles. Male flowers enclosed in spathe; seeds purplish, areoles ladder-like............. *N. minor*

Najas graminea Del., Descr. Egypte, Hist. Nat. 2: 282. 1813; Boiss., Fl. Or. 5: 28. 1882; Hook. *f.*, Fl. Brit. Ind. 6: 569. 1893; Parker, in Duthie, Fl. Gangetic Plain 3: 317. 1920; Subram., Aquat. Ang. in Bot. Mongr. 3: 101. 1974. *N. alagensis* PolL. Fl. Veron. 111: 49. 1824. *N. seminuda* Griff. ex Voigt, Hort. Suburb. Calcutt. 694. 1845.

Submerged aquatic herb, short grasslike, plumose. Stem slender, fragile, with densely foliose branchlets, upto 35 - 40 cm long, upto 0.7 mm thick, without spines, muricated, rooting at lower nodes; lower internodes up to 10 cm long. Leaves narrowly linear, 2.5 - 3.5 cm long and 0.5 - 1 mm broad, acute; with 2 - 3 teeth on the apex, flexible, minutely serrulate, ca 30 or more teeth on each margin, midrib without spines. Sheath wider than the laminae, prolonged on each side, narrowly triangular or deeply auriculate with 8 - 15 teeth on each side. Flowers without spathe, solitary or 1 - 2 per axil, staminate in the upper axil. Male flowers: 2 - 3 mm long, anther 4-

thecous. Female flower: style with 2-3 stigmas; pistil ca 2 mm long. Fruits ellipsoid oblong subcylindric, 1.75 - 2.25 mm long. Areoles minute, quadragular or polyhedral, irregularly diposed or in distinct longitudinal lines.

Field note: Common in stagnant, as well as slow running fresh water.

Fl. & Fr.: October - April.

Specimen examined: Baisrodgarh 1548, Mainal pond 1911 (MLVGCB Herb).

Distribution: Tropics of Africa, Asia and Australia. Introduced into some parts of Europe and North America

Najas minor All. Auct. Synop, Meth. Strip. Hort. Taur. 3. 1773 & All. Fl. Pedem. 2: 221. 1785; Boiss., Fl. Or. 5: 28. 1882; Hook. *f.,* Fl. Brit. Ind. 6: 569. 1893; Parker, in Duthie, Fl. Gangetic Plain 3: 317. 1929; Subram., Aquat. Ang. in Bot. Mongr. 3: 101. 1974. *Caulinia fragilis* Willd. in Mem. Acad. Sc. Berl. 94. 1798; Cl. Phil. Exper. 8. 1801. *Najas fragilis* (Willd.) Delile, Descr. Egypte, Hist. Nat. 2: 175. 1813. *N. indica* Cham. in Linnaea. 4: 501. 1829. *N. sternata* Roxb. *ex* Griff. Not. Pl. Asiat. 3: 183. 1851.

Submerged, dichotomously branched herb, monoecious, dark green; stem very slender, much branched, fragile, up to 25 - 30 cm long and up to 1 mm broad, profusely branched particularly in upper part, internodes without spines. Leaves linear, 15 - 40 x 0.7 - 1.5 mm, serrulate to spinose-dentate with 6 - 10 teeth on each side, apex acute with 1 - 2 teeth, sometimes with some spines on the back of the midrib. Sheath 1.5 - 3 mm long as well as broad, short rounded-truncate or spinulose, with 5 - 8 teeth on each side. Male flowers: solitary, enclosed in a spathe, neck conical or cylindrical, anther 1-thecous. Female flowers: without spathe, often 2 - 3 together, style with 2-stigmas; ovary ellipsoid. Achens brown, ellipsoid. Seeds slightly recurved at apex; areoles broader than long and arranged ladder like pits.

Field note: Common in stagnant, as well as slow running fresh water.

Fl. & Fr.: August - December.

Specimen examined: Rawatbhata 3649, Mainal 1931 (MLVGCB Herb).

Distribution: Europe to Asia through Africa, America (wide spread).

APONOGETONACEAE

APONOGETON L. f., *non. cons.*

Aponogeton natans (L.) Engl. & K. Krause in Engl. Pflanzenr. 24: 22. 1960; Subram. Aquat. Angiosp. 92. t. 5. f. 54. 1962; Raizada, Suppl. Duthie, Fl. Gangetic Plain 288. 1976; Pandey, in Shetty & Singh, Fl. Rajasthan 3: 875. 1993. *Saururus natans* L. Mant. Pl. 2: 227. 1771. *Aponogeton monostachyon* L. f. Suppl. Pl. 214. 1781; Hook. *f.,* Fl. Brit. Ind. 6: 564. 1892.

An aquatic, floating, scapigerous herb, with stoloniferous root-stock. Leaves 7.5 - 9.5 x 2 - 2.5 cm, floating, oblong to linear-oblong, acute or rounded at base and apex narrowly cuneate, with 2 - 3 parallel nerves; floating leaves upto 15 x 3 cm, 5-7 nerved. Petiole slender, much longer than the blade. Flowers in solitary spike, white or bright pink or purple blue. Spathe upto 15 mm long, caducous. Tepals 2, obovate or suborbicular, ca 1.5 x 0.75 mm, 1-nerved. Stamens 6; anthers bluish purple. Ovaries 3. Fruits upto 3.5 x 2.0 mm, smooth, beaked. Seeds ca 1.5 x 0.7 mm, oblong, winged, transparent and reticulately veined.

Field note: Common in stagnant shallow water, in ponds, tanks and along margins of river where water is stagnant.

Fl. & Fr.: August - November.

Specimen examined: Triveni 185, Mainal tank 1846 (MLVGCB Herb).

Distribution: Asia, Africa and Australia.

POTAMOGETONACEAE

POTAMOGETON L.

KEY TO THE SPECIES

1a. Leaves all alike, submerged; cordate-amplexicaul, ovate to lanceolate .... *P. perfoliatus*

1b. Leaves mostly dimorphic. Floating leaves coriaceous; submerged leaves petiolate, lanceolate or broadly lanceolate-oblong............................ *P. nodosus*

Potamogeton nodosus Poir. in Lam. Encycl. Meth. Bot. Suppl. 4: 535. 1816; Subram. Aquat. Angiosp. 95. 1962; Dandy, in Rech. f. Fl. Iran. 83 : 3. 1971; Parker, in Duthie, Fl. Gangetic Plain 3: 314. 1920. *P. indicus* Roxb., (Carey & Wallich ed.) Fl. Ind. 1: 452. 1820. non Roth ex Roem. & Schult. 1818; Hook. *f.*, Fl. Brit. Ind. 6: 565. 1892. *P. natans auct.* non L., Rech. f. Fl. Lowland Iraq. 27. 1964. *P. oblongus auct.* non Viv., Aitch. in J. L. Soc. Bot. 18: 99. 1880.

Perennial, rhizomatous, an aquatic herb, submerged. Stem creeping below, smooth, branched, leafy, terete, simple. Leaves heterophyllous, submerged leaves petiolate, lanceolate or broadly lanceolate-oblong, with thin veins 9 or more, thin translucent, 6 - 10.5 cm long; floating leaves broadly ovate, elliptic, coriaceous, of firm texture, 3 - 6 x 2 - 5 cm, entire; many nerved. Stipules free, keeled, lanceolate 2 - 15 cm x 8 - 10 mm. Spikes generally above the water surface, erect, reddish brown, cylindric, 4.5 - 11 cm long, 1.5 - 2.5 mm broad; peduncle axillary or leaf opposed, 7 - 11 cm long. Flowers in whorls, sessile, small. Perianth segments small, obtuse, entire, ca 1.5 mm long. Stigma short. Fruitlets 3 - 4 x 2 - 3 mm, 3-ribbed on the back, obliquely obovoid, spongy, ventral margin convex, shortly beaked and keeled. Seeds subreniform.

Field note: Common in ponds and ditches.

Fl. & Fr.: May - October.

Specimen examined: Bassi 110 (MLVGCB Herb).

Distribution: Warmer regions of Europe, America, Africa and Asia.

Notes: Sometimes confused with *Potamogeton natans* L. but in *P. nodosus* submerged leaves not reduced to narrowly linear phyllodes, and petioles without discoloured joints at the tips. Common from hills to the plains in the area.

Potamogeton perfoliatus L., Sp. Pl. 126. 1753; Hook. *f.*, Fl. Brit. Ind. 6: 566. 1892; Parker, in Duthie, Fl. Gangetic Plain 3: 314. 1920; Subram. Aquat. Angiosp. 95. f. 55 (4-5). 1962.

Submerged, rhizomatous aquatic herb. Stem terete, slightly branched or simple. Leaves submerged, sessile, ovate-lanceolate, cordate-amplexicaul, variable in size and shape, 2-4 x 1.5-2.5 cm., entire or minutely denticulate, rounded to obtuse at the apex, usually alternate below, opposite above, 5-9 nerved, thin, translucent. Stipules free, small, length variable, less than 1 cm long, soon evanescent. Spike 4-5 cm long, densely flowered. Tepals elliptic-obovate, wih long claw. Drupe 2-3.5 x 1.5-2 mm, ventral area convex, irregular or straight, usually curved beak; beak more or less keeled.

Field note: Common in stagnant, as well as slow running fresh water.

Fl. & Fr.: December - April.

Specimen examined: Kotri 487, Gurala pond 1412 (MLVGCB Herb).

Distribution: Europe, Asia, Africa, N. America and Australia.

Notes: Fruitlet polymorphism has been observed in this species.

ERIOCAULACEAE

ERIOCAULON L.

Eriocaulon cinereum R. Br. var. ***sieboldianum*** (Sieb. & Zucc.) T. Koyama ex Huang, Pollen Fl. Taiwan. 260. 1972. *E. cinereum* R. Br. var. *sieboldianum* Huang, Pollen. Fl. Taiwan. Pl. 171. f. 4 & 5. 1972. *E. heteranthum* Benth., Fl. Hong Kong. 382. 1861. Maxim., Diagn. Pl. As. 8: 12. 1893. *E. sexangulare* Mart. in WalL. Pl. As. rar. 3: 28. 1823. non. L. *E. sieboldianum* Sieb. & Zucc. ex Steudel, Syn. Pl. Glum. 2: 272. 1855. *"Sieboldtianum"* Hook. *f.*, Fl. Brit. Ind. 6: 577. 1893; Collett, Fl. Siml. 549. 1902; Ruhland, (Eriocaulaceae) in Engl. Das. Pflanzenr. 13 (IV): 3. 111. 1903. p. 112. f. 15 (Plate 314)

Monoecious annual, up to 11 cm high, glabrous. Leaves forming a rosette at the base, 2 - 5 x 0.1 - 0.2 cm, subulate to narrowly linear, acute, glabrous, brownish-green. Peduncles many, exceeding the leaves, 4-12 cm, 5-angled. Sheath 7-25 mm, shorter than the peduncles; apex oblique. Heads black, hemispherical, 2-4 mm across, Involucral bracts obovate-oblong, straw coloured, glabrous. Male flowers: sepals 3, united basally to form a cupular structure, glabrous, apically tridentate, teeth unequal, black; petals 3, minute, glabrous, or sparsely hairy, united into a tube, tridentate, teeth unequal, each ending in a black gland. Stamens 6, anthers globose, white. Female flower: sepals 1 or 2, free, linear, acuminate; petals absent. Ovary glabrous, 3-locular, style long; stigmas 3, filiform. Seeds ovoid-ellipsoid, 0.3 - 0.35 x 0.2 - 0.3 mm; cells of seed coat transversely elongated; appendages absent.

Field note: Common in margins of streams and ponds.

Fl. & Fr.: August-November.

Specimen examined: Bijoliya 1602, Kot 1848 (MLVGCB Herb).

Distribution: Afghanistan, Tien-Shan, Pamir-alai, Africa, India, Burma, Assam, Sri Lanka, Japan and China.

Notes: The type of *Eriocaulon cinereum* var. *cinereum* from Australia is a dwarf plant (3-5) cm tall with many slender peduncles, more or less of equal length. The present material differs from the typical variety in its larger stature (plants up to 11 cm tall), thicker and usually fewer peduncles, fairly unequal in size. The leaves in *Eriocaulon cinereum* var. *cinereum* are very linear, almost needle-like, while in *Eriocaulon cinereum* var. *sieboldianum* they are considerably broader. The differences are not strong enough to recognize present plants as a separate species (*Eriocaulon sieboldianum* Sieb. & Zucc.) as previous authors have done with the Asiatic material, but at the same time, while comparing with the type of *Eriocaulon cinereum*, it is difficult to consider the Asiatic material conspecific with the typical Australian material.

CYPERACEAE

KEY TO THE GENERA

1a. Style jointed with the ovary, style-base dilated or spongy-thickened, clearly demarcated.. 2

2a. Leaves absent. Inflorescence a simple spike .... *Eleocharis*

2b. Leaves present. Inflorescence with several to numerous spikes...................... *Fimbristylis*

1b. Style continuing with the ovary, withought demarcated border, hence, achens neither crowned by nor jointed with the style-base.... 3

3a. Hypogynous bristles and scales present. Glumes spirally imbricated............................ 4

4a. Leaves with a collar around stem. Inflorescence terminal. Bracts more than one, Glumes pubescent on the back, awned... *Fuirena*

4b. Leaves without a collar around stem. Inflorescence pseudo lateral. Bracts one, Glumes glabrous on the back, not awned................................ *Schoenoplectus*

3b. Hypogynous bristles and scales absent. Glumes spirally imbricated or 2 - ranked 5

5a. Nuts with flat face towards spike axis .... *Cyperus*

5b. Nuts with edge facing spike axis.................... 6

6a. Inflorescence a multiple spike or anthelodium. Spikes with more than 10 glumes .... *Pycreus*

6b. Inflorescence formed by 1 - 3 heads. Spikes with less than 6 glumes .... *Kyllinga*

CYPERUS L.

KEY TO THE SPECIES

1a. Stigma 2. Nuts dorsally compressed................................. *C. laevigatus*

1b. Stigmas 3. Nuts trigonous or triquetrous 2

2a. Spikelets digitately or stellately arranged 3

3a. Annuals. Roots fibrous 4

4a. Glumes with recurved cuspidate apex .... *C. cuspidatus*

4b. Glumes obtuse... 5

5a. Spikelts many, digitate, in compact heads. Rachilla slender. Glumes orbicular, 0.5 - 0.8 mm long, emrginate at epex, without mucro. Nuts punctate .... *C. difformis*

5b. Spikelts few digitate, in twice compound umbels. Rachilla flexuous. Glumes muticous. Nuts minutely verruculose .... *C. tenuispica*

3b. Perennials. Roots stock woody 6

6a. Rhizomes very short. Rootlets wiry. Stem slender .... *C. niveus*

6b. Rhizomes creeping. Rootlets thick .... *C. arenarius*

2b. Spikelets distinctly spicately or rarely sub-racemosely arranged 7

7a. Rachilla of spikelets distinctly winged with decurrent base of glumes 8

8a. Leaves short, rarely half the length of stem or absent .... *C. corymbosus*

8b. Leaves long, more than half the length of stem or longer 9

9a. Rhizomes not stoloniferous .... *C. exaltatus*

9b. Rhizomes stoloniferous 10

10a. Stolons thin (less than 0.5 mm, resembling roots), white, short-lived, ending to black bulbil .... *C. bulbosus*

10b. Stolons ca 2 mm diam. or more persistent, bearing ovoid black tubers 11

11a. Stem not nodosely thickened at base glumes obtuse .... *C. esculentus*

11b. Stem nodosely thickened at base Glumes apiculate 12

12a. Nuts oblong. Glumes frequently slightly recurved at apex, slenderly 5-nerved......................... *C. rotundus*

12b. Nuts oblong-ovoid. Glumes straight apex, conspicuously 7-nerved .... *C. tuberosus*

7b. Rachilla of spikelets not or hardly winged. Glumes without decurrent base 13

13a. Perennials, with woody rhizomes and often with long stolons .... *C. nutans*

13b. Annuals with tufted stems. Roots fibrous only 14

14a. Slender herbs. Spikes 8 - 20 mm long, strongly compressed.................. *C. compressus*

14b. Stout herbs. Spikelets 0.3 - 1.0 cm long, not compressed.. 15

15a. Spikes suberect. Rachis of spike smooth, keel of glume 3 - 5 nerved, glumes 1-1.5 mm .... *C. iria*

15b.Spikes finally reflexed, Rachis of spike scabridon angles, keel of glume 7 nerved glumes 1.5 - 2 mm .... *C. alulatus*

Cyperus alulatus Kern, Reinwardtia. 1: 463. f. 1. 1952; Raizada, Suppl. Duthie, Fl. Gangetic Plain 293. 1976; Pandey *et al.*, in Bull. Bot. Surv. India 24: 139. 1982; Parmar, in Shetty & Singh, Fl. Rajasthan 3: 897. 1993; Bhandari, Fl. Ind. Desert. 329. 1995. *C. iria* L. var. *rectangularis* Kük., in Engler, Pflanzenr. IV. 20, 101: 152. 1935; *C. iria* auct. non L. 1753; Clarke, in Ill Cyper. t. 14. 1909. *C. iria* L. var. *paniciformis* Clarke, in Hook. *f.*, Fl. Brit. Ind. 6: 607. 1893; Kern in Reinwardtia 1: 465, f. 1. 1952; Ahuja, Illus. of Indian plants, *Cyperus* I, Bull. Nat. Bot. Gardens 85: pl. 6 (as *C. iria* L.) 1963. (Fig. 160 & Plate 315)

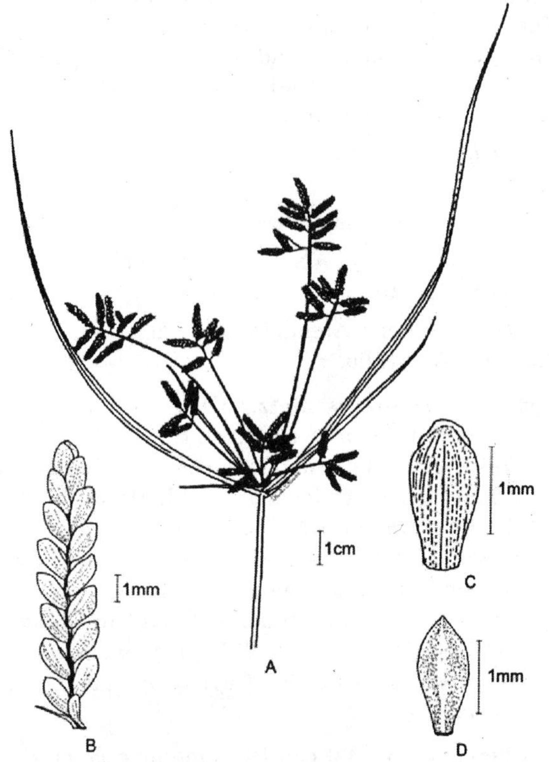

Fig. 160. *Cyperus alulatus* Kem.
A. Inflorescence, B. Spike, C. Glume, D. Nut.

Annual, tufted, 12 - 50 cm. Roots numerous, glabrous. Stem 1-2.5 mm diam., sharply trigonous, smooth. Leaves shorter than to equalling stem length, rather rigid; sheaths 25 - 45 mm, yellowish or grey, wide, mouth margin deeply concave; blades up to 15 cm, 2 - 5 mm wide, flat or slightly keeled, with brown dots and stripes, margins smooth, apex acute, scabrous. Inflorescence a simple or compound anthelodium lax, 4 - 15 cm across, primary branches 6 - 10, up to 13 cm, secondary anthelodium to 35 mm, tubular prophyll loosely sheathing, to 15 mm; lowest 4 - 6 bracts foliose, to more than 20 cm. Cluster of 6 - 16 spikes in loose spirals; spikes 4.3 - 10.8 x 2 - 2.5 mm, spreading at almost right angles, glume-like bract 1 - 1.5 mm, glume-like prophyll ca 1 mm, blunt, bi-nerved; rachis 4-angled, compressed, reddish brown, internodes ca 1.5 mm. Glumes 1.5 - 2 mm, loosely imbricating, broadly ovate or suborbicular, keel scabrous or smooth, sides prominently nerved, margins widely leathery, hard, yellow, keel 7 nerved, with narrow, serrate wing towards top. Stamens 2; anthers ca 1 mm; stigmas 3; style 3-fid almost to the base. Nut ca 1.7 x 1.0 mm, obovoid, sharply triangular with concave smooth sides, broadly stipitate, apiculate, brown to dark-brown, finely reticulate, with papilla in each areole.

Field note: Common in moist waterloddged places.

Fl. & Fr.: August - December.

Specimen examined: Fuliyakalan 134 (MLVGCB Herb).

Distribution: Afghanistan, Pakistan, NW India and Australia.

Cyperus arenarius Retz., Observ. Bot. 4: 9. 1786; Clarke, in Hook. *f.*, Fl. Brit. Ind. 6: 602. 1893; Raizada, Suppl. Duthie, Fl. Gangetic Plain 294. 1976; Parmar, in Shetty & Singh, Fl. Rajasthan 3: 897. 1993; Bhandari, Fl. Ind. Desert. 329. 1995. *C. persicus* Boiss. & Hohen., in Boiss., Diagn. Pl. Or. Nov. 13: 39. 1854. *C. persicus* Steud. & Boiss., in Steud., Syn. Pl. Cyperac.: 28. 1855; Koyama, in Dassanayake & Fosb., Fl. Ceylon 5: 210, f. 13. 1985.

Perennial, 6 - 30 cm. Rhizome long creeping, deep in sand, ca 1 mm diam. Roots thin, not tomentose. Stem solitary, ca 1 mm diam., almost terete, smooth. Leaves ca equalling the stem; sheaths 20 - 40 mm, strongly nerved, rigid, with brown veins, disintegrating into fibres with age, mouth margin straight; ligule 0; blades ca 1 mm wide, linear, stiff, curved, greyish green, semicircular in transverse section, margins slightly scabrous, apex flat, blunt or acute, slightly scabrous. Inflorescence ca 20 mm diam., capitate, with 5 to more than 20 spikes, seldom with some elongated branches; bracts 1 - 2 foliose, up to 50 (100) mm, rigid, sheath-less, one often as continuation of stem; spikes 8 - 20 mm, ovoid, compressed, glumelike bracts ca 5 mm, glumelike prophyll 4 mm, widely ovate, with scarious margins, blunt. Glumes 14 - 24, closely imbricating, caducous, ca 0.4 cm long, ovate-oblong, pale brown, obtuse, sides with ca 6 obscure nerves, which often reddish brown, margins scarious; rachis zigzagging, ca 1 mm wide, internodes ca 0.5 mm, margins sharp, hardly winged. Stamens 3. Nuts 0.18 - 0.2 x 0.1 cm, obovoid, rounded at the apex, trigonous, compressed, black or dark brown, almost smooth, glossy.

Field note: Common in sandy grounds and sandy river beds. The culms and roots are deep into the sand.

Fl. & Fr.: September - December.

Specimen examined: Gulabpura 1664, Bassi 186 (MLVGCB Herb).

Distribution: From W coasts of Persian Gulf to W and E Coasts of India and Sri Lanka; (E.G.Camus in M.H.Lecomte, Fl. générale Indoch. 7: 19-20. 1912); Vietnam, China, Mekong delta; obviously introduced to S. Australia and Durban, S. Africa

Cyperus bulbosus Vahl, Enum. Pl. 2: 342. 1805; Clarke, in Hook.f. Fl. Brit. Ind. 6: 611. 1893; Duthie, Fl. Gangetic Plain 3: 336. 1929; Parmar, in Shetty & Singh, Fl. Rajasthan 3: 899. 1993; Bhandari, Fl. Ind. Desert. 331. 1995. *Cyperus jeminicus* Retz., Obs. Bot. 4: 11. 1786, non Rottb. 1773; Kükenthal, in Engler, Pflanzenr., IV. 20.101: t. 15. f. C-E. 1935; Valckenier Suringar, Het Geslacht *Cyperus*: tab. IV, f. 28: t. V, f. 8. 1898. (Fig. 161)

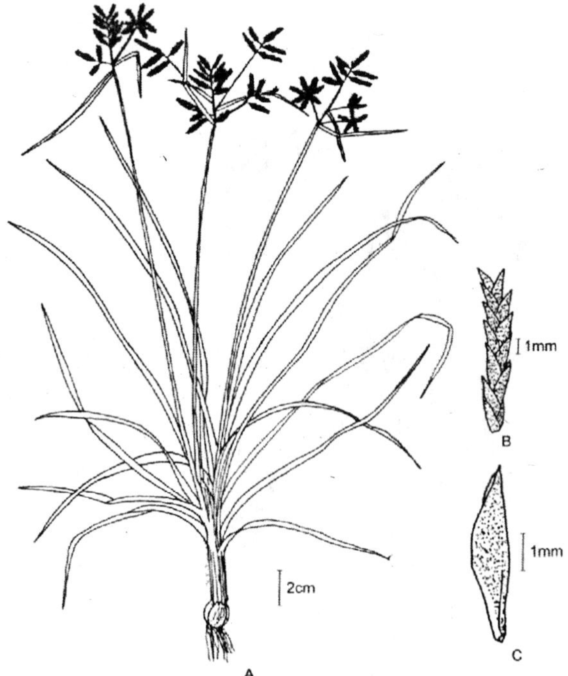

Fig. 161. *Cyperus bulbosus* Vahl.
A. Habit, B. Spike, C. Glume.

Perennial, 10 - 30 cm high, often forming small tufts; stolons thin, ephemeral. Roots fibrous. Stem solitary, 0.5 - 1.5 mm diam., sharply trigonous, smooth; base swollen, often with remains of bulbil, emitting stolons 0.5 mm diam., ending into bulbils, 8 - 12 mm, with dark black-brown scales. Leaves numerous, shorter than the stem; sheaths to 80 mm, yellowish, grey or brownish, soft, mouth margin slightly concave; blades to more than 20 cm, 1.5 - 3 mm wide, keeled, revolute, often curved, green or greyish-green, margins smooth, apex trigonous or flat, very long attenuate, acute, scabrous. Inflorescence a much reduced anthelodium or multiple spikes, 30 - 90 mm, primary branches few and short, each ending with compact cluster of 3 - 10 spirally arranged spikes, or basal part often two spikes close together (on a short branch) and supported by thin foliose bract longer than the spike; at base of branches tubular prophyll up to 7 mm; bracts up to 12 cm, several foliose, thin, clearly spaced. Spike 12 - 15 x 1.5 mm, slightly compressed, brown; rachis winged. Glumes 3 - 4 x 1.5 - 2.0 mm, oblong-lanceolate, back rounded, mid-nerve reddish-brown, ending into mucro just above rounded glume tip, sides with 3 - 4 prominent nerves, margins scarious. Nut 1.3 - 1.7 x ca 0.7 mm, ellipsoid-obovoid, trigonous, apiculate at the apex, obovoid, dark brown or blackish, finely reticulate.

Field note: Common in moist grasslands, riverbeds and waste places as weed throughout the area.

Fl. & Fr.: August - November.

Specimen examined: Rajsamand 4184 (MLVGCB Herb).

Distribution: Cap Verde Islands., Africa, Iran, Pakistan, India, Sri Lanka, Vietnam, Malaysia and Australia.

Uses: Bulbs are consumd as food during scarcity.

Cyperus compressus L. Sp. Pl. 1 : 46. 1753; Boiss., Fl. Or. 5: 372. 1882; Clarke in Hook. *f.*, Fl. Brit. Ind. 6: 605. 1893; Duthie, Fl. Gangetic Plain 3: 326. 1929; Mahesh. Illus. Fl. Delhi f. 217. 1966; Parmar, in Shetty & Singh, Fl. Rajasthan 3: 900. 1993; Bhandari, Fl. Ind. Desert. 332. 1995. *Chlorocyperus compressus* Palla, in K. Rech., Denkschr. Math.-Naturw. Kl. Akad. Wiss. 84 : 451. 1908; Britton & Brown, Illus. Fl. north. U.S., ed. 2, 1: f. 729. 1913; Matthew, Fl. Tamilnadu Carnatic 4: pl. 632. 1988. (Fig. 162 & Plate 316)

Annual, 3 - 27 cm high, forming small tufts. Stem 0.5-2 mm diam., sharply trigonous, smooth. Leaves greyish green, 1.5-4 mm broad, subcoriaceous from half to slightly shorter than stem; sheaths up to 50 mm, lowest blade-less, greenish grey, yellowish or often slightly brownish, soft, mouth margin wide, slightly concave; blades up to 16 cm, 2.5 mm wide, flat or folded, keeled, margins slightly revolute, smooth, apex acute, scabrous. Inflorescence a simple lax anthelodium, to 10 cm or single cluster of spikes, primary branches 0 - 4, up to 85 mm, tubular prophyll up to 8 mm; bracts 3 - 7, lowest 3 - 5 bracts foliose, 3 - 20 cm; cluster of spikes 10 - 20 cm, with 3 - 9 digitately arranged spikes; spikes 8 - 20 x ca 4 mm, compressed, linear-oblong, green to yellowish brown, glume-like bract ca 2.7 mm, acute, glume-like prophyll ca 1.7 mm, blunt; rachis compressed, ca 0.7 mm wide, narrowly winged, grey, brown-dotted, internodes 0.5 - 1 mm; glumes ca 4 mm, arista 0.5 - 0.7 mm long, sides with several prominent nerves, pale yellow or brownish, brown-dotted, margins scarious. Stamens 2 - 3; stigmas 3. Nut glossy, 1.5 - 1.7 x ca 1.2 mm, obovoid, rather sharply trigonous, apiculate, shortly stipitate, yellowish brown, smooth.

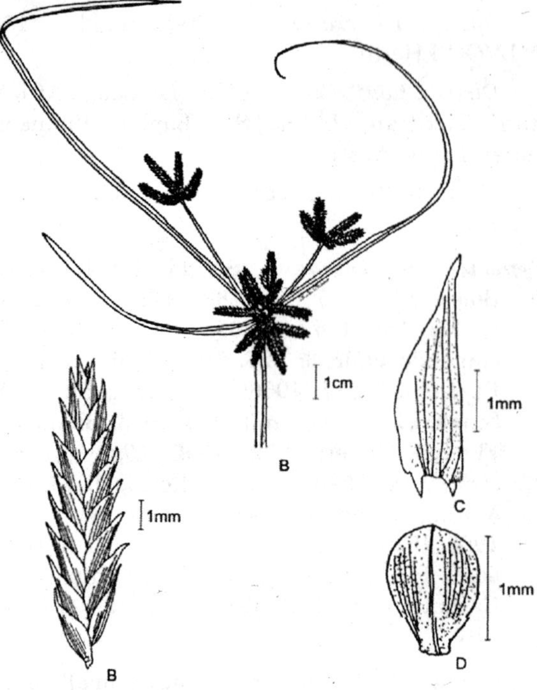

Fig. 162. *Cyperus compressus* L.
A. Inforescence B. Spike, C. Glume, D. Nut.

Field note: Weed in grasslands, road sides, waste places and fallow stream beds.

Fl. & Fr.: August - November.

Specimen examined: Bhunas 1978, Kunwara 1998 (MLVGCB Herb).

Distribution: Pantropical.

Cyperus corymbosus Rottb., Descr. et Ic. Rar. Nov. Pl. 42. t. 7. f. 4. 1773; Clarke, in Hook.f. Fl. Brit.Ind. 6: 612. 1893; in Duthie, Fl. Gangetic Plain 3: 336. 1929; Parmar, in Shetty & Singh, Fl. Rajasthan 3: 901. 1993. *C. diphyllus* Retz., Observ. bot. 5: 11. 1789. *C. koenigii* Vahl, Enum. Pl. 2: 302. 1805; Haines & Lye, Sedges and Rushes E. Afr. : f.. 354. 1983. (Fig. 163)

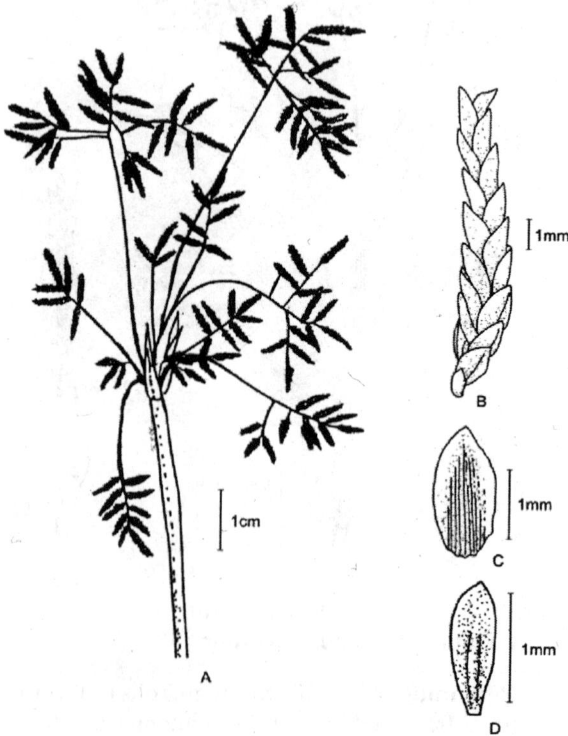

Fig. 163. *Cyperus corymbosus* Rottb.
A. Inflorescence, B. Spike, C. Glume, D. Nut.

Perennial, 0.5 - 1 m high. Rhizome creeping, 4 - 5 mm diam., woody. Stem spongy, 3 - 8 mm diam., terete below, trigonous above, smooth. Leaves absent or solitary, upto 13 cm long, 4 - 6 mm broad; sheaths to 30 cm, greyish green or brownish, often with reddish tint, mouth margin oblique. Inflorescence a conical compound anthelodium, 10 - 20 cm; bracts 3 - 5 foliose, lowest 2 - 10 cm long, flat or margins slightly recurved, keel scabrous, apex smooth or slightly scabrous, rather obtuse; primary branches 5 - 12, up to 130 mm; tubular prophyll up to 20 mm; secondary anthelodia to 80 mm, bracts up to 12 mm, c. as long as secondary tubular prophyll, secondary branches up to 45 mm, tertiary anthelodia occasional, tertiary branches 1 - 3, to 15 mm; cluster of spikes up to 20 x 30 mm, with 5-15 spikes, spikes reflexed; spikes 5 - 25 x ca 1 mm, with 8 - 22 glumes, fusiform, terete. Glumes stramineous, 3 - 5 x 2.0 - 2.5 mm, glume-like prophyll ca 1.5 mm, with spongy basis; rachis quadrangular, ca 0.2 mm wide, internodes ca 0.8 mm, winged; glumes 2 - 2.5 mm, loosely imbricating, cymbiform, straw coloured to reddish brown, mid-nerve not pronounced, mucronate, apex rounded, margins widely scarious. Anthers ca 1 mm. Stigma 3. Nuts 1.0 - 1.5 mm, narrowly obovoid-ellipsoid, trigonous, blackish, apiculate.

Field note: Rare in marshy habitats river banks etc.

Fl. & Fr.: August - November.

Specimen examined: Triveni 1604 (MLVGCB Herb).

Distribution: Pakistan; in tropical and subtropical regions from Myanmar and Sri Lanka west to Africa and the Americas.

Cyperus cuspidatus Kunth, in H. B. K. Nov. Gen. Sp. Pl. 1: 204. 1815; Clarke, in Hook. *f.*, Fl. Brit. Ind. 6: 598. 1893; Kern, in Van Steenis, Fl. Males. 7. 3: f. 59. 1974; Haines & Lye, Sedges and rushes E. Afr.: f. 508. 1983; Parmar, in Shetty & Singh, Fl. Rajasthan 3: 901. 1993. *C. uncinatus* Clarke, in Durand & Schinz., Consp. Fl. Africa 5: 580. 1895; Duthie, Fl. Gangetic Plain 3: 325. 1929.

Annual, 4 - 20 cm, forming small tufts, with fibrous roots. Stem slender, upto 1 mm diam., obtusely trigonous, smooth. Leaves to as long as stems; sheaths up to 15 mm, yellowish or reddish brown, mouth margin slightly concave; ligule 0; blades up to 80 mm, less than 1 mm wide, margins inrolled, smooth, apex slightly scabrous. Anthela simple or compound, contracted or evolute; primary branches 0 - 5, to 35 mm; lowest 2 - 3 bracts foliose, 80 x 1.5 mm, spreading. Cluster of spikes 5 - 20 mm diam., with 4 - 22 digitately arranged spikes; spikes 4 - 15 x 1 - 1.3 mm, flat, with from 4 to more than 30 glumes. Bracts ca 1.5 mm, glume-like prophyll ca 1 mm, scarious, bi-nerved; rachis ca 0.2 mm wide, compressed, 4-angled, reddish brown or greenish, very narrowly winged, often re-branching at axil of prophyll. Glumes 1.0 - 1.75 x 0.6 - 1.0 mm, obliquely spreading, usually fucous ferrugineous, sometimes hyaline, ovate, 3-nerved, nerves prominent, ending in a reflexed arista, ca 2 mm, mid-nerve area green with some reddish dots, nerveless sides wide, yellowish or reddish-brown. Stamens 3. Nut 0.5 - 0.8 x 0.2 - 0.4 mm, obovoid, trigonous, chestnut brown, apiculate, puncticulate.

Field note: Frequent in low-lying areas, open moist places, beds of ponds, rivers. It prefers sandy soil.

Fl. & Fr.: August - December.

Specimen examined: Mainal 1940, Gundali 1995 (MLVGCB Herb).

Distribution: Pantropical

Cyperus difformis L. Cent. Pl. 2: 6. 1756; Boiss. Fl. Or. 5: 370. 1882; Clarke, in Hook. *f.*, Fl. Brit. Ind. 6: 599. 1893; Duthie, Fl. Gangetic Plain 3: 325. 1929; Parmar, in Shetty & Singh, Fl. Rajasthan 3: 902. 1993; Bhandari, Fl. Ind. Desert. 333. 1995. (Fig. 164 & Plate 317)

Annual, 20 - 65 cm. Roots reddish-brown. Stem ca 5 mm diam., compressed triquetrous, smooth. Leaves shorter than stem; sheaths to ca 10 cm, green or yellow-brown, sometimes with reddish tint, mouth oblique; ligule ca 1 mm, papery, arch of attachment higher than wide; blades up to 50 cm, 3 - 5 mm wide, keeled, margins smooth, apex trigonous, scabrous. Inflorescence of 1-7 globose partial inflorescences, nearly sessile or peduncles to 30 mm; partial inflorescences 7-15 mm diam., axis tightly digitately branching and rebranching, with ca 100 or more sessile spikes. Bracts 2 - 3, lowest two foliose, up to 25 cm or more long, sheath-less. Spikes 2 - 8 x 1.0 - 1.25 mm, linear-lanceolate, 5 - 28 flowers; glumes 0.5 - 0.8 x 0.7 - 1.0 mm, broadly obovate-suborbicular, blunt, wider than long, midnerve strong, sides brown or dark brown, margins narrowly scarious; rachis compressed, slightly zigzag, internodes ca 0.8 mm. Stamens 1 - 2. Stigmas 3. Nut ca 0.5 x 0.3 mm, obovoid, trigonous, apiculate, sessile, yellowish brown.

Field note: Common in ditches, irrigation channels, grassy swamps and marshy habitats.

Fl. & Fr.: August - December.

Specimen examined: Mangrop 136, Charbhuja (Rawatbhata) 261 (MLVGCB Herb).

Distribution: Tropical and subtropical areas of all continents; from S. Europe to Turkey, Iraq, Caucasus, Uzbekistan, Tadjikistan, Pakistan and India.

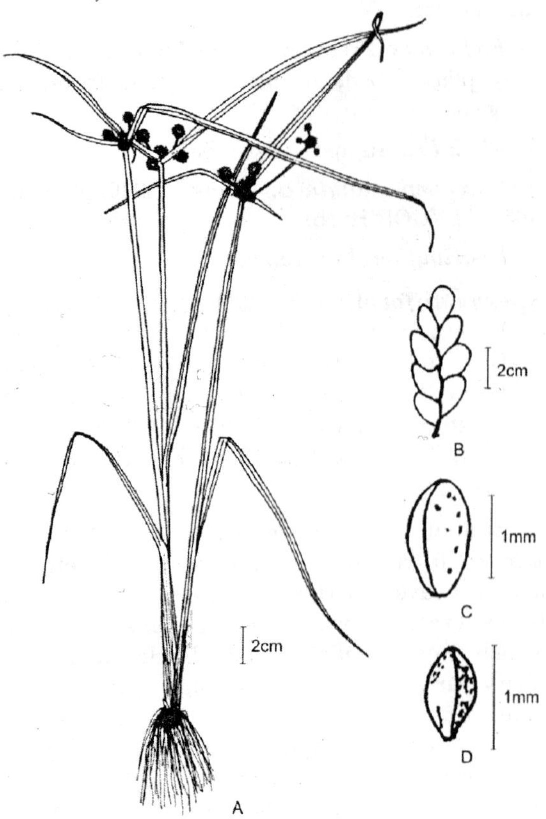

Fig. 164. *Cyperus difformis* L.
A. Habit. B. Inforescence - spike, C. Glume, D. Nut.

Cyperus esculentus L., Sp. Pl. 67. 1753; Boiss., Fl. Or. 5: 377. 1882; Clarke, in Hook. *f.,* Fl. Brit. Ind. 6: 616. 1893; Duthie, Fl. Gangetic Plain 3: 333. 1929; Parmar, in Shetty & Singh, Fl. Rajasthan 3: 903. 1993. *C. aureus* Tenore, Fl. Napol. Prodr. I. (1811) VIII; Holm & al., World's Worst Weeds: f. 46. 1977; Haines & Lye, Sedges and Rushes E. Afr.: f. 372 & 373. 1983.

Perennial, 10 - 50 cm. Base of stem slightly swollen, with stolons to ca 20 cm, tuber up to 15 mm diam., round or elongate, rooting, covered by greyish-brown scales. Stem solitary, 2 - 3 mm diam., trigonous, smooth. Leaves almost equalling the stem, 3 - 6 mm broad; sheaths 50 mm, yellowish or greyish brown, sometimes with reddish tint, mouth margin nearly straight or lingulate; blades to 35 cm, to 5 mm wide, yellow-green, keeled, margins slightly revolute, smooth or towards apex scabrous, apex acute, trigonous, scabrous. Inflorescence a terminal, lax, compound umbel, primary rays to more than 10, to 80 mm; lowest 2 - 3 bracts leaf-like, up to 30 cm, 5 mm wide, sheathless; primary branches end with often more than 20 spreading spikes in irregular spiral, but small secondary anthelodia frequent. Spikes up to 15 x ca 1.5 mm, elongating after flowering, with 5 - 15 flowers; glume-like bract ca 2 mm, glume-like prophyll two-veined, ca 1.5 mm, base swollen; rachis compressed, slightly zigzagging, internodes ca 1 mm, winged. Glumes rather laxy imbricating, 3 x 1.5 - 2.5 mm, ovate, obtuse, plicate, yellow-brown, with 5 - 9 conspicuous veins, mid-vein strong, margins scarious. Nut ca 1.5 mm, obovate-ellipsoid, obtusely trigonous, greyish black, obtuse, surface reticulate-papillose.

Field note: Common in sandy river-beds, irrigated fields, tanks, ponds etc.

Fl. & Fr.: August - December.

Specimen examined: Jaleshwar tank Mandalgarh 1608 (MLVGCB Herb).

Distribution: Tropical and subtropical Africa, but a weed in all continents; formerly cultivated, especially in the Mediterranean, for edible tubers rich in carbohydrates and oil (earth almond, chufa).

Notes: Often mixed with pale coloured races of the polymorphic *C.rotundus*, but stolons more slender and tubers more regularly zoned, glumes less closely imbricating and glume side with several distinct, pale nerves.

Cyperus exaltatus Retz., Obs. Bot. 5: 11. 1789; Clarke, in Hook. f., Fl. Brit. Ind. 6: 617. 1893; Duthie, Fl. Gangetic Plain 3: 331. 1929; Parmar, in Shetty & Singh, Fl. Rajasthan 3: 904. 1993.

var. ***exaltatus***

Perennial, loosely tufted, to more than 1 m high. Rhizome short, horizontal or ascending. Culms 3 - 5 mm diam., trigonous, smooth. Leaves many nearly equaling the culms; sheaths to 30 cm, brown, green or yellowish green, soft, mouth margin slightly concave; blades 50 cm or longer, 3 - 10 mm wide, flat or folded, keeled, margins recurved, smooth or scabrous, apex trigonous, barbed. Inflorescence a terminal, compound or decompounds umbel of spikes; bracts up to 6, leafy, 2 - 4, up to more than 50 cm long, up to 10 mm wide; primary branches 8 - 12, up to 20 cm, tubular prophyll up to 30 mm, bi-nerved, often with two aristas up to 10 mm; secondary anthelodia up to 10 cm, with up to more than 10 secondary branches, angle between primary and secondary branches sharp, with several foliose bracts, shorter than anthelodium; some primary and most secondary branches ending with cluster of spikes, sometimes with small tertiary anthelodia; spikes 2 - 5 x 0.7 - 1.5 cm, compressed, with 8 - 34 glumes. Spikelets 4 - 10 x 1.2 - 1.5 mm, patent, sub-distichous, linear to lanceolate – oblong, 6 - 20 - flowered. Rachila straight, with pale caduceus wings. Glums 1.5 - 2.0 x 1.2 - 1.5 mm, ovate, membranous, greenish to straw – coloured, tinged with brown. Keel 3 - 5 nerved. Nut ca 0.5 x 0.3 mm, ellipsoid or ovoid, trigonous, whitish or yellowish brown, finely reticulate or almost smooth.

Field note: Common in aquatic and marshy habitats usually in ditches, irrigation channels, etc.

Fl. & Fr.: September - November.

Specimen examined: Chhotisadari 4301 (MLVGCB Herb).

Distribution: Tropical Africa, SE Asia from India to Japan and southwards to southern Australia.

Cyperus iria L., Sp. Pl. 1 : 45. 1753, (excl. Rheede. cit.); Roxb., Fl. Ind. ed. Carey & Wall. 1 : 204. 1820; Boiss., Fl. Or. 5: 370. 1882; Clarke, in Hook.f. Fl. Brit. Ind. 6: 606. 1893; Duthie, Fl. Gangetic Plain 3: 329. 1929; Parmar, in Shetty & Singh, Fl. Rajasthan 3: 906. 1993; Bhandari, Fl. Ind. Desert. 333. 1995. *Chlorocyperus iria* (L.) Rikli, Jahrb. Wiss. Bot. 27 : 564. 1895; Holm & al., World's worst weeds: f. 95. 1977; Matthew, Fl. Tamilnadu Carnatic 4: pl. 640. 1988. (Fig. 165 & Plate 318)

var. ***iria***

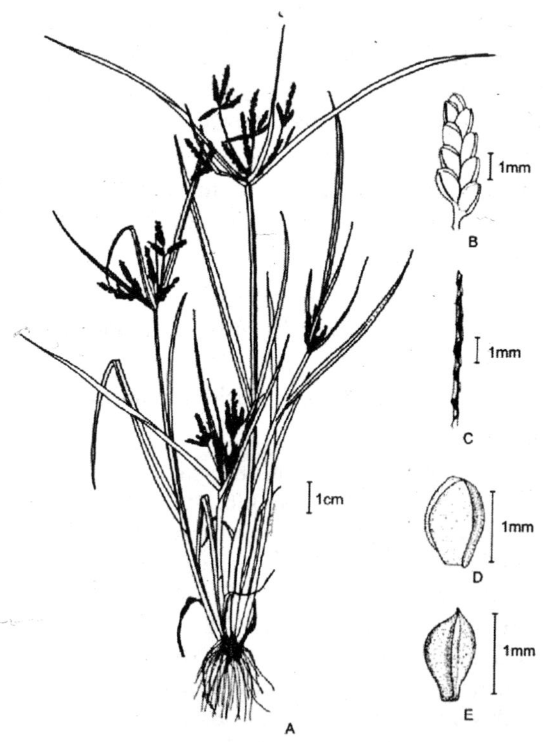

Fig. 165. *Cyperus iria* L. var. *iria* : A. Habit. B. Inflorescence C. Rachis, D. Glume, E. Nut.

Annual, 20 - 70 cm high, solitary or forming small tussocks. Rhizome small. Roots fibrous. Stem compressed, trigonous, smooth. Leaves up to equalling stem; sheaths to 20 cm, soft, yellow brown, sometimes with reddish tint, finely brown-dotted, mouth margin straight; ligule 0; blades to more than 30 cm, 2 - 5 mm wide, flat, keeled, green or greyish green, margins slightly revolute, margins and keel towards the apex scabrous, apex acute, scabrous. Inflorescence a terminal compound, diffuse umbel (anthelodium), 5 - 20 cm; 4 - 6 bracts foliose, to more than 25 cm, much longer than inflorescence; primary branches up to 15 cm, rigid; secondary anthelodium 20 - 50 mm; tertiary anthelodium 20 - 30 mm; cluster of spikes 10 - 40 mm, spikate, erect, lax, with ca 40 spikes; spikes narrow, 3.5 - 13 x 1.7 - 2.0 mm, compressed, lax, with 5 - 20 flowers, glume-like bracts ca 1 mm, glume-like prophyll scarious, bi-nerved; rachis ca 0.4 mm wide, 4-angled, flat, brown, narrowly winged, internodes ca 0.6 mm. Glumes 1.2 - 1.6 x 1.0 - 1.6 mm, broadly obovate - suborbicular, keeled, muticous or mucronate, sides nerveless, yellow-brown, shiny, margins scarious. Stamens 2; anthers 0.3 mm long. Style short; stigmas 3. Nuts glossy, ca 1.5 x 0.6 mm, rather sharply trigonous, obovoid-ellipsoid, dark brown, finely papillose.

Field note: Common in marshy habitats, particularly in fields, borders of irrigation ditches, ponds etc.

Fl. & Fr.: August - November.

Specimen examined: Rawatbhata, Gowtadam 1850 (MLVGCB Herb).

Distribution: Old World tropics and subtropics.

Cyperus laevigatus L. Mantissa. Pl. Alt. 179. 1771; Boiss., Fl. Or. 5: 366. 1882; Parmar, in Shetty & Singh, Fl. Rajasthan 3: 906. 1993; Bhandari, Fl. Ind. Desert. 334. 1995. *Juncellus laevigatus* (L.) Clarke, in Hook. *f.*, Fl. Brit. Ind. 6: 596. 1893; Duthie, Fl. Gangetic Plain 3: 341. 1929. *Acorellus laevigatus* (L.) Palla in Koch, Syn. deut. schweiz. Fl. ed. 3, 2 : 2558. 1905; Rottboell, Descr. et Ic. : t. 16, f. 1. 1773; Rateeb & Sherif, Fl. Libya 120 : f. 5. 1985; Feinbrun-Dothan, Fl. Palaestina 4 : Pl. 480. 1986.

Perennial, 10 - 65 cm, forming small tufts or stems in rows. Rhizome horizontal, creeping, 1.0 - 3 mm diam., covered with dark brown or reddish scales. Stem 0.5-5 mm diam., terete or obtusely trigonous, yellow green or grey green, smooth. Leaves long, short or absent; sheaths up to 20 cm, brown to dark brown, sometimes reddish, mouth margin concave; ligule 0; blade up to 10 or reduced to a point. Inflorescence capitate, pseudolateral, 7 - 25 mm, one cluster of 2 - 10 (-more than 30) spikes; bract usually one, as a continuation of stem, 5-50 mm, subterete, obtuse (sometimes another short, second bract can be distinguished); spikes 6 - 20 x 2 - 3.5 mm, often curved, turgid, with 10 - 40 closely imbricate glumes, glume-like bract, ca 2 mm, glume-like prophyll ca 2 mm; rachis 4-angled, concave side 1 - 1.2 mm wide, convex side 0.5 - 0.8 mm wide, internodes 0.7 - 1 mm, winged. Glumes 1.8 - 3 mm, membranaceous, broadly ovate, obtuse or mucronate, mid-nerve at apex slightly keeled, from yellowish or grey to dark reddish brown, glossy, margins narrowly scarious. Flowers on concave side of rachis. Stamens 3; stigma 2. Nut 1.3 - 1.7 x ca 1 mm, obovoid, plano-convex, grey-brown to black-brown, adaxial side flat, pressed against the rachis, abaxial side rounded, glossy, smooth or more or less clearly reticulate.

Field note: Rare in saline depressions, "*Usar ditches*".

Fl. & Fr.: August - November.

Specimen examined: Kachhola 1196, Kahjuri 1930 (MLVGCB Herb).

Distribution: From C. Asia, India and Pakistan W to southern N. America and S. America

Cyperus niveus Retz., Observ. bot. 5: 12. 1789; Clarke Hook. *f.*, Fl. Brit. Ind. 6: 601. 1893; Duthie, Fl. Gangetic Plain 3: 327. 1929; Parmar, in Shetty & Singh, Fl. Rajasthan 3: 907. 1993; Bhandari, Fl. Ind. Desert. 334. 1995. (Fig. 166)

Perennial, 10-80 cm. Rhizome short, horizontal, with tillers. Stem smooth, grey-green, upper part trigonous, lower part almost terete, base swollen. Leaves up to nearly equalling stem, somewhat flexuous; sheaths 20 - 40 mm, brown or greyish-brown, not shiny, rather soft, mouth wide, slightly concave; ligule 0; blades to ca 20 cm, 1 - 2 mm wide, channelled or folded, often curved, grey-green, margins scabrous, apex long, acute, trigonous, scabrous. Inflorescence capitate, 1 - 4 cm broad; 2 - 3 bracts foliose, to ca 60 mm, channelled or folded, scabrous. Spikes 10 - 20 x ca 5 mm, narrowly ovoid, compressed, with up to 40 glumes, glume-like bract ca 4 mm, acute, glume-like prophyll ca 2.5 mm, blunt; rachis quadrangular, angles rounded, not winged, internodes ca 0.8 mm. Glumes 4 - 5 mm long, tightly imbricating, slightly keeled, ovate-lanceolate, pale brown, mucronate, sides with 4 - 5 rather inconspicuous veins, margins scarious. Stamens 3, anthers ca 1.5 mm. Stigmas 3. Nuts ca 1.5 mm, obovoid, trigonous, brown, shiny, very weakly papillose, subsessile, densely puncticulate.

Field note: Common in marshy habitats along river, ponds etc.

Fl. & Fr.: August - December.

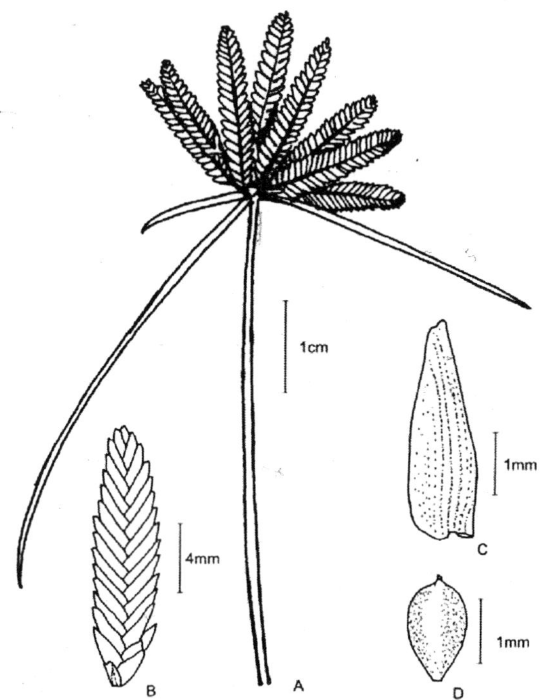

Fig. 166. *Cyperus niveus* Retz. A. Inflorescence B. Spike, C. Glum. D. Nut.

Specimen examined: Jahajpur 1237 (MLVGCB Herb).

Distribution: From E Iran to Myanmar; also in E Africa (R.W. Haines & Lye, Sedges and Rushes of E. Africa, 1983).

CYPERUS NUTANS Vahl

KEY TO THE SUBSPECIES

1a. Spikelets in cylindrical spike.................. *C. nutans* subsp. *eleusinoides*

1b. Spikelets in spicate - racemes....... *C. nutans* subsp. *nutans*

Cyperus nutans Vahl, subsp. ***eleusinoides*** (Kunth) Koyama in Gard. Bull. Singapore. 30 : 136. 1977; Parmar, in Shetty & Singh, Fl. Rajasthan 3: 909. 1993. *C. eleusinoides* Kunth, Enum. Pl. 2: 39. 1837; Clarke Hook. *f.*, Fl. Brit. Ind. 6: 608. 1892; Duthie, Fl. Gangetic Plain 3: 330. 1929. *C. nutans* Vahl var. *eleusinoides* (Kunth) Haines, Bot. Bihar Orissa 5: 898. 1924; Lecomte, Fl. Indo-Chine 7: f. 6, 7-11. 1912; Matthew, Fl. Tamilnadu Carnatic 4: pl. 641. 1988. (Fig. 167)

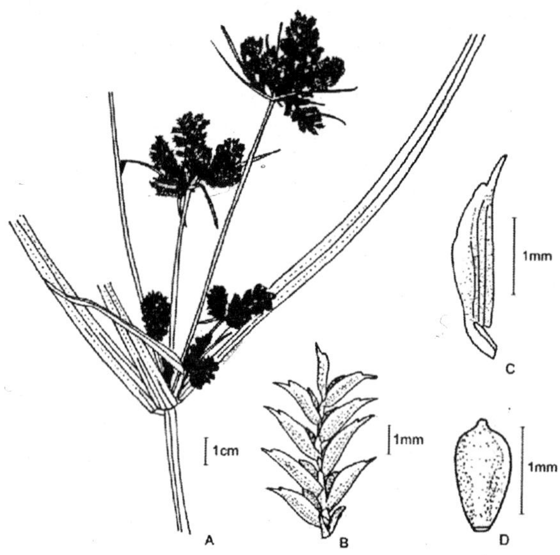

Fig. 167. *Cyperus nutans* Vahl ssp. *eleusinoides* (Kunth) Koyama. A. Infloresence B. Spike, C. Glume, D. Nut

Characteristic for subsp. *eleusinoides* is similar to subsp. *nutans*, but can be separated on the basis of primary rays of anthela being less patent or narrow anthelodium (not wider than long as in subsp. *nutans*), primary branches erect (more spreading). Spikelets few 4 - 8-flowered, 1.5 - 2.0 mm wide. Rachilla - internodes ca 0.5 mm long, suberect (more spreading). Glumes elliptic, more patent, upto 2.5 mm long, margins broadly scarious in the upper half, rather closely imbricating, with mid-nerve ending into mucro, ca 0.3 mm (obtuse or mucronate). Stamens 3. Styles 3-fid. Nuts brown, narrowly obovate-oblong, 1.0 - 1.3 x 0.5 - 0.7 mm, triquetrous, more suddenly narrowed to an apiculate tip.

Field note: Common in streams, along canal banks, river margins, in marshy habitats throughout the area.

Fl. & Fr.: August - December.

Specimen examined: Triveni 82 (MLVGCB Herb).

Distribution: Tropical Africa, India, Sri Lanka, NE Australia, Malaysia, S China and Taiwan.

Cyperus nutans Vahl, Enum. Pl. 2: 363. 1805; Clarke Hook. *f.*, Fl. Brit. Ind. 6: 607. 1892; Duthie, Fl. Gangetic Plain 3: 330. 1929; Parmar, in Shetty & Singh, Fl. Rajasthan 3: 908. 1993; Bhandari, Fl. Ind. Desert. 335. 1995. *C. jacquini* Schrad. in Linnaea 11 (Litt. -Bericht) : 87. 1837; *Chlorocyperus jacquini* (Schrad.) Rikli in Jahrb. Wiss. Bot. 27: 563. 1895.

subsp.: ***nutans***

Perennial, 40 - 100 cm high. Rhizome short, with tillers. Stems 2 - 6 cm diam., trigonous below, triquetrous above, smooth. Leaves shorter than stem; sheaths up to ca 15 cm, pale green or yellowish brown, soft, mouth margin straight or slightly ligulate; blades to 30 cm or more, 4 - 9 mm wide, grey-green, scabrous. Inflorescence simple or compound spicate umbel (anthelodia), 8 - 30 cm; lowermost bracts foliose, to more than 30 cm; primary branches 5 - 9, to more than 20 cm, secondary anthelodia 15 - 75 mm; tertiary anthelodia occasional, up to 35 mm; cluster of spikes sessile or pedunculate, 15 - 30 mm, with 8 - 24 spirally arranged spikes. Spikes dense, penicilliform, often somewhat nodding, 3 - 4.5 x 0.2 - 1 cm, with ca 10 glumes; glume-like bract ca 1.5 mm, acute, glume-like prophyll bi-nerved, ca 1 mm; rachis flat, reddish brown, internodes ca 1 mm, winged. Glumes 2 - 2.5 x 1 mm, elliptic, obtuse, acute with green mid-nerve ending into mucro up to 0.3 mm, sides with 3 - 4 prominent, 7-nerved, nerves reddish-brown. Stamens 3. Style 3-fid, almost to the base. Nut 1.3 - 1.5 x 1.5 - 0.5 mm, obovoid-oblong, trigonous, brown, apiculate densely and finely papillose.

Field note: Common in marshy habitats.

Fl. & Fr.: September - February.

Specimen examined: Mandalpond 1999 (MLVGCB Herb).

Distribution: Tropical Africa, India, Sri Lanka, NE Australia, Malaysia to S. China and Ryukyu Islands.

Cyperus rotundus L., Sp. Pl. 1: 45. 1753; Boiss. Fl. Or. 5: 376. 1882; Clarke, in Hook. *f.*, Fl. Brit. Ind. 6: 614. 1893; Duthie, Fl. Gangetic Plain 3: 332. 1929; Parmar, in Shetty & Singh, Fl. Rajasthan 3: 911. 1993; Bhandari, Fl. Ind. Desert. 336. 1995. *Chlorocyperus rotundus* (L.) Palla, Allg. Bot. Zeitschr. 6: 61. 1900; *Schoenus tuberosus* Burm. f., Fl. Ind.: 19. 1768; Rateeb & Sherif in Fl. Libya 120: f. 10. 1985. (Fig. 168 & Plate 319)

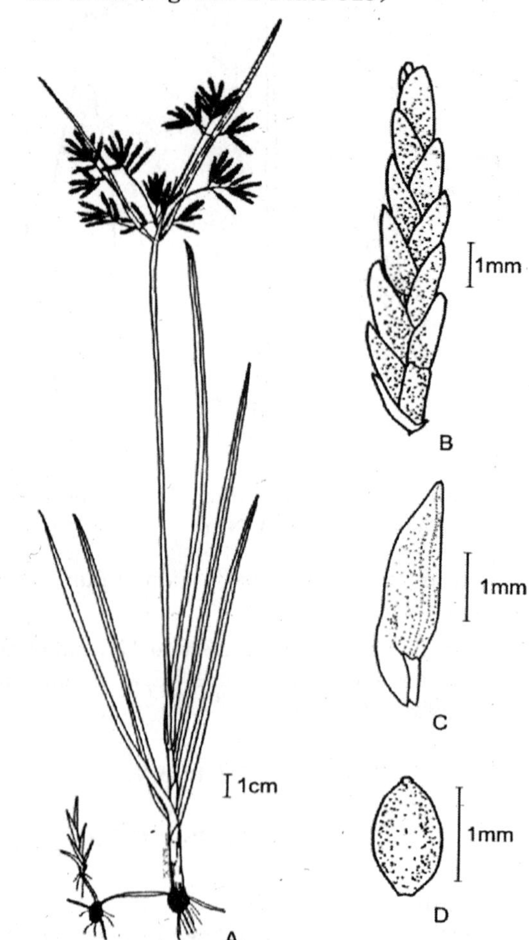

Fig. 168. *Cyperus rotundus* L.
A. Habit, B. Spike, C. Glume, D. Nut.

Perennial, 20 - 60 cm high. Rhizome dark brown, woody; stolons numerous, long-creeping, strong, 1 - 1.5 mm diam., producing ellipsoid aromatic tubers, ca 20 x 9 mm, dark brown. Stem 1 - 2.5 mm diam., trigonous, sometimes tortuous, green, smooth. Leaves shorter than stem; sheaths to 10 cm, wide, acute, grey-brown or brown, sometimes with reddish tint, mouth margin lingulate; blades to 30 cm, 1 - 4.5 mm wide, grey-green or green, flat or folded, slightly keeled, margins smooth or towards the apex barbed, apex long-attenuate, acute, trigonous, scabrous. Inflorescence simple or compound or terminal compound (an anthelodium), 1.5 - 4.5 cm; 2 - 3 lowermost bracts foliose, longer than inflorescence; primary branches 5 - 8, to more than 50 mm long; secondary anthelodia 25 - 5 0 mm; clusters of spikes 15 - 40 mm, of 5 - 7 spikes on slightly elongated axis, sometimes nearly digitately. Spikes 10 - 40 x 1.5 - 2 mm, with 11 - 51 glumes, glume-like bract ca 2 mm, glume-like prophyll 1.9 - 2.4 mm, bi-nerved, sterile; spike rachis 4-angled, ca 0.5 mm wide, internodes 0.7 - 1 mm, winged. Glumes 2.7 - 3.7 mm, ovate-ovate lanceolate, rather closely imbricate, obtuse, dark reddish brown, or grey-brown, sides with two distinct nerves, nerveless area wide, margin narrowly or sometimes widely scarious, towards apex embracing nut. Stamens 3. Styles long exserted; stigma 3. Nut 1.3-1.8 x 0.7 - 1 mm, obovate-oblong, trigonous, brown to blackish brown, clearly or obscurely reticulate papillose.

Field note: Common in wet meadows, ditches, moist road sides, waste places and marshy habitats throughout the area.

Fl. & Fr.: Almost throughout the year.

Specimen examined: Chhaprel 1319 (MLVGCB Herb).

Distribution: Tropical and subtropical areas of all continents.

Notes: Sometimes called the "Worlds worst weed". *Cyperus rotundus* is a variable species readily propagating vegetatively, but frequently producing seeds in S.W. Asia. Three subspecies, subsp. *rotundus,* subsp. *retzii* (Nees) Kük. and subsp. *tuberosus* (Rottb.) Kük. are commonly recognized (Kern, in Fl. Malesiana 1974; Koyama in Fl. of Taiwan, 1979; Haines & Lye, Sedges and Rushes E. Afr. 1983; Bhandari, 1990), and it is easier to identify them than in Fl. Iranica material (Kukkonen 1998).

Cyperus tenuispica Steud. Syn. Pl. Glum. 2: 11. 1855; Kern, in Van Steenis, Flora Malesiana 7, 3: f. 58. 1974; Haines & Lye, Sedges and Rushes East Afr.: f. 313-314. 1983; Parmar, in Shetty & Singh, Fl. Rajasthan 3: 913. 1993. *C. flavidus sensu* Clarke, in Hook. *f.*, Fl. Brit. Ind. 6: 600. 1893; Duthie, Fl. Gangetic Plain 3: 326. 1929.

Annual, erect, 10 - 17 cm, forming small tufts. Stem 0.5 - 1.5 mm diam., 3 - 5-angular, trigonous, smooth. Leaves shorter than stem, soon withering; sheaths green or red, soft, lower blade-less, mouth oblique or, when leaf blade present, rather straight; ligule 0; blades up to 60 mm, 1 - 2 mm wide, yellow-green, flat or folded, soft, rather acute, margins and keel smooth, apex barbed. Inflorescence a terminal, compound or decompound, lax umbel (anthelodium), 20 - 70 mm; bracts few, one or two foliose, up to more than 10 cm; primary branches ca 10, up to 70 mm; secondary anthelodium 10 - 20 mm, almost spherical, secondary branches 5 - 15 mm; cluster of spikes at the tip of all of the secondary branches, with 2 - 10 sessile spikes; spikes 2 - 8 x ca 1 mm, compressed, with 10 - 30 glumes; glume-like bracts ca 0.8 mm, linear-oblong, glume-like prophyll clasping, ca 0.6 mm, scarious, sterile; rachis 4-angled, ca 0.4 mm wide, green or reddish, internodes ca 0.5 mm, hardly winged. Glumes 0.8-1.0 x 0.5 - 0.75 mm, closely imbricate, ovate-oblong, recurving, obtuse, mid-nerve area green, slightly keeled near the apex, sides reddish brown, margins narrowly scarious. Stamens 1 or 2, connective tip smooth. Stigma 3. Nut 0.2 - 0.4 mm long, almost spherical to obovoid, obtusely trigonous, white or yellowish, papillose or tuberculate.

Field note: Common in marshlands, river banks, weed of waterloddged fields etc.

Fl. & Fr.: April - December.

Specimen examined: Triveni 825, Pahaola-jhar mahadev (R. Bhata) 574 (MLVGCB Herb).

Distribution: Tropical Africa, E. Africa, Kashmir, India, from Tadjikistan and Himalayas to Sri Lanka and eastwards.

Cyperus tuberosus Rottb. Descr. et Ic. Rar. Nov. Pl. 28. t. 7. f. 1. 1773; Clarke, in Hook. *f.,* Fl. Brit. Ind. 6: 616. 1893; Parmar, in Shetty & Singh, Fl. Rajasthan 3: 913. 1993. *C. rotundus* subsp. *tuberosus* (Rottb.) Kuk. in Engl. Pflanzenr. 4 (20), Heft 101: 113. 1936;

Raizada, Suppl. Duthie, Fl. Gangetic Plain 302. 1976.

Perennial, with stoloniferous creeping rhizomes. Culms trigonous, smooth. Leaves shorter then culms, 3-6 mm wide, linear, weakly folded; sheaths, plale stained with reddish-brown. Inflorescence to partial compound corymb. Spikes turbinate, losely bearing 2 - 11 spikelets, patent, 8 - 20 x 1.5 - 3.0 mm, linear, 8 to 20-flowered, yellow - brown. Rachilla with broad, white - hyaline wings. Glumes ca 3.5 mm long, elliptic to ovate, rounded at apex, yellowish - brown, conspicuously 7-nerved. Nuts obovate- oblong, trigonous, half the length of glume, puncticulate, brown.

Field note: Common in aquatic and marshy habitats.

Fl. & Fr.: August - December.

Specimen examined: Larkidam 511, Karoi 1214 (MLVGCB Herb).

Distribution: Africa and India.

ELEOCHARIS R. Br.

KEY TO THE SPECIES

1a. Glumes keeled, bristles equaling the nut, glistening white, scabrulous. Nuts ca 0.5 mm long .... *E. atropurpurea*

1b. Glumes scarcely keeled, bristles longer than the nut, reddish-brown, retrorsely scabrid. Nuts ca 1 mm long .... *E. geniculata*

Eleocharis atropurpurea (Retz.) J. Presl & K. Presl, Reliq. Haenk. 1: 196. 1828; Clarke, in Hook. *f.,* Fl. Brit. Ind. 6: 627. 1893 & Illus. Cyper. t. 36. f. 6 - 9. 1909; Parmar, in Shetty & Singh, Fl. Rajasthan 3: 915. 1993; Bhandari, Fl. Ind. Desert. 337. 1995. *Scirpus atropurpureus* Retz. Obs. Bot. Pl. 5: 14. 1789; *Heliocharis atropurpurea* (Retz.) Kunth, Enum. 2: 151. 1837; Duthie, Fl. Gangetic Plain 3: 348. 1929. *Eleocharis lereschii* Shuttlew., Flora 16: 241. 1837; Reichenb. Ic. Fl. Germ.8: Pl. 295, f. 699. 1846; Blake in Proc. Roy. Soc. Queensland 50: Pl. 10, f. 32 - 33. 1939. (Fig. 169)

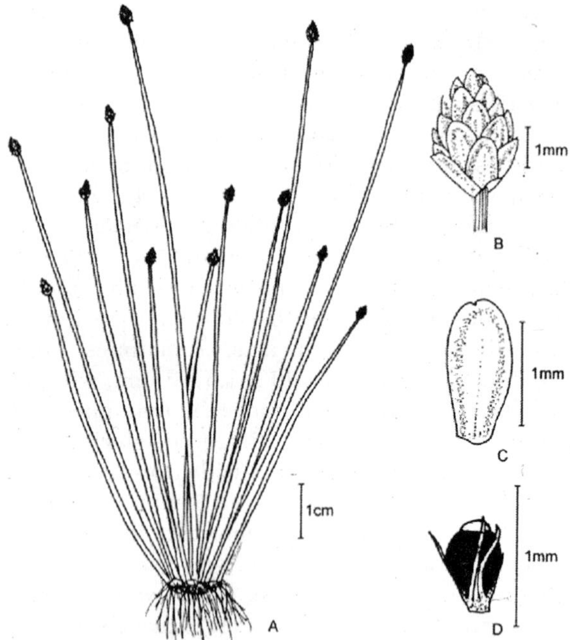

Fig. 169. *Eleocharis atropurpurea* (Retz) Presl. A. Habit. B. Inflorescence, C. Glume, D. Nut

Annual, 2 - 20 cm high, forming small tufts. Roots fibrous. Stem almost capillary, 0.3 - 0.5 mm diam., with 5 - 6 obtuse angles separated by deep grooves. Lower closed sheath 3 - 5 mm, brown or reddish, upper to 10 mm, green or grey green, base often reddish, with 5 - 6 green nerves, mouth oblique; blades 0 or tip ca 1 mm. Spikes 2 - 5 x 1 - 3 mm, ovoid-oblong, subterete, with ca 30 or more glumes in five spiral rows, all fertile; basal glume not distinct or slightly larger and with more pronounced mid-nerve than in other glumes. Glumes loosely imbricating, 1.1 - 1.6 mm, elliptic-oblong, with prominent, green mid-nerve not quite reaching obtuse apex, sides brown or dark reddish brown, nerves obscure or missing, margins more or less scarious. Perianth bristles rising from white collar, 4 - 7 or 0, to as long as nut, white or pale brownish; stamens 1 - 2, anthers 0.3 - 0.5 mm. Stigmas 2. Nuts glossy, 0.6 - 0.8 x 0.5 - 0.6 mm, obovoid, bi-convex, apex rounded, basal part gradually tapering, obscurely or finely reticulate, finally black, glossy; stylopodium much wider than long, ca 0.1 x 0.3 mm, white, clearly constricted from nut.

Field note: Common in marshy meadows throughout the area.

Fl. & Fr.: August - March.

Specimen examined: Triveni 1134 (MLVGCB Herb).

Distribution: Pantropical and extending into S. Europe, Turkey, Iraq, Afghanistan, Pakistan, Tadjikistan, Uzbekistan, China, Japan; tropical Africa, Madagascar, India, Malaysia, Australia, N. and S. America

Eleocharis geniculata (L.) Roem. & Schult., Syst. Veg. 2: 150. 1817; Parmar, in Shetty & Singh, Fl. Rajasthan 3: 916. 1993; Bhandari, Fl. Ind. Desert. 338. 1995. *Scirpus geniculatus* L., Sp. Pl. 1: 48. 1753. *Eleocharis caribaea* (Rottb.) Blake in Rhodora 20: 24. 1918. *Scirpus caribaeus* Rottb., Descr. Pl. Rar. 24. 1772. *Eleocharis capitata* R. Br. Prodr. Fl. Nov. Holl. 1: 225. 1810; Clarke, in Hook. *f.*, Fl. Brit. Ind. 6: 627. 1893, non *Scirpus capitatus* L.; *Eleocharis caduca* (Del.) Schultes, Mantissa 2: 88. 1824; *Scirpus caducus* Del., Descr. Egypte, Hist. nat. 9 (1813 n.v.); Matthew, Fl. Tamilnadu Carnatic 4: pl. 654. 1988. (Fig. 170)

Tufted annual, 12 - 25 cm high. Roots white, fibrous. Stem slender to rather rigid, 0.5 - 1.5 mm diam., green, with 5 - 6 obtuse angles separated by deep grooves. Open sheaths ca 5 mm, wide, scarious; lower closed sheath to 15 mm, reddish brown or green, mouth oblique, upper to 50 mm, green or basal parts reddish, with 5 nerves, mouth oblique, margin narrowly scarious; blades 0 or minute tip. Spikes 3 - 6 x 3 - 4 mm, ovoid or globular, with 50 or more very closely imbricating glumes; basal glumes fertile, not distinct from other glumes or its mid-nerve more pronounced and stronger; glumes 1.7 - 2.2 mm, not in protruding spiral rows, cymbiform, rounded, mid-nerve not very distinct, not reaching apex, sides grey brown, margins narrowly scarious. Perianth bristles 4 - 8, as long as or longer than nut, reddish brown; stamens 2 - 3, anthers 0.5 - 0.7 mm; stigmas 2. Nut 0.8 - 1 x 0.6 - 0.8 mm (excl. stylopodium), obovoid, bi-convex, apex rounded, basal part gradually tapering, obscurely reticulate, glossy, finally black; stylopodium 0.1 - 0.2 x 0.3 - 0.4 mm, much wider than long, white, clearly constricted from nut.

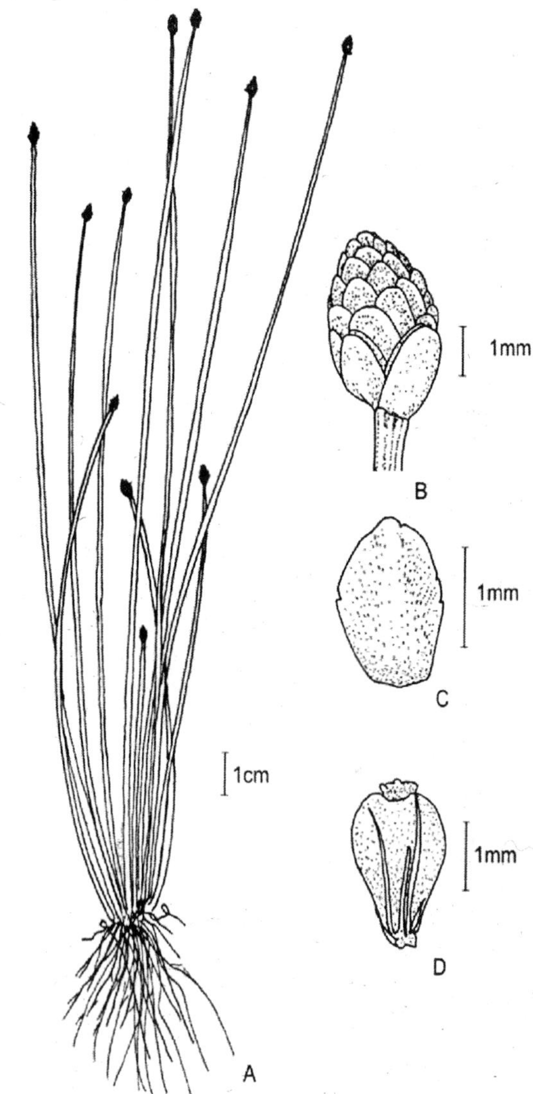

Fig. 170. *Eleocharis geniculata* (L.) Roem & Schult. A. Habit. B. Inflorescence, C. Glume, D. Nut

Field note: Common in wet and marshy places throughout the area.

Fl. & Fr.: August - February.

Specimen examined: Triveni 1217 (MLVGCB Herb).

Distribution: Pantropical.

FIMBRISTYLIS Vahl. *nom. cons.*

KEY TO THE SPECIES

1a. Stigma 2. Nuts biconvex or oblong cylindrical 2

2a. Glumes hairy at least the apical part, ferrugineuos .... *F. bisumbellata*

2b. Glumes glabrous, inflorescence open corymbose .... *F. cymosa*

1b. Stigma usually 3. Nuts triquetrous or trigonous 3

3a. Glumes 2-ranked, at least in the lower parts of a spikelet. Spikelets compressed below, terete or angular above........................ *F. ovata*

3b. Glumes spirally imbricate. Leaf blades laterally flattened, falcate. Spikelets throughout terete or angular.......... *F. miliacea*

Fimbristylis bisumbellata (Forsk.) Bubani, Dodecanthia. 30. 1850; Parmar, in Shetty & Singh, Fl. Rajasthan 3: 921. 1993; Bhandari, Fl. Ind. Desert. 339. 1995. *Scirpus bisumbellatus* Forsk., Fl. Aegypt-Arab. 15. 1775. *Fimbristylis dichotoma auct.* non (L.) Vahl, Enum. Pl. 2: 287. 1805; Nees in Wight, Contr. Bot. India 101. 1834; Clarke, in Hook. *f.,* Fl. Brit. Ind. 6: 635. 1893; Duthie, Fl. Gangetic Plain 3: 351. 1929; Mahesh. Ill. Fl. Delhi f. 223. 1966. (Plate 320)

Greyish-green, small tufts forming annual, 10 - 35 cm. Stem compresed-trigonous, glabrous. Leaves usually falcate, ca 1/2 of stem length; sheaths open, villous in upper parts; lowest bladeless, bicarinate, with scarious margins; ligule compact fringe of hairs; blades 0.5 - 2.5 mm wide, flat, adaxial side almost glabrous, abaxial side villous especially in proximal parts, margins towards apex scabrous. Inflorescence to 5 cm diam., with 15 - 30 spikes, compound to decompound, occasionally a few sessile, grouped together; lowest bract usually shorter than inflorescence. Peduncles grooved. Spikes 3.5 - 4.5 x 1.3 - 1.5 mm, narrowly ellipsoid, more or less angular, acute, light brown to greyish brown; rachis brown to dark brown, with scarious wings. Glumes 1.0 - 1.8 x 0.7 - 1.0 mm, broadly ovate, glabrous or occasionally, slightly hirsute, keeled, with prominent mid-nerve, arista conspicuous, margins widely scarious, slightly ciliate, yellowish brown. Stamens 1, sometimes 0; filaments scarious, compressed; anthers ca 1 mm; style brown, 0.7 - 1.2 mm long, flat, ciliate, stylopodium whitish, ca 0.25 x 0.25 mm, stigmas 2, ca 1 mm long. Nut 0.5 x 0.8 mm, lenticular, widely obovoid, trabeculate, yellowish to white.

Field note: Common in moist, marshy places along rivers, ditches, irrigation channels, etc.

Fl. & Fr.: July - December. But may be found in flowering stage almost throughout the year.

Specimen examined: Kotri 1240 (MLVGCB Herb).

Distribution: Western tropical and S.E. Africa to southern Europe and eastwards to Madagascar, Pakistan, India, Malaysia, tropical Australia and New Zealand.

Notes: The name *Fimbristylis dichotoma* is sometimes still used for this taxon, although C.E.C. Fischer (*Notes* on the Flora of Madras: Part X. Kew Bull. p. 149. 1935) corrected its usage. It is also to be noted that according to Stewart (1972), Clarke (1898) used *F. diphylla* Vahl for *F. dichotoma*, although Clarke clearly aimed for two distinct species.

Fimbristylis cymosa R. Br. subsp. ***spathacea*** (Roth) Koyama in Micronesica 1 : 83. 1964; Parmar, in Shetty & Singh, Fl. Rajasthan 3: 922. 1993. *F. spathacea* Roth, Nov. Pl. Sp.: 24. 1821; Clarke, in Hook. *f.,* Fl. Brit. Ind. 6: 640. 1893; Duthie, Fl. Gangetic Plain 3: 352. 1929. *F. glomerata* (Schrad.) Nees, fide Goetgheb. & Goudijzer, Bull. Jard. Bot. Nat. Belg. 54. 1984. *Isolepis glomerata* Schrad. in Roem. & Schult. 1824; Kral in Sida 4(2) : Pl. 43a (as *F. spathacea*) 1971; Matthew, Fl. Tamilnadu Carnatic 4 : Pl. 662. 1988.

Greyish green perennial, forming tight tufts, 10 - 20 cm. Stem rigid, terete to triangular, deeply grooved, smooth. Leaves less than half of stem length; sheaths 1 - 2.5 cm, pale brown, lustrous, with wide scarious margins; blades 1-1.5 mm wide, stiff, slightly turgid, frequently falcate, flat, margins barbed towards apex, apex short, frequently mucronate. Inflorescence compact or diffuse with elongate peduncles, of 10 or more solitary spikes, or, in compact inflorescences. Spikes sessile and clustered. Bracts usually short, margins scabrous; branches compressed or from triangular to terete. Spikes 2 - 5 x 1 - 2.5 mm, ovoid, more or less terete, greyish brown to brown; rachis articulate, castaneous brown, with prominent brown or colourless wings. Glumes 1.0 - 2 x 0.5 - 1.2 mm, broadly ovate, keeled, light brown, with green or pale mid-rib not reaching apex, margins widely scarious. Stamens 2; anthers 1.7 - 1.8 mm, with small basal lobes; style 1 - 1.2 mm, brown; stigmas 2 - 3, ciliate, 0.4 - 0.5 mm. Nut 0.1 - 0.7 x 0.5 cm, obovate, lenticular, slightly rugulose, black.

Field note: Common in marshy, aquatic places.

Fl. & Fr.: July - February.

Specimen examined: Kunwara 926 (MLVGCB Herb).

Distribution: Probably pantropical.

Notes: *F. cymosa* is interpreted here collectively following Kern, in Fl. Malesiana (1974). The bulk of Malaysian plants and specimens from present area also have, besides rather wide inflorescence, distigmatic flowers. Distigmatic plants were called by Kral (*l.c.*, 1971) and Bhandari (Flora of the Indian Desert, 1990) *F. spathacea* (type - B. Heyne from India), which is in accordance with the protologue. Both *F. cymosa* and *F. obtusifolia* (Lam.) Kunth are, according to their protologue, tristigmatic.

Fimbristylis miliacea (L.) Vahl, Enum. Pl. 2: 287. 1805; Clarke, in Hook. *f.*, Fl. Brit. Ind. 6: 644. 1893; Duthie, Fl. Gangetic Plain 3: 356. 1929; Parmar, in Shetty & Singh, Fl. Rajasthan 3: 925. 1993. *Scirpus miliaceus* L. Syst. Veg. Nat. ed. 10. 868. 1759; *Fimbristylis littoralis* Gaud. Voy. Uranie 10: 413. 1829; Kral in Sida 4 (2): Pl. 29. 1971; Haines & Lye, Sedges and Rushes E. Afr., f. 126. 1983. *Fimbristylis quinquangularis* Kunth, Enum. Pl. 2: 229. 1837; Clarke, in Hook. *f.*, Fl. Brit. India 6: 644. 1893. (Fig. 171)

Tufted, glabrous annual or perennial, 10 - 90 cm high, with many flowering culms and few sterile leafy culms. Flowering stems leafless or with solitary leaf, acutely 5-angled, 1.5 - 2.5 mm thick, prominently grooved. Leaves as long as stem or longer; lower sheaths open, prominently nerved, with widely scarious margins, upper 2 closed, 2 - 8 cm, mouth oblique, one side green, many-nerved, nerves ending in acute apex, mouth margin scarious; blades 1.5 - 2 mm wide, laterally compressed, adaxial side narrow groove at one margin, margins towards apex slightly barbed. Inflorescence corymbs lax, coumpound or supra-decoumpound, up to 50 mm diam., taller than wide, with 15-50 spikes; bracts up to 20 mm, involute; branches triangular or compressed, much longer than spikes, smooth or distal parts barbed. Spikes 1.5 - 3 mm, pedunculate or sessile, globular or ellipsoid, obtuse, pale yellow to brown, with from few to more than 30 glumes; spike rachis castaneous to almost black, with narrow hyaline wings. Glumes 1 - 2 x 0.6 - 1.0 mm, spiral, ovate, keeled, apiculate or sometimes obtuse, brown to almost hyaline. Stamens 1-2, anthers ca 1 x 0.3 mm, mucronate. Style 0.3-0.4 mm, brown, base mostly bulbous, deciduous; stigmas (-2) 3. Nut 0.4-0.8 x 0.3-0.5 mm, obovoid, or suborbicular, trigonous, shortly stipitate, verruculose, faintly trabeculate, papillose, grey to pale yellowish or white.

Field note: Common in wet ground, river beds etc.

Fl. & Fr.: July - December.

Specimen examined: Bari sadari 1148 (MLVGCB Herb).

Distribution: Pantropical.

Notes: It must be made clear, that here the well founded proposal by Strong and Kral (l.c.) to conserve *Scirpus miliaceus* L. with the type is followed and the name *F. miliacea* (L.) Vahl is being used in its traditional sense. In Fl. Iranica (Kukkonen 1998) the name *F. littoralis* Gaud. was applied for this species. S.T. Blake (Journ. Arnold Arbor. 35: 217, 1954) lectotypified *S. miliaceus* L. with LINN 71.41, which represents *F. quinquangularis* (Vahl) Kunth, the *F. miliacea* (L.) Vahl species.

Fig. 171. *Fimbristylis miliacea* (L.) Vahl.
A. Habit, B. Inflorescence

Fimbristylis ovata (Burm. f.) Kern, Blumea 15: 126. 1967; Parmar, in Shetty & Singh, Fl. Rajasthan 3: 925. 1993. *Carex ovata* Burm. f. Fl. Ind. 194. 1968. *Cyperus monostachyos* L. Mantissa Pl. 2: 180. 1771; *Fimbristylis monostachyos* (L.) Hassk., Pl. Jav. Rar. 61. 1848 '*monostachya*'; Clarke, in Hook. *f.,* Fl. Brit. Ind. 6: 649. 1893; Duthie, Fl. Gangetic Plain 3: 355. 1929.

Perennial, 10 - 30 cm high, with short, knotty rhizome. Stem erect, slender or filiform, slightly swollen at the base, spinulose - scabrous on upper angles. Leaves 2/5 - 2/3 of stem length; sheaths to 30 mm, basal parts strongly nerved, upper parts widely scarious, mouth concave to lingulate; ligule 0; blades 0.5 - 1 mm wide, canaliculate, margins involute, scabrous, apex flat, mucronate. Inflorescence terminal of 1 - 2 (-3) spikes. Peduncles ca 15 mm, with tubular prophyll. Bracts 2 or 3, scale - like, 8.2 mm, with barbed arista of 5 mm. Spike 7 - 10 mm, flat or angular, with 10 - 15 glumes, which distichously or irregularly spirally arranged. Glumes ovate-oblong, 3 - 6.2 x 2 -4 mm, coriaceous, acuminate, keeled, sides nerve-less, yellowish brown, lowest 2 - 3 sterile, with barbed arista to 1.9 mm; rachis ca 1 mm wide, deeply notched, internodes ca 0.8 mm, glumes caducous leaving basal corners on rachis, which thus winged. Stamens 2 - 3, anthers ca 1.3 mm. Style ca 1.5 mm, flat or shallowly trigonous, edges ciliate, stylopodium caducous, 0.8 - 1.2 x 0.6 - 0.7 mm, trigonous, triangular, edges sharp; stigmas 3. Nut 2.0 - 3 x 1.2 - 2 mm (excl. style base), pyriform, obovoid, trigonous, apiculate, stipitate, with ca 0.6 mm long stipe, tuberculate, white or brownish.

Field note: Rare, in dry grasslands and moist habitats.

Fl. & Fr.: August - October.

Specimen examined: Pander 1341, Kankoriliya 1607 (MLVGCB Herb).

Distribution: Pantropical.

FUIRENA Rottb.

1b. Annual. Leaves pubescent.
Inner perianth subquadrate .... *F. ciliaris*

1a. Perennial. Leaves glabrous.
Inner perianth linear or
absent....................................... *F. wallichiana*

Fuirena ciliaris (L.) Roxb., Fl. Ind. (Carey & Walliched) 1: 184. 1820; Blatter in Journ. Bombay Nat. Hist. Soc. 37 : 772. 1934; Parmar, in Shetty & Singh, Fl. Rajasthan 3: 930. 1993. *Scirpus ciliaris* L. Mant. Pl. 2: 182. 1771. *Fuirena glomerata* Lam. Tabl. Encycl. Meth. Bot. 1: 150. 1791; Clarke, in Hook. *f.,* Fl. Brit. Ind. 6: 666. 1893; Duthie, Fl. Gangetic Plain 3: 365. 1929.

Annual tufted, whole plant pubescent. Culms 5 - 40 cm tall. Leaves 5 - 15 x 0.4 - 0.6 cm, linear - lanceolate, acuminate, 3 to 5 - nerved, hairy. Sheaths more or less patently hairy. Spikelets in clusters 3 - 10, ovoid-oblong, elliptic, bracts leaf-like, surpassing the inflorescence. Glumes 1.5 - 1.7 mm long, awened, obovate to oblong elliptic, hypogynous bristles 3; awns hairy, usually recurved. Inner scales subquadrate, apiculate, base cordate with a slender claw, back with 3 raised nerves, adpressed to and concealing the nut. Nuts 0.25 - 1 mm long, stipitate, apiculate, trigonous, globosely obovoid, pale - brown.

Field note: Common in aquatic and marshy places.

Fl. & Fr.: September - February.

Specimen examined: Mandalgarh 335, Chittorgarh Fort 1622 (MLVGCB Herb).

Distribution: Warmer regions of the World.

Fuirena wallichiana Kunth, Enum. Pl. 2 : 182. 1837; Clarke, in Hook. *f.*, Fl. Brit. Ind. 6: 665. 1893, incl. var. *evoluta* Clarke, in Hook. *f., l.c.* & Illus. Cyper. t. 58. f. 8. 1909; Duthie, Fl. Gangetic Plain 3: 364. 1929; Parmar, in Shetty & Singh, Fl. Rajasthan 3: 931. 1993. *F. cuspidata* Kunth, *l. c.* 2: 187. 1837.

Perennial. Rizomes creeping. Culms triquetrous, up to 60 cm, slightly pubescent above. Leaves lenear-lanceolate, 5 - 20 x 0.4 - 0.8 cm, with prominent mid-rib. Sheaths long, with pubescent ligule. Spikelets 6 - 8 mm long, in terminal corymbose clusters, oblong, up to 1 x 0.3 cm, multi-fid. Bracts small. Rays small. Glumes 3 mm long excluding the stout scarbid mucro, broadly - oblong, rounded at the tip below the mucro, pubescent and ciliate, lined with green-red streaks. Bristles 3; outer bristles shorter than the nut, reddish-brown, scabrid; inner bristles much longer than the nut, sometimes small or abscent. Style 3 fid. Nut ca 1.2 mm long, pale yellow, turbinate, trigonous, globose, pyramidally pointed at each end, finely longitudinally ribbed.

Field note: Rare, in marshy habitats around temporary seasonal ponds.

Fl. & Fr.: September - January.

Specimen examined: Bari sadari 1148 (MLVGCB Herb).

Distribution: India, America

KYLLINGA Rottb. *nom. cons.*

KEY TO THE SPECIES

1a. Rhizome long-creeping, not tufted. Inflorescence as a rule single head, seldom 2 - 3 heads, 4 - 8 mm; spikes 3 - 3.5 mm. Keel of glumes spinulose at least towards the apex .... *K. brevifolia*

1b. Rhizome short.Iinflorescence usually formed by 3 sessile heads, seldom 1 or 4-5 heads, 6 -10 mm; spikes 1.6 - 2 mm. Keel of glumes smooth .... *K. bulbosa*

Kyllinga brevifolia Rottb., Descr. Ic. Rar. Nov. Pl. 13. t. 4. f. 3. 1773; Clarke, in Hook. *f.*, Fl. Brit. Ind. 6: 588. 1893; Duthie, Fl. Gangetic Plain 3: 343. 1929; Parmar, in Shetty & Singh, Fl. Rajasthan 3: 932. 1993. *Cyperus brevifolius* (Rottb.) Hassk. Cat. Hort. Bot. Bogor.: 24. 1844; Koyama, Fl. Taiwan: pl. 1345. 1979; Noltie, Fl. Bhutan 3. 1: f. 30, a-b. 1994. (Fig. 172)

Perennial, 5 - 35 cm high. Rhizome long-creeping, ca 2 mm diam., aerial stems arising from every second node. Stem 0.5 - 1.5 mm diam., sharply trigonous, smooth, green. Leaves much shorter than stem; sheaths 10 - 65 mm, grey or brown, sometimes with reddish tint, scarious; blades up to 10 cm, 1.5 - 3 mm wide, flat, slightly keeled, sometimes folded, margins smooth, apex flat, short, acute, scabrous along margins and midrib below. Inflorescence a terminal, single, round cluster of ca 100 spikes, 5 - 7 mm; 2 - 3 bracts foliose, spreading, up to 80 mm long. Spike 2.5 x 1 mm, compressed, pedicel 0.2 mm, glume-like bract scarious, glume-like prophyll scarious; glumes 2, opposite, ca 2.5 mm, keeled, folded, mid-nerve green, smooth or scabrous, acuminate, sides scarious, with 2 - 3 prominent nerves, white or yellowish and often with brown dots. Stamens 3. Nut 1.0 - 1.5 x 0.5 - 0.7 mm, compressed, globosely ovoid, brownish, apiculate, surface papillose.

Field note: Common in marshy habitats, on edges of ponds, dams etc.

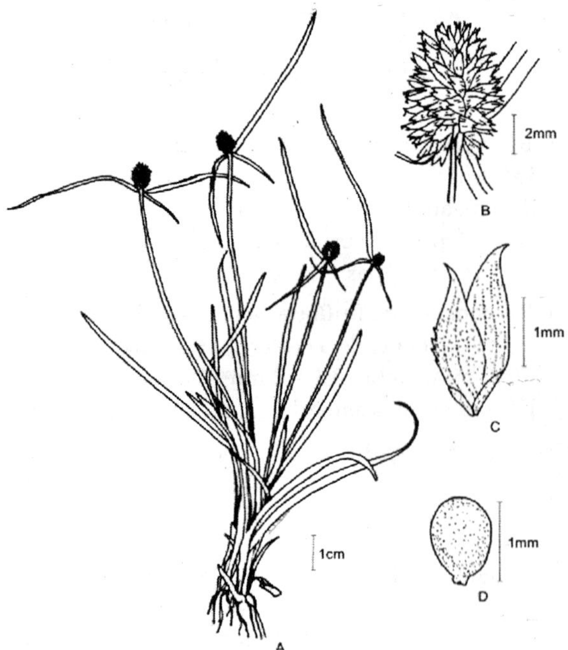

Fig. 172. *Kyllinga brevifolia* Rottb.
A. Habit. B. Inflorescence, C. Spike, D. Nut.

Fl. & Fr.: August - December.

Specimen examined: Kotri 183 (MLVGCB Herb).

Distribution: Pantropical; often introduced , e.g., to Caucasus; Azores and Portugal (in Fl. Europ.); in Africa South of Sahara; Russian Far East, Japan, China, Afghanistan, Pakistan, India, Taiwan, Malaysia, Australia, N and S America.

Kyllinga bulbosa Beauv. Fl. d'Oware & Benin 1: 11. t. 8. f. 1. 1804; Koyama, in Dassanayake & Fosberg, Fl. Ceylon 5: 245. 1985; Parmar, in Shetty & Singh, Fl. Rajasthan 3: 933. 1993. *K. triceps* Rottb. Descr. Ic. Rar. Nov. Pl. 14. t. 4. f. 6. 1773, *nom. illegit*; Clarke, in Hook. *f.,* Fl. Brit. Ind. 6: 587. 1893; Duthie, Fl. Gangetic Plain 3: 343. 1929; Kukkonen in Taxon 44: 26. 1995. *Schaenoides triceps* Rottb. Descr. Pl. Rar. Progr. 15. 1772. *invalid name. Cyperus triceps* (Rottb.) Endl., Cat. Hort. Acad. Vinodb. 1: 94. 1842. *Kyllinga tenuifolia* Steud., Syn. Pl. Cyperac.: 69. 1855. *Cyperus tenuifolius* (Steud.) Dandy, in Exell, Catal. Vasc. Pl. S. Tomé: 363. 1944; Koyama, in Acta Phytotax. Geobot. 16: Pl. 4, C. 1955; Haines & Lye, Sedges and Rushes E Afr.: fig. 466. 1983 (Plate 321).

Densely tufted, perennial herb, 5 - 35 cm, stem base bulbous, with tillers. Stem trigonous, smooth. Leaves blades up to 20 cm, 1 - 3 mm wide, flat, greyish green, margins smooth or margins and midnerve below towards apex scabrous, apex long-attenuate, acute, trigonous, scabrous, ca half of stem length. Sheaths stout, pale, lower part tinged with light brown. Inflorescence a terminal head of usually, sessile, compact, globose, greenish-white heads (clusters of spikes), 6 - 10 mm, central head usually larger and longer than others; 50 - 100 spikes in cluster, spirally arranged. Bracts foliose, to 10 cm, spreading or reflexed. Spikes 2 - 2.5 x *c.* 0.8 mm, oblong or oblong - lanceolate. Glumes 4, ovate - oblong to oblong - lanceolate, acute, keeled, folded, keel smooth, prominent, sides scarious with 3 prominent yellow-brown nerves. Stamens 2. Stigmas 2. Nut ca 1.2 mm, oblong, compressed, brown, papillose.

Field note: Common in marshy and wet meadows.

Fl. & Fr.: July - November.

Specimen examined: Atun 82, Shahpura 1188 (MLVGCB Herb).

Distribution: Africa of Sahara extending to Angola and Mozambique; Pakistan, Nepal, India and Sri Lanka, China, Vietnam, Australia.

Notes: *K. triceps* Rottb. is a superfluous illegitimate name. Kukkonen's proposal to conserve failed. Also "Taxon 44: 26." is an error for "Taxon 44: 626."

PYCREUS P. Beauv.

Pycreus flavidus (Retz.) T. Koyama, in J. Jap. Bot. 51: 316. 1976; Clarke, in Hook. *f.,* Fl. Brit. Ind. 6: 589. 1893; Parmar, in Shetty & Singh, Fl. Rajasthan 3: 941. 1993; *Cyperus flavidus* L., Sp. Pl. 46. 1753; Retz., Observ. bot. 5: 13. 1788; Clarke, in Hook. *f., l. c.* 6: 600. 1893; *C. globosus* (All.) Reichb. Fl. Germ. Excurs. 2: 140. 1830. *Pycreus globosus* (All.) Reichenb., Fl. Germ. Excurs.: 140. 1830. *P. cupillaris* (Koen. ex Roxb.) Nees Clarke, in

Hook. *f.,* Fl. Brit. India 6: 591. 1893. *Cyperus nilagiricus* Hochst. ex Steud. Syn. Pl. Cyperac.: 2. 1855; Hooper, in Townsend & Guest, Fl. Iraq 8: pl. 81. 1985. (Fig. 173)

var.: ***flavidus***

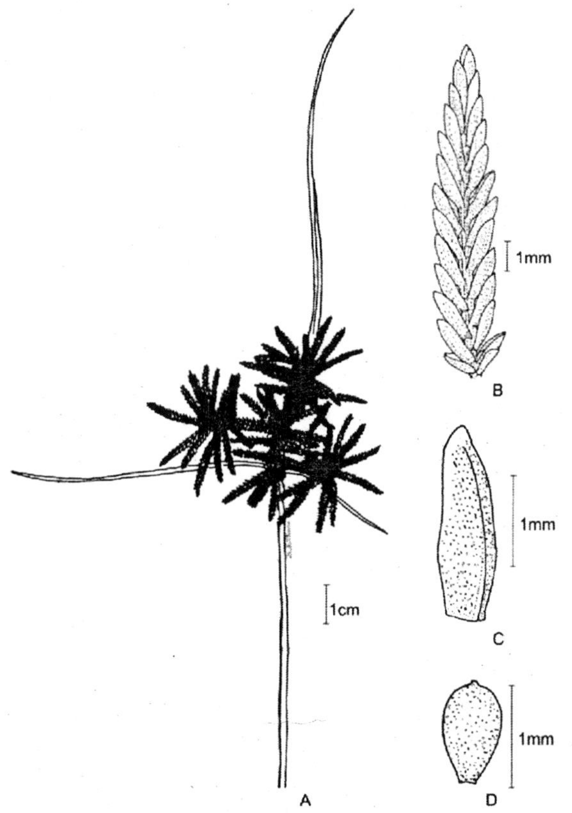

Fig. 173. *Pycreus flavidus* (Retz.) Koyama var. *flavidus.* A. Inflorescence, B. Spike, C. Glume, D. Nut

Tufted, annual 15 - 50 cm. Stem trigonous, basal parts almost terete, grooved, otherwise smooth. Leaves shorter than stem; sheaths 20 - 80 mm, brown or yellow-brown, mouth margin straight; blades up to 15 cm, 1 - 3 mm wide, flat or folded, margins involute, smooth, apex long, trigonous, acute, scabrous. Inflorescence simple umbel of contracted heads, group of up to ca 50 subsessile spikes or irregular anthelodium, up to ca 50 mm diam. Bracts 1-2, foliose. Spikes 8 - 30 mm long, 2 - 2.5 mm wide, compressed, with up to ca 50 glumes, glume-like bract ca 2.5 mm long, glume-like prophyll ca 2.5 mm, blunt, basal part adaxially swollen; rachis quadrangular, almost straight, brown, internodes ca 0.8 mm, lateral sides deeply concave, margins sharp. Glumes ca 2.0 mm, cymbiform, blunt, mid-vein area green, sides reddish brown, margins widely scarious. Keel 3 to 5 - nerved. Stamens 2. Nut 0.8 - 1 mm, ellipsoid or obovoid, bi-convex, with a short stipe, apiculate, brown, minutely tubercled.

Field note: Rare along rivers, also in forests and grasslands.

Fl. & Fr.: September - December.

Specimen examined: Chanderiya 1423 (MLVGCB Herb).

Distribution: S. Europe, Algeria, Israel, Iraq, Turkey, east to Pakistan, NW India, China, Japan, Taiwan, Malaysia and Australia.

SCHOENOPLECTUS (Reichb.) Palla *nom. cons.*

KEY TO THE SPECIES

1a. Glumes broadly ovate, acute. Nuts dark brown .... *S. articulatus*

1b. Glumes broadly ovate, mucronate. Nuts black.. 2

2a. Culms septate when dry; obtuse... *S. roylei*

2b. Culms not septate when dry; mucronate.. *S. supinus*

Schoenoplectus articulatus (L.) Palla in Bot. Jahrb. Syst. 10: 229. 1889; Parmar, in Shetty & Singh, Fl. Rajasthan 3: 946. 1993. *Scirpus articulatus* L., Sp. Pl. 1: 47. 1753; Clarke, in Hook. *f.,* Fl. Brit. Ind. 6: 656. 1893; Parker, in Duthie, Fl. Gangetic Plain 3: 360. 1929. (Fig. 174 & Plate 322)

An annual or pernnial with tufted stem. Stem terete, spongy, 5 - 15 (-40) cm long, smooth, dark green, clothed with 2 or 3 bladeless sheaths, circular, tapering towards the apex, transversely septed, upto 80 mm high. Leaves absent or reduced to brownish sheaths; sheaths 1 - 3, laxy enveloping the stem, sometimes bearing a short lamina. Inflorescence capitate, pseudolateral, 1 - 3 cm across, bearing 15 - 60 spikelets, born laterally near the base of the stem. Spikelets 7 -

17 x 3 - 6 mm, ovoid-oblong, terete, subacute, densely many-flowered. Glumes tightly spirally imbricated, 3-5.0 mm long, ovate or suborbicular, membranous, finely many-viened, pale green, mucronate. Perianth absent. Stamens 3. Stigma 3. Nuts trigonous, obvoid, about 2 mm long, dark brown or ultimately black, apiculate, sessile.

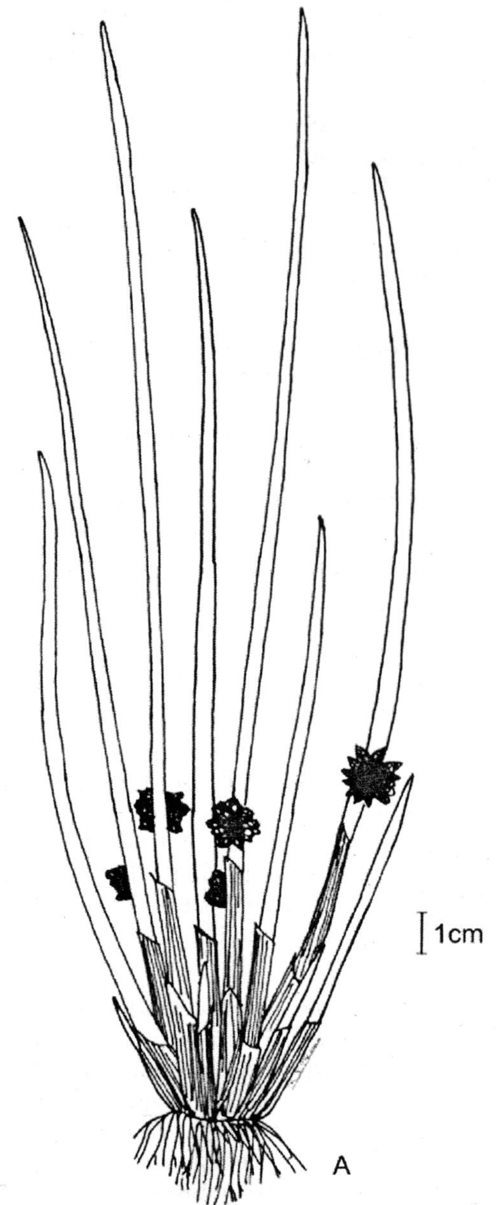

Fig. 174. *Schoenoplectus articulatus* (L.) Palla. A. Habit.

Field note: Common in ditches or near standing water.

Fl. & Fr.: September - February.

Specimen examined: Shripura 1308, Bhaisarorgarh 104 (MLVGCB Herb).

Distribution: Sudan to the Cape, in Madagascar and intermittently throughout the Old World tropics.

Schoenoplectus roylei (Nees) Lye. in Bot. Notiser 124: 290. 1971; Parmar, in Shetty & Singh, Fl. Rajasthan 3: 950. 1993. *Isolepis lupulina* Nees, in Wight, Contrib. Bot. India: 107. 1834. *Scirpus lupulinus* (Nees) Roshev. FL. USSR, 3: 53. 1935, non Spreng. *Isolepis roylei* Nees, in Wight, Contrib. Bot. Ind: 107. 1834. *Scirpus roylei* (Nees) Parker in Duthie, Fl. Gangetic Plain 3: 361. 1929; Bhandari, Fl. Ind. Desert. 343. 1995. *Scirpus quinquefarius* Buch. - Ham. ex Boeck., in Linnaea 36: 701. 1870; Clarke, in Hook. *f.*, Fl. Brit. Ind. 6: 657. 1893. (Fig. 175)

Annual, forming small tufts, 8 - 50 cm high. Root fibrous. Stem narrow, terete, grooved, transverslely septate, smooth, green. Leaves absent. Sheaths 2 - 3, lower ca 10 mm, upper to 50 mm, in both one side green, mouth oblique, margin widely scarious; blades reduced to mucro, ca equalling ligule in length. Inflorescence pseudo lateral capitate, yellowish spikelets, located above the middle of the stem, group of 2 - 12 sessile spikes, 8 - 12 mm across; lowest bract up to 13 mm, terete, 0.5 - 1 mm diam., slightly nodose, grooved, smooth, obtusely pointed. Spikes 5 - 12 x 3 - 7 mm, ovoid or ellipsoid. Glumes finally reflexed, 2.8 - 3.6 mm, oblong-lanceolate, midnerve area green, narrow, smooth, extending into short, slightly recurved mucro, sides many-nerved, golden brown, slightly glossy, apex gently tapering, margins hardly scarious. Perianth absent; stamens 3, anthers ca 0.5 mm; stigmas 3. Nut 1.2 - 1.3 x 0.8 - 1 mm, narrowly obovoid, apiculate, trigonous, almost plano-convex, sides concave, transversely way-wrinkled, black.

Field note: Common in marshy habitats and mudflates of monsoonal pools.

Fl. & Fr.: August - December.

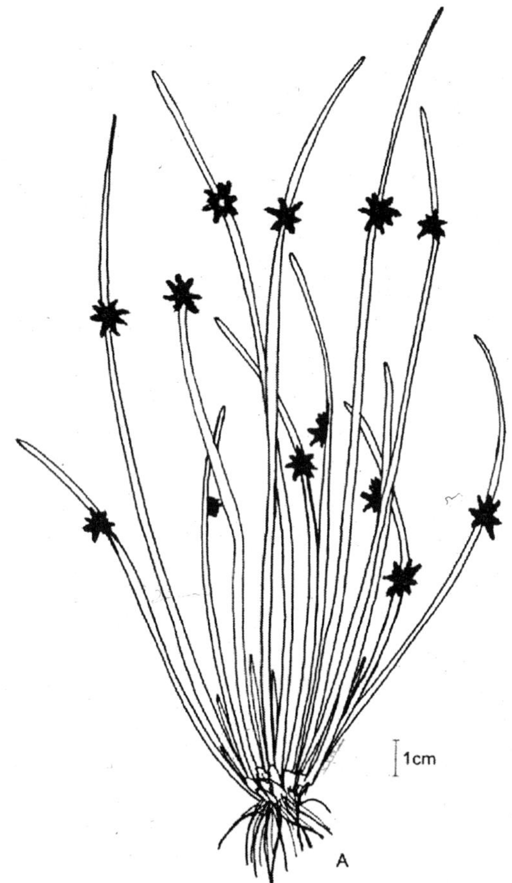

Fig. 175. *Schoenoplectus roylei* (Nees) Lye.
A. Habit.

Specimen examined: Agucha 1320 (MLVGCB Herb).

Distribution: Iraq, SE Uzbekistan, Tadjikistan, Afghanistan, Pakistan, India and Africa.

Notes: Names *Isolepis lupulinus* and *I. roylei* were published simultaneously, on same page by Nees (*l.c.* 1834), however, the former was combined under *Schoenoplectus*.

Schoenoplectus supinus (L.) Palla, subsp. *lateriflorus* (Gmelin) Koyama Bot. Jahrb. 10: 299. (1888) "1889"; Kukkonen, in Rech. f., *l.c.* 37; Parmar, in Shetty & Singh, Fl. Rajasthan 3: 950. 1993. *Scirpus lateralis* Retz. Obs. Bot. 4: 12. 1786, non Forsk. 1775. *Scirpus lateriflorus* Gmelin, Syst. Veg. 127. 1791. *Scirpus supinus* L., Sp. Pl. 1: 49. 1753; Clarke, in Hook. *f.,* Fl. Brit. Ind. 6: 655. 1893; Parker in Duthie, Fl. Gangetic Plain 3: 360. 1929; Bhandari, Fl. Ind. Desert. 344. 1995. *Isolepis supina* (L.) R. Br., Prodr. Fl. Nov. Holl. 1: 77. 1810; *Schoenoplectus melanospermus* (C.A. Mey.) Grossh., Fl. Kavk., ed. 1. 1: 146. 1928. *Scirpus melanospermus* C.A. Mey., in Mém. Acad. Imp. Sci. St.-Petersbourg, sér. 6, 1:199. tab.2. 1831; Madalski, Fl. Polon. Terr. Adiac. Iconogr. 3(1): 257. 1979. (Fig. 176 & Plate 323)

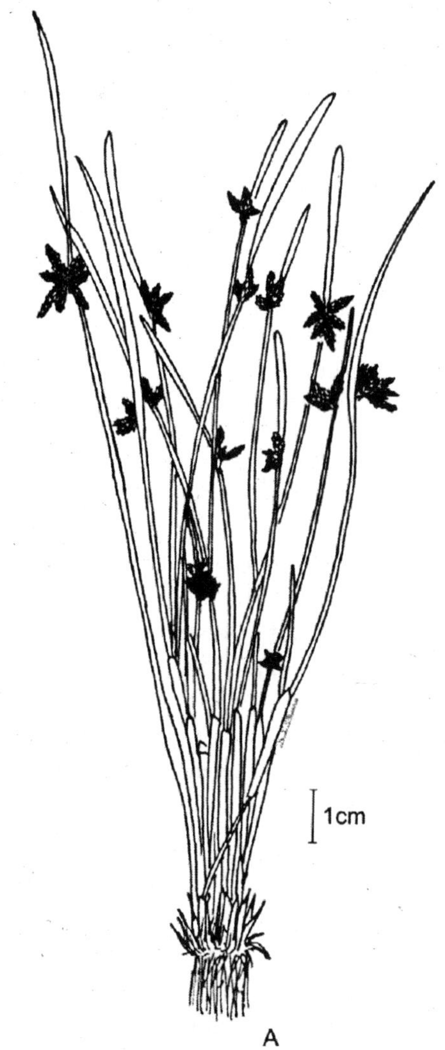

Fig. 176. *Schoenoplectus supinus* (L.) Palla subsp. *lateriflorus* (Gmelin) Koyama
A. Habit

Annual, 5 - 25 cm, tufted sedge. Roots fibrous, small. Culms 0.8 - 2 mm diam., terete or distal parts obscurely trigonous, grooved, green or greyish green, smooth. Sheaths membranous, up to 65 mm long, lowest short, ephemeral, bearing an occasional flower in its axil, upper longer, mouth oblique, margin widely scarious; ligule up to 0.8 mm, scarious; blades reduced to mucro or, in uppermost sheath, most often green, up to 50 mm long. Inflorescence pseulateral, head like or corymbose, sessile or occasionally one or a few spikes, pedunculate; lowest bract 2 - 16 cm long, terete, apex rather obtuse, smooth, other bracts seldom green and longer than their spike. Spikelets 3 - 20 to an inflorescence, congested or clustered in groups of 2 - 7 at the apices of rays, 4 - 8 x 2 - 3 mm, ovate lanceolate to oblong, pale - green, many flowered. Glumes rather tightly imbricating, 1.8 - 2.5 mm long, oblong-lanceolate, towards apex slightly keeled, smooth, sides obscurely nerved, brown or colourless, scarious, margins smooth or minutely fringed. Perianth bristles reduced, 3 - 4 occasionally elongated, unequal, longest shorter than nut. Stamens 3, anther 0.5 - 0.7 mm; stigmas 3. Nut 1.0 - 1.6 x 1 - 1.3 mm, widely obovoid, apiculate, trigonous, surface conspicuously transversely rugulose with very fine longitudinal striations, glossy, black brown.

Field note: Common in marshy habitats and mudflates of monsoonal pools.

Fl. & Fr.: August - December.

Specimen examined: Titoda 1663 (MLVGCB Herb).

Distribution: Europe, India, Africa, Turkey, Caucasus, Siberia, Kazakhstan, Iran and Pakistan.

POACEAE (GRAMINEAE *nom. alt.*)

KEY TO THE GENERA

1a. Arborescent, thorny. Lemmas many, subequal. Flowerring and fruiting rare .... *Bambusa*

1b. Herbaceous. Lemmas few, sub-equal. Flowering and fruiting annually 2

2a. Male and female spikelets in separate inflorescences or in different part of the same inflorescence. Male spikelets terminal on the same inflorescence .... *Coix*

2b. All spikelets bisexual or with male, barren and bisexual spikelet intermixed 3

3a. Tall reed-like. Panicle large feathery or fan-shaped 4

4a. Rachis fragile, breaking up at maturity, spikelets 2-flowered........ .*Saccharum*

4b. Rachis tough or taretily breaking up. Spikelets more than 2-flowered. lemma hairy........ *Arundo*

3b. Slender grass. Panicle not as above 5

5a. Spikelets 2-flowered, usually dorsally compressed, falling entire at maturity........ 6

6a. Spikelets solitary or paired, usually similar. Glumes dis-similar, lowers usually smaller or sometimes reduced 7

7a. Spikelets subtended by 1-many bristles or involucre of bristles 8

8a. Upper lemma transversely rugose. Bristles persistent .... *Setaria*

8b. Upper lemma smooth. Bristles falling with the spikelets 9

9a. Bristles numerous. Plant terrestrial........ 10

10a. Bristles rigid, united at the base into a hard cup-shaped involucre........ *Cenchrus*

10b. Bristles filiform, free to the base .... *Pennisetum*

9b. Bristles solitary. Plant aquatic.. *Pseudoraphis*

7b. Spikelets not subtended by bristles or involucre of bristles 11

11a. Spikelets in contracted or open panicles. Panicle open. Pedicels more than 2 mm long........ *Panicum*

11b. Spikelets in 1-sided spikes or spike-like racemes which are either digitate or arranged along a central axis 12

12a. Spikelet with a bead like swelling at the base........ *Eriochloa*

12b. Spikelet without a bead like swelling at the base........ 13

13a. Lower glume absent or as a small hyaline scale 14

14a. Spikelet solitary. Lemma without verrucose hairs. Rachis flattened... .*Paspalum*

14b. Spikelets binate or ternate, if solitary then lemma with verrucose hairs. Rachis teret or triangular .... *Digitaria*

13b. Lower glume usually well developed 15

15a. Racemes digitate. Upper glumes with globular-tipped purple hairs.......... *Alloteropsis*

15b. Racemes arranged along a central axis. Upper glumes without such hairs 16

16a. Glumes at least the upper, acuminate or awned, rarely acute...................................... 17

17a. Leaves ovate-lanceolate, base oblique. Ligule membranous .... *Oplismenus*

17b. Leaves linear, base not oblique. Ligule absent or a fringe of stiff hairs ..... *Echinochloa*

16b. Glumes not awned, if acuminate then the upper lemma mucronate................ 18

18a. Upper lemma obtuse, abruptly mucronate or aristate at the apex....................... *Urochloa*

18b. Upper lemma acute or acuminate; lower glume turned towards the rachis ... *Brachiaria*

6b. Spikelets paired or in groups of threes, dissimilar with one sessile and the other pedicelled. Glumes nearly similar............... .19

19a. Joints of rachis and pedicels swollen, 3-angeled, rounded or flattened...................... .20

20a. Upper lemma not awned. Racemes usually solitary. Pedicelled spikelets well developed. Rhizomes absent .... *Rottboellia*

20b. Upper lemma awned. Racemes 2-many 21

21a. Racemes 1-noded, spatheate. Spikelets in threes, 1-sessile and other 2-pedicelled. Fertile spikelets 1-flowered. Awn slender... *Apluda*

21b. Racemes many-noded, espatheate. Spikelets paired, 1-sessile and other pedicelled. Fertile spikelets 2-flowered ....*Ischaemum*

19b. Joints of rachis and pedicels narrow, not swollen .... .22

22a. Racemes solitary, terminating culm and its branches 23

23a. Upper glume of the sessile spikelets with 6-12 cm long awn. Lower glume of the fertile sessile spikelets rounded on the back....................................... *Heteropogon*

23b. Upper glume of the sessile spikelets with 7-12 mm long awn. Lower glumes channelled on the back... *Sehima*

22b. Racemes 2-many flowered, digitate or combined into a panicle................................ 24

24a. Inflorescence interrupted by spathes or spatheoles 25

25a. Aromatic. Lower glume of sessile spikelets 2-keeled; margins inflexed .... *Cymbopogon*

25b. Non-aromatic. Lower glume of sessile spikelets not keeled or keeled at top only; margins inflexed............................... *Themeda*

24b. Inflorescence not interrupted by spathes or spatheoles 26

26a. Racemes digitate or sub-digitate. Upper lemma of sessile spikelets entire. Lower pair of spikelets similar other dissimilar.................................... *Dichanthium*

26b. Racemes whorled along a long axis of panicle. Upper lemma of sessile spikelets 2-lobed .... .27

27a. Spikelets dorsally compressed. Lower glume flattened on back. Upper lemma awned from the sinus........... *Sorghum*

27b. Spikelets laterally compressed or terete. Lower glume concave on back. Spikelets in threes. 1-sessile and 2-pedicelled; glume equal lower glume smooth .... *Chrysopogon*

5b. Spikelets 1 to many flowered, usually laterally compressed or terete, breaking up at maturity above the persistent glumes or if falling entire, then not 2-flowered 28

28a. Inflorescence of a single, terminal, spike or spike-like racemes or distant group of spikelet on a 1-sided racemes 29

29a. Spikelets breaking up at maturity above the persistent glumes ... *Melanocenchris*

29b. Spikelets breaking up at maturity below the glumes 30

30a. Glumes not awned; upper glume with 5 rows of hooked spines .... *Tragus*

30b. Glumes awned; upper glume not as above.. ..*Perotis*

28b. Inflorescence a digitate spikes or an open or contracted spike-like panicle or a spicate head 31

31a. Inflorescence a digitate spike or spike-like racemes.. 32

32a. Spikelets only with 1-fertile florets............ 33

33a. Spikelets 1-flowered, usually without sterile florets above. Lemma not widened above, glabrous.. .34

34a. Lemmas long awned ....*Schoenefeldia*

34b. Lemmas awnless............................ .*Cynodon*

33b. Spikelets obviously several flowered, the lower 1-2 (-4), bisexual. Lemmas widened above, hairy.. 35

35a. Spikes 1to 3. Fertile florets 2 - 4.................................... .*Tetrapogon*

35b. Spikes 4 or more. Fertile floret only one. .... *Chloris*

32b. Spikelets containing 2 or more fertile florets.. 36

36a. Axis of spike ending into a sharp mucro ....*Dactyloctenium*

36b. Axis of spike ending in a spikelet .... *Eleusine*

31b. Inflorescence an open contracted or spike-like panicle or spicate head .... .37

37a. Lemma 3-awend from the tip. Awn simple or branched.................... *Aristida*

37b. Lemma 1-awend or awnless....................... 38

38a. Glumes absent or rudimentary. Palea similar to lemma. Stamens 6 .... .39

39a. Leaf-blades ovate-oblong, obtuse; sheath inflated. Lemma solitary ..... *Hygroryza*

39b. Leaf-blades linear, acute-acuminate; sheath not inflated. Spikelets straight. Lemma with 2 empty scales ....*Oryza*

38b. Glumes atleast the upper well developed. Palea not similar to lemma. Stamens 3 or less 40

40a. Spikelets usually with 2 - or more fertile florets, if one, then with sterile florets above it 41

41a. Glumes as long as or longer than the lowest floret. Lemma with a geniculate awn.... *Avena*

41b. Glumes usually shorter than the lowest floret. Lemma not awned or if awned then awn not geniculate 42

42a. Lemma emarginate or 2-4 toothed. 43

43a. Racemes with few sub-terete spikelets. Lemma rounded on the back............ *Diplanche*

43b. Racemes with many lateraly compressed spikelets. Lemma keeled. .... *Leptochloa*

42b. Lemma entire at apex. .... .44

44a. Glumes acuminate-aristate. Racemes deflexed on a flattened rachis .... *Dinebra*

44b. Glumes not as above. Racemes erect to spreading. Inflorescence an open or contracted panicle........................... *Eragrostis*

40b. Spikelets with 1 - fertile florets, with or without 1 or 2 male or barren florets below it.. .45

45a. Spikelets with 3-florets, lower 2-florets reduced to lemmas............................. *Phalaris*

45b. Spikelets with 1 or 2-florets........................ 46

46a. Spikelets with 1-floret............................... .47

47a. Lemma 3-5 nerved, awned. Glumes longer than lemmas. Panicle compact, interrupted........... *Polypogon*

47b. Lemma 1-3 nerved, not awned. Glumes shorter than lemmas. Panicle diffuse .... *Sporobolus*

4b. Spikelets with 2-florets................................ 48

48a. Inflorescence a contracted panicle..... *Isachne*

48b. Inflorescence an diffuse panicle. Awn setaceous .... *Arundinella*

ALLOTEROPSIS J.S. Presl ex C.B. Presl

Alloteropsis cimicina (L.) Stapf., In Prain, Fl. Trop. Africa 9: 487. 1919; Bor., Grass. Burm. Ceyl. Ind. Pak. 276. 1960; Raizada & Jain, in Ind. For. Rec. (n.s.) Bot. 5: 179. pl. 7. f. 1. 1965; Pandey, in Shetty & Singh Fl. Rajasthan 3: 970. 1993. *Milium cimicinum* L. Mant. Alt. 2: 184. 1771. *Panicum cimicinum* (L.) Retz. Obs. Bot. 3: 9. 1783. *Axonopus cimicinus* (L.) P. Beauv. Ess. Agrost. 120. 1812; Hook. *f.*, Fl. Brit. India 7: 64. 1896.

Tufted annual; culms upto 110 cm high, erect or ascending, nodes hairy. Leaves narrowly lanceolate, 2.5 - 15 x 5 - 20 mm, amplexicaul, tuberculate-ciliate on the margins. Inflorescence of 2 - 11 racemes, usually in 1 whorl. Racemes 7 - 25 cm long, the lowest third to half bare of spikelets. Spikelets narrowly elliptic-ovate, 3 - 5.5 mm long; lower glume acuminate, half to three-quarters as long as the spikelet, 3 - 5 nerved; upper glume subulate-acuminate, cartilaginous, smooth and shining, the margins ciliate with appressed or eventually spreading hairs 1 - 2 mm long; upper lemma with an awn 2 - 4.5 mm long, the palea pappillose with globular hairs. Caryopsis sub-orbicular to elliptic oblong.

Field Note: Common in open fields as well as grasslands.

Fl. & Fr.: July - October.

Specimen examined: Bhemali 178 (MLVGCB Herb).

Distribution: Throughout the tropical Old World.

APLUDA L.

Apluda mutica L., Sp. Pl. 82. 1753; Bor, Fl. Assam 5: 432. 1940; Bor., Grass. Burm. Ceyl. Ind. Pak. 93. 1960; Raizada *et al.* in Ind. For. Rec. (n.s.) Bot. 4: 203. 1961; Bor, in Towns., Guest & Al-Rawi, Fl. Iraq 9:508. 1968; Murty in Bull. Bot. Surv. India 14 (1-4): 150. 1975; Pandey, in Shetty & Singh Fl. Rajasthan 3: 972. 1993; Bhandari, Fl. Indian Desert 350. 1995. *Apluda geniculata* Roxb., Fl. Ind. 1: 3 27. 1820. *A. communis* Nees in Nova Acta Acad. Caesar. Leop. Carol. 19, Suppl. 1: 194. 1843. *A. aristita* L. Amoen. Acad. 4: 303. 1956. *A. varia* Hack., in DC. Monogr. Androp. 6: 197. 1889, include. subsp. *aristata* & *mutica*; Hook. *f.*, Fl. Brit. India 7: 150. 1896. *A. mutica* L. var. *aristata* (L.) Hack. ex Backer, Handb. Fl. Java 2: 54. 1928. (Fig. 177 & Flate 324)

Vern.: *Gundo.*

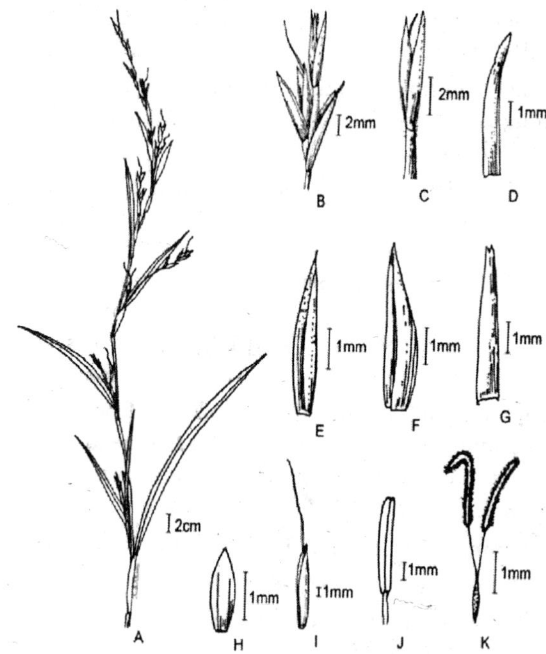

Fig. 177. *Apluda mutica* L.
A. Habit. B. Flowering twig, C. Developed pedicelled spikelet, D. Rudimentary pedicelled spikelet, E. Lower lemma, F. Upper glume, G. Lower glume, H. Upper palea, I. Upper lemma, J. Stamen, K. Pistil

Rambling perennial. Culms up to about 1.5 m long, rooting from the lower nodes. Leaves flat, 5-25 cm x 0.25 - 13 mm, linear - lanceolate, attenuate at the tip; ligule membrounous, rounded. Panicles of 2 to many, simple racemes terminating the culms, enclosed in a peduncled spathe. False panicle linear, interrupted, 3.40 cm long; spatheole narrowly ovate in side view, 3.5 - 10 mm long, acuminate. Sessile spikelet 2 - 6 mm long, hermaphrodite. Glumes equal, chartaceous; lower glume narrowly elliptic-lanceolate; upper lemma deeply bifid with an awn 4 - 12 mm long or entire to emarginate and awnless. Pedicelled spikelets broadly lanceolate, the larger 2 - 5 mm long; pedicels narrowly oblong, 24 mm long.

Field Note: Common in grasslands, boundaries of cultivated fields and waste lands throughout the area. It often constitutes a large part of the undergrowth in forests.

Fl. & Fr.: August - February.

Specimen examined: Rajasamand 1308 (MLVGCB Herb).

Distribution: India; Arabia and Socotra; throughout tropical Asia and in Australia and New Caledonia.

Uses: This is considered to be a fairly good fodder grass.

ARISTIDA L.

KEY TO THE SPECIES

1a. Lemma (or column) not articulated at its summit. Panicle ± contracted; awns usually less than 2 cm long; glumes never awned .... *A. adscensionis*

1b. Lemma or column articulated at the summit and breaking cleanly across when mature 2

2a. Articulation at the top of the column, just below the awns........................ *A. mutabilis*

2b. Articulation at the base of the column. Lower glume always longer than the upper; base of the callus pointed .... *A. funiculata*

Aristida adscensionis L., Sp. Pl. 82. 1753; Hook. *f.,* Fl. Brit. India 7: 224. 1896; Bor., Grass. Burm. Ceyl. Ind. Pak. 407. 1960; Mahesh. Illus. Fl. Delhi f. 271. 1966; Pandey, in Shetty & Singh Fl. Rajasthan 3: 973. 1993; Bhandari, Fl. Indian Desert 351. 1995. *A. adscensionis* L. var. *ehrenbergii* (Trio. & Rupr.) Henr. in Meded. Rijks-Herb. No. 54: 158. 1926. *A. adscensionis* L. var. *pumila* (Dcne.) Coss. & Dur., Fl. Alger. Glum. 84. 1855; Bor., *l.c.* 409. 1960. *A. caerulescens* Desf., Fl. Atlant. 1: 109, t.21/2. 1798; Boiss., Fl. Or. 5: 491. 1844; Bor in Towns., Guest & Al-Rawi, Fl. Iraq 9: 384. 1968; Bor in Rech. f., Fl. Iran. 70 : 367. 1970. *A. pumila* Dcne. in Ann. Sci. Nat., ser. 2, 4:85. 1835; Boiss., Fl. Or. 5: 491. 1884. *Aristida depressa* Retz., Obs. Bot. 4: 22. 1986. Duthie, Fodder Grasses 47. 1888; Bor., *l.c.* 409. 1960. (Fig. 178 & Plate 325)

var. ***adscensionis***

Vern.: *Lampriyo, Lappa, Lamba.*

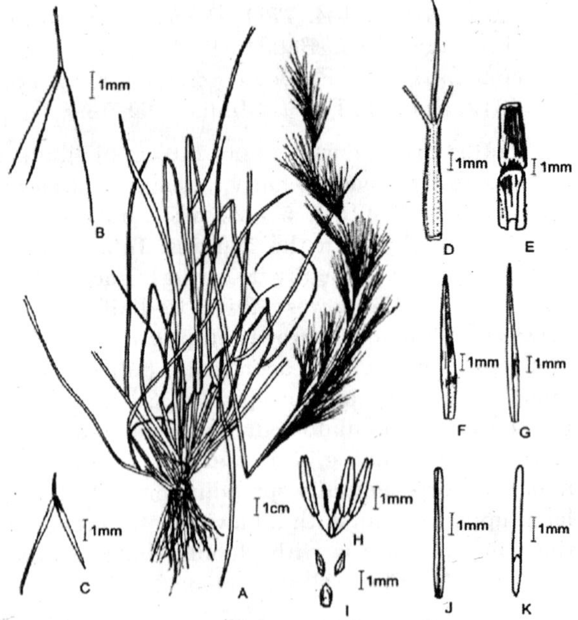

Fig. 178. *Aristida adscensionis* L. var. *adscenisonis* A. Habit. B. & C. Spikelets with flowers detached from glume, D. Lemma, E. Ligule, F. flower glume, G. Upper glume, H. Flower, I. Palea, J. & K. Grains.

Annual or short-lived perennial, forming erect or sprawling tufts upto 100 cm high. Leaves linear, up to 20 cm long and 3 mm wide, expanded or folded. Panicle up to 30 cm long, occasionally lax, usually ± contracted about the main branches, sometimes narrow and dense. Spikelets 6 - 8 mm, one - flowered, green or purple. Glumes unequal, linear-lanceolate to lanceolate, scaberulous on the keel, emarginate and mucronate or acute but never awned, the upper 5 - 10 mm long, the lower 1-3 mm shorter. Lemma 5 - 14 mm long, sometimes no longer than the glumes but usually exceeding the upper by 1 - 2 mm or more, laterally compressed, convolute, scabrid on the keel or sometimes generally scaberulous on the upper part, passing into the awn without constriction or articulation; callus narrowly oblong, 0.5 mm long, obtuse; central branch of the awn upto 3 cm long, the laterals subequal or occasionally conspicuously shorter. Caryopsis linear - oblong.

Field Note: Commonly found in sandy to rocky as well as wet habitats. Frequent occuring in pure formation.

Fl. & Fr.: August - December.

Specimen examined: Atun 336, Bhim 264 (MLVGCB Herb).

Distribution: Pantropical.

Notes: *Aristida adscensionis* is a pan-tropical weed and morphologically extremely variable. There are no discontinuities between var. *adscensionis* and either var. *pumila* or var. *ehrenbergii*. The distinctions between them are hardly worth maintaining. The same may be said of *Aristida depressa,* an extreme form completely intergrading with *Aristida adscensionis.* The distinction between annuals and perennials in the tropics and subtropics is not always very clear and often appears to be a facultative difference of little taxonomic consequence. This is particularly true in *Aristida adscensionis* where the separation of perennial plants as *Aristida caerulescens* cannot be justified. Indeed, perennial plants, otherwise indistinguishable from typical var. *ehrenbergii* which are seen in the Middle East.

Uses: Green as well as in dried form it is a good fodder.

Aristida funiculata Trin. & Rupr., Sp. Gram. Stip. 159. 1842 & in Mem. Acad. Sci. Petersb., Ser. 6, 7 (2) : 159. 1843; Boiss., Fl. Or. 5: 492. 1884; Duthie, Fodder Grasses 47. 1888; Hook. *f.,* Fl. Brit. Ind. 7: 226.1896; Blatter & McCann, Bombay Grasses 214. 1935; Bor, Grasses Burma Ceyl. Ind. Pak. 410. 1960; Bor, in Rech. f., Fl. Iran. 70: 365. 1970. Pandey, in Shetty & Singh Fl. Rajasthan 3: 974. 1993; Bhandari, Fl. Indian Desert 352. 1995. *A. funiculata* Trin. & Rupr. var. *mallica* (Edgew.) Henr. in Meded. Rijks-Herb. No. 54 A: 388. 1927; Bor., *l.c.* 410. 1960. *A. funiculata* Trin. & Rupr. var. *royleana* (Trin. & Rupr.) Hook. *f.,* l.c. 7: 227. 1896. *A. kotschyi* Hochst. ex Steud., Syn. Pl. Glum. 1: 142. 1854. *A. macrathera* A. Rich., Tent. Fl. Abyss. 2: 393. 1851. *A. mallica* Edgew., in J. Linn. Soc., Bot. 6:209. 1862; Duthie, *l.c.* 47. 1888. *A. royleana* Trin. & Rupr., *l.c.* 160; Duthie, *l.c.* 48. 1888; Bor, *l.c.* 412. 1960; Bor, in Rech.f., Fl. Iran. 70: 365. 1970.

var. ***funiculata***

Vern.: *Lampriyo.*

Tufted annual with wiry culms 15 - 50 cm high. Leaves flat or folded. 4 - 10 cm long and up to 2.5 mm wide. Panicle linear, up to 10 cm long, few-flowered. Spikelets upto 2.25 cm long, green, tinged with purple. Glumes linear, membranous, the lower 12 - 22 mm long, 1-nerved, awned; Upper 12 - 20 mm long, both finely attenuate. Lemma, cylindrical, about 4 mm long, scabrid above, articulated at the apex; callus about 2 mm long, conical, pungent; column 2 - 4.5 cm long, the base as wide as the summit of the lemma, straight for a short distance, thereafter strongly twisted; awn-branches subequal, 2.5 - 4.5 cm long. Lateral awns upto 5 cm long. Caryopsis cylindric, black.

Field Note: Commonly found in sandy to rocky habitats.

Fl. & Fr.: August - December.

Specimen examined: Amarvasi 1603, Nadadi 2140 (MLVGCB Herb).

Distribution: Pakistan (Sind, Baluchistan, Punjab & Kashmir); tropical East Africa, through Arabia to the Middle East and India.

Aristida mutabilis Trin. & Rupr., Sp. Gram. Stip. 150. 1842 & in Mem. Acad. Sci. Petersb., Ser. 6, 7(2): 150. 1843; Hook. *f.,* Fl. Brit. Ind. 7 : 226. 1896; Blatter & McCann, Bombay Grasses 212. 1935; Bor, Grasses Burma Ceyl. Ind. Pak. 411. 1960; Pandey, in Shetty & Singh Fl. Rajasthan 3: 974. 1993; Bhandari, Fl. Indian Desert 352. 1995. *A. articulata* Edgew. in J. Linn. Soc., Bot. 6: 209. 1862; Duthie, Grass. N.W. Ind. 26. 1883 et in Fodder Grass. N. Ind. 47. 1888.

Loosely tufted annual. Culms ascending, slender, rooting at lower nodes, 30 - 70 cm high.

Leaves, 6 - 15 cm long, linear, narrow or convolute. Panicle narrowly ovate, 5 - 20 cm long, loose or contracted, the spikelets in narrow spikes at the tips of the main branches. Spikelets upto 2.25 cm long, green with purplish ting. Glumes typically subequal, lanceolate, scaberulous on the keel and sometimes also on the flanks; lower glume 4 - 5 mm long; upper glume 6 - 7 mm long, acute or apiculate, less often with a short awn-point. Lemma terete, 4-5 mm long, scabrid upwards. Callus linear, 1-1.5 mm long, finely obtuse at the tip. Column 2-5 mm long, articulated just below the branching point of the awn. Awn-branches 10-30 mm long, slender, parallel or divergent.

Field Note: Commonly found in sandy to rocky habitats.

Fl. & Fr.: August - November.

Specimen examined: Jahajpur 1373, Rawatbhata (MLVGCB Herb).

Distribution: India and tropical East Africa.

ARUNDINELLA Reddi

KEY TO SPECIES

1a. Upper lemma not awned .... *A. leptochloa*

1b. Upper lemma with geniculate awned .... *A. pumila*

Arundinella leptochloa (Nees ex Steud.) Hook. *f.*, Fl. Brit. India 7: 76. 1896; Bor., Grass. Burm. Ceyl. Ind. Pak. 423. 1960; Pandey, in Shetty & Singh Fl. Rajasthan 3: 981. 1993. *Panicum leptochloa* Nees ex Steud. Syn. Pl. Glum. 1: 62. 1854. *Arundinella gigantea* Dalz. in Dalz. & Gibs. Bombay Fl. 293. 1861; Hook. *f.*, *l. c.* 7: 76. 1896. *A. lawsonii* Hook. *f.*, l.c. 7: 76. 1896.

Perennial, tufted. Sheaths glabrous. Culms upto 200 cm long, nodes glabrous. Leaves 25-50 x 2.0-3.5 cm, linear - lanceolate, acuminate, base rounded or cordate, coriaceous, margins scabrous; ligule an eciliate membrane. Infloresce-nce a panicle; bearing juvenile spikelets at emerg-ence. Peduncle glabrous. Panicle open; elliptic, or oblong. Panicle branches bearing congested spikelets. Spikelets in pairs. Fertile spikelets pedicelled; 2 in the cluster, comprising 1 basal sterile floret; 1 fertile floret; without rachilla extension. Spikelets elliptic; laterally compressed; 2.5 - 3.5 mm long; breaking up at maturity; disarticulating below each fertile floret. Glumes persistent; dissimilar; reaching apex of florets; thinner than fertile lemma. Lower glume ovate; 2 mm long, 1-keeled; 3 -veined, apex acute. Upper glume ovate; 2.5 - 3 mm long; 2.5 length of adjacent fertile lemma; membranous; 1-keeled; 5 -veined, apex acute. Florets basal sterile florets male, or barren; with palea; persisting on inflorescence. Lemma of lower sterile floret similar to upper glume; elliptic; 2.5 - 3.5 mm long; membranous; 1-keeled; 5 -veined; acute. Fertile lemma elliptic, or oblong, 1 mm long; chartaceous, light brown; without keel; 3 -veined. Lemma involute, keeled, apex entire; obtuse; muticous. Palea with auriculate. Lodicules 2; cuneate. Anthers 3. Stigmas 2. Caryopsis with adherent pericarp; 1 mm long.

Field Note: Common in moist habitats.

Fl. & Fr.: July - November.

Specimen examined: Mahua 1932 (MLVGCB Herb).

Distribution: Asia-tropical: India.

Arundinella pumila (Hochst. ex A. Rich.) Steud., Syn. Pl. Glum. 1: 114. 1854; Bor., Grass. Burm. Ceyl. Ind. Pak. 423. 1960; Pandey, in Shetty & Singh Fl. Rajasthan 3: 982. 1993. *Acrantherum pumilum* Hochst. ex A. Rich. Tent. Fl. Abyss 2; 414. t. 100. 1851. *Arundinella tenella* Nees ex Steud. *l.c.* 1: 115. 1854; Hook. *f.*, Fl. Brit. India 7: 71. 1896.

Annual. Culms erect, or geniculately ascending, 7 - 50 cm long; 1 - 3 -noded. Culm-nodes pubescent. Leaf-sheaths outer margin hairy. Ligule an eciliate membrane. Leaves upto 7.5 x 2.0 cm, linear-lanceolate, surface glabrous, or pilose; sparsely hairy. Inflorescence a panicle, open; obovate; diffuse. Spikelets solitary. Fertile spikelets pedicelled. comprising 1 basal sterile floret; 1 fertile floret; without rachilla extension. Spikelets ovate; laterally compressed; 1.5 - 2.5 mm long; breaking up at maturity; disarticulating below each fertile floret. Glumes persistent; dissimilar; exceeding apex of florets; thinner than fertile lemma; gaping. Lower glume ovate; 1 - 1.5 mm long; 0.5 - 0.75 length of upper glume;

membranous; 1-keeled; 3 -veined, apex acute, or acuminate. Upper glume ovate, membranous, surface glabrous, or pilose, apex acute or acuminate. Floret; basal sterile male, or barren; with palea; persisting on inflorescence. Lemma of lower sterile floret similar to upper glume; ovate; 1.3 length of fertile lemma; membranous; 1-keeled; 3 -veined; truncate, or obtuse. Fertile lemma ovate; 1 mm long; coriaceous; without keel; 1 -veined, asperulous, margins involute; interlocking with palea, keels, dentate; 2 -fid; awned; 1 -awned. Principal lemma awn from a sinus; geniculate; 2 - 3 mm long overall with twisted column. Column of lemma awn 1 - 1.5 mm long. Palea with auriculate flaps; 1 length of lemma; coriaceous; 2 -veined. Palea surface asperulous. Anthers 3. Caryopsis with adherent pericarp; 1 mm long.

Field Note: Common in moist, shady habitats.

Fl. & Fr.: August - October.

Specimen examined: Bigod 1979 (MLVGCB Herb).

Distribution: Africa: west tropical, west-central tropical, and northeast tropical. Asia-temperate: Arabia. Asia-tropical: India, Indo-China, and Malesia.

ARUNDO L.

Arundo donax L., Sp. Pl. 81. 1753; Hook. *f.*, Fl. Brit. India 7: 302. 1896; Bor., Grass. Burm. Ceyl. Ind. Pak. 413. f. 44. 1960; Pandey, in Shetty & Singh Fl. Rajasthan 3: 982. 1993. (Fig. 179 & Plate 326)

Vern.: *Bru.*

Erect, perennial, tufted, upto 2.5 m high. Culm fistular. Leaves 15 - 50 x 2 - 5 cm, distichous, linear - lanceolate, cordate or rounded at the base, margins scabrid. Panicles plumose, upto 70 cm long. Spikelets 3 - 4-flowered or more, terete, 1.0 - 1.5 cm long, light brown, 2 - flowered. Glumes subequal, lanceolate, siliceous apex with short awn or acute, glabrous; the lower glume 3 - 5-nerved, rarely 4-nerved, covered with a few silky hairs on the backside; upper glume 3-5-nerved. Lemma 7.5 - 12.5 mm long, 5-nerved, middle nerve prolonged into a 3 mm long awn, 2 lateral awn slightly cuspidate, basal part densely covered with long silky hairs which are nearly as long as the lemma. Palea 2-keeled, margins ciliate, apex truncate. Lodicules 2, truncate, nerved, stamens 2-3. Caryopsis with linear hilum.

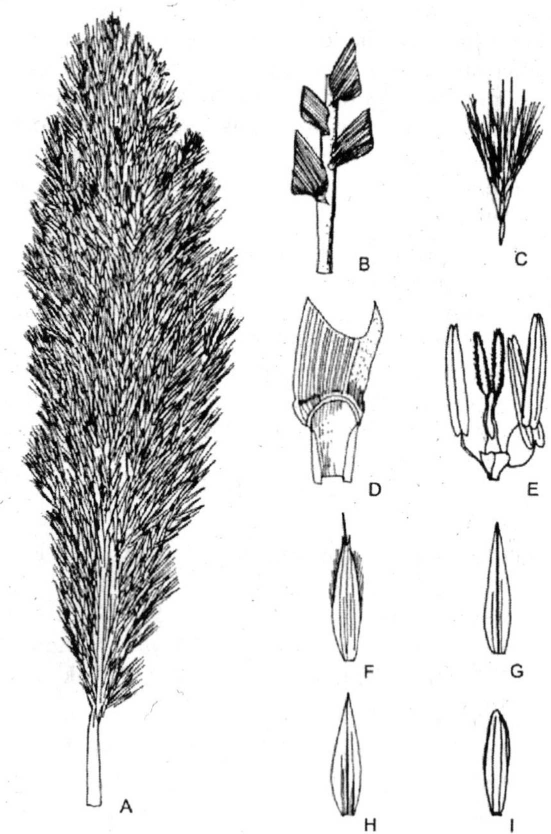

Fig. 179. *Arundo donax* L.
A. Inflorescence B. Part of stem showing leaf bases, C. Spikelet D. Ligule, E. Flower, F. Lemma abaxial, G. Lower glume, H. Upper glume I. Palea

Field Note: Rare, along rivers and canals in the area.

Fl. & Fr.: August - October.

Specimen examined: Ajmer 591 (MLVGCB Herb).

Distribution: Mediterranean region eastwards to Burma; North Africa; introduced into many parts of the World.

Notes: *A. donax* is a species of wet habitats but Bor (Grasses Burma Ceyl. Ind. Pak. 415) stated that it grows in dry places when once established. Considerable difficulty may be experienced in distinguishing immature plants of *Arundo, Neyraudia* Hook.f. and *Phragmites* Adanson and dissecting the spikelets will be of little use. *Phragmites* can be distinguished by the silky beard at the bases of the lowest panicle branches which is absent in the other two genera. The ligule of *Arundo* is membranous while that of *Phragmites* and *Neyraudia* is a fringe of hairs. Leaves of *Arundo* are very much broader than in the other genera and are conspicuously cordate or rounded at the base.

Uses: Cattle browse its young leaves but it is not of much account as a fodder grass. In Europe it is extensively cut to make mats, trays and baskets. The Romans use the stems for pens. It is sometimes used for making paper but is commercially of less value than *Phragmites australis* (Cav.) Trin. ex Steud.

AVENA L.

Avena sterilis L. var. ***culta*** Raizada in Ind. For. 80 : 36. 1854; Mahesh., Illus. Fl. Delhi f. 272. 1966; Pandey, in Shetty & Singh Fl. Rajasthan 3: 983. 1993. *A. sativa auct.* non L. 1753; Bor., Grass. Burm. Ceyl. Ind. Pak. 434. 1960.

Tufted erect, annual, 25 - 125 cm high. Clums slender to stout, glabrous, fistular. Leaves 20 - 40 x 1 - 2 cm, linear - lanceolate, acute, glabrous, rough; ligules 3.5 - 5.5 mm long. Panicle up to 22 cm long, and 8 cm wide, loose with scabrid branches, lax, terminal, hanging from one side, Spikelets 2. 25 - 2.75 cm long, 2 - 3-flowered; rachilla articulated above the glumes but not between the florets; glumes lanceolate, whitish on maturity, finely acute. Lemmas 1.5-2.1 cm long, densely bearded around the callus with hairs up to 5 mm long, densely hairy with long stiff hairs and finally dark brown in the lower two-thirds, scaberulous all over, narrowed upwards and acuminately 2-toothed at the apex. Awn 3 - 4 cm long, geniculate, the column dark brown to blackish, asperulous. Caryopsis silky.

Field Note: Common weed of *Wheat* and *Foeniculam* crops during winter.

Fl. & Fr.: January - March.

Specimen examined: Karoi 1929, Puskar 1671 (MLVGCB Herb).

Distribution: Mediterranean region and the Middle East eastwards to Northwest India.

Uses: Commonly used as fodder.

BAMBUSA Schreb. *nom. cons.*

Bambusa arundinacea (Retz.) Roxb. Corom. Pl. 1 : 56. f. 79. 1796; Gamble in Hook. *f.,* Fl. Brit. India 7: 395. 1896; Pandey, in Shetty & Singh Fl. Rajasthan 3: 984. 1993; *Bombos arundinacea* Retz., Obs. Bot. 5: 24. 1789. *Bambusa bambos* Voss in Vilm. Bluemeng. 1: 1189. 1896. *B. spinosa* Roxb, ex Buch.-Ham. in Trans. Linn. Soc. 13: 480. 1822.

Vern.: *Bans, Vando.*

Tall, arborescent, shrubby, thorny, upto 20 m tall, with 2 - 3 stout, recurved spines at nodes. Culms arising from a thick rhizome, usually caespitose, occasionally stoloniferous. Culms-sheaths variable, usually broad; imperfect blade broad, triangular. Leaves 5 - 20 cm long, base rounded, acuminate, margins thickly ciliate; ligule narrow. Panicles often occupying the entire stem. Spikelets 1.25 - 2.5 mm long, ovate - lanceolate, mucronate, many - nerved. Glume 5 - 8 mm long, ovate - lanceolate, mucronate, many - nerved. Upper lemma 1 -3, male or neuter. Lodicules 3. stamens 6; style 1; stigmas 1-3. Caryopsis 5 - 8 mm long, oblong, acute or obtuse, furrowed on one side, with persistent style base.

Field Note: Commonly cultivated for its valuable stems, sometimes occurs wild in forests.

Fl. & Fr.: Very rare; February - May.

Specimen examined: Bijoliya 1997, Bojunda 1861 (MLVGCB Herb).

Distribution: the tropics and sub-tropics of Asia, Africa and America.

Uses: Culms are extensively used for building purposes, agricultural equipments etc.

BRACHIARIA (Trin.) Griseb.

KEY TO THE SPECIES

1a. Rachis more or less flat, ribbon-like, narrowly winged, annuals. Spikelets upto 3 mm long. Upper lemma rugulose................. *B. distachya*

1b. Rachis solid, triquetrous, or crescentic, sometimes with very narrow wings, filiform..........2

2a. Spikelets 1.5 - 2.0 mm long, without a stipe; lower glume up to a quarter the length of the spikelet, rarely a little longer, truncate .... *B. reptans*

2b. Spikelets 3 mm long. Upper glume and lower lemma membranous, glabrous or pubescent. Lower glume acute or subacute. .... *B. ramosa*

Brachiaria distachya (L.) Stapf., in Prain, Fl. Trop. Africa 9: 565. 1919; Bor., Grass. Burm. Ceyl. Ind. Pak. 281. 1960; Raizada & Jain, in Ind. For. Rec. (n.s.) Bot. 5: 164. pl. 2. f. 3. 1965; Pandey, in Shetty & Singh Fl. Rajasthan 3: 987. 1993. *Digitaria distachya* (L.) Pers., Syn. Pl. 1: 85. 1805. *Panicum distachyon* L. Mant. Pl. 2: 183. 1771; Duthie, Fodder Grasses 6. 1888; Hook. *f.*, Fl. Brit. India 7: 37. 1896, *'distachyum'*. *Urochloa distachya* (L.) Nguyen in Nov. Sist. Vyssh. Rast. 1966: 13. 1966.

Creeping annual; culms 5-20 cm high, ascending, rooting at the lower nodes. Leaves 2 - 7 x 0.25 - 0.5 cm, broadly linear to narrowly lanceolate, acuminate, amplexicaul at base, pubescent; ligule a hairy ring. Inflorescence of 2 - 3 racemes on an axis 0.5 - 2 cm long; racemes 1 - 3 cm long, bearing the spikelets singly on a narrowly winged rachis. Spikelets 2.4 - 3.0 mm long, narrowly elliptic, glabrous, acute. Lower glume ca 1.5 mm long, 5-7 nerved, clasping. Upper glume separated from the lower by a short internode. Upper lemma ca 2.5 mm long, rugulose, subcoriaceous, 3-nerved, bluntly acute.

Field Note: Common in moist and shady places.

Fl. & Fr.: August - November .

Specimen examined: Bhilwara 1934 (MLVGCB Herb).

Distribution: India to Sri Lanka, Burma and Thailand; introduced in Africa.

Brachiaria ramosa (L.) Stapf., in Prain, Fl. Trop. Africa 9: 542. 1919; Bor., Grass. Burm. Ceyl. Ind. Pak. 284. 1960; Raizada & Jain, in Ind. For. Rec. (n.s.) Bot. 5: 164. pl. 2. f. 4. 1965; Pandey, in Shetty & Singh Fl. Rajasthan 3: 990. 1993; Bhandari, Fl. Indian Desert 355. 1995. *Panicum petiveri* Trin., Gram. Pan. 144. 1826. *Panicum ramosum* L. Mant. Pl. 1: 29. 1767; Hook. *f.*, Fl. Brit. India 7: 36. 1896, *pro parte*. *Urochloa ramosa* (L.) Nguyen, in Nov. Sist. Vyssh. Rast. 1966:13. 1966.

var.: ***ramosa***

Loosely tufted, annual. Culms upto 70 cm high, often rooting at the base. Leaves 2.5 - 15 cm long, 4 - 15 mm wide, broadly linear, acuminate, base rounded, glabrescent; ligule a line of hairs. Inflorescence of 3 racemes, borne on an axis 3 - 10 cm long; racemes 1 - 8 cm long, simple or the longest with branchlets at the base, bearing mostly paired loosely contiguous spikelets, appressed to the triquetrous rachis; pedicels shorter than the spikelets, 1 - 2 mm long. Spikelets ovoid, apiculate, 2.5 - 3.5 mm long, glabrous or pubescent, acute to cuspidate, with or without a stipe up to 0.5 mm long. Lower glume upto half the length of the spikelet; upper glume and lower lemma membranous, or rarely the latter coriaceous. Upper lemma transversely rugose, subacute to acute. Caryopsis 1.5 - 2.0 mm long, flattened, broadly elliptic.

Field Note: Common in mixed habitats, throughout the area.

Fl. & Fr.: July - November

Specimen examined: Bhilwara 1994 (MLVGCB Herb).

Distribution: Senegal to Yemen and southwards to Malawi, Rhodesia and South Africa; tropical Asia.

Brachiaria reptans (L.) Gardner & Hubbard in Hook Ic. Pl. 34. t. 3363. p. 3. 1938; Bor., Grass. Burm. Ceyl. Ind. Pak. 285. 1960; Raizada & Jain, in Ind. For. Rec. (n.s.) Bot. 5: 162. pl. 2. f. 1. 1965; Pandey, in Shetty & Singh Fl. Rajasthan 3: 991. 1993; Bhandari, Fl. Indian Desert 355. 1995. *B. prostrata* (Lam.) Griseb. in Abhandl. Ges. Wiss. Gott. 7 : 263. 1857. *Panicum reptans* L. Syst. Nat. ed. 10. 870. 1759. *P. prostratum* Lam. Tab. Encycl. 1 : 171. 1791; Duthie, Fodder Grasses 11. 1888; Hook. *f.*, Fl. Brit. India 7: 33. 1896. *Urochloa reptans* (L.) Stapf., in Prain, Fl. Trop. Africa 9: 601. 1919. Blatter & McCann, Bombay Grasses 144. 1935. (Plate 327)

Annual, tufted, erect or creeping, usually rooting at the nodes, upto 60 cm high. Leaves 2-7 cm long, 3 - 15 mm wide, linear-lanceolate, acuminate, subamplexicaul, glabrescent; ligule a densely ciliate rim. Inflorescence of 3 - 15 racemes on an axis 1 - 8 cm long; racemes 4 cm long, bearing paired spikelets crowded on a triquetrous rachis with hirsute pedicels. Spikelets upto 2 mm long, oblong or broadly elliptic, subacute, glabrscent. Lower glume reduced to a hyaline, minute scale; upper one faintly 5-nerved. Upper lemma crustaceous, shining, rugose, subacute, mucronulate. Caryopsis ca 1.25 mm long, oblong - elliptic, flattened.

Field Note: Common in mixed habitats throughout the area.

Fl. & Fr.: July - October.

Specimen examined: Asind 2009 (MLVGCB Herb).

Distribution: Tropical Asia; introduced throughout the tropics.

Uses: A good fodder grass whose grains have been used as food during of famine.

CENCHRUS L.

KEY TO THE SPECIES

1a. Annuals. Bristles of the involucel retrorsely scabrid, tenaciously prickly.............. *C. biflorus*

1b. Perennials. Bristles of the involucel antrorsely scabrid, not prickly 2

2a. Bristles united only at the base to form a shallow disc. Base of the involucre small, elliptic. Inner bristles not very stouter at the base.. *C. ciliaris*

2b. Bristles connate above the base into a 1 - 4 mm long cup. Rachis scabrid. None of the inner bristles longer and stouter than the rest, all similar, acute.............. *C. setigerus*

Cenchrus biflorus Roxb., Fl. Ind. 1: 238. 1820; Hook. *f.*, Fl. Brit. India 7: 89. 1896; Bor., Grass. Burm. Ceyl. Ind. Pak. 287. 1960; Raizada & Jain, in Ind. For. Rec. (n.s.) Bot. 5: 215. pl. 10. f. 4. 1965; Pandey, in Shetty & Singh Fl. Rajasthan 3: 993. 1993; Bhandari, Fl. Indian Desert 356. 1995. *C. barbatus* Schumach., Beskr. Guin. Pl. 43. 1827. *C. catharticus* Delile, Cat. Hort. Monsp. 1838: 4. 1839; Hook. *f., l. c.* 7: 90. 1896; Blatt. & McC., Bombay Grasses 186. pl. 119. 1935.

Vern.: *Bhurat.*

Annual, erect or geniculately ascending. Culms 5 - 90 cm high, ascending. Leaves 2 - 25 cm x 2 - 8 mm, acuminate, scarbid; ligule a ciliate rim. Inflorescence a cylindric, spiciform, dense panicle, 2 - 15 cm long, tinged with purple. Involucres ovoid, 4-11 mm long; inner spines flattened, connate at the base to form an ovoid or diamond-shaped disc 2 - 4 mm across, bearing 1 - 3 shallow grooves on the outer face, ciliate below, retrorsely barbellate and mostly pungent at the tip; outer spines numerous, acicular, shorter than the inner, commonly divergent, rarely suppressed. Spikelets 1 - 3 in each involucre, 3.5 - 6 mm. long, sessile. Glumes 4, obtuse or acute, hyaline, the upper 4 mm ovate or broadly ovate, 1 - 5 - nerved; lower floral glume ovate, acuminate, 3 - 5-nerved; upper floral glume as long as the lower one, 5 - nerved. Caryopsis dorsally compressed, elliptic or elliptic-oblong, rugulose.

Field Note: Common in sandy habitats

Fl. & Fr.: August - December

Specimen examined: Ajmer 1631, Badnor 1937 (MLVGCB Herb).

Distribution: Tropical Africa, extending through Arabia to India.

Uses: Used as fodder.

Cenchrus ciliaris L. Mant. Alt. 302. 1771; Bor., Grass. Burm. Ceyl. Ind. Pak. 287. f. 33. 1960; Raizada & Jain, in Ind. For. Rec. (n.s.) Bot. 5: 213. 1965; Pandey, in Shetty & Singh Fl. Rajasthan 3: 994. 1993; Bhandari, Fl. Indian Desert 355. 1995. *C. longifolius* Hochst. ex Steud., Syn. Pl. Glum. 1: 109. 1854. *Pennisetum cenchroides* Rich., in Pers. Syn. 1: 72. 1805; Hook. *f.*, Fl. Brit. India 7: 88. 1896. *P. ciliare* (L.) Link, Hort. Berol. 1: 213. 1827.

Vern.: *Dhaman.*

Perennial, forming clumps, erect or ascending. Culms 5 - 120 cm high, compressed, rigid. Leaves 2 - 25 cm x 2 - 7 mm wide, linear, tapering to a fine point, glabrous or loosely hairy; ligule a densely ciliate rim. Panicle 2 - 12 cm, purplish, cylindrical to ovoid; Rachis angular, puberulous. Involucres subsessile, cup-shaped, 3 - 7 mm; inner spines short, flattened, connate for 1/4 - 1/2 their length to form a cup, glabrous or obscurely puberulous and grooved on the face, occasionally sparsely ciliate on the margins, green to dark purple, antrorsely scaberulous, narrowly triangular at the tip; outer spines few, short, often suppressed, rarely almost as long as the inner. Spikelets 1-3 in each involucel, upto 5 mm long, oblong lanceolate. Glume ovate to broadly ovate-oblong, acute or acuminate; lower involucral glume ovate-lanceolate 1-nerved, acute; upper involucral glume 3 mm long, ovate, acuminate, 1-3-nerved, hyaline; lower floral glume male or barren, rarely hermaphrodite. Caryopsis 2 mm long, dorsally compressed.

Field Note: Common in dry sandy habitats.

Fl. & Fr.: August - March.

Specimen examined: Kishangarh 1814 (MLVGCB Herb).

Distribution: Throughout Africa, extending through Arabia and the Middle East to India; widely introduced else-where in the Old World.

Uses: Used as fodder.

Cenchrus setigerus Vahl, Enum. Pl. 2: 395. 1805; Bor., Grass. Burm. Ceyl. Ind. Pak. 290. 1960; Raizada & Jain, in Ind. For. Rec. (n.s.) Bot. 5: 215. 1965; Pandey, in Shetty & Singh Fl. Rajasthan 3: 996. 1993; Bhandari, Fl. Indian Desert 358. 1995. *C. biflorus auct.* Non Roxb. 1820; Hook. *f.,* Fl. Brit. India 7: 89. 1896; Blatter & McCann, Bombay Grasses 185. 1935, *non* Roxb. 1820. *C. bulbifer* Hochst. ex Boiss., Fl. Or. 5: 448. 1884. in syn. *C. montanus* Nees ex Royle, Ill. Bot. Him. 416. 1840. *nom. nud.*; Steud. Syn. Pl. Glum. 1: 111. 1854; Duthie, Fodder Grasses 16. 1888. *Pennisetum vahlii* Kunth, Rev, gram. 1 : 49. 1829.

Vern.: *Dhaman*.

Perennial, forming clumps from a bulbous base. culms 5-80 cm. high, geniculately ascending. Leaves 5-22 cm long, 2-7 mm wide, linear - lanceolate, acuminate; ligule a shortly ciliate rim. Panicle 2-12 cm long, greenish with purple tinge, solitary, cylindric; rachis angular. Involucres cup-shaped, sub-sessile, 3-7 mm. long; inner spines short, flattened, connate for 1/4 - 1/2 their length to form a cup, glabrous or obscurely puberulous and grooved on the face, occasionally sparsely ciliate on the margins, green to dark purple, antrorsely scaberulous, narrowly triangular at the tip; outer spines few, short, often suppressed, rarely almost as long as the inner. Spikelets 1-3 in each involucel, upto 4 mm long, ovate, acute or obtuse. Glume ovate, acute hyaline, 1-nerved, puberulous; lower floral glume male or barren, as long as upper glume; palea 3 mm long, oblong, obtuse. Caryopsis 2 mm long, dorsally compressed.

Field Note: Common in sandy habitats, often gregarious in tufts.

Fl. & Fr.: August - December

Specimen examined: Raipur 79 (MLVGCB Herb).

Distribution: Sudan and Ethiopia, through Arabia to India; introduced to several other tropical countries.

Uses: Used as fodder.

Notes: Bhandari 1995, noted this species as annual in the Indian Desert, but in south eastern Rajasthan this has been recorded as a perennial species by Pandey (1993).

CHLORIS Sw.

Chloris virgata Sw. Fl. Ind. Occ. 1: 203. 1797; Hook. *f.* Fl. Brit. Ind. 7: 291. 1896; Blatt. & Mc Cann, Bombay Grasses 255. pl. 170. 1935; Bor, Fl. Assam 5: 123. 1940; Bor, Grasses Burma Ceyl. Ind. Pak. 468. 1960; Pandey, in Shetty & Singh Fl. Rajasthan 3: 1001. 1993; Bhandari, Fl. Indian Desert 360. 1995. *C. alba* Presl Rel. Haenk. 1: 289. 1830. *C. compressa* DC. Cat. Hort. Monsp. 94. 1813. *C. decora* Nees ex Steud., Syn. Pl. Glum. 1: 205. 1854. *C. elegans* H.B.K., Nov. Gen Sp. 1: 166, t.49. 1816. *C. meccana* Hochst. & Steud. ex Schlechtend. in Ind. Sem. Hort. Hal. 1843. and in Linnaea 18: 158. 1844; Boiss. Fl. Or. 5: 554. 1884. *C. penicellata* Willd. ex Steud., Nom. Bot., ed. 2, 1: 353. 1840. *non* Pers. 1805.

C. pubescens Lag., Var. Cienc. 4: 143. 1805. *C. virgata* Sw. var. *elegans* (H.B.K.) Stapf in Thiselt: Dyer, Fl. Cap. 7: 642. 1900. *Rhabdochloa virgata* (Sw.) P. Beauv. Ess. Agrost. 84, 158. 1812. (Fig. 180 & Plate 328)

Fig. 180. *Chloris virgata* Sw.
A. Habot. B. Ligule, C. Palea, D. Spikelet, E. Flower, F. Lower glume, G. Upper glume, H. Embryo & lemma, I. Fertile lemma, J. Grains

Tufted, erect, annual. Culms up to 60 cm high, occasionally rooting at the lower nodes, glabrous below the inflorescence. Leaves 5 - 25 cm long x 2 - 6 mm wide, flat, linear, acute, glabrous tapering at the apex; basal sheaths strongly keeled and often flabellate; ligule a narrow, membranous ridge. Inflorescence of 4 - 12 digitate, spreading, feathery spikes 2 - 10 cm long. Spikelets (2-)3-flowered, 2-awned; lower glume 1.5 - 2.5 mm long; upper glume 2.5 - 4.5 mm long including the short awn-point if present; lowest lemma obliquely obovate in sideview, 254 mm long, pallid or dark, ciliate along the margins, keel and flanks, with a crown of hairs 1.5 - 4 mm long, the awn 5 - 15 mm long; callus rounded, ciliate; 2nd lemma slightly projecting from the side of the lowest lemma, oblong, 2 - 2.5 mm long, glabrous, with an awn 5 - 12 mm long; 3rd lemma an awnless clavate scale 0.5 - 1.2 mm long. Caryopsis fusiform.

Field Note: Feather finger-grass is common in mixed habitats. It is one of the characteristic grasses of the saline or usar tracts.

Fl. & Fr.: July - November

Specimen examined: Bijoliya 78, Beawar 1522 (MLVGCB Herb).

Distribution: Widely distributed throughout the tropics of both hemispheres.

Uses: Known as a good fodder grass.

CHRYSOPOGON Trin. *nom. cons.*

Chrysopogon serrulatus Trin. in Mem. Acad. Sci. Petersb. ser. 6. 2 : 318. 1832; Bor, Grasses Burma Ceyl. Ind. Pak. 118. 1960; Bor in Rech. f., Fl. Iran. 70 : 533. 1970; Pandey, in Shetty & Singh Fl. Rajasthan 3: 1004. 1993. *Andropogon monticola* Schult,. var. *trinii* (Steud.) Hook. *f.* Fl. Brit. Ind. 7: 193. 1896. *Andropogon trinii* Steud. Syn. Pl. Glum. 1: 395. 1854. *Chrysopogon ciliolatus* (Nees ex Steud.) Boiss. Fl. Or. 5: 458. 1884. *Chrysopogon fulvus* (Spreng.) var. *serrulatus* (Trin.) R. R. Stewart in Brittonia. 5: 466. 1945.

Vern.: *Dhaulu, Sabba.*

Perennial. Culms 30 - 100 cm high, erect, robust. Leaves 25.0 - 30.0 cm x 5 mm, but sometimes basal and much shorter, bluntly acute to acuminate, glabrous or sometimes with a few tubercle-based hairs. Panicle ovate, 3 - 12 cm long, with delicate capillary branches fulvously bearded at the tip. Sessile spikelet narrowly oblong; lower glume 4 - 7 mm long, laterally compressed to a rounded keel, hispidulous at the tip; upper glume glabrous or with a few white hairs on the keel, bearing an awn 6 - 10 mm long; upper lemma minutely bidentate, with a glabrous or pubescent awn 2 - 3 cm long. Pedicelled spikelets 4.5-8 mm long, lower glume bearing an awn 3.5-6.5 mm long. Pedicels fulvously villous, a quarter to a third the length of the sessile spikelet.

Field Note: Common in aquatic habitats, along bank of rivers, prefer running water.

Fl. & Fr.: August - December

Specimen examined: Triveni 509, Mandalgarh 1031 (MLVGCB Herb).

Distribution: Afghanistan to northern India, Nepal, Pakistan and Burma; Madagascar and tropical Africa.

Uses: Used as a fodder.

Notes: *Chrysopogon serrulatus* Trin. is one of a cluster of closely related species which includes the *Chrysopogon aucheri* (Bios.) Stapf. and two species from southern India, *Chrysopogon fulvus* (Spreng.) Chiov. and *Chrysopogon polyphyllus* (Hack.) Blatter & McCann. *Chrysopogon aucheri* is distinguished by its puberulous leaves while in *Chrysopogon polyphyllus* (Hack. ex Hook.f.) Blatt. & McC. the leaves are distinctly distichous. The distinction between *Chrysopogon fulvus* (*Chrysopogon montanus* Trin.) and *Chrysopogon serrulatus* is based on more or less arbitrary characters which however, seem to be correlated with different geographical territories. The spikelets of *Chrysopogon fulvus* are smaller than those of *Chrysopogon serrulatus* and the upper glume of the sessile spikelet has long golden-brown hairs along the keel. In both these respects there is continual intergradation and but for the fact that a little is known of the cytogenetics and morphological variations within the group, as a whole it might have been better to regard them as subspecies.

COIX L.

Coix lacryma-jobi L., Sp. Pl. 972. 1753; Hook. *f.*, Fl. Brit. India 7: 100. 1896; Bor., Grass. Burm. Ceyl. Ind. Pak. 264. 1960; Raizada *et al.* in Ind. For. Rec. (n.s.) Bot. 4 : 176. pl. 1. f. 2. 1965; Pandey, in Shetty & Singh Fl. Rajasthan 3: 1006. 1993. *C. lachryma* L. Syst. Nat. ed. 10. 1261. 1759. (Plate 329)

Vern.: *Motingo.*

Annual, robust, spongy, glabrous, upto 2.5 m tall. Culms rooting at lower nodes; branching erect or ascending, Leaves 7.5 - 40.5 x 2 -5 cm, linear - lanceolate, cordate at the base, acuminate at the apex; ligule glabrous membrane. Inflorescence a false raceme, consisting of solitary, basal female spikelet. Spikelets terminal, and in the upper axils, unisexual, staminate spikelets two-flowered, in twos or threes on the continuous rachis; pistillate spikelets three together, one fertile, and two sterile. Glumes of the fertile spikelet several-nerved, all enclosed finally in a bony beadlike involucre. Caryopsis white to bluish white, globular ovoid, 6 - 12 mm long.

Field Note: Common in aquatic habitats, particularly at the banks of river, it prefers shallow water.

Fl. & Fr.: August - December

Specimen examined: Triveni 509, Mandalgarh 1031 (MLVGCB Herb).

Distribution: Native perhaps to Southeast Asia. Pantropical

Uses: Used as a fodder.

CYMBOPOGON Spreng.

Cymbopogon martinii (Roxb.) Watson, in Atkins., Gaz. N.W. Prov. India. 10: 392. 1882; Stapf, in Kew Bull. 1906: 335. 1906; Burkill, in J. Proc. Asiat. Soc. Beng. 5: 89-93. 1909; Blatter & McCann, Bombay Grasses 104. 1935; Bor, in J. Bombay nat. Hist. Soc. 51: 908. 1953; Morton, in Taxon 8: 13. 1959; Bor, Grasses Burma Ceyl. Ind. Pak. 129. 1960; Raizada *et al.,* in Ind. For. Rec. (n.s.) Bot. 4: 244. pl. 11. f. 1. 1961; Pandey, in Shetty & Singh Fl. Rajasthan 3: 1010. 1993; Bhandari, Fl. Indian Desert 363. 1995. *Andropogon martinii* Roxb., Fl. Ind. 1: 280. 1820. *A. pachnodes* Trin., in Mem. Acad. Sci. Petersb., ser. 6, 2(3): 284. 1832. *A. schoenanthus* L. var. *genuinus* Hack., in DC. Mongor. Phan. 6: 609. 1889. *A. schoenanthus* L. var. *martinii* (Roxb.) Hook. *f.* Fl. Brit. Ind. 7: 204. 1896. *Cymbopogon martinianus* Schult. Mant. 2: 459. 1824, based on *Andropogon martinii* Roxb., *C. motia* Gupta in Proc. Indian Acad. Sci., Sect. B, 71: 92. 1970. *C. pachnodes* (Trin.) Wats. in Atkins., Gaz. NW. Ind. 10: 392. 1882.

Vern.: *Rhauns.*

Tufted perennial arising from a short, stout, woody rootstock. Culms up to 3 m high, the lower nodes often swollen. Leaves, linear-lanceolate or lanceolate, up to 50 cm long, 10-30 mm wide, glaucous or pruinose below, usually dark green above, cordate at the base and often amplexicaul, tapering to a filiform tip. False panicle linear-oblong, up to 30 cm long, erect; spatheoles narrowly elliptic to narrowly lanceolate, 2 - 4 cm long, green, becoming orange or reddish at maturity. Racemes 15 - 20 mm long, the lowermost pedicles swollen and barrel-shaped; internodes and pedicels densely ciliate along the margins, sparsely pilose on the back. Sessile spikelet elliptic-oblong or oblong, 4 - 4.5 mm long. Lower glume flat on the back in the upper half and with a deep V-shaped groove in the lower, the keels winged above. Upper lemma deeply bifid, with an awn 12 - 18 mm long.

Field Note: Common in moist habitats.

Fl. & Fr.: September-November.

Specimen examined: Luhari kalan 2141 (MLVGCB Herb).

Distribution: Pakistan, India; cultivated in the tropics.

CYNODON L. C. Rich. *nom. cons.*

Cynodon dactylon (L.) Pers., Syn. Pl. l: 85. 1805; Hook. *f.*, Fl. Brit. India 7: 288. 1896; Blatt. & McC. Bombay Grasses 250. pl. 166. 1935; Bor., Grass. Burm. Ceyl. Ind. Pak. 469. f. 52. 1960; Pandey, in Shetty & Singh Fl. Rajasthan 3: 1011. 1993; Bhandari, Fl. Indian Desert 364. 1995. *Panicum dactylon* L., Sp. Pl. ed. 1: 58. 1753; Bor *l.c.* 496. 1960.

Perennial. Culms decumbent, finally erect, slender. Leaves green or glaucous, 2 - 10 x 1 - 3 mm, norrowly linear or lanceolate, finally acute, more or less glaucous, soft, smooth, usually conspicuously distichous in the barren shoots and at the base of clums; sheaths tight, glabrous or hairy, bearded at the mouth; ligule a fine, ciliated rim. Spikes 2 - 6, upto 6.5 cm long. Spikelets 2.0 - 2.5 mm long, crowded, secund. Glumes more or less persistent, equal or slightly unequal, narrow, pointed, keeled, 1-nerved, 1.5 mm long, membranous. Lemma as long as the spikelet, boat-shaped, blunt, keeled, 3-nerved, firm, densely and minutely hairy on the keel. Palea as long as the lemma, 2-keeled. Anthers 3. Caryopsis ca 1 mm, oblong, tightly enclosed between the hardened lemma and palea.

Field Note: Common throughout the area.

Fl. & Fr.: Throughout the year.

Specimen examined: Bhilwara 902, Jawaja 1164 (MLVGCB Herb).

Distribution: Cosmopolitan.

Uses: Widely used as a lawn grass.

DACTYLOCTENIUM Willd.

KEY TO THE SPECIES

1a. Annual. Spikes 1.2 - 6.5 cm long. Anthers 0.25 - 0.8 mm long..........*D. aegyptium*

1b. Perennial. Spikes 0.8 - 2 cm long. Anthers 1.1 - 2 mm long...............*D. scindicum*

Dactyloctenium aegyptium (L.) Willd. Enum. Hort. Berol. 1029. 1809; Boiss., H. Or. 5: 556. 1884; Blatter & McCann, Bombay Grasses 262. pl. 176. 1935; Bor, Grasses Burma Ceyl. Ind. Pak. 489. f. 54. 1960; Pandey, in Shetty & Singh Fl. Rajasthan 3: 1012. 1993; Bhandari, Fl. Indian Desert 364. 1995. *Chloris mucronata* Michx., Fl. Bor.-Amer. 1: 59. 1803. *Cynosurus aegyptius* L., Sp. Pl. 1: 72. 1753. *Dactyloctenium mucronatum* (Michx.) Willd., Enum. Hort. Berol. 1029. 1809. *Eleusine aegyptia* (L.) Desf., Fl. Atlant. 1: 85. 1798. Duthie, Fodder Grasses 56. 1888 (as *E. aegyptiaca* Pers.); Hook. *f.*, Fl. Brit. Ind. 7: 295. 1896. (Fig. 181)

Vern.: *Madhana*.

Annual. Culms up to 70 cm high, usually geniculately ascending and rooting at the lower nodes, frequently shortly stoloniferous and mat-forming, less often erect. Leaves 3 - 25 cm long, 2.5 - 7.5 mm wide, flat, linear, acute, margins and nerves covered with bulbous - based hairs; ligule a ring of white hairs. Inflorescence composed of (1)3 - 9 linear to narrowly oblong spikes 1.2 - 6.5 cm long. Spikelets 3 - 4-flowered, broadly ovate, 3.5 - 4.5 mm long; glumes subequal, 1.5 - 2.2 mm long, the lower lanceolate in profile with a thick scabrid keel, the upper elliptic to narrowly obovate in profile, the smooth keel extended into a stout divergent scabrid awn half to twice as long as the glume. Lemmas narrowly ovate to ovate in profile, 2.6 - 4 mm long, keel gibbous, concave and scabrid above the middle and often extended into a stout cusp or mucro up to 1 mm long; palea-keels winged or wingless. Anthers 0.25 -0.8 mm long. Caryopsis about 1 mm long, broadly obovate to obtriangular in profile, transversely rugose.

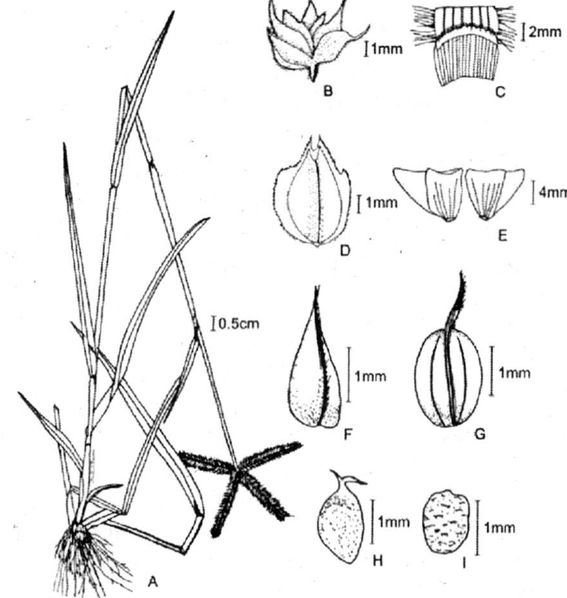

Fig. 181. *Dactyloctenium aegyptium* (L.) Willd. A. Habit. B. Spikelet, C. Ligule, D. Palea, E. Lodicules, F. Lower glume, G. Upper glume, H. Fruit, I. Nut.

Field Note: Common in waste lands and in moist habitats.

Fl. & Fr.: July-October.

Specimen examined: Badnor 1028 (MLVGCB Herb).

Distribution: Widely distributed in tropical and warm temperate regions of the Old World; introduced in America.

Notes: An exceedingly variable pantropical annual weed, typically with sprawling geniculately ascending culms and linear spikes radiating in a star-like manner from the culm-tip, but also including plants with compact oblong spikes, occasionally very similar to those of *Dactyloctenium aristatum* Link, from which it is most easily distinguished by its rugose grains.

Uses: it is a very nutritious fodder grass.

Dactyloctenium scindicum Boiss. Diagn. Pl. Orient. ser. 2. 4: 131. 1859; Boiss., Fl. Or. 5: 557. 1884; Blatter & McCann, Bombay Grasses 264. 1935; Bor, Grasses Burma Ceyl. Ind. Pak. 489. 1960; Pandey, in Shetty & Singh Fl. Rajasthan 3: 1013. 1993; Bhandari, Fl. Indian Desert 365. 1995. *D. glaucophyllum* Courb., in Annls. Sci. Nat., set. 4. 18: 133. 1862. *Eleusine glaucophylla* (Courb.) Benth., in J. Linn. Soc., Bot. 19: 107. 1881. *E. aristata* Ehrenb. ex Bioss. Fl. Orient. 5: 557. 1884; Hook. *f.,* Fl. Brit. India 7: 296. 1896. *E. scindica* (Boiss.) Duthie, Illus. Indig. Fodd. Grass. pl. 36. 1886 & Fodder Grasses. N. India 58. 1888. (Plate 330)

Stoloniferous perennial forming extensive spreading mats. Culms slender with swollen bases, 7 - 45 cm high, erect. Leaves 1 - 11 cm long, 1.5 - 3 mm wide, flat or loosely folded, tough and rather glaucous, scattered papillose-hispid especially along the margins. Inflorescence composed of 34(5) slightly falcate oblong spikes, 0.8 - 2 cm long forming a compact head, the spikes readily disarticulating at maturity from the top of the culm. Spikelets 3 - 9-flowered, broadly lanceolate to ovate, 4-8 mm long. Glumes broadly elliptic in profile, the lower 1.7-2.5 mm long, keel narrowly winged, the upper 1.5 - 2.3 mm long, prominent scabrid keel extended into an awn half to as long as the glume. Lemmas lanceolate in profile, 3 - 3.8 mm long, obtuse to acute with the scabrid keel extended into a mucro up to 0.8 mm long. Palea-keels finely scabrid, unwinged. Anthers 1.1-2 mm long. Caryopsis 0.7- 1 mm long, ovoid, transversely rugose, brownish.

Field Note: Common in sandy and gravelly soils forming thick tufts on the ground. This is a well defined species readily recognised by its small tufts of often rather glaucous leaves on spreading stolons, by its long anthers, and by the compact inflorescence of short falcate spikes.

Fl. & Fr.: September - January.

Specimen examined: Badnor 827, (MLVGCB Herb).

Distribution: Kenya north to the Sudan and eastward to Northwest India and Pakistan.

Uses: It is cosidered to be a good fodder grass.

DESMOSTACHYA Stapf

Desmostachya bipinnata (L.) Stapf in Thiselt.-Dyer, Fl. Cap. 7: 632. 1900. Blatter & McCann, Bombay Grasses 244. 1935; Bor, Grasses Burma Ceyl. Ind. Pak. 491. 1960; Pandey, in Shetty & Singh Fl. Rajasthan 3: 1015. 1993. *Briza bipinnata* L., Syst. Nat. ed. 10. 2: 875. 1759. *Eragrostis bipinnata* (L.) K. Schum. in Engl., Pflanzenw. Ost. Afr. C : 113. 1895. *E. cynosuroides* (Retz.) P. Beauv., Ess. Agrost. 162. 1812; Boiss., Fl. Or. 5: 583. 1884; Duthie, Fodder Grasses 62. 1888; Hook. *f.*, Fl. Brit. Ind. 7: 324. 1896. *Poa cynosuroides* Retz., Obs. Bot. 4: 20. 1786. *Stapfiola bipinnata* (L.) O. Ktze., in Post & O. Ktze, Lexic. Gen. Phan. 532. 1903. *Uniola bipinnata* (L.) L., Sp. Pl. ed. 2. 1: 104. 1762.

Vern.: *Kando.*

Harsh tussocky rihozomatous perennial; culms stout, up to 1.2 m high. Leaves up to 65 cm long, 3.8 - 10.5 mm wide when unrolled; lower leaf-sheaths leathery, often densely flabellate towards the base of the culm. Inflorescence up to 60 cm long; spikes clustered or spaced, 14 cm long. Spikelets 3 - 17-flowered, narrowly ovate to linear-oblong, 3 - 10 mm long; lower glume 0.7 - 1.5 mm long, upper glume 1.1 - 2.0 mm long; lemmas straw-coloured or suffused with purple, 1.8 - 2.7 mm long.

Field Note: Common in grasslands, saline-moist habitats throughout the area.

Fl. & Fr.: Rainy season.

Specimen examined: Bhilwara 2215 (MLVGCB Herb).

Distribution: Pakistan, India: throughout the Middle East to Indo-China; North and tropical Africa.

Notes: On account of the existence of the genus *Desmostachys* Planch. (Olacinaceae, 1882) the new generic name *Stapfiola* Kuntze. was proposed by Kuntze (1903) for *Desmostachya* Stapf. This new name has not been generally accepted and it is expressly stated in the International Code of Botanical Nomenclature (1961 edition, Art. 75) that *Desmostachya* is not likely to be confused with *Desmostachys* and is therefore to be maintained as a valid generic name. Opinions about the economic usefulness of this grass vary. Duthie reports that cattle not normally eat it although buffalo do, while Blatter & McCann pointed out that in Afghanistan it is considered to be a good fodder grass.

DICHANTHIUM Willemet

KEY TO THE SPECIES

1a. Inflorescence digitate or subdigitate racemes. Lower glume of sessile spikelet not pitted.................................... *D.annulatum*

1b. Inflorescence a solitary raceme. Lower glume of sessile spikelet pitted ....*D. foveolatum*

Dichanthium annulatum (Forsk.) Stapf, in Prain Fl. Trop. Afr. 9: 178. 1917; Blatter & McCann, Bombay Grasses 94. 1935; Bor, Grasses Burma Ceyl. Ind. Pak. 133. 1960; Deshpande, Fl. India Fasc. 15: 5. f. 1 - 2. 1984; Pandey, in Shetty & Singh Fl. Rajasthan 3: 1017. 1993; Bhandari, Fl. Indian Desert 366. 1995. *Andropogon annulatus* Forssk., Fl. Aegypt.-Arab. 173. 1775; Hook. *f.*, Fl. Brit. Ind. 7: 196. 1896. *A. nodosus* (Willemet) Nash N. Amer. Fl. 17: 122. 1912. *Dichanthium nodosum* Willemet in Usteri, Ann. Bot. 18: 11. 1796. *nom. superfl.*, based on *Andropogon annulatus* Forssk., but the description is that of *Dichanthium aristatum* (Poir.) C.E. Hubbard. (Fig. 182)

Vern.: *Palwan.*

Tufted, erect or ascending perennial. Culms 25 - 100 cm high, slender, terete, smooth. Leaves 3 - 30 cm long, 2 - 7 mm wide, linear - lanceolate, acuminate; ligule about 1 mm. Inflorescence composed of (1-) 2-15 subdigitate shortly peduncled racemes, the peduncles glabrous. Racemes 3 - 7 cm long, spikelets subimbricate with 0 - 6 smaller homogamous pairs at the base; internodes and pedicels solid. Sessile spikelet narrowly oblong, 2 - 6 mm long. Lower glume firmly cartilaginous, slightly concave, pubescent to villous below the middle with long bulbous-based hairs on the margins above, obtuse to subacute; awn 8 - 25 mm long. Caryopsis broadly elliptic, brown, pidicelled spikelets oblanceolate, spathulate, reduced to empty glumes.

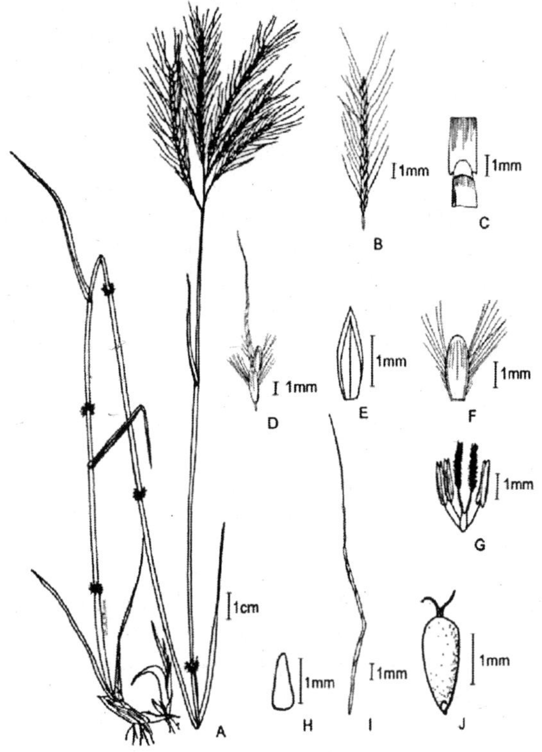

Fig. 182. *Dichanthium annulatum (*Forsk.) Stapf. A. Habit. B. Spike, C. Ligule, D. Spikelet, E. Upper glume, F. Lower glume of sessile spikelet, G. Flower, H. Lower lemma, I. Upper awened lemma, J. Fruit

Field Note: Common in grasslands, moist habitats throughout the area.

Fl. & Fr.: Throughout the year.

Specimen examined: Bhilwara 944, Keli 1851 (MLVGCB Herb).

Distribution: India, Pakistan, Kenya, Tanzania and Senegal, through the Middle East to Indonesia; introduced to southern Africa, Tropical America and Australia.

Notes: *Dichanthium annulatum* is morphologically and cytologically very variable. For details of its variation see Mehra & Celarier in Proc. Okla. Acad. Sci. 38: 22-25.1958; Mehra in Fyton 17 : 157-166. 1961; and Mehra in Caryologia 17: 545-556. 1964

Uses: It is an excellent fodder grass eagerly sought by stock.

Dichanthium foveolatum (Del.) Roberty in Boissiera. 9: 170. 1960; Pandey, in Shetty & Singh Fl. Rajasthan 3: 1019. 1993. *Andropogon foveolatus* Del. Fl. Egypt 16. t. 8 f. 2. 1812; Hook. *f.,* Fl. Brit. Ind. 7: 168. 1896. *A. strictus* Roxb., Fl. Ind. 1: 260. 1832. *Eremopogon foveolatus* (Del.) Stapf, in Prain Fl. Trop. Afr. 9: 183. 1917; Blatter & McCann, Bombay Grasses 96. 1935; Bor, Grasses Burma Ceyl. Ind. Pak. 148. 1960; *E. strictus* (Roxb.) A. Camus in Annls Soc. Linn. Lyon. (n.s.) 68: 208. 1921.

Vern.: *Ghandel.*

Tufted, ascending, perennial with silky hairy basal sheaths. Culms 15 - 80 cm high, wiry. Leaves 1 - 15 cm x 1-3 mm, flat or rolled, green or glaucous; ligule truncate, ciliate. Inflorescence a solitary raceme subtended by a narrow spatheole, these loosely aggregated on the culms. Raceme 1.5 - 4.5 cm long, with 0 - 1 smaller homogamous pairs of spikelets at the base. Sessile spikelet narrowly elliptic, 2.5 - 4 mm long. Lower glume cartilaginous, glabrous, shining, with a circular depression in the upper third, acute. Upper lemma entire or minutely bidentate, its awn 12 - 18 mm long. Pedicelled spikelet with or without a pit.

Field Note: Common in dried riverbeds and in moist habitats.

Fl. & Fr.: August - December.

Specimen examined: Chileshwar 184 (MLVGCB Herb).

Distribution: India, Pakistan, Kenya, Mali and North Africa, eastwards through the Middle East and Somalia to India and Sri Lanka.

Notes: The form with an unpitted pedicelled spikelet has been treated as a distinct species (*Eremopogon strictus*) but Bor (Grasses Burma Ceyl. Ind. Pak. 149) has shown that it cannot be satisfactorily distinguished from *Dichanthium foveolatum.*

Uses: It is an excellent fodder grass eagerly sought by stock.

DIGITARIA Haller nom. cons.

KEY TO THE SPECIES

1a. Racemes mostly 2. Spikelets of each pair heteromorphus; lower glume present..*D. bicornis*

1b. Racemes more than 3. Spikelets of each pair similar; lower glume absent ......*D. ciliaris*

Digitaria bicornis (Lam.) Roem. & Schult. ex Loud. Hort. Brit. 24 : n. 1758. 1830; Bor, in Webbia 11: 323. 1955 & Grasses Burma Ceyl. Ind. Pak. 299. 1960; Raizada & Jain, in Ind. For. Rec. (n.s.) Bot. 5: 185. pl. 5. f. 7. 1965; Pandey, in Shetty & Singh Fl. Rajasthan 3: 1024. 1993; Bhandari, Fl. Indian Desert 367. 1995. *Paspalum bicorne* Lam. Tabl. Encycl. Meth. Bot. 1 : 176. 1791. *P. heteranthum* Hook. *f.*, Fl. Brit. India 7: 16. 1896, non Link 1820, non Nees & Meyen 1843.

Tufted, erect or decumbent, annual, 10 - 20 cm high. Culms decumbent, rooting at lower nodes. Leaf sheaths glabrous or lower sheaths papillose-pilose. Leaves 2.0 - 11 × 0.2 - 0.8 mm, linear, usually scabrid on both surfaces, rarely sparsely pilose; ligule 1 - 3 mm long, truncate. Inflorescence digitate or subdigitate; racemes 2 - 5, 4-14 cm. Spikelets paired, imbricate, heteromorphous, elliptic-lanceolate; the sessile ones glabrous or scarsely hairy and pedicellate ones with spreading hairs. Lower glume very small, almost absent or up to 0.4 mm, triangular or bifid, veinless; upper glume lanceolate, ca 2/3 length of spikelet, 3-veined, pilose. Lower lemma of pedicellate spikelets hairy, 7 - nerved; upper lemma acute, yellowish, slightly shorter than spikelet. Stamens 3. Caryopsis ca 2.25 mm long, obtusely elliptic, hilum a circular dot.

Field Note: Common in waste lands, often on sand, including wet and humid habitats.

Fl. & Fr.: August - October.

Specimen examined: Badnor 929 (MLVGCB Herb).

Distribution: China, India, Indonesia, Malaysia, Myanmar, New Guinea, Sri Lanka, Thailand; Africa, Australia and introduced in America.

Digitaria ciliaris (Retz.) Koel., Descr. Gram. 27. 1802; Blake, in Proc. R. Soc. Queensland 81: 10. 1969; Bor, in Rech. f., Fl. Iran. 70: 490. 1970; Pandey, in Shetty & Singh Fl. Rajasthan 3: 1025. 1993; Bhandari, Fl. Indian Desert 367. 1995. *Panicum ciliare* Retz. Obs. Bot. 4: 16. 1786. *P. adscendens* H.B. & Kunth, Nov. Gen. Sp. Pl. 1: 97. 1816. *Digitaria fimbriata* Link, Hort. Berol. 1: 226. 1827. *D. marginata* Link var. *fimbriata* (Link) Stapf, in Prain Fl. Trop. Afr. 9: 440. 1919. *D. adscendens* (H.B & Kunth) Henr., in Blumea. 1: 92. 1934 & Monogr. *Digitaria* 9. 1950, incl. vars. & subsp. *chrysoblephara*, *marginata* & *criniformis* Bor, in Webbia 11: 350. 1955 & Grasses Burma Ceyl. Ind. Pak. 298. 1960. *D. biformis* Willd. subsp. *willdenowii* Henr. *l.c.* 978. 1950. *Digitaria adscendens* (H.B. & Kunth) var. *fimbriata* (Link) Cuf., in Bull. Jard. Bot. Etat. Brux. 39: suppl. 1327. 1969.

Annual. Culms 20 - 100 cm high, decumbent at the base and geniculately ascending. Leaves 3 - 25 cm x 3 - 10 mm, linear-lanceolate, acuminate, rounded at the base; ligule 1.5 - 3.0 mm long, membranous, more or less lacerate. Inflorescence digitate or subdigitate, composed of 2 - 12 racemes. Racemes stiff, 5 - 22 cm long, sometimes adorned with sparse long white hairs, the spikelets binate and overlapping by about two-thirds their length on a winged rachis with triquetrous midrib. Pedicels triquetrous, scabrid, obtuse or slightly expanded at the tip. Spikelets narrowly elliptic, 2.5 - 3.5 mm long, acute. Lower glume distinct, 0.2 - 0.4 mm long and triangular; upper glume at least half as long as the spikelet, typically two-thirds to three-quarters as long, 3-nerved; lower lemma as long as the spikelet, 5 - 7-nerved, the nerves smooth and evenly spaced or with a wide interspace flanking the midrib appressed puberulous, silky pubescent or rarely shortly villous, often encompassed by a ciliate frill, occasionally glabrous, one or both spikelets often beset with stiff glassy bristles. Stamens 3. Caryopsis ca 2 - 7.5 mm long, ellipsoid, grey to light brown.

Field Note: Common in bushes and hedges, in waste lands and neglected places throughout the district.

Fl. & Fr.: August - December.

Specimen examined: Hurda 2211 (MLVGCB Herb).

Distribution: Throughout the tropics.

Notes: *Paspalum sanguinale* (L.) Lam. var. *rottleri* Hook. *f.* is typified by two sheets, *Wight* 3032 and *Wight* 3033, both at Kew. Of these, the former is *Digitaria ciliaris* and the latter *Digitaria bicornis* (Lam.) Loud. In the protologue to the variety, Hooker described the wings of the rachis as being "two or three times broader than the midrib." Since this character does not agree with *Digitaria bicornis* (including Wight 3033), the specimen *Wight* 3032 is selected as the lectotype of var. *rottleri*. This is included in the synomymy of *Digitaria ciliaris* because the spikelets are narrow and apparently lack the spiny protuberences on the nerves of the lemma.

Digitaria ciliaris is a polymorphic species showing at least four parallel series based on quite independent sources of variation. The rachis may have long white hairs (*rachiseta*) or not; in terms of indumentum the spikelets of a pair may be the same or different (*biformis*); the spikelets (independently in each pair) may be glabrous (*nubica*), long or short hairy (*ciliaris, desvauxii*) or conspicuously fringed (*criniformis, fimbriata*); one or both spikelets may be beset with stiff glassy bristles (*chrysoblephara, willdenowii*). None of these variants shows any indication of geographical segregation, and since there are at least these four independent sources of variation the number of potential varieties is enormous, but none would be of any taxonomic significance. The sporadic occurence of the siliceous spines characteristic of *Digitaria sanguinalis* (L.) Scop. in spikelets otherwise typical of *Digitaria ciliaris* suggests that there may be some gene-flow between these two species in Pakistan.

Digitaria bicornis (Lam.) Loud. is often included in subsp. *chrysoblephara* (or *vice versa*, see Veldkamp 1973) by virtue of its possession of glassy bristles, but like *Digitaria ciliaris* it too shows several types of indumentum and quite frequently is perfectly glabrous (even to the exclusion of the glassy bristles). It is in fact a distinct species differing in a number of important characters. There are normally only 2, sometimes 3 racemes each with a slightly swollen base. The midrib is thick and the wings are usually extremely narrow. The pedicels are stout and straight and this, combined with the thick midrib, gives the spikelets the appearance of being sunk in the rachis. The lower lemma is horny with much thickened, contiguous or closely spaced nerves. The first pair of interspaces, and the second where present, are reduced to deep V-shaped grooves. Lemma characters are best seen in the subsessile spikelets towards the base of the raceme.

DINEBRA Jacq.

Dinebra retroflexa (Vahl) Panzer, in Denkschr. Acad. Wiss. Munch. 1813 : 270. t. 12. 1814; Boiss., Fl. Or. 5: 557. 1884; Blatter & McCann, Bombay Grasses 264. pl. 177. 1935; Bor, Grasses Burma Ceyl. Ind. Pak. 491. 1960; Pandey, in Shetty & Singh Fl. Rajasthan 3: 1031. 1993. *Cynosurus retroflexus* Vahl, Symb. Bot. 2: 20. 1791. *Dactylis paspaloides* Willd., Enum. Hort. Berol. 111. 1809. *Dinebra arabica* Jacq. Fragm. 77. t. 121 f. 1. 1807; Duthie, Fodder Grasses 55. 1888; Hook. *f.*, Fl. Brit. Ind. 7: 297. 1896. *Leptochloa arabica* (Jacq.) Kunth, Rev. Gram. 1: 91. 1829.

Vern.: *Bada sarpot.*

Tufted annual. Culms up to 112 cm high, slender, usually straggling and ascending from a decumbent base, often rooting at the lower nodes. Leaves glandular especially on the sheaths; leaf-blades 4.5-28 cm x 4-8 mm wide, linear-lanceolate, acute, finely pointed. Inflorescence 8 - 34 cm long, linear or elliptic-oblong to pyramidal. Spikes 0.6 - 5 cm long, stiff, ascending at first, reflexing and deciduous at maturity. Spikelets 1-3-flowered, narrowly wedge-shaped, 5-9 mm long, closely over-lapping on the flattened, narrowly green-winged rachis; glumes narrowly elliptic with caudate curving tips, 6 - 8 mm long, coriaceous, asymmetric, usually over-lapping on the abaxial side and obscuring the florets, glandular along the keel. Lemmas narrowly ovate, 2 - 3 mm long, appressed pilose along the lateral nerves and on the lower half of the back around the central nerve, acute to emarginate, mucronulate. Palea appressed pilose on the flaps alongside the keels.

Field Note: A very common weed of cultivation.

Fl. & Fr.: July - October.

Specimen examined: Singoli 1204, Triveni 1980 (MLVGCB Herb).

Distribution: South Africa, through tropical Africa to Egypt; Iraq to India.

Uses: Used as a fodder grass.

DIPLACHNE P. Beauv.

Diplachne fusca (L.) P. Beauv. ex. Roem. & Schult. Syst. Veg. 2: 615. 1817; Hook. *f.,* Fl. Brit. India 7: 329. 1896; Blatter & McCann, Bombay Grasses 246. pl. 163. 1935; Bor., Grass. Burm. Ceyl. Ind. Pak. 492. 1960; Pandey, in Shetty & Singh Fl. Rajasthan 3: 1031. 1993. *Festuca fusca* L. Syst. Nat., ed. 10, 2: 876. 1759 & Sp. Pl. ed. 2. 1: 109. 1762.

Aquatic or semi-aquatic rhizomatous perennial. Culms moderately stout, 60 - 150 cm high, rooting and branching from the lower nodes. Leaves 15 - 50 x 0.15 - 0.5 cm, flat or convolute, filiform, acuminate; ligule 3 - 8 mm long, membranous, lacerate, pointed. Inflorescence 10 - 30 cm long, fairly compact at first with erect racemes, then after spreading. Spikelets 6 - 11-flowered, narrowly elliptic to elliptic, 8 - 15 mm long, slightly overlapping. Glumes keeled, scabrid on the keels, lower glume lanceolate to narrowly lanceolate, 2.1 - 4.6 mm long, acute or acuminate, upper glume narrowly oblong, 3.3 - 7.4 mm. long, acute or obtuse and minutely mucronate. Lemmas narrowly oblong, 3.2 - 5.9 mm. long, pilose on the lower part of the nerves, 2- or more toothed and mucronate or shortly awned from the sinus (awn 0.3-1.6 mm. long). Anthers 1.3 - 2.7 mm long. Caryopsis elliptic-oblong, 1.0 - 2.0 mm long, dorso-ventrally flattened.

Field Note: Common in water logged habitats.

Fl. & Fr.: March - December.

Specimen examined: Gowata 2212 (MLVGCB Herb).

Distribution: Tropical and subtropical Old World, including Australia

Notes: A polymorphic species, varying considerably in length of the rhizome (some plants are more or less tufted), height and thickness of the culm, compactness of the inflorescence, and the lemma-tip. The extension of the central nerve of the lemma varies from a barely apparent mucro to a clearly visible awn. The lateral nerves are also sometimes shortly excurrent.

ECHINOCHLOA P. Beauv. *nom. cons.*

Echinochloa colona (L.) Link, Hort. Berol. 2 : 209. 1833; Bor., Grass. Burm. Ceyl. Ind. Pak. 308. f. 34. 1960; Pandey, in Shetty & Singh Fl. Rajasthan 3: 1032. 1993; Bhandari, Fl. Indian Desert 369. 1995. *Panicum colonum* L. Syst. Nat. ed. 10. 2 : 870. 1759; Hook. *f.,* Fl. Brit. India 7: 32. 1896. (Plate 331)

Annual. Culms erect or ascending, hairy at nodes, often rooted at lower nodes, upto 80 cm tall. Leaves 4 - 20 × 0.2 - 0.7 cm,, linear, flat, glabrous, sometimes with transverse purple bands, margins slightly scabrous, apex acute; sheaths upto 5 cm long, striate, keeled; ligule absent. Inflorescence narrow, 5 - 10 cm; racemes 1 - 2 cm, erect or sometimes stiffly diverging, simple, separated or overlapping by up to half their length or more, rachis usually without long, tubercle-based hairs, spikelets tightly congested in 2 - nate rows. Spikelets plumply ovate-oblong, 2 - 3 mm, ovoid-ovate or obovate-elliptic, sharply acute, more or less pubescent, secund. lower glume ca. 1 mm long, ovate or suborbicular, membranous, 5-nerved; upper glume 1.5 - 2 mm long, broadly ovate, concave, 5-nerved, hairy. Lower lemma similar, staminate or sterile, with hyaline palea; upper lemma broadly ovate, turgid, coriaceous, whitish at maturity. Caryopsis 1 - 1.5 mm long, plano-convex, elliptic, reddish brown.

Field Note: A weed of damp places and irrigated fields. It is a widespread, weedy species, distinguished by its short, neat, usually rather openly spaced racemes of rounded, awnless spikelets.

Fl. & Fr.: August - November.

Specimen examined: Bhilwara 2153, Mandalgarh 1132 (MLVGCB Herb).

Distribution: Warmer regions throughout the world.

Uses: It is good nutritive grass for cattle. The grains are eaten at the time of scarcity.

Notes: Bhandari 1995, published this species as *'colonum'*, but references from other worker cited this species as *'colona'*. authors follow the species name *colona*.

ELEUSINE Gaertn.

Eleusine coracana (L.) Gaertn. Fruct. 1: 8. t. 1. f. 11. 1789; Hook. *f.*, Fl. Brit. India 7: 294. 1896; Blatter & McCann, Bombay Grasses 260. pl. 173. 1935; Bor., Grass. Burm. Ceyl. Ind. Pak. 492. 1960; Pandey, in Shetty & Singh Fl. Rajasthan 3: 1035. 1993. *Cynosurus coracanus* L. Syst. Nat. ed. 10. 2: 875. 1759.

Vern.: *Kangni*.

Tall, erect, stout, annual. Culms tufted, robust, erect or ascending, usually branched, 50 - 150 cm tall. Leaves 2 - 20 × 0.5 - 0.8 cm, flat, acute, pilose or glabrous; sheaths glabrous; ligule 1 - 2 mm. Inflorescence subdigitate, racemes 5 - 20, stout, often incurved at maturity, 5 - 10 × 0.8 - 1.5 cm, hairy at base. Spikelets very closely imbricate, ovate, 3 - 4 mm, florets 3 - 6, not disarticulating at maturity. Glumes lanceolate-oblong, scabrid along the winged keel; lower glume 3-veined, 1 - 3 mm; upper glume with additional veins in keel, 1 - 5 mm. Lemmas triangular-ovate, 2 - 5 mm, keel 3-veined, scabrid and narrowly winged, subacute. Palea narrowly ovate, keels scabrid, winged. Caryopsis smooth or rugose, with depressed black hilum and slightly flattened on one side, dark brown or yellowish brown.

Field Note: Cultivated cereal crop.

Fl. & Fr.: August - November.

Specimen examined: Jahajpur 2155 (MLVGCB Herb).

Distribution: Widely cultivated in tropical and subtropical regions of the Old World

Uses: This species is used for cereal, forage, papermaking, and soil-conservation.

ERAGROSTIS N.M.Wolf

KEY TO THE SPECIES

1a. Spikelets shedding their florets from the apex downwards at maturity, the rachilla internodes falling with the florets 2

2a. Lemmas ciliate on the keels. Panicle woolly, spike-like or lobed and interrupted, the spikelets densely clustered. Stamens 2 .... *E. ciliaris*

2b. Lemmas smooth or slightly scaberulous on the keel. Panicles open, neither woolly nor with the spikelets in dense clusters. Stamens 3 .... *E. tenella*

1b. Spikelets shedding their florets from below upwards, the rachilla remaining upon the pedicel (or eventually breaking up) 3

3a. Erect with disagreeable odour. Pedicels without crateriform glands. Spikelets 2 - 4 mm wide. Caryopsis subglobose, 0.5 mm long .... *E. cilianensis*

3b. Erect or acsending withiout specific smell. Pedicels of spikelets with crateriform glands. Spikelets 1.5 - 2.0 mm wide. Caryopsis broadly oblong, 0.7 - 0.8 mm long .... *E. minor*

Eragrostis cilianensis (All.) Vig. - Lut. ex F.T. Hubbard, in Philipp. J. Sci. C. Bot. 8: 159. 1913. Blatter & McCann, Bombay Grasses 237. pl. 157. 1935; Bor, Grasses Burma Ceyl. Ind. Pak. 503. 1960; Roy, in Bull. Bot. Surv. India 18: 106. 1979; Pandey, in Shetty & Singh Fl. Rajasthan 3: 1045. 1993; Bhandari, Fl. Indian Desert 373. 1995. *Briza eragrostis* L., Sp. Pl. 1: 70. 1753. *non Poa eragrostis* L. *Eragrostis major* Host, Gram. Austr. 4 : 14. t. 24. 1809; Stapf in Hook. *f.*, Fl. Brit. Ind. 7: 320. 1896. *Eragrostis megastachya* (Koel.) Link, Hort. Berol. 1: 187. 1827; Boiss., Fl. Or. 5: 580. 1884; Duthie, Fodder Grasses 63. 1888. *Poa cilianesis* All., Fl. Pedem. 2: 246. t. f. 2. 1785. *P. megastachya* Koel. Descr. Gram.

181. 1802. based on *Briza eragrostis* L. (Fig. 183 & Plate 332)

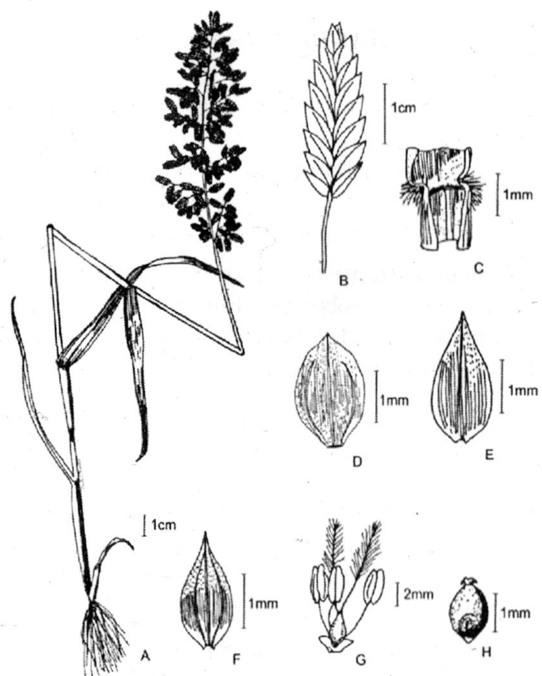

Fig. 183. *Eragrostis cilianensis* (All.) Vig. - Lut. ex F. T. Hubb. A. Habit. B. Spikelet, C. Ligule, D. Lemma, E. Upper glume, F. Lower glume, G. Flower, H. Grain

Loosely tufted annual. Culms 10 - 80 cm high, erect or ascending. Leaves up to 15 cm long and 8 mm wide, flat or convolute, linear, acute, mostly glabrous and usually with a row of warty glands along the margins; ligule a ring of hairs. Panicle 4 - 20 cm long, ovate, fairly dense, contracted, stiffly branched, usually with glands on pedicels and branchlets. Spikelets 5 - 50-flowered, narrowly ovate to narrowly oblong, rarely linear, 3 - 20 mm long, 2 - 4 mm wide, pallid yellowish green to leaden grey, breaking up from the base, the rachilla often breaking off above the glumes before all the lemmas have fallen. Glumes subequal, oval, boat-shaped, 1.5 - 2.5 mm long, 1-3-nerved, often glandular on the keel, acute. Lemmas broadly ovate to subrotund, 2 - 2.8 mm long, chartaceous, often glandular on the keel, the lateral nerves distinct, emarginate or obtuse (occasionally bluntly acute) when flattened, usually obtuse in side-view; palea scabrid on the keels, persistent. Anthers 3, 0.3 mm long. Caryopsis typically subglobose, ca 0.5 (0.7) mm long, rarely oblong, dark reddish brown.

Field Note: Common in open waste lands and cultivated fields.

Fl. & Fr.: August - December.

Specimen examined: Pancha ka bara 1939, Hajari khera 2142 (MLVGCB Herb).

Distribution: Tropical and warm temperate regions of the Old World; introduced to the New World.

Notes: Henrard (in Blumea 3 : 420.1940) has pointed out that Allioni's description of *Poa cilianensis* mentions 3 - 4 branches per node of the inflorescence, and has suggested that he was in fact describing *Poa trivialis* L. However, the rest of the description, the figure (which shows only 1-2 branches per node) and the isotypes all contradict this view. Clayton (in Fl. Trop. E. Afr. 234. 1974) has accepted Lutati's conclusion that *Poa cilianensis* is synonymous with the plant at one time known as *Eragrostis major* Host. and his treatment is adopted here. Lutati, however, doubted his own conclusion and would not formally make the new combination although it is often ascribed to him. *Eragrostis major,* often applied to this species, is a new name dating from 1809 and is predated by *Poa cilianensis* (1785), the basis of Lutati's combination. Sprague & Hubbard (in Kew Bull. 1933: 17. 1933) have listed the full synonymy.

This species is best recognised by the warty, crateriform glands on the leaf-margins and the rather stiff panicles of yellowish green or leaden grey spikelets. It does, however, intergrade with *Eragrostis minor* Host.

Uses: Occasionally used as fodder.

Eragrostis ciliaris (L.) R. Br., in Tuckey, Narr. Exp. Congo App. 478. 1818; Duthie, Fodder Grasses 62. 1888; Stapf, in Hook. *f.,* Fl. Brit. Ind. 7: 314. 1896; Blatter & McCann, Bombay Grasses 231. pl. 154. 1935; Bor, Grasses Burma Ceyl. Ind. Pak. 506. 1960; Roy in Bull. Bot. Surv. India 18: 103. 1979; Pandey, in Shetty & Singh Fl. Rajasthan 3: 1045. 1993; Bhandari, Fl. Indian Desert 374. 1995. *E. ciliaris* (L.) R. Br. var. *brachystachya* Boiss.,

Fl. Orient. 5 : 582. 1884; Stapf, in Hook. *f., l. c.* 7: 315. 1896; Bor, *l.c.* 506. 1960. *E. pulchella* Parl, in Atti Riun. scient. Ital. 8: 586. 1847. *Megastachya ciliaris* (L.) P. Beauv., Ess. Agrost. 74, 167. 1812. *Poa ciliaris* L. Syst. Nat. ed. 10. 2: 875. 1759.

var. ***ciliaris***

Tufted annual. Culms 5 - 60 cm high, erect. Leaves 2-10 x 0.15 - 0.5 cm, flat, linear-lanceolate, acute; ligule ciliate. Panicle spike-like, but often ± lobed or interrupted, 1 - 20 cm long, woolly, the spikelets densely clustered. Spikelets 5 - 12-flowered, ovate, 2 - 5 mm long, fluffy, often purplish, breaking up from the apex, the rachilla fragile. Glumes lanceolate, subequal, 0.7 - 1.2 mm long, acute. Lemmas oblong-elliptic, 0.9 - 1.5 mm long, keels (at least in the upper lemmas) bearing a few short stiff hairs, otherwise smooth, broadly obtuse, more or less obscurely mucronate. Palea equal to lemmas; keels with tuberculate-ciliate, hairs 0.6 - 0.7 mm long, exceeding the width of the adjacent floret. Anthers 2, 0.2 mm long. Caryopsis ellipsoid, 0.3 - 0.6 mm long, across, ellipsoid.

Specimens with the inflorescence reduced to an ovoid head have been separated as var. *brachystachya,* but they seem to be no more than a dwarfed form of arid habitats.

Field Note: Common in sandy soils.

Fl. & Fr.: Throughout the year.

Specimen examined: Tikar 1201, Bhilwara 1665 (MLVGCB Herb).

Distribution: Tropical and South Africa, extending through Arabia and the Mascarene Islands to India; tropical America.

Notes: The woolly spike is usually distinctive, but sometimes the panicle shape may approach extreme forms of *Eragrostis tenella* (L.) P. Beauv ex Roem. & Schult.; the cilia on the lemma-keel, though rather obscure, are then the surest means of determination. *Eragrostis ciliaris* also differs in the possession of only 2 stamens.

Uses: Used as a fodder grass.

Eragrostis minor Host, Icon. Descr. Gram. Austr.: 15. 1809; Stapf in Hook. *f.,* Fl. Brit. Ind. 7: 321. 1896; Blatter & McCann, Bombay Grasses 238. pl. 158. 1935; Pandey, in Shetty & Singh Fl. Rajasthan 3: 1048. 1993; Bhandari, Fl. Indian Desert 375. 1995. *Poa eragrostis* L., Sp. Pl. 1: 68. 1753. *Eragrostis poaeoides* P. Beauv. Ess. Agrost. 162. 1812; Bor, Grasses Burma Ceyl. Ind. Pak. 512. 1960; Roy in Bull. Bot. Surv. India 18: 106. 1979.

Tufted annual. Culms 6 - 60 cm high, ascending. Leaves up to 12 cm long and 5 mm wide, flat, linear, acute, usually with a row of warty glands along the margins; sheath glandular and bearded at the mouth; ligule a pubescent ridge. Panicle 4 - 20 cm long, ovate, fairly dense to open, stiffly branched with short pedicels (lateral pedicels 1 - 3 mm), usually with glands on pedicels and branchlets. Spikelets 6 - 16 (40)-flowered, narrowly oblong or almost linear, 3 - 9(15) mm long, 1.3 - 2 mm wide, yellowish green, leaden grey or purplish, breaking up from the base. Glumes subequal, ovate, boat-shaped, 1 - 1.7 mm long, 1 - 3-nerved, often glandular on the keel, acute. Lemmas broadly ovate to subrotund, 1.5 - 2 mm long, chartaceous, often glandular on the keel, the lateral nerves distinct, obtuse. Palea scabrid on the keels, persistent. Anthers 3, 0.3 mm long. Caryopsis broadly oblong, 0.7 - 0.8 mm long, dorsally flattened, dark brown.

Field Note: It occurs as a weed in gardens, irrigated fields and ditches.

Fl. & Fr.: September - March.

Specimen examined: Hamirgarh 79, Bhilwara 2154 (MLVGCB Herb).

Distribution: Warm temperate and subtropical regions of the Old World; occasionally found as an introduction in the tropics and the New World.

Notes: *Eragrostis minor* intergrades with *Eragrostis cilianensis,* being distinguished by the narrower oblong (rather than ovate) spikelets, shorter lemmas, more open panicle and oblong (rather than globose) grains. No single character can be relied upon to separate the species. For the present purpose, however, grain shape has been taken as decisive in doubtful cases. The characteristic glands are occasionally absent from the leaves giving rise to confusion with *Eragrostis nutans* (Retz.) Steud. (a perennial) which has not yet been recorded from Pakistan. A diligent search will usually reveal at least a few glands on the panicle branches and pedicels. Annual plants completely devoid of these crateriform glands (although glandular dots are present on the panicle branches) and without the beard at the mouth of the sheath have been separated as *Eragrostis rottleri* Stapf, a little known species collected only from Madras over 100 years ago.

Eragrostis tenella (L.) P. Beauv. ex Roem. & Schult. Syst. Veg. 2: 576. 1817; Duthie, Fodder Grasses 65. 1888; Stapf in Hook. *f.*, Fl. Brit. Ind. 7: 315. 1896; Blatter & McCann, Bombay Grasses 232. 1935; Bor, Grasses Burma Ceyl. Ind. Pak. 513. 1960; Roy in Bull. Bot. Surv. India 18 : 104. 1979; Pandey, in Shetty & Singh Fl. Rajasthan 3: 1051. 1993; Bhandari, Fl. Indian Desert 376. 1995. *Poa tenella* L., Sp. Pl. 1: 69. 1753. *Eragrostis amabilis* Wight & Arn. ex Nees in Hook. & Arn. Bot. Beechy Voy. 251. 1838. *E. plumosa* (Retz.) Link, Hort. Berol. 1: 192. 1827; Boiss. Fl. Or. 5: 582. 1884. *E. tenella* (L.) P. Beauv. ex Roem. & Schult. var. *plumosa* (Retz.) Stapf in Hook. *f., l.c.* 7: 315. 1896. *Poa amabilis* L. *l.c.* 1: 68. 1753. *P. plumosa* Retz., Obs. Bot. 4: 20. 1786.

var. ***tenella***

Vern.: *Bharbhuri.*

Tufted, annual. Culms 6 - 50 cm high, erect or geniculately ascending. Leaves up to 9 cm long and 4 mm wide, flat, linear-lanceolate, acuminate; mouth of the sheath hairy; ligule a narrow rim of hairs. Panicle elliptic or pyramidal, 2-15 cm long, open, hairy in the axils or not, the branches spreading and bearing oblong yellowish glands but not sticky. Spikelets 4 - 9-flowered, oblong-ovate, 1.5 - 2.5 mm long, breaking up from the apex, the rachilla fragile. Glumes ovate to narrowly ovate, subequal, 0.5 - 1 mm long, acute. Lemmas ovate-oblong, 0.7 - 1 mm long, smooth or scaberulous, broadly obtuse. Palea-keels ciliate with hairs 0.1 - 0.3 mm long, normally shorter than the width of the adjacent floret. Anthers 3, 0.2 mm long. Caryopsis ovoid-ellipsoid, 0.5 mm long, brown.

Field Note: Common in wet places.

Fl. & Fr.: August - November.

Specimen examined: Gowta dam 2212 (MLVGCB Herb).

Distribution: Throughout the tropics.

Uses: Used as a fodder grass. The grains are said to be very nutritious.

ERIOCHLOA H. B. Kunth

KEY TO THE SPECIES

1a. Spikelets more than 4 mm long, distincatly awned. Pedicel commonly setose.. *E. fatmensis*

1b. Spikelets less than 3.5 mm long, acute to acuminate. Pedicel commonly glabrous .... *E. procera*

Eriochloa fatmensis (Hochst. & Steud.) W.D. Clayton, in Kew Bull. 30: 108. 1975; Pandey, in Shetty & Singh Fl. Rajasthan 3: 1054. 1993. *E. acrotricha* (Steud.) Hack. ex Thell. in Vjschr. Naturf. Ges. Zurich. 52: 435. 1907. *E. nubica* (Steud.) Hack. & Stapf ex Thell. *l.c.* 64: 697. 1919; Bor, Grasses Burma Ceyl. Ind. Pak. 312. 1960. *Helopus nubicus* Steud. Syn. Pl. Glum. 1: 100. 1854. *Panicum annulatum* A. Rich., Tent. Fl. Abyss. 2: 370. 1851.

Annual. Culms 10 - 120 cm high, erect or geniculately ascending. Leaves 3 - 25 cm long, 2 - 12 mm wide, convolute, acute. Inflorescence 3 - 20 cm long. Racemes 1 - 5 cm long, the rachis puberulous (occasionally pilose), triquetrous or narrowly winged, bearing paired spikelets. Pedicels commonly setose, those of a pair usually free. Spikelets lanceolate, 3 - 5 mm long, thinly pubescent. Lower glume absent; upper glume attenuate to an awnlet 0.5 - 4 mm long. Lower floret represented only by a lemma; upper lemma with a mucro 0.3 - 1 mm long.

Field Note: Common in moist habitats.

Fl. & Fr.: September - November.

Specimen examined: Mangrop 1342, Badnor 1938 (MLVGCB Herb).

Distribution: Throughout tropical and South Africa; also in Arabia and a few places in India and Thailand.

Notes: It is a grass that can adapt itself to wet or dry conditions, and is eaten by all stock. It is a probable introduction in Asia.

Eriochloa procera (Retz.) C. E. Hubbard, in Kew Bull. 1930: 256. 1930; Bor, Grasses Burma Ceyl. Ind. Pak. 312. 1960; Raizada & Jain, in Ind. For. Rec. (n.s.) Bot. 5: 196. pl. 6. f. 2. 1965; Pandey, in Shetty & Singh Fl. Rajasthan 3: 1055. 1993. *Agrostis procera* Retz., Obs. Bot. 4: 19. 1786. *Eriochloa annulata* (Flugge) Kunth, Rev. Gram. 1: 30. 1829. *E. polystachya sensu* Duthie, Fodder Grasses. 2. 1888; Hook. *f.,* Fl. Brit. Ind. 7: 20. 1896, *non* Kunth 1816. *E. ramosa* (Retz.) O. Ktze., Rev. Gen. Pl. 2: 775. 1891; Blatter & McCann, Bombay Grasses 131. 1935. *Helopus annulatus* (Flugge) Nees, Agrost. Bras. 17. 1829. *Milium ramosum* Retz., Obs. Bot. 6: 22. 1791. *Paspalum annulatum* Flugge, Gram. Monogr. 133. 1810.

Annual. Culms 20 - 100 cm high, erect or geniculately ascending. Leaves 2 - 30 cm long, 2 - 4 mm wide, glabrous, convolute when dry. Inflorescence 4 - 20 cm long. Racemes 2 - 10 cm long, the rachis puberulous, triquetrous, bearing paired spikelets. Pedicels commonly glabrous, those of a pair often fused below. Spikelets lanceolate, (2.5) 3.4 mm long, thinly pubescent. Lower glume and palea absent; upper glume acute to acuminate. Lower - floret represented only by a lemma; upper lemma with a mucro 0.3 - 0.5 mm long. Caryopsis free within the hardened lemma, lenticular, compressed, smooth.

Field Note: Widely distributed in damp places, ditches and the edges of waterloged fields.

Fl. & Fr.: August - November.

Specimen examined: Vindhyabhata 342 (MLVGCB Herb).

Distribution: Throughout Southeast Asia; introduced in tropical America.

Uses: It is a good fodder.

HETEROPOGON Pers.

Heteropogon contortus (L.) P. Beauv. ex Roem. & Schult., Syst. Veg. 2: 836. 1817; Duthie, Fodder Grasses 32. 1888; Blatter & McCann, Bombay Grasses 109. pl. 71. 1935; Bor, Grasses Burma Ceyl. Ind. Pak. 163. f. 6. 1960; Raizada *et al.*, in Ind. For. Rec. (n.s.) Bot. 4: 247. pl. 11. f. 5. 1961; Pandey, in Shetty & Singh Fl. Rajasthan 3: 1059. 1993; Bhandari, Fl. Indian Desert 379. 1995. *Andropogon allionii* DC. in Lam. & DC. Fl. Franc. ed. 3. 3 : 97. 1805. *A. contortus* L., Sp. Pl. 2: 1045. 1753; Hook. *f.,* Fl. Brit. Ind. 7: 199. 1896. *Heteropogon allionii* (DC) Roem & Schult., Syst. Veg. 2: 835. 1817. *H. hirsutus* P. Beauv. Ess. Agrost. 134. 1812. (in error for *Heteropogon hirtus* Pers.). *Heteropogon hirtus* Pers. Syn. Pl. 2: 533. 1807. *Heteropogon hispidissimus* Hochst. ex A. Rich., Tent. Fl. Abyss. 2: 453. 1851. (Fig. 184)

Vern.: *Haru*

Tufted perennial, the basal sheaths laterally compressed. Culms 30-100 cm high, erect. Leaves 3-30 cm long, 2-8 mm wide, flat, linear, abruptly narrowed at the tip; sheaths compressed, keeled, glabrous; ligule a ciliate, scarious rim. Racemes 3-12 cm long, mostly long exserted, solitary or aggregated into a scanty false panicle, the awns forming a twisted spire. Homogamous spikelets 3-17 pairs, resembling the pedicelled spikelets. Sessile spikelet 5.5-10 mm long including a ferociously pungent and rufously bearded callus 2 - 3 mm long; lower glume ca 6.5 mm long elliptic-oblong, brown, hispidulous; awn 5-8 cm long, hirtellous. Pedicelled spikelet 5-15 mm long; lower glume eglandular, green with yellowish membranous margins; the hairiness varying from glabrous to tuberculate-villous; callus 2-3 mm long. Lower lemma empty, ciliate. Caryopsis oblongish, terete.

Field Note: Fairly common in waste lands and fallow fields throughout the district.

Fl. & Fr.: August - November.

Specimen examined: Atun 1416, Raipur 1935 (MLVGCB Herb).

Distribution: Tropical and warm temperate regions.

Fig. 184. *Heteropogon contortus* (L.) P. Beaux. ex Roem. & Schuff. A. Habit. B. Pair of sessile & pedicelled spikelets

Uses: It is considered to be a good fodder grass before it reaches the flowering stage, and hay made from it is said to keep for many years. After flowering it is avoided by stock. Animals can be seriously injured by the sharp callus which, aided by the hygroscopic awns, can burrow into the flesh causing septic sores, lameness or more serious harm. Apart from the irritation having an adverse effect on the animals health, the commercial value of their hides is much reduced by these injuries.

HYGRORYZA Nees

Hygroryza aristata (Retz.) Nees ex Wight & Arn. in Edinb. New Phil. Journ. 15: 380. 1833; Hook. *f.,* Fl. Brit. Ind. 7: 95. 1896; Blatter & McCann, Bombay Grasses 271. pl. 184. 1935; Bor, Grasses Burma Ceyl. Ind. Pak. 597. 1960; Pandey, in Shetty & Singh Fl. Rajasthan 3: 1061. 1993. *Pharus aristatus* Retz., Obs. Bot. 5: 23. 1789. *Leersia aristata* (Retz.) Roxb., Fl. Ind. 2: 207. 1832.

Floating aquatic perennial herb, with inflated sheaths. culm soft, rooting at nodes; Leaves 2 - 7 x 1 - 2 cm, ovate-oblong, obtuse, base rounded or subcordate, upper surface pappilate; ligules hyaline, apex truncate, 0.8 mm long. Panicle upto 5.5 cm long, triangular, much branched. Spikelets 1.8 - 2 cm long, including long awn, sessile or shortly pedicellate. Glumes absent. Lemma ca 1 cm long, lance-elliptic, suddenly narrowed at both ends, scaberulous awn, 5-nerved; callus stipe-like; palea as long as lemma, 3-nerved. Caryopsis 5 - 6 mm long lanceolate.

Field Note: Occasional in ponds.

Fl. & Fr.: October - March.

Specimen examined: Gambhiridam 1161 (MLVGCB Herb).

Distribution: Tropical Asia, India, Ceylon, Indo-China, southern China and Taiwan to Malesia.

ISACHNE R. Br.

Isachne globosa (Thunb.) O. Kuntze, Rev. Gen. Pl. 2: 778. 1891; Bor, Grasses Burma Ceyl. Ind. Pak. 580. 1960; Ved Prakash & Jain in Fl. India Fasc. 14: 22. f. 1 - 13. 1984; Pandey, in Shetty & Singh Fl. Rajasthan 3: 1063. 1993. *Milium globosum* Thumb. Fl. Jap. 49. 1784. *Ischane australis* R. Br. Prodr. Fl. Nov. Holl. 1: 196. 1810; Hook. *f.,* Fl. Brit. India 7 : 24. 1896. *I. australis* R. Br. var. *effusa* Trim. ex Hook. *f., l.c.* 7: 25. 1896. *I. globosa* (Thunb.) O. Ktze. var. *effusa* (Trim ex Hook. *f.*) Senaratna, Grass. Ceyl. 109. 1956; Bor, *l.c.* 580. 1960. *I. dispar auct.* non Trin. 1828; Hook. *f., l.c.* 7: 26. 1896, *pro parte*; Bor, *l.c.* 580. 1960, *pro parte*.

Perennial, tufted. Culms slender, erect or decumbent and rooting at lower nodes, up to 80 cm tall, nodes glabrous, eglandular or with glandular ring. Leaves 3 - 10 × 0.2 - 1.0 cm, linear-lanceolate or ovate-lanceolate, glabrous, scabrid, base rounded, apex acute; Leaf sheaths shorter than internodes, glabrous except for ciliate outer margins; ligule 1 - 2 mm. Panicle open, ovate in outline, 4 - 11 cm, glandular, many-spiculate; branches and pedicels filiform, flexuose. Pedicels variable in length, shorter or longer than spikelets. Spikelets elliptic-globose, 1.5 - 2.5 mm long, greenish or purplish brown; florets slightly to clearly dissimilar; lower floret male, upper floret female; glumes subequal, as long as or shorter than florets, broadly elliptic, 5 - 7 veined, usually glabrous, rarely hispidulous or scaberulous above middle, apex broadly rounded. Lower lemma oblong, glabrous. Anthers 0.8 - 1.3 mm. Upper lemma shorter and more convex, slightly rough, back glabrous or puberulous, upper margins ciliate.

Field Note: Rare, in wet places, and as a weed in water-logged fields

Fl. & Fr.:Almost throughout the year.

Specimen examined: Kachhola 1908 (MLVGCB Herb).

Distribution: Bangladesh, Bhutan, China, India, Indonesia, Japan, Korea, Malaysia, Nepal, New Guinea, Philippines, Sri Lanka, Thailand, Vietnam; Australia and Pacific Islands.

ISCHAEMUM L.

KEY TO THE SPECIES

1a. Leaves cordate to rounded at the base. Lower glume of sessile spikelet smooth across the back or faintly rugose, usually nodular on the margins ....*I. molle*

1b. Leaves tapering at the base. Lower glume of sessile spikelet with deep ridges across the back in the lower part.....................*I. rugosum*

Ischaemum molle Hook. *f.,* Fl. Brit. Ind. 7: 128. 1896; Blatter & McCann, Bombay Grasses 13. 1935; Bor, Grasses Burma Ceyl. Ind. Pak. 183. 1960; Raizada *et al.*, in Ind. For. Rec. (n.s.) Bot. 4: 201. 1961; Pandey, in Shetty & Singh Fl. Rajasthan 3: 1067. 1993.

Tufted perennial. Culms erect or geniculately ascending, 35 - 110 cm high, often robust, usually branched. Leaves 10 - 20 cm long, 5 - 13 mm wide, linear-lanceolate, acute, sparsely hairy on both surfaces. Sheaths ribbed, glabrous or sparsely hairy. Inflorescences terminal and axillary, of paired racemes each 4 - 14 cm long; internodes and pedicels clavate, ciliate. Sessile spikelet broadly ovate, 5.6 mm long; callus short, bearded. Lower glume coriaceous and nodular on the margins below, the back smooth and shining or faintly rugose, membranous above, densely ciliate to quite glabrous, 2-keeled along its length. Upper lemma with an awn 12 - 20 mm long. Pedicelled spikelet resembling the sessile or often much smoother, and on a pedicel about one-third the length of the internode.

Field Note: Rare, in marshy habitats.

Fl. & Fr.: September - December.

Specimen examined: Sareri 1982 (MLVGCB Herb).

Distribution: India and Pakistan.

Notes: This species is easily confused with *Ischaemum rugosum,* form which it differs by the faintly rugose rather than transversely ridged, and often ciliate lower glume of the sessile spikelet.

Ischaemum rugosum Salisb., Ic. Stirp. Rar. 1. t. 1. 1791; Hook. *f.,* Fl. Brit. Ind. 7: 127. 1896; Blatter & McCann, Bombay Grasses 12. pl. 7. 1935; Bor, Grasses Burma Ceyl. Ind. Pak. 184. 1960; Raizada *et al.* in Ind. For. Rec. (n.s.) Bot. 4: 200. 1961; Pandey, in Shetty & Singh Fl. Rajasthan 3: 1068. 1993. *Ischaemum rugosum* Salib. var. *segetum* (Trin.) Hack. in DC. Monogr. Phan. 6: 208. 1889. *I. segetum* Trin. in Mem. Acad. Sci. Petersb., ser. 6, 2(3): 294. 1832. *Meoschium arnottianum* Nees, in Nova Acta Acad. Caesar. Leop. Carol. 19, suppl. 1: 198. 1843. *M. elegans* Nees & Arn., in Nova Acta Acad. Caesar. Leop. Carol. 19, suppl. 1: 199. 1843. *M. griffithii* Nees & Arn., in Ann. Mag. Nat. Hist. 1: 284. 1838. *M. royleanum* Nees ex Steud., Syn. Pl. Glum. 1: 375. 1854. (Fig. 185)

Vern.: *Bharda*.

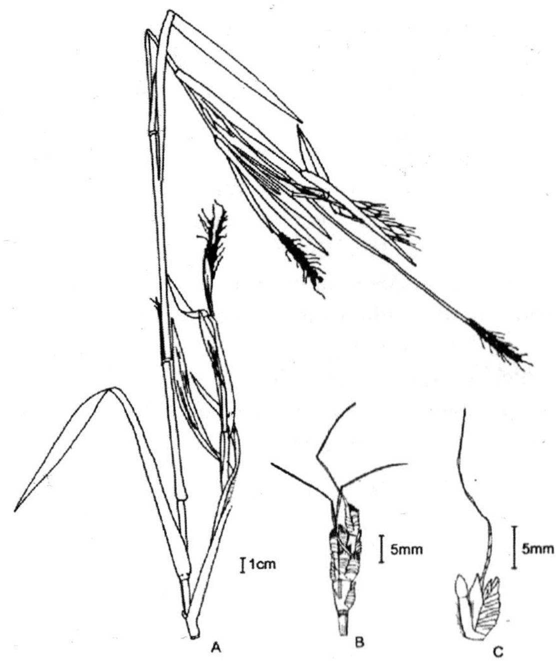

Fig. 185. *Ischaemum rugosum* Salisb.
A. Habit, B. Spike, C. Spikelets.

Straggling annual. Culms 10 - 60 cm high. Leaves 5 - 30 cm long, 3 - 15 mm wide, linear-lanceolate, acuminate, with scabrid margins; ligule truncate. Inflorescences terminal and axillary, of paired racemes each 3 - 10 cm long, internodes and pedicels clavate, ciliate. Sessile spikelet oblong-ovate, 4 - 6 mm long. Lower glume coriaceous and transversely ridged below, membranous above, 2-keeled along its length. Upper lemma with an awn 15 - 20 mm long. Pedicelled spikelet resembling the sessile and on a pedicel one-third the length of the internode, or much reduced and on a pedicel equalling the internode.

Field Note: Common in moist-marshy habitats.

Fl. & Fr.: September - December.

Specimen examined: Dhunwala 1922 (MLVGCB Herb).

Distribution: Tropical Asia and tropical Africa; introduced to most tropical countries.

Notes: It is a very variable species in which the pedicelled spikelet may be large and on a short pedicel (var. *rugosum*) or reduced and on a long pedicel (var. *segetum*).

Uses: It is grazed by cattle when young.

LEPTOCHLOA P. Beauv.

Leptochloa panicea (Retz.) Ohwi, in Bot. Mag. Tokyo. 55: 311. 1941; Bor, Grasses Burma Ceyl. Ind. Pak. 517. 1960; Mahesh. Illus. Fl. Delhi f. 269. 1966; Pandey, in Shetty & Singh Fl. Rajasthan 3: 1072. 1993. *Leptochloa filiformis sensu* Hook. *f.*, Fl. Brit. Ind. 7: 298. 1896. *non* P. Beauv. 1812. *Poa panicea* Retz., Obs. Bot. 3: 11. 1783.

Tufted annual. Culms 25-100 cm high, slender, geniculately ascending. Leaves up to 20 cm long, 7.5 mm wide, linear, flat, long-attenuate; leaf-sheaths and both surfaces of the blades papillate-pilose, the hairs spreading. Inflorescence elliptic to oblong, often purplish, 20 - 30 cm long, composed of numerous straight slender ascending racemes 4 - 11 cm long scattered along the central axis. Spikelets 2-5 (usually 3)-flowered, narrowly elliptic, 1.9-25 mm long, overlapping. Glumes subequal, shorter than the lemmas, lower glume lanceolate, 0.7-1.5 mm long, acute, upper glume narrowly oblong, 0.9-1.6 mm long, obtuse and mucronate; lemmas elliptic oblong, 0.8 - 1.2 mm long, minutely hairy on the back, bluntly 2-toothed. Caryopsis broadly elliptic, 0.5 mm long, brown, obtusely trigonous in cross-section.

Field Note: A weed of cultivated fields.

Fl. & Fr.: August - October.

Specimen examined: Potalan 1917, Shivrati 1993 (MLVGCB Herb).

Distribution: Sudan to Transvaal and Natal; West Africa and tropical Asia (India and Pakistan).

Notes: It is often confused with *Leptochloa filiformis* (Lam.) P. Beauv. a native of warm temperate and tropical America which has long acuminate glumes only slightly shorter than the spikelet and glabrous lemmas. *Leptochloa contracta* (Retz.) Blatter & McCann, Bombay Grasses 243. 1935 is based on *Poa contracta* Retz., Obs. Bot. 3: 11. 1783 which, according to

Bor, Grasses Burma Ceyl. Ind. Pak. 517, is *Diplachne fusca* (Linn.) P. Beauv., *Leptochloa chinensis* (L.) Nees in Syll. Ratisb. 1: 4. 1824, differs from *Leptochloa panicea* in having 3 - 6, rather than 2-3 florets per spikelet and longer glumes. It is found over much of Southeast Asia.

MELANOCENCHRIS Nees

Melanocenchris jacquemontii Jaub. & Spach, Ill. Pl. Or. 4: 36. t. 325. 1851; Bor, Grasses Burma Ceyl. Ind. Pak. 473. 1960; Blatter & McCann, Bombay Grasses 248. 1935; Nair & Nayar, in Bull. Bot. Surv. India 16: 142. f. 1 A-F. 1977; Pandey, in Shetty & Singh Fl. Rajasthan 3: 1074. 1993; Bhandari, Fl. Indian Desert 382. 1995. *M. royleana* Nees ex Steud. Syn. Pl. Glum. 1: 218. 1854; Duthie, Fodder Grasses 54. 1888. *Gracilea royleana* (Nees ex Steud.) Hook. *f.*, Fl. Brit. India 7: 284. 1896.

Annual. Culms fasciculate, rarely solitary, geniculately ascending, 6 - 25 cm high. Leaves 1 - 6 cm long, 1 - 2 mm wide, filiform, scaberulous above and on the margins, and with sparse whitish tubercle-based hairs at the base; sheaths with scattered tubercle-based hairs; ligule a hairy ridge. Spikes 5 - 10 (-12) on a rachis 24 cm long, with triquetrous, smooth, flexuous rachis. Clusters of spikelets including the awns upto 8 cm long, green with purplish tinge, 2-flowered. Glumes, including the awns, up to 8 mm long, rarely longer, usually with white or occasionally purple hairs, the lower a little longer than the upper. Lowest lemma 3-lobed, the lobes aristate, the central longer and narrower at the base than the laterals, 4 - 4.5 mm long. Palea equalling the lemma or slightly shorter than it. Caryopsis upto 2.5 mm long, oblong-elliptic, brownish.

Field Note: A small tufted grass common in stony and barren places.

Fl. & Fr.: August - November.

Specimen examined: Kot 1927, Bhilwara 2008 (MLVGCB Herb).

Distribution: India and Pakistan.

Uses: It is a good grazing grass when young, though rather too small to be of much account.

OPLISMENUS P. Beauv. *nom. cons.*

Oplismenus burmannii (Retz.) P. Beauv., Ess. Agrost. 54, 168, 169. 1812; Duthie, Fodder Grasses 13. 1888; Hook. *f.*, Fl. Brit. Ind. 7: 68. 1896; Blatter & McCann, Bombay Grasses 154. 1935; Bor, Grasses Burma Ceyl. Ind. Pak. 317. 1960; Raizada & Jain, in Ind. For. Rec. (n.s.) Bot. 5: 181. pl. 4. f. 2. 1965; Pandey, in Shetty & Singh Fl. Rajasthan 3: 1076. 1993. *Oplismenus multisetus* A. Rich., Tent. Fl. Abyss. 2: 377. 1851. *Orthopogon burmanii* (Retz.) Trin., Fund. Agrost. 181. 1820. *Panicum bromoides* Lam., Tab. Encycl. 1: 170. 1791. *P. burmanii* Retz., Obs. Bot. 3: 10. 1783.

Prostrate or trailing annul. Culms 10 - 50 cm long. Leaves 1 - 9 cm long, 5 - 20 mm wide, lanceolate to narrowly ovate, acuminate, sparsely pilose; sheaths glabrous or pubescent; ligule a ciliate rim. Inflorescence 2 - 11 cm long racemes, racemes 0.5 - 2.5 cm long, their rachis hirsute, the spikelets contiguous. Spikelets lanceolate, 2.5 - 5.0 mm long, pubescent, often with a transverse band of hairs across the middle of the lower lemma, rarely glabrous. Lower glume thinly membranous, 5-nerved, dorsally pubescent, margins ciliate; awn 2.5 - 10 mm long, pubescent; upper glume similar to the lower, 5 - 7- nerved, awn 2.5 - 7 mm long; lower lemma empty, lanceolate, 2-keeled; awn 1 - 1.5 mm long; palea absent. Upper lemma chartaceous, oblong, 5 - 7- nerved. Palea sub-hyaline, 2-keeled. Caryopsis oblong.

Field Note: This species often forms pure mats in the shaded as well as in open areas.

Fl. & Fr.: July - December.

Specimen examined: Kanechhan kalan 508 (MLVGCB Herb).

Distribution: Throughout tropical Africa, Ttropical America and Asia.

Uses: Grazing grass.

ORYZA L.

Oryza rufipogon Griff., Notul. 3: 5. 1851 & Icon. Pl. Asiat. 3: t. 144. f. 2. 1851; Bor, Grasses Burma Ceyl. Ind. Pak. 605. 1960; Pandey, in

Shetty & Singh Fl. Rajasthan 3: 1079. 1993. *O. nivara* Sharma & Shastri, in Ind. Journ. Gen. & Pl. Breed. 25: 161. 1965. (Plate 333)

Vern.: *Chawal*

Aquatic or marshy, annual or perennial, tufted or stoloniferous. Culms erect, 50 cm - 1 m or more tall. Leaves up to 40 × 1 - 2 cm, linear-lanceolate margins and midrib scabrid, apex acuminate; leaf sheaths slightly inflated below, upper sheaths tight, glabrous, auricles conspicuous, glabrous or ciliate ligule up to 17 mm; ligule membranous, 10 - 15 mm long, truncate or splitting at tip. Panicle effuse. Spikelets oblong, 8 - 24 mm, yellowish green with reddish apex, deciduous; sterile lemmas lanceolate, ca 2.5 mm, keeled, 1-nerved, apex acuminate; fertile lemma with a conical callus, coriaceous, sparsely hispid, apex glandular and terminating into about 6 cm long, scabrid awn with jointed base. Caryopsis elliptic-oblong, reddish brown, 5 - 7 mm.

Field Note: Occasional along the margins of ponds, dams etc.

Fl. & Fr.: September - December.

Specimen examined: Mainal 402 (MLVGCB Herb).

Distribution: Moist Tropical and sub-tropical regions of the world.

Uses: It is consumed as a food by tribals.

PANICUM L.

KEY TO THE SPECIES

1a. Leaves not distichous. Culms soft, spongy. Spikelets narrowly lanceolate. Leaf-blades broad, flat .... *P. paludosum*

1b. Leaves distichous. Culms firm, not spongy. Spikelets lanceolate or oblong-lanceolate. Leaf-blades narrow, often involute .... *P. repens*

Panicum paludosum Roxb., Fl. Ind. 1: 310. 1820; Duthie, Fodder Grasses 11. 1888; Blatter & McCann, Bombay Grasses 162. 1935; Bor, Grasses Burma Ceyl. Ind. Pak. 329. 1960; Raizada & Jain, in Ind. For. Rec. (n.s.) Bot. 5: 157. pl. 1. f. 4. 1965; Majumdar in Bull. Bot. Soc. Bengal 27: 51. f. 19. 1973; Pandey, in Shetty & Singh Fl. Rajasthan 3: 1084. 1993. *P. natans* Koen. ex Trin. Gram. Pan. 237. 1826. *P. proliferum auct.* plur. non Lam. 1797; Hook. *f.*, Fl. Brit. Ind. 7: 50. 1896. *P. proliferum* Roxb. var. *paludosum* (Roxb.) Stapf in Thiselt.-Dyer, Fl. Cap. 7: 407. 1899. *P. repens* L. var. *paludosum* (Roxb.) O. Ktze., Rev. Gen. Pl. 3: 363. 1898.

Aquatic perennial. Culms 25 - 150 cm high, spongy, erect or ascending from the creeping lower portion, this usually rooting at the nodes. Leaves linear to narrowly lanceolate, 5 - 30 cm long, 5 - 14 mm wide, subcordate at the base, flat, glabrous, blunt or sharply acute; sheaths glabrous and loose; ligule a rim of hairs. Panicle ovate or oblong, 10-30 cm long, the branches naked below the lowermost whorled. Spikelets narrowly lanceolate, 3.24 mm long, glabrous, acute or shortly acuminate, appressed to the branches. Lower glume ovate-orbicular, hyaline, up to a quarter the length of the spikelet, obtuse or truncate, 3-5 nerved; upper glume 9-11-nerved. Lower lemma 9-nerved, its palea well developed or absent; upper lemma pallid, smooth and shining.

Field Note: Common creeper in marshy habitats.

Fl. & Fr.: September - December.

Specimen examined: Antali 1921 (MLVGCB Herb).

Distribution: Nepal, India, Pakistan and Ceylon, through Burma to Southeast Asia and tropical Australia

Uses: A vigorous creeping water-grass grazed by animals.

Panicum repens L., Sp. Pl., ed. 2, 1: 87. 1762; Hook. *f.*, Fl. Brit. Ind. 7: 49. 1896; Bor, Grasses Burma Ceyl. Ind. Pak. 330. 1960; Raizada & Jain, in Ind. For. Rec. (n.s.) Bot. 5: 157. pl. 1. f. 5. 1965; Majumdar, in Bull. Bot. Soc. Bengal 27: 51. f. 18. 1973; Pandey, in Shetty & Singh Fl. Rajasthan 3: 1085. 1993.

Perennial, rhizomatous. Culms tough, erect or decumbent, 30-125 cm tall. Leaves cauline, linear-lanceolate, often stiff and pungent, markedly distichous, ascending close to the culm, 5 - 20 × 0.5 - 1.25 cm, apex acute or acuminate;

leaf sheaths glabrous, striate, puberulous to ciliate on margins, especially toward throat; ligule 0.5 - 1.5 mm, a ciliolate membrane. Panicle terminal, narrowly oblong in outline, 5 - 20 cm, sparsely to moderately branched; branches glabrous, scabrid, ascending. Spikelets ovate, 2.5 - 4 mm, acute. Lower glume 0.5 - 1 mm long, ovate-orbicular, obscurely 1-5-nerved; upper glume elliptic - lanceolate, 5-9-nerved, acute. Lower floret staminate, lemma similar to upper glume, palea well developed; upper floret almost as long as spikelet, pale yellow, shiny. Caryopsis elliptic.

Field Note: Rare, in marshy places.

Fl. & Fr.: August - December.

Specimen examined: Amarwasi 2214 (MLVGCB Herb).

Distribution: Tropical and subtropical locations worldwide

PASPALUM L.

KEY TO THE SPECIES

1a. Spikelets ovate, acute at apex. Upper glume herbaceous, appressed puberulent .... *P. paspaloides*

1b. Spikelets broadly-elliptic, obovate or suborbicular, obtuse at apex. Upper glume papery, glabrous .... *P. scrobiculatum*

Paspalum paspaloides (Michaux) Scribner in Mem. Torrey Bot. Club. 5: 29. 1894; Jain & Das in Ind. For. 99: 575. 1973; Pandey, in Shetty & Singh Fl. Rajasthan 3: 1090. 1993; Bhandari, Fl. Indian Desert 387. 1995. *Digitaria paspaloides* Michx. Fl. Bor. Amer. 1: 46. 1803. *Paspalum distichum auct.* non L. Syst. Nat. ed. 10, 2: 855. 1759; Hook. *f.,* Fl. Brit. India 7: 12. 1896; Bor., Grass. Burm. Ceyl. Ind. Pak. 338. 1960; Raizada & Jain, in Ind. For. Rec. (n.s.) Bot. 5: 198 1965.

Creeping stoloniferous perennial. Culms 6 - 50 cm high, rooting at nodes. Leaves 3 - 20 cm long, 2.5 - 7.5 mm wide, flat, linear-lanceolate, glabrous, acute at apex; ligule membrenous. Inflorescence composed of 2-3, one sided, divericate racemes, each raceme 1.5 - 8 cm long, the spikelets single in 2 rows on a flattened rachis 1 - 2 mm wide. Spikelets ovate, 2.5 - 3.5 mm long, plano-convex, relatively plump, pale-green. Lower glume often present as a small scale. Upper glume appressed-puberulent, herbaceous, with a distinct mid-vein. Lower lemma similar but glabrous: upper lemma smooth, pallid at maturity. Caryopsis elliptic-oblong.

Field Note: Common along ditches and irrigation channels, as a weed in fields, and is gregarious in swampy places and moist habitats.

Fl. & Fr.: September -December.

Specimen examined: Shahpura 1981, Triveni 2207 (MLVGCB Herb).

Distribution: Tropics and sub-tropics throughout the world.

Notes: *Paspalum paspaloides* closely resembles the more tropical *Paspalum vaginatum* Sw. The latter can be recognised by its narrowly ovate-elliptic spikelets, the papery upper glume and lower lemma, and the glabrous upper glume.

Paspalum scrobiculatum L., Mant. 1: 29. 1767; Blatter & McCann, Bombay Grasses 136. 1935; Bor, Grasses Burma Ceyl. Ind. Pak. 340. 1960; Raizada & Jain, in Ind. For. Rec. (n.s.) Bot. 5: 198. pl. 7. f. 2. 1965; Pandey, in Shetty & Singh Fl. Rajasthan 3: 1091. 1993; Bhandari, Fl. Indian Desert 387. 1995. *P. cartilagineum* Presl, Rel. Haenk. 1: 216. 1830; Bor, *l.c.* 335. 1960. *P. comm-ersonii* Lam. Tab. Encycl. Meth. Bot. 1 : 175, t. 43. f. 1. 1791; Bor, *l.c.* 335. 1960. *Paspalum kora* Willd., Sp. Pl. 1 (1): 332. 1798. *P. orbiculare* Forst., Fl. Ins. Austr. Prodr. 7. 1786; Bor, *l.c.* 340. 1960. *P. polystachyum* R.Br. Prodr. 188. 1810. *P. scrobiculatum* L. var. *commersonii* (Lam.) Stapf, in Prain Fl. Trop. Afr. 9: 573. 1919.

Vern.: *Khoda*.

Perennial. Culms 10 - 120 cm high, erect or ascending from a procumbent base and rooting at the lower nodes. Leaves 5 - 20 cm long, 3 - 15 mm wide, linear, tapering to a filiform tip. Inflorescence composed of 1 - 20 racemes, these digitate or borne on an axis up to 8 cm long, the lowest raceme 4 - 15 cm long, with spikelets borne singly on a ribbon-like rachis 1 - 2.5 mm wide. Spikelets broadly elliptic, obovate or suborbicular, 1.4 - 3 mm long, green, becoming brown. Lower glume absent; upper glume papery. Lower lemma similar or rarely coriaceous, 3 - 5-nerved; upper lemma finely striate, brown at maturity. Caryopsis subglobose.

Field Note: Common in wet places, along ditches and water streams.

Fl. & Fr.: August - December.

Specimen examined: Pander 398, Raipur 2143 (MLVGCB Herb).

Distribution: Throughout the Old World tropics.

Notes: This is a highly polymorphic species, but with the variation apparently quite continuous; possibly it is a swarm of apomicts (see Clayton 1975).

PENNISETUM L. C. Rich.

KEY TO THE SPECIES

1a. Spikelets solitary, sessile. Lower glume minute or small .... *P. hohenackeri*

1b. Spikelets solitary and pdicelled or in groups of 2 - 5, with one sessile and the other pedicelled. Lower glume atleast half the length of the lower lemma..... ..*P. pedicellatum*

Pennisetum hohenackeri Hochst. ex Steud., Syn. Pl. Glum. 1: 103. 1854; Bor, Grasses Burma Ceyl. Ind. Pak. 344. 1960; Raizada & Jain, in Ind. For. Rec. (n.s.) Bot. 5: 209. pl. 9. f. 4. 1965; Pandey, in Shetty & Singh Fl. Rajasthan 3: 1092. 1993. *P. alopecuros* Nees ex Steud., Syn. Pl. Glum. 1: 102. 1854, non Jacq. 1844; Hook. *f.,* Fl. Brit. Ind. 7: 84. 1896; Blatter & McCann, Bombay Grasses 178. pl. 113. 1935.

Perennial, forming large tussocks with thick tough roots. Culms 60 - 150 (-200) cm high, erect. Leaves 15 - 40 cm long, 1 - 5 mm wide, narrowly linear, rigid, tightly folded, ribbed on the outside, the ribs smoothly flat-topped on the inside; sheaths markedly keeled and imbricate; ligule hairy. Panicle linear, 10 - 25 cm long; rachis with shallow angular ribs below the cupular scars, scaberulous; involucre enclosing 1 sessile spikelet, the base with a short oblong stipe 0.3 - 0.5 mm long; bristles glabrous, the longest 6-30 mm long. Spikelets narrowly lanceolate, 7 - 11 mm long. Lower glume up to a quarter the length of the spikelet, acute, rotund or truncate; upper glume one-third to half as long as the spikelet, acute to emarginate. Lemmas similar, the lower male or barren, the upper firmer in texture; anther tips with or without a minute tuft of hairs. Caryopsis ca 6 mm long, oblong, brownish.

Field Note: Common along water streams, river beds, dam sides etc.

Fl. & Fr.: August - February.

Specimen examined: Nagdidam 1367 (MLVGCB Herb).

Distribution: Tropical Africa, Madagascar and Western India.

Notes: Asian specimens of *Pennisetum hohenackeri* often have a truncate lower glume, but never have penicillate anthers; in Africa the reverse is true. However, many specimens from both regions are otherwise identical, and it seems impracticable to recognise more than one taxon.

Pennisetum pedicellatum Trin. in Mem. Acad. Sci. Petersb. ser. 6. 3: 184. 1834, *pro parte,* Hook. *f.,* Fl. Brit. Ind. 7: 86.1896, *pro parte,* Bor, Grasses Burma Ceyl. Ind. Pak. 346. 1960, *pro parte,* Raizada & Jain, in Ind. For. Rec. (n.s.) Bot. 5: 209. pl. 9. f. 1. 1965; Pandey, in Shetty & Singh Fl. Rajasthan 3: 1094. 1993.

subsp. ***pedicellatum***

Tufted, annual. Culms 30 - 80 cm high, much branched. Leaves 5 - 25 cm long, 4 - 15 mm wide, flat, linear-lanceolate, acuminate, truncate or rounded at the apex. Panicle cylindrical, 5 - 15 cm long; rachis with sharp decurrent wings below the involucral scars, glabrous; involucre fluffy, ovate, 0.5 - 1 cm long, enclosing 1 - 5 spikelets at least one of them upon a pedicel 0.5-3.5 mm. long; bristles densely woolly. Spikelets lanceolate, 3.5 - 6 mm. long; lower glume lanceolate to oblong, 1 - 3 mm long; upper glume as long as the spikelet, apiculate; lower lemma similar but tridentate, Male or barren; upper lemma coriaceous, shining, readily deciduous at maturity, obtuse.

Field Note: Common in grasslands and in rocky habitats.

Fl. & Fr.: August - December.

Specimen examined: Sahara 1376 (MLVGCB Herb).

Distribution: Tropical Africa and Asia.

PEROTIS W. Ait.

Perotis hordeiformis Nees, in Hook. & Arn. Bot. Beechey Voy. 247, 248. 1838; Bor., Grass. Burm. Ceyl. Ind. Pak. 611. 1960; Pandey, in Shetty & Singh Fl. Rajasthan 3: 1096. 1993; Bhandari, Fl. Indian Desert 388. 1995. *P. latifolia* Ait., Hortus Kew. 1: 85. 1789, *pro parte*; Hook. *f.*, Fl. Brit. India 7: 98. 1896, *pro parte*.

Annual or short-lived perennial. Culms loosely tufted, erect or decumbent at base, 25 - 50 cm tall. Leaves lanceolate to narrowly ovate, 2 - 5 cm, 4 - 7 mm wide, glabrous, margins spiny-scabrous or loosely ciliate near base; sheaths loose, glabrous; ligule membranous, ca 0.5 mm. Inflorescence up to 20 cm long, delicate, like bottle brush, slender, purplish-green. Spikelets usually densely arranged, 1.5 - 2.5 mm, ascending or horizontally spreading, basal callus 0.1 - 0.2 mm, obtuse; rachis scabrous. Glumes narrowly lanceolate, back scaberulous-hirtellous in rather indistinct close rows, midvein pectinate-ciliolate, apex abruptly acute, clearly demarcated from awn, 1-nerved, awn 0.5 - 1.5 cm. Lemmas ca 1.5 mm long, solitary. Anthers ca 0.3 mm. Caryopsis ca 1.5 mm long, subcylindric, pale brown.

Field Note: Common in sandy habitats.

Fl. & Fr.: August - November.

Specimen examined: Parsoli 379 (MLVGCB Herb).

Distribution: China, India, Indonesia, Malaysia, Nepal, Myanmar, Pakistan, Sri Lanka, and Thailand.

Notes: It is very close to *Perotis indica* (L.) O. Ktze. and is sometimes included within it. No single character by itself is reliable for separating the two, but the combination of characters given in the key will usually suffice.

PHALARIS L.

Phalaris minor Retz. Obs. Bot. 3: 8. 1783; Boiss. Fl. Or. 5: 472. 1884; Duthie, Fodder Grasses 45. 1888; Hook. *f.*, Fl. Brit. Ind. 7: 221. 1896; Bor, Grasses Burma Ceyl. Ind. Pak. 616. 1960; Mahesh. Illus. Fl. Delhi f. 246. 1966; Pandey, in Shetty & Singh Fl. Rajasthan 3: 1097. 1993. (Plate 334)

Annual. Culms 20-100 cm high. Leaves flat, linear-lanceolate, scabrid along margins; ligule membranous 5-10 mm long. Panicle contracted, 1-6 x 1 - 2 cm, ovate-oblong. Spikelets ca 5 mm long, oblong, acute laterally strongly compressed. Glumes 4 - 6.5 mm long, equal, broadly winged, the wing margins usually erose-denticulate, occasionally entire. Sterile floret, 1 - 2 mm long or very short and only 0.2 - 0.3 mm long, glabrous; fertile floret broadly lanceolate-ovate, 2.7 - 4 mm long, grey-brown and shiny at maturity, pubescent. Caryopsis closely invested by the indurated lammas and paleas.

Field Note: It is a winter season weed of in cultivated fields and waste places.

Fl. & Fr.: January - March.

Specimen examined: Jahajpur 75 (MLVGCB Herb).

Distribution: Throughout the world, but apparently native only in the Mediterranean region and eastwards to Baluchistan and the Northwest Himalayas.

Uses: It (*"Lesser Canary Grass"*) is considered as a useful forage grass for sheep and cattle.

POLYPOGON Adanson

Polypogon monspeliensis (L.) Desf. Fl. Atlant. 1: 67. 1798; Boiss. Fl. Or. 5: 520. 1884; Hook. *f.*, Fl. Brit. Ind. 7: 245. 1896; Blatter & McCann, Bombay Grasses 207. 1935; Bor, Grasses Burma Ceyl. Ind. Pak. 403. 1960;

Mahesh. Illus. Fl. Delhi f. 274. 1966; Pandey, in Shetty & Singh Fl. Rajasthan 3: 1099. 1993; Bhandari, Fl. Indian Desert 389. 1995. *Alopecurus monspeliensis* L., Sp. Pl. 1: 61. 1753.

Vern.: *Malar*.

Tufted annual. Culms 6 - 80 cm high, erect or geniculately ascending. Leaves 2.5-20 cm long; 2-8 mm wide, flat, broadly lanceolate, acute, rough; ligule membranous, 2-12 mm long. Panicle narrowly ovate to narrowly oblong, cylindrical or lobed, 1.5-16 cm long, 1-3.5 cm wide, very dense and bristly, pale green or yellowish. Spikelets 1.2 - 3 mm long, pale greenish -yellow. Lower glume oblong to obovate-oblong, scabrid on keel, ciliate along margins, awn delicate, 2-4.5 mm long; upper glume little longer. Lemma hyaline, obovate-oblong, about half the length of the glumes, smooth, nerveless, apex 2-fid, with an awn up to 2 mm long or awnless. Palea delicate, notched at the tip. Caryopsis ovoid,

Field Note: Common in open gravelly-sandy waste lands, along water streams and fallow fields.

Fl. & Fr.: July - February.

Specimen examined: Choudhriyas 2213 (MLVGCB Herb).

Distribution: Northeast tropical and South Africa; Mediterranean region northwards to the British Isles and eastwards to India, Pakistan and China; introduced and naturalised in most warm temperate countries.

Uses: Little value for fodder.

PSEUDORAPHIS Griff.

Pseudoraphis spinescens (R. Br.) Vickery, Proc. Roy. Soc. Queensland. 62 (7): 69. 1952; Bor., Grass. Burm. Ceyl. Ind. Pak. 353. 1960; Pandey, in Shetty & Singh Fl. Rajasthan 3: 1100. 1993. *Panicum asperum* Koenig in Naturf. 23: 209. 1788, non Lam. 1778. *P. spinescens* R. Br. Prod. Fl. Nov. Holl. 193. 1810. *Chamaeraphis spinescens* (R. Br.) Poir., in Lam. Encycl. Meth. Bot. Suppl. 2: 189. 1811; Hook. *f.*, Fl. Brit. India 7: 62. 1896. *Pseudoraphis aspera* (Koenig) Pilger., in Notizbl. Bot. Gard. Berlin 10: 210. 1928, *non illegit*; Blatter & McCann, Bombay Grasses 168. pl. 109. 1935; Raizada & Jain, in Ind. For. Rec. (n.s.) Bot. 5: 207. pl. 6. f. 4. 1965.

Perennial, floating, pubescent at nodes, 20 - 100 cm, much branched. Leaves 3 - 7 cm long, 2 - 4 mm wide, lanceolate-linear, with a petiole of ca 1 mm long, membranaceous, erose, acuminate; ligule a ridge of minute hairs. Panicle 6 - 12 cm long, branches stiff, extending beyond spikelet into a bristle of 4 - 10 mm long. Spikelets hispid, lanceolate, 4 - 7 mm long, solitary on pedicel of ca 0.5 mm long. Lower glume truncate or rounded, hyaline, ca 0.5 mm long; upper glume 5 - 7 mm long, lanceolate, acuminate, hispid on veins. Lower lemma slightly shorter than upper glume, veins ca 13, obscure, acuminate, enclosing a lower palea and a staminate flower; upper lemma pistillate 1 - 2 mm long, membranaceous. Anther ca 1 mm long. Caryopsis elliptic, much exceeding upper floret when mature.

Field Note: Growing in shallow water.

Fl. & Fr.: September - December.

Specimen examined: Triveni 504, Jaleshwar Tank (Mandalgarh) 1375 (MLVGCB Herb).

Distribution: India, Sri Lanka, Myanmar. Indochina, Malesia, northern Australia, China and Japan.

ROTTBOELLIA L. f. nom. cons.

Rottboellia exaltata L. *f.*, Nov. Gram. Gen. 40 t. 1. 1779 & Suppl. Pl. 114. 1781; Duthie, Fodder Grasses 28. 1888; Hook. *f.*, Fl. Brit. Ind. 7: 156. 1896; Blatter & McCann, Bombay Grasses 38. Pl. 27. 1935; Bor, Grasses Burma Ceyl. Ind. Pak. 206. f. 13. 1960; Raizada *et al.*, in Ind. For. Rec. (n.s.) Bot. 4: 208. 1961; Pandey, in Shetty & Singh Fl. Rajasthan 3: 1102. 1993. *Manisuris exaltata* (L. *f.*) O. Ktze., Rev. Gen. Pl. 2: 779. 1891. *Rottboellia arundinacea* Hochst. ex A. Rich., Tent. Fl. Abyss. 2: 444. 1851. (Plate 335)

Vern.: *Phallo*.

Tall, erect, annual supported below by stilt roots, the basal sheaths painfully hispid. Clums 30 cm to 3 m high. Leaves up to 45 cm long and 2 cm wide, broadly linear, acute, scabrid; sheaths with tubercle - based hairs; ligule membranous, rounded, ciliolate. Raceme 3-15 cm long, glabrous, terminating in a tail of reduced spikelets, gathered into a spatheate false panicle. Sessile spikelets 6-8 mm, oblong-elliptic, pallid. Lower glume, coriaceous, minutely 2-toothed, 3.5 - 5 mm long; upper glume with the keel narrowly winged towards the tip, many nerved. Pedicelled spikelet narrowly ovate, 3 - 8 mm long, herbaceous, green. Pedicel shorter then the internode.

Field Note: Occasional along water streams.

Fl. & Fr.: August - December.

Specimen examined: Thaggo ka khera 2205 (MLVGCB Herb).

Distribution: Throughout the Old World tropics and introduced to the Caribbean.

Uses: A good fodder grass but not common.

SACCHARUM L.

KEY TO THE SPECIES

1a. Leaves neither petiolate nor winged at the base. Lower glume of sessile spikelet hairy on the back. Upper glume of sessile spikelete glabrous .... *S. bengalense*

1b. Leaves petiolate. Lower glume of sessile spikelet glabrous on the back. Upper glume also glabrous or sparsely ciliate on the lateral nerves .... *S. spontaneum*

Saccharum bengalense Retz., Obs. Bot. 5: 16. 1789; Bor, Grasses Burma Ceyl. Ind. Pak. 211. 1960; Pandey, in Shetty & Singh Fl. Rajasthan 3: 1103. 1993. *Erianthus sara* (Roxb.) Rumke in Archf Suik. Ind. Ned.-Indie 1934, No. 7 : 223. t. 2. 1934. *Ripidium bengalense* (Retz.) Grassl in Proc. 14th Cong., ISSCT. 244. 1972. *Saccharum arundinaceum* Hook. *f.*, Fl. Brit. Ind. 7: 119. 1896, *non* Retz. 1786; Blatter & McCann, Bombay Grasses 47. 1935. *S. ciliare* Anderss., in Ofvers. K. VetenskAkad. Forh. 12: 155. 1855. *S. munja* Roxb., Fl. Ind. 1 : 246. 1820; Blatter & McCann, *l.c.* 48. 1935. *S. sara* Roxb., *l.c.* 249. 1820; Boiss., Fl. Or. 5: 453.1884.

Vern.: *Hanti*.

Tall caespitose perennial. Culms up to 4 m high, solid. Leaves up to 90 cm long, 3 - 24 mm wide, flat or markedly channelled, linear, acuminate, midrib occupying the greater part of the width, glaucous; sheaths very long, hairy at the base and margins, striate; ligule membranous. Panicle 20 - 75 cm long, the peduncle glabrous. Racemes 2 - 4 cm long, considerably shorter than the supporting branches, the internodes and pedicels hirsute with hairs up to 7 mm long. Spikelets slightly heteromorphous, 3.8 - 5.5 mm long, the callus bearded with whitish or greyish hairs up to 2.5 mm long. Glumes equal, membranous, lower glume of sessile spikelet hairy on the back, the upper glume glabrous, both glumes of pedicelled spikelet hairy, the hairs at least 4 mm long, often up to 9 mm. Lower lemma oblong-elliptic, hairy on the back; upper lemma ovate-lanceolate, ciliate on the margins, acute or very shortly awned, the awn not visible beyond the glumes. Stamens 3. Caryopsis oblong, brown.

Field Note: Common along the hedges, river beds and in moist habitats.

Fl. & Fr.: September - March.

Specimen examined: Bigod 183, Khajuri 907 (MLVGCB Herb).

Distribution: North and Northwest India, Pakistan and Afghanistan

Uses: It is a very large tufted grass which is of little account as a fodder plant since cattle grazed the tender young leaves only. Valuable fibres are obtained from the upper leaf-sheaths of the flowering culm.

Saccharum spontaneum L. Mant. 2: 183. 1771; Hook. *f.*, Fl. Brit. Ind. 7: 118. 1896; Panje in Indian J. Agric. Sci. 3: 1013. 1933; Blatter & McCann, Bombay Grasses 45. Pl. 31. 1935; Mukerjee in Bull. Bot. Soc. Bengal 8 : 145. 1949; Bor, Grasses Burma Ceyl. Ind. Pak. 214. 1960; Raizada *et al.* in Ind. For. Rec. (n.s.) Bot. 4: 185. pl. 2. f. 3. 1961; Pandey, in Shetty & Singh Fl. Rajasthan 3: 1104. 1993. *Imperata spontanea* (L.) P. Beauv. Ess. Agrost. 8. 1812. *Saccharum canaliculatum* Roxb., Fl. Ind. 1: 251. 1820. *S. propinquum* Steud. Syn. Pl. Glum. 1 : 406. 1855. *S. semidecumbens* Roxb. *l.c.* 1: 241. 1820. (Plate 336)

Vern.: *Kans.*

Tall, rhizomatous, perennial. Culms 1 - 2 m high or more, solid above and fistular below. Leaves 0.5 - 2 m long, up to 7.5 mm wide, rigid, coriaceous, folded, narrowly linear, margins scabrid; gradually tapered towards the base into a narrow wing on either side of the petiole, glaucous; sheaths auricled, galbrous; ligule membranous, triangular. Panicle 25 - 50 (-60) cm long, the peduncle usually hairy. Racemes 3 - 15 cm long, usually much longer than the supporting branches, the internodes and pedicels hirsute. Spikelets all alike, 2.5 - 5.5 (-7) mm long, the callus bearded with silky white hairs, 2 - 3 times as long as the spikelet. Glumes equal, subcoriaceous in the lower third, glabrous on the back, often ciliate on the margins above. Lower lemma lanceolate-elliptic, ciliate on the margins; upper lemma very narrow, very shortly awned. Stamens 3. Caryopsis oblong.

Field Note: Common in corners of rivers and in moist habitats

Fl. & Fr.: October - February.

Specimen examined: Chanderiya, Mangaol 157 (MLVGCB Herb).

Distribution: India and Pakistan; Widely distributed in the warmer regions of the Old World.

Notes: Asian plants with the leaves narrowed to the midrib towards the base and with a more or less triangular ligule are referable to subsp. *spontaneum*. In contrast, the African subsp. *aegyptiacum* (Wald.) Hack. has the leaves laminate to the base and the ligule is crescent-shaped. The latter also occurs in parts of the Middle East.

Uses: This species flowers and fruits at the end of the rains and is therefore capable of colonising areas such as soil and sand left bare by retreating floods. The root-system is extremely extensive and the grass acts as an effective sand-binder.

SCHOENEFELDIA Kunth

Schoenefeldia gracilis Kunth, Rev. Gram. 1: 283. t. 53. 1830; Boiss., Fl. Or. 5: 553. 1884; Duthie, Fodder Grasses 52. 1888; Bor, Grasses Burma Ceyl. Ind. Pak. 474. 1960; Pandey, in Shetty & Singh Fl. Rajasthan 3: 1105. 1993; Bhandari, Fl. Indian Desert 391. 1995. *Chloris myosuroides* Hook. *f.*, Fl. Brit. India 7: 290. 1896. *C. pallida* (Edgew.) Hook. *f.*, *l.c.* 7:289. 1896; Blatter & McCann, Bombay Grasses 252. pl. 167. 1935. *Schoenefeldia pallida* Edgew in J. Asiat. Soc. Beng. 21: 183. 1852.

Annual. Culms tufted, erect or geniculately ascending, 10 - 70 cm high, glabrous and smooth. Leaves up to 25 cm long, 3 mm broad, narrowly linear, acuminate, scaberulous on both sides or smooth beneath, glabrous or pilose towards the base; ligule hairy. Spikes 1-4, 3-20 cm long, straight or slightly flexuous, pale, slender, the rachis 0.4 - 0.6 mm wide. Spikelets 3 - 5 mm long, 1-flowered. Glumes 4, shortly aristate; lower 2.5 - 4.5 mm long, lanceolate, subulate, acuminate; upper glume 3 - 5 mm long, lanceolate, obscurely keeled, 1-nerved. Lower floral glume ovate-lanceolate, subacute, ciliate, hyaline, nerveless; upper floral glume very slender, ciliate. Palea minute ciliate. Caryopsis ca 1.25 mm long, linear, pericarp loose.

Field Note: Gregarious in dry sandy places and is easily recognised by the beautifully braided awns of its spikes.

Fl. & Fr.: August - December.

Specimen examined: Parsoli 1343 (MLVGCB Herb).

Distribution: tropical Africa, Arabia India and Pakistan.

SEHIMA Forsk.

Sehima nervosum (Rottl.) Stapf in Prain, Fl. Trop Africa. 9: 36. 1917; Blattet & McCann, Bombay Grasses 21. Pl. 14. 1935; Bor, Grasses Burma Ceyl. Ind. Pak. 218. 1960; Raizada *et al.* in Ind. For. Rec. (n.s.) Bot. 4: 198. 1961; Pandey, in Shetty & Singh Fl. Rajasthan 3: 1108. 1993. *Andropogon nervosus* Rottl. *apud* Willd. in Verh. Ges. Naturf. Freunde Berlin Neue Schr. 4 : 218. 1803. *A. striatus* Klein ex Willd. Sp. Pl. 4 (2) : 903. 1806. *Ischaemum laxrum* R. Br. Prodr. 205. 1810, *pro parte,* Duthie, Fodder Grasses 31.

1888; Hook. *f.*, Fl. Brit Ind. 7: 136. 1896. *I. nervosum* (Rottl.) Thw. Enum. Pl. Zeyl. 365. 1864. (Fig. 186)

Fig. 186. *Sehima nervosum* (Rottl.) Stapf.
A. Habit, B. Portion of raceme,
C. Pair of spikelet

Tufted perennial. Culms erect, 30-90 cm high, wiry. Leaves 5-20 cm long, 8 mm wide, flat, linear, glaucous, long acuminate to a filiform tip; sheaths terete, striate, glabrescent. Raceme 3-12 cm long, terminal, solitary, gently curved. Sessile spikelet 6-11 mm long, laterally compressed, acuminate. Lower glume narrowly oblong-elliptic, thinly coriaceous with a membranous tip up to a quarto its length, 2 keeled, concave between the keels, lyrately nerved with 2 - 3 closely spaced nerves adjacent to each keel. Upper lemma with an awn 2 - 4 cm long, minutely ciliolate along the spiral. Pedicelled spikelet lanceolate, 4 - 10 mm long, purplish, ciliate. Lower glume with a mid-nerve and 2 - 3 conspicuous lateral nerves adjacent to each keel.

Field Note: Common in open grasslands.

Fl. & Fr.: September - December.

Specimen examined: Piploond 76 (MLVGCB Herb).

Distribution: Tropical Africa and Arabia, extending through India to China, and southwards to Australia.

Uses: It is an excellent fodder grass.

SETARIA P. Beauv. *nom. cons.*

KEY TO THE SPECIES

1a. Bristles retrorsely barbed........... *S. verticillata*

1b. Bristles antrorsely barbed 2

2a. Upper glumes as long as upper lemma, the latter quite smooth........................*S. italica*

2b. Upper glumes shorter than the upper lemma, the latter distinctly rugose 3

3a. Spikelets ca 3 mm long, upper lemma coarsely rugose, cymbiform, slightly keeled upwards.................................. *S. glauca*

3b. Spikelets 2 - 2.5 mm long, upper lemma finely rugose, narrow, not keeled...... *S. pumila*

Setaria glauca (L.) P. Beauv., Es. Agrostogr. 51: 169. 178. 1812; Hook. *f.*, Fl. Brit Ind. 7: 78. 1896; Bor, Grasses Burma Ceyl. Ind. Pak. 360. 1960; Raizada & Jain, in Ind. For. Rec. (n.s.) Bot. 5: 204. pl. 8. f. 5. 1965; Pandey, in Shetty & Singh Fl. Rajasthan 3: 1111. 1993. *Panicum glaucum* L., Sp. Pl. 56. 1853 (var.) & Syst. Nat. ed. 10. 870. 1759 & Sp. Pl. ed. 2: 83. 1762. (Fig. 187 & Plate 337)

Vern.: *Kutta ghas.*

Annual, erect to geniculate. Culms up to 60 cm high. Leaves 5 - 25 x 0.2 - 1.5 cm, linear, acuminate, base rounded, rough on margins; sheaths keeled, glabrous; ligules ciliate. Inflorescence ca 10 cm long, dense, cylindrical, spike like panicle, yellow at maturity; axis densely pubescent; bristles 5 - 20 in a cluster, scabrous, very unequal, 6 - 10 mm long. Spikelets broadly elliptic, ca 3 mm long. Lower glume 1 - 1.5 mm long, hyaline, broadly ovate, 3-nerved, apiculate; upper glume ca 1.5 mm long, suborbicular, 5-nerved. Lower floret male or barren. Lower lemma membranous, 5 - nerved; upper ones coarsely rugose, boat-shaped. Anthers 3, ca 1.8 mm long. Upper floret cymbiform, convex on back, distinctly transversely rugose. Palea granular-punctate. Anthers 3, ca 1 mm long. Caryopsis rounded, ellipsoid, ca 2 mm long, plano-convex; hilum punctate; embryo more than half.

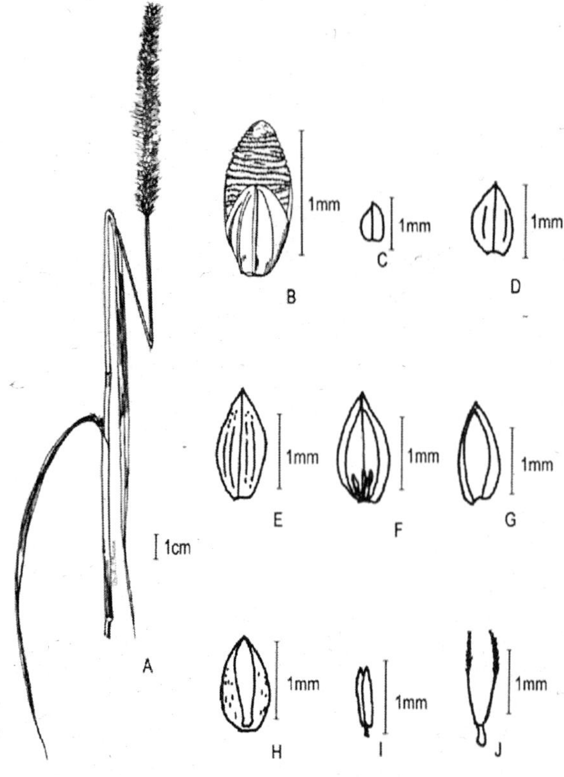

Fig. 187. *Setaria glauca* (L.) P. Beauv.
A. Habit, B. Spikelet, C. Lower involucral glume, D. Upper involucral glume, E. Lower floral glume, F. Palea of lower glume with stamens, G. & H. Upper floral glume & its palea. I. Stamen, J. Pistil

Field Note: Common in grasslands.

Fl. & Fr.: August - December.

Specimen examined: Jahajpur 76 (MLVGCB Herb).

Distribution: India, Burma, Pakistan and Sri Lanka; introduced to tropical Africa.

Uses: Used as a fodder grass.

Setaria italica (L.) P. Beauv., Ess. Agrost. 51. 170. 1812; Hook. *f.,* Fl. Brit Ind. 7: 78. 1896; Bor, Grasses Burma Ceyl. Ind. Pak. 362. 1960; Pandey, in Shetty & Singh Fl. Rajasthan 3: 1113. 1993. *Panicum italicum* L., Sp. Pl. 56. 1753. (Plate 338)

Annual. Culms robust, erect, up to 100 cm, nodes glabrous. Leaves soft, linear-lanceolate, 15 × 0.35 cm, usually glabrous, ciliate on margins; sheaths glabrous or pubescent, ciliate; ligule 1 - 3 mm, densely ciliate. Inflorescence a condensed, 5 - 10 cm long panicle; branches spirally arranged. Spikelets broadly elliptic to ovate or subglobose, 2 - 3.5 mm long, yellow, brown, orange or purple; lower glume ca 1 mm long; upper glume elliptic, 2.5-3.5 mm long, concave, 5-7 (-9)-nerved, obtuse; lower lemma equal to spikelet, 5-7-nerved. Anthers 3, yellow, 2 - 2.5 mm long. Caryopsis light yellow, ellipsoid.

Field Note: Common in shady areas of grass lands throughout the south central Rajasthan.

Fl. & Fr.: August - December.

Specimen examined: Raipur 180 (MLVGCB Herb).

Distribution: Pantropical

Uses: It is a useful forage grass.

Setaria pumila (Poir.) Roem. & Schult., Syst. Veg. 2: 891. 1817; Clayton, in Tutin *et al.*, Fl. Eur. 5: 263. 1980; Pandey, in Shetty & Singh Fl. Rajasthan 3: 1113. 1993. *Panicum pumilum* Poir., in Lam. Encycl. Meth. Bot. Suppl. 4. 273. 1816. *P. pallide-fuscum* K. Schum. Beskr. Guin. Pl. 58. 1827. *P. rubiginosum* Steud., Syn. Pl. Glum. 1: 50. 1854. *Setaria pallide-fusca* (K. Schum.) Stapf. & Hubbard. in Kew Bull. 1930: 259. 1930; Bor, Grasses Burma Ceyl. Ind. Pak. 363. 1960; Raizada & Jain, in Ind. For. Rec. (n.s.) Bot. 5: 204. pl. 8. f. 6. 1965. *S. rubiginosa* (Steud.) Miq., Fl. Ned. Ind. 3: 467. 1857.

Vern.: *Kangni*.

Loosely tufted annual. culms 5 - 70 cm high, ascending. Leaves 3-20 cm long, 2-12 mm wide, flat or convolute, linear-lanceolate; ligule membranous, truncate, ciliate. Panicle spiciform, cylindrical, 1-10 cm long, the rachis tomentellus; bristles 3-12 mm long, slender, commonly fulvous. Spikelets ovate, 1.5-3.5 mm long, solitary, enclosed in an antrorsely barbellate. Both glumes one-third to two-third the length of the spikelet.

Lower floret male or sterile, its palea almost as long as the lemma; upper lemma rugose to corrugate, rarely almost smooth. Caryopsis 1.5 - 1.75 mm long, plano - convex, elliptic-rotundate.

Field Note: Common in shady areas of grass lands throughout the area.

Fl. & Fr.: August - December.

Specimen examined: Dungala 1813 (MLVGCB Herb).

Distribution: Tropical and warm temperate regions of the Old World; introduced to North America.

Notes: *Panicum pumila* is a polymorphic weed segregating into regional populations. These overlap to such an extent that the recognition of more than one species cannot be uphold, though there may be a case for distinguishing between tropical and Mediterranean populations at infra-specific level (see Clayton in Kew Bull. 33 : 501. 1979). Much argument has been devoted to the nomenclature of this species (summarised by Terrell in Taxon 25 : 297-304. 1976), but it now seems that the commonly used name *Panicum glauca* L. should be referred to *Pennisetum*.

Setaria verticillata (L.) P. Beauv. Ess. Agrost. 51, 178. 1812; Boiss., Fl. Or. 5: 443. 1884; Duthie, Fodder Grasses 15. 1888; Hook. *f.*, Fl. Brit. Ind. 7: 80. 1896; Blatter & McCann, Bombay Grasses 174. 1935; Bor, Grasses Burma Ceyl. Ind. Pak. 365. 1960; Raizada & Jain, in Ind. For. Rec. (n.s.) Bot. 5: 203. pl. 8. f. 2. 1965; Pandey, in Shetty & Singh Fl. Rajasthan 3: 1114. 1993; Bhandari, Fl. Indian Desert 393. 1995. *Panicum verticillatum* L., Sp. Pl., ed. 2. 1: 82. 1762.

Vern.: *Kutta ghas*.

Loosely tufted annual. Culms 10 - 100 cm high or more, geniculately ascending. Leaves 5 - 30 cm long, 3 - 16 mm wide, linear - lanceolate, flaccid, glabrous to loosely pilose, acuminate; sheaths glabrous to pubescent; ligule truncate. Panicle spiciform, linear to untidily lobed, 2 - 15 cm long, often entangled, the rachis hispidulous; bristles few, 3 8 mm long, retrorsely barbed, tenaciously clinging. Spikelets ellipsoid, 1.5 - 2.5 mm long, enclosed in an involucre. Lower glume ca 1 mm long, hyaline, acute; upper glume as long as the spikelet, 5 - 7-nerved, obtuse. Lower floret barren, the palea minute; upper lemma finely rugose. Caryopsis elliptic.

Field Note: Common on old walls, waste lands, hedges of agriculture fields etc.

Fl. & Fr.: August - December.

Specimen examined: Bhilwara 77 (MLVGCB Herb).

Distribution: Tropical and warm temperate regions.

Uses: It is a good fodder grass.

SORGHUM Moench. nom. cons.

Sorghum halepense (L.) Pers., Syn. Pl. 1: 101. 1805; Blatter & McCann, Bombay Grasses 55. Pl. 37. 1935; Bor, Grasses Burma Ceyl. Ind. Pak. 222. 1960; Raizada *et al.* in Ind. For. Rec. (n.s.) Bot. 4: 215. 1961; Pandey, in Shetty & Singh Fl. Rajasthan 3: 1115. 1993; Bhandari, Fl. Indian Desert 394. 1995. *Holcus halepensis* L., Sp. Pl. 2: 1047. 1753. *Andropogon halepensis* (L.) Brot. Fl. Lusit. 1 : 89. 1804; Hook. *f.*, Fl. Brit. Ind. 7: 182.1896. *A. miliaceus* Roxb., Fl. Ind. 1: 276. 1820. *A. miliformis* Schult. Mant. 2: 448. 1824. *Sorghum giganteum* Edgew. in J. Asiat. Soc. Bengal. 21: 181. 1853. *Sorghum miliac-eum* (Roxb.) Snowden in J. Linn. Soc., Bot. 55 : 207. 1955; Bor, Grasses Burma Ceyl. Ind. Pak. 223. 1960. *Sorghum miliaceum* (Roxb.) Snowden var. *parvispiculum* Snowden, *l.c.* 209. 1955; Bor, *l.c.* 223. 1960. (Plate 339)

Vern.: *Baru*.

Rhizomatous perennial. Culms simple or branched, 0.5 - 3 m high or more, slender to rather stout. Leaves 20 - 90 cm long, 0.5 - 4 cm wide, broadly linear - lanceolate, base rounded or subamplexcaul, acuminate; sheath glabrous, striate; ligule short, membranous, ciliate. Panicle loose when in flower, sometimes somewhat contracted after flowering, lanceolate to pyramidal, mostly 10 - 55 cm long, 3 - 25 cm wide; primary branches compound, bare at the base, ultimately bearing racemes of 1 - 5 spikelet pairs. Sessile spikelet elliptic to subelliptic, 4.5 -5 (-5.5) mm long, at first ± hairy and cream to buff-yellow or twany, at the length often tawny, mahogany red, purplish to blackish brown or black and sometimes almost glabrous. Lower glume keeled above, the wings of the keels widening upwards to end in minute teeth, forming with the short apex a distinctly 3-toothed tip; upper lemma acute and minutely mucronate or 2-lobed with an awn 10 - 16 mm long. Pedicelled spikelet at length deciduous, often purplish, awnless, lower involucral glume herbaceous, 5-7-nerved, glabrous, keels ciliate; upper involucral glume similar, 4-5-nerved; lower floral glume as in the sessile spikelets; upper floral glume linear, oblong, hyaline.

Field Note: Common along boundaries of cultivated feilds and waste lands.

Fl. & Fr.: August - December.

Specimen examined: Mandal 78 (MLVGCB Herb).

Distribution: Throughout the Warm countries.

Notes: Snowden (in J. Linn. Soc., Bot. 55: 199. 1955) considers that *Sorghum halepense* reaches the eastern limit of its range in Kashmir and that it is not native of India. It is widely grown in temperate regions, especially North America, as Johnson Grass, and has been introduced into most warm countries.

Because of its tenacious rhizomatous habit *Sorghum halepense* has earned a bad reputation as a weed of cultivated fields. Worse than this, under certain circumstances its tissues contain hydrocyanic acid and is a positive danger to stock.

Endrezzi (in Bot Gaz. 119: 1. 1957) considers that *Sorghum halepense* arose as a segmental allotetraploid from a cross involving *Sorghum propinquum* (2n=20) and a member of Subsect. *Arundinacea* (2n=20). *Sorghum propinquum* (Kunth) Hitchc. extends from China to the Malay Peninsula, but does not occur in India, and is the only diploid *Halepensia*.

Uses: It is used as a fodder grass. The dried plants are used in making huts as well as shades for cattles.

SPOROBOLUS R. Br.

KEY TO THE SPECIES

1a. Spikelets 1.2 - 1.6 mm long, rather crowded. Glumes obtuse. Grains oblong, truncate at the tip .... *S. diander*

1b. Spikelets 1.7 - 2.1 mm long, scattered. Glumes acute. Grains ellipsoid .... *S. nervosus*

Sporobolus diander (Retz.) P. Beauv. Ess. Agrost. 26, 147, 178. 1821; Duthie, Fodder Grasses 48. 1888; Hook. *f.*, Fl. Brit. Ind. 7: 247.1896; Blatter & McCann, Bombay Grasses 221. pl. 148. 1935; Bor, Grasses Burma Ceyl. Ind. Pak. 629. 1960; Pandey, in Shetty & Singh Fl. Rajasthan 3: 1120. 1993; Bhandari, Fl. Indian Desert 395. 1995. *Agrostis diandra* Retz., Obs. Bot. 5: 19. 1789. *Vilfa retzii* Steud. Nom. Bot. ed. 2, 2: 768. 1841.

Tufted, rhizomatous, perennial. Culms 10 - 75 cm high, erect or geniculately ascending. Leaves up to 30 cm long, 3 - 4 mm wide, nearly always involute, occasionally flat or convolute, acute. Panicle pyramidal or narrowly lanceolate, loose, up to 20 cm long, the primary branches erect or somewhat spreading, 1.5 - 5 cm long, bare at the base. Spikelets 1.2 - 1.6 mm long. Glumes lanceolate to ovate-lanceolate, shorter than the spikelet, obtuse, the lower 0.3 - 0.5 mm long, the upper 0.6 - 0.9 mm long. Lemma narrowly ovate, as long as the spikelet, acute. Anthers 2, sometimes 3, 0.3 - 0.4 mm long. Caryopsis oblong or elliptic-oblong, 0.6 - 0.8 mm long, truncate at the tip.

Field Note: Common in the plains.

Fl. & Fr.: August-October.

Specimen examined: Eras 394, Raipur 1920 (MLVGCB Herb).

Distribution: India to Burma, Pakistan, Sri Lanka and Australia.

Sporobolus nervosus Hochst. in Flora, Jena. 38: 202. 1855; Pandey, in Shetty & Singh Fl. Rajasthan 3: 1123. 1993. *Vilfa nervosa* (Hochst.) Schweinf. Beitr. Fl. Aethiop. 303. 1867. *Sporobolus ioclados sensu* Hook. *f.*, Fl.

Brit. Ind. 7: 249. 1896. *non* (Nees ex Trin.) Nees 1841. *S. longibrachiatus* Stapf. in Kew Bull. 1907: 219. 1907. *S. stocksii* Bor. in Kew Bull. 3 : 45. 1948; Bor, Grasses Burma Ceyl. Ind. Pak. 633. 1960.

Densely caespitose perennial, or with the shoots contiguous from the closely spaced nodes of a short oblique rhizome; basal sheaths yellowish, indurated, often covered by the fibrous remains of decaying sheaths. Culms 10-70 cm high, erect. Leaves 4 - 10 cm long, 1-3.5 mm wide, flat, or more often convolute, the tip filiform. Panicle narrowly ovate, 7-20 cm long, secondarily branched, the branches bare below and the spikelets secund and clustered at the tips of the branchlets. Spikelets 1.7-2.1 mm long, scaberulous or smooth, pallid greyish green to dark green. Lower glume lanceolate, 0.8 - 1 mm long, acute; upper glume narrowly ovate, 1 - 2 mm long, two-thirds to four-fifths as long as the spikelet, acute. Lemma narrowly ovate, as long as the spikelet. Anthers 3, 0.8 - 1 mm long. Caryopsis ellipsoid, 0.6 mm long.

Field Note: Occasional in waste lands.

Fl. & Fr.: August - November.

Specimen examined: Hurda 1907 (MLVGCB Herb).

Distribution: India, Kenya, Pakistan, Tanzania and Mauritania through Ethiopia and Somalia to Arabia; also in South West Africa.

TETRAPOGON Desf.

KEY TO THE SPECIES

1a. Annual or short-lived perennial, loosely tufted. Lowest lemma shortly ciliate or subglabrous. .... *T. tenellus*

1b. Perennial, densely tufted. Lowest lemma long ciliate .... *T. villosus*

Tetrapogon tenellus (Koen. ex Roxb.) Chiov. in Ann. R. Ist. Bot. Roma. 8: 352. 1908; Bor, Grasses Burma Ceyl. Ind. Pak. 475. 1960; Pandey, in Shetty & Singh Fl. Rajasthan 3: 1127. 1993; Bhandari, Fl. Indian Desert 399. 1995. *Chloris tenella* Koen. ex Roxb., Fl. Ind. 1: 330. 1820; Hook. *f.,* Fl. Brit. Ind. 7: 290. 1896; Blatter & McCann, Bombay Grasses 254. pl. 169. 1935. *C. macrantha* Jaub. & Spach, Ill. Pl. Or. 4: 42, t.328. 1851. *non* Desv. 1831. *C. triangulata* Hochst. ex A. Rich., Tent. Fl. Abyss. 2: 409. 1851. (Plate 340)

Annual or short-lived perennial, loosely tufted, with erect or ascending 4 - 6-noded culms up to 60 cm high; nodes not prominent. Leaves 5 - 20 cm long, 2 - 5 mm wide, linear-lanceolate, acuminate, ciliate; sheaths loosely flabellate at the base of the culm. Spikes exserted, solitary or paired and divergent, 3 - 7 cm long, solitary or paired. Spikelets 4 - 6-flowered, the lower 3 - 5 florets fertile and 3-awned. Glumes ovate-lanceolate, acute, acuminate or rarely aristate; lower glume 3 - 5.5 mm long. Upper glume 4 - 6.5 mm long. Lowest lemma obovate, 4 - 6.5 mm long, obovate, shortly ciliate along the nerves or subglabrous, with an awn 3 - 15 mm long; upper 1 - 2 florets sterile and reduced to minute awnless lemmas. Caryopsis oblong.

Field Note: Found in desert areas and sometimes on limestones.

Fl. & Fr.: July - November.

Specimen examined: Hurda 501, Raipur 1666 (MLVGCB Herb).

Distribution: Tropical Africa eastwards to India, Pakistan and southwards to Rhodesia and Angola.

Uses: It is considered to be a good fodder grass.

Tetrapogon villosus Desf., Fl. Atlant. 2: 389. t. 255. 1799; Boiss., Fl. Or. 5 555. 1884; Duthie, Fodder Grasses 55. 1888; Bor, Grasses Burma Ceyl. Ind. Pak. 475. 1960; Pandey, in Shetty & Singh Fl. Rajasthan 3: 1128. 1993; Bhandari, Fl. Indian Desert 400. 1995. *Chloris villosa* (Desf.) Pers., Syn. Pl. 1: 87. 1805; Hook. *f.,* Fl. Brit. Ind. 7: 291. 1896; Blatter & McCann, Bombay Grasses 254. 1935.

Vern.: *Hanger.*

Densely tufted perennial with erect 34-noded culms up to 40 cm high; nodes black and prominent. Leaves 2 - 10 cm long, 1 - 2.5 mm wide, flat, acute or obtuse; sheaths 2.3 cm long, keeled, minute, very strongly flabellate at the base of the culm. Spikes exserted, paired, dorsally adpressed, rarely separating at maturity, 2 - 8 cm long, pallid or purple-tinged. Spikelets 4 - 6-flowered, all florets awned, the lowest 1 - 3 fertile. Glumes dissimilar, the lower ovate-lanceolate, 2 - 3 mm long, acuminate, the upper ovate or elliptic-oblong, 3 - 4 mm long, abruptly acute, obtuse or truncate with the nerve excurrent; lowest lemma ovate-elliptic, 2.5 - 11 mm long, with long spreading cilia on the lateral nerves and keel and with an awn 8 - 10 mm long; upper 2 - 3 florets sterile and reduced to glabrous lemmas. Caryopsis linear-oblong, brownish.

Field Note: A common grass of desert areas and mountain slopes.

Fl. & Fr.: August - November.

Specimen examined: Asind 503, Badnor 1667 (MLVGCB Herb).

Distribution: Tropical Africa eastwards to India, Pakistan and westwards to North Africa and the Canary Islands.

THEMEDA Forsk.

Themeda quadrivalvis (L.) Kuntze, Revis. Gen. Pl. 2: 794. 1891; Bor, Grasses Burma Ceyl. Ind. Pak. 252. 1960; Raizada *et al.* in Ind. For. Rec. (n.s.) Bot. 4: 251. pl. 12. f. 1. 1961; Pandey, in Shetty & Singh Fl. Rajasthan 3: 1130. 1993. *Andropogon quadrivalvis* L. in Murr. Syst. Veg. ed. 13. 758. 1774. *Anthistiria ciliata* L. f. Suppl. 113. 1781; Hook. *f.*, Fl. Brit. India 7: 213. 1896.

Annual. Culms moderately robust, erect or geniculate at base, ca. 1 m tall. Leaves up to 30 cm long, 0.3 - 0.5 cm wide, flat or folded, linear, glabrous, abruptly acute to acuminate; sheaths glabrous or with tubercle-based bristles at mouth; ligule *ca* 2 mm long, membranous. Compound panicle large, dense; spathes and spatheoles lanceolate-caudate, glabrous, innermost 1.3 - 1.7 cm. Raceme composed of a triad of 1 sessile and 2 pedicelled spikelets above the involucre of 2 homogamous pairs. Homogamous spikelets all sessile, arising at same level, barren, both glumes present, 4.5 - 6 mm, lanceolate, stiffly setose in upper half with 3 - 4 mm, tubercle-based bristles. Sessile spikelet 4 - 5 mm; callus 0.8 - 1 mm, subacute, brown bearded. Lower glume dorsally rounded, dark brown at maturity, pubescent, often thinly or glabrous on lower back; awn 3.5 - 4 cm. Pedicelled spikelet 4 - 6 mm, barren. Palea absent. Caryopsis subcylindric.

Field Note: Common along riverbanks and water streams.

Fl. & Fr.: August - December

Specimen examined: Triveni 1131, Swaroopganj 1216 (MLVGCB Herb).

Distribution: China, India, Indonesia, Myanmar, Nepal, Thailand, Vietnam and Australia.

Notes: This species has dense fascicles of racemes similar to those of *Themeda triandra* Forsk. but differs with smaller spikelet parts and conspicuously setose homogamous spikelets with long, patent bristles from very large tubercles. It can also generally be separated by its annual habit.

TRAGUS A. Haller nom. cons.

Tragus roxburghii Panigrahi, in Kew Bull. 29: 495. 1974; Pandey, in Shetty & Singh Fl. Rajasthan 3: 1131. 1993; Bhandari, Fl. Indian Desert 400. 1995. *Lappago biflora* Roxb. *sensu* Fl. Ind. 1: 284. 1820, *quoad descript., excl. typo. Tragus biflorus sensu* Schult. Syst. Veg. Mant. 2: 205. 1824, *quoad. descript.*; Bor, Grasses Burma Ceyl. Ind. Pak. 682. 1960. *T. racemosa sensu* Hook. *f.*, Fl. Brit. India 7: 97. 1896. non Scop. 1777. *Nazia racemosa auct.* non O. Ktze. 1891; Blatter & McCann, Bombay Grasses. 217. pl. 144. 1935. (Plate 341)

Radiately spreading annual. Culms 3 - 20 cm long, ascending from a prostrate base, rooting from the nodes, often matted. Leaves 0.5 - 5 cm long, 1 - 6 mm wide, linear-lanceolate, acuminate, margins undulate, with tubercle - based bristles; ligule a ciliate rim. Inflorescence 1 - 7 cm long, cylindrical, often loosely so; clusters straight or slightly gaping at the tip, consisting of 2 subequal and subopposite spikelets. Spikelets. in pairs, 2.5

- 4 mm long. Lower glume very minute, hyaline. Upper glume 5-nerved, the prickles terete in their lower half and hooked at the tip. Lemma solitary, shorter, oblong-lanceolate, apiculate. Caryopsis oblong-elliptic, 1.5 mm long, compressed.

Field Note: Common in wast lands.

Fl. & Fr.:August - November.

Specimen examined: Manpura 2195, Paroli 1172 (MLVGCB Herb).

Distribution: E. Africa, S. E. Asia, India and Pakistan.

Notes: It is commonly known as *Tragus biflorus* Schult., but for complex technical reasons, resolved by Panigrahi (*l.c.*) this name is superfluous.

Uses: A sporadic xerophytic species grazed during the rains and said to be nutritious.

UROCHLOA P. Beauv.

Urochloa panicoides P. Beauv. Ess. Agrost. 53. t. 11. f. 1. 1812; Bor, Grasses Burma Ceyl. Ind. Pak. 372. 1960; Raizada & Jain, in Ind. For. Rec. (n.s.) Bot. 5: 177. pl. 4. f. 4. 1965; Pandey, in Shetty & Singh Fl. Rajasthan 3: 1135. 1993; Bhandari, Fl. Indian Desert 401. 1995. *Panicum javanicum* Poir. in Lam. Encycl. Math. Bot. Suppl. 4: t. 274. 1816; Hook. *f.*, Fl. Brit. Ind. 7: 35. 1896. *P. rothii* Spreng. Syst. Veg. 1: 310. 1825. *P. panicoides* (P. Beauv.) Hitchc. in J. Wash. Acad. Sci. 9: 551. 1919. *P. hirsutum* Koen. ex Roxb., Fl. Ind. 1: 303. 1820, *non* Swartz 1797. *Urochloa helopus* (Trin.) Stapf, in Prain Fl. Trop. Afr. 9: 595. 1920; Blatter & McCann, Bombay Grasses 146. 1935. *Panicum helopus* Trin. in Spreng. Neue Entdeck. 2 : 84. 1821; Duthie, Fodder Grasses 8. 1888. *Urochloa panicoides* P. Beauv. var. *marathensis* (Henr.) Bor. *l.c.* 373. 1960. *U. panicoides* P. Beauv. var. *pubescens* (Kunth) Bor, *l.c.* 372. 1960. *U. panicoides* P. Beauv. var. *velutina* (Henr.) Bor, *l.c.* 373. 1960. *U. marathensis* Henr. in Meded. Rijks.-Herb. 43: 2. 1922; Blatter & McCann, *l.c.* 146. 1935. *U. marathensis* Henr. var. *velutina* Henr. *l.c.* 43: 2. 1922. *U. pubescens* Kunth, Rev. Gram. 1: 31. 1829.

var. ***panicoides***

Tufted annual. Culms 10 - 60 cm high, rooting at lower nodes. Leaves 2 - 15.5 cm long, 5-15 mm wide, linear to narrowly lanceolate, subamplexicaul, coarse, glabrous or pubescent, the margins tuberculate-ciliate at least near the base; sheaths striated, densely ciliated-pubescent; ligule a ciliate rim. Inflorescence of 2 - 8 (-10 or rarely more) racemes on a common axis 1 - 6 cm long; racemes 1 - 6 cm long, bearing single or sometimes paired spikelets on a narrowly winged rachis, the pedicels (sometimes also the rachis) with white hairs. Spikelets elliptic, 3.5 - 5 mm long, acute. Lower glume ovate, a quarter to almost half as long as the spikelet, very rarely more, 3-5-nerved, obtuse to subacute. Upper glume often with crossveins, glabrous or pubescent. Lower lemma sometimes with a setose fringe. Upper lemma rugulose with a mucro 0.3 - 1 mm long. Caryopsis upto 2.5 mm long, compressed, yellowish-brown.

Field Note: Common in cultivated fields.

Fl. & Fr.: August - November.

Specimen examined: Rupaheli 909, Pander 1202 (MLVGCB Herb).

Distribution: Sudan to Yemen and southwards to South Africa: India, Pakistan; introduced to Australia.

Notes: In common with several other species of the genus, *Urochloa panicoides* has forms with glabrous, pubescent or setosely fringed spikelets; they appear to be of no taxonomic importance.

REFERENCES

Bhandari, M.M. 1965. On *Coccinia grandis* (Linn.) Voigt. *Ann. Arid Zone* **4** : 227-229.

Bhandari, M.M. 1978. *Flora of the Indian desert.* Scientific Publishers (India), Jodhpur.

Bhandari, M.M. 1991. Evolution of Desert vegetation in Thar desert in Rajasthan. In Dhir, P.P. *et al.* (Eds.) *Land, man and environment*, pp. 73-78. Geological Society of India, Bangalore.

Bhandari, M.M. 1994. Floral diversity of the Indian Desert. In Frame, B., Victor, J. and Joshi, Y. (Eds.) *Biodiversity Conservation in Forest, Wetland and Desert*, pp. 75-90. TERI, New Delhi and British High Commission.

Bhandari, M.M. 1999. Assessment of the status, trends and option for conservation and sustainable use of biological diversity in the Indian desert. *WWF - India Biodiversity Technical Report* no. 221-65.

Bhandari, M.M. and Kumar, M. 1995. Desert floristics - need for updating. *J. Indian bot. Soc.* **74**A : 195-204.

Braun-Blanquet, J. 1932. *Plant Sociology.* McGraw Hill Book Company, New York.

Cabrera, A. 1952. Report on plant ecology of the arid and semi-arid areas of Latin America NS/A2 **80** : 1-69. UNESCO.

Cain, S. A. 1950. Life-forms and phyto-climate. *Bot. Rev.* **16** : 1-32.

Charan, A. K., Sen, D. N. and Rajpurohit, K. S. 1978. Biological spectrum of the vegetation of western Rajasthan desert. *Ind. J. For.* **1** (3) : 226-228.

Das, R.B. and Sarup, S. 1951. The Biological spectrum of Indian desert flora. *Univ. Raj. Stud. (Bio. Sci.)* **1** *:* 36-43.

Drude, O. 1897. *Manual de-geographie botanique.* Klincksieck, Paris.

Drude, O. 1913. Die Oekologie der Pflanze. Brunswick.

Duthie, J.F. 1903-1929. Flora of the Upper Gangetic Plains and of the adjascettt Siwalik and Sub - Himalayan tract. Calcutta.

Gena, C.B. 2006. Phytodiversity of the Thar desert. *Proc. Nat. Conf. Biod. Consev.*, M.L.V.G.C., Bhilwara. pp 1-4.

Gupta, R.S. 1965a. A contribution to the flora of Rajasthan. *Indian For.* **91** : 438-439.

Hooker, J.D. 1872-1897. Flora of British India. Vols. 1-7. Kent.

Hooker, J.D. and Thomson, T. 1855. Flora Indica. London, pp. 247.

Humboldt, A. 1805. Essai, Sar la geographic des plantes, Paris.

Jain, S.K. 1970. Floral composition of Rajasthan - A review. *Bull. bot. Surv. India* **12** . 176-187 (also in ed. Roonwal, M.L. The Natural Resources of Rajasthan. **1** : 47-64. 1977. University of Jodhpur, Jodhpur).

Linnaeus with his classic publication Genera plantarum (1737).

Maheshwari, J.K. and Singh, V. 1974. Aquatic and Marsh - land plants of Bundi dist. Rajasthan. *Journ. Bombay Nat. Hist. Soc.* **70** : 437 - 446.

Majumdar, R.B. 1969. Notes on the Flora of Rajasthan - II (Additions to the Flora of N.W. Rajasthan). *Bull. Bot. Soc. Bengal* **23** : 35 - 40.

Majumdar, R.B. 1976. Synoptic flora of Kota division (S.E. Rajasthan). *Bull. Bot. Surv. India* **18** : 49-79, 105 - 160.

Majumdar, R.B. 1977. Flora of Rajasthan East of Aravalli. *Bull. Bot. Surv. India* **19** : 23 - 24.

Meena, K.L. 2010. *Morinda coreia* Buch.-Ham.: A new record to the flora of Rajasthan. *J. Indian. Bot. Soc.* 89 (1 & 2):210-212.

Meena, K.L. and Yadav, B.L. 2006. Floristic diversity of Bhilwara tehsil, Rajasthan, India. *Proc. Nat. Conf. Biod. Cons.* MLV GC Bhilwara- PP 26-30.

Meena, K.L. and Yadav, B.L. 2009. *Senna uniflora* (Mill.) Irwin & Barneby (Leguminosae: Ceasalpinioideae) - A New Record for Rajasthan. *Natural Product Radiance.* **8** (5) : 525 - 527.

Meher-Homji, V. M. 1962b. The bioclimates of India in relation to the vegetational criteria. *Bull. Bot. Surv. India* **4** : 105-112.

Mertia, R. S. and Bhandari, M. M. 1978. Vegetational adaptations in the extremely arid regions of the Indian Desert. In Mann, H. S. (Edit.) *Arid Zone Research and Development,* pp. 155-164. Scientific Publishers (India), Jodhpur.

Pandey, R.P. and Parmar, P. J. 1993. An assessment of Biological spectrum of the flora of Rajasthan. *J. Econ. Taxon. Bot.* **17** (1) : 99-103.

Pandey, R.P., Parmar, P.J. and Singh. V, 1985. A reassessment of Biological spectrum of the flora of Rajasthan desert. *Trans. Isdt.* and *Ucds.* **10** (1) : 121-128.

Parmar, P.J. and Singh, A.N. 1982. A contribution to the flora of Bhilwara district, Rajasthan. *J. Econ. Taxon. Bot.* **3** (2) : 491-515.

Raizada, M.B. and Sharma, V.S. 1962. New plant records for Upper gangetic plains from Ajmer Merwara. *Bull. Bot. Surv. India* **88** : 356-359.

Raunkiaer, C. 1911. Liveformernes statistik som Grundlag for Biologisk Plantegeographi. Botanisk Tidsskr. 29 (1) : (see also, statistik der Lebensformeno als Grundslage fur die biologische geographic. *Beih. Bot. Zb*1. **27** : 171-206).

Raunkiaer, C. 1934. Life forms of plants and statistical plant geography. Clarendon Press, Oxford.

Roxburgh, W. 1820-1824. *Flora Indica* (Edited by W. Carey and N. Wallich). Serampore. 1(1820): 2(1824).

Sharma, K.C. 1978. Medicinally important wild plants of Ajmer. *Scientia* **21** : 71-80.

Sharma, N. 2002. *The flora of Rajasthan.* Aavishkar Publishers & Distributors, Jaipur.

Sharma, S. and Tiagi, B. 1979. *Flora of North-East Rajasthan.* Kalyani Publishers, New Delhi

Sharma. K.C., Gena, C.B. and Joshi, A.K. 1990. An enumeration of grasses and sedges of Ajmer district. *Geobios New Reports.* **9** : 52-54

Shetty, B. V. and Pandey, R. P. 1978. Vegetation and floral composition of the Rajasthan Desert. *Arid Zone Research in India, Silver Jubilee Souvenir* (1952-1977). pp. 25-31. Central Arid Zone Research Institute, Jodhpur.

Shetty, B. V. and Pandey, R. P. 1979. Studies on the flora of Rajasthan desert - A review. *Bull. Bot. Surv. India* **19** (1-4) : 5-22.

Shetty, B. V. and Singh, V. (Eds.) 1987. *Flora of Rajasthan.* Vol. **I.** BSI, Howrah.

Shetty, B. V. and Singh, V. (Eds.) 1991. *Flora of Rajasthan.* Vol. **II.** BSI, Howrah.

Shetty, B.V. and Pandey, R.P. 1977. Studies on the flora of Rajasthan desert - A review. *Bull. Bot. Surv. India* **19** : 5 - 22.

Shetty, B.V. and Pandey, R.P. 1983. Flora of Tonk district. *BSI*, Howrah.

Shetty, B.V. and Singh, V. (Eds.) 1993. *Flora of Rajasthan.* Vol. **III.** BSI, Howrah.

Singh, V. 1977. Addition to the flora of Rajasthan. *J. Bombay Nat. Hist. Soc.* **74** : 45-49.

Singh, V. 1980. Some interesting plant records for Rajasthan. *J. Bombay Nat. Hist. Soc.* **76** : 206-209.

Singh, V. 1983. Flora of Banswara District. *BSI*, Howrah.

Singh, V. and Singh, M. 2006. Biodiversity of Desert National Park Rajasthan. *BSI*, Kolkata.

Singh, V. and Srivastava, A.K. 2007. Biodiversity of Ranthambhore Tiger Reserve Rajasthan. Scientific Publishers (India), Jodhpur.

Tiagi, Y.D. & Aery, N.C. 2007. Flora of Rajasthan (South & South-East Region). Himanshu Publications, New Delhi.

Vyas, L.N. 1967. Contributions to the flora of North-East Rajasthan. *J. Bombay Nat. Hist. Soc.* **64** : 191-231

Vyas, L.N. and Ramdeo, K.D. 1965. Contribution to the flora of Rajasthan from Udaipur. *Indian For.* **91** : 672-675.

Whittaker, R.H. 1975. Communities and Ecosystems. New York (2nd ed.).

Yadav, B.L. and Meena, K.L. 2008. *Solanum sisymbriifolium* Lam. (Solanaceae): A new record for Rajasthan, India. *J. Econ. Taxon. Bot.* **32** (3): 749 - 753.

Yadav, B.L. and Meena, K.L. 2009. *Cleome burmannii* Wight & Arn. (Cleomaceae): A New Record to the flora of Rajasthan, India. *Rheedea.* **19**(1): 65 - 66.

INDEX TO FAMILIES, GENERA AND SPECIES

INDEX TO LOCAL NAMES